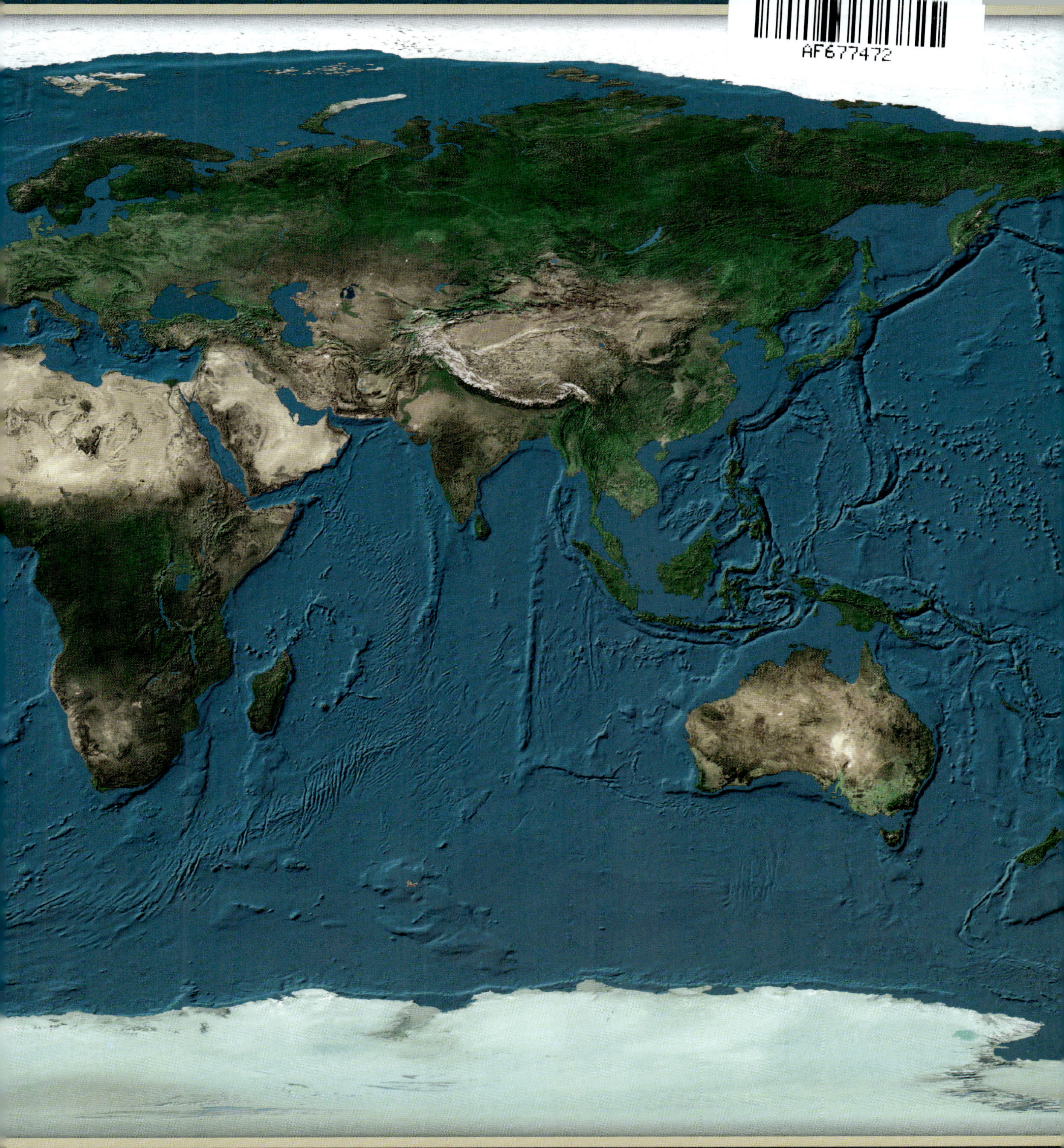

INTERNATIONAL PETROLEUM ENCYCLOPEDIA 2009

International Petroleum Encyclopedia is published annually by:
PennWell Corporation
1421 South Sheridan Road
Tulsa, Oklahoma 74112, USA
Phone: 918-835-3161
Fax: 918-831-9555
www.pennwellbooks.com

800.752.9764
+1.918.831.9421
sales@pennwell.com
www.pennwellbooks.com
www.pennwell.com

ISBN 9781593701901

Printed in the United States of America

1 2 3 4 5 13 12 11 10 09

PennWell

Contents

Guest Essay

Impact of the Financial Crisis on the Energy Industry

Samuel A. Van Vactor, Ph.D.

The four shockwaves of the financial crisis

The global economy was slammed by a four-wave economic tsunami in 2008.

The first shockwave concerned the U.S. sub-prime mortgage market and the housing bubble, where accumulating bad debt undermined a number of key banks.

In the second wave, the demise of Lehman Brothers and anxiety over the creditworthiness of banks and insurance companies led to a global bailout of the financial industry.

In the third wave, cash hoarding, frozen loans, and stock market panic led to a massive realignment of expectations that dooms the developed world to a serious recession.

In the final wave, the contagion has spread to emerging markets, leading to a global slide reminiscent of the 1970s oil shocks.

The first wave: U.S. sub-prime mortgages

In the first phase of the crisis it was discovered that many mortgages written during the housing boom were headed for default and that the mortgages themselves had been repackaged and resold as if they were secure assets. Lacking transparency, the "toxic" assets could not be valued and resold at any price. Moreover, the weak U.S. economy put commercial loans, credit cards, and other mortgages at risk.

To offset the macroeconomic impact, Congress authorized a tax rebate for U.S. consumers, and the resulting bump in cash propped up the "real" economy through the summer of 2008.

However, the illness was bound to spread. As housing prices declined, the toxicity of mortgages grew and the asset base of important international banks dwindled. Major financial institutions—such as The Bear Stearns Companies, Inc., and IndyMac Bank (Independent National Mortgage Corp.)—were forced into unappetizing mergers or dissolution, with the government accepting responsibility for much of the bad debt.

The broad scale of the problem became apparent on September 7, when the federal government was forced to seize control of federally sponsored mortgage companies, Fannie Mae and Freddie Mac.

The second wave: The growing risk of systemic default

As the crisis wore on and mortgage defaults mounted, a more serious problem emerged. It turned out that insurance companies and investment banks had sold trillions of dollars of credit default swaps (CDSs). These swaps were not limited to the housing market or to sub-prime mortgage loans, but covered a broad set of financial instruments and institutions.

Banks that had been weakened by sub-prime mortgage holdings were also involved in the CDS market, and the bankruptcy of Lehman Brothers Holdings Inc. on September 15 tripped a variety of credit obligations.

Within a few hours, the U.S. Treasury was forced to bail out major insurance company American International Group Inc. (AIG.) Almost immediately other dominos began falling around the globe. The resulting chaos set off a global panic in stock markets, which further weakened the assets of many financial institutions.

The U.S. was not the only country that enjoyed a significant increase in housing prices. In particular, the fastest growing European countries had experienced housing price surges even greater than those in U.S. Moreover, most European banks and insurance companies had been involved in the CDS market.

Over the weekend of September 27, the credit crunch hit Europe with full force. Within a few days:

- The Dutch banking and insurance activities of Fortis were acquired by the government of The Netherlands, and its other banking activities were acquired by the Belgian state.
- The UK government nationalized British bank, Bradford & Bingley
- The Iceland government took control of its third largest bank, Glitnir.

As the financial stress deepened, it became apparent that a piecemeal approach was unworkable. The problem crystallized on September 30, when the Irish government announced that it would guarantee all deposits in its banks, amounting to €400 billion—twice its annual GDP.

Although Ireland was a small country, it had adopted the euro as its currency and it did not take central bankers in neighboring countries long to realize they faced the threat of international bank runs if they did not follow the Irish example.

In the meantime, U.S. officials groped for a strategy to save its banks. They had focused on pulling out the banks' bad loans, rather than recapitalizing them by direct investment. The plan, however, looked too much like a bailout of Wall Street to Congress, and the House of Representatives quashed the idea on September 28. Finally, on October 3, a revised plan passed both houses of Congress.

Ironically, however, the Treasury later shifted strategies and began investing directly in banks. Treasury officials had little choice. Without a clear demonstration of government guarantees, the international market would quickly pull cash from any institution perceived to be under-capitalized.

The third wave: Recession in developed countries

Despite the central bank infusion, it was obvious that the economies of developed nations were in for a hard time. The IMF had pegged global economic growth at 3.8% for 2009, but a new forecast in early October dropped that figure to 3.0%.

Within a few days, however, pessimism deepened, signaling the third wave of growing concern over the health of the real economy. In response to the new pessimism, global stock and commodity markets plunged, as traders factored in the lowered expectations.

The depth and length of the developed countries' recession is unclear, but early indicators are decidedly negative. There are also features of this recession that are significantly different from anything experienced since the Great Depression.

Most importantly, there is an asset deflation, in real property, stocks, and commodities. This is in contrast to the oil price shocks of the 1970s. In that era stock prices dropped, but commodities and real estate values increased along with general price inflation. Most economists would agree that aggregate consumption is dependent on both income and perceived wealth.

The asset deflation is exacerbated by the fact that there is huge debt outstanding against these assets. Many home-owners are "under water," owing more on their mortgage than the value of their home. Government treasuries are also depleted.

Hopefully, much of the cash infusion will be recovered as banks regain solvency, but some will be permanently lost. Economic recovery will be burdened by excessive debt, which reduces flexibility for both governments and consumers.

The fourth wave: A global recession

As of late 2008, China was forecasting economic growth on the order of 8% in 2009, and its leaders have announced infrastructure investments intended to support it. However, legitimate questions can be raised about the viability of such plans when economies around the world are faltering. After all, China has prospered by selling manufactured goods to consumers in developed economies.

The swift rise in commodity prices, particularly crude oil, is partially responsible for the meltdown. It was learned in the oil

Dr. Samuel A. Van Vactor

Dr. Van Vactor is an energy economist who has analyzed oil, natural gas, and electricity markets since 1973.

Since 1981, he has been the president of Economic Insight, Inc. (EII), in Portland, OR.

EII provides consulting services on a range of topics related to the petroleum, natural gas, and electricity industries. The firm has undertaken a variety of strategic studies for energy companies and government agencies on topics from market development forecasting and analysis of regulatory changes to critique of public policies and evaluation of market and pricing structures.

His consulting clients have included the American Petroleum Institute, the U.S. Department of Justice, the Bonneville Power Administration, Pacific Gas & Electric Corporation, the Western Interstate Energy Board, the California Power Exchange, the Mitsubishi Research Institute, Tokyo Gas, Gaz de France, and the Government of India.

He also has advised major oil and gas companies, including Texaco, Chevron, Exxon, BP, ARCO, and Shell. He was an expert witness for the North Slope producers on Alaska tax and royalty issues.

Dr. Van Vactor holds a Ph.D. from Cambridge University; and both an M.A. (University of Washington) and a B.S. (University of Oregon) in economics. He also has done research at the London School of Economics.

In prior years, Dr. Van Vactor served as an economist at the U.S. Department of the Treasury, a senior economist at the International Energy Agency of the OECD, and as head of energy planning for the State of Oregon.

From 1983 to 1989, Dr. Van Vactor also was an associate of Dr. Arlon R. Tussing and vice president of Tussing's consulting firm, Arta Inc. (Seattle), regarding natural gas and Alaska energy issues.

During the 1980s, ARTA was deeply involved with issues regarding deregulation of the North American natural gas industry.

He has lectured on energy and economics at the University of Maryland (European Division), Portland State University, University of Southwest Australia, Cambridge University, Columbia University, and Victoria University of New Zealand.

He has also conducted seminars for the Singapore Centre for Management Technology on gas pricing and contracting in Asia.

Dr. Van Vactor is the author of numerous books and articles concerning energy industries.

price shocks of the 1970s that the sudden shift in cash from consumers to energy producers jolts the economy and provokes a combination of inflation and recession.

Oil is not as important to the overall energy market now as it was in the 1970s, but it is still significant, particularly for America. In 1974, the United States produced more oil than it imported, so much of the money associated with higher prices re-circulated. The situation is reversed today, and most of the money spent on oil flows out of the country. The rapid shift in funds withdrew liquidity from banks, making it difficult for them to recapitalize.

The drop in demand for manufactured goods and commodities will reverse the financial trends of the last few years. Countries with huge pots of cash now may find the funds rapidly depleted over the coming months, and the counter-shock could also be disruptive.

Perspective on derivatives

How could a relatively small set of low-grade mortgages cause a global recession and require public funding of trillions of dollars? The answer, of course, is that derivatives can be used as leverage, and leverage magnifies either profit or loss.

According to *The New York Times*, the derivative market at the end of 2008 totaled $531 trillion—or about ten times global GDP. The market for CDSs has been about $55 trillion.

When market values change quickly, and in unexpected ways, it disturbs normal commerce. The essence of finance is trust – to make a loan in the expectation of repayment. Trust in turn depends on confidence in the institutions and traders in the marketplace. Derivatives that are not transparent or not adequately backed by assets have the power to undermine confidence and trust.

What Wall Street should have learned from Enron's collapse

Some important lessons from the collapse of Enron Corp. in 2001 appear to have been lost in the din of good times.

First, Enron misused derivatives and engaged in other manipulations to move suspect assets from its balance sheet and cover up losses. It completed these transactions through a series of complex maneuvers, the purpose of which was to obfuscate, rather than illuminate, their true liabilities.

It seems that no one on Wall Street learned much from the Enron example. No one—including the bond rating agencies—questioned the veracity of the sub-prime mortgage bonds, despite the complexity of their packaging, their opaque nature, and the flimsiness of their credit credentials.

Second, sight was lost of the interrelationship between derivatives, mark-to-market accounting, and managerial bonuses. Most derivatives are traded in the over-the-counter (OTC) market, but the concepts that underlie them are extracted from futures exchanges. In a futures exchange, traders' positions are marked to market at the close of each trading day. There is no other way to treat a liquid and marketable asset; book value or the price paid is irrelevant. Problems arise, however, when the asset cannot be easily sold and its value has to be estimated.

Enron frequently tied bonuses to the present value of future profits from freshly negotiated deals. They even claimed a portion of the present value as a current profit in their balance sheet. This is akin to mark-to-market accounting, except that the contractual commitment is unlikely to be resold; thus, valuation is subjective.

This incentive structure created two distortions. The company's managers focused on short-term deal-making rather than that long-term planning. Moreover, the booking of theoretical profits was at variance with actual cash flow, so Enron had to constantly borrow money to make ends meet.

One of the most troubling aspects of the 2008 financial crisis is the way in which Wall Street financial firms followed the Enron paradigm and ignored its consequences.

Large bonuses were paid out even if the firm retained the asset or a residual liability associated with it. Most financial firms deal in paper and have limited physical assets. The basis for their profitability is having clever people who need to be compensated if they are to be retained. So, as all the bad paper rolled in, bonuses rolled out.

Banks simply did not keep the profits garnered in the mortgage-writing frenzy, and when the paper and other questionable assets turned toxic, liabilities grew, necessitating a public-sector bailout. Once again, the banking sector proved "too big to fail."

There is no question that the global system of derivatives trading must be carefully reviewed. There are two issues to be addressed: inadequate third-party clearing and asymmetrical risk.

Third-party clearinghouses for commodities, such as NYMEX's Clearport, are already being offered by the marketplace and volume is rising rapidly. In these arrangements an exchange manages and clears bilateral contracts between buyers and sellers, providing objective valuation and minimizing the risk of default.

Commodity trading has a natural symmetry in that there is active hedging by both buyers and sellers. There may not be a perfect match, but financial derivatives offered by a bank can be offset so that the net position is nearly balanced.

Unfortunately, there is no natural offset for CDSs; they are akin to insurance. That is, insurance buyers want to hedge against a catastrophe, but usually no one benefits from such an event, so the activity is typically one-sided. Thus, issuers need to be sure they have adequate capital to cover all eventualities. Obviously, the capital base has been inadequate for these types of derivatives.

The consequence of the financial crisis on the energy industry

Policy makers, traders, and investors are all trying to sort out the depth and duration of the coming recession, but the impact on the energy industry is immediately obvious. Oil prices have dropped by more than two-thirds since their peak in July 2008. With lower economic activity comes less demand for oil, and although the difference is not great, it is sufficient to completely reverse the market.

One striking feature of oil market developments in the last few years is the speed at which things can change. After the price run-ups of the 1970s, seven years passed between the peak of oil prices in 1980 and their collapse in 1987.

The price drop observed from mid-July through November 2008 took only about 100 business days.

Although the drop in oil prices provides consumers a well-deserved break, it raises a pointed question. Is the relief short- or long-term? The primary reason that economic growth has stalled is that investment has dried up. Many energy development projects are funded by cash flow. This is particularly the case for OPEC's National Oil Companies (NOCs,) which have the greatest potential for developing new supply. For that supply to be realized, however, investment must continue, and low oil prices will be a major inhibition to such investment.

A variety of financial reforms have already been suggested, including:

- Greater regulation of banks
- An enhanced role for the Commodity Futures Trading Commission (CFTC)
- A streamlining of regulatory bodies
- An international clearing house
- Registration of derivative contracts.

Restrictions on trading will reduce the risk of default, but that restriction will impose a cost: reducing the flexibility and liquidity of the market.

Officials in some countries are even seeking a complete overhaul of the international system of foreign exchange and trade—a "Bretton Woods II." It is important to recognize that there is a tradeoff between economic efficiency and stability, just as there is between risk and return, and it is hoped that the coming process will get the balance right.

Like it or not, the financial crisis will enhance the role of NOCs and government-sponsored enterprises (GSEs). The movement of oil around the globe, like the smooth workings of the international financial system, is crucial to economic growth and prosperity.

Many important private institutions collapsed in 2008, so it will be argued that the new reality necessitates a much larger role for government.

In contrast, consider Raymond Mikesell's summary of the goals of the 1944 Bretton Woods conference and the post-war plan to restructure the international economy: "...free and non-discriminatory markets for currencies, capital, and goods."

The Bretton Woods system bestowed more than 60 years of economic growth and prosperity following the Great Depression.

Viewed in that context, the 2008 financial crisis has been only a minor setback.

(Based on a presentation made by Dr. Van Vactor for a conference of the International Association for Energy Economics, Perth, Australia, November 6, 2008.)

Chronology

2008 In Brief

JANUARY

Exploration/Development

- Iran begins a prequalification process for an oil licensing round planned for later in the year.
- Brazil's Petrobras signs agreements with Cuba's Cupet for cooperation in oil and gas exploration and production offshore Cuba.

Production

- API reports that U.S. companies drilled a record 30,625 natural gas wells in 2007.
- Gas production begins from Starling field in the Central North Sea; peak production of 140 MMcfd is expected.
- Pemex reports that Mexico's 2007 oil production dropped by 5.3% in 2007 to an average 3.08 million b/d.

Natural Gas/LNG

- Uzbekistan approves creation of a JV for production of LNG by Russia's Stroytransgaz and Uzbekistan's Uzbekneftegaz.

Pipelines/Transportation

- South Korea says it will ban single-hull tankers in its waters starting in 2011, in reaction to a major December 2007 oil spill along its west coast.
- Alaska rejects a proposal from ConocoPhillips to build a pipeline to transport North Slope gas to the Lower 48. A similar proposal from TransCanada is still under review.
- Turkey and Syria agree to build a pipeline to move Egyptian gas from Aleppo to the Turkish town of Kilis.

> *"What we've seen so far from shale fields is just the tip of the iceberg."*
>
> *~ Devon Energy Corp. Senior Vice President Terrence L. Ruder*

- A Paris court rules that Total SA must pay 192 million euros in damages for action that led to the sinking of the tanker *Erika* off Brittany in 1999, polluting the coast.

Government

- A U.S. study commission says the government should be ready to spend $200 million/year over ten years for R&D on alternatives to petroleum-based motor fuels.

FEBRUARY

Exploration/Development

- U.S. MMS calls the federal lease sale in the Chukchi Sea offshore Alaska the most successful OCS lease sale in the state's history.
- U.S. and Indian researchers report finding one of the richest marine gas hydrate accumulations ever discovered, in India's Krishna-Godavari basin.
- BG Group boosts its estimate of total hydrocarbons in place on the Tupi discovery (offshore Brazil) to 12–30 billion boe.

Production

- ExxonMobil drills a record-breaking extended-reach well (Z-12) from the onshore Sakhalin-1 project in eastern Russia to a target area 7 miles offshore.

Natural Gas/LNG

- Excelerate Energy LLC begins commissioning its Northeast Gateway deepwater port, 18 mi offshore Boston, for accepting shipments of LNG.
- StatoilHydro delivers its first cargo of LNG from its Hammerfest liquefaction plant in Norway to the U.S. Atlantic seaboard.

Refining/Petrochemicals

- Trinidad and Tobago announces it will build a new 200,000-b/d refinery next to the existing 168,000-b/d facility at Pointe-a-Pierre.
- Abu Dhabi announces plans to build the country's third refinery, a 417,000-b/d unit in Ruwais.

Pipelines/Transportation

- Russia and Ukraine settle their disagreement over the supply, pricing and movement of Central Asian natural gas.

Government

- Venezuela says it will compensate StatoilHydro and Total SA $1.1 billion for its nationalization in May 2007 of an Orinoco heavy-oil project they previously controlled.

Markets

- For the first time, a front-month contract (March) for benchmark U.S. light, sweet crudes closed above $100/bbl on Feb. 19, after setting a new intraday record of $100.10/bbl on the NYMEX.

MARCH

Exploration/Development

- Federal sale of leases in the central and eastern Gulf of Mexico attract more than $3.7 billion in bids—a record since area-wide leasing began in 1983.
- Colombia offers 43 blocks for bid in three gas-prone areas and in one region prospective for heavy oil.

Production

- A strike by Gabon's petroleum workers shuts in production of some 90,000 b/d.
- The North West Shelf gas project decides to go ahead with development of the $5 billion (Aus.) North Rankin B platform, off Western Australia.

> *"The climate change and energy security issues are creating the economic opportunity of our lifetime."*
>
> *~ Deloitte LLP Senior Advisor on Energy and Resources Joseph A. Stanislaw*

Natural Gas/LNG

- Construction begins on Thailand's first LNG terminal, a 5-mtpy facility at Rayong.
- The first shipment to France of LNG from Snøhvit in Norway is delivered to the Montoir-de-Bretagne terminal in Brittany.

Refining/Petrochemicals

- Shortages in China of diesel and gasoline reflect a widening gap between rising international crude prices and centrally controlled fuel markets.
- The new Cienfuegos refinery in Cuba begins producing gasoline.

Markets

- Petrobras (Brazil) and Mitsui & Co. Ltd. (Japan) announce plans to form a JV to target ethanol production for use in Japanese markets.
- On March 3, the price of crude oil reached $103.95/bbl on the NYMEX, topping a record set nearly 30 years ago.
- For seven straight sessions, March 5–13, the front-month crude contract sets record high prices on the NYMEX.

APRIL

Exploration/Development

- USGS reports that the Bakken formation in North Dakota and Montana contains 25 times more oil than a 1989 estimate.
- The UN allows Australia to extend its continental shelf by about 2.5 million sq km (1 million sq mi) after 15 years of lobbying.
- Petrobras gets U.S. MMS approval for a major project in the deepwater Gulf of Mexico that will be the first to use an FPSO vessel in the gulf.

Production

- Iraq discusses with five major IOCs ways to increase production from several oil wells.

Natural Gas/LNG

- New York State rejects the Broadwater floating LNG terminal and regasification facility proposed for construction in Long Island Sound.
- Commissioning cargoes arrive at two new U.S. LNG terminals: Sabine Pass (Cameron Parish, Louisiana) and Quintana (south of Houston).

Pipelines/Transportation

- BP PLC and ConocoPhillips announce plans to build the Denali pipeline to bring natural gas from Alaska's North Slope to markets in Canada and the United States.
- Iran and Pakistan restart talks on construction of the 2,700-km (1,678-mi) Iran-Pakistan-India natural gas pipeline.

Government

- The U.S. Congress begins debating proposals for a cap-and-trade program to control carbon emissions.
- New UK policy takes effect, requiring that all fuels contain 2.5% biofuels.
- U.S. MMS designates five areas offshore New Jersey, Delaware, Georgia, Florida and California for research on wind, wave, and ocean-current energy development.

MAY

Exploration/Development

- Chile signs contracts with four IOCs to explore 8 blocks in the southern Magellanes region.
- Trinidad and Tobago offers five blocks for exploration: two off the eastern coast, three off the northern coast.

Production

- The Iranian oil and gas industry marks its 100th anniversary.
- Eni proposes to Kazakhstan officials that start of production from the giant Kashagan oil field be delayed to 2012–2013, from 2011.
- Gas production begins from Shamrock field in the North Sea. Capacity will average 120 MMscfd.
- Royal Dutch Shell PLC and Repsol YPF SA suspend plans to develop Phase 13 of South Pars gas field in Iran.

Natural Gas/LNG

- Northeast Gateway Deepwater LNG Port, 18 mi (29 km) offshore Boston, begins commercial operation.
- Italy's environment ministry approves plans for construction of an LNG terminal in Calabria.
- A government/industry agreement paves the way for engineering and design of an LNG project in Papua New Guinea.

Refining/Petrochemicals

- Work begins on a $1.2-billion upgrade of Sri Lanka's sole refinery, near Colombo.
- Construction starts on Vietnam's $6 billion Nghi Son refinery and petrochemical plant.

Pipelines/Transportation

- Offshore pipe-laying begins on the subsea Medgaz line from Beni Saf, Algeria, to Almeria, Spain.
- Construction starts on the Keystone pipeline to carry Canadian crude oil from Alberta to Illinois and Oklahoma.
- Pipeline delivery of gas begins from the Energia Costa Azul LNG regas terminal in Baja, California (Mexico) to markets in the U.S. southwest.

Government

- U.S. DOE decides not to add 10 million bbl of oil to the Strategic Petroleum Reserve July-December, reflecting public concern about oil prices.
- U.S. FTC starts process to determine whether a rule is needed to prohibit oil market manipulation, spurred by soaring gasoline prices.
- Indonesia says it will pull out of OPEC at yearend, citing low oil production.

Markets

- Truckers, farmers and others in France, Spain, Portugal, and Italy protest the soaring price of diesel fuel.
- Benchmark U.S. light sweet crudes hit a record intraday high of $135.09/bbl May 22.

JUNE

Exploration/Development

- UK awards 97 onshore licenses for exploration in mature basins in its 13th licensing round.
- Trinidad and Tobago awards licenses for exploration of one northern offshore block and for several southern onshore blocks.
- Norway launches its 20th licensing round, offering 79 blocks for exploration.

> *"Recent studies indicate that by 2025, we could see a gap of 15 to 20 bcfd between natural gas production and demand."*
>
> *–Royal Dutch Shell Executive Director-Gas and Power Linda Cook*

Production

- Oil and gas production begin from Saqqara field offshore Egypt in the Gulf of Suez.
- U.S. rig count for the week ended June 13 tops 1,900 working units for the first time in more than 22 years.
- Oil production begins from Alvheim field off Norway; output is expected to reach 75,000 boe/d by early 2009.
- Brazil's Petrobras signs an agreement with a U.S. firm and Japan's Mitsui & Co. to study development of oil shale projects in Utah.

Natural Gas/LNG

- Petrobras contracts with Britain's BG Group PLC for the first-ever importation of LNG into Brazil.

Pipelines/Transportation

- Petrobras announces plans to build its first ethanol pipeline, to be operational by yearend 2009.

Government

- U.S. Supreme Court overturns $2.5-billion punitive damage ruling against ExxonMobil Corp. for the 1989 *Exxon Valdez* oil spill in Alaska. It limits punitive damages to $507.5 million.

JULY

Exploration/Development

- U.S. SEC proposes changes to oil and gas reserves reporting requirements to aid investor decision-making.
- USGS estimates that the area north of the Arctic Circle contains 90 billion bbl of crude oil, and 1,670 tcf of gas.
- U.S. BLM proposes rules for commercial oil shale development, a resource it says could yield 800 billion bbl.

Production

- Oil and gas production starts from Yuzhno Khylchuyu field in northwestern Siberia, one of the largest in Russia's Timan Pechora oil and gas province.
- U.S. EPA releases proposals related to long-term underground carbon dioxide storage as an element in carbon capture and storage (CCS) strategies.
- Oil and gas production begin from the Neptune field in the deepwater Gulf of Mexico
- Oil production begins from the Bualuang field off Thailand.

Natural Gas/LNG

- Construction begins on Neptune deepwater LNG port, 10 mi (16 km) offshore Boston.
- Dutch government approves construction of the Netherlands' first LNG import terminal in Rotterdam.

Pipelines/Transportation

- Egypt begins natural gas exports to Syria as part of a larger project to export Egyptian gas to several Middle East countries.
- Hungary and Romania agree to connect their natural gas systems to improve supply security.

Markets

- NYMEX contract for light sweet crude for August delivery reaches $147.27/bbl in intraday trading July 11, a high for the year.

AUGUST

Exploration/Development

- U.S. Coast Guard begins a project to map the Arctic Ocean floor at the northern edge of the Beaufort Sea.
- Queensland, Australia, bans oil shale development for two years based on concerns about environmental impact.
- Canadian Superior reports a major natural gas discovery off eastern Trinidad and Tobago.
- MMS Lease Sale 207 in the Gulf of Mexico attracts $487 million in apparent high bids.

Production

- Shell Exploration & Production Co. installs the world's deepest production spar in 8,000 ft (2,438 m) of water for the Perdido development in the Gulf of Mexico.
- Oil production begins from the presalt layer of the big Jubarte field in the Campos Basin offshore Brazil.
- On August 14, the one hundred millionth barrel of crude oil is produced from the White Rose oil field offshore northeast Canada.

- Agreements are reached to develop the three fields in the Hebron complex off northeastern Canada.
- Pemex announces that oil output from Mexico's offshore Cantarell oil field dropped by 36% over the past year.
- Natural gas and condensate production starts from Chergui field in Tunisia.
- Oil production starts from offshore Agbami oil field, the largest deepwater discovery in Nigeria.

Natural Gas/LNG

- Fifth LNG train at the North West Shelf gas project on Australia's Burrup Peninsula comes on stream, boosting capacity by 4.4 Mtpy (to 16.3 Mtpy).
- Subsidiary of Japan's Mitsubishi Corp. abandons plans to build an LNG regasification terminal at Long Beach, California

Refining/Petrochemicals

- NPRA reports that U.S. refiners raised capacity 4.1% since 2003 despite soaring oil prices and shrinking margins.

Pipelines/Transportation

- Contract is awarded to build a pipeline to connect the Rubiales heavy-oil field in east-central Colombia to the OCENSA pipeline.
- Alaska approves a state license for Trans-Canada Corp. to pursue federal certification for a natural gas pipeline from the Alaska North Slope to the Alberta Hub.
- FERC approves construction of the Midcontinent Express gas pipeline.
- Bolivia's YPFB and Transredes start construction of the Gasoducto Carrasco-Cochabamba natural gas pipeline.
- Iran threatens to fire missiles at shipping in the Straits of Hormuz in a standoff with the UN Security Council over the Iranian nuclear program.

Government

- Fighting erupts between Georgian military units and Russian peacekeeping forces based in the Georgian provinces of South Ossetia and Abkhazia. Attacks on three major pipelines and a rail line through Georgia are reported.
- Nigeria cedes control over the potentially oil-rich Bakassi Peninsula to Cameroon.
- Beijing hosts 10,500 athletes at the Summer Olympic Games—the first time that any Olympic Games are hosted in China.

SEPTEMBER

Exploration/Development

- U.S. Congressional moratoria on oil and gas leasing on the OCS expire September 30 in a budget-resolution compromise by Democrats with the White House. Pres. George Bush in July lifted the presidential moratorium on such leasing.
- USGS announces a joint U.S.-Canadian program to map the unexplored Arctic seafloor.
- Three European institutes launch a consortium to pursue a program to map possible gas shale sites in Europe.

Production

- Hurricanes Gustav and Ike sweep through the Gulf of Mexico, prompting evacuations and damaging several rigs and platforms.
- OPEC members agree September 10 to action that implies a possible production reduction of 530,000 b/d.
- Saudi Aramco starts production from Khursaniyah oil field after a one-year delay.
- OMV AG starts gas production from Strasshof and Ebenthal fields in the Vienna basin.
- Oil production starts from 6 subsea wells in the Vincent field off Western Australia.

Natural Gas/LNG

- FERC approves the first new LNG terminal in the Pacific Northwest. The Bradwood Landing project is expected to deliver 1.3 bcfd in 2012.
- Iraq approves a Royal Dutch Shell plan to gather 500 MMcfd of associated natural gas in the Basra region.
- Origin Energy and ConocoPhillips announce plans to extract coalbed methane in Queensland, Australia, and convert it to LNG at a new facility in Gladstone.

Refining/Petrochemicals

- Iran signs an agreement to help upgrade Ecuador's refineries and jointly build a petrochemical production facility in Ecuador.

> *"The time is now to tackle our energy issues together. If we fail to do so now, America is at the very real risk of losing the global race for energy."*
>
> *~ Marathon Oil Corp. Chief Executive Officer Clarence Cazalot*

Markets

- On September 22, the October contract for benchmark light sweet crudes rises by $16.37/bbl, the largest one-day jump ever on the NYMEX.
- On September 29, crude plunges $10.52/bbl—the largest one-day drop since January 1991—as the U.S. House of Representatives rejects a proposed $700 billion financial markets rescue plan.

OCTOBER

Exploration/Development

- A subsidiary of CNPC announces a venture with Uzbekneftegaz to develop Mingbulak oil field in northeastern Uzbekistan.
- USGS reports that 85.4 tcf of undiscovered, technically recoverable natural gas resources are locked in gas hydrates on the Alaskan North Slope.
- Anadarko reports presalt Wahoo discovery off Brazil, southeast of Jubarte field.

Production

- Production begins from Galoc 3 well offshore northwest Palawan, Philippines.
- Gas production starts from Peru's Pagoreni field, part of the Camisea gas-condensate project.
- Petrobras inaugurates its first domestically produced semisub platform, the P-51. It is likely to be employed in the Campos basin.
- Ta'Kuntah FSO vessel in Cantarell field offshore Mexico logs its one billionth barrel of oil offloaded after 10 years of continuous operation.
- Gas production begins from Khor Mor field in Iraq's northern Kurdistan area.
- Angel gas field comes onstream at the North West Shelf project, off Western Australia.

Refining/Petrochemicals

- PDVSA breaks ground on 100,000-b/d Santa Ines refinery in Barinas state, Venezuela.

Markets

- Crude futures sales post the biggest monthly loss for front-month NYMEX crude prices (down 32.6% or $32.83) since that commodity began trading in 1983.
- U.S. consumer prices plunge by 1%, the largest amount since 1947, as gasoline pump prices drop by a record amount.
- November contract for benchmark light sweet crudes hits new low for the year, closing at $77.70 on the NYMEX.

NOVEMBER

Exploration/Development

- USGS estimates that 2.4 tcf of natural gas and 9.8 million bbl of NGLs lie beneath eastern Oregon and Washington.
- U.S. MMS takes initial steps toward sale of acreage for exploration offshore Virginia in 2011.
- Bahrain launches bid round for exploration of its onshore gas fields.

Production

- OPEC's announced cut of 1.5 million b/d in its oil output targets takes effect November 1. The action reflects concern that a global economic recession will drive oil prices down and dampen E&P efforts.
- Several firms announce cuts in planned spending for development of oil sands in Alberta.
- Oil production starts from Chevron's Blind Faith field in the deepwater Gulf of Mexico.
- U.S. BLM publishes final rules for development of oil shale on public lands in Colorado, Utah, and Wyoming.
- Oil production starts from Area 12 on North Duri field, Sumatra, one of the world's largest steamflood projects.
- Shell Oil Co. sets a world water-depth record in drilling and completing a subsea well 9,356 ft (2,852 m) below the water's surface in the Silvertip field in the Gulf of Mexico.

Natural Gas/LNG

- Centrica PLC begins commissioning the second expansion phase of the Isle of Grain LNG terminal near Kent, England.
- Shanghai Gas Group receives its first spot cargo of LNG from Malaysia's Petronas.
- Shell Eastern LNG agrees to sell up to 2 mtpy of LNG to PetroChina Ltd. from Australia's Gorgon project and other sources.

> *"An oil crisis is coming, and sooner than most people think."*
>
> *~ Hess Corp. Chairman and Chief Executive John B. Hess*

Pipelines/Transportation

- Oil flow resumes through the Baku-Supsa and the Baku-Tbilisi-Ceyhan pipelines after hostilities end between Russia and Georgia.
- Russia and China agree to build a spur from the East Siberian Pacific Ocean oil pipeline, running from Skovorodino to the Chinese border.
- Somali pirates hijack the Saudi-owned tanker *Sirius Star* November 15, carrying about 2 million bbl of oil, more than 450 mi (725 km) southeast of Mombasa, Kenya.
- Kurdish rebels attack a section of the Kirkuk-Ceyhan oil pipeline in southeast Turkey, shutting down the 400,000-b/d flow of Iraqi crude for about ten days.

Government

- Barack Obama is elected 44th president of the United States; the Democratic Party gains additional seats in both the Senate and the House of Representatives.
- UK conducts first auction of carbon permits. It will sell 4 million, each equivalent to a tonne of carbon dioxide, under the EU's greenhouse gas emissions trading scheme.
- Negotiations resume in Nice, France, on EU-Russia pacts on economic, energy, and security issues.
- Pres. Dmitry Medvedev makes the first visit ever by a Russian head of government to Venezuela.

Markets

- Benchmark crude falls as low as $48.50/bbl on November 20 on the NYMEX, a level last seen on May 18, 2005.
- U.S. MMS reports record disbursals of $23.4 billion to state, Native American, and federal accounts from onshore and offshore energy production in FY 2008.

DECEMBER

Exploration/Development

- China's CNPC signs agreement with Cuba on oil and gas development in the Gulf of Mexico.
- Iraq offers 10 gas and oil fields in its second postwar licensing round, including two southern fields with reserves of 12 billion bbl of crude each.
- China says it found a major gas field in Xinjiang Uygur Autonomous Region.

> *"We are going through what I call the 'great bubbling,' a high degree of innovation all across the energy spectrum."*
>
> *~ Cambridge Energy Research Associates Chairman Daniel Yergin*

Production

- Full oil production starts from ultradeep Thunder Horse field in the Gulf of Mexico.
- Oil production also starts from Pegasus field, also in the Gulf of Mexico.
- Year-round oil export begins from a new terminal in Aniva Bay in the south of Sakhalin Island.

Refining/Petrochemicals

- Venezuela and Cuba sign agreements to expand two refineries in Cuba and build a third.
- Kuwait cancels a $9-billion JV with Dow Chemical Co. The action threatens Dow's plan to acquire Rohm & Haas Co.

Government

- Iraqi officials approve a security pact under which U.S. forces will withdraw from the country by January 1, 2012. It will replace a UN mandate that expires December 31, 2008.
- EU nations agree on an ambitious plan to reduce greenhouse-gas emissions and spur renewables use, aimed at cutting carbon emissions by 20% by 2020.
- Attendees from 190 nations meet in Poznan, Poland, for the UN Framework Convention on Climate Change. They aim to devise a treaty to replace the Kyoto Protocol, which expires in 2012.
- The EU applies for permanent observer status on the Arctic Council, which monitors resource development in that region.
- Hostilities erupt again between Israel and Gaza militants as a mid-December ceasefire ends. Oil prices exceed $40/bbl as the fighting continues for several days.

Markets

- The WTI crude price sinks to $32.40 (a near five-year low) in intraday trading December 19 on the NYMEX before settling at $33.87.

World Energy Market: The Big Picture

The Turbulence of 2008

"Oil market heading for surprises in '08."

That was the headline on a major look-ahead story in the January 7, 2008, issue of *Oil & Gas Journal*.

OGJ's senior editor for economics was pretty close to the mark with the following projections:

- Commodity price volatility will continue, but it will moderate, and prices will move lower.
- The U.S. economy will grow at a much slower pace than during 2007.
- There is still some risk of entering a recession this year.

The environment in which the global oil and gas industry operated was shaped by two major, related developments: extreme volatility in oil prices, and a global financial crisis of a magnitude not seen since the 1930s.

CRUDE OIL FUTURES CONTRACT PRICES

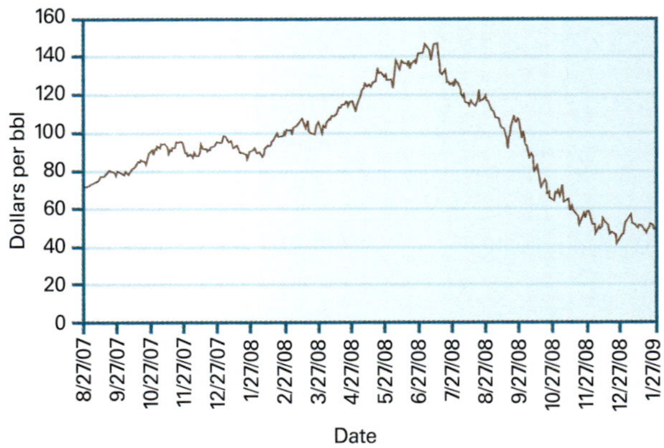

Oil price volatility

On January 2, 2008, a single trade for light, sweet U.S. crude was made at \$100/bbl (though the price did not stay above \$100 until late February).

The price rose to \$110 on March 12; \$125 on May 9; \$135 on May 22; and to \$145 on July 3.

On July 11, 2008, oil reached a peak for the year at \$147.27/bbl.

Cambridge Energy Research Associates (CERA; Cambridge, MA) noted in a November 2008 analysis that the tight balance between supply and demand was not the only factor pushing oil prices up. In addition, beginning in summer 2007, a weakening U.S. dollar sparked a "flight to commodities" (including oil).

As expected, rising oil prices drove oil demand down. For example, U.S. motorists drove 112 billion fewer miles from November 1, 2007 to November 30, 2008 compared with the year-prior period, according to the U.S. Department of Transportation.

The price shock also affected the price of food, the viability of alternative and renewable energy technologies, and the fortunes of companies involved in transporting people and cargo.

Trouble in the U.S. housing and financial markets at mid-year also began to slow overall economic activity, also reducing energy demand.

Oil prices settled around $125 on July 24, and by September 15, fell back below $100.

On October 11, oil fell almost $9/bbl to $77.70 as global equities tumbled, then to below $70 on October 16.

On December 21, oil traded at $33.87/bbl—less than one-fourth of the peak price reached July 11.

The front-month contract rebounded to settle at $44.60 on December 31 on the NYMEX.

In summary, it took more than four years for the price of oil to go from $35/bbl to $147.27 and less than six months to fall all the way back down.

Global economic downturn

In the third quarter of 2008, the global economic system began to weaken significantly, not just from high oil prices but also from serious problems that had started in 2007 in the U.S. housing and financial markets.

The Guest Essay at the front of this book by energy economist Sam Van Vactor, Ph.D., describes the four shocks that led to the global financial crisis.

In reaction to those shocks, governments around the world at yearend 2008 had proposed drastic plans to rescue financial institutions and some other businesses (e.g., automakers) judged too big or too important to fail.

Hundreds of billions in government funds—$700 billion by the United States—were expended or pledged in efforts to forestall bankruptcies, maintain liquidity, fund business operations, and spur the extension of credit.

At yearend, many governments (including the United States, the UK, France, Germany and China) also were also considering large economic stimulus programs to revive economic growth.

Some analysts called the economic crisis the worst since the 1930s. Several major banks told *Oil & Gas Journal* in December 2008 that the downturn could last into 2011.

The *Financial Times* (London) presented this yearend summary of 2008 economic indicators:

- S&P 500: down 39.3% (worst drop since 1931)
- Dow Jones Industrials Average: down 34.7% (worst drop since 1931)
- Japan Nikkei 225 Index: down 42.1% (worst drop since 1990)
- Korea Kospi Index: down 40.7%
- London FTSE 100: down 32.4% (worst drop since 2002)
- London FTSE Eurofirst 300: down 45.7% (worst since inception in 1986)
- MSCI World Index (emerging market equities): down 55%
- U.S. Consumer Confidence Index: 38 in December 2008 (record low)
- Case-Shiller Home Price Index (large U.S. cities): down 18% in October, a record drop.

Global View: 2008

EIA: Short-term outlook

In a monthly short-term energy outlook released July 8, the U.S. Energy Information Administration (EIA) said that worldwide, oil markets remained tight. Consumption continued to grow despite seven years of rising prices, and production was not rising as quickly as expected.

Preliminary data showed global oil consumption up in the first half of 2008 by roughly 520,000 b/d compared with year-earlier levels. The increase reflects a 170,000-b/d year-to-year gain in the first three months, followed by an 870,000 b/d increase in the second quarter.

A first-half drop of 760,000 b/d in consumption in OECD nations was more than offset by a 1.3-million b/d increase in non-OECD nations, led by China and the Middle East.

EIA expected global oil demand to climb by almost 1.2 million b/d during the second half of 2008, in response to higher prices, less economic growth, and growing pressure on China, India, and other countries to ease price subsidies.

Oil supplies were not growing as rapidly as first anticipated, especially outside OPEC, due chiefly to less production from Russia and the North Sea and lower expectations for Brazil.

Worldwide oil supply and demand

	2007 1st Qtr.	2007 2nd Qtr.	2007 3rd Qtr.	2007 4th Qtr.	2007 Year	2008 1st Qtr.	2008 2nd Qtr.	2008 3rd Qtr.	2008 4th Qtr.	2008 Year
	Million b/d									
DEMAND										
OECD										
North America	25.7	25.4	25.5	25.5	25.5	24.8	25.0	25.2	25.2	25.0
Europe	15.2	14.9	15.4	15.6	15.3	15.1	15.1	15.4	15.5	15.3
Asia/Pacific	8.8	7.8	7.8	8.6	8.3	8.8	7.9	7.9	8.7	8.3
Total OECD	**49.7**	**48.2**	**48.7**	**49.8**	**49.1**	**48.7**	**48.0**	**48.4**	**49.4**	**48.6**
Non-OECD										
FSU	4.1	3.9	4.2	4.3	4.1	4.1	4.0	4.3	4.4	4.2
Europe	0.8	0.7	0.7	0.7	0.7	0.8	0.8	0.7	0.8	0.8
China	7.3	7.7	7.5	7.6	7.5	7.9	8.1	7.9	8.0	8.0
Other Asia	9.3	9.4	9.1	9.4	9.3	9.6	9.5	9.1	9.4	9.4
Latin America	5.4	5.6	5.7	5.7	5.6	5.7	5.8	6.0	6.0	5.9
Middle East	6.4	6.5	6.7	6.4	6.5	6.7	6.8	7.1	6.8	6.8
Africa	3.1	3.1	3.0	3.1	3.1	3.1	3.1	3.0	3.2	3.1
Total Non-OECD	**36.4**	**36.9**	**36.9**	**37.3**	**36.9**	**37.9**	**38.0**	**38.1**	**38.6**	**38.1**
TOTAL DEMAND	**86.1**	**85.1**	**85.6**	**87.1**	**86.0**	**86.6**	**86.1**	**86.5**	**87.9**	**86.8**
SUPPLY										
OECD										
North America	14.4	14.4	14.2	14.1	14.3	14.2	13.9	14.1	14.4	14.1
Europe	5.2	4.9	4.7	5.0	5.0	4.9	4.5	4.3	4.6	4.6
Asia	0.6	0.6	0.6	0.6	0.6	0.6	0.7	0.8	0.8	0.7
Total OECD	**20.2**	**19.9**	**19.5**	**19.8**	**19.8**	**19.7**	**19.1**	**19.1**	**19.9**	**19.4**
Non-OECD										
FSU	12.8	12.7	12.8	12.8	12.8	12.8	12.9	13.1	13.5	13.1
Europe	0.1	0.1	0.1	0.1	0.1	0.1	0.1	0.1	0.1	0.1
China	3.7	3.8	3.7	3.7	3.7	3.8	3.8	3.9	3.9	3.8
Other Asia	2.7	2.7	2.6	2.7	2.7	2.7	2.7	2.7	2.8	2.7
Latin America	4.4	4.4	4.3	4.2	4.3	3.9	4.0	4.1	4.2	4.1
Middle East	1.7	1.7	1.7	1.6	1.7	1.6	1.6	1.6	1.6	1.6
Africa	2.5	2.5	2.5	2.5	2.5	2.5	2.6	2.6	2.6	2.6
Total Non-OECD	**27.9**	**27.8**	**27.7**	**27.7**	**27.8**	**27.5**	**27.6**	**28.1**	**28.6**	**28.0**
Processing gain	2.0	2.1	2.1	2.1	2.1	2.1	2.1	2.1	2.2	2.1
Other biofuels	0.3	0.3	0.3	0.4	0.3	0.4	0.5	0.5	0.6	0.5
Total Non-OPEC	**50.4**	**50.2**	**49.7**	**49.9**	**50.0**	**49.7**	**49.3**	**49.9**	**51.2**	**50.0**
OPEC[1]										
Crude	30.3	30.2	30.6	31.5	30.7	32.3	32.2	32.4	32.4	32.3
NGL	4.8	4.8	4.8	4.9	4.8	4.9	5.0	5.2	5.4	5.1
Total OPEC	**35.1**	**34.9**	**35.4**	**36.4**	**35.5**	**37.3**	**37.2**	**37.6**	**37.8**	**37.4**
Total SUPPLY	**85.5**	**85.1**	**85.1**	**86.4**	**85.5**	**87.0**	**86.5**	**87.5**	**89.0**	**87.4**
Stock change	(0.6)	—	(0.5)	(0.8)	(0.4)	0.4	0.4	1.0	1.1	0.6

[1]Includes Ecuador beginning Dec. 2007. Totals may not add due to rounding.
Source: International Energy Agency; OGJ estimates for OPEC 2nd, 3rd, and 4th quarter 2008 crude supply

OGJ view

Oil & Gas Journal offered a mid-2008 assessment of worldwide energy demand. (Its detailed view of the mid-year U.S. energy outlook is described in the United States section of the North America chapter.)

While U.S. energy demand in 2008 was expected to be limited by weak economic growth and improved fuel-use efficiency, China, Latin America, and the Middle East were expected to sustain worldwide oil demand growth despite high global prices and keep pressure on supply.

OGJ judged that the extent to which producers would be able to respond to the need for more oil in the short run would be inadequate to relieve high prices (though a sharp global economic downturn later in the year did, in fact, drive oil prices down sharply).

World crude prices[1]

Country	Type of crude and API gravity°	June 1, 2008, $/bbl	% change 6–08/1–08	In effect Jan. 1, 2008, $/bbl	% change 08/07	In effect June 1, 2007, $/bbl	In effect June 1, 2006, $/bbl	In effect June 1, 2005, $/bbl	In effect June 1, 2004, $/bbl	In effect June 1, 2003, $/bbl
OPEC										
Saudi Arabia	Arabian Light 34	120.79	29.9	93.02	89.7	63.68	45.96	33.77	23.52	21.57
Abu Dhabi	Murban 39	126.63	33.5	94.85	81.5	69.77	52.67	37.53	26.69	24.98
Algeria	Saharan 44	126.32	28.5	98.28	80.3	70.07	50.07	37.73	27.17	22.92
Nigeria	Bonny Light 37	129.82	31.8	98.52	81.6	71.48	51.61	37.87	27.44	23.32
Libya	Es Sider 37	121.80	25.8	96.79	80.3	67.55	49.88	37.34	26.57	22.67
Indonesia	Minas 34	127.97	30.1	98.34	83.2	69.85	50.81	38.06	28.24	24.51
Venezuela	Tia Juana 31	122.26	30.3	93.85	88.9	64.71	47.69	36.92	27.81	23.18
Total OPEC		**121.52**	**29.9**	**93.56**	**86.6**	**65.11**	**47.29**	**35.09**	**25.04**	**22.47**
OTHER										
UK	Brent Blend 38	125.47	27.5	98.42	80.7	69.43	51.21	37.78	27.58	23.26
Norway	Ekofisk 42	127.92	30.2	98.23	85.3	69.02	50.23	37.63	27.78	23.20
Mexico	Isthmus 33	122.15	30.3	93.74	89.1	64.60	47.58	36.81	27.70	23.07
Russia	Urals 32	119.69	27.4	93.98	84.9	64.74	47.61	35.78	26.16	21.99
Total World		**121.36**	**30.6**	**92.93**	**87.7**	**64.67**	**46.65**	**35.29**	**25.72**	**22.31**
US[2]		117.82	33.3	88.41	84.6	63.84	44.79	34.83	25.50	21.66

[1]Represents estimated contract prices based on government stated prices, netback deals and spot market quotations.
[2]Average prices (fob) weighted by estimated import volume.

Source: US Energy Information Administration

First quarter worldwide oil production

Country	First quarter 2008	First quarter 2007	Change	Change, %
	1,000 b/d			
OPEC	**32,314**	**30,225**	**2,089**	**6.9**
Non-OPEC	**41,254**	**42,222**	**–968**	**–2.3**
Argentina	631	631	—	—
Brazil	1,766	1,754	12	0.7
Canada	2,567	2,615	–48	–1.8
Colombia	560	519	41	7.9
Ecuador	500	499	1	0.2
Mexico	2,911	3,158	–247	–7.8
United States	5,101	5,174	–73	–1.4
Other	2,744	2,798	–54	–1.9
Western Hemisphere	16,780	17,148	–368	–2.1
Norway	2,205	2,426	–221	–9.1
United Kingdom	1,484	1,597	–113	–7.1
Other	583	616	–33	–5.4
Western Europe	4,272	4,639	–367	–7.9
FSU	12,481	12,203	278	2.3
Other	174	180	–6	–3.3
Eastern Europe & FSU	12,655	12,383	272	2.2
Egypt	630	660	–30	–4.5
Gabon	227	230	–3	–1.3
Other	8,788	8,536	252	3.0
Africa	9,418	9,196	222	2.4
Oman	727	723	4	0.6
Syria	390	393	–3	–0.8
Other	21,939	20,598	1,341	6.5
Middle East	23,056	21,714	1,342	6.2
Australia	409	440	–31	–7.0
China	3,771	3,755	16	0.4
India	678	695	–17	–2.4
Malaysia	777	753	24	3.2
Other	1,751	1,724	27	1.6
Asia-Pacific	7,386	7,367	19	0.3
Total world	**73,568**	**72,447**	**1,121**	**1.5**

Source: Oil & Gas Journal

Global View: 2009 and Beyond

General perspective

OGJ view

In the face of the global economic slowdown and expected continuing volatility of oil and gas prices, companies in 2009 are likely to slash capital spending, drill fewer wells, and delay or cancel investment decisions, said *OGJ* in an early-2009 assessment.

Reduced options for obtaining credit or capital and the extreme volatility of oil and gas prices hamper project development as well as ongoing operations.

Several companies—e.g., StatoilHydro (Stavanger, Norway), and Royal Dutch Shell PLC (The Hague, Netherlands)—postponed Canadian heavy oil projects in late 2008 because of price volatility.

Underinvestment, noted *OGJ*, could limit future supply and produce another violent price cycle.

EIA view

World oil consumption was projected to fall by 800,000 b/d in 2009, according to the *Short-Term Energy Outlook* released January 13, 2009, by EIA.

The agency said it expects a "fairly loose oil market balance over the next two years." The global economic downturn will likely reduce oil consumption in 2009, while additional production capacity from both OPEC and non-OPEC nations should boost surplus production capacity.

Total world oil consumption is expected to rebound modestly in 2010, rising by 880,000 b/d from 2008, assuming the start of economic recovery.

EIA projects that WTI crude will average about $43/bbl in 2009 and $55/bbl in 2010.

World oil consumption

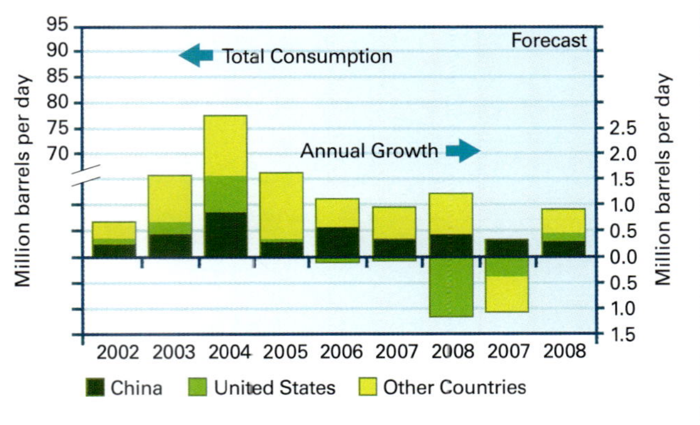

Oil supply, demand and prices

Oil prices up in early 2009

At the start of 2009, crude prices rose in the face of increased fighting between Israel and Hamas militants in Gaza and on concerns over supplies from Nigeria after militants there blew up part of a pipeline operated by a unit of Eni (Rome).

The February contract for WTI on the NYMEX settled at $48.81/bbl. ICE February Brent reached $49.62/bbl.

Various views

Because of the volatility of 2008 oil prices and the onset of the global economic crisis, no attempt is made in this book to document projections for future oil price, supply and demand that were made in the first three quarters of that year.

The tables below show selected data from analyses published in December 2008 and early January 2009 by various agencies.

		Expected Oil Prices ($/bbl, nominal)			
Agency	2009 Global Economic Growth (%)	2009	2010	2011	2012 and Beyond
IEA	2.1				2008–15: 100 2030: 200
EIA	0.5	WTI: 43.25	WTI: 54.50		
OPEC	1.5				
Deutsche Bank		47.50	55	80	
World Bank				75	
BMI		OB: 52 Brent: 55.65 WTI: 56.63 Urals: 52.48	OB: 58	OB: 65 Brent: 68.70 WTI: 69.60 Urals: 65.50	OB: 70

	Change in Oil Demand (% or b/d)					
	Global		Regional: 2009 vs. 2008			Global: Longer-term
Agency	2008 vs. 2007	2009 vs. 2008	OECD	Non-OECD	North America	2010 and beyond
CERA	-300,000	-660,000				
IEA		+0.5% (+440,000)	-1.4%	+2.9%	-330,000	2006–30: +1.6% per yr 2008–13: +0.97% per yr 2010: +1.2% vs. 2009 2011: +1.3% vs. 2010
EIA	+0.01% (+10,000)	-0.94% (-810,000)	-2.7% (-1,300,000)	+1.2% (+490,000)	-270,000	2010: +1.0% vs. 2009 (+880,000)
OPEC		-150,000	-980,000	+2.2%	-580,000	
WoodMac		-250,000				
BMI		-0.5% (-417,000)	-1.3%	+2.3%	-520,000	2007–13: +1.17% per yr 2013–18: +1.42% per yr 2010: +0.58% vs. 2009 2011: +1.42% vs. 2010

Saudi view: $75 oil needed

At a London conference of oil consumers and producers December 19, 2008, the Saudi oil minister warned that the world needed $75 oil to sustain investment in both conventional production and research into alternatives.

	Change in Oil Supply (% or b/d)			
	Global	Regional: 2009 vs. 2008		Longer-term
Agency	2009 vs. 2008	OPEC	Non-OPEC	2012 and beyond
IEA			Almost +1% (+480,000)	2008–13: 0.52% per yr (non-OPEC)
EIA	-0.62%	-1.99% (-713,000)	+0.37% (+183,000)	2010: +1.96% vs. 2009 (global)
OPEC			+640,000	
BMI	+0.4%	+0.1% (+265,000)	+0.5%	2007–13: +1.52% per yr 2010–13: +2.0% per yr 2013–18: +1.37% per yr (all global)

Abbreviations and shorthand used in all three tables:

- CERA: Cambridge Energy Research Associates, an IHS Company (Cambridge, MA)
- IEA: International Energy Association (Paris)
- EIA: Energy Information Administration (U.S. Dept. of Energy)
- OPEC: Organization of Oil Exporting Countries
- WoodMac: Wood Mackenzie Ltd. (Edinburgh)
- World Bank: The World Bank (Washington)
- BMI: Business Monitor International (London)
- OECD: Organization for Economic Cooperation and Development (Paris)
- OB: OPEC Basket of crude oils from 12 OPEC member countries
- Brent: Brent crude oil blend (North Sea origin)
- WTI: West Texas Intermediate crude oil (also called Texas Light Sweet crude)
- Urals: Urals crude oil (Russian origin)

EIA: Flat U.S. oil demand

A long-range analysis in mid-December 2008 by EIA projected virtually no growth in U.S. petroleum use through 2030.

For the first time in more than 20 years, said EIA, U.S. petroleum demand is projected to be essentially flat for years to come, halting steady annual petroleum demand increases dating back to the 1980s.

As a result, net U.S. petroleum imports are expected to decline sharply, with liquid fuel imports—primarily oil—accounting for only 40% of U.S. consumption by 2025, compared with 58% in 2007.

The analysis envisions a rebound in oil prices to $130/bbl in 2007 dollars by 2030 ($189/bbl real).

EIA: Global oil demand down

In its *Short Term Energy Outlook* of in mid-January 2009, EIA projected that global oil demand would be essentially flat in 2008 and then drop by about 800,000 b/d in 2009 (1.03%), to 85.102 million b/d.

The agency pointed to entrenched market perceptions that a prolonged global economic downturn was likely, which would put downward pressure on oil prices.

Such a drop in 2009 would mark the first time since 1983 that global oil demand dropped from the previous year.

Though oil demand fell in developed countries in 2008, it was masked by steady demand growth in emerging economies.

EIA: International Energy Outlook

Total world energy consumption could grow by 50% from 2005 to 2030, with most of the increase coming in developing countries, said EIA in its *International Energy Outlook* (*IEO*) of June 2008.

Demand could grow by 85% in non-OECD countries and by 19% within the OECD, said the *IEO* reference case.

Petroleum and other energy liquids are expected to provide only 33% of total consumption in 2030, down from 37% in 2005.

Energy price assumptions for the study were set in late 2007 and do not reflect carbon constraints.

The *IEO* projected that global natural gas use could reach 158 tcf in 2030, up from 104 tcf in 2005, displacing oil in some industrial and electricity markets.

Non-OECD nations will likely provide a major part of new gas production, much of it as LNG exports from the Middle East and Africa.

The agency said it expected no meaningful carbon capture and storage contributions to be available before 2030.

IEA: View to 2030

The annual *World Energy Outlook* released in November 2008 by the International Energy Agency (IEA; Paris) projected that oil prices would rise to $200/bbl by 2030. In 2007, the agency had projected a price of only $108/bbl by 2030.

The era of cheap oil is over, said IEA—likely averaging (in real terms) more than $100/bbl from 2008 to 2015—but prices are likely to be very volatile.

In view of these high oil prices, concerns about climate change, and the challenge of new production, said IEA, the "current global trends in energy supply and consumption are patently unsustainable – environmentally, economically, socially."

Other points from the report:

- Subsidies paid to fossil fuel energy generators should be used instead to support low-carbon energy resource development, including renewable sources, and carbon capture and storage.
- Governments should put a higher price on carbon dioxide emissions and encourage the growth of wind and solar energy.
- The world must invest $26.3 trillion by 2030 to ensure adequate energy supplies.
- Global oil demand in 2030 will be 106 million b/d—down more than 10 million b/d from a previous IEA projection. Demand in 2007 was about 85.22 million b/d.
- All of the growth in oil demand until 2030 comes from emerging countries.

IEA pointed out that even if oil demand remained flat to 2030, about 45 million b/d of gross capacity—roughly four times the current capacity of Saudi Arabia—would need to be developed by 2030 just to offset oil-field production declines.

World Bank: Commodities boom over

In December 2008, the *Global Economic Prospects* report of The World Bank said the commodities boom of the past five years—which drove prices up by 130%—had "come to an end."

Some other organizations see the ongoing price drops as a correction within an upward trend.

Commodity prices will return to neither the high levels of mid-2008, the bank said, nor to the depressed levels of the 1990s.

An oil price of about $75/bbl is expected by 2011 as slower population and income growth reduce world demand for raw materials.

The bank's chief economist said the current downturn was likely to see simultaneous recessions in most of the industrialized world, and that they were likely to last longer than those of the early 1980s.

ExxonMobil: *"Outlook for Energy: A View to 2030"*

ExxonMobil Corp. (Irving, TX) said in early 2009 that it expects global energy demand to increase by an average 1.2%/year during 2005–30, even assuming major energy efficiency gains.

The company sees demand increasing to 310 million boe/d in 2030, driven by growing populations and expanding economies. In 2005, total energy demand was 229 million boe/d.

The firm's annual *"Outlook for Energy: A View to 2030"* included an examination of improved energy efficiency, development of all viable forms of energy, climate risk technology, and public policy.

The outlook analyzed about 100 countries.

Though fossil fuels are expected to continue to provide about 80% of the world's energy needs through 2030, low-carbon fuels will receive growing emphasis and use of renewable resources will rise rapidly.

Natural gas supply, demand and price

IEA: Challenges for gas industry

A September 2008 report by IEA on the medium-term outlook for the global natural gas industry concluded that investment uncertainties, cost increases and delays continue to be a major problem in most gas markets and pose a threat to long-term supply security.

As with oil, the price of gas has been falling as demand forecasts have been reduced.

For both fuels, a future rebound in demand—coupled with limited supplies due to the delay or cancellation of development projects—could result in a tight supply/demand balance and higher prices.

GECF chartered

The Gas Exporters Countries Forum (GECF), a group of the world's largest natural gas suppliers, met in Moscow in December 2008, establishing a charter and choosing Doha as home for its permanent secretariat.

The actions formalize the group, which was loosely established in 2001, and strengthen its determination to shape the global gas market. Its members account for roughly two-thirds of world gas reserves.

GECF said that it has no plans to form a cartel similar to OPEC.

OPEC Actions

Production quotas

OPEC members acted several times in late 2008 in response to falling oil prices.

- They agreed to "strictly comply" with the official 28.8-million b/d production quota adopted in September 2007, which would have eliminated more than 500,000 b/d of overproduction.
- On October 24 (in Vienna), the cartel agreed to a cut of 1.5 million b/d starting November 1 (reducing global supply by about 2%).
- In early December, it appeared that some members were not complying fully with their lower quotas.
- On December 17 (in Algeria), OPEC pledged even deeper cuts that would reduce output by 4.2 million b/d compared to actual September levels of the 11 members that have quotas. (New target: 24.8 million b/d).

If achieved and sustained into the first half of 2009, this production cut could reverse a contra-seasonal build-up of crude stocks that began in October 2008.

Analysts say that Venezuela and Iran need crude at about $90/bbl to generate revenue for domestic spending needs.

Other OPEC members, such as Nigeria and Ecuador, also face budget problems that make them reluctant to undertake production cuts.

The International Monetary Fund estimated that Saudi Arabia needed a crude price of about $50/bbl for 2008 fiscal accounts to break even.

OPEC's desire to boost oil prices stood in contrast to its views of a year earlier. In autumn 2007, when the OPEC reference basket price reached $80/bbl, the cartel blamed much of the increase on speculators.

OPEC: World Oil Outlook

In July 2008, OPEC released its *World Oil Outlook* to 2030, citing as oil-industry challenges uncertainty of demand, rising costs, and environmental regulations.

The study said that the low oil prices of the 1980s and 1990s discouraged investment, leaving the world unprepared for a surge in demand at the turn of this century to support higher than expected global economic growth.

In its reference case, the OPEC study assumed a 50% rise in global energy from 2006 to 2030. This is based on world economic growth averaging 3.5%/yr and assuming no significant changes in policies and technologies.

Other key points from the study:

- Fossil fuels will continue to provide more than 85% of the world's energy needs.
- There is plenty of oil available to meet foreseeable needs, but deliverability will be a challenge.
- From 2007–30, in 2007 dollars, total upstream investments will be $2.8 trillion.
- Policies such as the U.S. Energy Security and Independence Act of 2007 and proposed EU climate change and renewable energy targets could reduce the call on OPEC oil substantially.
- To meet product demand in 2030, OPEC said a total of almost 20 million b/d of additional distillation capacity will likely be required.
- Total investment in refinery processing to 2030 will be close to $800 billion in 2007 dollars.

OPEC OIL PRODUCTION

Country	First quarter 2008	First quarter 2007	Change	Change, %	OPEC quota, 1,000 b/d
	1,000 b/d				
Angola[1]	1,891	1,618	—	—	1,900
Algeria	1,390	1,330	60	4.5	1,360
Ecuador[1]	500	499	1	0.2	520
Indonesia	859	850	9	1.1	870
Iran	4,023	3,903	120	3.1	3,820
Iraq[3]	2,383	1,893	490	25.9	—
Kuwait[2]	2,583	2,435	148	6.1	2,530
Libya	1,763	1,693	70	4.1	1,710
Nigeria	2,057	2,227	–170	–7.6	2,160
Qatar	847	797	50	6.3	830
Saudi Arabia[2]	8,993	8,475	518	6.1	8,940
United Arab Emirates	2,623	2,567	56	2.2	2,570
Venezuela	2,400	2,437	–37	–1.5	2,470
Total OPEC	**32,314**	**30,225**	**2,089**	**6.9**	**29,680**

[1]Angola and Ecuador became a member of OPEC beginning in 2007. [2]Kuwait and Saudi Arabia production each include half of Neutral Zone. [3]Not included in 2008 quota.
Source: Oil & Gas Journal

SHIFT IN WEALTH

The annual *World Energy Outlook* released November 2008 by IEA projected a $2-trillion shift in wealth from oil-consuming countries to members of OPEC as oil prices rise to $200/bbl by 2030.

The agency estimated that OPEC would control 51% of the world's oil supply by 2030, up from 44% in 2007.

IEA expects Saudi Arabia to remain the largest producer, with output rising to 15.6 million b/d in 2030 from the current 9.5 million.

In a January 2009 analysis, the U.S. EIA estimated the following OPEC net oil-export revenues:

- 2008: $972 billion
- 2009: $387 billion
- 2010: $526 billion

North America

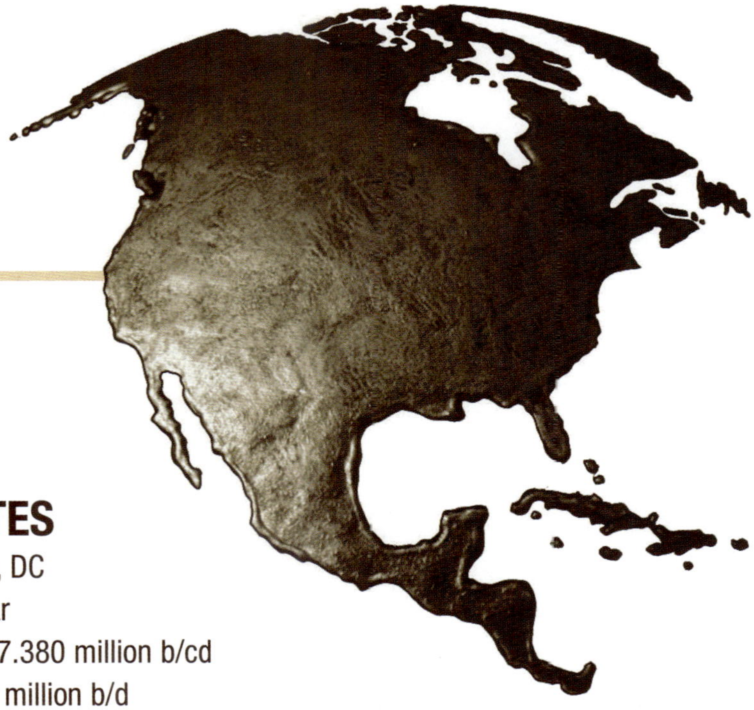

UNITED STATES

Capital: Washington, DC
Monetary unit: Dollar
Refining capacity: 17.380 million b/cd
Oil production: 4.90 million b/d
Oil reserves: 21.317 billion bbl
Gas reserves: 237.726 tcf

Overview

Major issues shaping markets in the United States in 2008 were a hard-fought national election campaign, extremely volatile oil prices (topping $147/bbl in July, then plunging to about $40 at yearend), and an unnerving economic downturn in the last half of the year that overshadowed virtually everything else. (See the Guest Essay at the front of this book for further details on the evolution and impact of the economic crisis.)

Energy picture: OGJ mid-2008 view

Oil & Gas Journal assessed U.S. energy demand in mid-2008, concluding that it would be limited by weak economic growth and improved fuel-use efficiency.

US ENERGY DEMAND

	2007 — Trillion btu —	*2008	Change, %	% share of total energy 2007	*2008
Oil	39,818	38,942	-2.2	39.2	38.2
Gas	23,641	24,470	3.5	23.3	24.0
Coal	22,767	23,000	1.0	22.4	22.5
Nuclear	8,415	8,400	-0.2	8.3	8.2
Hydro, other	6,962	7,240	4.0	6.9	7.1
Total	**101,603**	**102,052**	**0.4**	**100.0**	**100.0**

*OGJ estimate. Source: 2007 US Energy Information Administration

U.S. oil demand for 2008 was expected to contract, while demand for natural gas and coal would get a lift from their use in increased power consumption.

Hydroelectric power and other renewable sources of energy also were expected to be in higher demand, though they still represent a small share of the total energy market.

Production of crude, condensate, and liquids in the United States was expected to climb modestly, but with demand waning, imports were projected to decline.

OGJ said it expected U.S. natural gas production to increase in 2008.

(*Editor's note:* This assessment preceded the economic downturn that began in third quarter 2008. Falling oil prices, tight credit markets, a housing-market contraction, and financial-sector difficulties in late 2008 were expected to dramatically reshape global growth projections and energy use patterns.)

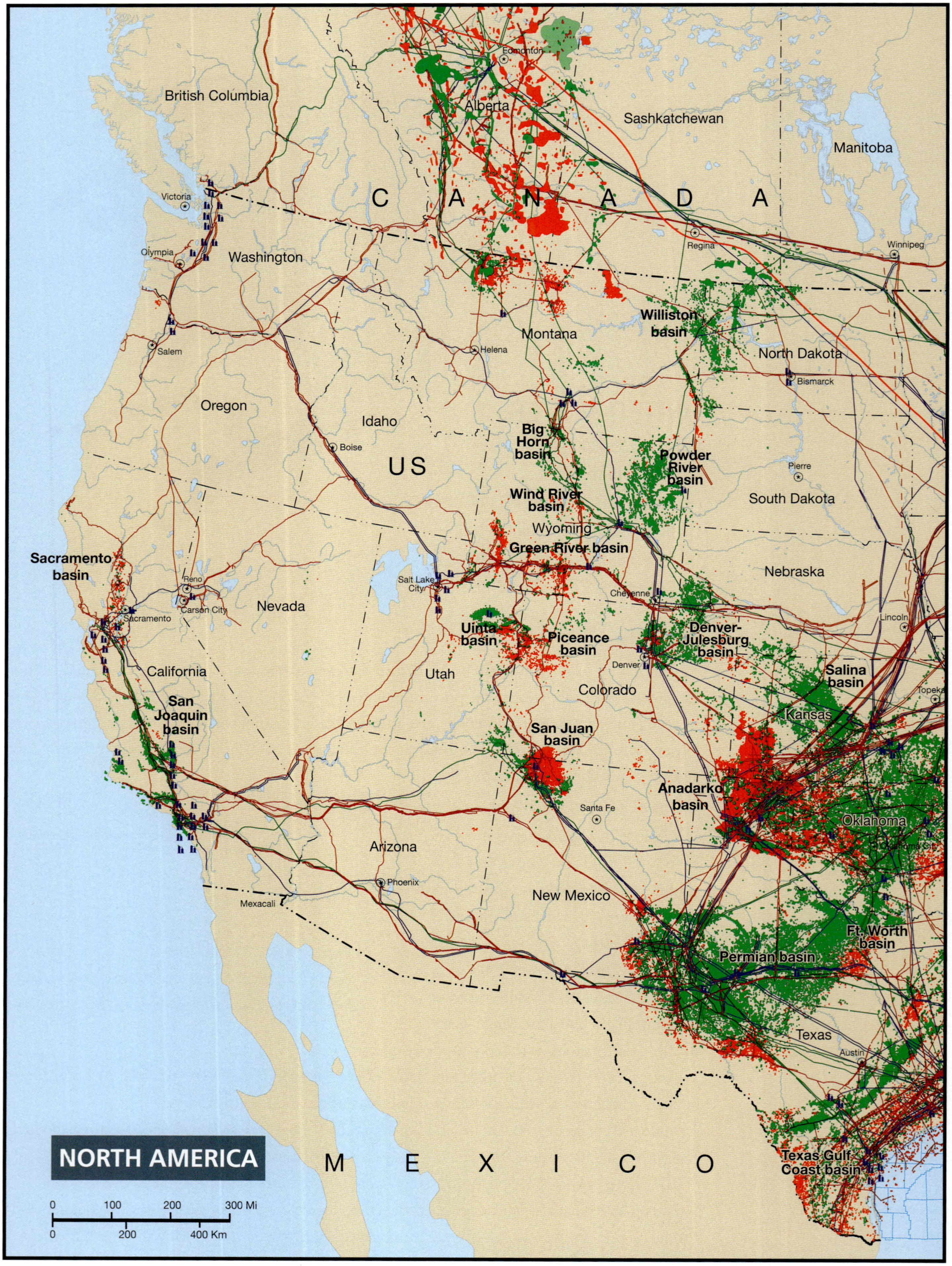
British Columbia
Alberta
Edmonton
Sashkatchewan
Manitoba
C A N A D A
Victoria
Regina
Winnipeg
Olympia
Washington
Williston basin
Salem
Montana
Helena
North Dakota
Bismarck
Oregon
Idaho
Big Horn basin
Boise
US
Powder River basin
Pierre
South Dakota
Wind River basin
Wyoming
Green River basin
Sacramento basin
Nebraska
Reno
Carson City
Salt Lake City
Cheyenne
Nevada
Sacramento
Uinta basin
Piceance basin
Denver-Julesburg basin
Lincoln
Denver
California
Utah
Salina basin
Colorado
Topeka
San Joaquin basin
Kansas
San Juan basin
Anadarko basin
Santa Fe
Oklahoma
Arizona
Phoenix
Mexacali
New Mexico
Ft. Worth basin
Permian basin
Texas
Austin
Texas Gulf Coast basin
M E X I C O
NORTH AMERICA
0 100 200 300 Mi
0 200 400 Km

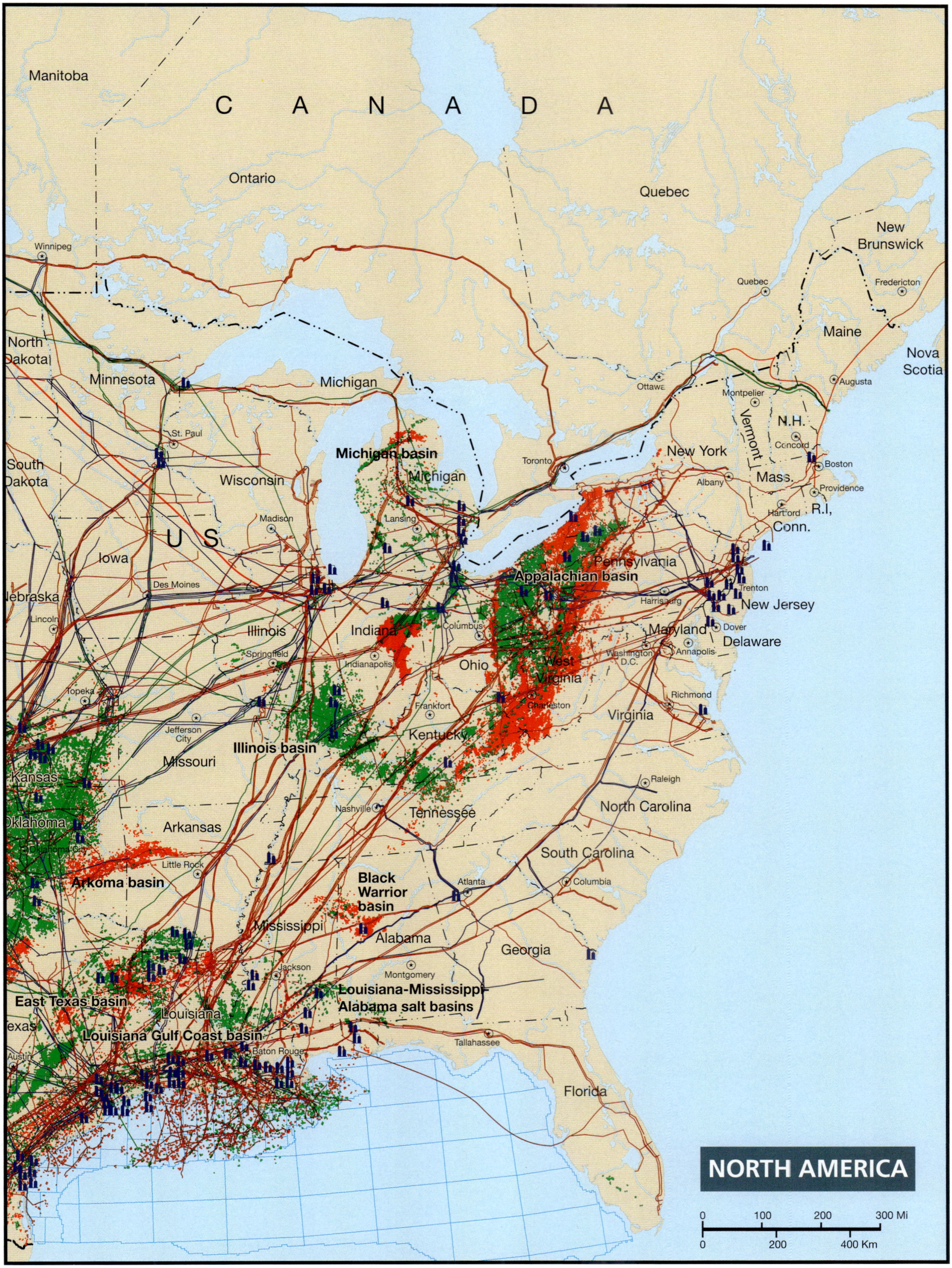
CANADA
Manitoba
Ontario
Quebec
New Brunswick
Nova Scotia
Winnipeg
Quebec
Fredericton
Maine
Augusta
Ottawa
Montpelier
Vermont
N.H.
Concord
Boston
Mass.
Providence
R.I.
Hartford
Conn.
New York
Albany
Toronto
North Dakota
South Dakota
Minnesota
St. Paul
Wisconsin
Madison
Michigan
Michigan basin
Lansing
US
Iowa
Des Moines
Nebraska
Lincoln
Illinois
Springfield
Indiana
Indianapolis
Columbus
Ohio
Pennsylvania
Appalachian basin
Harrisburg
Trenton
New Jersey
Dover
Delaware
Maryland
Annapolis
Washington D.C.
West Virginia
Charleston
Richmond
Virginia
Frankfort
Kentucky
Topeka
Jefferson City
Missouri
Illinois basin
Kansas
Oklahoma
Nashville
Tennessee
Raleigh
North Carolina
Arkansas
Little Rock
Arkoma basin
South Carolina
Columbia
Black Warrior basin
Atlanta
Mississippi
Alabama
Georgia
Jackson
Montgomery
East Texas basin
Louisiana-Mississippi-Alabama salt basins
Louisiana
Texas
Louisiana Gulf Coast basin
Austin
Baton Rouge
Tallahassee
Florida
NORTH AMERICA
0 100 200 300 Mi
0 200 400 Km

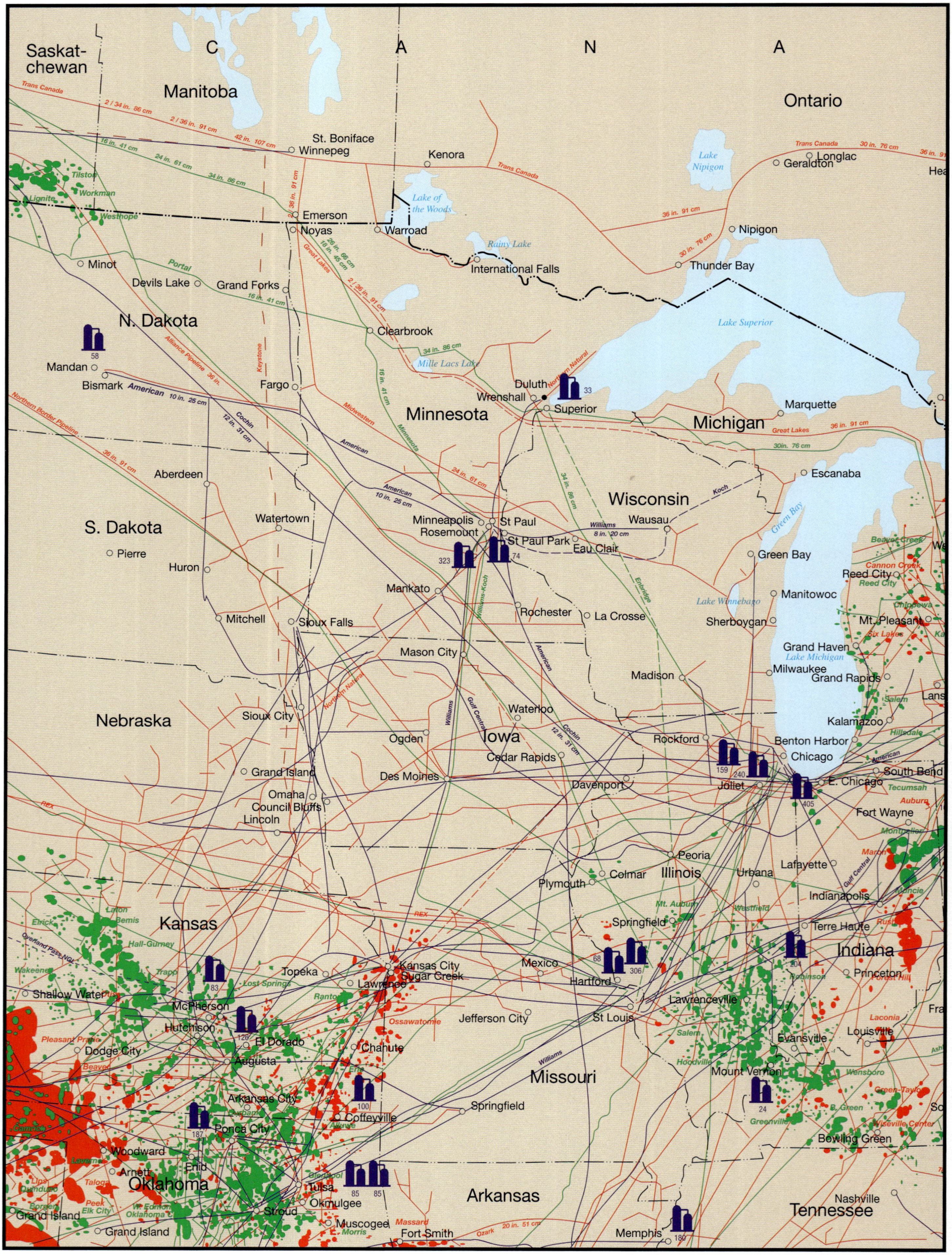

C A N A D A
Saskat-chewan
Manitoba
Ontario
Trans Canada
St. Boniface
Winnepeg
Kenora
Lake Nipigon
Geraldton
Longlac
Lake of the Woods
Emerson
Noyas
Warroad
Rainy Lake
International Falls
Nipigon
Thunder Bay
Lake Superior
Tilston
Workman
Lignite
Westhope
Minot
Portal
Devils Lake
Grand Forks
N. Dakota
Mandan
Bismark
American 10 in. 25 cm
Alliance Pipeline 36 in.
Keystone
Fargo
Clearbrook
Mille Lacs Lake
Duluth
Wrenshall
Superior
Northern Natural
Minnesota
Marquette
Michigan
Great Lakes
Northern Border Pipeline
Cochin 12 in. 31 cm
Midwestern
American
Aberdeen
Escanaba
Wisconsin
Koch
Green Bay
S. Dakota
Pierre
Watertown
Minneapolis
Rosemount
St Paul
St Paul Park
Williams 8 in. 20 cm
Wausau
Eau Clair
Huron
Mankato
Rochester
La Crosse
Enbridge
Lake Winnebago
Manitowoc
Sherboygan
Mitchell
Sioux Falls
Mason City
Madison
Milwaukee
Lake Michigan
Grand Haven
Grand Rapids
Mt. Pleasant
Reed City
Nebraska
Sioux City
Waterloo
Iowa
Ogden
Cedar Rapids
Rockford
Benton Harbor
Chicago
Kalamazoo
Grand Island
Des Moines
Davenport
Joliet
E. Chicago
South Bend
Omaha
Council Bluffs
Lincoln
Fort Wayne
REX
Peoria
Illinois
Colmar
Plymouth
Urbana
Lafayette
Indianapolis
Kansas
Springfield
Terre Haute
Indiana
Topeka
Kansas City
Sugar Creek
Lawrence
Mexico
Hartford
Shallow Water
McPherson
Hutchinson
El Dorado
Jefferson City
St Louis
Lawrenceville
Princeton
Dodge City
Augusta
Chanute
Evansville
Louisville
Missouri
Mount Vernon
Arkansas City
Springfield
Ponca City
Coffeyville
Bowling Green
Woodward
Arnett
Enid
Oklahoma
Tulsa
Okmulgee
Stroud
Arkansas
Nashville
Tennessee
Grand Island
Muscogee
Fort Smith
Memphis
58
33
323
74
159
240
405
83
126
100
187
85
85
68
306
204
24
180

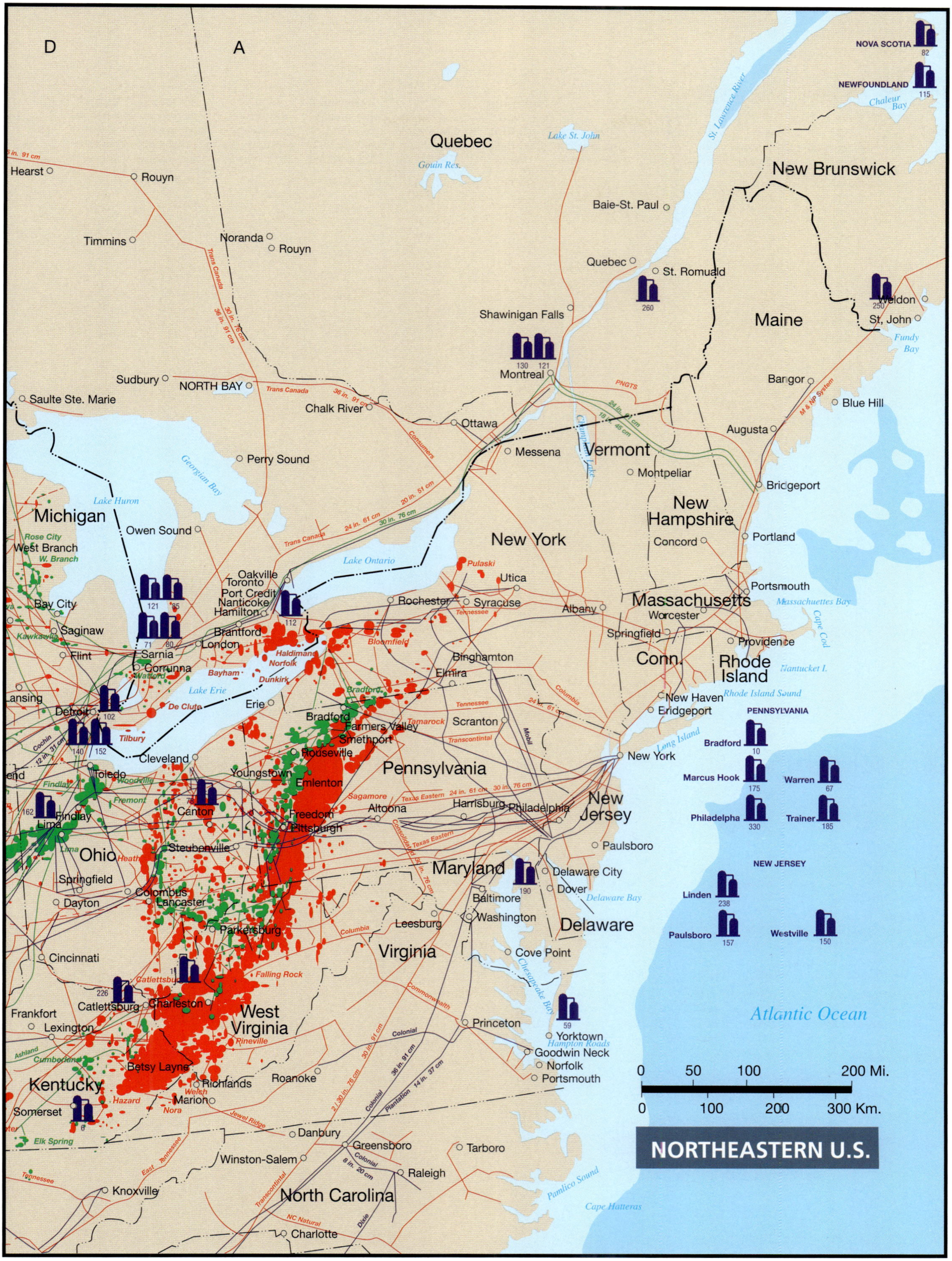
NORTHEASTERN U.S.
D
A
Quebec
New Brunswick
NOVA SCOTIA
82
NEWFOUNDLAND
115
Maine
Vermont
New Hampshire
Massachusetts
Rhode Island
Conn.
New York
Pennsylvania
New Jersey
Maryland
Delaware
Virginia
West Virginia
North Carolina
Kentucky
Ohio
Michigan
Atlantic Ocean
Lake Ontario
Lake Erie
Lake Huron
Georgian Bay
Lake St. John
Gouin Res.
PENNSYLVANIA
Bradford 10
Marcus Hook 175
Warren 67
Philadelphia 330
Trainer 185
NEW JERSEY
Linden 238
Paulsboro 157
Westville 150
0 50 100 200 Mi.
0 100 200 300 Km.

US ENERGY CONSUMPTION AND EFFICIENCY

	GDP (billion 2000 dollars)	Energy consumption (trillion btu)	Energy consumption per GDP, 2000 dollar (Mbtu)	Oil energy consumption (trillion btu)	Oil energy consumption per GDP, 2000 dollar (Mbtu)	Natural gas energy consumption (trillion btu)	Total natural gas energy consumption per GDP, 2000 dollar (Mbtu)	Total oil and natural gas energy consumption (trillion btu)	Oil and gas energy consumption per GDP, 2000 dollar (Mbtu)	Oil and natural gas energy % of total energy
1973.....	4,341.5	75,708	17.4	34,840	8.0	22,512	5.2	57,352	13.2	75.8
1974.....	4,319.6	73,991	17.1	33,455	7.7	21,732	5.0	55,187	12.8	74.6
1975.....	4,311.2	71,999	16.7	32,731	7.6	19,948	4.6	52,679	12.2	73.2
1976.....	4,540.9	76,012	16.7	35,175	7.7	20,345	4.5	55,520	12.2	73.0
1977.....	4,750.5	78,000	16.4	37,122	7.8	19,931	4.2	57,053	12.0	73.1
1978.....	5,015.0	79,986	15.9	37,965	7.6	20,000	4.0	57,965	11.6	72.5
1979.....	5,173.4	80,903	15.6	37,123	7.2	20,666	4.0	57,789	11.2	71.4
1980.....	5,161.7	78,280	15.2	34,202	6.6	20,394	4.0	54,596	10.6	69.7
1981.....	5,291.7	76,335	14.4	31,931	6.0	19,928	3.8	51,859	9.8	67.9
1982.....	5,189.3	73,234	14.1	30,231	5.8	18,505	3.6	48,736	9.4	66.5
1983.....	5,423.8	73,066	13.5	30,054	5.5	17,357	3.2	47,411	8.7	64.9
1984.....	5,813.6	76,693	13.2	31,051	5.3	18,507	3.2	49,558	8.5	64.6
1985.....	6,053.7	76,580	12.7	30,922	5.1	17,834	2.9	48,756	8.1	63.7
1986.....	6,263.6	76,722	12.2	32,196	5.1	16,708	2.7	48,904	7.8	63.7
1987.....	6,475.1	79,156	12.2	32,865	5.1	17,744	2.7	50,609	7.8	63.9
1988.....	6,742.7	82,774	12.3	34,222	5.1	18,552	2.8	52,774	7.8	63.8
1989.....	6,981.4	84,886	12.2	34,211	4.9	19,712	2.8	53,923	7.7	63.5
1990.....	7,112.5	84,730	11.9	33,553	4.7	19,730	2.8	53,283	7.5	62.9
1991.....	7,100.5	84,522	11.9	32,845	4.6	20,149	2.8	52,994	7.5	62.7
1992.....	7,336.6	85,866	11.7	33,527	4.6	20,835	2.8	54,362	7.4	63.3
1993.....	7,532.7	87,579	11.6	33,841	4.5	21,351	2.8	55,192	7.3	63.0
1994.....	7,835.5	89,248	11.4	34,670	4.4	21,842	2.8	56,512	7.2	63.3
1995.....	8,031.7	91,200	11.4	34,553	4.3	22,784	2.8	57,337	7.1	62.9
1996.....	8,328.9	94,226	11.3	35,757	4.3	23,197	2.8	58,954	7.1	62.6
1997.....	8,703.5	94,790	10.9	36,266	4.2	23,328	2.7	59,594	6.8	62.9
1998.....	9,066.9	95,200	10.5	36,934	4.1	22,936	2.5	59,870	6.6	62.9
1999.....	9,470.3	96,827	10.2	37,960	4.0	23,010	2.4	60,970	6.4	63.0
2000....	9,817.0	98,966	10.1	38,404	3.9	23,916	2.4	62,320	6.3	63.0
2001.....	9,890.7	96,304	9.7	38,333	3.9	22,861	2.3	61,194	6.2	63.5
2002....	10,048.8	97,793	9.7	38,401	3.8	23,628	2.4	62,029	6.2	63.4
2003....	10,301.0	98,103	9.5	39,074	3.8	22,967	2.2	62,041	6.0	63.2
2004....	10,703.5	100,199	9.4	40,594	3.8	22,993	2.1	63,587	5.9	63.5
2005....	11,048.6	100,505	9.1	40,735	3.7	22,886	2.1	63,621	5.8	63.3
2006....	11,415.3	99,661	8.7	40,217	3.5	22,518	2.0	62,735	5.5	62.9
2007....	11,566.8	101,603	8.8	39,818	3.4	23,641	2.0	63,459	5.5	62.5
*2008..	11,682.5	102,052	8.7	38,942	3.3	24,470	2.1	63,412	5.4	62.1

*Estimated. Source: US Energy Information Administration

US NATURAL GAS SUPPLY AND DEMAND

	2005	2006	2007	Change, %07/06	2008 bcf	Change, % 08/07
	bcf					
Production						
Texas	5,276	5,514	6,093	10.5	6,700	10.0
Louisiana	1,296	1,361	1,327	–2.5	1,393	5.0
Federal Gulf of Mexico	3,132	2,902	2,771	–4.5	2,855	3.0
Other states	9,223	9,605	9,960	3.7	10,212	2.5
Total production	**18,927**	**19,382**	**20,151**	**4.0**	**21,160**	**5.0**
Imports						
Canada	3,700	3,590	3,777	5.2	3,800	0.6
Mexico	9	13	54	315.4	10	–81.5
LNG	631	584	771	32.0	375	–51.4
Total imports	**4,340**	**4,186**	**4,602**	**9.9**	**4,185**	**–9.1**
Supplemental gas	64	66	61	–7.6	60	–1.6
Losses, etc.*	(643)	(821)	(1,111)	35.3	(850)	–23.5
Total new supply	**22,688**	**22,813**	**23,703**	**3.9**	**24,555**	**3.6**
Supply from storage	52	(436)	177	—	200	—
Total supply	**22,740**	**22,377**	**23,880**	**6.7**	**24,755**	**3.7**
Exports	729	724	822	13.5	900	9.5
Total consumption	**22,011**	**21,653**	**23,058**	**6.5**	**23,855**	**3.5**

*Extraction losses and unaccounted for gas. Source: 2005, 2006, 2007 EIA; 2008 OGJ

FIRST HALF US CRUDE, CONDENSATE PRODUCTION

	First half 2008*	First half 2007	Change, %
	1,000 b/d		
PAD District 1	18	21	–14.3
Florida	5	6	–16.7
Others	13	15	–13.3
PAD District 2	463	480	–3.5
Illinois	26	28	–7.1
Kansas	94	100	–6.0
Michigan	15	17	–11.8
North Dakota	113	118	–4.2
Oklahoma	172	175	–1.7
Others	43	42	2.4
PAD District 3	2,952	2,934	0.6
Alabama	15	19	–21.1
Arkansas	13	15	–13.3
Louisiana	1,349	1,326	1.7
Mississippi	50	54	–7.4
New Mexico	165	163	1.2
Texas	1,360	1,357	0.2
PAD District 4	325	357	–9.0
Colorado	45	41	9.8
Montana	92	102	–9.8
Utah	44	54	–18.5
Wyoming	144	160	–10.0
PAD District 5	1,362	1,394	–2.3
Alaska	708	725	–2.3
California	653	668	–2.2
Others	1	1	—
Total	**5,120**	**5,186**	**–1.3**

*OGJ estimate. Source: US Energy Information Administration

OGJ FORECAST OF US OIL SUPPLY AND DEMAND

	First half 2008		Year 2008	
	Volume 1,000 b/d	% change from 2007	Volume 1,000 b/d	% change from 2007
DEMAND				
Motor gasoline	9,125	–0.9	9,200	–0.9
Dist. 1–4	7,540	–0.9	7,606	–0.9
Dist. 5	1,585	–0.9	1,594	–1.0
Jet fuel	1,605	–1.0	1,583	–2.5
Dist. 1–4	1,119	–1.0	1,092	–2.5
Dist. 5	486	–1.2	491	–2.5
Distillate	4,250	–0.4	4,178	–1.1
Dist. 1–4	3,692	–0.4	3,620	–1.1
Dist. 5	558	–0.4	558	–1.1
Residual	665	–14.4	600	–18.3
Dist. 1–4	539	–14.4	472	–18.3
Dist. 5	126	–14.4	128	–18.3
LPG and ethane	2,100	–2.2	1,977	–5.1
Dist. 1–4	2,048	–2.2	1,928	–5.1
Dist. 5	52	–2.2	49	–5.1
Other products	2,605	–3.4	2,710	–1.4
Dist. 1–4	2,291	–3.4	2,387	–1.4
Dist. 5	314	–3.1	323	–1.4
Total domestic demand	**20,350**	**–1.8**	**20,248**	**–2.2**
Dist. 1–4	17,229	–1.8	17,105	–2.2
Dist. 5	3,121	–1.7	3,143	–2.2
Exports	**1,700**	**25.6**	**1,550**	**10.8**
Dist. 1–4	1,453	25.6	1,326	10.8
Dist. 5	247	25.6	224	10.8
Total Demand	**22,050**	**–0.1**	**21,798**	**–1.4**
Dist. 1–4	18,682	–0.1	18,432	–1.4
Dist. 5	3,368	–0.1	3,366	–1.4
SUPPLY				
Domestic production				
Crude and condensate	5,120	–1.3	5,120	0.4
Dist. 1–4	3,737	–1.3	3,735	0.4
Dist. 5	1,383	–1.3	1,385	0.4
NGL and LRG[1]	1,830	4.3	1,850	3.8
Dist. 1–4	1,765	4.3	1,784	3.8
Dist. 5	65	2.7	66	3.8
Total domestic production	6,950	0.1	6,970	1.2
Dist. 1–4	5,502	0.5	5,519	1.4
Dist. 5	1,448	–1.1	1,451	0.5
IMPORTS				
Crude oil	9,800	–2.1	9,700	–3.1
Dist. 1–4	8,736	–2.1	8,589	–3.1
Dist. 5	1,064	–2.1	1,111	–3.2
Products and unfinished oils	3,250	–8.0	3,250	–5.0
Dist. 1–4	2,848	–8.0	2,902	–5.0
Dist. 5	402	–8.0	348	–5.0
Total Imports	**13,050**	**–3.6**	**12,950**	**–3.6**
Dist. 1–4	11,584	–3.6	11,491	–3.6
Dist. 5	1,466	–3.8	1,459	–3.6
Processing gain, loss, etc.	1,700	62.2	1,919	88.3
Dist. 1–4	1,419	62.2	1,537	88.3
Dist. 5	281	62.2	382	88.3
Total new supply	**21,700**	**0.8**	**21,839**	**2.4**
Dist. 1–4	18,506	0.7	18,546	2.1
Dist. 5	3,194	1.0	3,293	4.2
Stock change	**–350**	—	**41**	—
Dist. 1–4	–176	—	114	—
Dist. 5	–174	—	–73	—
Crude runs to stills	**14,845**	**–0.9**	**15,045**	**–0.6**
Total Input to stills	**15,115**	**–1.2**	**15,250**	**–1.2**
Total refining capacity	**17,594**	**0.8**	**17,650**	**1.2**
Refining utilization, %	**85.9**	**–2.0**	**86.4**	**–2.4**
Total industry stocks[2]	**970**	**–6.6**	**980**	**1.6**
Refined products	670	–2.0	690	1.6
Crude oil	300	–15.5	290	1.4
SPR crude oil stocks	706	2.3	710	1.9
IMPORT DEPENDENCY				
Total imports % domestic demand	64.1	—	64.0	—
Net imports % domestic demand	55.8	—	56.3	—

[1]Liquefied refinery gases. [2]Million bbl at end of period.

Petroleum: API view

In early 2009, the American Petroleum Institute (API; Washington, DC) reported that U.S. petroleum demand fell by 6% in 2008. It was the most rapid year-over-year decline since 1980.

Reasons included a slowing economy, plunging oil and natural gas prices, and the credit crisis.

API also said that U.S. crude oil production in 2008 fell below 5 million b/d for the first time since 1946, as the result of lower Alaskan production and hurricane-related shutdowns in the Gulf of Mexico.

Petroleum: EIA view

DOE's Energy Information Administration (EIA), in a long-range analysis released in December 2008, projected virtually no growth in the nation's petroleum use through 2030.

The scenario envisions a rebound in oil prices after current economic problems subside, reaching $130/bbl in 2007 dollars by 2030 ($189/bbl real).

That price trend, coupled with growing interest in alternative fuels (3.3% growth/yr), will lead to:

- wider use of ethanol and biodiesel
- a push for greater automobile fuel efficiency, including wider use of gas-electric hybrid vehicles.

For the first time in more than 20 years, said EIA, U.S. petroleum demand is projected to be essentially flat for years to come. As a result, net U.S. petroleum imports are expected to decline sharply.

EIA predicted that liquid fuel imports—chiefly oil—will account for only 41% of U.S. consumption by 2025, compared with 58% in 2007.

But the agency also said that coal, oil and natural gas will still provide nearly 80% of U.S. energy in 2030, barring mandatory limits on carbon dioxide emissions.

US CRUDE, PRODUCTS, AND NATURAL GAS PRICES

Year	Average wellhead crude price, $/bbl	Refiner's acquistion cost of crude, $/bbl	Retail motor gasoline, all types, ¢/gal	Residential heating oil, ¢/gal	Average wellhead natural gas price, $/Mcf
1976	8.19	10.89	59.5	40.6	0.58
1977	8.57	11.96	63.1	46.0	0.79
1978	9.00	12.46	65.2	49.0	0.91
1979	12.64	17.72	88.2	70.4	1.18
1980	21.59	28.07	122.1	97.4	1.59
1981	31.77	35.24	135.3	119.4	1.98
1982	28.52	31.87	128.1	116.0	2.46
1983	26.19	28.99	122.5	107.8	2.59
1984	25.88	28.63	119.8	109.1	2.66
1985	24.09	26.75	119.6	105.3	2.51
1986	12.51	14.55	93.1	83.6	1.94
1987	15.40	17.90	95.7	80.3	1.67
1988	12.58	14.67	96.3	81.3	1.69
1989	15.86	17.97	106.0	90.0	1.69
1990	20.03	22.22	121.7	106.3	1.71
1991	16.54	19.06	119.6	101.9	1.64
1992	15.99	18.43	119.0	93.4	1.74
1993	14.25	16.41	117.3	91.1	2.04
1994	13.19	15.59	117.4	88.4	1.85
1995	14.62	17.23	120.5	86.7	1.55
1996	18.46	20.71	128.8	98.9	2.17
1997	17.23	19.04	129.1	98.4	2.32
1998	10.87	12.52	111.5	85.2	1.96
1999	15.56	17.51	122.1	87.6	2.19
2000	26.72	28.26	156.3	131.0	3.68
2001	21.84	22.95	153.1	125.0	4.00
2002	22.51	24.10	144.1	112.9	2.95
2003	27.56	28.53	163.8	135.5	4.88
2004	36.77	36.98	192.3	154.8	5.46
2005	50.28	50.24	233.8	205.2	7.33
2006	59.69	60.24	263.5	236.5	6.40
2007	66.52	67.93	284.9	259.0	6.39
2008*	110.00	108.00	367.0	360.0	10.00

*OGJ estimate. Source: US Energy Information Administration, for 1976-2007 data

US REFINERY UTILIZATION

	Crude runs	Input to distillation units	Operable capacity	Utilization rate, %
	1,000 b/d			
1987	12,851	12,999	15,643	83.1
1988	13,246	13,447	15,927	84.4
1989	13,401	13,551	15,701	86.3
1990	13,409	13,610	15,623	87.1
1991	13,301	13,508	15,707	86.0
1992	13,411	13,600	15,460	87.9
1993	13,613	13,851	15,143	91.5
1994	13,866	14,032	15,150	92.6
1995	13,973	14,119	15,346	92.0
1996	14,195	14,337	15,239	94.1
1997	14,662	14,831	15,594	95.2
1998	14,889	15,108	15,802	95.6
1999	14,804	15,078	16,282	92.6
2000	15,067	15,296	16,525	92.6
2001	15,086	15,351	16,512	92.6
2002	14,947	15,180	16,700	90.7
2003	15,304	15,503	16,747	92.6
2004	15,517	15,781	16,982	93.0
2005	15,220	15,479	17,128	90.4
2006	15,240	15,598	17,400	89.7
2007	15,143	15,443	17,447	88.5
*2008	15,045	15,250	17,650	86.4
1990-2008 change				
Volume	1,636	1,640	2,027	—
Percent	12.2	12.0	13.0	—
2007-08 change				
Volume	(98)	(193)	203	—
Percent	−0.6	−1.2	1.2	—

Source: US Energy Information Administration

SPR REFILL

DOE said in late 2008 that it will start buying oil to replenish the Strategic Petroleum Reserve. The SPR was drawn down following Hurricanes Katrina and Rita in 2005.

Congress ordered DOE to suspend SPR purchases in May 2008 as oil prices topped $100/bbl, but the ban expired at yearend.

Planned 2009 acquisitions will restore the SPR's 727-million bbl storage capacity and provide about 70 days of net import protection.

Natural gas perspectives

U.S. natural gas is expected to trade at $5–6/MMbtu over the next five years, said Wood Mackenzie Ltd. (Edinburgh) at a late-2008 energy forum it sponsored in Houston.

WoodMac said that shale-gas development could well lead to "significant potential over-supply."

The firm also saw declining demand due to "a prolonged recession" through the end of 2010, new wind and coal-fired power generation capacity coming online, and a major drop in drilling because of a lower commodity price.

EIA expected the use of natural gas to grow from 2009–2030, in a December 2008 energy forecast.

This growth will be supported by robust gains in domestic production, largely from unconventional reservoirs. As a result, said EIA, imports will likely provide only about 3% of total supply in 2030, compared to 16% in 2008.

New leadership

U.S. Sen. Barack H. Obama (D-IL) was elected 44th president of the United States on November 4, 2008, besting Republican nominee, Arizona Sen. John McCain. Mr. Obama was the first African-American to be elected president.

U.S. Sen. Joseph R. Biden, Jr. (D-DE) was elected vice president.

By January 2009, Pres. Obama had announced several major appointments that will shape U.S. energy policy and initiatives.

- Steven Chu, Ph.D.—winner of the 1997 Nobel Prize in physics and head of the Lawrence Berkeley National Laboratory of the U.S. Dept. of Energy—to be Secretary of DOE.
- U.S. Sen. Kenneth L. Salazar (D-CO) to lead the Interior Department.
- Carol Browner, former administrator of the EPA in the Clinton Administration, to serve in the new post of Energy Coordinator. Observers expect her job will be to promote cooperation among the different federal energy and climate entities.

US OIL IMPORTS

Sources of crude imports[1]

Country	Share of total 2008 %	First quarter average 2008, 1,000 b/d	Change 08/07, %	Annual average 2007, 1,000 b/d	Change 07/06, %	Annual average 2006, 1,000 b/d
Angola[2]	4.4	433	–12.7	496	–3.3	513
Ecuador[2]	2.2	217	9.6	198	–27.2	272
Indonesia	0.1	14	–6.7	15	–6.3	16
Nigeria	11.3	1,102	1.8	1,082	4.3	1,037
Saudi Arabia	15.8	1,541	6.1	1,453	2.1	1,423
Venezuela	10.1	980	–14.8	1,150	0.7	1,142
Other OPEC	13.0	1,268	26.8	1,000	163.2	380
Total OPEC	**57.0**	**5,555**	**3.0**	**5,394**	**12.8**	**4,783**
Canada	19.4	1,886	1.2	1,864	3.4	1,802
China	0.1	10	66.7	6	–68.4	19
Colombia	1.8	174	27.0	137	–2.8	141
Gabon	0.6	54	–14.3	63	5.0	60
Mexico	12.5	1,220	–13.5	1,410	–10.6	1,577
Norway	0.1	14	–75.0	56	–42.9	98
Trinidad and Tobago	0.5	49	2.1	48	–28.4	67
UK	0.8	80	–21.6	102	–21.5	130
Other Non–OPEC	7.2	702	–25.1	937	–35.0	1,441
Total imports	**100.0**	**9,744**	**–2.7**	**10,017**	**–1.0**	**10,118**

Sources of refined product imports

Country	Share of total 2008 %	First quarter average 2008, 1,000 b/d	Change 08/07, %	Annual average 2007, 1,000 b/d	Change 07/06, %	Annual average 2006, 1,000 b/d
Algeria	7.0	220	–3.1	227	–23.3	296
Saudi Arabia	0.5	15	–58.3	36	–12.2	41
Venezuela	5.5	172	–18.5	211	–23.8	277
Other OPEC	3.5	109	–6.0	116	–2.5	119
Total OPEC	**16.4**	**516**	**–12.5**	**590**	**–19.5**	**733**
Canada	20.5	646	15.2	561	1.8	551
Colombia	0.8	26	52.9	17	30.8	13
Italy	1.5	48	–12.7	55	–5.2	58
Mexico	3.5	111	–10.5	124	–3.9	129
Netherlands Antilles	0.5	17	88.9	9	–71.9	32
Virgin Islands	10.8	340	–1.7	346	5.5	328
Other Non–OPEC	45.8	1,440	–16.3	1,720	–1.4	1,745
Total imports	**100.0**	**3,144**	**–8.1**	**3,422**	**–4.7**	**3,589**

[1]Includes imports for the Strategic Petroleum Reserve. [2]Angola and Ecuador became a member of OPEC in 2007.Source: US Energy Information Administration

FIRST HALF US CRUDE AND PRODUCTS STOCKS

	[1]2008, 1,000 bbl	Change 08/07, %	[1]2007, 1,000 bbl	Change 07/06, %	[1]2006, 1,000 bbl
Motor gasoline	210,900	2.0	204,913	–3.9	213,292
Jet fuel	39,600	–1.2	41,185	4.6	39,388
Distillate fuel oil	120,700	–2.6	123,449	–5.0	129,912
Residual fuel oil	40,000	11.9	36,118	–15.5	42,726
Unfinished oils	85,400	–3.9	88,850	–2.5	91,158
Other	173,400	2.5	189,428	0.9	187,667
Total products stocks	**670,000**	**0.9**	**683,943**	**–2.9**	**704,143**
Crude stocks[2]	300,000	–18.3	354,791	5.4	336,735
Total	**970,000**	**–5.7**	**1,038,734**	**–0.2**	**1,040,878**

[1]At end of June. [2]Excludes Strategic Petroleum Reserve.
Source: US Energy Information Administration, 2006-2007. 2008, OGJ estimate

Obama energy plans

Early in his campaign, then-Sen. Obama identified three major energy-related goals: reduced dependence on imported oil, reduction of greenhouse gas emissions implicated in climate change, and creation of new jobs associated with wider use of alternative energy resources.

Among proposed specific actions:

- Revise regulations to discourage excessive speculation in energy trading.
- Institute a cap-and-trade program to cut greenhouse gas emissions.
- Use proceeds from the auction of emission credits to fund clean energy technology.
- Create "green jobs" in such fields as electric power production, improved energy efficiency, weatherization, and clean-coal technology.
- Tighten vehicle fuel-efficiency standards.
- Promote U.S. production of plug-in hybrid electric vehicles.
- Promote oil and gas production from unconventional resources in areas already producing
- Construct an Alaskan gas pipeline
- Pursue enhanced oil recovery and carbon dioxide sequestration.
- Diversify energy sources with a renewable fuel standard for electricity.

Oil & Gas Journal noted that Obama outlined his energy agenda while energy prices were very high and while the economy seemed reasonably healthy. Those conditions changed sharply toward the end of 2008, making uncertain how much of the agenda will be supported by Congress.

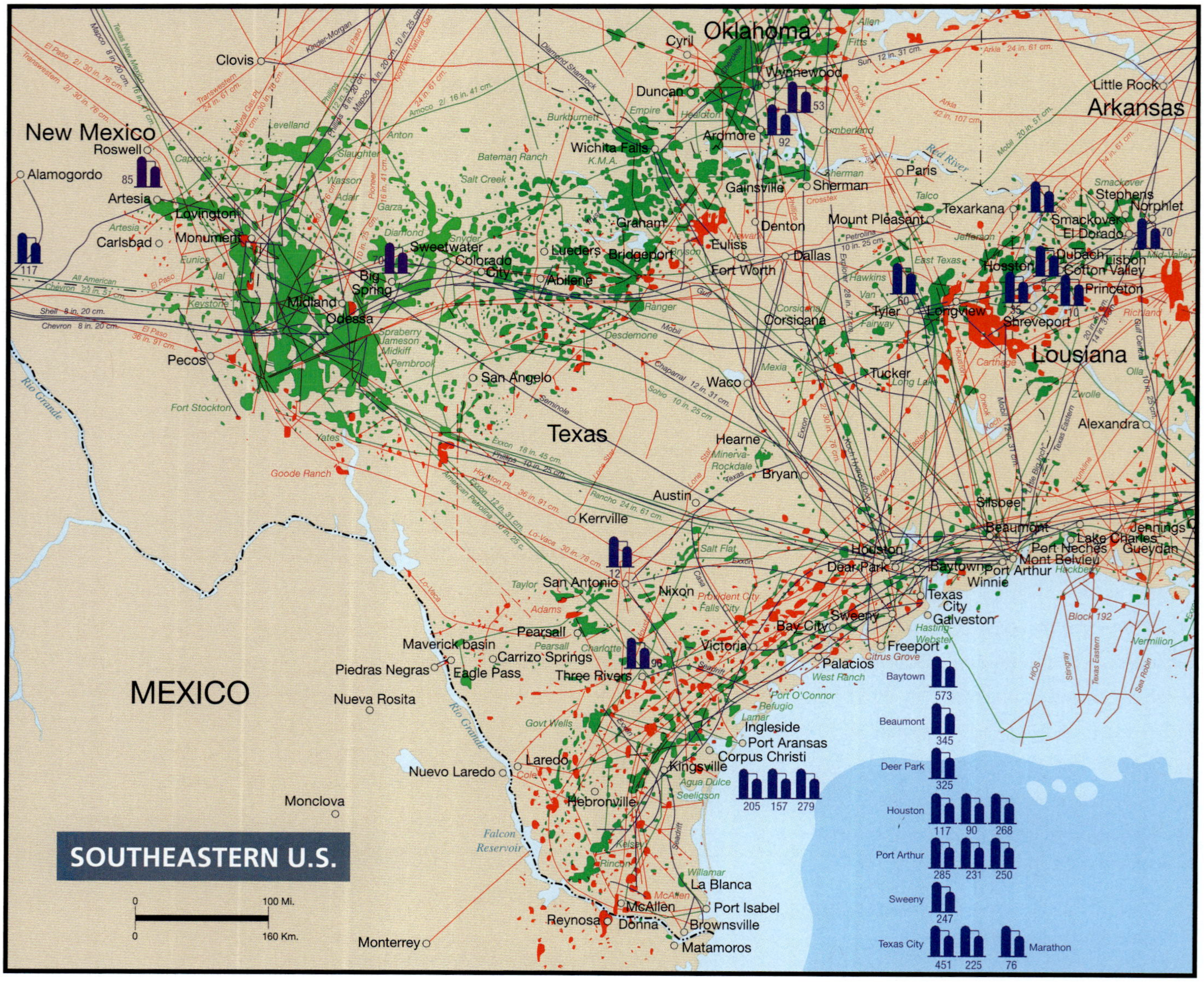

Exploration and Development

Relook at offshore Florida?

Rising crude oil prices in mid-2008 prompted producers as well as politicians to look again at drilling for oil and natural gas in the deep waters off Florida.

Reconsideration was under way even before then-Pres. Bush urged Congress in June to lift its moratoria, enacted for environmental reasons, that barred new offshore drilling since 1981.

A 2006 Congressional compromise allows drilling on 8.3 million acres that lie more than 125 mi off the Panhandle, an area that had been covered by the moratorium.

The Interior Department estimates there could be 18 billion bbl of oil and 77 tcf of natural gas beneath the 574 million acres of federal coastal waters that are now off-limits.

In March 2008, BHP Billiton Petroleum Deepwater Inc. (Melbourne, Australia), Anadarko E&P Co. (Houston), Shell Offshore Inc. (Houston), and Eni SPA (Milan) bought leases on 36 Gulf of Mexico tracts under the 2006 compromise.

BLM: Utah plans

New resource management plans (RMPs) for five of the six Utah field offices of the U.S. Bureau of Land Management (BLM) received final approval in October 2008. The sixth was still under review.

In the new plans, 53% of the acres open to leasing are subject to stricter environmental controls, and 18% of the lands within the planning areas are not available at all.

BLM wants to encourage use of innovative methods such as directional drilling, well placement, and sound muffling to reduce the impact of production on public lands.

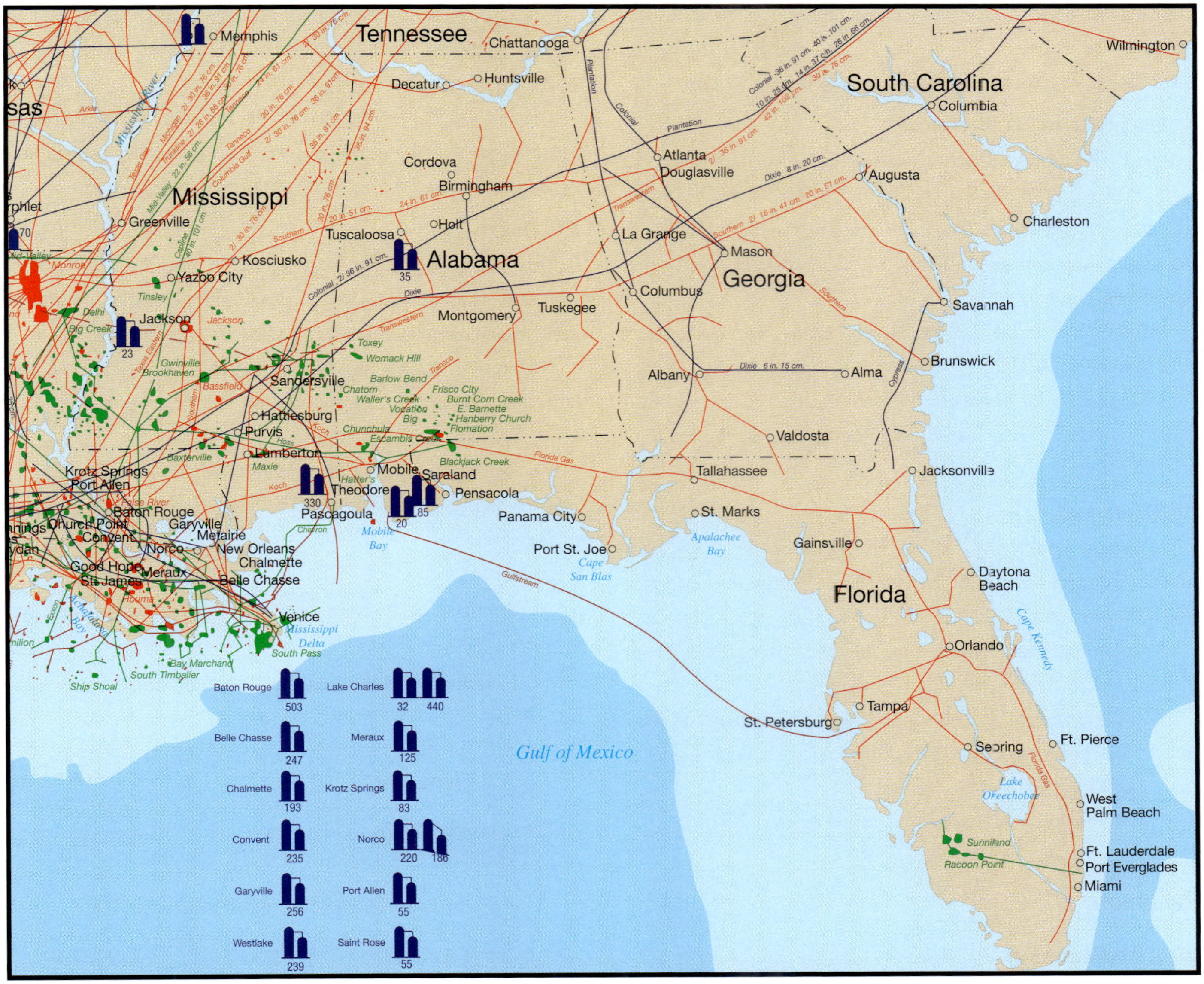

BLM: Oil shale

The BLM published final regulations in November 2008 to establish a commercial oil shale development program on public lands in Colorado, Utah, and Wyoming.

However, officials said that lease sales probably will not occur for another 5–10 years, and that certain environmental documents would need to be developed.

The new regulations will guide private investors considering future financial commitments for prospective oil shale projects.

Oil shale development on public land in the three states potentially could add 800 billion bbl of oil to U.S. reserves, according to BLM—enough to meet U.S. oil demand for 110 years.

Tuscaloosa oil shale

Encore Acquisition Co. (Fort Worth, TX) was exploring for oil in the highly overpressured Cretaceous Tuscaloosa marine shale in late 2008. The firm controls 210,000 net acres along the Louisiana-Mississippi line east of the Mississippi River.

It drilled the Weyerhaeuser-1H, in section 60-1s-4e, in the northwestern corner of St. Helena Parish, Louisiana, and was planning a 4,100-ft lateral.

Mapping the Arctic Seabed

The U.S. Coast Guard Cutter *Healy* conducted two arctic cruises in 2008 to collect data about the extended continental shelf in the Arctic Ocean.

Details are presented in the Alaska section of this chapter.

Louisiana State University (Baton Rouge) estimated that the shale could hold 7 billion bbl of recoverable oil in an area extending across south Louisiana and southern Mississippi.

Petrobras interest: Utah oil shale

Brazil's Petrobras (Rio de Janeiro) signed an agreement in mid-2008 with Oil Shale Exploration Co. LLC (OSEC; Mobile, AL) and Japanese investment and trading company Mitsui & Co. Ltd. (Tokyo) to jointly study the development of oil shale projects in Utah.

Petrobras has its own proprietary oil-shale retort technology that may be applicable.

OSEC reportedly has access to more than 30,000 acres of oil shale property, with a discovered resource base in the range of 3 billion bbl.

RPSEA program

In January 2009, the U.S. DOE approved the 2009 plan of the Research Partnership to Secure Energy for America (RPSEA; Sugar Land, TX).

The plan outlines collaborative research to be carried out under RPSEA's Ultra-Deepwater Program.

The program aims to develop and apply technology needed to find and produce oil and natural gas in water depths greater than 15,000 ft (4,572 m).

Requests for proposals are to be issued by mid-2009.

Reserves

USGS: Northwest gas

An estimated 2.4 tcf of natural gas and 9.8 million bbl of natural gas liquids lie beneath eastern Oregon and Washington, said USGS in a November 2008 report.

The figures are mean estimates of undiscovered gas resources beneath 60,000 sq mi of the two states.

The hypothetical Columbia basin assessment unit (AU), holding an estimated 2.1 tcf of gas and 9.2 million bbl of NGL, covered more than 4 million acres (6,250 sq mi).

The hypothetical Eastern Oregon and Washington conventional gas AU covered more than 22.2 million acres (34,685 sq mi) and estimated 300 bcf of conventional gas and 610,000 bbl of NGL.

BLM: Hydrocarbons inventory

U.S. onshore public lands contain an estimated 31 billion bbl of crude oil and 231 tcf of natural gas, reported the BLM in its June 2008 inventory.

It also found that 60% of the onshore federal acreage with oil and gas potential is closed to leasing, effectively making 62% of the oil and 41% of the gas inaccessible.

Another 30% of federal onshore oil and 49% of federal onshore gas may be developed subject to seasonal timing and other restrictions over and above standard lease terms.

E&Y report

Ernst & Young (London) released in mid-2008 an evaluation of oil and gas reserves and production, based on data from benchmark companies in the U.S. E&P industry.

The study included integrated companies, independents, and large independents and was based on annual reports filed with the U.S. SEC.

U.S. EIA estimated that the benchmark companies hold 74% of estimated total proved U.S. oil reserves and 68% of gas reserves.

Among key findings:

- U.S. proved oil reserves for 40 exploration and production companies increased by 2% during 2003–07.
- Oil production held steady at 1.2 billion bbl/yr since 2004 (including condensate and NGLs).
- Estimated proved natural gas reserves increased 47% during 2003–07.
- Benchmark companies had 138.6 tcf of gas reserves in 2007 and 129.5 tcf in 2006.
- Gas production reached 10.2 tcf in 2007, up 7% from 2006.
- Exploration costs increased to $12.8 billion in 2007, up 165% from 2003.
- Development costs at yearend 2007 had grown 180% from $18.4 billion in 2003.

Permian basin estimates

In a March 2008 report, the USGS estimated that there is about 40.58 tcf of undiscovered natural gas, 1.02 billion bbl of NGLs, and 1.26 billion bbl of oil in the Permian basin.

Most of the gas (about 35.38 tcf) and gas liquids (about 785 million bbl) is in unconventional formations in the Delaware and Midland basins.

USGS cautioned, "given that few wells have produced from these assessment units, there is significant geological uncertainty in these estimates."

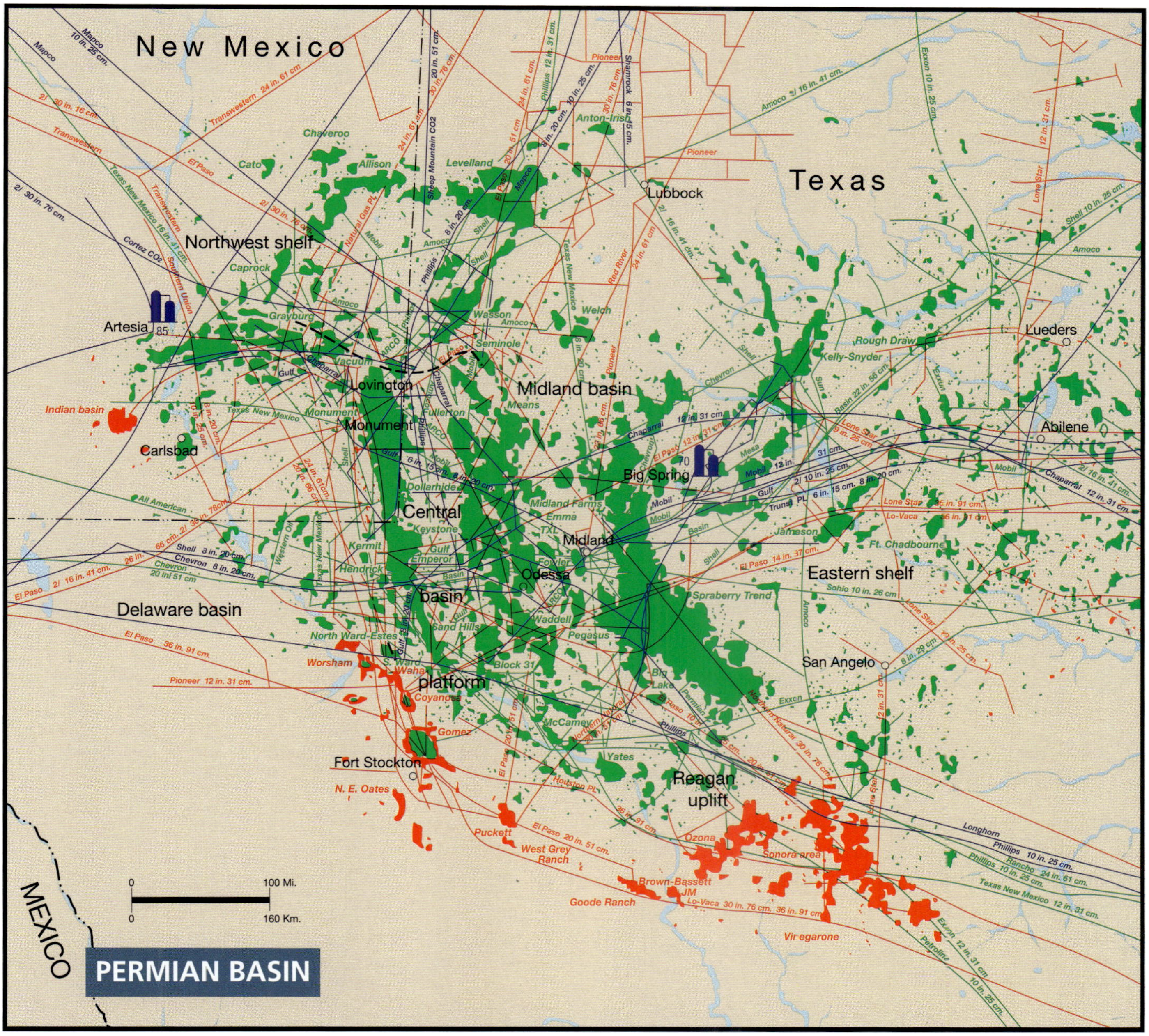

The study assessed the potential for technically recoverable resources in new field discoveries only; it did not include field or reserve growth in conventional fields.

Reserves reporting

In mid-2008, the U.S. Securities and Exchange Commission (SEC) proposed several revised oil and natural gas reserves reporting requirements, plus changes on disclosure of estimated probable and possible reserves.

Currently, oil and gas companies report only estimated proved reserves to the SEC.

Proposed rule changes include:

- Defining "reasonable certainty" to mean "much more likely to be achieved than not."
- Allowing use of both deterministic and probabilistic methods to estimate reserves.
- Reporting estimated oil and gas reserves using an average price based on the previous 12 months.
- Permitting use of proven new technologies to determine proved reserves.
- Classifying previously excluded resources (e.g., oil sands) as oil and gas reserves.

Drilling and Production: General

Gas shales boost production

In an October 2008 report, EIA said that U.S. natural gas production has undergone "an historic shift."

After nine years of little change, gas output rose by 4% in 2007. And in the first seven months of 2008, it increased by 9% year-over-year.

Independent producers have pointed to shale gas as a major contributor to the increase.

The Barnett shale of North Texas produced about 4.4 bcfd in 2008, compared to 1 bcfd in 2004.

Horizontal drilling was seen as a key technology in boosting shale-gas production.

Mid-year drilling outlook

In a July 2008 analysis, *OGJ* noted that oil and gas drilling were on the upswing in the United States and that drilling in Canada was coming back after slumping in 2007.

OGJ expected the drilling of 7% more wells in the United States in 2008 compared to 2007, and a slight gain in Canada.

Highlights from *OGJ*'s midyear drilling forecast for North America in 2008:

- U.S. operators will drill 50,475 wells, up from an estimated 47,057 in 2007.
- All operators will drill 5,246 exploratory wells of all types, up from an estimated 3,833 in 2007.
- The Baker Hughes Inc. count of active U.S. rotary rigs will average 1,840 rigs/week in 2008, compared to 1,768 in 2007 and 1,649 in 2006.
- Shale and other unconventional gas plays were claiming large shares of the drilling in Texas and numerous other states. Of note was activity in East Texas Districts 5 and 6, the Texas Panhandle counties, and the Jurassic Haynesville.
- The Bakken shale oil play moved eastward. North Dakota added 147 wells producing from the Bakken in 2007.

Key U.S. plays

Plays likely to see increased activity for 2008, said *OGJ*, include:

- The Texas Barnett shale (noncore counties beyond Denton, Wise, and Tarrant)
- East Texas (from Tarrant and Robertson counties to the Oklahoma and Louisiana state lines)
- Arkansas Fayetteville shale
- Arkoma basin Woodford shale
- Spraberry Trend of West Texas
- Raton basin in southeastern Colorado.

How US, Canada drilling compare

Average number of active rigs per week, year to date

	Jan. 1– June 29, 2007	Jan. 1– June 27, 2008
Alabama	4.1	4.1
Alaska	9.4	7.4
Arkansas	43.3	44.0
California	33.3	38.1
Land	31.6	36.9
Offshore	1.7	1.2
Colorado	102.6	114.0
Florida	0.4	0.1
Kansas	13.4	10.9
Kentucky	9.0	10.0
Louisiana	185.9	149.5
N. land	56.7	52.2
S. inland waters	23.9	20.0
S. land	38.0	25.5
Offshore	67.3	51.9
Michigan	1.4	0.8
Mississippi	16.4	11.3
Montana	19.8	11.3
Nebraska	0.1	0.1
Nevada	1.9	3.2
New Mexico	80.2	73.6
New York	6.8	6.3
North Dakota	33.7	58.7
Ohio	12.9	11.8
Oklahoma	183.1	203.1
Pennsylvania	14.2	20.1
South Dakota	1.9	1.7
Tennessee	3.9	4.8
Texas	820.2	886.8
Offshore	10.2	8.8
Inland waters	0.9	2.5
Dist. 1	21.8	23.9
Dist. 2	29.5	34.3
Dist. 3	58.0	62.2
Dist. 4	92.1	91.6
Dist. 5	164.6	182.8
Dist. 6	123.9	121.1
Dist. 7B	38.8	32.0
Dist. 7C	54.0	60.2
Dist. 8	107.9	128.3
Dist. 8A	25.9	22.5
Dist. 9	34.6	40.5
Dist. 10	58.1	76.2
Utah	41.7	40.3
Virginia	2.4	4.2
West Virginia	30.8	26.4
Wyoming	74.8	71.0
US total	**1,749.9**	**1,817.0**
US land	**1,645.5**	**1,732.0**
US offshore	**79.9**	**62.4**
Canada total	**332.7**	**341.2**

Source: Baker Hughes Inc.

Oil & Gas Journal well forecast for 2008

State	First half 2008 Total wells	First half 2008 Exploratory wells	First half 2008 Field wells	Full year 2008 Total wells	Full year 2008 Exploratory wells	Full year 2008 Field wells	Full year 2008 Total footage (1,000)
Alabama	230	12	218	471	24	447	2,033
Alaska	77	14	63	160	30	130	861
Arizona	2	2	0	5	5	0	9
Arkansas	397	67	330	810	138	672	5,422
California land	1,470	49	1,421	2,905	96	2,809	13,415
California offshore	3	0	3	8	0	8	49
Colorado	1,902	510	1,392	3,825	1,025	2,800	23,053
Florida	0	0	0	1	0	1	13
Illinois	194	60	134	380	117	263	914
Indiana	68	22	46	130	42	88	196
Kansas	1,240	102	1,138	2,505	205	2,300	9,662
Kentucky	474	19	455	965	39	926	2,679
Louisiana	1,049	96	953	2,125	196	1,929	20,075
North	568	49	519	1,155	100	1,055	10,277
South	237	13	224	475	27	448	4,849
Offshore	244	34	210	495	69	426	4,949
Michigan	239	47	192	490	97	393	801
Mississippi	119	12	107	250	25	225	2,242
Montana	346	47	299	709	97	612	4,480
Nebraska	28	7	21	57	15	42	280
Nevada	3	2	1	5	4	1	25
New Mexico - East	577	41	536	1,145	81	1,064	9,090
New Mexico - West	449	6	443	915	13	902	5,409
New York	52	2	50	105	4	101	387
North Dakota	298	64	234	620	133	487	6,597
Ohio	567	56	511	1,155	114	1,041	4,813
Oklahoma	1,980	93	1,887	3,975	187	3,788	28,878
Pennsylvania	2,062	184	1,878	4,100	365	3,735	10,090
South Dakota	27	3	24	55	7	48	210
Tennessee	93	34	59	188	70	118	422
Texas	7,735	712	7,023	15,565	1,436	14,129	130,614
Dist. 1	298	26	272	605	53	552	3,802
Dist. 2	404	65	339	820	133	687	7,191
Dist. 3	478	65	413	950	130	820	7,914
Dist. 4	712	67	645	1,440	135	1,305	14,141
Dist. 5	852	34	818	1,680	67	1,613	18,394
Dist. 6	863	106	757	1,740	214	1,526	17,920
Dist. 7-B	607	25	582	1,240	52	1,188	6,217
Dist. 7-C	784	44	740	1,555	87	1,468	12,485
Dist. 8	1,030	64	966	2,080	129	1,951	16,742
Dist. 8-A	446	44	402	905	89	816	4,997
Dist. 9	600	17	583	1,225	36	1,189	6,148
Dist. 10	620	140	480	1,235	279	956	13,795
Offshore	41	15	26	90	32	58	867
Utah	569	114	455	1,160	232	928	9,103
Virginia	270	38	232	555	78	477	1,476
Washington	1	1	0	1	1	0	15
West Virginia	992	144	848	2,010	291	1,719	8,729
Wyoming	1,559	51	1,508	3,150	104	3,046	13,580
US total	**25,072**	**2,611**	**22,461**	**50,500**	**5,271**	**45,229**	**315,622**
Western Canada	**8,279**	**1,689**	**6,590**	**18,605**	**3,791**	**14,814**	**77,509**
Alberta	5,952	1,184	4,768	13,525	2,691	10,834	53,153
Saskatchewan	1,743	333	1,410	3,765	719	3,046	14,710
Brit. Columbia	427	153	274	935	335	600	8,317
Manitoba	157	19	138	380	46	334	1,329
NWT + Yukon Terr.	4	3	1	12	10	2	80
Eastern offshore	3	1	2	6	2	4	40
Eastern land	67	7	60	135	15	120	271

Canadian outlook

- Drilling in Canada is showing initial signs of a turnaround from weakness the past few years, especially in gas drilling.
- Factors in the resurgence: Stronger gas and oil prices; expectations for newer shale oil and shale gas plays (e.g., Bakken, Montney, Muskwa); and returning to CBM and other gas programs.
- Operators will drill 18,605 wells in western Canada, up from an estimated 18,391 wells in 2007.
- Baker Hughes counted an average 349 rigs in Canada January-May, up 3 from same-period 2007.

A Look at 30 Years of US Well Completions

Year	Total wells[1]	Total footage	Total exploratory wells
[2]2008	49,012	302,906,000	4,001
[2]2007	47,057	279,001,000	3,833
2006	49,375	289,959,000	3,696
2005	44,679	254,844,000	3,727
2004	39,051	213,908,000	3,192
2003	30,487	158,221,000	2,593
2002	27,794	145,055,000	2,271
2001	36,061	184,462,000	3,181
2000	31,261	149,848,000	2,517
1999	22,107	109,854,000	2,141
1998	25,822	143,625,000	2,723
1997	30,208	165,480,000	3,353
1996	25,724	138,588,000	3,364
1995	23,061	124,426,000	3,406
1994	23,324	130,654,000	3,788
1993	26,032	138,509,000	3,604
1992	23,921	123,456,000	3,494
1991	28,417	141,391,000	4,399
1990	30,615	149,518,000	5,074
1989	28,363	134,901,000	5,251
1988	32,238	155,164,000	6,350
1987	36,253	163,848,000	6,903
1986	39,015	177,641,000	7,156
1985	70,806	316,778,000	12,208
1984	84,983	368,796,000	15,138
1983	75,738	316,617,464	13,845
1982	83,889	375,382,919	15,882
1981	89,234	406,520,453	17,430
1980	69,486	311,444,837	12,870
1979	51,890	243,685,430	10,735

[1]Well counts in most recent years subject to reporting lag. [2]Estimated.
Source: 1975-2006 American Petroleum Institute

Rig Activity: New 22-Year High

The U.S. rig count during the week ended June 13, 2008, topped 1,900 working units for the first time in more than 22 years.

Baker Hughes Inc. reported 1,901 rotary rigs drilling in the United States, 15 more than the previous week and up from 1,773 during the same period in 2007.

The last time the rig count topped that level was in the week ended January 3, 1986, when 1,915 units were working.

As usual, land operations accounted for the bulk of the increase.

Of the rigs working, 1,504 were drilling for natural gas and 389 for crude. Eight were unclassified.

API: Gas drilling record in 2007

U.S. oil and gas exploration and production activity remained robust during 2007, API reported in its January 2008 quarterly report. Some highlights:

- Producers drilled a record number of natural gas wells in 2007 (30,625).
- Oil well completion activity was the most active since 1990; total estimated footage was the highest in two decades.
- An estimated 52,731 oil wells, gas wells, and dry holes were completed domestically in 2007.
- A resurgence in oil well completion activity continued in 2007, with an estimated 17,223 wells drilled.
- Exploratory gas well estimates rose for a fifth consecutive year to a record 3,613.

2007 drilling outlays

U.S. oil and gas drilling expenditures reached a record $226.4 billion in 2007, more than double the previous record of $109.8 billion set in 2006, the API said in January 2008.

API said that records also were set in 2007 in average costs per well and per foot.

Average costs per U.S. oil well grew to $4 million in 2007 from $2.2 million in 2006; per-foot costs climbed to $717 from $412 year-to-year.

Average costs per domestic natural gas well rose to $3.9 million in 2007 from $1.9 million in 2006; average costs per foot grew to $604 from $348.

The number of wells drilled in 2007 went up by 4% from 2006; average well depth increased by 9%.

Spending in 2007 was greater for U.S. gas wells than for oil wells for a 20th consecutive year.

MMS: record disbursements

The U.S. Minerals Management Service (MMS) said it disbursed a record $23.4 billion to state, American Indian, and federal accounts from onshore and offshore energy production during fiscal 2008.

Disbursements from royalties, rents, and bonuses from production on federal land broke the previous record of $12.8 billion set in fiscal 2006.

MMS attributed the record disbursements to higher energy prices in 2008, and the more than $10 billion in bonus bids paid by companies to lease tracts in the Gulf of Mexico and offshore Alaska.

Thirty-five States received a record $2.59 billion as their share of federal revenue collected.

Land drilling: mid-2008

Oil & Gas Journal reported in August 2008 a rapid recovery of the land drilling market in North America due to solid natural gas prices and a growing focus on deep shale gas.

New land rigs, offshore rigs, and related equipment, such as marine drilling risers, were under construction worldwide.

The new rush to drill new, deep shale gas plays will require land rigs of 1,500 hp and higher.

3Q-2008 completions up

According to API, U.S. oil and gas drilling continued to outpace year-ago activity during third quarter 2008 and was nearly twice the levels seen during the 1990s.

In the three months ended September 30, an estimated 16,379 oil wells, natural gas wells, and dry holes were completed, 16% more than the comparable 2007 period.

A resurgence in U.S. oil well drilling that began in 2000 continued during 2008.

The group also pointed out that gas continued to be the primary drilling target.

Construction costs up

In a mid-2008 report, IHS-Global Inc.-Cambridge Energy Research Associates (CERA; Cambridge, MA) noted that construction costs for upstream oil and gas facilities rose by 6% during the previous six months, and had doubled since 2005.

The group said that rising costs had become "one of the new fundamentals" driving the price of oil.

Increased demand for raw materials and transportation was driving costs, which contributed to the delay of many projects.

Independents cut budgets

In October 2008, Chesapeake Energy Corp. (Oklahoma City) as well as Petrohawk Energy Corp. and ATP Oil & Gas Corp. (both of Houston) announced that they were cutting drilling budgets to focus on their best prospects, in view of the deteriorating economy and declining demand.

Chesapeake announced a drilling capital expenditure budget cut of $3.2 billion, or 17%, for the second half of 2008 through 2010.

Company officials blamed a 50% drop in natural gas prices since June and a possible gas surplus.

Petrohawk said it will cut its 2009 capital budget by one-third and focus on development of the Haynesville and Fayetteville shales.

ATP Oil & Gas announced plans to cut more than $200 million from its capital expenditure budget for the rest of 2008 and 2009.

Top 12 US contractors, large land rigs[1]

Drilling contractor	2007 rank[2]	1,500 hp	2,000 hp	Total
Helmerich & Payne International Drilling Co.	3	136	10	146
Nabors Drilling USA	2	91	21	112
Grey Wolf Drilling Co. LP	5	29	25	54
Unit Drilling Co.	6	28	8	36
Patterson-UTI Drilling Co. LP	1	26	10	36
Rowan Cos.	85	3	24	27
Cactus Drilling Co. LLC	13	16	6	22
Trinidad Drilling LP	10	15	—	15
Pioneer Drilling Co.	8	9	—	9
Cyclone Drilling Inc.	18	8	—	8
Scandrll Inc.	15	7	—	7
Bronco Drilling Co. Inc.	12	6	1	7
Total		**374**	**105**	**479**

[1]Land Rig Newsletter, June 2008.
[2]Land Rig Newsletter, May 2008; based on US footage drilled.

US drillers, August 2008[1]

Drilling contractor	Market capitalization, $ billion	Debt/ capital, %	Rank[2]	US land fleet
Land				
Patterson-UTI	4.505	0	1	3350
Nabors Industries Ltd.	10.702	44	2	336
Helmerich & Payne	6.542	17	3	181
Unit Drilling Co.	43.02		4	131
Grey Wolf Drilling	1.901	27	5	120
Total	**26.67**			**1,118**
Offshore			**Fleet size**	**Rigs under construction**
Transocean Inc.	44.173	49	145	10
Diamond Offshore Inc.	16.729	14	47	—
Noble Corp.	14.170	13	64	5
Nabors	10.702	44	73	—
ENSCO International	10.057	6	52	6
Pride International Inc.	6.994	24	47	3
Rowan Cos.	4.484	15	30	9
Hercules Offshore	2.173	36	66	—
Parker Drilling Inc.	0.946	39	18	—
Total	**110.428**		**542**	**33**

[1]Company reports and Lehman Bros. Equity Research, Original Oil Patch Weekly, Aug. 4, 2008. [2]Based on US land footage drilled, 2007. [3]Includes western Canada. [4]For Unit Corp., of which Unit Drilling is a subsidiary.

Piceance basin

Chevron Corp. (San Ramon, CA) started gas production in August 2008 in a major project in the Piceance basin of northwestern Colorado.

The $7.3 billion project aims to develop an estimated 3 tcf of gas over several decades.

Initial production was 5 MMcfd from the Cretaceous Williams Fork formation.

Chevron had drilled 82 wells and was laying plans to drill, fracture and complete a total of 2,000 to 3,000 gas-condensate wells in Garfield County.

Production capacity is intended to build to 400–450 MMcfd in several years.

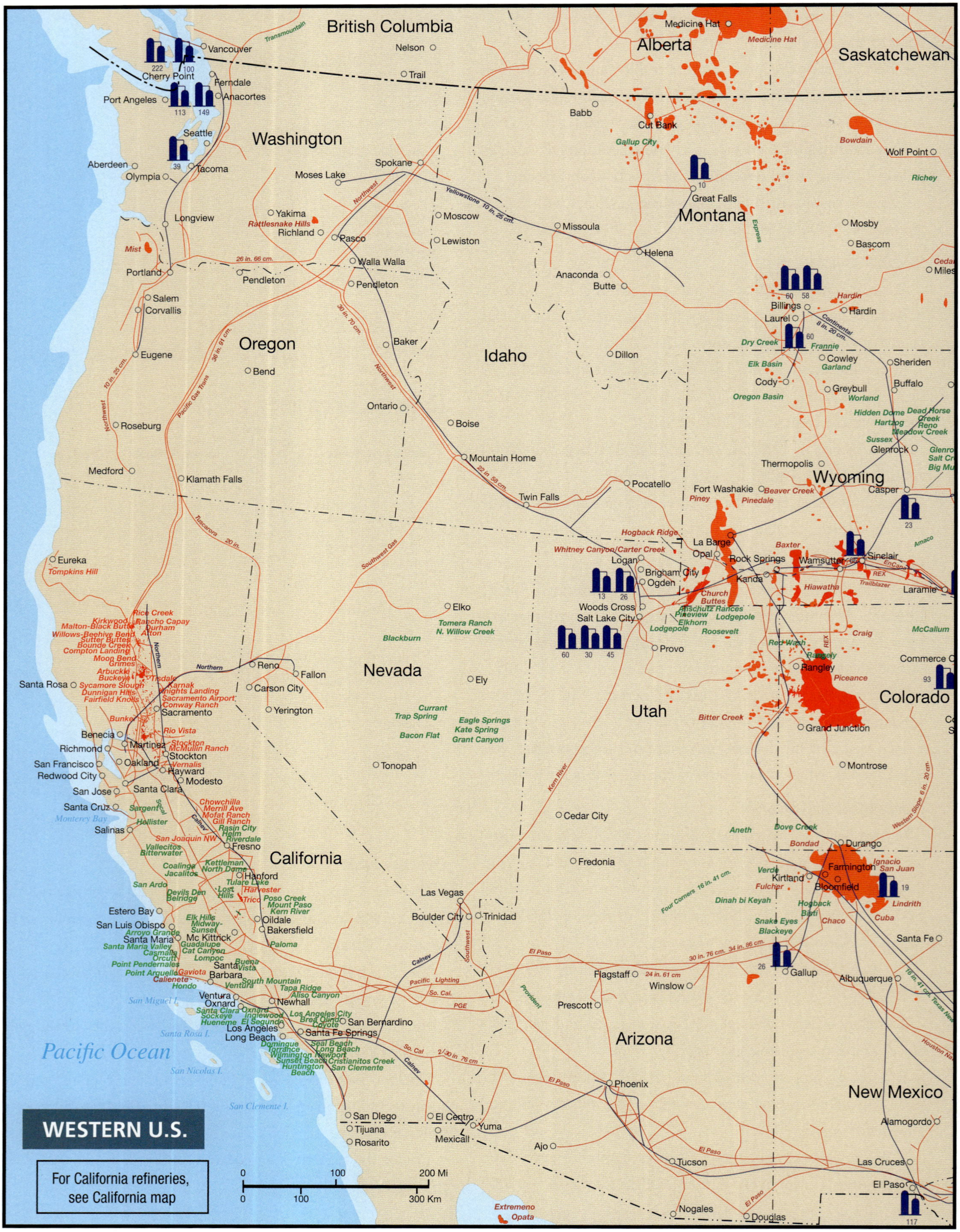
WESTERN U.S.
For California refineries, see California map
British Columbia
Alberta
Saskatchewan
Washington
Oregon
Idaho
Montana
Wyoming
Nevada
Utah
Colorado
California
Arizona
New Mexico
Pacific Ocean
Vancouver
Nelson
Trail
Cherry Point
Ferndale
Anacortes
Port Angeles
Seattle
Tacoma
Aberdeen
Olympia
Longview
Portland
Mist
Salem
Corvallis
Eugene
Roseburg
Medford
Klamath Falls
Bend
Pendleton
Yakima
Rattlesnake Hills
Richland
Pasco
Walla Walla
Moses Lake
Spokane
Moscow
Lewiston
Baker
Ontario
Boise
Mountain Home
Twin Falls
Pocatello
Missoula
Helena
Anaconda
Butte
Dillon
Great Falls
Babb
Cut Bank
Gallup City
Medicine Hat
Bowdain
Wolf Point
Richey
Mosby
Bascom
Billings
Laurel
Hardin
Dry Creek
Elk Basin
Frannie
Cowley
Garland
Cody
Oregon Basin
Greybull
Worland
Sheriden
Buffalo
Hidden Dome
Dead Horse Creek
Sussex
Glenrock
Thermopolis
Casper
Fort Washakie
Beaver Creek
Piney
Pinedale
La Barge
Opal
Rock Springs
Wamsutter
Sinclair
Laramie
Baxter
Hiawatha
Kanda
Hogback Ridge
Whitney Canyon/Carter Creek
Logan
Brigham City
Ogden
Woods Cross
Salt Lake City
Church Buttes
Lodgepole
Elkhorn
Roosevelt
Provo
Craig
McCallum
Red Wash
Rangely
Piceance
Commerce City
Grand Junction
Bitter Creek
Montrose
Eureka
Tompkins Hill
Elko
Tomera Ranch
N. Willow Creek
Blackburn
Reno
Fallon
Carson City
Yerington
Ely
Currant
Trap Spring
Eagle Springs
Kate Spring
Grant Canyon
Bacon Flat
Tonopah
Cedar City
Fredonia
Aneth
Dove Creek
Bondad
Durango
Farmington
Kirtland
Bloomfield
Ignacio
San Juan
Lindrith
Cuba
Chaco
Dinah bi Keyah
Snake Eyes
Blackeye
Gallup
Santa Fe
Albuquerque
Flagstaff
Winslow
Prescott
Phoenix
Ajo
Tucson
Nogales
Douglas
Alamogordo
Las Cruces
El Paso
Las Vegas
Boulder City
Trinidad
Santa Rosa
Benecia
Martinez
Richmond
Oakland
San Francisco
Redwood City
Hayward
San Jose
Santa Clara
Santa Cruz
Monterey Bay
Salinas
Sacramento
Stockton
Modesto
Fresno
Hanford
Oildale
Bakersfield
Estero Bay
San Luis Obispo
Santa Maria
Santa Barbara
Ventura
Oxnard
Newhall
Los Angeles
Long Beach
Santa Fe Springs
San Bernardino
San Miguel I.
Santa Rosa I.
San Nicolas I.
San Clemente I.
San Diego
Tijuana
Rosarito
El Centro
Yuma
Mexicali
Extremeno
Opata
0 100 200 Mi
0 100 300 Km

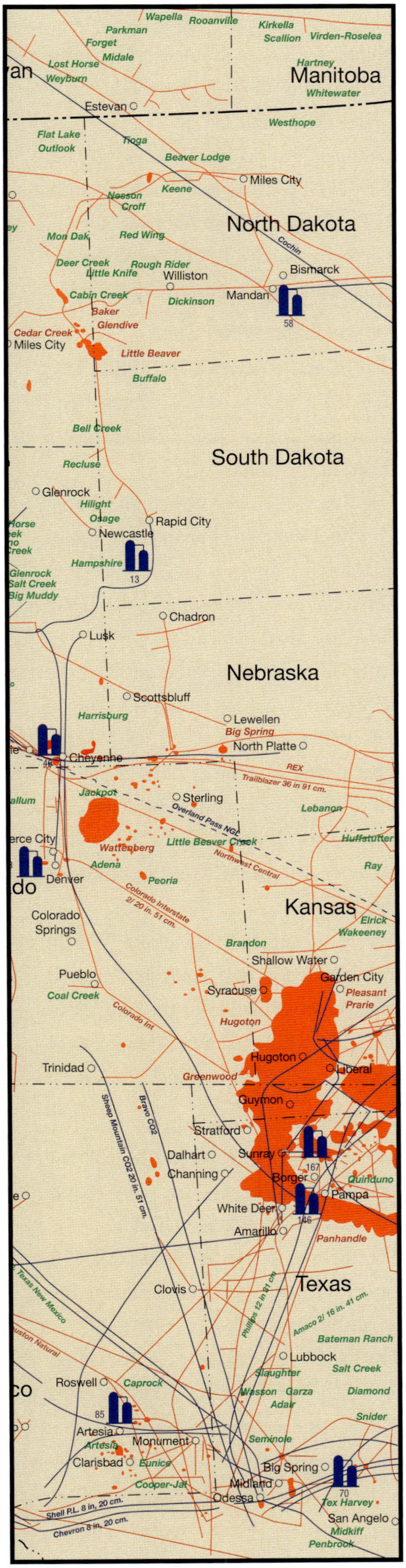

Amoruso field

EnCana Corp. (Calgary) drilled 11 net wells in its Amoruso field deep Jurassic Bossier gas play in East Texas in first quarter 2008 and was planning to drill a total of 35 wells there for the year.

Five wells brought on line in the quarter averaged 25 MMcfd; one exceeded 60 MMcfd.

EnCana said it expected production from the Amoruso to average 315–355 MMcfd in 2008, from approximately 200 drilling locations.

Colorado North Park basin

EOG Resources Inc. (Houston) was planning to drill seven wells in 2008 and to shoot 3D seismic survey in a horizontal play for oil in Cretaceous Niobrara in Colorado's remote North Park basin.

EOG estimated that the Niobrara has a reserve potential of 10 to 80 million boe at an implied recovery efficiency of 1–5%.

The firm's first well, Buffalo Ditch 1-32H, some 15 mi south-southwest of Walden, was drilled to 7,500 ft true vertical depth (plus a 4,000-ft single lateral) and fraced in late 2007.

The company expects meaningful production in 2009.

Vermillion Creek

Devon Energy Corp. (Oklahoma City) started production in late 2008 at the 5-3 Horseshoe Basin Unit well in the Vermillion Creek area of the Greater Green River basin, in Wyoming.

After initial flow, volumes from Cretaceous Baxter shale settled at 2.19 MMcfd and 412 b/d of condensate. TD is 13,534 ft.

Three wells had been drilled, and Devon was acquiring 3D seismic, aiming to pursue horizontal drilling in 2009 in the Baxter.

CO_2: Permian basin

In mid-2008, Occidental Petroleum Corp. (Los Angeles) said it will work with SandRidge Energy to increase substantially the amount of carbon dioxide (CO_2) available in the Permian basin for use in enhanced oil recovery projects.

The CO_2 will come from Oxy's planned gas processing plant to be built in Pecos County, Texas, which could be operational in 2011.

CO_2: New Mexico

Enhanced Oil Resources Inc. (Houston) August 2008 started CO_2 injection at Milnesand (San Andres) Unit in Roosevelt County, New Mexico.

The firm estimated that unit oil production could grow to 5,000 b/d when fully flooded from the present 70 b/d. The cumulative oil production target is 18 million bbl.

Full development awaits construction of a pipeline to deliver CO_2 from western New Mexico and eastern Arizona.

CO_2: Texas, Mississippi

Denbury Resources Inc. (Plano, TX) announced plans to acquire giant Conroe field north of Houston, where it intends to begin injecting CO_2 by 2013.

Potential at Conroe from CO_2 injection is estimated at 130 million boe, depending on the ultimate recovery factor. The properties contain 18.2 million boe of proved conventional reserves.

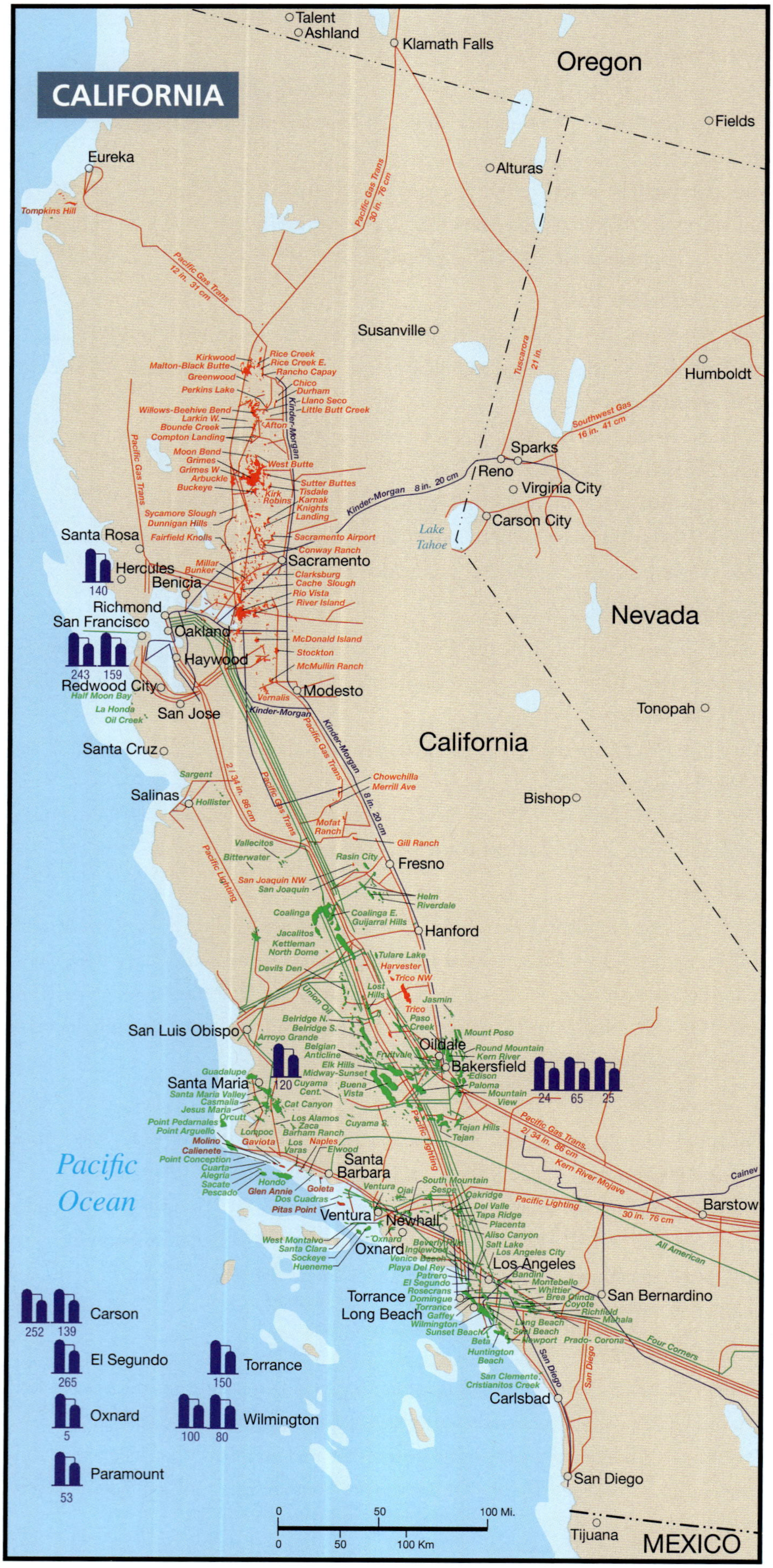

Shipping CO_2 to Conroe will require construction of an 80-mi pipeline spur, to be in service by 2012.

Denbury also booked 29.8 million bbl of proved oil reserves as of June 30, 2008, at giant Tinsley field in Mississippi.

Tinsley responded to CO_2 injection and became Denbury's highest valued field.

Three Forks/Sanish formation

Continental Resources Inc. (Enid, OK) completed a well on the Williston basin Nesson anticline in Dunn County, North Dakota, in mid-2008.

The Bice 1-29H well flowed oil and gas from the Devonian Three Forks/Sanish formation.

Seven-day average flows were 618 b/d of oil and 543 Mcfd of gas.

Only 52 wells have produced from the little-known formation in the U.S. part of the basin.

Continental and others are trying to learn if the formation may be a separate reservoir not drained by a horizontal completion in the Middle Bakken zone above it.

Encore Acquisition Co. (Fort Worth) also completed in mid-2008 its first well in the Sanish formation of the Bakken in the Williston basin. It was aiming to drill five more by yearend.

New ERD record

ExxonMobil Corp. (Irving, TX) reported in early 2008 the drilling of a record-depth-setting, extended-reach well.

The drilling platform was in 1,075 ft of water off southern California on the Spanish Bay prospect.

The well was drilled to a measured TD of 33,435 ft (TVD of 7,663 ft), with 29,720-ft displacement.

The company said that the well established a new North American extended reach drilling (ERD) record and a

new global ERD record from an offshore location.

Long Beach drilling

State officials approved legislation in October 2008 enabling further oil exploration and drilling in the western portion of Wilmington oil field in Long Beach, CA.

The field has produced more than 2.5 billion bbl since 1932.

The city and Occidental Petroleum Corp. (Oxy; Los Angeles) will negotiate a contract for the work, including evaluation of tidelands mineral deposits.

Oxy is expected to pursue enhanced recovery from some of the field's 500–700 existing wells, and could also drill up to 200 new wells.

Analysts say that the deal could boost reserves by 22 million bbl from the current 35 million bbl in the West Wilmington unit.

Cleveland formation

Infill horizontal wells helped redevelop the Cleveland formation that lies at 6,500–8,200 ft depths in the northeastern Texas Panhandle and Oklahoma.

The formation is an interbedded sequence of sand with thin shale laminations and fine-grained sandstone. Staff with BP PLC (London) described the success they and others had using this approach.

The success of the infill wells resulted from use of subsurface mapping, core analysis, openhole logging, and microseismic during fracturing to optimize horizontal well configuration and completion.

CLEVELAND HORIZONTAL PLAYS

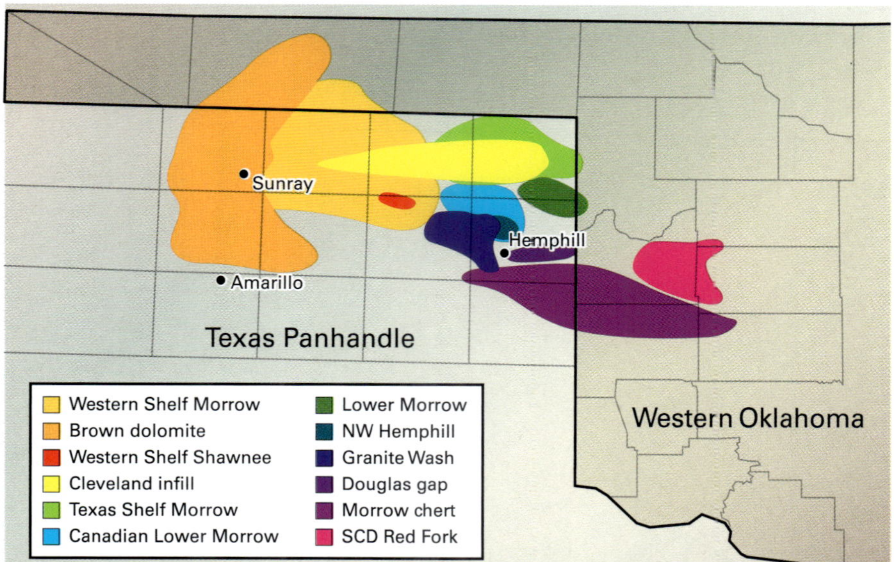

Source: Melvin, J., "BP Coiled Tubing Drilling Experiences," PTTP Workshop, Aug. 17, 2005, Houston

Studies showed that a fraced 1,600-ft horizontal well may provide a 1.4-MMscfd initial production rate and an ultimate recovery of about 1.4 bcf.

Bowdoin field

A decision to drill new and replacement wells at a rate equal to the retirement of no-longer-productive wells will keep the Bowdoin natural gas field in Montana active for another 35–50 years, said the U.S. BLM.

In December 2008, BLM officials approved a project by Fidelity Exploration & Production Co. (Bismarck, ND) and five other parties who want to drill up to 635 wells in the field.

Some replacement wells are to be drilled from the original drilling pads of the 1940s. Also, areas previously passed over will be tapped.

Drilling and Production: Oil Shales

Oil shales: Promising, challenging

In an October 2008 assessment, *OGJ* noted the promise and problems posed by U.S. oil shale resources.

On one hand, researchers say oil shale ultimately could prove the most practical option among unconventional hydrocarbon resources to add large quantities of proved U.S. reserves.

Yet, it could take another 10–15 years to determine whether more energy can be extracted from oil shale than is used to produce it.

Concentrated in northwestern Colorado, eastern Utah, and southwestern Wyoming, about 70% of oil shale is on federal land.

Government reports suggest the resource could add as much as 800 billion bbl of oil to U.S. reserves.

Companies typically have used heat to try to unlock the oil from the shale's kerogen. Researchers are now exploring use of carbon dioxide and special additives to "pull" the oil out of the shale.

BLM has approved leases—all in Colorado—for field-based research by a unit of Shell Exploration & Production Co. (Houston), and by Chevron Corp. (San Ramon, CA) and American Shale Oil LLC (Rifle, CO).

The University of Utah and three national laboratories are participating with industry in the research.

US OIL SHALE RESOURCES

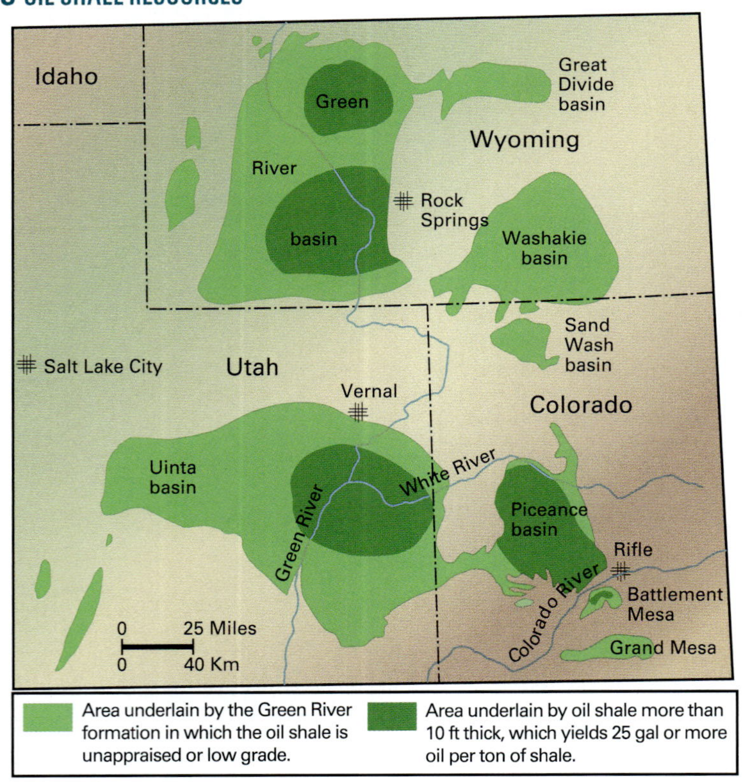

Source: Oil & Gas Journal

Bakken: New assessment

In April 2008, the USGS published an updated assessment of the Bakken oil formation.

The Bakken underlies about 25,000 sq mi of North Dakota, eastern Montana, northwestern South Dakota, and southern Saskatchewan and Manitoba, Canada.

The agency estimated that between 3.0 and 4.3 billion barrels of oil are undiscovered and technically recoverable from the Bakken shale formation—a 25-fold increase compared to its initial assessment in 1995.

USGS called the Bakken the largest "continuous" oil accumulation it has ever assessed.

The U.S. EIA said that the success of horizontal drilling and fracturing technology in Montana prompted the reevaluation of a 1995 USGS assessment that had estimated technically recoverable oil at only 151 million bbl.

Continental Resources Inc. (Enid, OK), the Bakken's largest oil producing company, has worked there since the 1980s.

USGS BAKKEN ASSESSMENT UNITS

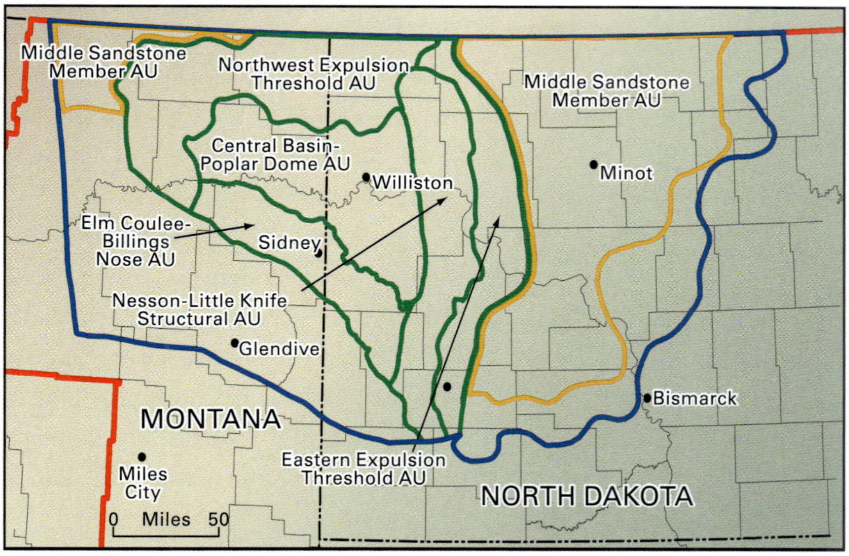

Map showing boundary of Bakken-Lodgepole Total Petroleum System (TPS) in blue, five continuous assessment units in green, and one conventional assessment unit in orange defined for the assessment of undiscovered oil resources in the Upper Devonian-Lower Mississippian Bakken formation in the US portion of the TPS. The outermost green line defines the area of oil generation for the upper shale member of the formation.

MONTANA BAKKEN EVOLUTION

	Wells	Initial production, bo/d	Estimated ultimate recovery, 1,000 bbl	Comment
2000	1	100	200	—
2001	12	160	220	Without refrac
2002	15	200	260	—
2003	42	330	450	Success is achieved
2004	100	280	400+	Includes infill wells
2005	150	300	400+	—

Source: Continental Resources Inc.

Five continuous assessment units (AUs) in the southern Williston basin were identified and evaluated. They are Elm Coulee-Billings Nose; Central Basin-Poplar Dome; Nesson-Little Knife Structural; Eastern Expulsion Threshold; and Northwest Expulsion Threshold.

USGS did not assess the Canadian portion of the basin.

Other active operators in the U.S. Bakken play besides Continental Resources are: Headington Oil Co., Lyco Energy Corp., and Petro-Hunt Corp. (all in Dallas); EOG Resources, ConocoPhillips, and Marathon Oil (all in Houston); Whiting Petroleum Corp. (Denver), and Hess Corp. (New York City).

USGS said about 105 million bbl of oil were produced from the Bakken through yearend 2007, with about 65 million bbl of that from the Elm Coulee field.

Bakken: Industry activity

Newfield Exploration Co. (Houston) accumulated 170,000 net acres in the Williston basin Bakken shale oil play from 2006–08 and was making plans to operate at least two rigs there in 2009.

The acreage includes areas along and west of the Nesson anticline, and in Elm Coulee field in Montana.

Continental Resources Inc. (Enid, OK) holds 577,000 net acres in the Bakken shale play, in Montana and North Dakota.

It is operating 10 rigs in North Dakota. Six target the Three Forks/Sanish formation, aiming to determine if that zone is separate from the overlying Middle Bakken over the majority of its acreage.

Marathon Oil Corp. was making plans to drill 65 company-operated wells in the Williston basin Bakken shale oil play in 2008.

The company, which already has some Bakken oil and gas production, has seven rigs working and expected to have 100 wells in the play by yearend.

Drilling and Production: Gas Shales

Gas shales snapshot

The fastest growing sources of U.S. gas production have been shale plays, through the advancement and expanding use of horizontal drilling and hydraulic fracturing, said *OGJ* in early 2009.

There are more than 20 shale basins spread across as many states.

The Barnett in North Texas was the first gas shale formation to receive wide attention. Production there rose from virtually nothing in 1999 to more than 5% of total Lower 48 gas production in 2007.

Now, however, emerging plays are drawing increased interest. These include the Fayetteville shale of north-central Arkansas, the Haynesville shale of northwestern Louisiana, and the Marcellus shale in the Appalachians.

Barnett Shale Wells in 19 Texas Counties

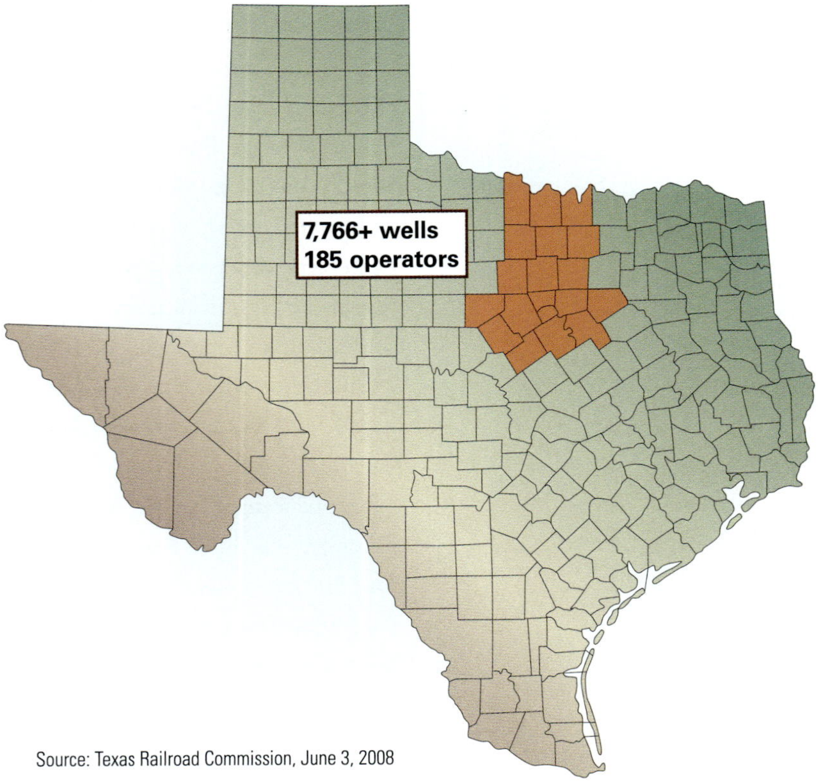

Source: Texas Railroad Commission, June 3, 2008

Others also drawing attention are the Woodford in eastern Oklahoma, and the Green River Piceance basin play in Colorado.

The Natural Gas Supply Association (NGSA; Washington, DC) said production from U.S. gas shale plays could double to 15–20 bcfd in the next 10 years and provide 25% of U.S. supply, compared to about 10% in 2008.

Gas shales impact

The American Clean Skies Foundation (Washington, DC) released a study in mid-2008 that found the United States has 2,247 tcf of gas reserves—a 118-yr supply at 2007 demand levels.

Navigant Consulting Inc. (Chicago) conducted the study for the foundation.

Navigant said that horizontal drilling and other new technologies have opened supplies in shale formations over the last several years, boosting reserve estimates.

Gas production from shale grew from less than 1 bcfd in 1998 to about 5 bcfd in 2008.

Some observers thought the study might dampen calls to open more of the OCS for exploration.

Haynesville

A Jurassic Haynesville shale play in Louisiana, with just a few wells producing gas so far, could eventually be worth 7.5–20 tcf of recovery net to Chesapeake Energy Corp. (Oklahoma City), said the company in March 2008.

The 20-tcf figure depends on securing 500,000 net acres in the play, compared with the 200,000 acres the firm held in March.

Haynesville could have "a larger impact on the company than any other play in which it has participated to date," said Chesapeake.

Others also are venturing into the Haynesville.

Penn Virginia Corp. (Radnor, PA) has drilled and completed 17 vertical wells on its East Texas leasehold and tested four reservoirs below the Jurassic Cotton Valley. Five more wells were planned for 2008.

Goodrich Petroleum Corp. (Houston) was testing early wells, acquiring acreage, and forming up a drilling program in Bethany-Longstreet and Longwood fields in Northwest Louisiana.

Three operators reported new horizontal completions in late 2008 in the Jurassic Haynesville: Petrohawk Energy Corp. (Houston); and Comstock Resources Inc. and EXCO Resources Inc. (both of the Dallas area) plan much more activity in the Haynesville in East Texas and Northwest Louisiana.

Petrohawk reported a flow rate of 28.2 MMcfd rate at its Sample 9-1 well about 12 mi south of Elm Grove gas field, in Louisiana.

Marcellus: Atlas Energy

As of late 2008, Atlas Energy Resources (Moon Township, near Pittsburgh)—largest gas producer from Devonian Marcellus shale in the Appalachian basin—reported 90 wells total in Pennsylvania, producing a combined 25 MMcfd from the Devonian Marcellus.

Atlas said cumulative production topped 4 bcf.

The company was making plans to drill 32 vertical wells and 12 horizontal wells by April 2009, plus 75 more vertical and 12 more horizontal wells in the rest of 2009.

Atlas said it was moving 60 MMcfd, and was planning to expand system capacity to 250 MMcfd by yearend 2009.

Atlas Energy's North Appalachian Operating Area

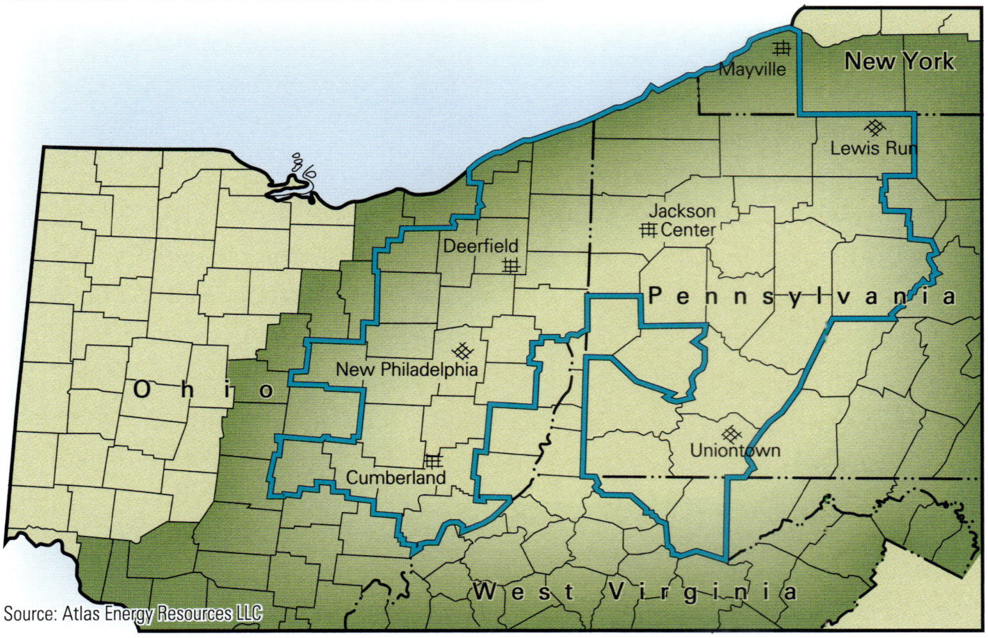

Source: Atlas Energy Resources LLC

Extent of Marcellus, Devonian Shales

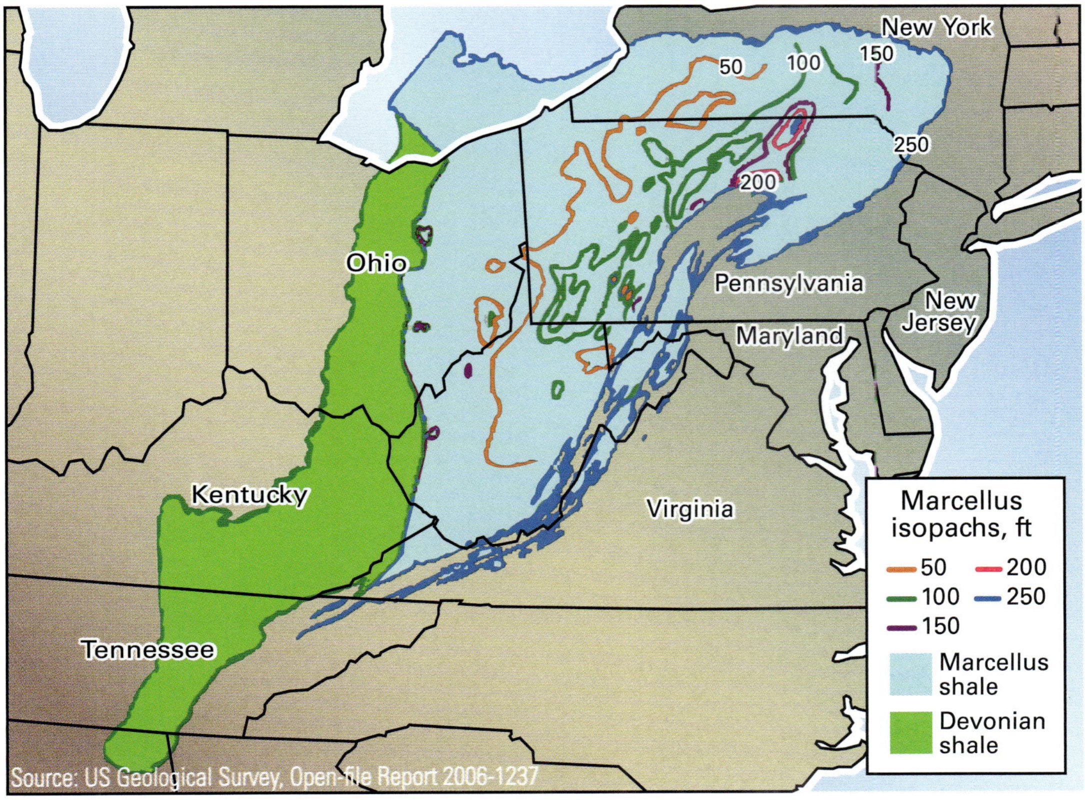

Source: US Geological Survey, Open-file Report 2006-1237

The firm previously estimated that it could ultimately recover 4–6 tcf from the Marcellus on its properties mostly in southwestern Pennsylvania.

Marcellus: Range Resources

Range Resources Corp. (Fort Worth) said seven of its wells totaling 30 MMcfd from the Marcellus shale are connected to Pennsylvania's first large-scale gas processing plant, operated by MarkWest Energy Partners LP (Denver).

Marcellus challenges

The growing potential of the Marcellus shale deep-gas formation is prompting action by producers, landowners, state and local officials, and others to address development issues.

Of particular concern are:

- General education about industry procedures in regions that have never had oil and gas production.
- Special aspects of larger-scale production, in regions (e.g., parts of Pennsylvania) where the historical focus has been on smaller wells.
- Water use, conservation and produced-water management
- Environmental impacts of hydraulic fracturing operations
- General environmental impacts of industry operations.

Major Marcellus player Chesapeake Energy Corp. (Oklahoma City) said the rapid ramp-up of gas production from the Marcellus would be slower than that seen in the Barnett shale.

The firm does not expect a meaningful contribution from the Marcellus until 2013–2015.

The firm plans to begin flowing more wells when two more gas processing plants are completed in 2009.

Plans call for having six horizontal drilling rigs operational by yearend 2009, aiming to reach net production of 80–100 MMcfed.

Delaware basin

The Mississippian Barnett shale has the potential to be a prolific gas producer in the Delaware basin, but it will take time, said geoscientists in the AAPG Bulletin in late 2008.

One well is reported to have an estimated ultimate recovery of 9 bcf.

Study authors noted that drilling and stimulation costs to as deep as 18,000 ft are now about $8 million, down from more than $18 million in earlier years.

Both the Barnett and Devonian Woodford shales should be evaluated, said the study.

The first Delaware basin shale gas wells were drilled in southwestern Reeves County in 2002.

Maverick basin

Early gas flow rates in 2008 from a horizontal Cretaceous Pearsall shale well in the Maverick basin of southwest Texas were encouraging, said TXCO Resources Inc. (San Antonio, TX).

Marcellus shale wells, New York State*

Operator	Well permits applied for	Well permits granted	Wells drilled
Chesapeake Appalachia LLC	7	5	2
Houghton College	1	1	1
Petroleum Development Corp.	1	—	—
Texas Keystone Inc.	1	1	—
East Resources Inc.	2	1	1
Nornew Inc.	11	1	—
Frontier-Kempner/Flatiron JV	6	6	6
EOG Resources Inc.	13	12	5
Fortuna Energy Inc.	19	12	11
East Resources Inc.	3	1	—
Eastern American Energy Corp.	1	1	1
PetroEdge Resources LLC	9	6	3
Glenn O. Tiffany	1	1	1
MegaEnergy Operating Inc.	2	—	—
Total	**77**	**48**	**31**

*Based on New York State Dept. of Environmental Conservation wells database search, July 28, 2008.

Chesapeake Appalachia's Marcellus shale wells*

Well name	County	Type	Proposed TD, ft	Permit application filed	Permit application approved	Well spud	Confidental?
Beagell 2-B	Broome	Vertical	4,800	Aug. 17, 2007	Oct. 17, 2007	Oct. 25, 2007	No
Butkowsky 1-B	Broome	Vertical	4,800	Aug. 17, 2007	Oct. 17, 2007	—	No
Trimber 1-B	Chemung	Vertical	3,900	July 13, 2007	July 23, 2007	Aug. 20, 2007	Yes
Bartlett 1-A	Chemung	Horizontal	3,702	Feb. 15, 2008	—	—	Yes
Bartlett 2	Chemung	Horizontal	3,785	Feb. 15, 2008	—	—	Yes
Manwaring 1-B	Tioga	Vertical	7,850	Feb. 5, 2008	Feb. 28, 2008	—	Yes
TGS Holdings	Tioga	Vertical	4,050	May 7, 2008	May 13, 2008	—	Yes

*Data from New York State Dept. of Environmental Conservation, wells database search, July 28, 2008.

The Pearsall shale—for decades a marginal producer from vertical wells—flowed 3.5 MMcfd with 2,500 b/d of frac fluid after a mid-2008 five-stage frac job by TXCO at a horizontal well.

The rate later improved to 3.8 MMcfd.

Myers 2-683H was the company's third horizontal Pearsall well to be frac-treated in 2008.

Two earlier horizontal wells gave up about 1 MMcfd each after open hole completions.

Black Warrior basin

Two emerging gas shale plays in Alabama's Black Warrior basin and geologic provinces to its east drew increased producer attention in 2008.

One play is for gas in Middle to Late Cambrian Conasauga. The other is for gas in Mississippian Floyd shale and Devonian Chattanooga shale.

Conasauga play

The Conasauga play is more advanced and has the only production, Big Canoe Creek field in St. Clair County.

The top three Conasauga players are the 50-50 combine of Energen Resources Corp. (Birmingham) and Chesapeake Energy Corp. (Oklahoma City); and HighMount Black Warrior Basin LLC (Birmingham).

Virtually all of the Conasauga exploration has taken place in St. Clair County. Other areas of interest are in Etowah, Cullman, Bibb, Greene, and Cherokee counties.

ALABAMA SHALE PLAY AREAS

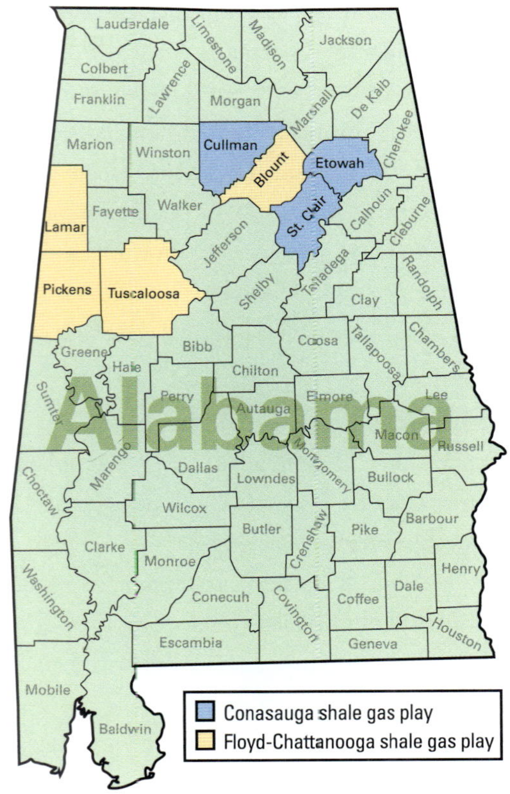

MAVERICK BASIN PEARSALL-EAGLEFORD ACTIVITY

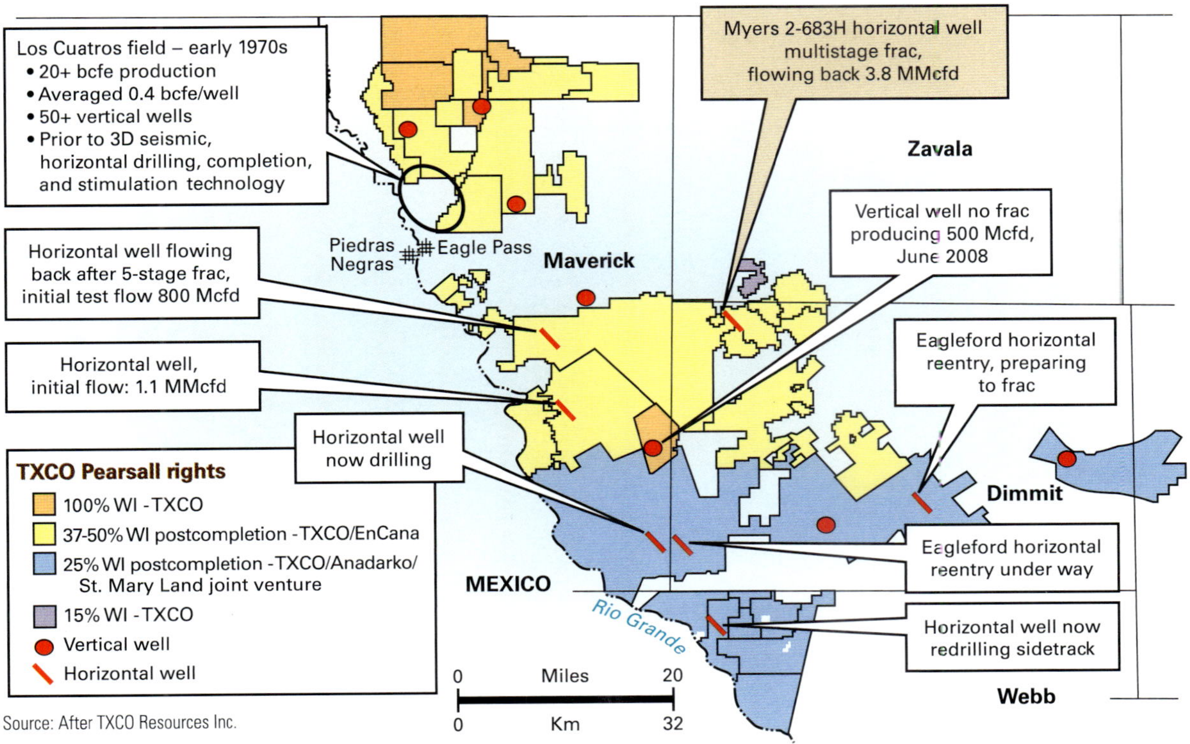

Source: After TXCO Resources Inc.

Floyd-Chattanooga play

Exploration for Floyd-Chattanooga gas has occurred across a wider area than the Conasauga, but has resulted in little gas recovery. Most drilling has been directed at the Floyd shale, in Lamar, Pickens, Tuscaloosa, and Blount counties.

Active companies include Denbury Offshore LLC (Plano, TX), Samson Resources Co. (Tulsa), Murphy Exploration & Production Co.-USA (Houston), Wagner & Brown Ltd. (Midland, TX), Elysium Energy LLC (Denver), Jim Walter Resources Inc. (Brookwood, AL), Geomet (Houston), and Shannon J. Flicklinger (West Salem, IL).

Conasauga, Chattanooga gas shales

Energen Corp. (Birmingham, AL) was planning further exploratory tests for 2009 of the Conasauga and Chattanooga gas shales in Alabama.

The Conasauga is a carbonate-dominated sequence of interbedded shales and carbonates alternating every few feet, posing unique development challenges. Work will focus on prospects on south of the Appalachian thrust.

The Chattanooga shale appears rich in organics, usually in deposits 25–40 ft thick.

However, Energen said is has identified a subbasin along the Appalachian thrust in which intervals of the Devonian age Chattanooga shale are 100 ft thick or more.

Attempts at hydraulic fracturing yielded disappointing results from the Chattanooga, but the company saw promise in an unusually thick Floyd shale formation encountered during related operations.

Raton basin Pierre shale

Pioneer Natural Resources Co. (Dallas) estimated in early 2008 that a shale gas play on the Colorado side of the Raton basin had recovery potential exceeding 2 tcf of gas net to the firm.

Planned Raton basin drilling of 175 wells in 2008 was to include 15 Pierre shale wells.

Pioneer said five wells were producing 2 MMcfd of gas from one of five prospective zones in Cretaceous Pierre laminated shale, and five more delineation wells were in early stages of completion.

The company expects its Pierre proved reserves to exceed 200 bcf by yearend 2010.

Energen's Alabama shale plays

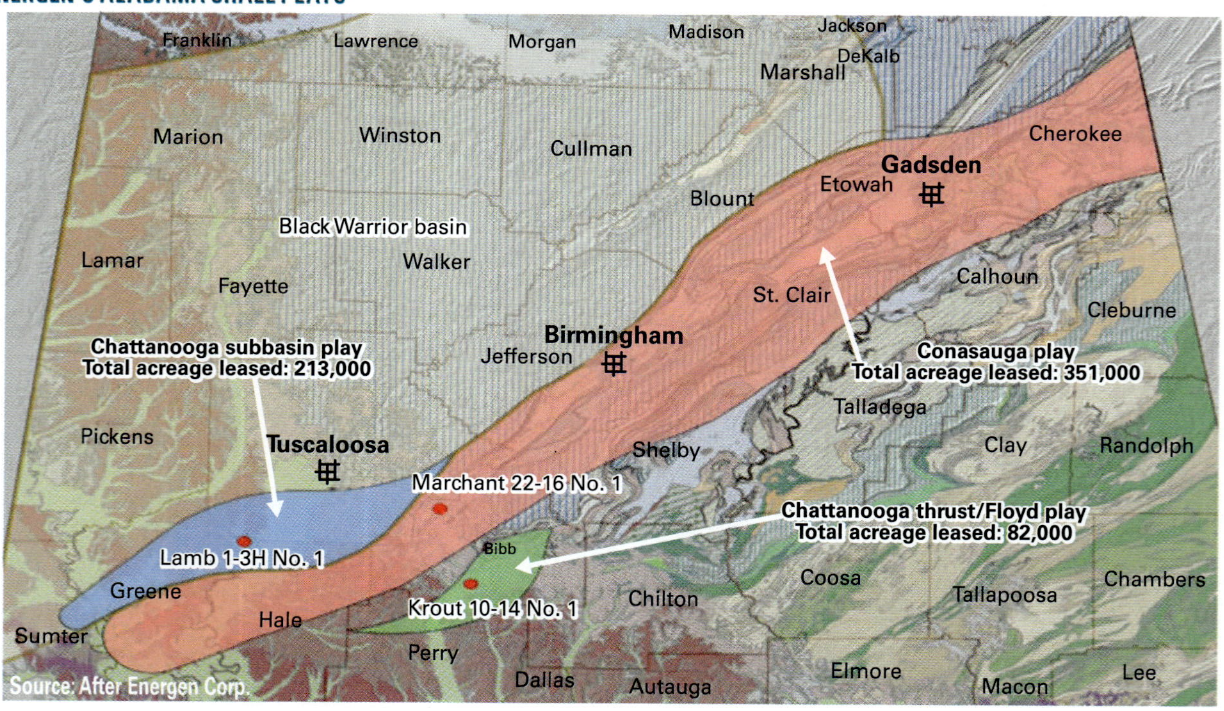

Where Newfield is pursuing Woodford shale gas

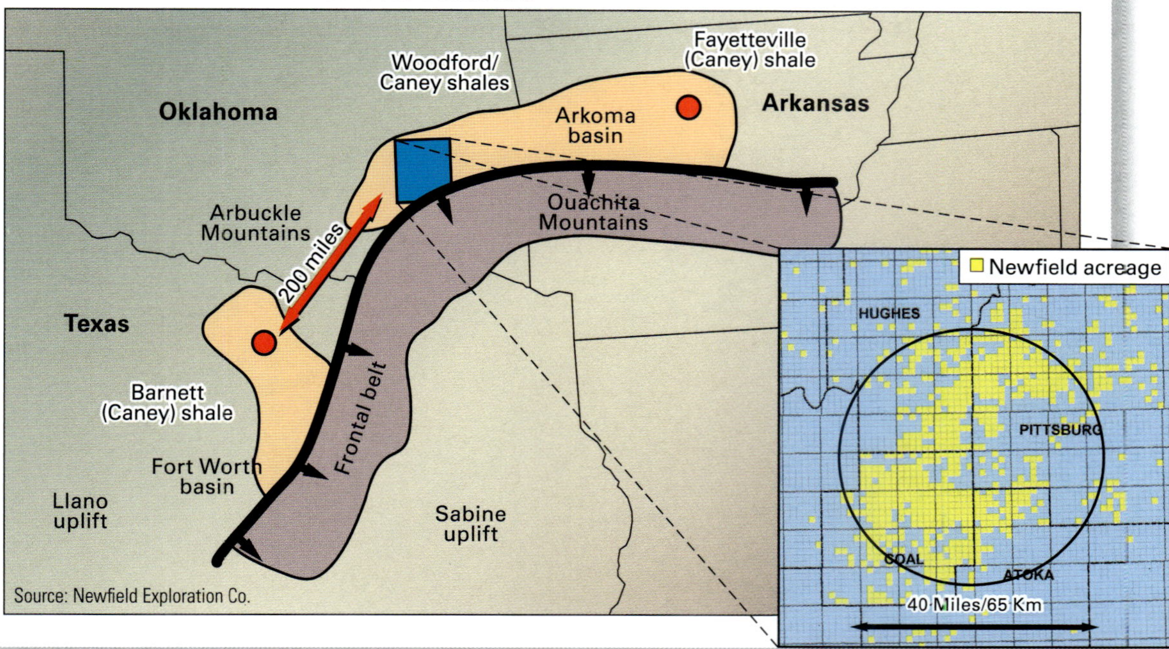

Source: Newfield Exploration Co.

Arkoma basin Woodford shale

In the Arkoma basin Woodford shale gas play, Newfield Exploration Co. (Houston) was finding value in longer horizontal lateral drilling in 2008. The firm also was planning to try multilateral drilling.

Some analysts think that the Woodford acreage in southeastern Oklahoma could be as prolific as the Barnett shale in North Texas.

Newfield was planning to drill about 100 operated horizontal wells and participate in 60–70 others in 2008, with half likely to involve laterals of more than 3,000 ft.

Ardmore basin Woodford shale

Operators actively pursued gas and liquid hydrocarbons in 2008 in Mississippian-Devonian Woodford shale in the northwestern Ardmore basin in southern Oklahoma.

Bankers Petroleum Ltd. (Vancouver, BC), which calls the field Tishomingo, planned to drill 30 wells by yearend.

The wells produce from the Woodford, which averages twice the formation's thickness in the Arkoma basin.

Five Bankers wells on production in Tishomingo in May were producing a combined 3.1 MMcfd of gas equivalent, 64% gas.

Ardmore-Arkoma Woodford comparison

	Ardmore	Arkoma
Depth range, ft	6,000-11,000	6,000-12,000
Gross thickness, ft	320-380 (345 avg.)	120-220 (150 avg.)
Avg. %Ro	[1]0.9-1.01	1.1-3
Total organic carbon, %	6.04-6.54	3-10
Silica content, %	52.5-66.5	60-80
Gas in place, bcf/sq mile	80-300	40-120
Drilling costs, million $	3.5	[2]7
Per well recoveries, bcfe	[3]3-5	[2]3-4.5
Gas content, btu/Mcf	~1,300	[4]1,050

[1]Suppression corrected. [2]Newfield 4.5 bcfe from laterals greater than 3,000 ft, Feb. 4, 2008, press release. [3]Low case MHA reserves report undeveloped case, high probability case on 10 proved undeveloped locations. [4]Newfield March 2008 Raymond James presentation.
Source: Bankers Petroleum Ltd.

Other operators in the area include Wagner & Brown (Houston), Chesapeake Energy Corp. (Oklahoma City), Range Resources Corp. (Fort Worth), and Walter Oil & Gas Corp. (Houston).

Anadarko basin Woodford shale

Cimarex Energy Co. (Denver) was pursuing gas and oil in Devonian Woodford shale with horizontal wells in the Anadarko basin in western Oklahoma.

The company had participated in 28 wells by October 2008, of which 16 were completed and 12 were still drilling or being completed.

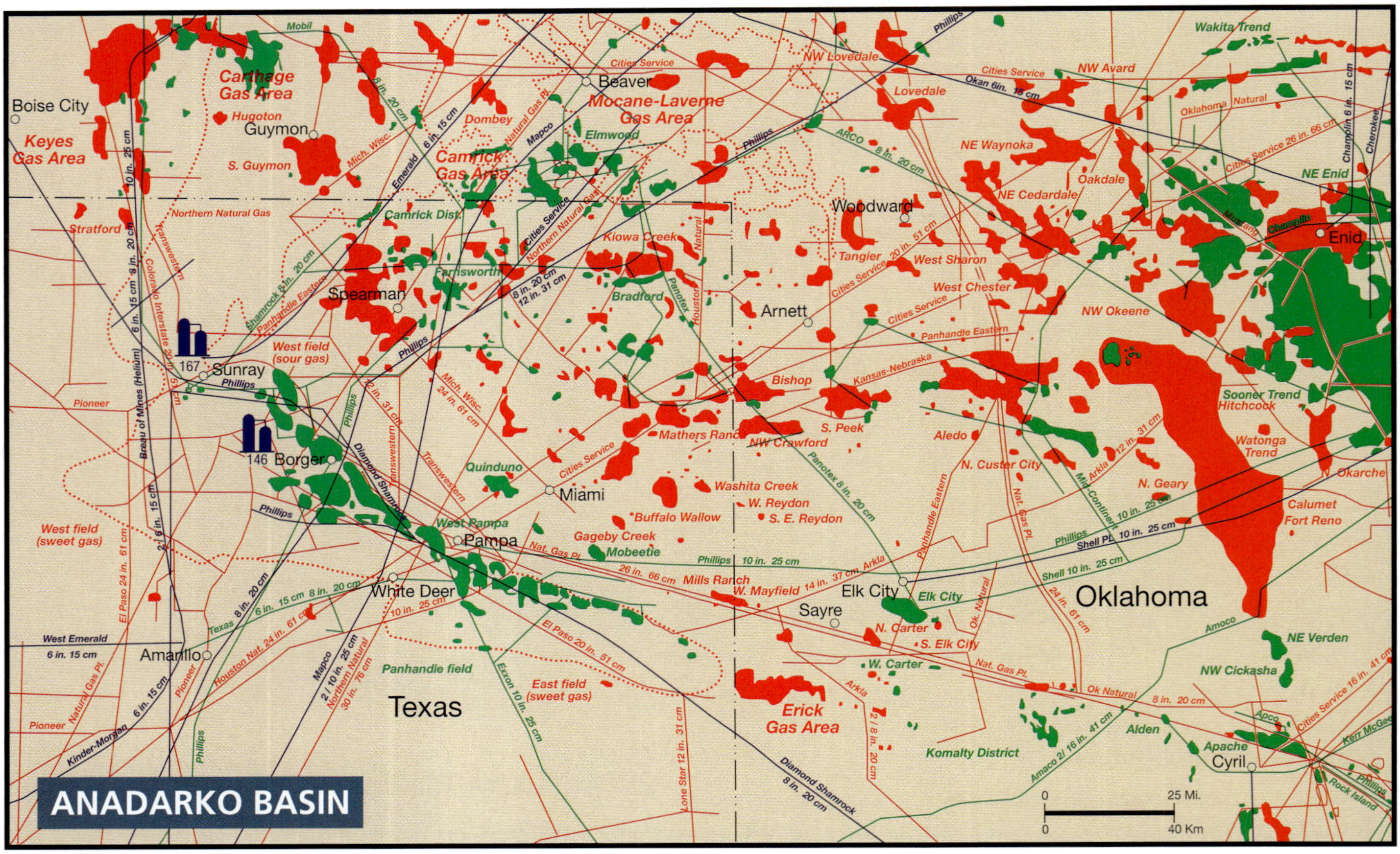

Cimarex said it expected to be running 9 to 11 rigs in the spring of 2009.

The firm estimated a potential of 1.5 to 2 tcf recoverable to the company from the play.

Other interested operators are Linn Energy LLC (Houston); Continental Resources Inc. (Enid, OK); and Devon Energy and Western Oil & Gas Development (both in Oklahoma City).

Nora field, Huron shale

Range Resources Corp. (Fort Worth) completed drilling its fifth horizontal well to Devonian Huron shale in Nora field in southwestern Virginia in late 2008.

The company, which says the Huron produces gas from 107 vertical wells in the field, estimated net reserve potential at Nora from horizontal drilling at 8–1.5 tcf.

Plans called for drilling five more Huron shale wells by yearend 2008.

New Albany shale research

GTI (Des Plaines, IL) took initial steps to lead a research consortium aiming to turn noncommercial wells into commercial producers in the Devonian New Albany shale in the Illinois basin.

GTI signed a contract for a multiyear program with the Research Partnership to Secure Energy for America (Sugar Land, TX).

RPSEA supports research to increase U.S. gas supply and lower costs for consumers. GTI is a research, development, and training organization.

The consortium, involving GTI and 14 participants, will target the 10.5 tcf of technically recoverable gas in the New Albany shale formation.

Fayetteville output report

Southwestern Energy Co. (Houston) said in mid-2008 that its gross operated production from the Fayetteville shale in Arkansas was 500 MMcfd of natural gas.

This included 12 MMcfd from 14 wells producing from conventional reservoirs.

The production rate a year ago was 200 MMcfd.

The company projected that estimated ultimate recovery could be improved by 15% using closer perforation cluster spacing, which it has tested.

The company had drilled and completed 619 operated wells, 554 of them horizontal, as of June 30.

Others with substantial acreage or operations in the play include Chesapeake Energy Corp. (Oklahoma City), XTO Energy Inc. (Fort Worth), Petrohawk Energy Inc. (Houston), and Edge Petroleum Corp. (Houston).

Paradox basin: Colorado, Utah

Bill Barrett Corp. (Denver) was spudding in mid-2008 its next well in the Yellow Jacket gas shale project, an emerging two-formation play in the Paradox basin of Colorado and Utah.

Targets are the Pennsylvanian Hovenweep shale and Gothic shale, atop and below the Lower Ismay formation. Hovenweep covers 1,300 sq mi in the basin and Gothic 1,850 sq mi.

In San Juan County, Utah, Barrett and partners were making plans to drill one well in 2008 in its Green Jacket and Pine Ridge projects.

Green Jacket targets the Hovenweep shale. Pine Ridge objectives are structural salt flank plays.

Lease Sales

OCS bans lifted

On July 14, 2008, Pres. George W. Bush lifted a presidential withholding of oil and gas acreage leasing on the Outer Continental Shelf.

He took the action in the face of sharply rising prices for crude oil and gasoline.

At the time, Congressional leasing moratoria still were in place on 85% of the OCS.

The president's action prompted two months of debate about their impact on U.S. consumers.

In the end, Democratic leaders in the House chose not to try to reimpose the Congressional moratoria when they expired at the end of September.

The issue likely will be revisited in 2009 by the Obama Administration.

New OCS leasing program

In September 2008, the MMS began to take steps for developing a new five-year oil and gas leasing program for the U.S. OCS.

The current program, which will end in 2012, includes 21 lease sales in eight of the 26 OCS planning areas in the Gulf of Mexico, Alaska, and the Atlantic.

Utah parcels

The Utah branch of the BLM agreed to defer from a scheduled December 2008 lease sale several parcels located near national parks in southeastern and eastern Utah.

The National Park Service (NPS) had expressed concern about the environmental impact of development of the parcels.

The two agencies also agreed to renew a prior MOU regarding how they will work together on prospective oil and gas lease offerings.

Virginia OCS

In November 2008, MMS took a first step toward holding a possible lease sale for acreage offshore Virginia. Lease Sale 220 is proposed for 2012.

MMS published notice of its intent to prepare an environmental impact statement and sought public comments by January 13, 2009.

The sale could lead to the first new exploration and production leases off the U.S. East Coast in three decades.

The proposed lease area includes 2.9 million acres beginning about 50 mi offshore. MMS said the area could contain 130 million bbl of oil and 1.14 tcf of natural gas.

PROPOSED LEASE SALE 220 AREA OFFSHORE VIRGINIA

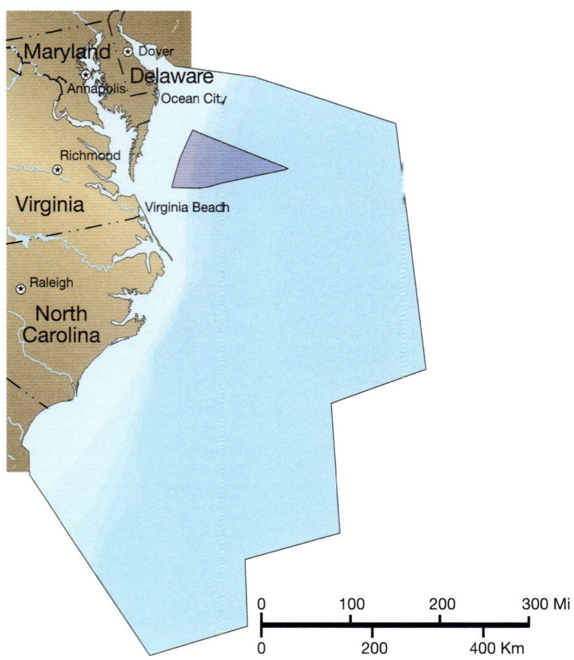

Roan Plateau

The long-anticipated Roan Plateau lease sale in August 2008 netted $113.9 million, a record for the U.S. BLM in the Lower 48, said the agency.

But the amount fell short of the $2 billion that many politicians and groups in Colorado anticipated.

The Colorado governor and several other Colorado Democrats in the Congress advocated leasing federal land on the western Colorado plateau in phases instead of holding a full lease sale.

First OCS revenue shares

The U.S. MMS proposed regulations in May 2008 to distribute qualified federal OCS oil and gas revenues to four Gulf Coast states and their eligible coastal governmental subdivisions.

The distributions will take place under the 2006 Gulf of Mexico Energy Security Act, which established federal OCS revenue sharing for affected coastal states and communities.

The law authorized that 37.5% of all federal OCS revenue from new leases in the gulf be shared with Alabama, Mississippi, Louisiana, and Texas and their coastal communities and counties/parishes.

The first sale with immediate revenue-sharing leases, OCS Lease Sale 224, was held March 19.

MMS calculated payments for fiscal 2008 to be 30% for Alabama, 27% for Mississippi, 32% for Louisiana, and 11% for Texas.

LNG

OceanWay: Quake threat?

Pipelines from a proposed deepwater LNG terminal off Southern California face a 16–48% probability of a magnitude-6.5 earthquake within 30 mi of their route, the USGS reported in December 2008.

The planned pipeline for the OceanWay Secure Energy project would run from 23 mi offshore in Santa Monica Bay to a terminus near Los Angeles International Airport.

The proposed deepwater LNG project would be sited in 3,000 ft of water and would be connected to onshore systems by twin 24-in pipelines to onshore systems 35 mi away.

Sparrows Point

The U.S. FERC identified more adverse environmental impacts as it issued a final environmental impact statement (EIS) in early December 2008 for a proposed LNG project near Baltimore, MD.

AES Corp. (Arlington, VA) plans to build the terminal at Sparrows Point. The facility would store up to cu ft (480,000 cu m) of LNG and send out 1.5 bcfd of revaporized natural gas.

Maryland state and local government officials have expressed concerns about safety, security and environmental impact, as well as potential disruption of Baltimore Harbor traffic by LNG tankers.

Bradwood Landing

The FERC approved in September 2008 the first new LNG terminal and related send-out pipeline in the Pacific Northwest, after a review lasting more than three years.

Developers of Bradwood Landing, on the Columbia River about 20 mi east of Astoria, OR, must implement 109 mitigation measures as a condition of the approval, FERC said.

The facility would deliver up to 1.3 bcfd of gas. It will have a single marine berth, two insulated storage tanks, and related facilities.

A 36-mi pipeline would connect it with the Williams Northwest pipeline north of Kelso, WA.

Quintana, Sabine Pass

Terminal commissioning was ongoing in the first half of 2008 at two Texas Gulf Coast LNG terminals.

The Quintana terminal of Freeport LNG Development LP, about 70 mi south of Houston, began receiving its first cool-down shipment in April. Start-up was planned to occur by yearend.

In August 2008, Freeport LNG became the first U.S. firm to request permission to export LNG imported into its terminal. It sought "blanket authorization" for short-term exports totaling 24 bcf—about eight tanker cargoes—over two years.

The Sabine Pass terminal of Cheniere Energy—in Cameron Parish, LA, along the Sabine River border near Port Arthur, TX—also received its first cargo for commissioning in April.

Two other terminals in the area were scheduled to begin commissioning by yearend 2008 or in first quarter 2009. They are:

- ExxonMobil's Golden Pass terminal, across the Sabine River from the Cheniere facility.
- The Cameron terminal of Sempra LNG, east of Sabine Pass, in Hackberry, LA, on the Calcasieu Channel about 18 mi from the Gulf of Mexico.

Cove Point expansion

A federal district court in July 2008 overturned the approval that had been granted by the FERC for a proposed expansion of the Cove Point LNG facility.

US GULF COAST LNG TERMINALS

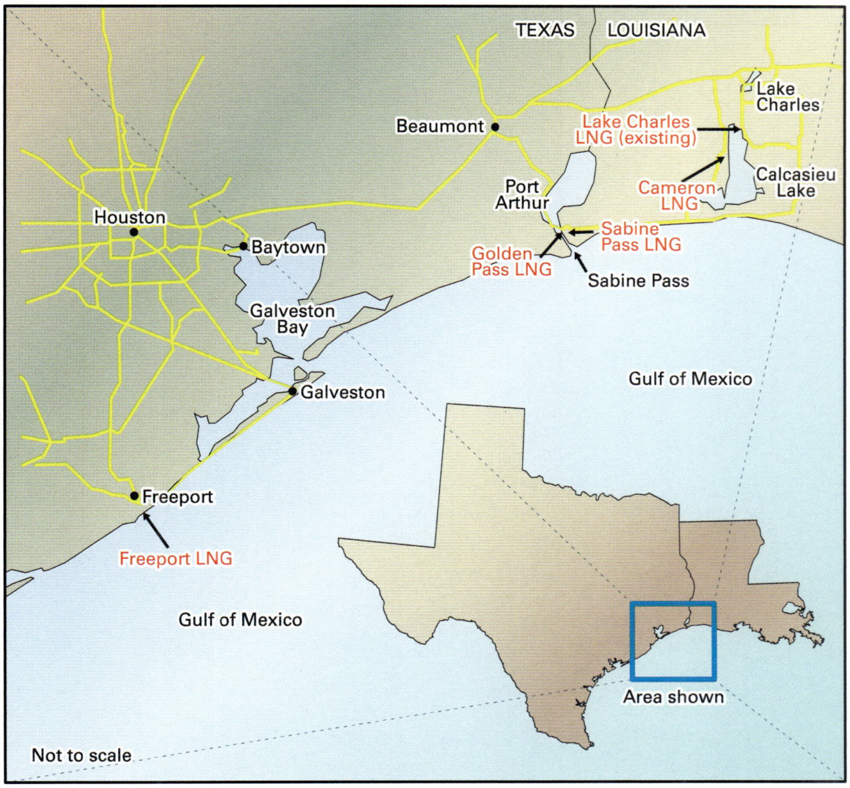

The terminal, owned by Dominion Resources (Richmond, VA) is on Chesapeake Bay, south of Baltimore, MD.

The court rejected FERC's conclusion that WGL could address safety concerns before the project's scheduled November 2008 in-service date.

WGL's petition said that expansion would introduce more LNG into the utility's system and that doing so would likely increase leakage from the WGL distribution network

Valdez project

Mitsubishi Corp. (Tokyo) signed in June 2008 an agreement with the Alaska Gasline Port Authority (AGPA) to develop an LNG project in southern Alaska.

A spur pipeline would carry North Slope gas to a liquefaction plant and export facility at Valdez, following the existing trans-Alaska oil pipeline from a point near Delta Junction.

Sempra LNG also had joined the Alaska project, said AGPA. Sempra owns the Energia Costa Azul regasification facilities just below the U.S.-Mexico border.

Mitsubishi had decided in 2008 to abandon plans for a regasification terminal at Long Beach, California, that drew strong protest from local residents.

Neptune project

Construction began in July 2008 on the Neptune LNG project, offshore Gloucester, MA.

Suez Energy NA said that the first construction phase includes installation of a 13-mi subsea pipeline from the Neptune facility to the existing onshore Spectra Energy HubLineSM.

The second phase, slated for completion by late 2009, will connect the pipeline to HubLineSM and install two offshore offloading buoys.

Weaver's Cove

The U.S. Department of Commerce (DOC) upheld an objection by Massachusetts to a proposal by Weaver's Cove Energy LLC and Mill River Pipeline LLC to build an LNG terminal and pipeline near Fall River.

The state had determined that the project was not consistent with its plan under the Coastal Zone Management Act.

DOC upheld the state's objection because it determined that adverse coastal effects outweighed the project's national interest.

In particular, navigation-related concerns were raised regarding movement of LNG tankers on the Taunton River.

Northeast Gateway

In June 2008, the Northeast Gateway (NEG) Deepwater LNG Port, 18 mi off Boston, MA, began commercial operation.

Owner Excelerate Energy LLC (Houston) said its Excellence regasification vessel off-loaded its natural gas cargo into the HubLineSM pipeline system, operated by Spectra Energy (Houston).

The system is designed to handle peak deliveries of 800 MMcfd through two turret buoys. Under normal operations it will deliver about 500 MMcfd.

Offshore New Jersey terminal

Excalibur Energy (USA) Inc.—a 50-50 JV of Canadian Superior Energy Inc. (Calgary) and Global LNG Inc (New York City)—announced plans to build a 2.4-bcfd LNG terminal off New Jersey, as well as subsea pipelines to shore.

U.S. LNG terminals

As of May 2008, five import terminals and one export plant were operating in the United States, under the jurisdiction of the U.S. FERC.

- Everett, MA
- Cove Point, MD
- Elba Island, GA
- Lake Charles, LA
- Penuelas, Puerto Rico
- Sabine, LA
- Freeport, TX
- Kenai, AK

All are import terminals except the Kenai facility, which exports LNG to Japan.

In addition, the Gulf Gateway Energy Bridge was operating in the Gulf of Mexico, under jurisdiction of the U.S. Maritime Administration.

FERC listed 47 other North American proposed or expanded terminals.

US LNG terminals, FERC-regulated

Source: US Energy Information Administration

The proposed facility would include:

- a submerged gas importation turret anchored to the seabed in 100 ft of water, 15 mi offshore
- 50 mi of subsea pipeline
- a directionally drilled underground shoreline approach near South Amboy, NJ
- 11 miles of onshore pipeline to interconnections in Linden, NJ.

Excalibur said it will submit U.S. permit applications in 2009, and targets late 2011 for start up.

Broadwater

In April 2008, a floating LNG terminal and regasification facility proposed for Long Island Sound was rejected by New York's state government.

Officials said that the Broadwater LNG project is not consistent with several policies under the state's coastal zone management plan.

The installation would be designed to deliver up to 1.25 bcfd of gas.

The FERC had conditionally approved the project in March.

Concerns regarding the project include its size, potential disruption of commercial and recreational fishing, and alleged failure to guarantee low-cost gas to Long Island consumers.

2007 LNG import data

Pan EurAsian Enterprises Inc. (Raleigh, NC) conducted a yearend-2007 review of U.S. LNG imports. Highlights include the following:

- The United States was the world's fourth-largest LNG importer in 2007
- U.S. imports were 770.8 bcf (not including 26.3 bcf imported into Puerto Rico)
- Imports were 583.5 bcf in 2006
- Weighted average cost in 2007 was \$6.66/MMbtu versus \$6.82/MMbtu in 2006.
- An import surge during March-September 2007 offset a drop during fourth quarter of 2007.
- Largest importers of LNG into the United States in 2007 were BG Group (Reading, UK) and Suez Energy NA (Houston).

BlueOcean Energy

ExxonMobil Corp. (Irving, TX) announced plans in early 2008 for construction of BlueOcean Energy, a floating LNG receiving terminal off Long Island.

The facility would supply up to 1.2 bcfd to New Jersey and New York.

The terminal would be moored in 150 ft of water, away from shipping lanes, ports, and recreational areas. Earliest projected start-up is 2015.

The proposal calls for a subsea pipeline to deliver gas into New Jersey's Raritan Bay.

Pipelines

Midcontinent Express

Midcontinent Express Pipeline, a JV of Kinder Morgan Energy Partners LP (KMP) and Energy Transfer Partners LP (ETP), both in Houston, received final approval from the FERC in August 2008 to build the $1.27-billion Midcontinent Express gas pipeline.

The 506-mi system will deliver as much as 1.5 bcfd of gas to customers in the southern and eastern United States through transmission pipelines in Oklahoma, northeastern Texas, northern Louisiana, central Mississippi, and Alabama.

The pipeline is scheduled to be in service by early March 2009.

Rockies Express East

In mid-2008, the FERC authorized Rockies Express Pipeline LLC (Houston) to build and operate its planned 1.8-bcfd REX-East interstate natural gas pipeline.

REX-East will run 639 mi from the eastern terminus of Rockies Express' REX-West pipeline in Audrain County, MO, to an interconnection with the Dominion Transmission Inc., Dominion East Ohio, and Texas Eastern Transmission LP pipelines at the Clarington Hub in Monroe County, OH.

REX-East is the third leg of a project to bring Rockies gas to Midwestern and Eastern markets.

- REX-Entrega extends 327 mi from Colorado and Wyoming to the Cheyenne Hub in Weld County, CO. Carrying up to 1.1 bcfd it entered service in early 2007.
- REX-West extends 717 mi from the Cheyenne Hub to Audrain County. With a capacity is 1.5 bcfd, the line entered full service in mid-2008.

Ruby Pipeline

The BLM began public hearings in early 2008 on a proposed $2-billion natural gas pipeline that would run 680 mi between southwestern Wyoming and southern Oregon near the California state line.

The Ruby Pipeline system initially would transport 1.2 bcf/day of gas from the Opal Hub across northern Utah and Nevada to an interconnection near Malin, OR.

BLM and the U.S. Forest Service are participating agencies in the preparation by the FERC of an environmental impact statement on the project.

Pathfinder Pipeline

TransCanada Corp. (Calgary) was gauging support in early 2008 for a proposed pipeline to move natural gas from U.S. Rocky Mountains basins to Midwest markets.

The proposed 500-mi Pathfinder Pipeline will extend northeastward from Wamsutter, Wyoming, through Montana and North Dakota to a point on the Northern Border Pipeline still to be determined.

Initial capacity is to be 1.2 bcfd, with an ultimate capacity of 2 bcfd. Anticipated in-service target is late 2010.

Sunstone Pipeline

Williams Cos. Inc. (Tulsa) and TransCanada Corp. (Calgary) were evaluating joint development of the Sunstone Pipeline to move natural gas from the Rockies to the U.S. West.

The proposed 618-mi line would have a capacity of 1.2 bcfd. A tentative in-service date is 2011.

The line would run parallel to the existing Williams Northwest Pipeline system between the Opal Hub in Wyoming and Stanfield, OR.

The Northwest system interconnects at Stanfield with TransCanada's Gas Transmission Northwest system.

Bakken gas pipeline

Williston Basin Interstate Pipeline Co. (Bismarck, ND) was holding open season in mid-2008 for a proposed 100-mi pipeline in northwestern North Dakota that would transport associated gas from the Bakken oil play in the Williston basin.

Its capacity would be 100 MMcfd, expandable to double that volume, with a tentative in-service target of mid-2010.

The Bakken line would originate at an interconnect with Williston Basin's existing pipeline near Tioga, ND, and extend to the Alliance Pipeline in Bottineau County.

Fayetteville pipeline

Kinder Morgan Energy Partners LP (KMP) and Energy Transfer Partners LP (ETP), both in Houston, announced a 50:50 JV venture, Fayetteville Express Pipeline LLC.

The venture aims to build a 187-mi, 2-bcfd pipeline to increase takeaway capacity from the Fayetteville shale.

The line, from Conway County, AR, to an interconnection with Trunkline Gas Co., in Quitman County, MS, could be in service by early 2011.

Ethanol pipeline

Kinder Morgan Energy Partners LP (KMP said in October 2008 that it successfully tested what could become the nation's first transmarket ethanol pipeline.

KMP batched denatured ethanol with natural gas for the trip through the 106-mi Central Florida Pipeline System from Tampa to Orlando, FL.

The company also is testing the movement of B-5 blended biodiesel on the Plantation Pipe Line from Mississippi to South Carolina, and from Portland to Eugene, OR.

Pipeline business stats

Oil & Gas Journal published in September 2008 a comprehensive special report on U.S. oil and gas pipeline companies.

- Natural gas pipeline operators saw net profits surge nearly 19% in 2007 to $4.8 billion.

US INTERSTATE PIPELINE MILEAGE

Year	Miles: Gas[1] [2]	Miles: Oil	Miles: Total[1]
1998	190,250	157,234	347,484
1999	180,489	155,904	336,393
2000	186,151	152,823	338,974
2001	180,961	154,877	335,838
2002	190,899	149,619	340,518
2003	188,178	139,901	328,079
2004	190,117	142,200	332,317
2005	188,847	131,334	320,181
2006	189,012	140,407	329,419
2007	192,189	147,235	339,424

[1]FERC-defined major gas pipelines only; transmission mileage. See GAS COMPANIES table for definition of major and nonmajor companies and details of companies reporting mileage for 2006. [2]Totals revised from initial publication.
Source: US FERC annual reports: Form 6, oil pipelines; Forms 2 & 2A, gas pipelines

TOP 10 US INTERSTATE OIL PIPELINE COMPANIES—2007

	Company	Mileage	Company	Trunkline traffic, million bbl-miles	Company	Income, $1,000
1	Magellan Pipeline Co. LP	8,564	Colonial Pipeline Co.	708,334	Kinder Morgan Operating LP "A"	393,994
2	Plains Pipeline LP	8,387	Enbridge Energy LP	408,038	BP Pipelines North America Inc.	271,959
3	Mid-America Pipeline Co.	7,833	Marathon Pipeline LLC	169,388	ExxonMobil Pipeline Co.	259,971
4	ConocoPhillips Pipe Line Co.	6,693	Explorer Pipeline Co.	153,386	Shell Pipeline Co. LP	241,089
5	Colonial Pipeline Co.	5,591	TE Products Pipeline Co. LP	122,108	Colonial Pipeline Co.	222,024
6	Sunoco Pipeline LP	5,432	Plantation Pipe Line Co.	115,445	Magellan Pipeline Co. LP	191,591
7	TE Products Pipeline Co. LP	4,676	Plains Pipeline LP	110,222	Marathon Pipeline LLC	175,994
8	ExxonMobil Pipeline Co.	4,559	Laclede Pipeline Co.	93,586	TE Products Pipeline Co. LP	128,201
9	TEPPCO Crude Pipeline LP	3,967	Mid-America Pipeline Co. LLC	91,562	Whiting Oil & Gas Corp.	124,147
10	NuStar Logistics LP	3,746	ConocoPhillips Transportation Alaska Inc.	86,751	Enbridge Energy LP	122,491
	Top 10 total—2007	**59,448**		**2,058,820**		**$2,131,461**
	Part of all companies	**40.38%**		**59.52%**		**56.74%**
	Top 10 total—2006	60,447		2,001,551		$2,087,825

Source: US FERC Form 6: Annual Report of Oil Pipeline Companies, Dec. 31, 2007

TOP 10 US INTERSTATE GAS PIPELINE COMPANIES—2007

	Company*	Transmission mileage	Company*	Volumes moved for fee, MMcf	Company*	Net income, $1,000
1	Northern Natural Gas Co.	15,487	Transcontinental Gas Pipe Line Corp.	2,669,937	Natural Gas Pipeline Co. of America	309,104
2	Tennessee Gas Pipeline Co.	14,463	ANR Pipeline Co.	2,043,743	Texas Eastern Transmission Corp.	266,984
3	Columbia Gas Transmission Corp.	10,339	Columbia Gas Transmission Corp.	1,849,756	Duke Energy Ohio Inc.	263,544
4	Transcontinental Gas Pipe Line Corp.	10,325	Tennessee Gas Pipeline Co.	1,801,283	Distrigas of Massachusetts LLC	215,092
5	El Paso Natural Gas Co.	10,240	Natural Gas Pipeline Co. of America	1,783,189	Southern Natural Gas Co.	208,736
6	ANR Pipeline Co.	9,587	El Paso Natural Gas Co.	1,673,647	Transcontinental Gas Pipe Line Co.	193,989
7	Texas Eastern Transmission LP	9,115	Texas Eastern Transmission Corp.	1,438,099	Columbia Gas Transmission Corp.	190,394
8	Natural Gas Pipeline Co. of America	9,045	Columbia Gulf Tranmission Co.	1,112,959	Tennessee Gas Pipeline Co.	174,639
9	Southern Natural Gas Co.	7,636	Northern Natural Gas Co.	1,055,067	Dominion Transmission Inc.	163,778
10	Gulf South Pipeline Co. LP	6,499	CenterPoint Energy Gas Transmission Co.	959,905	Northern Natural Gas Co.	161,088
	Total—2007	**102,736**	**Total—2007**	**16,387,585**	**Total—2007**	**$2,147,348**
	Part of majors	**53.46%**		**44.94%**		**46.09%**
	Part of all companies	**51.80%**		**44.13%**		**45.06%**
	Total—2006 top 10	102,810		15,710,325		$1,862,092

*All FERC-classified as "major."

Source: US FERC Forms 2 & 2A: annual reports for natural gas companies, Dec. 31, 2007

- Oil pipeline operators' profits were nearly unchanged.
- Operators began to rein in plans to expand capacity.
- Proposed mileage fell by more than 50%, and only four of 18 projects called for new or additional compression of 20,000 hp or greater.
- Estimated $/mile pipeline costs rose nearly 22% to more than $3.38 million/mile.
- Labor prices rose by nearly $300,000/mile.

Balloon-based remote CP monitoring

Space Data Corp. (Chandler, AZ) described an innovative system for maintaining cathodic protection of pipeline networks without the need to travel to remote locations.

Standard weather balloons provide a platform for a patented GPS navigation system that maintains a continuous monitoring presence above the southwestern United States and northwestern Gulf of Mexico.

Pipeline company revenues, incomes

	Gas		Oil	
	Operating revenues, $1,000	Net income, $1,000	Operating revenues, $1,000	Net income, $1,000
1998	13,584,783	3,010,821	6,890,083	2,050,982
1999	14,616,949	2,545,043	7,219,500	2,928,460
2000	14,980,925	2,910,835	7,483,100	2,705,463
2001	14,407,467	2,246,109	7,729,972	3,006,898
2002	14,015,308	2,734,182	7,811,951	3,408,753
2003	15,082,011	3,260,797	7,703,998	3,469,996
2004	15,781,445	3,588,344	8,019,554	3,322,738
2005	16,375,921	3,863,331	7,917,176	3,076,476
2006	$17,122,586	$4,015,253	$8,516,563	$3,743,115
2007	$21,736,725	$4,765,815	$8,996,329	$3,756,749

Source: US FERC annual reports (Forms 2, 2A, and 6) by regulated interstate natural gas and oil pipeline companies

Major cost components—10 years

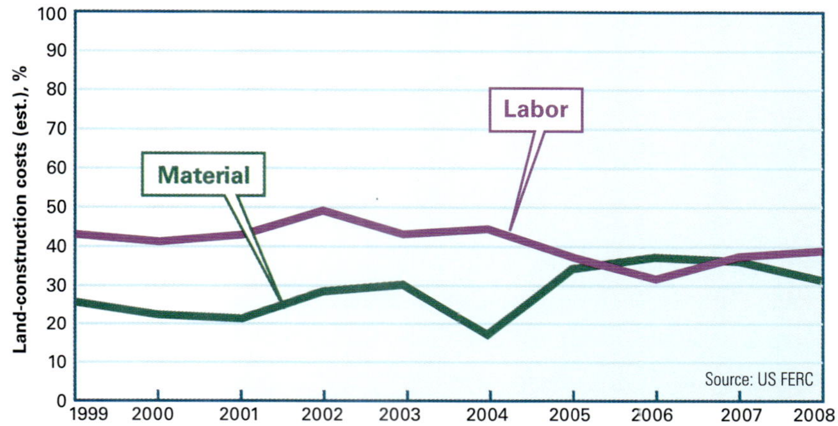

Oil pipeline investment

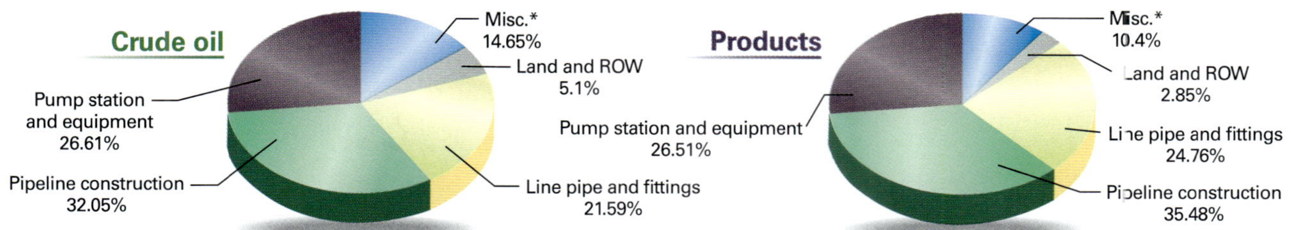

*Generally includes delivery systems, communications, office furniture and equipment, vehicles and other work equipment, and other property.
Source: US oil pipeline company annual reports (Form 6) to FERC for 2007

Pipeline construction costs—estimated

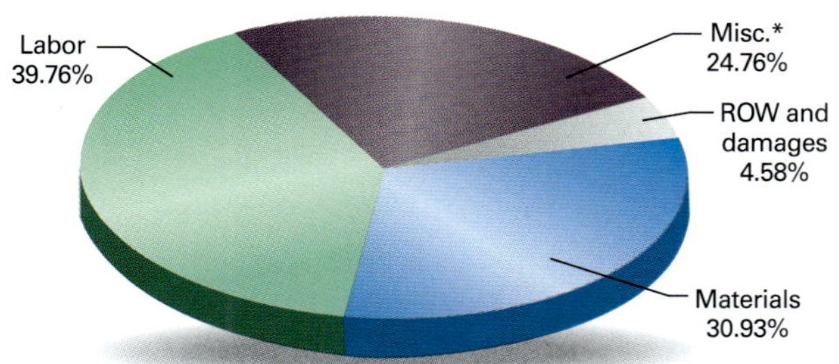

*Generally includes surveying, engineering, supervision, administration and overhead, interest, contingencies and allowances for funds used during construction (AFUDC), and regulatory filing fees.
Source: US FERC construction-permit filings July 1, 2007, to June 30, 2008

Compressor construction costs—estimated[1]

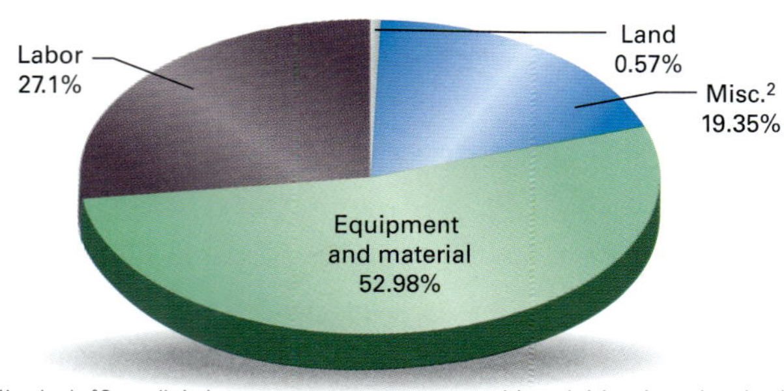

[1]Land only. [2]Generally includes surveying, engineering, supervision, administration and overhead, interest, contingencies, allowances for funds used during construction (AFUDC), and regulatory filing fees.
Source: US FERC construction-permit filings, July 1, 2007, to June 30, 2008

The company said its SkySite systems operate at an altitude of 20 mi.

It also enables general supervisory control and data acquisition; pump controller monitoring; alarm monitoring of storage tanks; field communications; and flow monitoring.

Platform location

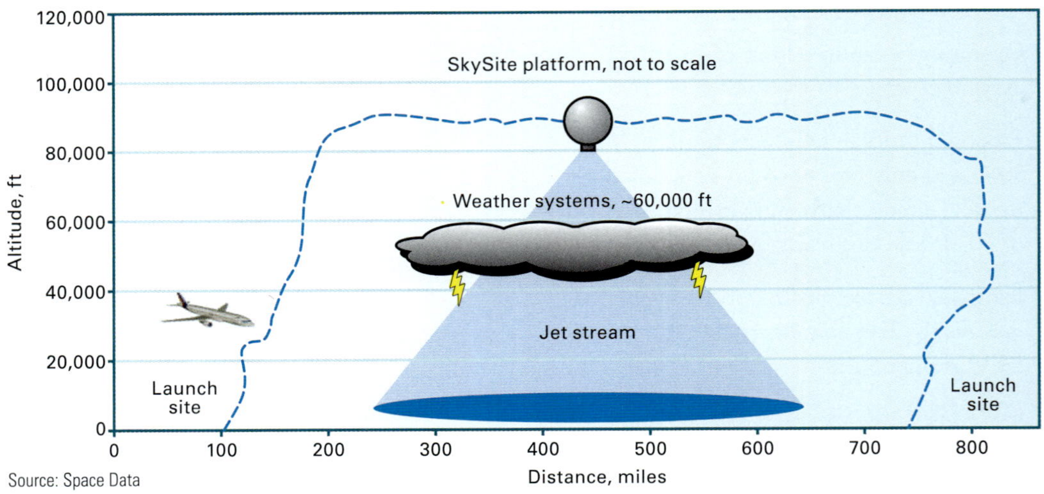

Source: Space Data

Refining and Products

Refining capacity up in 2007

According to the National Petrochemical & Refiners Association (NPRA; Washington, DC), U.S. refiners continued to add capacity despite soaring oil prices and shrinking margins in 2007.

In an August 2008 annual report, the group said that domestic refining capacity increased by 0.86% during 2007, by 4.1% over the last five years, and by 8.2% over the last nine years.

As of January 1, 2008, there were 149 operable U.S. refineries, excluding Puerto Rico and the Virgin Islands, with a total crude distillation capacity of 17.6 million b/cd and 18.6 million b/sd.

ExxonMobil plans

ExxonMobil Refining & Supply (Irving, TX) announced in early 2009 its plan to invest more than $1 billion in three refineries to boost production of ultralow-sulfur diesel (ULSD) by about 6 million gpd.

By 2010, the firm aims to add new units and modifying existing facilities at its refineries in Baytown, TX, Baton Rouge, LA, and Antwerp, Belgium.

Sunoco: Tulsa

Reuters reported in November 2008 that Sunoco Inc. (Philadelphia) abandoned plans to upgrade its refinery in Tulsa, OK, and instead was seeking to sell the facility.

Marathon: Detroit

Marathon Oil Corp. (Houston) started construction in mid-2008 of the $1.9-billion heavy oil upgrading project at its refinery in Detroit, MI.

The project will include an additional 80,000 b/d of heavy oil capacity and increase the total refining capacity to 115,000 b/d from 102,000 b/d.

The increased capacity will yield more than 400,000 gpd of additional transportation fuels.

The project is expected to be completed in late 2010.

Total: Port Arthur

Total SA (Paris) announced plans in early 2008 to build a 50,000-b/d coker and desulfurization, vacuum distillation, and related units at its 231,000-b/d refinery in Port Arthur, TX.

The $2.2-billion modifications, to be completed in 2011, will boost output of ultralow-sulfur automotive diesel by about 60,250 b/d (3 million tpy).

Canadian oil sands products

In an August 2008 analysis, Wood Mackenzie (Edinburgh) projected that the supply of Canadian oil sands products will increase by 2 million b/d between 2007 and 2015, and that half of this growth will be in Canadian heavy crude blends.

The firm expects that the export pipeline infrastructure and U.S. refinery capacity will handle the additional Canadian crude during 2007–15, but that more export capacity may be needed soon thereafter.

Currently disclosed plans indicate that pipeline companies and U.S. refiners plan to invest more than $31 billion by 2015 to export and distribute oil sands products and to process them in the U.S. refining system.

That figure excludes investments in pipelines within Alberta and in the Canadian refining and upgrading system.

Existing, Proposed Pipelines for Canadian Crude

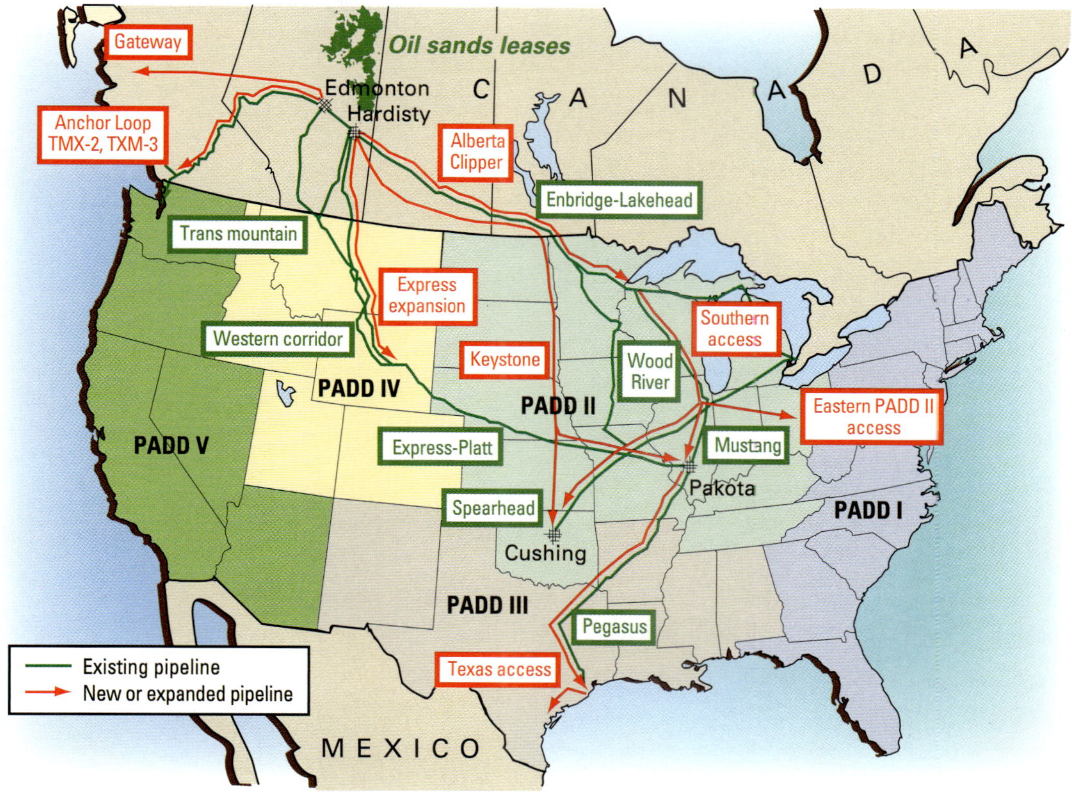

Allocation of Additional Products, 2015

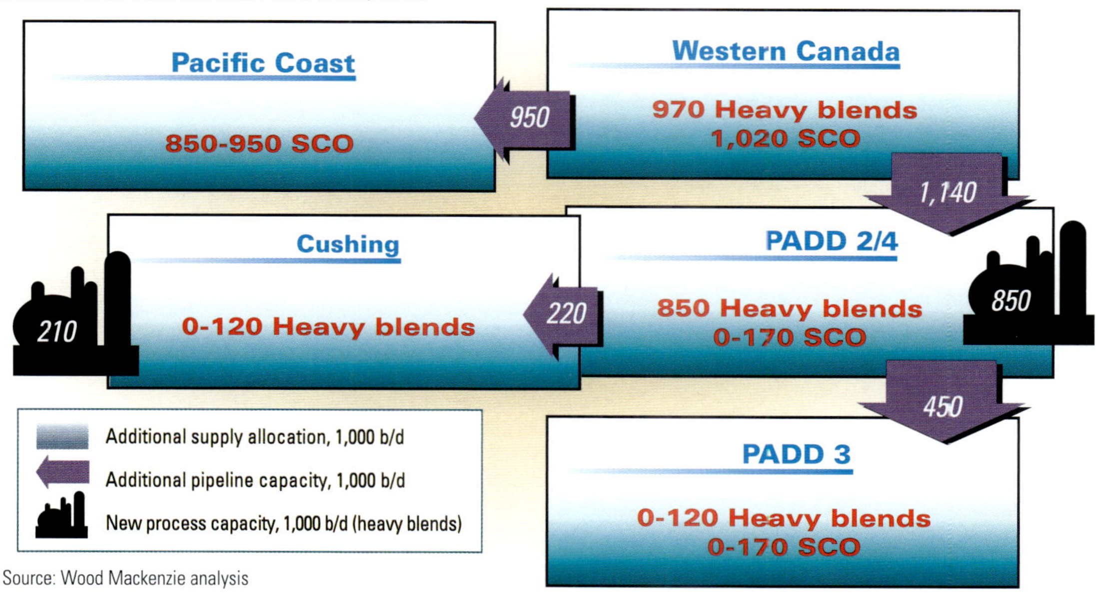

Source: Wood Mackenzie analysis

Planned Refinery Investments

Company	Refinery location	Project	Additional heavy blends, 1,000 b/d	Investment, $ billion	Start-up
Flint Hills Resources	Rosemount, Minn.	Expansion 2007 (oil pipeline 2008)	50	0.2	2007
Sinclair Oil	Sinclair, Wyo.	Upgrade	35	N/A	2008
Holly	Woods Cross, Utah	Capacity expansion, upgrade	10	0.1	2008
Frontier Oil	El Dorado, Kan.	Capacity expansion, upgrade	10	0.4	2009
Sinclair Oil	Tulsa, Okla.	Capacity expansion, upgrade	60	1.0	2010
Marathon Petroleum	Detroit, Ill.	Capacity expansion, upgrade	80	1.9	2010
WRB Refining	Wood River, Ill.	Capacity expansion, upgrade	130	1.9	2010
BP	Whiting, Ind.	Upgrade	260	3.8	2011
Holly	Artesia, NM	Capacity expansion, upgrade	40	N/A	2011
WRB Refining	Borger, Tex.	Debottleneck	40	0.3	>2012
WRB Refining	Wood River, Ill.	Capacity expansion, upgrade	210	2.0	2014
BP/Husky Energy	Toledo, Ohio	Capacity expansion, upgrade	60	2.5	2015
WRB Refining	Borger, Tex.	Capacity expansion, upgrade	70	0.6	2015
Total			**1,055**	**14.7+**	

Source: Wood Mackenzie refinery evaluation model

Refinery Safety

Over the last decade, according to data from the U.S. Department of Labor, the U.S. refining industry had much better safety performance than all of private industry.

US Injury, Illness Rates

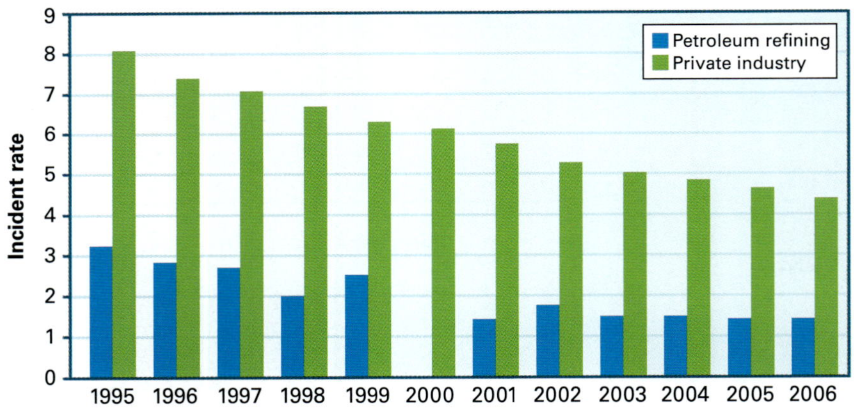

Source: US Department of Labor, Bureau of Labor Statistics

Planned facilities

New gas-processing capacity totaling 920 MMcfd is slated to come online in 2009, according to *OGJ*'s November 2008 *Worldwide Construction Update.*

- Anadarko Petroleum Corp. (The Woodlands, TX) plans to expand its Chapita gas processing plant in eastern Utah from 250 MMcfd to 500 MMcfd, with planned start-up in first quarter 2009.
- Williams Companies (Tulsa) was building a 450-MMcfd processing plant in Rio Blanco County, in western Colorado's Piceance basin. Start-up is slated for third-quarter 2009.
- MarkWest Energy Partners LP (Denver) announced in late 2007 that it will invest $60 million to expand nearly all of its plants in the Appalachian region. That includes replacing its Boldman and Cobb plants with cryogenic processing facilities.
- Enogex LLC [subsidiary of OGE Energy Corp. (Oklahoma City)] announced that it will provide gas processing and transportation services beginning in 2009 for Chesapeake Energy Corp. (Oklahoma City) in the Colony Granite Wash play in western Oklahoma.
- Quicksilver Resources LP (Fort Worth) was making plans for a third gas processing plant to serve Barnett shale gas producers. Start-up is scheduled for first quarter 2009.

Gas Processing

Pioneer plant

The Pioneer natural gas processing plant of Enterprise Products Partners LP (Houston) began operation in February 2008 in Sublette County, WY.

It will process up to 750 MMcfd of gas and extract as much as 30,000 b/d of NGLs.

Gas Storage

Caverns at Spindletop

Golden Triangle Storage Inc., a subsidiary of AGL Resources (Atlanta, GA), began building in late 2008 a 12-bcf working-gas-capacity salt dome natural gas storage facility at Spindletop in Beaumont, TX.

Nine miles of pipeline will connect the facility to six interstate and intrastate transmission pipelines.

The project will offer 600 MMcfd of deliverable capacity and 300 MMcfd of injection capacity.

The company says the first cavern will be ready for operation by early 2011, the second by 2013.

Storage along Gulf Coast

A mid-2008 study by Bentek Energy LLC (Boulder, CO) indicated that planned storage capacity increases in the Southeast U.S.-Gulf Coast region would add nearly 30% to working-gas storage capacity and have an even greater effect on withdrawal capacity.

Thirty-one storage projects have been announced for 2008–10 in the region,adding 304.9 bcf of working-gas capacity and 17.5 bcfd of withdrawal capacity.

Southeast US-Gulf Coast Gas Storage Projects

	— Storage capacity, bcf —				—Withdrawal capacity, bcfd —			
Location	2008	2009	2010	Total	2008	2009	2010	Total
Louisiana	28.8	13.2	32.3	74.3	3.1	1.0	1.0	5.1
Texas	13.4	38.7	35.5	87.6	0.9	2.5	1.6	4.9
Alabama	5.4	60.0	—	65.4	0.6	2.2	—	2.8
Mississippi	20.0	24.6	33.1	77.7	1.3	1.3	2.1	4.7
Total	**67.6**	**136.5**	**100.9**	**304.9**	**5.8**	**7.0**	**4.6**	**17.5**

Additional Storage Deliverability, 2008-2011

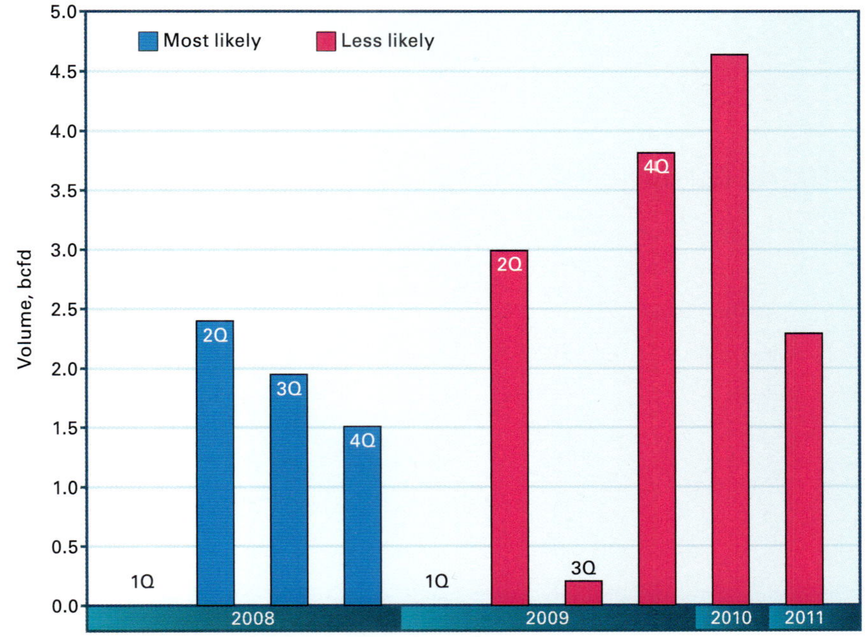

Planned Storage Projects, 2008*

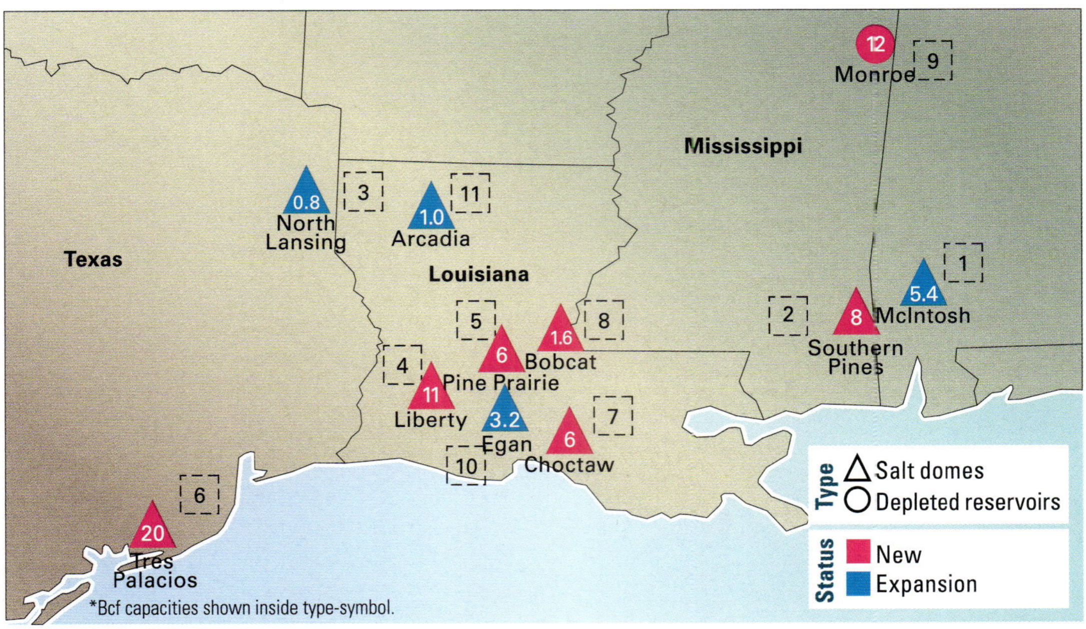

Southeast US-Gulf Coast Storage Projects, 2008

Project	In-service date	State	County	Status	Type	Storage, MMcf	Withdrawal, MMcfd
McIntosh-Bay Gas	Apr. 1	Alabama	Washington	Expansion	Salt dome	5,400	600
Southern Pines Energy Center	May 1	Mississippi	Greene	New	Depleted reservoir	8,000	800
North Lansing	May 12	Texas	Harrison	Expansion	Depleted reservoir	824	0
Liberty Gas Storage	May 15	Louisiana	Calcasieu	New	Salt dome	11,000	1,000
Pine Prairie Energy Center	July 1	Louisiana	Evangeline	New	Salt dome	6,000	800
Tres Palacios Gas Storage	July 1	Texas	Matagorda, Wharton	New	Salt dome	12,580	850
Choctaw	Sept. 1	Louisiana	Iberia	New	Salt dome	6,000	300
Bobcat	Oct. 1	Louisiana	St. Landry	New	Salt dome	1,600	1,000
Monroe Gas Storage	Oct. 1	Mississippi	Monroe	New	Depleted reservoir	12,000	465
Egan	Nov. 1	Louisiana	Acadia	Expansion	Salt dome	3,150	0
Arcadia	Dec. 15	Louisiana	Bienville	Expansion	Salt dome	1,000	20
		Louisiana				28,750	3,120
		Texas				13,404	850
		Alabama				5,400	600
		Mississippi				20,000	1,265
		Total				**67,554**	**5,835**

Southeast US-Gulf Coast Storage Projects, 2009

Project	In-service date	State	County	Status	Type	Storage, MMcf	Withdrawal, MMcfd
North Dayton	Apr. 1	Texas	Liberty	Expansion	Salt dome	5,500	425
Southern Pines Energy Center[1]	Apr. 1	Mississippi	Greene	Expansion	Salt dome	8,000	400
Tarpon Whitetail	Apr. 1	Mississippi	Monroe	New	Depleted reservoir	8,600	300
Tres Palacios Gas Storage[1]	Apr. 1	Texas	Matagorda, Wharton	Expansion	Salt dome	13,900	825
Pine Prairie Energy Center[1]	June 1	Louisiana	Evangeline	Expansion	Salt dome	7,000	800
Moss Bluff	June 1	Texas	Liberty	Expansion	Salt dome	3,250	250
Bobcat[1]	Aug. 1	Louisiana	St. Landry	Expansion	Salt dome	6,200	200
Mont Belvieu[2]	Cancelled	Texas	Chambers	New	Salt dome	0	0
MoBay	Oct. 1	Alabama	Mobile	New	Depleted reservoir	50,000	1,000
Leaf River	Nov. 1	Mississippi	Smith	New	Salt dome	8,000	625
McIntosh-Bay Gas[1]	Nov. 1	Alabama	Washington	Expansion	Salt dome	10,000	1,200
Houston Storage Hub	Nov. 1	Texas	Liberty	New	Salt dome	16,000	1,000
		Louisiana				13,200	1,000
		Texas				38,650	2,500
		Alabama				60,000	2,200
		Mississippi				24,600	1,325
		Total				**136,450**	**7,025**

[1]Next phase of existing project. [2]Mont Belvieu had a planned 10 bcf capacity.

Planned Storage Projects, 2009*

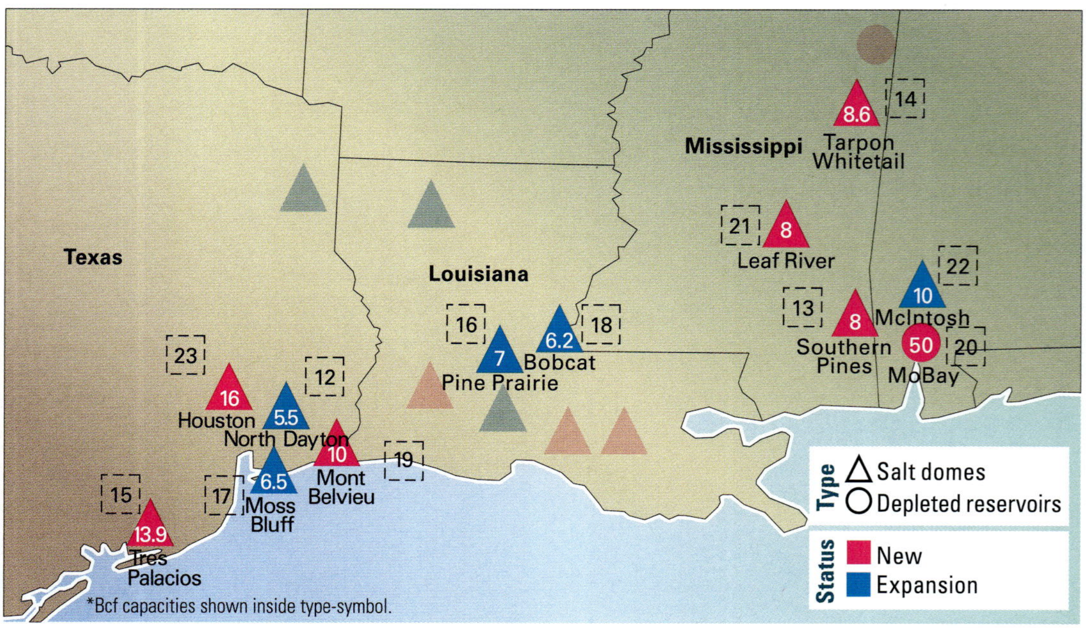

Alternative Energy

Investment outlook

The credit crunch and falling oil prices could dampen investment in alternate energy companies, said speakers at an October 2008 energy forum hosted by Ernst & Young (New York City).

E&Y noted that initial public offerings for alternative energy companies essentially stopped in July.

However, U.S. government subsidies and tax credits for renewals will continue.

An October 2008 law extended tax credits for eight years for investments in solar energy and one year for investments in wind energy.

And the $787-billion federal stimulus package signed into law in February 2009 contains almost $20 billion for renewable energy.

Ethanol rush slows

According to an analysis by the *Financial Times*, six of the largest publicly traded U.S. ethanol producers had lost more than $8.7 billion in market value since mid-2006 (the peak of the ethanol boom) and early October 2008.

The rush to ethanol followed a 2005 law requiring refiners to mix billions of gallons of the biofuel with gasoline as a way to cut both oil imports and greenhouse gas emissions.

The losses came despite billions of dollars of government support for the industry, including more than $11 billion spent since 2005 on tax breaks for companies that blend ethanol into gasoline.

Billions more were spent on state and federal subsidies for U.S. ethanol production.

However, a growing number of critics said that ethanol was contributing to rising food prices.

BP: Cellulosic ethanol plans

BP PLC (London) announced in August 2008 plans to invest $90 million in Verenium Corp. (Cambridge, MA) to accelerate development and commercialization of cellulosic ethanol.

BP said Verenium already has demonstrated an advanced technology for cellulosic ethanol production. Verenium opened a demonstration plant in Jennings, LA, in mid-2008.

RFS biofuels mandate

Debate continued in 2008 regarding the technical feasibility, economic viability, and net environmental impact of wider use of renewable biofuels.

The federal Renewable Fuels Standards (RFS)—part of the Energy Independence and Security Act of 2007 (EISA)—is a legislative mandate for renewable fuels such as bioethanol and biodiesel.

The RFS requires that national production of renewable biofuels be increased from 9 billion gal/yr in 2008 to 36 billion gal/yr by 2022.

An analysis by Sklar & Associates (Murrells Inlet, SC) in early 2008 concluded that there may be a chance that the RFS targets can be met, but only with extensive government support.

EISA RFS* FOR BIOFUELS PRODUCTION

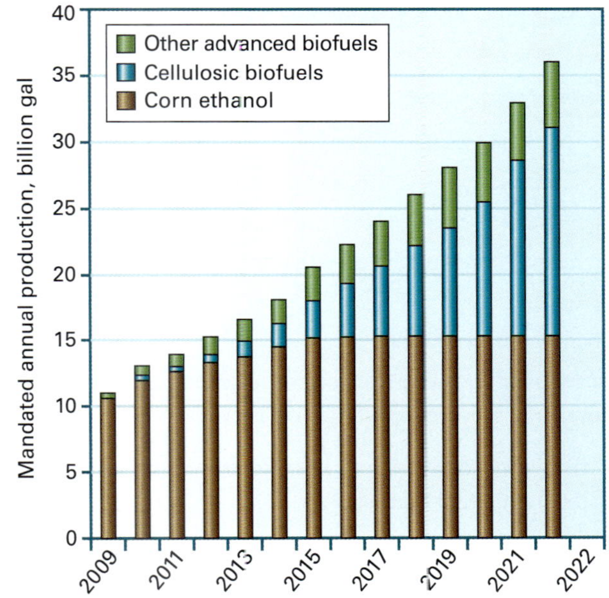

*Renewable Fuels Standards.

EISA RFS REQUIREMENTS FOR BIOFUELS TO 2022

	2008	2012	2017	2022
	Mandated amounts, %			
Corn ethanol	100.0	86.8	62.5	41.7
Cellulosic biofuel	0.0	3.3	22.9	44.4
Other advanced biofuels	0.0	9.9	14.6	13.9
	RFS annual mandate, billion gal			
Corn ethanol	9.0	13.2	15.0	15.0
Cellulosic biofuel	0.0	0.5	5.5	16.0
Other advanced biofuels	0.0	1.5	3.5	5.0
	No. of 55-million gal/year biofuels plants operating			
Corn ethanol	164	240	273	273
Cellulosic biofuel	0	9	100	291
Other advanced biofuels	0	27	64	91
Total No. of plants for RFS	**164**	**276**	**436**	**655**
	Plant investment, billion $ (@ $5/installed gal)			
Corn ethanol	45	66	75	75
Cellulosic biofuel	0	3	28	80
Other advanced biofuels	0	8	18	25
Total	**45**	**76**	**120**	**180**

Questions about biofuels

Two scientific reports in early 2008 questioned whether substituting biofuels for gasoline will reduce greenhouse gases.

The first report was in *Science*, the weekly magazine of the American Academy for the Advance of Science.

It found that, on a global basis, corn-based ethanol, instead of producing a 20% savings, nearly doubles greenhouse emissions over 30 years and increases greenhouse gases for 167 years.

The second study, by the Nature Conservancy (Arlington, VA) and the University of Minnesota, looked at the impact of clearing land to plant crops for biofuels.

Results suggested that biofuels produced on converted lands could be greater net emitters of greenhouse gases than the fossil fuels they typically displace.

More alternative-fuel R&D

The National Surface Transportation Policy and Revenue Study Commission recommended in January 2008 that the U.S. government spend $200 million/yr over 10 years for more research and development of alternatives to petroleum-based motor fuels.

Other recommendations included:

- Incentives for the private sector to speed development of an infrastructure for alternative fuels
- Tax credits and revolving loans to encourage transition of fleets to alternative fuels.

Its report, *Transportation for Tomorrow*, said that new programs could be partially financed by increasing the federal fuel tax to 8¢/gal and then indexing it to inflation.

LNG from landfill gas

Linde North America (Murray Hill, NJ) and Waste Management Inc. (Houston) announced in mid-2008 their plans to convert landfill gas into clean vehicle fuel.

The venture plans to build an LNG production plant at the Altamont Landfill near Livermore, CA. Start-up is planned for 2009.

The aim is to fuel hundreds of waste collection trucks with clean fuel produced from garbage.

Environmental Issues

Climate-change deliberations

In early 2008, Congress entered a new phase of its global climate change deliberations as it considered whether to enact a carbon cap-and-trade program.

Congress had passed three major laws (the Energy Policy Act of 2005; the America Competes Act of 2007; the Energy Independence and Security Act of 2007) that began to address global climate change.

The Senate had before it in early 2008 the proposed Lieberman-Warner cap-and-trade bill.

Other committees were discussing whether elements of the successful sulfur dioxide control program from the 1980s could be used in the more difficult carbon dioxide situation.

One key question is whether a carbon cap-and-trade program would use State Implementation Plans, New Source Review, or an entirely new approach.

Many parties have expressed the concern that U.S. carbon emissions reduction programs will have no meaningful impact unless other countries—notably China and India—participate.

Rules for carbon dioxide storage

The U.S. EPA released proposals in July 2008 for a category of injection wells for long-term underground carbon dioxide storage.

The move paralleled energy-industry discussion about carbon capture and storage (CCS) technology.

EPA hopes to ensure that stored CO_2 does not move or leach heavy metals or other contaminants out of surrounding rock into underground water.

The new rules represent the first federal regulations for long-term CO_2 storage. Regulations are expected by early 2011.

Rules already exist for CO_2 injected for enhanced oil recovery.

CO_2 sequestration tests

The Southwest Regional Partnership (SWP) of the U.S. DOE continued its Phase 2 and 3 carbon dioxide projects in 2008 evaluating the sequestration of CO_2 in saline aquifers, coalbeds, and oil fields.

As part of Phase 2, SWP started injecting CO_2 in July in the Fruitland coalbeds of the San Juan basin in New Mexico.

The San Juan Pump Canyon test aims to demonstrate whether CO_2 can enhance methane production from coalbeds (ECBM) as well as the potential of coalbeds to sequester injected CO_2.

Other parts of Phase 2 involve CO_2 injection in oil fields.

- Tests in Aneth oil field in the Paradox basin of Utah will test CO_2 injection in a saline zone.
- Work at the SACROC Unit in the Permian basin of West Texas will test CO_2-enhanced oil recovery concurrent with CO_2 storage in the reservoir.

Pump Canyon Site

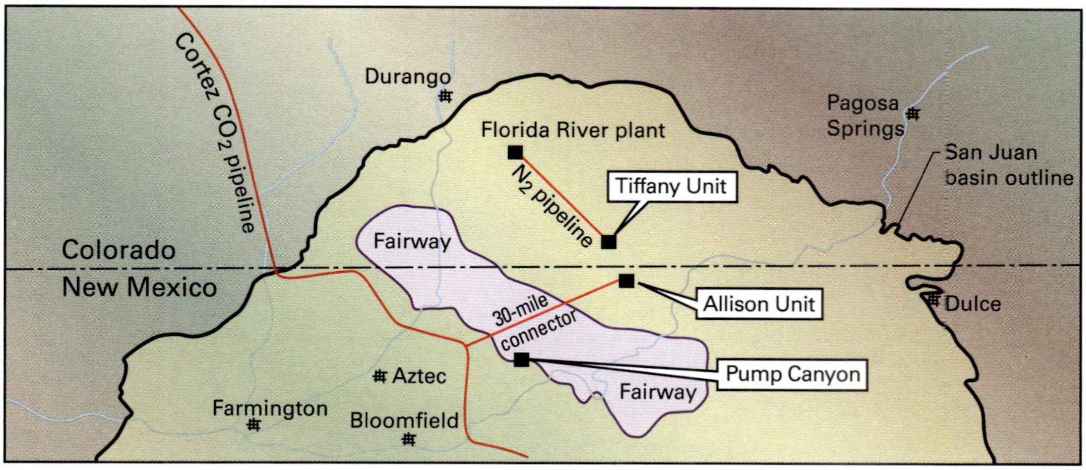

Southwest Regional Partnership Area

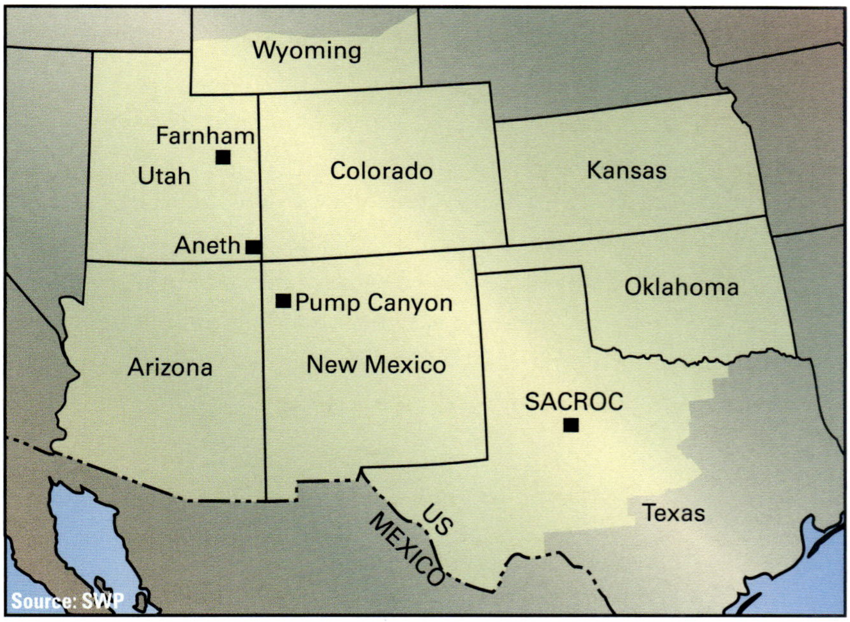

SWP also started Phase 3, injecting CO_2 into saline water zones in the Farnham dome in Utah. It was making plans to drill the injection well by yearend 2008.

SWP is one of seven regional partnerships organized by DOE in 2003 to validate and demonstrate the feasibility of CO_2 sequestration underground.

FutureGen restructured

In December 2007, the FutureGen Alliance selected a site in Mattoon, IL, to build a $1.5-billion, 275-MW power plant to research and test technology to burn coal and control carbon dioxide emissions.

With funding from the U.S. DOE, it would evaluate methods to gasify coal, strip out harmful emissions, then burn the gas to produce electricity and hydrogen.

It also would seek to capture carbon dioxide for storage underground.

However, in January 2008, DOE restructured FutureGen to focus more on carbon capture and storage. DOE also cited rising costs for the Mattoon plant and improvements in competing technologies.

The FutureGen Alliance includes U.S. utilities and coal producers plus three international companies involved in mining and power generation.

Polar bears

The Department of the Interior in May 2008 listed the polar bear as a threatened species under the federal Endangered Species Act (ESA).

The agency said the decision was based on scientific evidence that loss of sea ice—not oil and gas development or native populations' subsistence activities—threatens polar bear habitat.

The vulnerability of the bear was an issue in discussion about oil and gas development in the Arctic.

The agency noted that the bears already are protected under the Marine Mammal Protection Act, which has provisions more stringent than those in the ESA.

Refinery air emissions

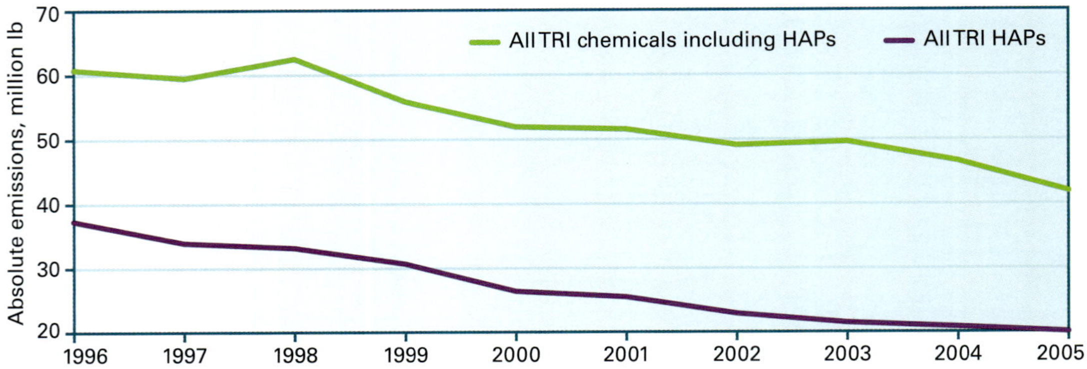

Refinery emissions: 1996–2005

From 1996–2005, U.S. refiners significantly reduced the amount of hazardous air emissions from their operations, according to a September 2008 report from the U.S. Environmental Protection Agency (EPA).

In 2005, the refiners reported 42.2 million lb of absolute air emissions to EPA's annual Toxics Release Inventory (TRI). This was a 31% decrease from the level in 1996.

TRI chemicals in water discharges, however, increased during the reporting period.

The total amount of waste managed decreased during 1996–2005.

Tougher air-quality standard

The U.S. EPA in March 2008 signed its most stringent standard for ground-level ozone, revising the standards for the first time in more than a decade.

The new National Ambient Air Quality Standard (NAAQS) is 0.075 ppm, compared to the previous standard of 0.08 ppm.

API and NPRA criticized the change, calling it unnecessary.

EPA said the change was based on the most recent scientific evidence about the effects of ozone.

Texas RFS waiver request

The U.S. EPA in August 2008 denied a request from Texas officials for a 50% waiver from the federal renewable fuel standard (RFS) mandate for corn-based ethanol.

Increased demand for ethanol was said to be contributing to escalating corn prices, higher food prices and also higher costs for livestock feed.

EPA found no compelling evidence that the mandate was causing severe economic harm during the time period specified by Texas.

California GHG waiver

In March 2008, the U.S. EPA denied a request from California officials for the state to set its own standards on greenhouse gas emissions from motor vehicles.

EPA has authority to allow waivers for California and had done so in the past.

However, EPA said there was a difference in scope in the latest request compared to previous waiver requests it had granted.

In early 2009, EPA officials in the Obama Administration likely will reconsider the decision.

Company News

M&A activity in 2008

OGJ noted in an early 2009 analysis that most notable merger and acquisition (M&A) transactions initiated in 2008 in the United States involved independents. Among the largest deals:

- Independent XTO Energy Inc. (Fort Worth) obtained Fayetteville shale leasehold acreage from Southwestern Energy Co. (Houston) and also acquired producing properties, pipeline, and leasehold acreage from Linn Energy LLC (Houston).
- Stone Energy Corp. (Lafayette, LA) acquired and then merged with Bois d'Arc Energy Inc. (Houston).
- XTO purchased producing properties and undeveloped acreage in the Williston basin Bakken shale from private Headington Oil Co. LP (Dallas).

- XTO also acquired private Hunt Petroleum Corp. (Dallas).
- Plains Exploration & Production Co. (Houston) agreed to acquire a 20% interest in the Haynesville shale leasehold of Chesapeake Energy Corp. (Oklahoma City) in a new JV.
- Quicksilver Resources Inc. (Fort Worth) entered into agreements with various private parties to acquire significant Barnett shale assets.
- BP America (Warrenville, IL) and Chesapeake Energy established a JV in which BP obtained 25% interest in Chesapeake's Fayetteville shale assets.

Independent E&P industry

Moody's Investors Service (New York City) downgraded to negative its outlook in early 2009 for the independent exploration and production industry.

The firm cited the "precipitous decline" in oil and natural gas prices to levels likely to result in abnormally low cash margins and fundamental credit deterioration.

Moody's also expressed concern about credit availability to the speculative grade companies.

In Moody's view, there is significant risk that E&Ps have entered a prolonged period of abnormally low cash margins and returns due to persistent demand-driven price declines.

Collapse of K-Dow deal

The government of Kuwait in December 2008 cancelled a $17.4-billion petrochemicals JV with Dow Chemical (Midland, MI).

Details are given in the Kuwait section of the Middle East chapter of this book.

Marathon separation decision

In December 2008, Marathon Oil Corp. (Houston) delayed a decision on whether or not to split into two separate companies. The company gave no new timeframe for a decision.

GULF OF MEXICO

Exploration and Development

Eugene Island Blocks

Leed Petroleum PLC (London) hit pay in late 2008 with its A-8 well on the Eugene Island Blocks.

The well reached TD of 15,470 ft (4,715 m) and encountered commercial hydrocarbons in the two primary targets.

The blocks are 50 mi (80 km) south of Morgan City, Louisiana, in approximately 80 ft (24 m) of water.

Electric line logs confirmed three pay sands with a total of 155 ft (47 m) of true vertical thickness.

First production from the deeper pay zone is expected in first quarter 2009.

South Timbalier: Blackbeard

Operator McMoRan Exploration Co. (New Orleans) reported in late 2008 that the ultradeep Blackbeard No. 1 well on the Gulf of Mexico shelf logged four potential hydrocarbon-bearing zones below 30,067 ft (9,165 m).

Having reached 32,997 ft (10,058 m) in October, McMoRan and partners were pondering whether to halt drilling and test the zones encountered, or continue drilling through another 2,000 ft (610 m) of sediment to basement [expected to be below the permit depth of 35,000 ft (10,668 m)].

The four zones are in the Miocene Rob-L formation. Bottomhole pressure exceeded 25,000 psi

The company said that test results have implications out to a 50-mi (80-km) radius from the well and that additional drilling on the flanks could result in significant reserve potential.

DeSoto Canyon: Block 48

A well targeting the Dalmatian prospect on De Soto Canyon block 48 hit high quality, dry natural gas pay in late 2008, according to Murphy Exploration and Production Co. (Houston).

The well, in water nearly 5,900 ft (1,798 m) deep, encountered 120 ft (37 m) net measured depth of gas pay.

DeSoto and Mississippi Canyons

Operator Shell Oil Co. (Houston) made an oil discovery at the Vicksburg prospect in 7,500 ft (2,286 m) of water on DeSoto Canyon Blocks 353 and 397 and Mississippi Canyon Block 393, in the eastern gulf

The well was drilled to 25,400 ft (7,742 m) and found a 300-ft (91-m) hydrocarbon column.

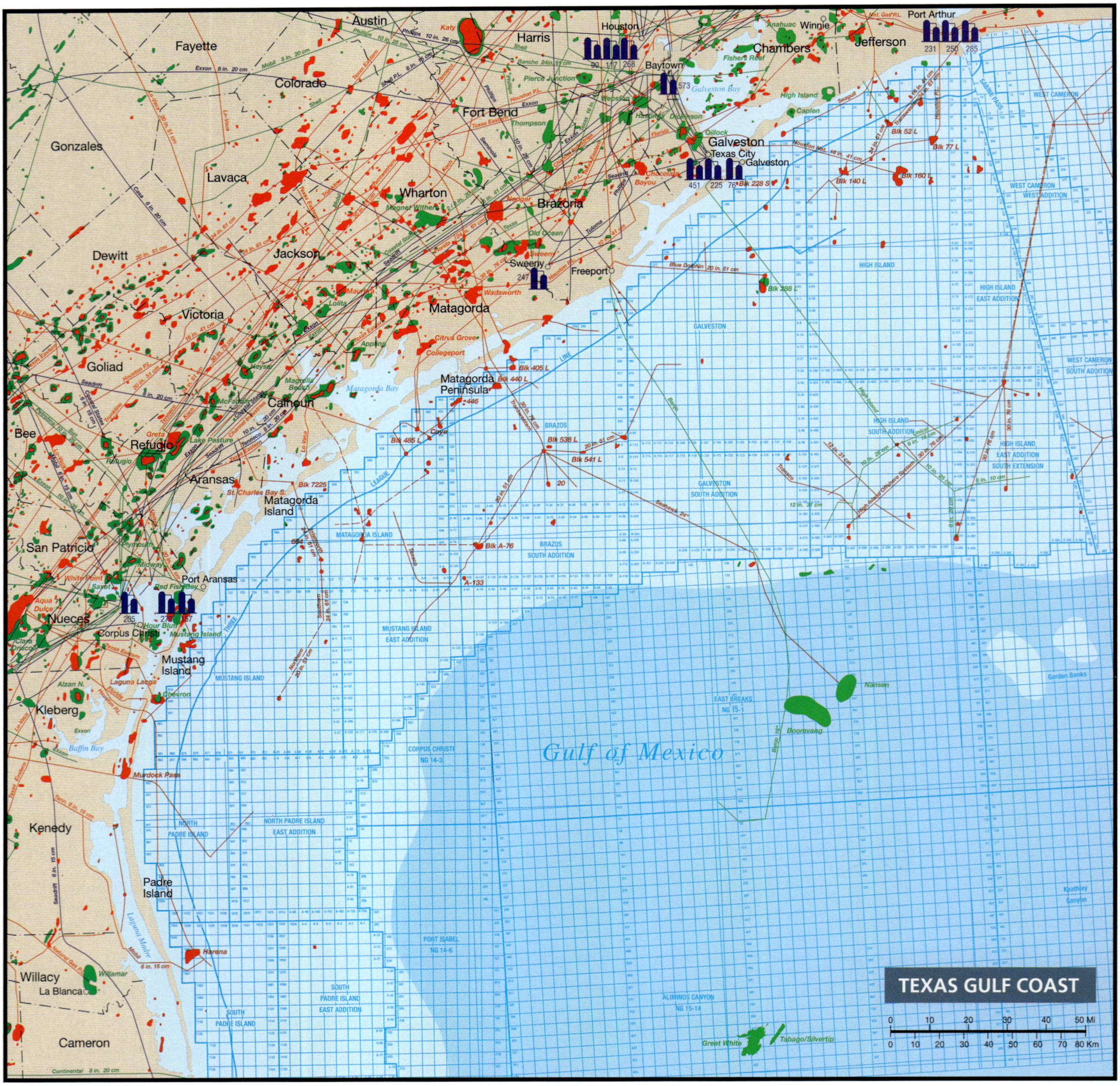

Mississippi Canyon: Block 948

The E&P subsidiary of BP America Inc. (Warrenville, Illinois) discovered oil for the third time in the deepwater gulf. The Freedom discovery, 70 mi (113 km) southeast of Louisiana, is on Mississippi Canyon Block 948 in 6,100 ft (1,859 m) of water.

The well reached 29,280 ft (8,925 m) TD and encountered more than 550 net ft (168 m) of hydrocarbon-bearing sands in Middle and Lower Miocene reservoirs.

BP had earlier finds in nearby Tubular Bells and Kodiak fields.

Partners with BP are Noble Energy Inc. (Houston); Samson Offshore Co. (Tulsa); and Marathon Oil Co. (Houston).

Mississippi Canyon: Blocks 800, 754

ATP Oil & Gas Corp. (Houston) participated in drilling two successful exploration wells; one is an oil discovery, and the other a gas-condensate discovery. Newfield Exploration Co. (Houston) operates both discoveries.

The Gladden prospect on Mississippi Canyon Block 800 found 80 ft (24 m) of net oil pay. The discovery well was drilled in 3,116 ft (950 m) of water, and an updip well was being drilled in mid-2008.

The Anduin West prospect on Mississippi Canyon Block 754 found 30 ft (9 m) of net gas-condensate pay. The well was drilled in 2,696 ft (822 m) of water.

Production from the discoveries is expected to begin in late 2009.

Mississippi Canyon: Block 771

The exploration and production subsidiary of BP PLC (London) made an oil find in early 2008 at its Kodiak prospect on Mississippi Canyon Block 771, about 60 mi (97 km) southeast of Louisiana.

The discovery well, drilled to TD 31,150 ft (9,494 m) in 5,000 ft (1,524 m) of water, hit 500 net ft (152 m) of hydrocarbon-bearing sands in Middle and Lower Miocene reservoirs.

Garden Banks: Block 462

Mariner Energy Inc. (Houston) made a deepwater discovery on its Geauxpher prospect in Garden Banks block 462 in mid-2008. The well, in water 2,700 ft (823 m) deep, was drilled to 23,156 ft (7,058 m) TD.

Mariner estimates the field's gross proved and probable reserves on the block to be approximately 100 bcfe (2.83 bcme) of natural gas. Initial production is expected by yearend 2008.

A second well is planned to delineate the extent of the discovery. Mariner also said it identified additional potential on the block.

South Marsh Island: Block 212

Flatrock field No. 4 well, drilled to 15,315 ft (4,668 m) in mid-2008, indicated hydrocarbon- bearing sands in the Rob-L section approximating 135 net ft (41 m), said McMoRan Exploration (New Orleans).

The field is on South Marsh Island block 212 in 10 ft (3 m) of water. The well will be deepened to 18,500 ft (5,639 m) TD to evaluate additional targets in the Rob-L and Operc sections.

The Flatrock No. 3 delineation well was drilled to 18,587 ft (5,665 m) and the well will be completed in the Operc section. The well encountered 256 net ft (78 m) of hydrocarbon pay.

McMoRan previously drilled the successful Flatrock No. 2 well. The initial discovery well came online in January 2008 at a gross rate of 50 MMcfe/d. That well was tested in a Rob-L sand at a gross rate of 114 MMcfe/d and was expected to begin production in mid-2008.

Loop current study

The MMS awarded a $5.5 million contract in third quarter 2008 to Science Applications International Corp. (SAIC; San Diego) for a major study of the Gulf of Mexico loop current.

That current—the upstream portion of the Gulf Stream—is the principal ocean current that transports energy, mass, heat, momentum, and salt from the eastern to the western half of the Gulf.

The five-year study will evaluate the dynamics of the loop current in the Eastern Gulf through observations and numerical modeling. Findings will provide information on how energetic currents may interrupt oil production and affect the movement of oil spills, including natural seeps from the sea floor.

Scientists from Princeton University, the University of Rhode Island, and the University of Colorado are part of the SAIC team. The Atlantic Oceanographic and Meteorological Laboratory also will work with SAIC to explore the role of the thermal structure of the loop current in hurricane intensification.

All study data will be available to other organizations, including the National Hurricane Center, the U.S. Navy, and the oil and gas industry.

Green Canyon: Block 468

A 33,362-ft (10,169-m) sidetrack well successfully appraised the 2006 Pony discovery of Hess Corp. (New York) in Green Canyon Block 468, just north of the Knotty Head area in 2,623–5,246 ft (800–1,599 m) of water.

The Pony-2 sidetrack, drilled from a surface location 7,400 ft (2,255 m) northwest of the discovery well, encountered the same objective Miocene sands as the Pony-1 and Pony-2 wells.

Hess said the Pony acreage has 200 million boe recoverable; it had earlier estimated 100–500 million boe.

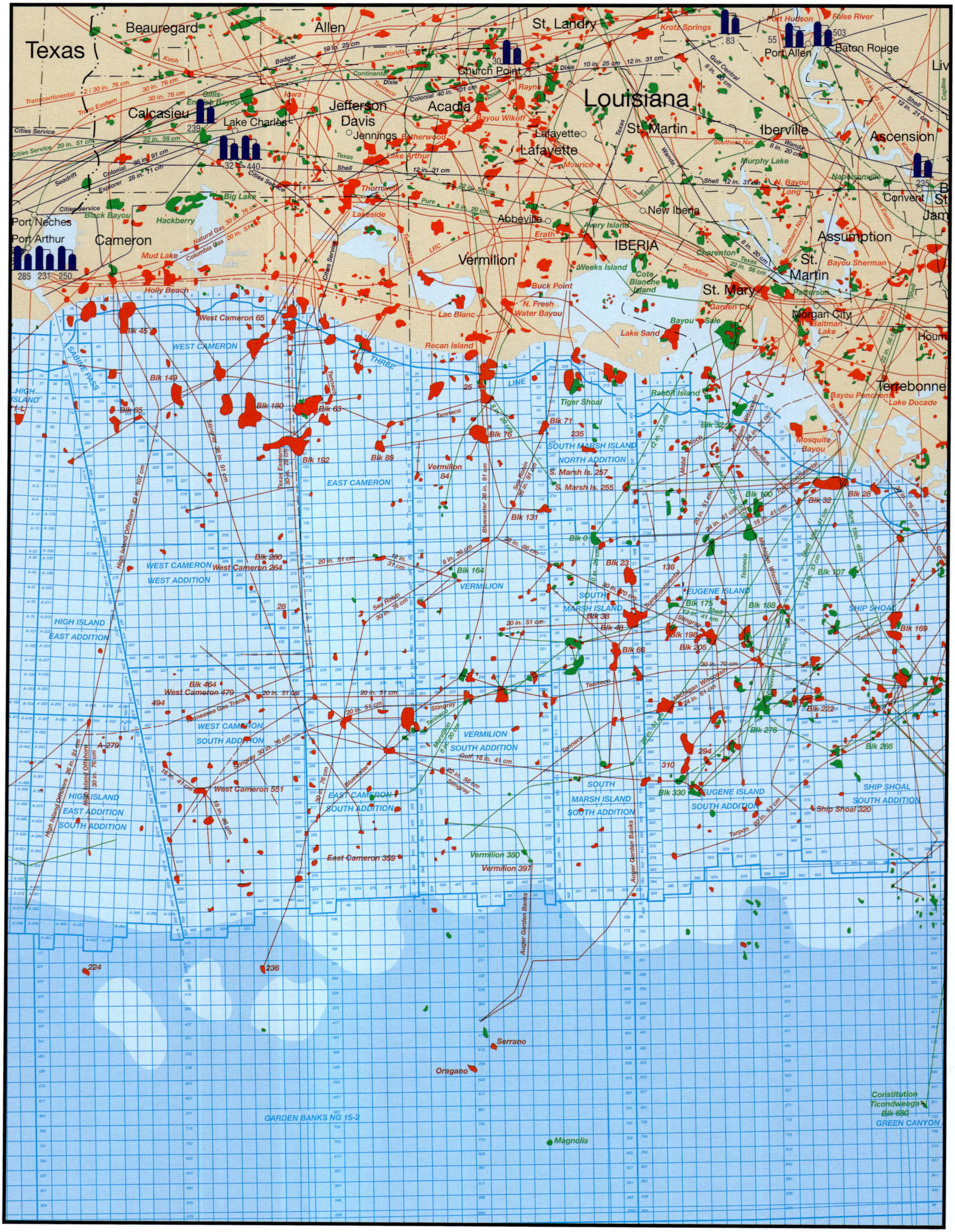

Texas
Beauregard
Allen
St. Landry
Krotz Springs
Port Hudson
False River
Baton Rouge
Port Allen
Church Point
Louisiana
Calcasieu
Lake Charles
Jefferson Davis
Jennings
Acadia
Lafayette
St. Martin
Iberville
Ascension
Convent
Port Neches
Port Arthur
Cameron
Big Lake
Hackberry
Black Bayou
Abbeville
New Iberia
Vermilion
IBERIA
Assumption
St. Mary
Morgan City
Weeks Island
Avery Island
Cote Blanche Island
Holly Beach
Mud Lake
Lac Blanc
Recan Island
Lake Sand
Terrebonne
Lake Docade
West Cameron 65
WEST CAMERON
SABINE PASS
THREE
LINE
Tiger Shoal
Rabbit Island
SOUTH MARSH ISLAND
NORTH ADDITION
S. Marsh Is. 257
S. Marsh Is. 255
EAST CAMERON
WEST CAMERON
WEST ADDITION
West Cameron 264
HIGH ISLAND
EAST ADDITION
VERMILION
SOUTH MARSH ISLAND
EUGENE ISLAND
SHIP SHOAL
West Cameron 479
WEST CAMERON
SOUTH ADDITION
VERMILION
SOUTH ADDITION
West Cameron 551
HIGH ISLAND
EAST ADDITION
SOUTH ADDITION
EAST CAMERON
SOUTH ADDITION
SOUTH
MARSH ISLAND
SOUTH ADDITION
EUGENE ISLAND
SOUTH ADDITION
SHIP SHOAL
SOUTH ADDITION
Ship Shoal 320
East Cameron 359
Vermilion 380
Vermilion 397
Auger Garden Banks
Serrano
Oragano
GARDEN BANKS NG 15-2
Magnolia
Constitution
Ticondewoga
Blk 680
GREEN CANYON

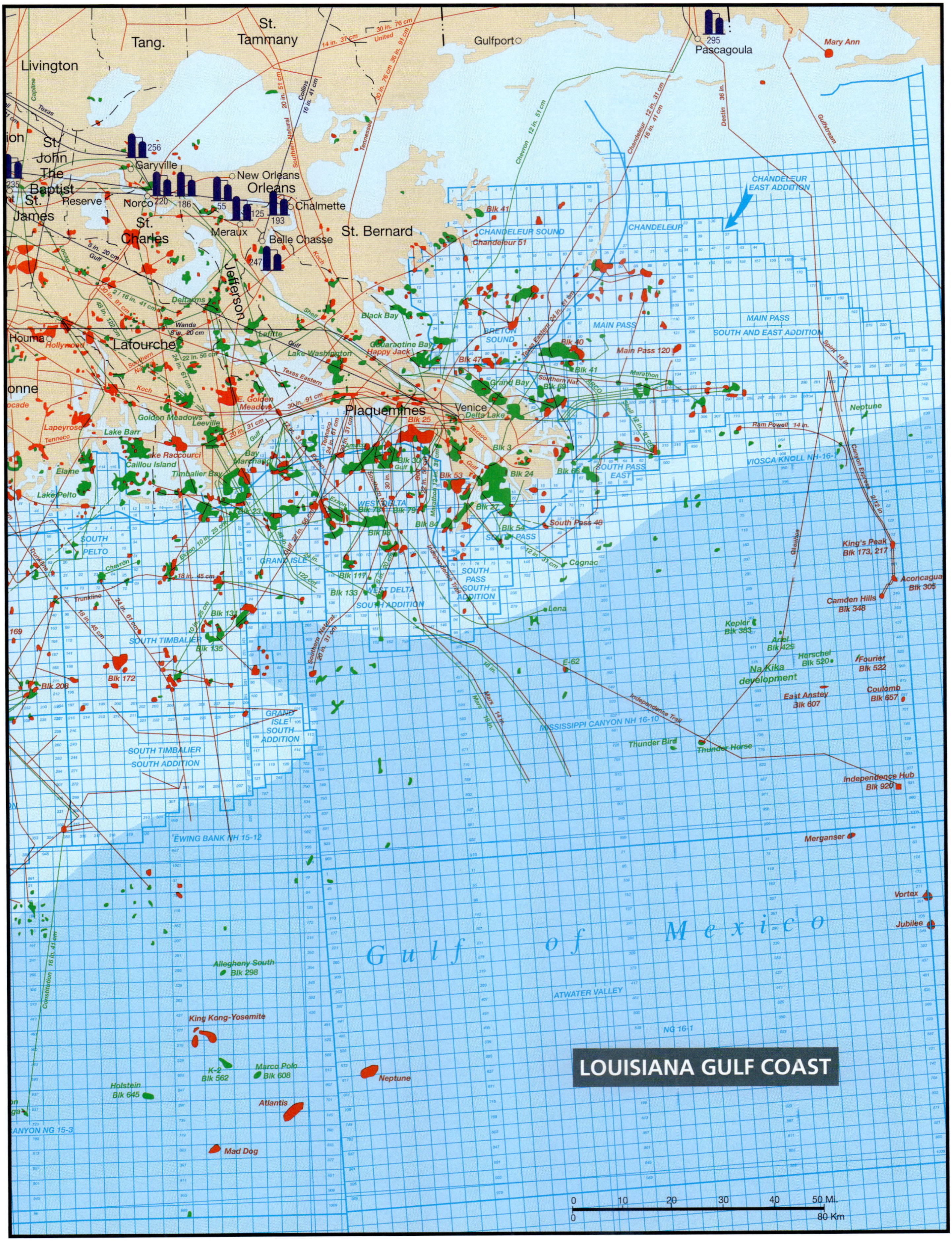
LOUISIANA GULF COAST
Livingston
Tang.
St. Tammany
Gulfport
Pascagoula
Mary Ann
St. John The Baptist
St. James
Reserve
Garyville
Norco
St. Charles
New Orleans
Orleans
Chalmette
Meraux
Belle Chasse
St. Bernard
Jefferson
Houma
Lafourche
Hollywood
Plaquemines
Venice
CHANDELEUR SOUND
CHANDELEUR
CHANDELEUR EAST ADDITION
BRETON SOUND
MAIN PASS
MAIN PASS SOUTH AND EAST ADDITION
SOUTH PASS
SOUTH PASS EAST
WEST DELTA
GRAND ISLE
SOUTH PELTO
SOUTH TIMBALIER
SOUTH TIMBALIER SOUTH ADDITION
GRAND ISLE SOUTH ADDITION
WEST DELTA SOUTH ADDITION
SOUTH PASS SOUTH ADDITION
EWING BANK NH 15-12
MISSISSIPPI CANYON NH 16-10
VIOSCA KNOLL NH 16-7
ATWATER VALLEY
NG 16-1
Neptune
King's Peak Blk 173, 217
Aconcagua Blk 305
Camden Hills Blk 348
Kepler Blk 383
Ariel Blk 429
Herschel Blk 520
Na Kika development
Fourier Blk 522
Coulomb Blk 657
East Anstey Blk 607
Thunder Bird
Thunder Horse
Independence Hub Blk 920
Merganser
Vortex
Jubilee
Cognac
Lena
Allegheny South Blk 298
King Kong-Yosemite
K-2 Blk 562
Marco Polo Blk 608
Holstein Blk 645
Atlantis
Mad Dog
Gulf of Mexico
0 10 20 30 40 50 Mi.
0 80 Km

Louisiana State Lease 18860

Contango Oil & Gas Co. (Houston) drilled a successful well at its Mary Rose prospect in Louisiana waters on State Lease No. 18860. It was the fourth well the company has drilled on the prospect.

The company expected to complete the well and begin flow to the platform at Eugene Island block 11 by August 2008.

Walker Ridge: Block 29

Chevron Corp. reported successful completion in early 2008 of an appraisal well at its Big Foot prospect. It lies in more than 5,000 ft (1,523 m) of water on Walker Ridge Block 29, some 225 mi (362 km) south of New Orleans and 180 miles offshore.

The appraisal well confirmed the same pay intervals of the previously announced discovery and sidetrack wells.

The appraisal well reached a measured depth of 25,113 ft (7,654 m), including water.

New Gulf Coast coalition

A coalition called America's Energy Coast (AEC) was formed in mid-2008 to ask the nation to invest in restoring the degraded landscapes of the western Gulf Coast and bolster oil production.

Members include major oil companies, business and transportation interests, environmental organizations, sporting and recreational groups, and state and local officials from Louisiana, Mississippi, Alabama, and Texas.

The coalition thinks the Department of Homeland Security should do more to secure refineries, offshore platforms and oil ports and should consider protecting energy infrastructure in danger from rising sea levels.

They note that about 90% of the U.S. Gulf of Mexico's oil and gas production passes through the four-state area. A huge pipeline network is anchored on barrier islands protected by wetlands there.

The Council's Web site is: www.americasenergycoast.org.

Galveston Island

Providence Resources (Dublin) and partners made a gas discovery in early 2008 on Galveston Island block A-155 in the Gulf of Mexico.

Initial evaluation suggested reserves of at least 13 bcf (368 MMcm) of recoverable reserves.

Lease Sales

Sale 205: Central Gulf

As of early 2008, MMS accepted bids valued at $2.83 billion and awarded 683 leases in connection with Central Gulf of Mexico oil and gas Lease Sale 205, which was held in October 2007.

Eighty-four companies submitted 1,428 bids on 723 tracts in the sale. The total for high bids submitted on all tracts was $2.90 billion.

The highest bid accepted on a tract was $90.5 million, made by Shell Offshore Inc. for Walker Ridge block 7 in water 5,240–6,562 ft (1,600–2,000 m) deep.

Sales 206 and 224: Central, Eastern Gulf

Two lease sales conducted by the MMS in March 2008 attracted more than $3.7 billion in high bids.

Central Gulf of Mexico Sale 206 attracted 1,057 bids from 85 companies on 615 tracts resulting in a record $3.68 billion in high bids.

For Eastern Gulf of Mexico Sale 224, MMS received 58 bids from six companies on 36 tracts resulting in $ 64.7 million in high bids.

Lease Sale 224 encompasses 118 whole or partial unleased blocks covering 546,971 acres (855 sq mi; 2,215 sq km) in the Eastern Planning Area, in water depths ranging from 2,657–10,213 ft (810–3,113 m).

Beginning with Sale 224, Louisiana, Mississippi, Alabama, and Texas will receive a greater share of all revenues, including bids, rental payments, and royalties, said MMS.

The Central Sale 206 offered 5,569 tracts comprising about 29.8 million acres (46,563 sq mi; 17,978 sq km) in federal areas offshore Louisiana, Mississippi, and Alabama. The acreage is located from 3–230 mi (5–370 km) offshore in water depths of about 10 ft (3 m) to more than 11,200 ft (3,400 m).

About 34% of the tracts receiving bids in this sale are in ultradeep water of more than 5,249 ft (1,600 m).

Results of Central Gulf of Mexico Lease Sale 206

Top 10 companies based on total number of high bids submitted

Company	Total high bids	Sum of high bids, $
1. BP Exploration & Production Inc.	63	336,575,445
2. Chevron USA Inc.	49	240,987,863
3. BHP Billiton Petroleum (Deepwater) Inc.	42	95,652,498
4. Cobalt International Energy LP	36	389,056,079
5. Devon Energy Production Co. LP	35	94,795,020
6. Repsol E&P USA Inc.	32	95,477,260
7. Eni Petroleum US LLC	31	113,125,970
8. Hess Corp.	25	437,541,152
9. LLOG Exploration Offshore Inc.	25	104,854,780
10. Stone Energy Corp.	25	43,019,844

Companies submitting Top 10 single highest bids

Company	Block	Water depth, m	High bid amount, $
1. Anadarko E&P Co. LP* Murphy E&P Co.-USA Samson Offshore Co.	Green Canyon 432	800-<1,600	105,600,789
2. Marathon Oil Co.* Hess Corp	Walker Ridge 226	>2,000	93,024,910
3. Cobalt International Energy LP	Green Canyon 858	1,600-2,000	85,418,889
4. Chevron USA Inc.	Green Canyon 945	1,600-2,000	81,063,073
5. Cobalt International Energy LP	Keathley Canyon 163	1,600-2,000	74,418,889
6. ConocoPhillips Co.	Keathley Canyon 654	>2,000	73,448,446
7. Marathon Oil Co.* Hess Corp	Walker Ridge 269	>2,000	73,424,910
8. Hess Corp	Keathley Canyon 469	1,600-2,000	71,489,800
9. Hess Corp	Garden Banks 829	800-1,600	71,489,800
10. Statoil Gulf of Mexico LLC	Green Canyon 480	800-1,600	68,525,300

*Denotes the submitter.

Source: US Minerals Management Service

The top five companies submitting the highest dollar amount of high bids for Sale 224 included:

- BHP Billiton Petroleum (Deepwater) Inc., 27 bids, $47,858,420
- Anadarko E&P, 7 bids, $12,754,728
- Murphy Exploration & Production Co., 7 bids, $3,188,682
- Eni Petroleum US LLC, 1 bid, $527,006
- Shell Offshore Inc., 1 bid, $384,377.

Top 10 Tracts by High Bid

Rank, company	Block	Water depth, m	High bid, $
1. Statoil Gulf of Mexico LLC	Alaminos Canyon 380	>2,000	61,110,000
2. Chevron USA Inc.	Garden Banks 973	800->2,000	52,107,556
3. Chevron USA Inc.	Garden Banks 972	800-<1,600	34,607,556
4. Statoil Gulf of Mexico LLC	Alaminos Canyon 424	>2,000	22,332,000
5. Chevron USA Inc.	Alaminos Canyon 775	>2,000	20,121,217
6. Shell Gulf of Mexico Inc.	Alaminos Canyon 771	>2,000	11,999,999
7. Eni Petroleum US LLC	Keathley Canyon 580	>2,000	8,103,115
8. LLOG Exploration Offshore Inc.	High Island Area 169	0-<200	8,012,000
9. ConocoPhillips Co.	Keathley Canyon 146	800-<1,600	6,200,110
10. LLOG Exploration Offshore Inc.	High Island Area A7	0-<200	6,012,000

Source: US Minerals Management Service

Sale 207: Western Gulf

The Western Gulf of Mexico Oil and Gas Lease Sale 207 in August 2008 attracted $487.3 million in high bids, according to MMS. Fifty-three companies submitted 423 bids on 319 tracts comprising over 1.8 million acres offshore Texas in the western gulf's OCS planning area.

The sum of all bids received totaled $607 million.

Approximately 17% of the tracts receiving bids are in ultra deepwater of more than 5,249 ft (1,600 m). The deepest tract bid on is Alaminos Canyon, block 783 in 9,767 ft (2,977 m) of water.

The highest bid received on a tract was $61.1 million, from Statoil Gulf of Mexico LLC for Alaminos Canyon block 380.

Top 10 Companies by Sum of Apparent High Bids

Rank, company	Total high bids	Sum of high bids, $
1. ExxonMobil Corp	130	127,331,000
2. Chevron USA Inc.	20	127,281,322
3. Statoil Gulf of Mexico LLC	5	87,352,000
4. LLOG Exploration Offshore Inc.	11	23,172,800
5. Shell Gulf of		12,341,000
Mexico Inc.	15	20,199,985
6. Anadarko E&P Co. LP	19	14,806,839
7. Hess Corp.	22	14,162,407
8. Eni Petroleum US LLC	5	11,127,512
9. ConocoPhillips Co.	2	10,504,410
10. Devon Energy Production Co. LP	20	6,671,461

Source: US Minerals Management Service

Sale 208: CPA

The U.S. MMS announced plans to hold Lease Sale 208 for the Central Planning Area (CPA) in the Gulf of Mexico on March 18, 2009.

They estimate that the proposed sale could result in production of 0.8–1.3 billion bbl of oil and 3.4–5.4 tcf of natural gas. The acreage lies 3–230 mi (5–370 km) offshore in water depths ranging from 10 to 11,150 ft (3 to 3,400 m).

The sale encompasses 6,200 unleased blocks covering more than 33.5 million acres (52,343 sq mi; 20,210 sq km) off Louisiana, Mississippi, and Alabama.

This area includes a large section in the southeastern part of the CPA known as the 181 South Area, which will be offered for lease for the first time since 1988.

RIK sales

In early 2008, MMS sold more than 4.2 million bbl of Royalty in Kind (RIK) crude oil and condensate from federal leases in the Gulf of Mexico and Pacific Ocean to six companies. The sale is expected to bring in more than $362 million.

Companies submitting winning bids include Chevron Products Co., ConocoPhillips, ExxonMobil Oil Corp., Sempra Trading, Shell Trading US Co., and Plains Marketing LP.

Drilling and Production

Thunder Horse

The ultra-deep, high-pressure, high-temperature Thunder Horse field began full production in mid-December 2008, with its third and fourth wells flowing, according to BP PLC (London).

The field was pumping more than 200,000 boe/d.

Thunder Horse, which first began producing in June 2008, was slated to come online more than three years ago but was plagued by technical problems.

Located 150 mi (241 km) southeast of New Orleans, the platform has a daily production capacity of 250,000 barrels of oil and 200 MMcf of natural gas.

Nearby Thunder Horse North was expected to begin producing by mid-2009.

BP is the operator and owns three-fourths of the project.

Pegasus field

Eni (Milan) started production from the Pegasus field in Green Canyon block 385 in late 2008.

The field tied back to the Eni-operated Allegheny TLP. Peak oil production of approximately 5,000 b/d of oil is expected in 2009.

Perdido completion

Shell Oil Co. (Houston) set a world water depth record in drilling and completing a subsea well 9,356 (2,852 m) below the water's surface.

The SA001 well was drilled to 16,300 ft (4,968 m) TD in Silvertip field at the Perdido regional development project, 200 mi (322 km) from Houston. Silvertip field is on Alaminos Canyon Block 815 in OCS lease G19409 in the Gulf of Mexico.

The Silvertip record is 35% deeper than the previous oil well record of 6,950 ft (2,118 m), also set by Shell, in Fourier field in the gulf.

At Perdido, Shell intends to drill an even deeper well in Tobago field at 9,627 ft (2,934 m) on Alaminos Canyon Block 859. Shell operates Perdido on behalf of partners Chevron Corp. and BP PLC.

PERDIDO DEVELOPMENT

Deepwater Gulf of Mexico discoveries

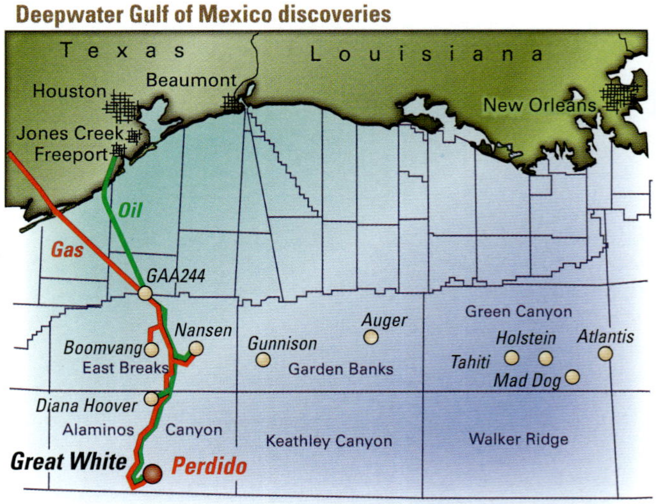

Alaminos Canyon wells

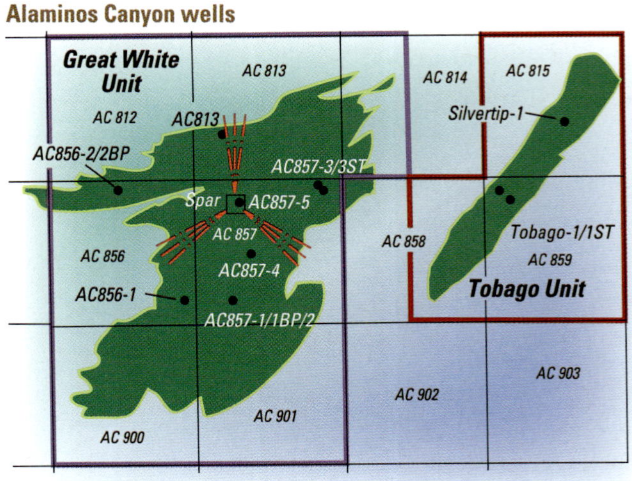

Perdido Regional Development Truss Spar

Spar dimensions	555 ft (173.3 m) x 118 ft (35.97 m)
Spar draft	505 ft
Spar total displacement	186,270 kips (186 million lb_f)
Spar weight	20,956 tonnes
Mooring system	Nine taut lines (3 x 3); 9.68-in. polyester rope main section with short lengths of 5.28-in. chain at each end; 18-ft OD suction piles
Risers (6)	• Five 14-in. diameter production risers with subsea separation and boosting systems • One drilling and completion riser
Topsides dimensions	150 ft x 210 ft
Operating payload	37,000 kips (37 million lbf), including rig and equipment
Accommodation	150

Main contractors	Providing
Noble Drilling Corp.	Delineation drilling
Technip SA	Spar design, fabrication; umbilicals; mooring system
Alliance Engineering	Topsides design
Kiewit Offshore Services Ltd.	Topsides fabrication; spar completion
FMC Technologies Inc.	Subsea completion and processing systems
Oceaneering International Inc.	Oil pipeline tie-in; subsea jumpers
Heerema Marine Contractors	Offshore installation of host, subsea flowlines, and equipment
Dockwise Shipping BV	Finland to Texas spar tow
Otto Candies LLC	Gulf of Mexico spar, topside tows
American Bureau of Shipping	Certification

Shell and its partners will drill 35 wells at Perdido—22 direct vertical access and 13 remote—in the Great White, Tobago, and Silvertip fields in Alaminos Canyon.

Moored in 8,000 ft (2,438 m) of water, the drilling and production facility will be the deepest in the world. The floating structure will weigh 50,000 tons (45,359-metric tons) and will be nearly as tall as the Eiffel Tower.

First production from Perdido is expected in early 2010. The facility can handle 130,000 boe/d.

Blind Faith field

Chevron Corp. (San Ramon, CA) reported the start of oil production from its Blind Faith field in the deepwater Gulf of Mexico.

Initial production is 30,000 b/d of oil and 30 MMcfd of natural gas. Daily production is expected to ramp up to 65,000 b/d of oil and 55 MMcfd of gas in 3 months.

Blind Faith uses a deep-draft semisub hull moored about 160 mi (250 km) southeast of New Orleans.

Chevron's deepest offshore production facility, Blind Faith is in 6,500 ft (1,981 m) of water, with subsea systems in 7,000 ft (2,134 m) of water in Mississippi Canyon blocks 695 and 696.

The field has an estimated gross resource potential exceeding 100 MMboe.

Droshky, Ozona projects

Marathon Oil (Houston) announced plans in October 2008 to proceed with two Gulf of Mexico development projects, Droshky ($1.3 billion) and Ozona ($300 million).

The company expects to initially book proved reserves of approximately 29 MMboe for Droshky and Ozona.

The Droshky discovery lies in 2,900 ft (884 m) of water in Green Canyon block 244, about 18 mi (29 km) southeast of Shell's Bullwinkle platform. Four development wells will be tied back to Bullwinkle. Drilling is slated to start in 2009, and first production is targeted for 2010. Expected net peak production is about 45,000 b/d of oil and 43 MMcf/d of natural gas, after royalties.

The Ozona discovery is on Garden Banks block 515, in 3,000 ft (914 m) of water, about 6 mi (10 km) from Shell's Auger platform. First production is expected in 2011, with an anticipated net peak rate of about 6,000 b/d of oil and 13 MMcf/d of natural gas.

Big Foot project

Chevron USA Inc. (San Ramon, CA) was evaluating development options for the Big Foot project in the Gulf of Mexico in late 2008.

Big Foot lies in more than 5,000 ft (1,524 m) of water on Walker Ridge Block 29, about 225 mi (362 km) south of New Orleans and 180 mi (290 km) offshore.

Ursa/Princess waterflood

Royal Dutch Shell PLC (The Hague, Netherlands) started water injection in mid-2008 at the Ursa/Princess Waterflood project, aiming to extend the life of the field by 10 years.

Producing wells will include three Princess subsea wells and up to six Ursa wells.

The Ursa-Princess complex has produced nearly 400 MMboe.

The waterflood initiative is one of the largest construction projects on an existing platform in the Gulf of Mexico. The facility is 100 mi (161 km) offshore in 3,800 ft (1,158 m) of water and encompasses Mississippi Canyon blocks 808, 809, 810, 852, 853 and 854.

Structure count by production category

Structure type*	Year	<100	100-500	500-1,000	>1,000	Total
		Production, bo/d				
Oil	2000	8	18	7	5	38
	2001	8	18	6	7	39
	2002	9	15	8	6	38
	2003	6	17	7	5	35
	2004	7	16	9	3	35
	2005	13	15	1	3	32
		Production, Mcfd				
		<600	600-3,000	3,000-6,000	>6,000	Total
Gas	2000	12	19	9	8	48
	2001	13	14	9	11	47
	2002	14	17	7	7	45
	2003	15	18	7	7	47
	2004	18	17	6	9	50
	2005	14	21	9	2	46
		Production, boe/d				
		<100	100-500	500-1,000	>1,000	Total
All	2000	14	37	17	18	86
	2001	26	18	14	28	86
	2002	12	34	21	16	83
	2003	14	28	23	17	82
	2004	18	35	15	17	85
	2005	21	35	16	6	78

*Structures are classified as oil or gas producers according to their cumulative GOR measured in cu ft/bbl. Structures with GOR ≤5,000 are classified as primarily oil producers; structures with GOR >5,000 are primarily gas producers.

Share of gulf production from destroyed structures

Fluid type	2000	2001	2002	2003	2004	2005
Gas	1.7	1.9	2.0	2.3	2.1	1.2
Oil	2.2	3.0	3.7	2.9	2.7	1.5
BOE*	1.7	2.3	2.7	2.6	2.4	1.3
Water	3.9	3.7	4.2	4.3	4.3	3.4

*BOE = barrels of oil equivalent, computed on a heat equivalent basis, where 6,040 cu ft of gas provides 1 BOE. The BOE stream is the combined oil and gas production output.

Structure and rig damage by storm

	Ivan	Katrina	Rita
Platforms destroyed	7	46	69
Platforms extensively damaged	24	20	32
Rigs adrift	5	6	13
Rigs destroyed	1	4	4
Rigs extensively damaged	4	9	10

Source: US Minerals Management Service

Operators with three or more structures destroyed*

Operator	Structures destroyed	Total wells	Active wells, 2005
Chevron	15	179	38
Apache	11	131	39
BP America	11	109	24
Forest Oil	11	53	11
Energy XXI	9	17	6
Stone Energy	7	37	11
Energy Resources Technology	5	29	10
Newfield Exploration Co.	5	21	10
Anglo-Suisse Offshore Partners	4	72	13
Noble	4	101	18
Mariner Energy	3	4	1
Maritech	3	28	2

*Structures destroyed in 2004-05 hurricane seasons.

2008 hurricane damage report

Hurricanes Gustav and Ike moved through the Gulf of Mexico in September 2008. Gustav made landfall September 2 in Louisiana; Ike came ashore September 13 in Galveston, Texas.

In its December 2008 final storm damage assessment, the U.S. MMS said a total of 60 offshore platforms was destroyed.

The destroyed platforms produced 13,657 b/d of oil and 96.5 MMcfd of natural gas, or 1.05% of the oil and 1.3% of the gas produced daily in the gulf.

MMS estimated that 127 of the more than 3,800 oil and gas production platforms in the gulf were exposed to hurricane conditions, with winds greater than 74 mph during the storms.

The agency also said that 93 platforms suffered "moderate" damage (needing as long as three months for repairs) and another 31 had "extensive" damage (requiring up to 6 months to repair).

Ivan, Katrina, Rita: Economic impact

Researchers at Louisiana State University (Baton Rouge) revisited the damage caused by Hurricanes Ivan, Katrina and Rita (during the 2004 and 2005 storm seasons) to evaluate the financial impacts of decommissioning and the potential value of the lost oil and gas production.

Their report in mid-2008 noted that 122 structures were destroyed and 76 others severely damaged in the three hurricanes.

Structures destroyed in hurricanes Ivan, Katrina, and Rita

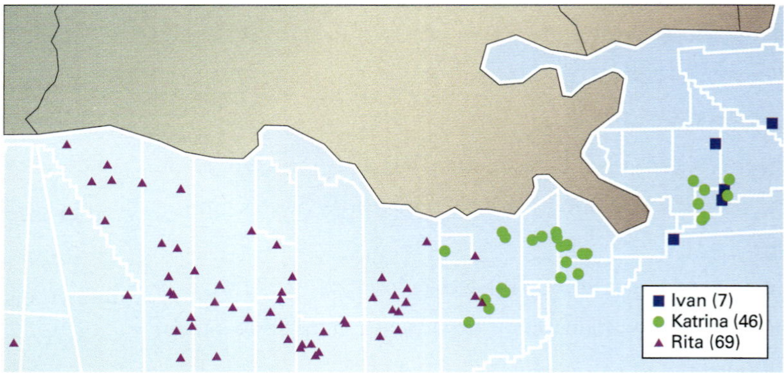

Structure damage rate and physical damage estimates

Hurricane	Year	Structures in storm path	Structures destroyed and with major damage	Damage* rate,%	Physical damage, billion $
Andrew	1992	700	87	12	0.9
Lili	2002	800	10	1	0.4
Ivan	2004	150	31	21	1.5
Katrina	2005	2,068	66	3	6.4
Rita	2005	793	101	13	3.7

*Damage rate expressed as percent exposed, computed as the number of structures destroyed and with major damage divided by the number of structures in a 50-mile envelope centered on the storm path.

Shut-in oil production in Gulf of Mexico

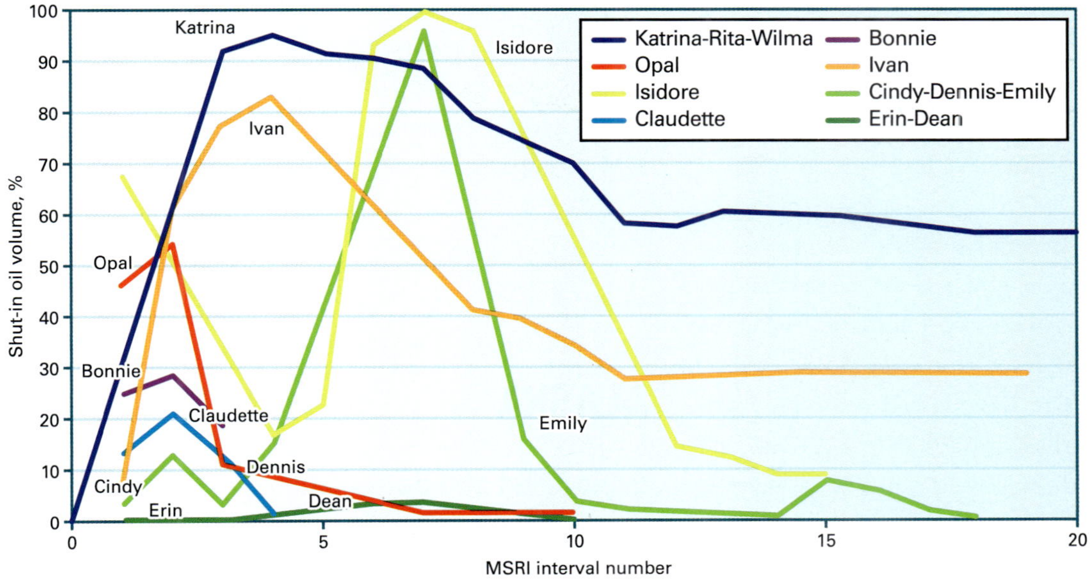

Destruction by age group*

Age group, years	Number
<10	24
11-20	9
21-30	17
31-40	49
>40	23
Total	**122**

*Structures destroyed by hurricanes Ivan, Katrina, and Rita.

Shut-in gas production in Gulf of Mexico

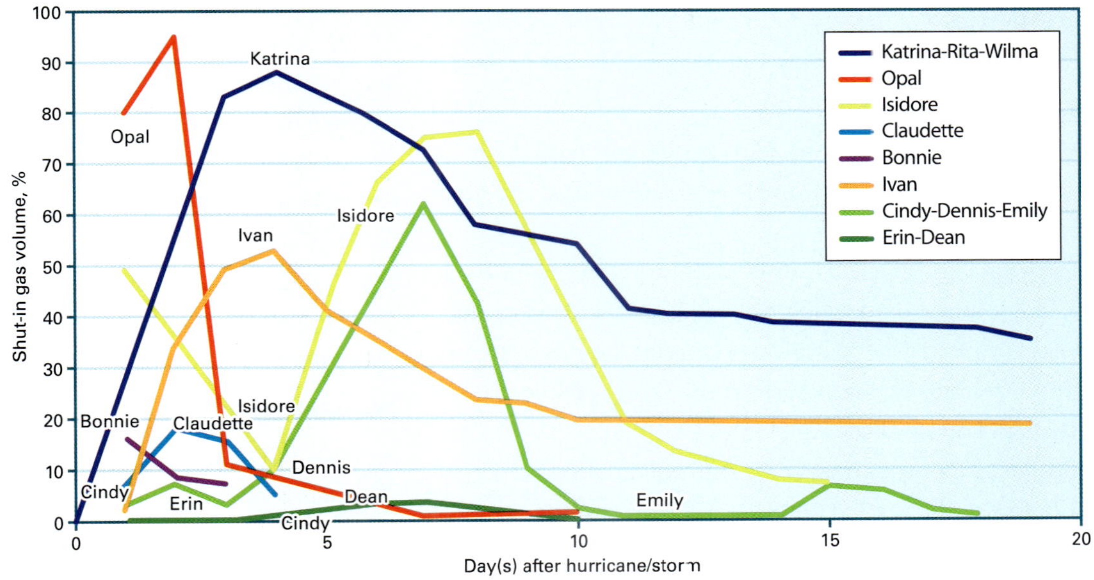

Historic and forecast production from destroyed structures

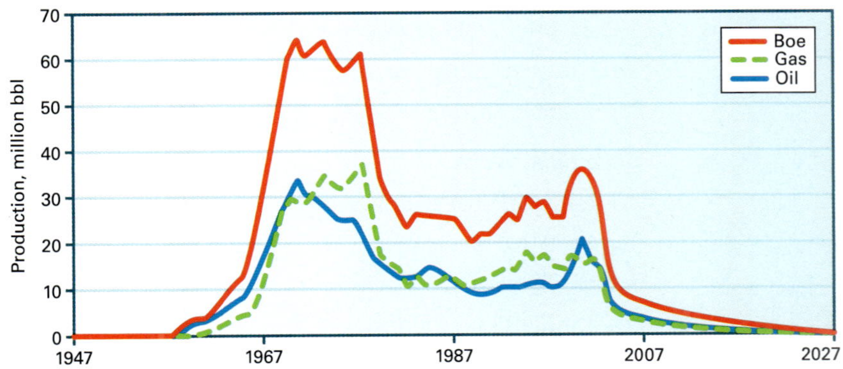

Abandonment cost estimates

Operator	Structures destroyed	Estimated cost, million $	Source[1]
Maritech	3	72-96	A
Taylor Energy	[illegible]	500	T
BP America	1[illegible]	500	T
Pioneer Natural Resources	[illegible]	138	T
Mariner Energy[2]	3	35	A
Apache	1[illegible]	724	A
Stone Energy	7	50	A

[1] A = 10K annual report, T = trade press.
[2] Costs pertain to assets acquired from Forest Oil.

Cascade-Chinook: First Gulf FPSO

In mid-2008, the MMS approved development plans for the Cascade-Chinook oil and natural gas project in block 425 of the Walker Ridge area, 165 miles off Louisiana in 8,200 ft (2,499 m) of water.

The project, operated by Brazil's Petrobras, will involve the first use of an FPSO vessel in the gulf, and will be the first production from deep discoveries in the Lower Tertiary trend of the Walker Ridge and Keathley Canyon areas, said MMS.

At least two subsea wells in Cascade and one subsea well in Chinook will each be drilled to 27,000 ft (8,229 m) and tied back to the FPSO. Additional wells could be included in the future.

Pending MMS approval, the FPSO is to be installed in early 2010 with production to start soon thereafter. The vessel will have a storage capacity of 600,000 bbl of oil, a processing capacity of 80,000 b/d of oil, and natural gas export facilities of 16 MMscfd.

Petrobras will build a pipeline to transport the gas to shore.

Atwater Valley: Block 425

Mariner Energy Inc. (Houston) started gas production in February 2008 from the two-well Bass Lite field in the deepwater gulf.

The wells, in 6,750 ft (2,057 m) of water on Atwater Valley Block 425, represent one of the gulf's deepest deepwater developments.

Initial production of 60 MMcfd of biogenic gas flows through a 56-mi (90-km) subsea tieback to the Devil's Tower spar on Mississippi Canyon Block 773.

Operator Mariner estimated the field will attain peak output of 125 MMcfd by yearend 2008.

Refining

Terminal and storage: Port Arthur

Sempra Energy (San Diego) proposed in early 2008 to build a marine petroleum terminal and storage facility in Port Arthur, Texas, to improve tanker access to refineries in that city and Beaumont.

An initial phase of the proposed terminal would handle oil, LP gas, and related products.

The terminal would allow 24-hr marine terminal access with first-phase throughput of up to 500,000 b/d. That phase would require about 120 of the 2,900 acres Sempra Energy owns near Port Arthur.

The property also is the proposed site of a larger energy complex that includes the planned development of a LNG project capable of storing and regasifying up to 3 bcfd (85 MMcm) of gas.

ALASKA

Exploration and Development

North Slope reserves potential

U.S. oil reserves could more than double by 2050 if commodity prices remain high and leasing restrictions ease on Alaska's North Slope (ANS), according to an early-2008 federal government study.

The U.S. DOE's Office of Fossil Energy study said exploration and development might add 28 billion bbl of economically recoverable oil and 125 tcf (3.54 tcm) of gas during 2015–50. Reserves for all of the US now are 21 billion bbl of oil and 211 tcf (5.98 tcm) of gas.

The forecast assumes continued high oil and gas prices, stable fiscal policies, and access by the producing industry to all areas of the North Slope, including the coastal plain of the Arctic National Wildlife Refuge.

Kuparak formation

The Tofkat-1 well struck oil-bearing sandstone in the Kuparak formation on Alaska's North Slope in early 2008, said partner Bow Valley Energy Ltd. (Calgary).

Tofkat-1, east of the Colville River, awaits evaluation to determine its commerciality. Log analysis confirmed an interval of 10 ft (3 m) of gross pay and 6 ft (2) of net pay. No production test was attempted.

Data from a 3D seismic survey were to be analyzed by late 2008 in preparation for an appraisal drilling program in the 2009 winter drilling season.

White Hills blocks

Total SA (Paris) said it planned to join Chevron Corp. (San Ramon, CA) in exploring 772 sq mi (2,000 sq km) of the onshore White Hills blocks in Alaska in late 2008. The blocks lie 25 mi (40 km) southwest of Prudhoe Bay field in the ANS.

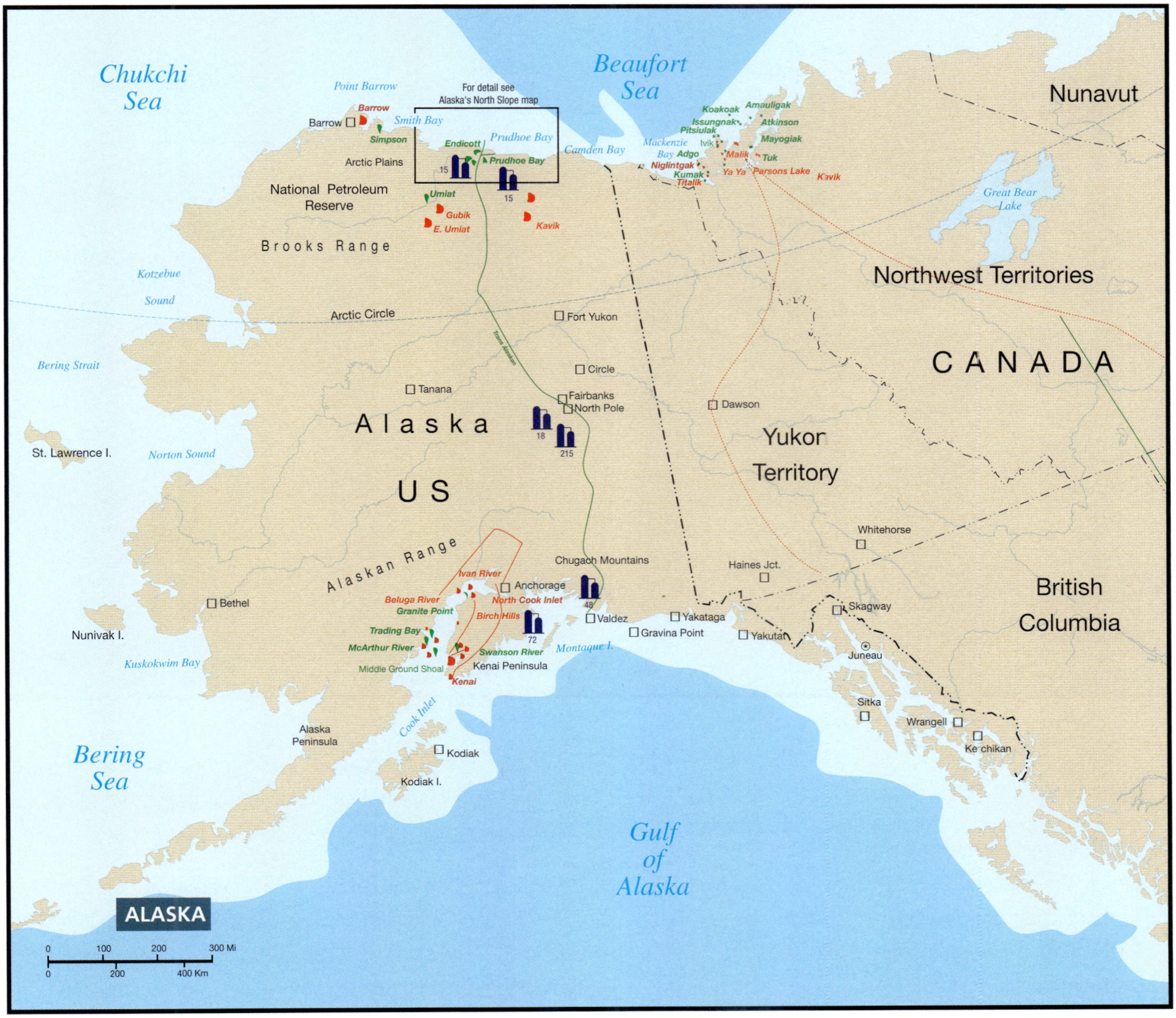

Gas hydrates assessment

In an October 2008 assessment, the USGS found that Alaska holds 85.4 tcf (2.42 tcm) of natural gas hydrates recoverable with current technology.

Scientists assessed a 55,894-sq mi (144,765-sq km) area. It extends from the National Petroleum Reserve-Alaska (NPR-A) on the west through the Arctic National Wildlife Refuge (ANWR) on the east, and from the Brooks Range on the south, to the state-federal offshore boundary about 3 mi (5 km) into the Arctic Ocean off the northern coast.

Gas hydrates are naturally occurring, ice-like solids in which water molecules trap gas molecules in a cage-like structure.

The evaluation assumed that the gas hydrate resources can be produced by existing conventional technology, and that depressurization appears to be the most promising method.

Of the 85.4 tcf, about 24%, is in the Sagavanirktok formation; 33%, is in the Tulavak-Schrader Bluff-Prince Creek formations, and 43%, is in the Nanushuk formation.

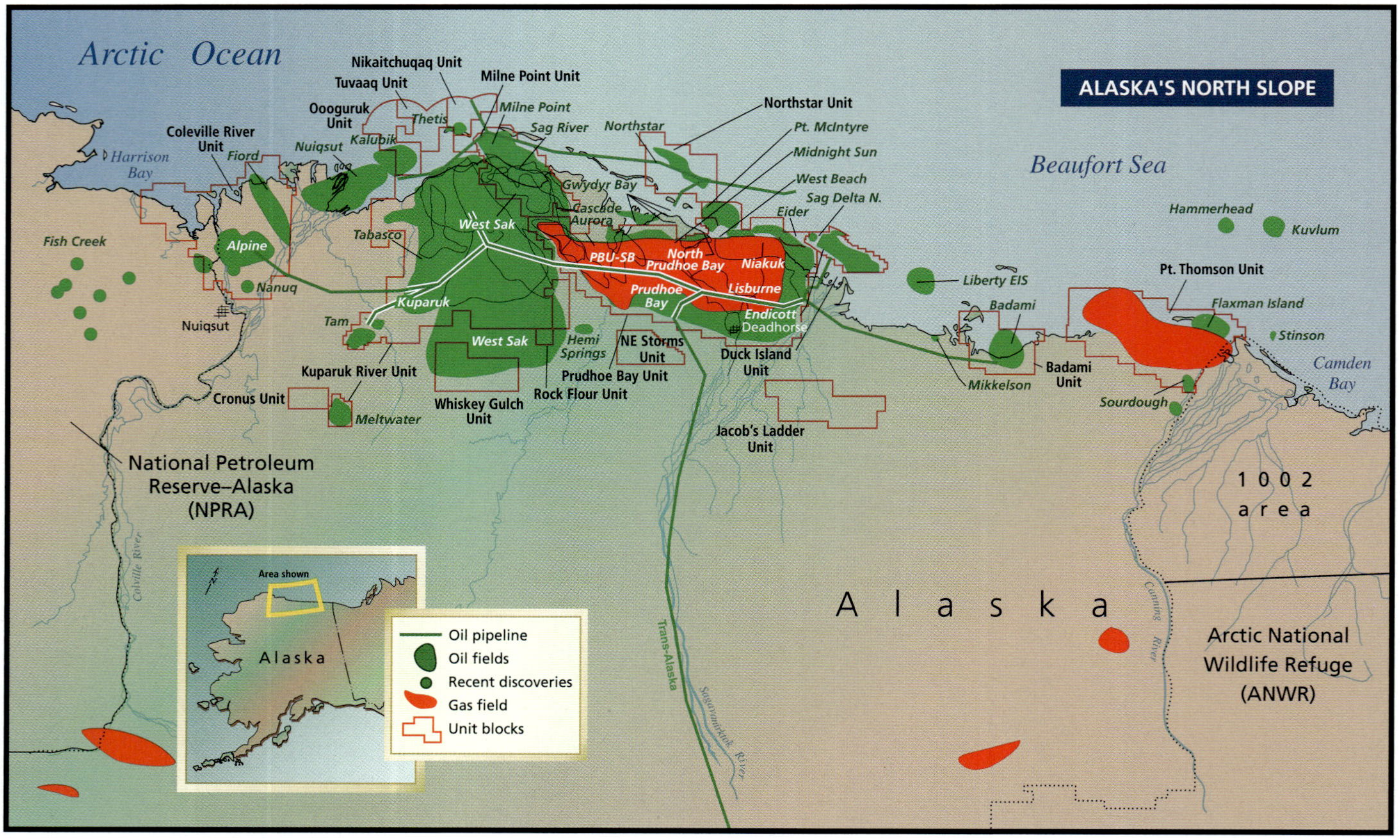

Point Thomson Unit

Alaska state officials rejected in April 2008 the latest plan of ExxonMobil Corp. (Irving, TX) to develop Point Thomson Unit (PTU), east of Prudhoe Bay on Alaska's North Slope. The gas-condensate area currently has no production.

The Alaska Department of Natural Resources said the proposed plan was not in the state's best interest because it would have involved development on the 106,200-acre (166 sq mi; 430 sq km) PTU without any commitment to produce gas.

ExxonMobil, PTU operator, has estimated that the high-pressure reservoir has reserves of more than 8 tcf (226 bcm) of gas and 200 million bbl of condensate.

ExxonMobil along with BP Exploration (Alaska) Inc., Chevron USA Inc., and ConocoPhillips Alaska Inc. hold working interests in PTU, covering 45 state oil and gas leases. The oil companies have said production hinges upon construction of an Alaska gas pipeline to the Lower 48.

PTU development has been under discussion for years, and ExxonMobil has submitted numerous plans and modified plans. The latest plan ($1.3 billion) called for PTU production by yearend 2014.

Arctic seafloor exploration

The USGS announced a joint program with Canada to jointly map the unexplored Arctic seafloor where the two countries may have sovereign rights over oil, gas, and other resources.

One ship from each nation's icebreaker Coast Guard fleet was scheduled to collect data in fourth quarter 2008.

Both countries will share resulting data to establish the continental shelf's outer limits under criteria in the UN Law of the Sea Convention. The extended continental shelf, the seafloor, and subsoil beyond 200 nautical miles from shore that meet those criteria could be an area for economic development, said USGS.

Then-Pres. Bill Clinton signed the convention in 1994, but it has not been ratified by the Senate.

About 150 nations have already ratified the treaty, including every major industrialized nation other than the United States.

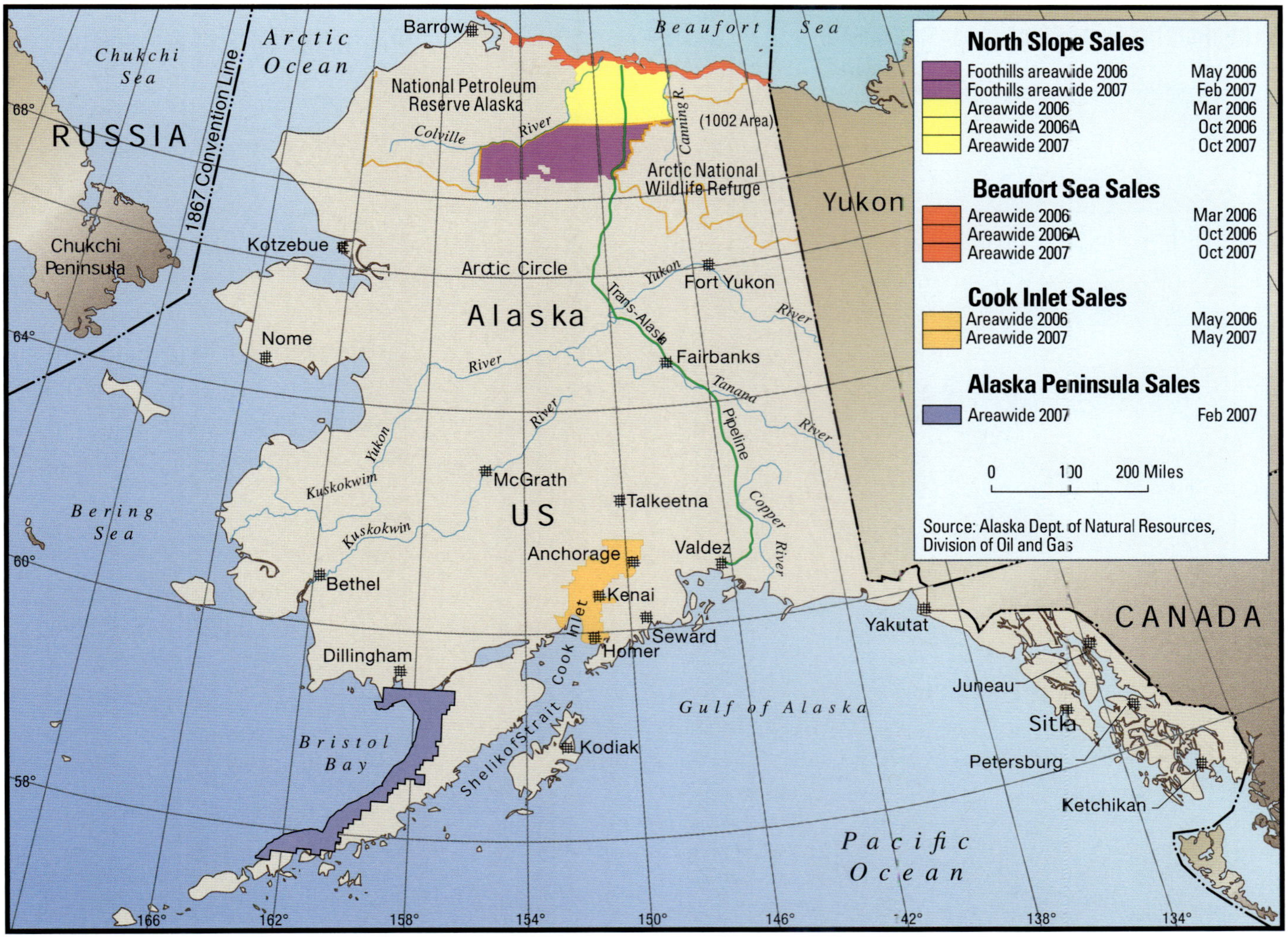

Proponents of ratification argue that the United States would have no legal claim to much of the oil and gas under the Arctic Ocean unless the nation becomes a party to the treaty.

Lease Sales

Lease Sale 193: Chukchi basin

The U.S. MMS was evaluating in April 2008 the $2.66 billion in high bids offered in February for leases in the Chukchi basin offshore northwestern Alaska.

Industry submitted 677 bids for 448 blocks in the area of more than 29 million acres (45,313 sq mi; 117,360 sq km) where only five exploration wells have been drilled previously. Previous leasing was in 1988 and 1991.

The 677 bids made it the most successful OCS lease sale in Alaska's history, said MMS.

The leases offered extend from 25 to 200 mi offshore (16 to 124 km).

Shell Gulf of Mexico Inc. (Houston) submitted the single largest bid—more than $105.3 million. Shell also led in total bids, with 302 totaling nearly $2.2 billion.

Other participants included ConocoPhillips Co.; Royal Dutch Shell PLC; Repsol E&P USA Inc.; Eni Petroleum US Inc.; StatoilHydro USA E&P Inc.; and Iona Energy Co. (US) Ltd.

The Chukchi Sea is home to 20% of the world's 25,000 remaining polar bears, and the sale took place despite protests that oil and gas activity could threaten them. The protection issue was resolved in December 2008 by the U.S. Department of the Interior.

NPR-A

The U.S. Bureau of Land Management (BLM) issued a record of decision that sets the stage for a major lease sale in the National Petroleum Reserve-Alaska's northeast portion (NPR-A).

Land in the NPR-A's northwest section also will be offered in the sale, planned for fourth quarter 2008.

Where industry bid for Chukchi basin acreage

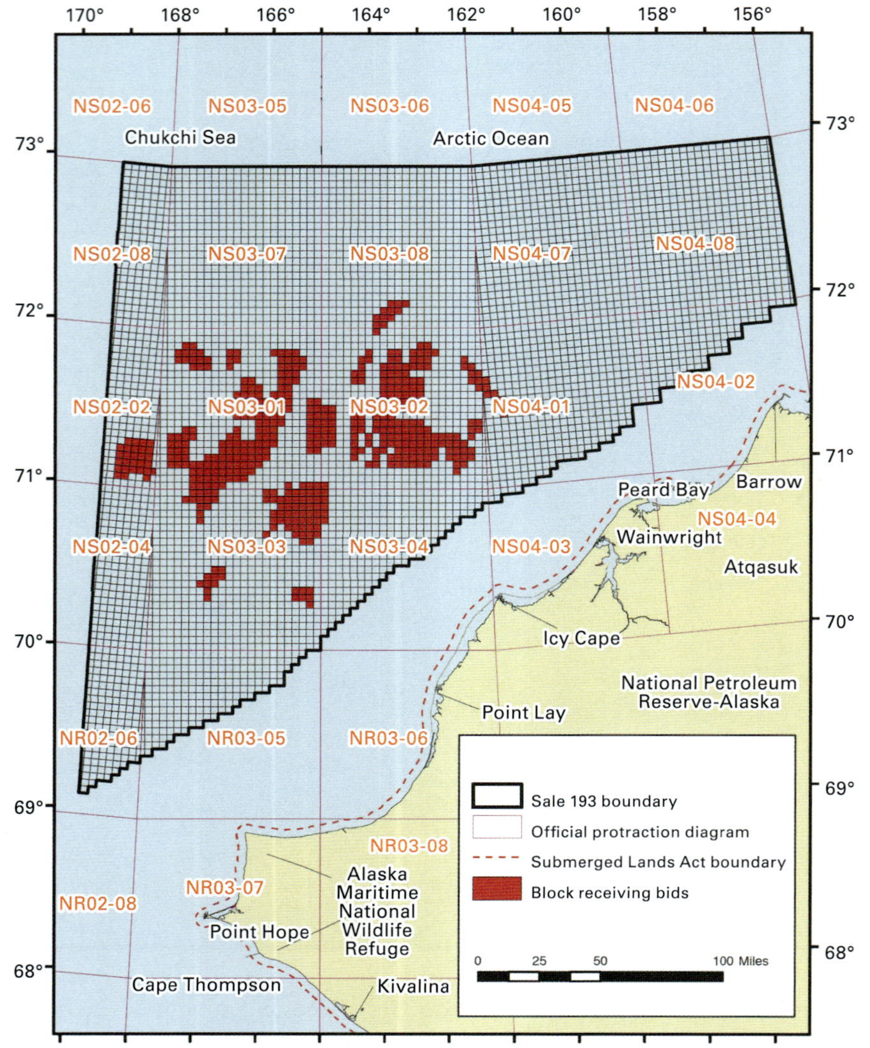

Source: US Minerals Management Service

BLM said that the acreage could result in development of as much as 8.4 billion bbl of oil and several trillion cubic feet of natural gas.

BLM noted that the decision provides for protection of polar bears, caribou and water fowl, and meets North Slope residents' subsistence needs while making land available for exploratory leasing.

Oil produced from this area is expected to move south through the Trans-Alaska Pipeline System. In mid-2008, TAPS throughput was around 700,000 b/d, down from its full 2.1 million b/d capacity.

BLM likely will not open 219,000 acres (342 sq mi; 886 sq km) of Teshepuk Lake and its island to oil and gas activity.

Protecting Chukchi polar bears

The U.S. MMS estimates that the Chukchi Sea might contain 15 billion bbl of recoverable oil and 2.2 trillion cu m of natural gas.

However, the agency also estimates that new offshore drilling would introduce a 33–51% chance of a major oil spill, which could have a severe impact on polar bears and other marine mammals.

In the run-up to the Chukchi Lease Sale, environmentalists and Alaska Native groups also linked the future of the polar bear to climate change. Noting reports of shrinking Arctic Ocean ice fields, they urged Congress to list the bears under the Endangered Species Act (ESA), calling the diminishing ice one of the impacts of climate change.

In December 2008, the U.S. Interior Department finalized a special rule for the polar bear. It allows oil and gas exploration in areas where the bears live, as long as the companies comply with the Marine Mammal Protection Act.

North Slope and Beaufort Sea

The Alaska Department of Natural Resources in October 2008 opened the bidding for competitive oil and gas leasing in North Slope Areawide (NS) 2008 and Beaufort Sea Areawide (BS) 2008, oil and gas lease sales.

NS has 1,225 tracts and lies between the Arctic National Wildlife Refuge and the National Petroleum Reserve Alaska. The southern boundary of the NS sale area is the Umiat baseline.

BS has 573 tracts consisting of state-owned tide and submerged lands from Point Barrow to the Canadian border.

Lease Sale 214: OCS

The Aleutians East Borough and the U.S. MMS signed a memorandum of agreement (MOA) in mid-2008 that outlines a cooperative arrangement to ensure local interests are considered during the evaluation of proposed North Aleutian basin oil and gas Lease Sale 214 in Alaska's Outer Continental Shelf (OCS).

The MOA provides a framework for coordination, communication, collaboration, and the exchange of ideas and information.

The area included in the proposal, believed to be natural gas-prone, is located in the southeastern Bering Sea offshore the Borough and the western end of the Aleutian Peninsula. The area under consideration in the North Aleutian area proposed for the sale covers about 5.6 million acres (8,750 sq mi; 22,663 sq km). The sale is proposed for 2011.

The Aleutians East Borough stretches from the Alaska Peninsula to the easternmost Aleutian Islands.

Drilling and Production

Oooguruk and Nikaitchuq fields

Eni SPA (Rome) said it expected production to begin by mid-2008 from Oooguruk field on Alaska's North Slope about 25 km (16 mi) west of Nikaitchuq field.

Looking to 2009, the company also said that it will spend $1.45 billion developing Nikaitchuq and that field production was scheduled to begin at yearend 2009.

Nikaitchuq, in about 10 ft (3 m) of water, will have 35 production wells and 35 injection wells. Production will go to a new 40,000-b/d processing facility onshore near the field and then transported some 14 mi (22 km) to the Kuparuk network, which is linked to the Trans-Alaska Pipeline System.

Nikaitchuq has reserves estimated at 180 million bbl of oil.

Advanced drill rig

BP Exploration Alaska Inc. (BPXA), a division of BP PLC (London) signed a contract in mid-2008 with Parker Drilling Co. (Houston) to build a new ultraextended-reach drilling rig for the Alaskan Beaufort Sea.

The rig will be able to drill wells nearly two miles deep (10,600 ft; 3,219 m) and with offsets as long as eight miles (42,240 ft; 12,875 m).

BP will own the land-based rig and intends to use it to drill as many as six extended-reach wells into offshore Liberty field beginning in 2010. BPXA expects oil production to begin in 2011, ramping up to 40,000 b/d, with an ultimate recovery of 100 million bbl oil.

BP drilled the Liberty discovery well about 20 mi (32 km) east of Prudhoe Bay. It will drill the development wells from an existing satellite pad at its Endicott oil field.

Cook Inlet basin

GeoPetro Resources Co. (San Francisco) said it plans to spud an exploratory well in Alaska's Cook Inlet basin in May 2009.

The planned 8,000-ft (2,438-m) Frontier Spirit-1 well is to evaluate a 100% owned, 11,500-acre (18 sq mi; 47 sq km) prospect that is six miles (4 km) north of Anchorage.

Refining

North Pole refinery

Alaska's state government and Flint Hill Resources said in December 2008 they will begin a joint effort to position the company's refinery at North Pole for long-term success. Options to be studied will be aimed at improving the plant's ability to respond to volatile energy costs, varying product demands and volatile refining margins as well as facilitating upgrades.

Flint Hill had announced in May that it was reviewing alternatives for the refinery due to financial challenges.

The plant is Alaska's largest refinery with a 240,000 b/d processing capacity. About 60% of its products are sold to the aviation market.

Environment

Coastal impact aid funds

Alaska officials learned in October 2008 that the state will receive as much as $80 million of coastal impact assistance authorized under the Energy Policy Act of 2005. The state and eight eligible boroughs will share the funds.

The law, including the Coastal Impact Assistance Program (CIAP), authorizes the U.S. Department of the Interior to distribute $250 million/year to six OCS oil and gas-producing states in fiscal 2007–10, said the U.S. MMS.

Alabama, Mississippi, Louisiana, Texas, and California are the other five states.

CIAP funding can be used for mitigation of the impact of OCS activities. MMS said Alaska will receive $2.5 million/year each year for FY 2007 and FY 2008. The annual payout will climb to $30–40 million in FY 2009 and 2010 as a result of the February 2008 federal OCS Lease Sale 193 in the Chukchi Sea.

Alaska has identified 51 projects to be supported by CIAP funds and may add more.

Polar bears: Lawsuit

Five U.S. industry groups filed a federal lawsuit in third quarter 2008 against the U.S. Department of Interior over a rule designed to protect the polar bear.

The groups say the rule discriminates against business activities in Alaska that might harm the bear, designated in May 2008 by the department as a threatened species.

Plaintiffs are the American Petroleum Institute, the U.S. Chamber of Commerce, the National Association of Manufacturers, the National Mining Association, and the American Iron & Steel Institute.

The groups want to block government plans to review projects in Alaska that might harm the polar bear by contributing to global warming.

Environmental groups and university researchers pointed to recent news reports about polar bears struggling to survive in the Arctic sea after the decay of sea ice from which the bears hunt seals. The loss of ice is said to be related to warming due to greenhouse gas emissions.

Transportation and Pipelines

Major pipeline projects

In August 2008, the FERC reported to Congress that of the two major projects competing to deliver natural gas from Alaska's North Slope (ANS), only one likely will be built, and it urged the sponsors work together to build that single line.

Denali

Alaska Gas Pipeline LLC (Denali) got approval in June from the state's Office of Energy Projects to begin the prefiling process. The company already had begun field work. The 4-bcfd system is sponsored by BP PLC and ConocoPhillips.

Project partners call it the "largest private-sector construction project ever built in North America."

Denali plans a 750-mi (1,207-km), 2,500-psi pipeline from ANS, generally following the Trans-Alaska oil pipeline to Fairbanks, where it would branch off, following the Alaska Highway to the Alaska-Yukon border.

Another 1,000 mi (1,609 km) of transmission line, to be built by Canadian affiliates, would extend from the Alaska border to Alberta.

Denali says if additional capacity is needed to transport these volumes to the Lower 48, the affiliates would build an extension through Alberta to the U.S. border, and Denali would construct a mainline across parts of North Dakota, Minnesota, Iowa, and Illinois to the Chicago area. These sections would add another 1,500 mi (2,414 km) to the system.

Denali also would build a gas treatment plant on the North Slope, several off-take points in Alaska for in-state distribution, and gathering lines from production areas to connections with other sections of its system.

Initial pipeline routing surveys and environmental resource studies began in mid-2008. The company plans to apply for a FERC certificate in fall 2011. Its construction schedule calls for bringing gas to market by mid-2018.

TC Alaska

Alaska Gov. Sarah Palin in August signed a bill designating TC Alaska the state's preferred applicant under its Alaska Gasline Inducement Act (AGIA) and granting it an exclusive AGIA license.

Project partners are TransCanada Corp. affiliates TransCanada Alaska Co. LLC and Foothills Pipe Lines Ltd.

The award authorizes the state to contribute as much as $500 million toward TC Alaska expenditures for planning and preparation leading to federal and other permit applications. It also represents an agreement to settle various project issues between the state and the applicant.

The route of the proposed 5-bcfd line is similar to Denali's, with 750 mi (1,207 km) of pipe extending from the existing gas treatment plant on the North Slope, following the same route in Alaska, and 1,000 mi (1,609 km) of pipeline from the Alaska-Yukon border to Alberta's existing gas distribution hub.

TC Alaska said it would build a gas treatment plant on the North Slope if necessary.

FERC said the company could request that many of its prefiling activities be combined with existing prefiling activities for Denali.

TC Alaska originally planned to conduct open season in 2009, to begin the FERC prefiling process in June 2010, and apply by December 2011 for a FERC certificate, but this schedule is likely to slip.

Alaska Gasline Port Authority

The AGPA project contemplates delivering Prudhoe Bay gas by pipeline 800 mi (1,288 km) to Valdez, where it could be liquefied and shipped on carriers to the Asian market, Hawaii, and the west coasts of the United States and Mexico.

AGPA has added new partners—Mitsubishi Corp. (Tokyo) and Sempra LNG (San Diego, California) —and Alaska is assisting AGPA with project development. FERC would also regulate this gas export project.

Intrastate gas line

In July 2008, Alaska formed a public-private partnership to build an intrastate gas pipeline to serve south-central and interior Alaska. Construction is slated to start in 2011.

The $16-billion system would begin service in 2013 with a capacity of 460 MMcfd of gas. The supply would mainly come from undiscovered supplies in Cook Inlet, interior basins along the pipeline, and the North Slope foothills, officials said.

Construction would start at Cook Inlet and progress north along the Richardson Highway for about 400 mi (644 km), reaching Fairbanks and interior Alaska.

If sufficient supplies fail to materialize from the Cook Inlet and Copper River basins and exploration along the pipeline, a second phase would involve a leg to bring gas from the North Slope foothills.

Project partners are the Alaska Natural Gas Development Authority, Enstar Natural Gas Co. (Anchorage), and the state.

The $16 billion system would be regulated by Canada's National Energy Board, not the FERC.

LNG

Kenai export license extension

Alaska officials said in January 2008 that they support an application by Marathon Oil Co. (Houston) and ConocoPhillips Alaska Natural Gas Corp. for a two-year federal extension of an export license for their Kenai LNG plant at Nikiski, on the Kenai Peninsula.

The application was filed with the U.S. Department of Energy in January 2007. If granted, the extension would run through March 2011. If not granted, the plant would close in March 2009.

Both Marathon and ConocoPhillips said they would drill more Cook Inlet gas wells if the extension is approved and also agreed to make data available to other potential oil and gas explorers in the Cook Inlet region to encourage exploration and development.

The plant has the ability to divert gas from export to Alaska markets for local heating and electricity generation during peak demand.

Exxon Valdez lawsuit

In June 2008, the U.S. Supreme Court overturned the $2.5-billion punitive damages judgment against ExxonMobil Corp. (Irving, Texas) for an oil spill from the tanker Exxon Valdez in Alaska's Prince William Sound in 1989.

The court said punitive damages should be limited to $507.5 million, not exceeding what the company had already paid to plaintiffs for economic losses.

ExxonMobil Chairman and CEO Rex W. Tillerson said the company had spent more than $3.4 billion, including compensatory payments, cleanup payments, settlements, and fines.

The punitive damages case arose from a 1994 lawsuit in which more than 32,000 plaintiffs sued the vessel's captain, Joseph J. Hazelwood, and ExxonMobil as the ship's owner and Hazelwood's employer.

The Supreme Court's ruling overturned punitive damages set by the U.S. Ninth Circuit Court of Appeals.

CANADA

Capital: Ottawa
Monetary unit: Canadian Dollar
Refining capacity: 2,029,450 b/cd
Oil production: 2.570 million b/d
Oil reserves: 178.092 billion bbl
Gas reserves: 57.906 tcf

Overview

Canada has considerable natural resources and is one of the world's largest producers and exporters of energy.

According to the Energy Information Administration (EIA) of the U.S. Department of Energy, almost all of Canada's energy exports go to the United States, making it the largest foreign source of U.S. energy imports.

Oil

According to *Oil & Gas Journal* (*OGJ*), Canada had 178.1 billion barrels of proven oil reserves as of January 2009, second only to Saudi Arabia. More than 95% of these reserves are oil sands deposits in Alberta, which are much more difficult to extract and process than conventional crude oil.

Canada is a net exporter of oil, with 2007 net exports of 1.0 million b/d.

Canada's oil production has steadily increased as new oil sands and offshore projects have come on-stream to replace aging fields in the western province.

Canadian oil production comes mainly from three sources:

- the Western Canada Sedimentary Basin (WCSB)
- the oil sands deposits of northern Alberta
- offshore fields in the Atlantic Ocean.

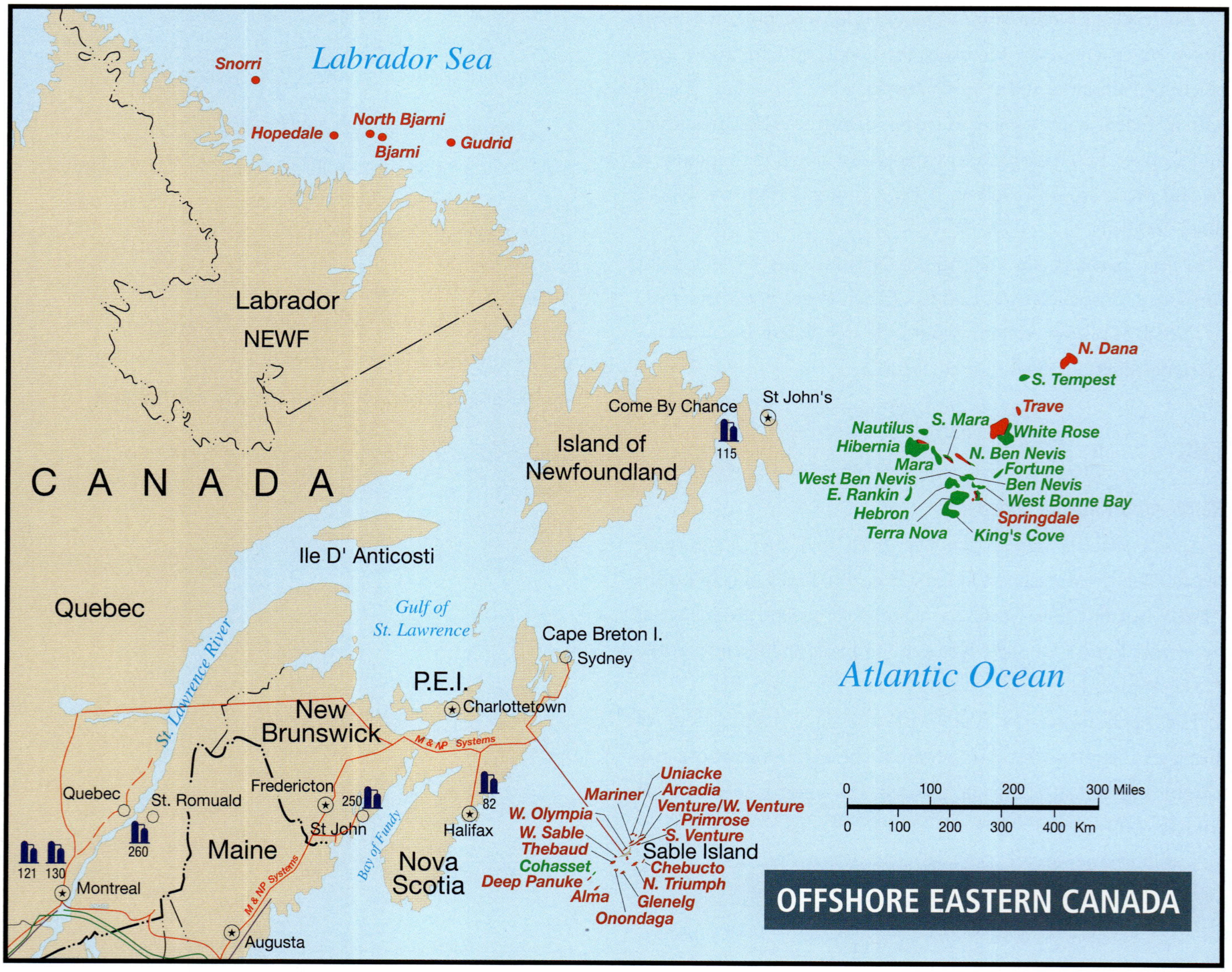

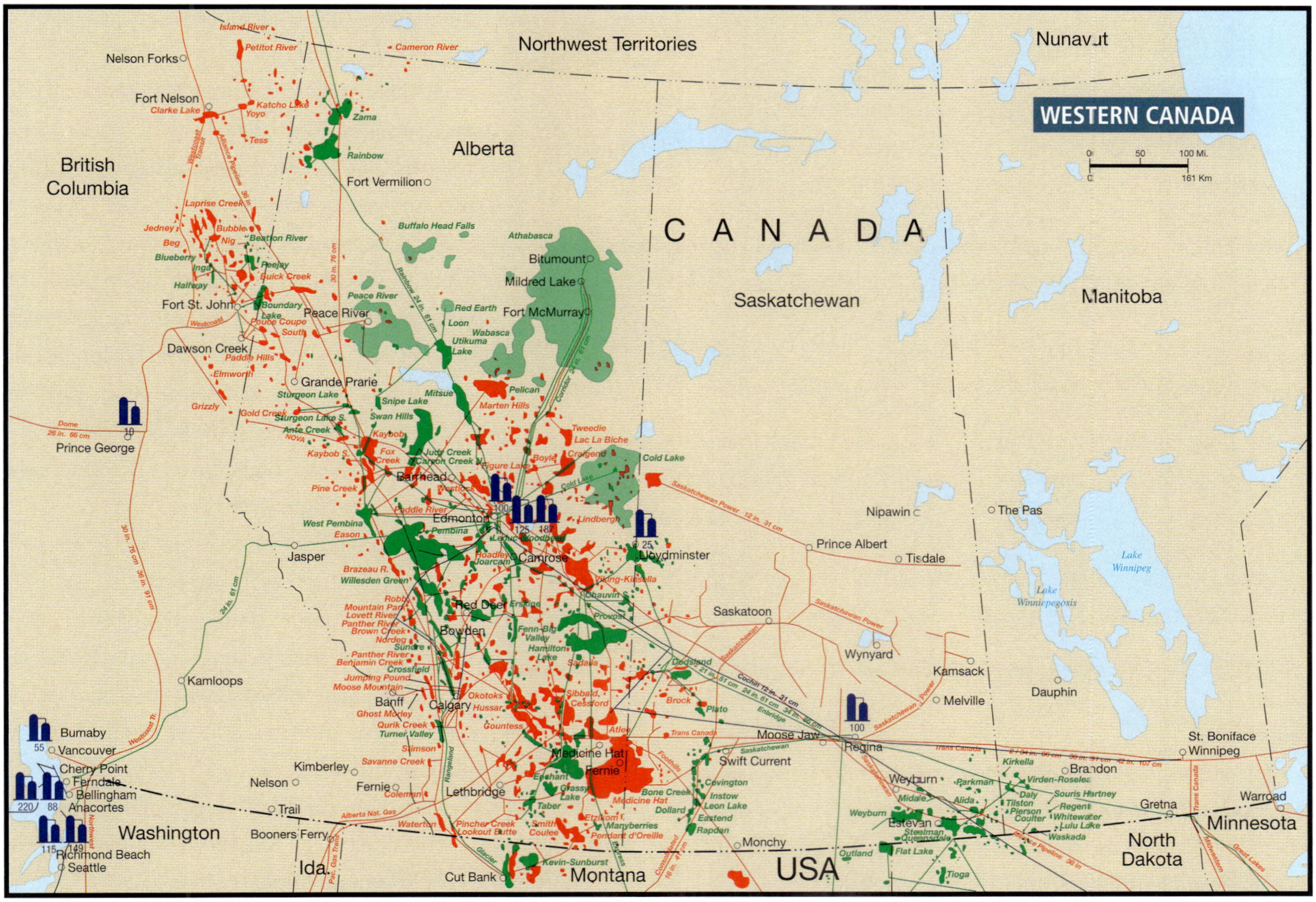

The Canadian Association of Petroleum Producers (CAPP; Calgary) in mid-2008 projected that total Canadian crude production—including conventional, oil sands, and Atlantic offshore—would increase to almost 4.5 million b/d in 2020 from 2.7 million b/d in 2007.

Alberta produces most of Canada's oil because it holds the majority of oil sands deposits and the bulk of the WCSB. According to Statistics Canada, Alberta accounted for 68% of Canada's national oil production in 2007.

In 2007, oil sands production represented approximately half of Canada's total crude production. The Athabasca oil sands deposit in northern Alberta is one of largest oil sands deposits in the world. There are also sizable oil sands deposits on Melville Island in the Canadian Arctic, and two smaller deposits in northern Alberta near Cold Lake and Peace River.

Phased development of the Alberta oil sands continues. A combination of projects is expected to add about 2 million b/d by 2016 to the current 1.2 million b/d being produced from the oil sands.

However, falling oil prices and the deepening global financial crisis in late 2008 likely will slow or halt progress on some oil sands projects.

Canada has three oil projects off its Atlantic coastline, all in the Jeanne d'Arc basin:

- Hibernia producing 135,000 b/d (Petro-Canada; Calgary)
- Terra Nova producing 116,000 b/d (Petro-Canada)
- White Rose producing 117,000 b/d (Husky Energy; Calgary).

Investment in the basin has increased in recent years, and both White Rose and Hibernia have announced plans to develop satellite fields.

Outside of the Jeanne d'Arc basin, StatoilHydro (Stavanger, Norway) announced in 2008 that it would start drilling the Mizzen field in the Flemish Pass basin. Chevron Corp. (San Ramon, CA) signed an agreement with the provincial government in 2007 to develop the Hebron-Ben Nevis field, which could hold recoverable reserves of 700 million bbl.

Parliament suspended

At the request of Prime Minister Stephen Harper, leader of the minority Conservative government, the Governor General Michaelle Jean on December 4, 2008, suspended Parliament until January 26, 2009.

The action avoided a confidence vote scheduled for mid-December that likely would have defeated the Conservatives.

Opposition parties had called for such a vote, following a provocative economic statement tabled by the Conservatives.

The New Democratic Party (NDP) and the Liberal Party of Canada had agreed to form a coalition to govern Canada—with support from the separatist Bloc Quebecois—as an alternative to the Conservatives.

Prime Minister Harper promised to present a revised budget in January to address some of the concerns that triggered the confrontation.

Natural Gas

Canada had 57.9 tcf of proven natural gas reserves in January 2009, according to OGJ. The country produced 6.5 tcf of natural gas in 2006 and is the second largest producer of natural gas in the Western Hemisphere, after the United States.

In 2006, Canada exported 3.6 tcf of natural gas to the United States, representing 86% of total U.S. natural gas imports that year.

Natural gas production is concentrated in the WCSB, particularly in Alberta. Even though there have been some new conventional natural gas finds in the WCSB, many analysts predict that conventional natural gas production in the WCSB has reached its zenith.

Future natural gas production could center on coal bed methane (CBM) deposits in the WCSB, Arctic frontier natural gas deposits, the Deep Basin area, and offshore natural gas fields.

Offshore Nova Scotia is the center of Canadian gas production on the Atlantic coast. The Sable Offshore Energy Project (SOEP), led by ExxonMobil Corp. (Irving, Texas) and Shell Canada Ltd. (Calgary), began production in 1999. The Alma and South Venture fields are the latest brought on stream by the SOEP.

Offshore Newfoundland, the Hibernia and White Rose fields contain a combined 4 tcf in recoverable natural gas reserves. There is no current natural gas production at either site, but both ExxonMobil (Hibernia) and Husky Energy (White Rose) have stated that they could commence natural gas production in the future.

Offshore basins in British Colombia contain an estimated 43.4 tcf of total natural gas reserves, but a federal moratorium on drilling prevents production activity.

The Mackenzie Delta in the Northwest Territories holds an estimated 5–6 tcf of recoverable natural gas reserves. Moving gas from the region to southern markets hinges on completion of the Mackenzie Gas Pipeline. There are three large, proven gas fields in the Mackenzie Delta: Taglu (3 tcf); Parsons Lake (1.8 tcf); and Niglintgak (1 tcf).

CBM production is still in its infancy in Canada; the first wells were drilled in 1997. But CBM output could eventually replace some of the decline in conventional natural gas production. The Alberta Geological Service has estimated as much as 500 tcf of CBM gas in place in Alberta alone.

Shale gas also is still in the early stages of development. According to industry sources, the Montney shale formation in British Columbia could contain 50 tcf of shale gas.

Exploration and Development

NL: West

Vulcan Minerals Inc. (St. John's, NL) said it expected to pursue several exploration plays in Newfoundland and Labrador in 2009.

- A planned 3,600-m (11,810-ft) exploration well, Robinson-1, onshore in the Bay St. George basin would be the basin's deepest well ever.
- Two other more shallow planned wells 1,500–2,500 m (4,921–8,202 ft) deep onshore in the basin.
- Work with operator Leprechaun Resources Ltd. (Calgary) on three onshore permits at Parsons Pond, in which Vulcan holds partial interests.
- Possible work with NWest Energy Inc. (St. John's, NL) in blocks that cover a Cambrian-Ordovician-Silurian platform and foreland basin rocks with petroleum potential.
- Exploration plans with Investcan Energy Corp. (St. John's, NL) to explore License 1107 off Labrador, which offsets two undeveloped gas discoveries and has several seismic leads.

NB: McCully field

Corridor Resources Inc. (Halifax, NS) was preparing in late 2008 to appraise an oil discovery 3 km (2 mi) southeast of its McCully gas field in southern New Brunswick.

The South Branch G-36 well flowed 59 b/d without water after hydraulic fracturing. The producing interval was the Hiram Brook member of the Mississippian Albert formation at 1,574–1,612 m (5,164–5,289 ft).

Corridor also ran a frac in Hiram Brook at 1,757–1,840 m (5,764–6,037 ft), resulting in a show of oil and no measurable gas.

Corridor said it expected 2008 yearend production of 35–40 MMscfd of gas from McCully field.

NT: Ellice Island, Langley Island

MGM Energy Corp. (Calgary) reduced its 2008–09 winter drilling program to three wells from four in November, citing lack of regulatory progress on the Mackenzie Valley gas pipeline project.

The three wells to be drilled are Ellice J-27 on Ellice Island to test the Taglu formation; Ellice J-17 to test a stratigraphic play on the west side of Langley Island; and Ellice A-25 to test a footwall of the Ellice Island anticline.

The first well was to spud by yearend 2008.

Maritimes: Flemish Pass, Jeanne d'Arc basins

Canada granted StatoilHydro (Stavanger, Norway) operatorship of one license and a 50% interest in another, both off Newfoundland, under land sale awards in November 2008.

One license is in the Flemish Pass basin, near StatoilHydro's Mizzen well, which was to be drilled by early 2009. StatoilHydro's partner will be Husky Energy Inc. (Calgary).

The second license is in the Jeanne d'Arc basin, where StatoilHydro will partner with operator Petro-Canada Ltd. (Calgary).

StatoilHydro also is a partner in development of the Hibernia, Terra Nova, and Hebron fields offshore eastern Canada.

QC: Gaspe Peninsula

Quebec independents Petrolia, Junex, and Gastem completed 3D seismic and soil geochemistry surveys in late 2008 around the Petrolia Haldimand-1 oil and gas discovery well.

The well is near the tip of the Gaspe Peninsula.

The surveys are part of a $5 million exploration effort to better understand the area's geology.

Petrolia (Rimouski, QC) also completed surface geochemical and geo-microbial surveys on its holdings on Anticosti Island in the Gulf of St. Lawrence off Quebec, to prioritize a number of structures already identified on seismic surveys.

NL: North Amethyst oil field

Husky Energy Inc. (Calgary) received approval in early 2008 for subsea development of North Amethyst oil field, the first of three White Rose field satellites planned for development off Newfoundland.

Husky said it expects to drill 11 North Amethyst wells for tie-back to the SeaRose FPSO vessel moored in 120 m (394 ft) of water in White Rose field. The production estimate for the satellite is 62,900–75,500 b/d.

The company said White Rose, in the Jeanne d'Arc basin 350 km (217 mi) east of St. John's, NL, was producing an average of 90,000–95,000 b/d in January 2008.

AB: Peace River arch

Redcliffe Exploration Inc. (Calgary) tested a Triassic Halfway sour oil and gas discovery in early 2008 at Wapiti in the Peace River arch area of west-central Alberta.

Operator Redcliffe expected that the well would go on production in the second quarter.

On initial production test, the well stabilized at 750 b/d of oil and 2.6 MMcfd of gas.

The firm was planning to drill at least one more well into the pool by yearend 2008.

AB: Marten Mountain

Exall Energy Corp. (Calgary) said its Marten Mountain horizontal oil discovery, 224 km (140 mi) north-northwest of Edmonton, flowed 1,585 b/d of oil and 575 Mcfd of gas.

The well penetrated 420 m (1,378 ft) of Devonian Gilwood sand.

Operator Exall planned to drill a second horizontal well from the same pad and a vertical well nearby by yearend 2008.

Mapping the unexplored Arctic seafloor

The United States and Canada announced a joint program to map the unexplored Arctic seafloor in a program to begin in late 2008.

The study area will encompass a region where the two countries may have sovereign rights over oil, gas, and other resources, said the USGS.

Further details are presented in the "Alaska" section of this chapter.

Energy research funds

The Nova Scotia government said in mid-2008 it would use about $50 million of an estimated $870 million in Canadian government settlement funds to support offshore E&P research.

The settlement concerns the "Crown share dispute" from 1986, when Nova Scotia was promised compensation for giving up its ownership interest in offshore oil and gas resources, but never received the funds.

More than 90% of the settlement funds are slated to go against the province's $12.4 billion debt.

About $50 million will be set aside for offshore energy research and university infrastructure, and $25 million to protect public land.

Policy

AB: New royalty framework

A new royalty framework was set to take effect in Alberta on January 1, 2009, for production from new natural gas wells and conventional oil wells.

Spurred by then-high commodity prices, Alberta officials announced the "New Royalty Framework" in October 2007, saying it would bring in an additional $1.4 billion (Can.) in 2010, 20% more than the prior royalty regime.

The new royalty structure did not include oil sands, and the Alberta government spent 2008 negotiating with Syncrude and Suncor, large joint-venture projects that had royalty agreements in place through 2016. Syncrude partners, the last to come to terms, announced their acceptance in November 2008.

In response to the unforeseen drop in natural gas and oil prices and credit restrictions from the global economic crisis in the last half of 2008, however, the government introduced optional, five-year transitional royalty rates for new wells begun from mid-November to December 31, 2008.

Operators were given the option to pay royalties under the transitional scheme or the new royalty framework scheme.

Other adjustments had been made first quarter 2008 easing terms for production from deep wells.

AB: Royalty hike gas impact

In early 2008, Ziff Energy Group Ltd. (Calgary) said the planned Alberta royalty increase was aggravating a "recession" in the province's conventional oil and gas industry, particularly in gas drilling.

For natural gas, the new framework raises the maximum royalty to 50% from 35% and eliminates tiers, starting in 2009. It retains and will revamp programs for special production categories such as deep formations and marginal wells.

Ziff Energy said that the new framework—coupled with taxes, producer return requirements, and rising costs to find new gas reserves—was hampering new gas exploration efforts.

The firm estimated that at least 80% of companies active in Alberta reduced their planned conventional spending since the new royalty announcement.

Most affected by the spending cuts are deep gas projects in western and northwestern Alberta and shallow gas and coalbed methane in the central and southeastern areas, said Ziff Energy.

Drilling and Production

NL: Hebron heavy oil fields

Chevron Canada Ltd. (Vancouver, BC) and its co-venturers finalized in mid-2008 legal agreements with Newfoundland and Labrador to develop the Hebron heavy oil fields off northeastern Canada.

The complex consists of Hebron, Ben Nevis, and West Ben Nevis fields, all in the Jeanne d'Arc basin between the Hibernia and Terra Nova production areas on the Grand Banks.

The complex is 338 km (210 mi) off Newfoundland and Labrador in 91 m (300 ft) of water. Hebron was discovered in 1981, but the project was put on hold in 2006 when the operator could not reach an agreement with Newfoundland and Labrador.

The Canada-Newfoundland and Labrador Offshore Petroleum Board (C-NLOPB) in 2006 estimated that the Hebron complex contains 731 million bbl of proved and probable oil resources, about 30 million bbl of natural gas liquids, and 12 bcm (429 bcf) of natural gas.

Chevron estimates recoverable oil to be in the 400–700 million bbl range.

Partners with Chevron Canada are ExxonMobil Canada Properties, Petro-Canada, StatoilHydro Canada Oil & Gas Inc., and Oil & Gas Corp. of Newfoundland and Labrador.

Jeanne d'Arc basin discoveries

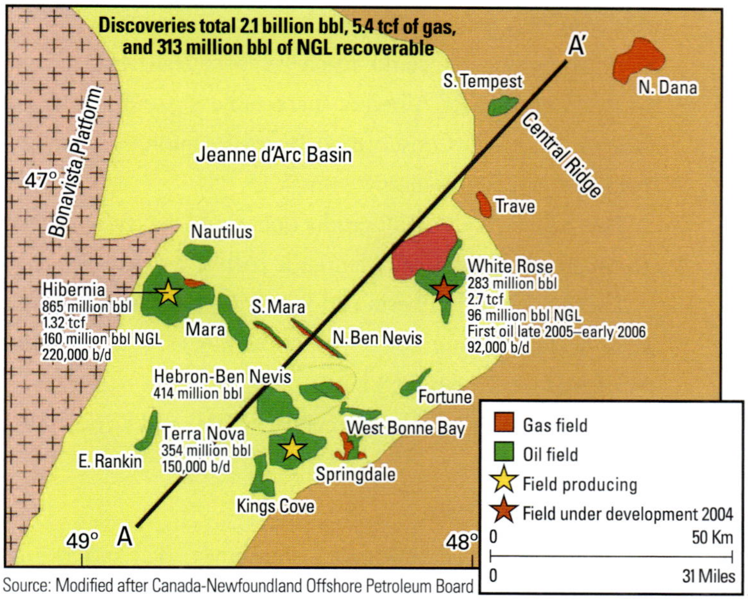

Source: Modified after Canada-Newfoundland Offshore Petroleum Board

Provincial officials estimate that the 20- to 25-year project could generate $20 billion for the province and that the federal Government and other provinces could receive more than $8 billion in revenues.

First oil is expected no earlier than 2016, with peak production of 150,000 b/d two years after startup.

White Rose field: Production milestone

Husky Energy Ltd. (Calgary) reported that on August 14, 2008, the one hundred millionth barrel of crude oil was produced from the White Rose oil field.

Reservoir production capacity at the White Rose field is 120,000 b/d. An infill production well was expected to come onstream in third quarter 2008.

NL: North Amethyst, White Rose satellites

Husky Energy Inc. (Calgary) said that North Amethyst will be the first of several White Rose field satellite discoveries that it expects to develop off Newfoundland.

Husky's August 2007 development plan was under review in early 2008 by the Canada-Newfoundland and Labrador Offshore Petroleum Board (CNLOPB). The start of development drilling was projected for mid- to late 2008.

Oil production from White Rose began in late 2005. It lies on the eastern margin of the Jeanne d'Arc basin, 350 km (217 mi) east of St. John's, NL. Both Husky and Petro-Canada Ltd. (Calgary) have interests in White Rose.

White Rose field

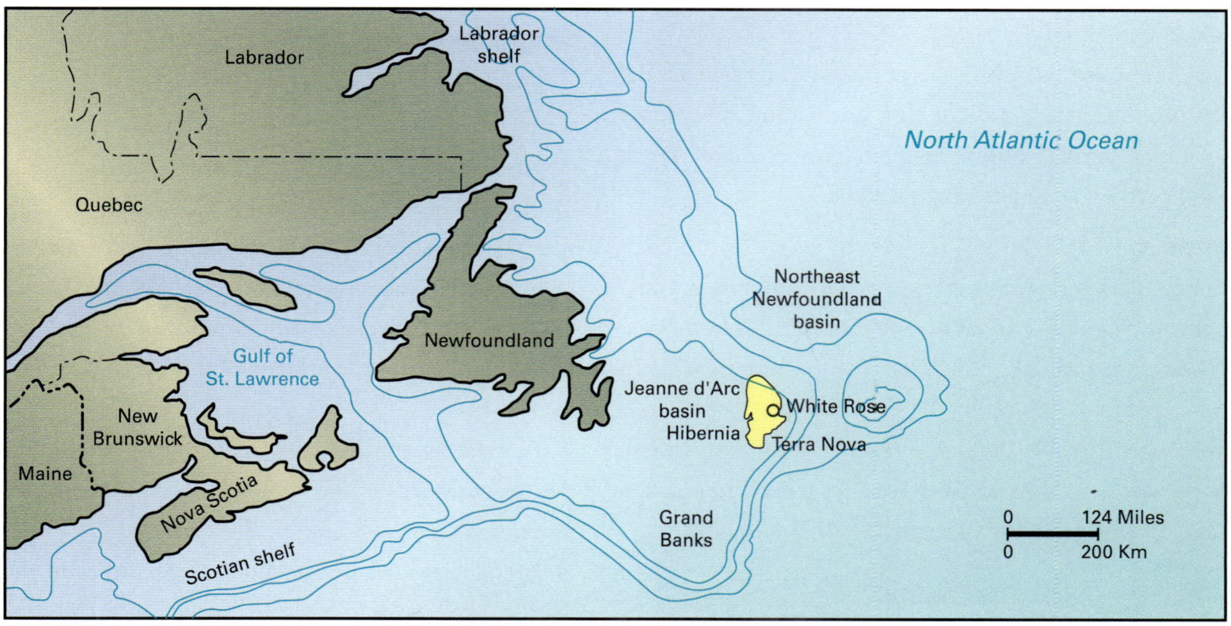

Source: Husky Report SR-SRT-RP-0002, North Amethyst Satellite Tie-Back to SeaRose FPSO Development Plan, August 2007

Husky was partnering with StatoilHydro (Stavanger, Norway) and Petro-Canada Ltd. (Calgary) on a drilling program to drill infill producers and injectors to enhance production at the existing White Rose oil pool and to advance the North Amethyst and West White Rose development programs.

Husky estimated P50 recoverable oil from North Amethyst at 70 million bbl out of the 256 million bbl of oil in place and that the field's wells will have a maximum oil production of 62,900–75,500 b/d.

The company said its White Rose satellite fields could come onstream by early 2010.

MT: Sinclair field waterflood

Tundra Oil & Gas Ltd. (Winnipeg) said it planned to begin a waterflood program in mid-2008 at the Sinclair oil field in the Williston basin in southwestern Manitoba and might inject carbon dioxide later.

The waterflood program would inject water at 192 wells over a 31-sq km (12-sq mi) area.

Recovery is expected to grow from 10% of OOIP by primary means to as much as 20% with waterflooding and possibly to 30% or more with CO2 and waterflooding.

AB: Crossfire area

Great Plains Exploration Inc. (Calgary) completed and tested its first exploration well in early 2008 in the Crossfire area of Pembina, AB, as a Devonian Nisku zone discovery.

The 11-12-50-05w5 well encountered a 10-m (34-ft) hydrocarbon column and flowed at rates up to 1,390 b/d of oil and 470 Mcfd of gas.

The well suggests a significant northeast extension of the Pembina Nisku trend. It was the first of as many as six more wells targeting similar Nisku light oil prospects that Great Plains and partners were planning for 2008.

BC: Horn River

EOG Resources Inc. (Houston) expected to begin natural gas production in mid-2008 from British Columbia's Horn River basin. It estimated initial production potential from the basin to be 3.5–5 MMcfd.

With leaseholds on 140,000 net acres (219 sq mi) near British Columbia's northern boundary, EOG said it expects production to be ramped up significantly starting in 2010.

Coalbed Methane

AB: CBM view from ERCB

CBM production could reach 18 bcm (640 bcf) per year in 2017, or 16% of Alberta's marketable gas output, said the Energy Resources Conservation Board in mid-2008. The comparable CBM figure in 2007 was 5%.

The ERCB said CBM production could be even higher if commercial output from Mannville coals is accelerated.

The agency said Alberta had 9,339 wells producing 2.2 bcm (77.7 bcf) of CBM in 2007, with remaining established reserves estimated at 24 bcm (858 bcf).

The province's first commercial CBM production began in 2002.

BC: Coalbed methane

Canada Energy Partners Inc. (Vancouver, BC) and an undisclosed partner hoped to start production in late 2008 from a coalbed methane project in Northeast British Columbia.

They drilled and ran fracs on five wells and connected them and three previously shut-in wells to the Spectra Energy pipeline that crosses the lease northwest of Dawson Creek, BC.

The Peace River project property has room for 315 wells on 160-acre spacing.

Consulting engineers estimate a discovered resource of more than 2.3 tcf of gas.

Shale Gas

NS and NB: Shale gas development

Several firms were working to develop shale gas in 2008 in eastern Canada.

The Canadian unit of Triangle Petroleum Corp. (Calgary) was drilling in Nova Scotia's Windsor Block, 40 km (25 mi) from the Halifax lateral of the Maritimes & Northeast pipeline.

In 2007, Triangle acquired new seismic over the block, and drilled, cored, and completed two vertical wells in the Kennetcook area of Hants County, about 70 km (43 mi) north of Halifax.

Triangle also drilled two more vertical wells in mid-2008, on the north side of the Windsor Block, plus another in fourth quarter 2008 to test Horton Bluff shales.

Triangle also has been evaluating gas shales in the Moncton-Sackville subbasin in New Brunswick since 2007, in the Beech Hill Block.

Triangle Petroleum's Shale Gas Blocks

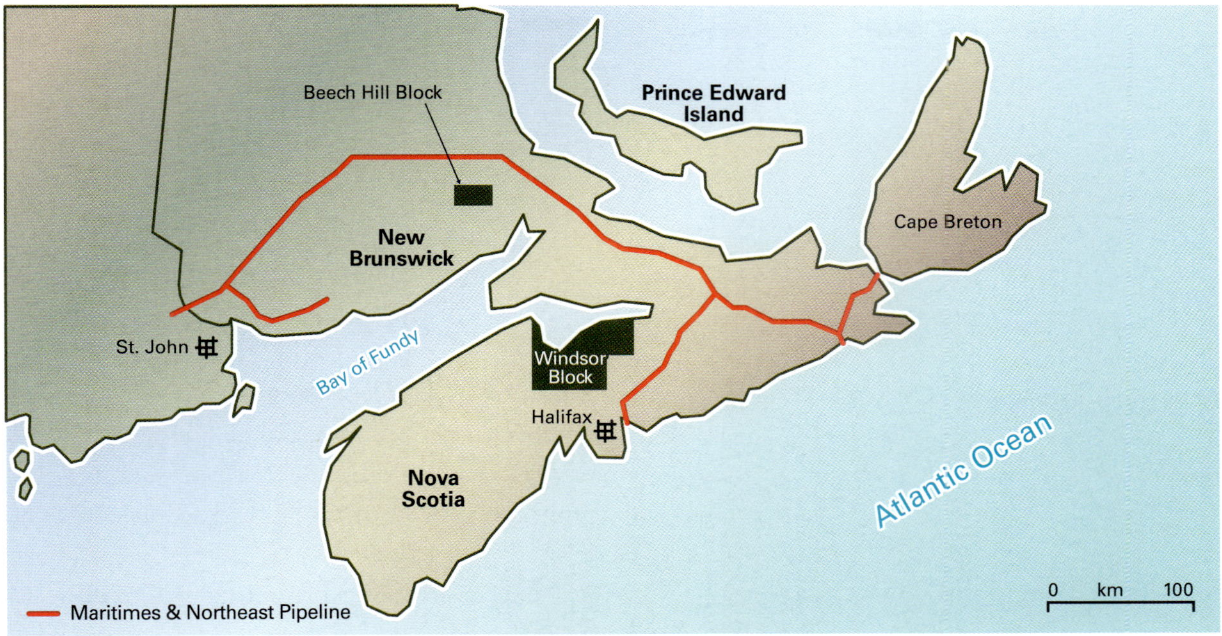

Corridor Resources Inc. (Halifax) said it would spend $32 million in 2008–09 to assess the Mississippian Frederick Brook shale in an area near Elgin, NB.

Corridor will drill vertical wells to obtain log and core data, followed by a horizontal well with a lateral as long as 1,000 m (3,281ft) in the most prospective part of the shale. It plans a 65-sq km (25-sq mi) 3D seismic program over the most promising area.

Corridor's Elgin licenses, excluding McCully field area holdings, cover 118,000 acres (185 sq mi).

QC: Utica, Lorraine shale gas

Talisman Energy Inc. (Calgary) announced results in September 2008 from its first well in Ordovician Utica shale in Quebec's St. Lawrence Lowlands.

The firm says the overlying Lorraine shale—not yet tested by Talisman—has the potential to be just as prospective as the Utica but is "unproven."

Talisman was planning to drill three more wells by the end of 2008.

Gas Shales in Quebec's St. Lawrence Lowlands

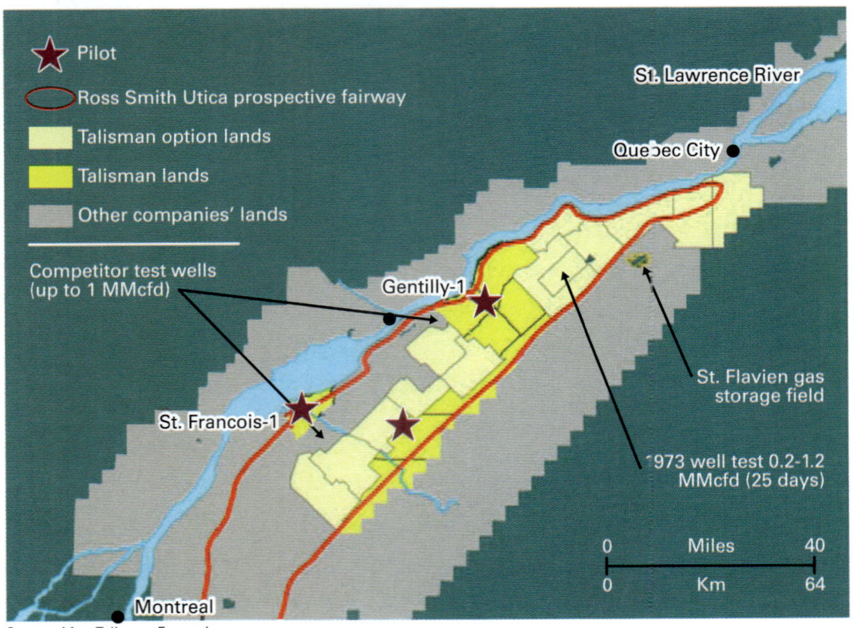

Source: After Talisman Energy Inc.

Geology of unconventional plays

	Outer Foothills	Montney	Bakken	Marcellus	Quebec
Reservoir lithology	Sandstone	Sandstone/siltstone/shale	Sandstone/siltstone	Shale	Siltstone/shale
Drill depth, ft	11,500-13,000	3,000-13,000	5,000-7,000	2,500-6,000	4,900-8,800
Thickness, ft	100-2,000	160-1,000	32-40	50-250	300-6,500
TOC weight, %	n/a	1.5-6.0	7.0-12.0	1.0-12.0	0.1-2.5
Ro, %	n/a	0.8-2.5	Immature	1.1-3.0	1.1-4.0
Silica weight, %	n/a	20-60	90-95	20-50	*30-35
Clay weight, %	n/a	<30	<5	20-45	*30-38
Gas filled porosity, %	2.0-12.0	1.0-6.0	10.0-14.0 (oil)	1.6-7.0	1.2-3.5
Pressure gradient, psi/ft	0.42-0.78	0.44-0.70	0.43	0.45-0.60	0.60-0.76
OGIP, bcf/section	55-180	70-300	3.4-4.5 MMboe	20-100	75-350
Relative to NYMEX $	–1.10	–1.30	+5.00/bbl	+0.40	+0.60

*Refers to Lorraine only.

Source: Talisman Energy Inc.

Utica, Lorraine shale characteristics

Parameter	Lorraine	Utica
Prospective depth, ft	1,500-10,000	1,500-11,000
Thickness, ft	1,500-6,500	300-1,000
TOC weight, %	01.-1.5	0.3-2.5
Ro, %	1.1-4.0	1.1-4.0
Silica weight, %	30-35	12-51
Clay weight, %	30-38	8-66
Gas-filled porosity, %	1.2-3.2	2.2-3.5
Pressure gradient, psi/ft	0.6	0.6
OGIP, bcf/section	50-190	25-160

Source: Talisman Energy Inc.

The company reported a sustained, 18-day test rate of 800 Mcfd from a single completed Utica interval at Gentilly-1, about 100 km (62 mi) south of Quebec City.

Plans also were being made to test the shallower Basal Lorraine and Lorraine shale intervals in the same well.

In mid-2008, Talisman refocused its efforts on several North American onshore gas and oil shale plays.

The firm said in May that it would spend $1.1–1.3 billion through the end of 2009 evaluating a material part of the gas potential in its 2.5 million net acres (3,906 sq mi) of unconventional land in North America.

Of that total, it will budget $900 million for development in areas where the company experienced early success, including 200 wells in the Western Canada Outer Foothills and Devonian Montney shale gas plays, and the U.S. Williston basin Bakken shale oil play.

AB: Gas shale at Wildmere

Stealth Ventures Ltd. (Calgary) was pursuing in early 2008 what it called Alberta's first commercial shale gas development, at Wildmere, 193 km (120 mi) east-southeast of Edmonton. The company had completed 10 wells of a planned 80-well program for biogenic gas in Upper Cretaceous Colorado Group shales.

Stealth said it booked 350 MMcm (12.4 bcf) of proved and probable reserves in 2007 from 39 gross (29 net) wells across 15 townships.

BC: Muskwa-Evie, Montney shale gas

Petrobank Energy & Resources Ltd. (Calgary) in mid-2008 acquired 25 sections of land in Northeast British Columbia to pursue gas in the Devonian Muskwa and Evie shales of the Horn River basin.

The firm plans to drill its first vertical well in the region in early 2009.

Petrobank also was also moving to acquire 15 sections with strong horizontal development potential in the Triassic Montney formation in the Monias area, northwest of Dawson Creek. Consulting engineers assessed the recoverable potential at 4 bcm (148 bcf).

Petrobank was planning to drill two horizontal Montney wells on the lands in by yearend 2008.

BC: Beaver River shale gas

The life of northeast British Columbia's Beaver River field may be extended if initial flows from shale and siltstone formations prove out in the Liard basin field.

Questerre Energy Corp. (Calgary) and Transeuro Energy Corp. (Vancouver, BC) reported an initial flow of 10 MMcfd of gas on a three-day test at the field's A-5 well. The well was to go on production by fourth quarter 2008.

Beaver River field is 150 km (93 mi) northwest of Fort Nelson, BC, near the provincial border with Yukon and 100 km (62 mi) east of the Horn River basin, which has an emerging gas play in the Devonian Muskwa and Evie shale.

A-5 is the third well to be put on production from the shale-siltstone intervals at Beaver River field, said Transeuro. The intervals are collectively more than 2,000 m (6,562 ft) thick with three major intervals:

- Mississippian Mattson shale at 1,200–2,100 m (3,937–9,842 ft), producing in the A-7 well
- Devonian Besa River shale at 2,000–3,000 m (6,562–9,842 ft), producing in the A-2 well

Muskwa shale gas play

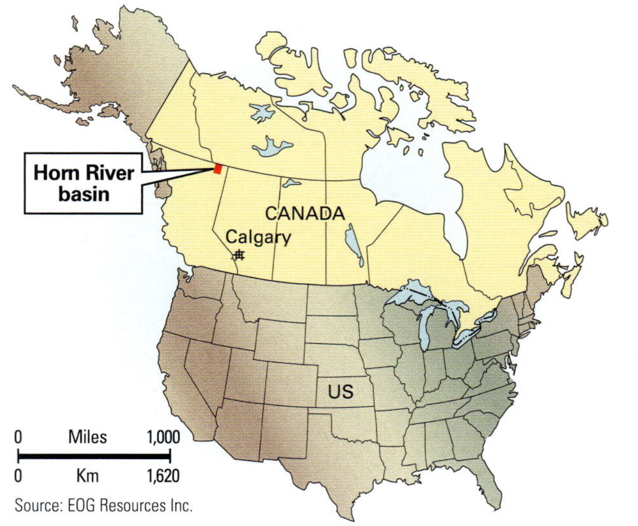

Source: EOG Resources Inc.

- Devonian Golata and Muskwa shales at 3,000–4,000 m (9,842–13,124 ft).

BC: Muskwa shale gas

O&GJ reported in early 2008 that an Upper Devonian shale gas play, emerging over the past 2–3 years in remote northern Northeast British Columbia, could become one of North America's larger gas fields.

The play is for gas in the Muskwa shale in the Horn River basin, 1,247 km (775 mi) northwest of Calgary.

EOG Resources Inc. (Houston), with 140,000 acres (219 sq mi) in the play, estimates 70 tcf of gas in place in the shale, of which a net 6 tcf seems recoverable.

Gas in place in the Muskwa on EOG lands averages 2.5 times that of the Barnett shale in Johnson County, Texas (in the Fort Worth basin south).

Others with acreage in the play are EnCana Corp. (Calgary) and Apache Corp. (Houston) and several other independents. The total area under lease is estimated at 600,000–700,000 acres (938–1,094 sq mi).

EOG estimated that industry would drill a total of 17–21 horizontal wells in the play by the end of 2008. EnCana-Apache will drill nine horizontal wells and EOG two vertical and four to eight horizontal wells.

The Muskwa shale, which lies at about the same depth as the Barnett, is thicker and has higher pressure than the Barnett in the Northeast BC area south of Fort Liard, NWT.

Shale Rock Properties

	Muskwa, Horn River basin, BC	Barnett, Johnson County, Tex.
Thickness, ft	530	355
Permeability, nd	230	250
Gas-filled porosity, %	4	4.5
Maturity, Ro	2.8	2.2
Silica content, %	65	55
Gas in place, bcf/sq mile	265	193

Source: EOG Resources Inc.

EOG was making plans in mid-2008 to begin production from its first two wells, but said it did not expect major contributions of production until 2010–11.

Oil Sands

Financial crisis impact

In the last third of 2008, deteriorating financial conditions, falling oil prices, and high construction and operating costs led to spending cutbacks and delays on several oil sands projects in Canada. A summary by Thomson-Reuters business information service in mid-November 2008 included the following:

November

- Petro-Canada (Calgary) deferred a final investment decision on construction of an upgrader for its C$21 billion ($17 billion) Fort Hills oil sands project. It will not make decision on the mine until 2009 as it expects costs to fall as the pace of oil sands development falls. It had previously planned a go-ahead decision in December 2008.

- Canadian Natural Resources Ltd. (CNRL; Calgary) slowed spending on the second phase of the Horizon oil sands project for 2009 after first phase costs rose to C$9.7 billion. It also scrapped timelines for the second phase, which would lift output to 250,000 bpd from 110,000. The company said it was still targeting first synthetic crude oil production from Horizon by the end of 2008.

October

- Royal Dutch Shell PLC (The Hague, Netherlands) said it will delay its investment decision on second expansion of Athabasca oil sands project. It also postponed the 100,000 b/d Carmon Creek oil sands project, near Peace River.
- Suncor Energy Inc. (Calgary) delayed an oil sands upgrader project for the Voyageur expansion by one year to 2013. The expansion is to boost production from Suncor's oil sands operations near Fort McMurray, Alberta, to 550,000 bpd from 350,000.
- Nexen Inc. and Opti Canada Inc. (both in Calgary) delayed the second phase of the Long Lake project to 2009. Expansion would double production of synthetic crude to 120,000 bpd. The first phase—cost: C$6.1 billion (Can.)—was just starting up in late 2008.
- Value Creation Group (Dallas) reportedly halted its Heartland upgrader near Edmonton. The first phase would have processed 77,500 bpd of bitumen into synthetic crude. The private firm has regulatory approval for a plant with a 260,000-bpd capacity.

Other sources reported that in December, StatoilHydro ASA (Stavanger, Norway) dropped plans to develop a $16 billion (Can.) bitumen upgrader in Alberta. In May, the project was initially delayed for two years to 2016.

Leismer project

StatoilHydro Canada Ltd. let a $41-million construction contract in late 2008 to Flint Energy Services Ltd. (Calgary) to develop its Leismer Oilsands steam-assisted gravity drainage (SAGD) demonstration project, 160 km (99 mi) south of Fort McMurray.

The facility will have the capacity to produce from oil sands 10,000 b/d of bitumen from 22 horizontal well pairs–one for injection, the other for extraction–linked to four well pads.

Production is slated to start by early 2010 and is expected to increase to more than 200,000 b/d by 2020 from an initial 20,000 b/d in 2012 from all of StatoilHydro's oil sands leases.

CAPP: Oil sands outlook

According to a November 2008 assessment by the Canadian Association of Petroleum Producers (CAPP; Calgary), current oil sands production is 1 million b/d and is expected to increase to 4 million b/d by 2020.

The largest of three oil sands deposits in Alberta is at Fort McMurray; the other two are at Peace River and Cold Lake. There are more than 20 active mining and in-situ oil sands projects in those areas.

Some analysts anticipate a 10–15% drop in capital spending in western Canada in 2009 as producers try to manage cash flow.

Many observers see this as a sign that low oil prices are starting to discourage new investment.

By some estimates, integrated oil sands projects require long-term prices of $85/bbl for a solid rate of return. Meanwhile, the difference between bitumen and synthetic crude has narrowed, leaving less value for the facilities to capture.

On the other hand, CAPP noted that postponements may end spiraling costs due to the scarcity of workers and material in Alberta.

Pipes, refineries ready for oil sands

A mid-2008 analysis by Wood Mackenzie (Edinburgh) indicated that supply of Canadian oil sands products will increase by 2 million b/d between 2007 and 2015 and that half of this growth will be in Canadian heavy crude blends.

The firm also concluded that the export pipeline infrastructure and U.S. refinery capacity will be able to handle the forecast additional supplies of Canadian crude during 2007–15.

Further details are presented in the "United States" section of this chapter.

Long Lake SAGD: Phase 1 progress

Nexen Inc. (Calgary) continued in mid-2008 to ramp up Phase 1 of its Long Lake steam-assisted gravity drainage (SAGD) and upgrading project in the Athabasca region.

Nexen and Opti Canada Inc. (both in Calgary) are jointly developing the project. Nexen operates the SAGD portion; Opti is the builder and operator of the upgrader.

Long Lake production, upgrading

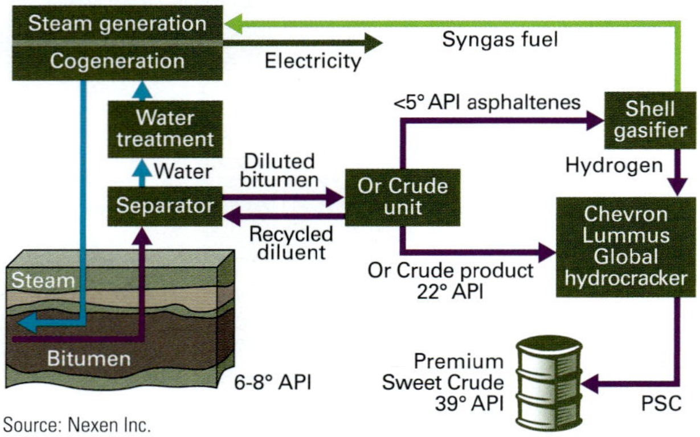

Source: Nexen Inc.

Reservoir quality

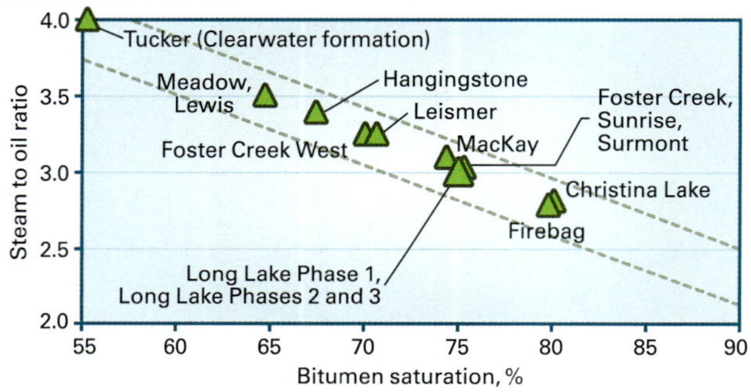

Source: Nexen Inc. Note: all projects within McMurray formation except Tucker.

Product comparison

	West Texas intermediate (WTI)	2005 Syncrude (SSP)	Long Lake (PSC)
Gravity, °API	38.7	34	39
Sulfur, ppm	4,500	1,040	<10
Transportation fuel, %	61.4	57	70.4
Gas-oil, %	24.1	40.4	27.6
Diesel cetane No.	56	40	47

Source: Nexen

The Long Lake project initiated steam injection in mid-April 2007, and first production of synthetic crude through its upgrader was anticipated for early fourth quarter 2008.

The project—40 km (25 mi) southeast of Fort McMurray—is expected to reach the 72,000 b/d Phase 1 full-design bitumen production capacity in 2009.

When operational, the Long Lake upgrader will produce about 60,000 b/d of synthetic crude. Nexen expected first commercial production in late summer 2008 with full production 12–18 months later.

The company says Long Lake has more than 2 billion bbl of recoverable bitumen, translating to a production life of about 40 years once all phases are complete.

Phase 2 work (delayed to 2009) will double the upgrading capacity to 144,000 b/d.

The project may produce as much as 288,000 b/d of bitumen with completion of additional phases.

Nexen said it wants more clarity on the regulations for carbon capture so that it can include these rules in the design of the expansion for Phase 2.

Sulfur challenge

A specialist at a Calgary conference in July 2008 said that sulfur might be a bigger problem for producers of bitumen and heavy oil in Alberta than water and energy supplies, royalty rates, and air-quality regulations.

The former expert with Shell Canada Ltd. (Calgary) expected sulfur supply to double in the next 5–10 years, creating "a market meltdown and transportation gridlock."

He sees sulfur supply jumping from bitumen upgraders in northern Alberta, refineries responding to gasoline and diesel desulfurization regulations in the United States and Europe, and large sour-gas projects in China, Kazakhstan, and the Middle East.

In a sulfur-market collapse, he said, Alberta (and other high-cost producers) will be unable to sell the byproduct and may be forced to spend money to store, inject, or bury it.

Reservoir comparison

Project	Operator	Pay thickness, m	Bitumen saturation, %	Porosity, Porosity, %	Reservoir depth, m	No. of cores analyzed	Reservoir pressure, kPa	Injection pressure, kPA
Long Lake	Nexen	40	74.9	34.9	253	155	1,200	2,750
Christina Lake	Encana	25	80.4	34.7	399	32	2,000	5,000
Firebag	Suncor	37	80.0	33.8	285	164	800	3,000
Surmont	ConocoPhillips	30	74.7	36.1	402	12	1,700	1,700
Foster Creek	Encana	25	73.7	33.9	500	73	2,700	2,700
Mackay River	PetroCanada	15-35	73.5	34.6	137	90	500	1,700
Tucker Lake	Husky	30	54.0	36.0	480	48	—	—

Note: The core analysis is from public EUB data.

Source: Nexen

Taiwan interest in oil sands

Taiwan's state-owned CPC Corp. (Taipei) signed a memorandum of understanding in mid-2008 with Indian Oilsands Corp. (Saskatchewan) regarding a possible partnership on oil sands development in Alberta.

CPC said it plans to budget $792 million for a five-year exploration program as part of an effort to diversify energy sources and stabilize domestic energy supplies as global energy prices keep rising.

Formosa Petrochemical Corp. (Taipei), part of Formosa Plastics Group, also is studying a plan to tap Canada's oil sands and said it plans to send a delegation to Alberta.

Evolving bitumen, SCO markets

Purvin & Gertz Inc. (Calgary) described in mid-2008 the strategies Canadian oil sands producers are adopting in marketing bitumen blends and synthetic crude oil (SCO) as production increases.

Some seek to minimize capital cost and accept a low price and possible market risk with bitumen production. Others are willing to upgrade or refine bitumen to add value and perhaps reduce market risk.

Expected growth in bitumen and SCO production will require growth in markets, as well as more pipeline capacity.

Market expansion will depend on growth in bitumen supply. The U.S. Gulf Coast market is of interest due to its large size and complexity. Other potential markets include the East Coast, West Coast, and Asia.

Bitumen from oil sands generally is available to the market as a diluted bitumen blend or SCO.

Oil sands environmental impact

Major discussion at a July 2008 conference in Calgary focused on actions taken to minimize the environmental impact of developing Alberta's oil sands resources.

The Alberta Energy Research Institute (AERI) reported that in 2006, Alberta GHG emissions broke down this way: power plants 51.7 million tonnes; oil sands operations 24 million tonnes; and heavy oil operations 6.8 million tonnes.

Earlier in 2008, Alberta released its climate change action plan for cutting GHG emissions (see box elsewhere in this chapter).

It was noted that although the province's oil sands lie beneath a region about the size of Florida (roughly 140,000 sq km; 54,000 sq mi), only about 500 sq km (162 sq mi) have been disturbed by oil sands activity to date.

Alberta's strict reclamation standards require that all land disturbed must be reclaimed. The province also strictly limits industry water use.

Also, ambient air is monitored continuously for levels of carbon monoxide, nitrogen dioxide, ozone, fine particulate matter, sulfur dioxide, and hydrogen sulfide.

In 2007, officials said air quality near Fort McMurray was rated good or better 98% of the time.

Oil sands crude supply

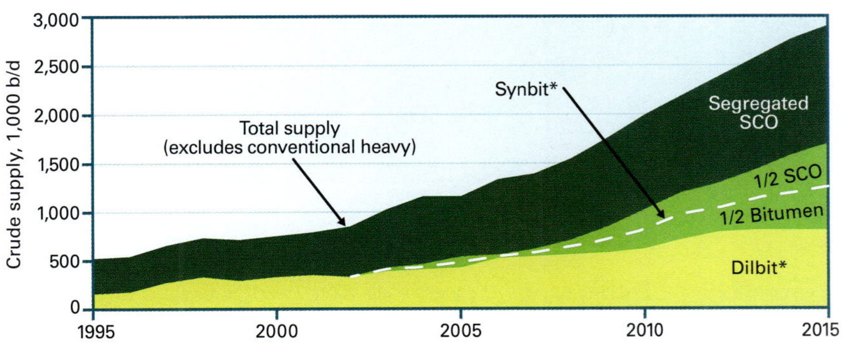

*Some bitumen blend in Western Canadian Select (WCS) or other blends.

Oil sands market outlets

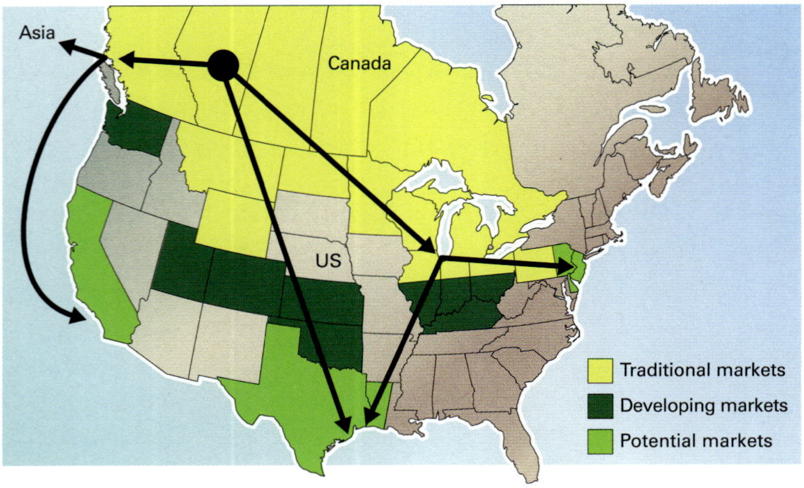

AB: Regulatory board reorg

The Alberta Energy and Utilities Board (EUB) split on January 1, 2008, into two bodies.

The Energy Resources Conservation Board (ERCB) now regulates oil and gas companies and the Alberta Utilities Commission (AUC) regulates utilities.

Sweet SCO disposition

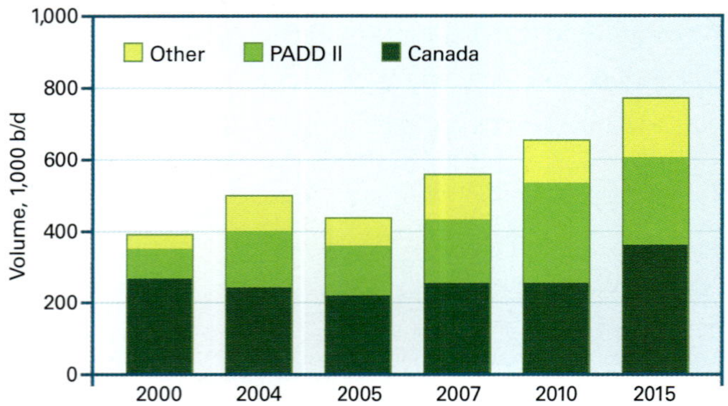

Bitumen blend distillation

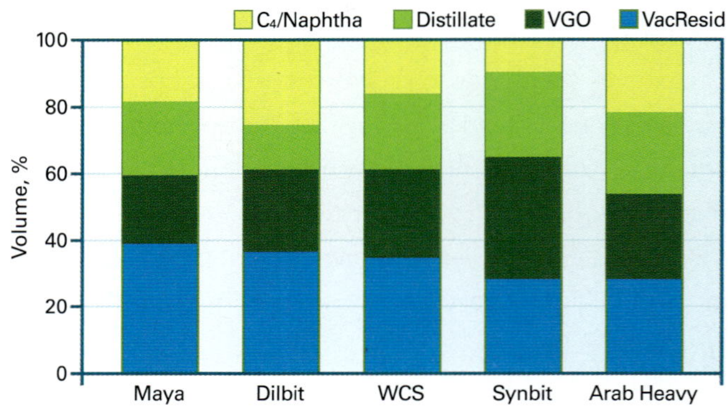

SCO vs. WTI distillation

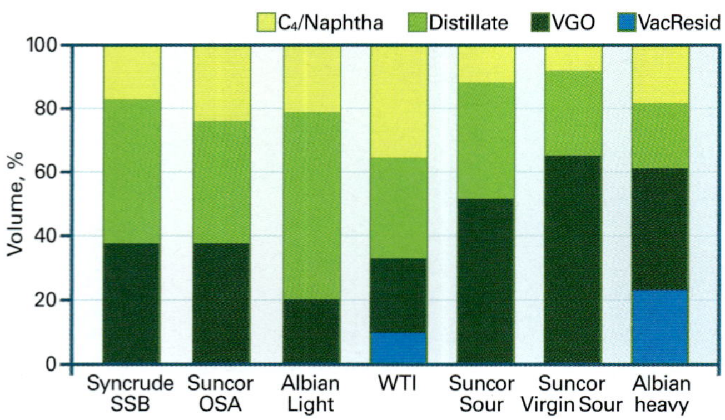

Western Canadian heavy crude disposition

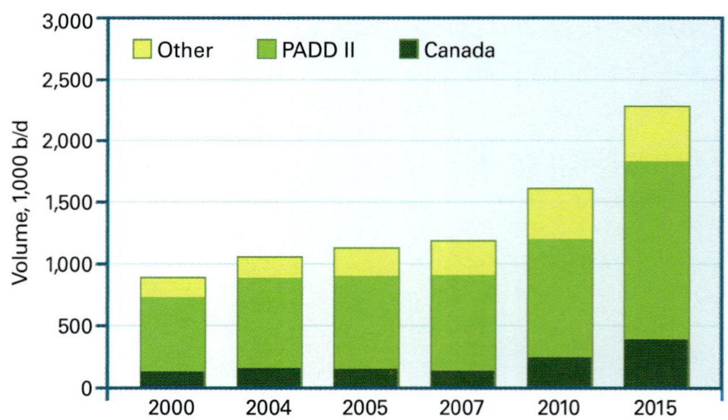

Some refiners in the traditional markets have announced plans for facilities to use more oil sands crude, especially diluted bitumen.

Historically, producers have diluted bitumen for shipment in pipelines using condensate, in a blend commonly referred to as DilBit. In some cases, they have used SCO as diluent, yielding an SCO blend called SynBit.

With growing production, SCO exports from Canada likely will rise. Main U.S. importers of SCO have been Sunoco (Philadelphia) , Marathon Oil Co. (Houston), and Sinclair Oil Corp. (Salt Lake City, Utah).

The largest market for Canadian heavy crudes has been the U.S. Petroleum Administration for Defense District II (PADD II), including refineries near Minneapolis, Chicago, and Wood River, Illinois.

Other markets include Western Canada, Ontario, Montana, and California. Following completion of pipeline reversals in early 2006, Canadian heavy crude has been moving to Cushing, Oklahoma, and the U.S. Gulf Coast.

U.S. GHG rules impact oil sands

Analysts with Purvin & Gertz (Calgary) note that oil sands crude oil faces a threat from the U.S. market due to its greenhouse gas (GHG) footprint.

The 2007 U.S. Energy Independence and Security Act appears to restrict use of oil sands crudes by federal agencies; that includes the U.S. military.

"Alternative or synthetic fuels including fuel produced from non-conventional petroleum resources" are restricted if the life cycle GHG emissions exceed those from using conventional fuels. Uncertainty remains regarding interpretation of this directive.

Some states (e.g., California) also may have environmental rules that affect oil sands crude imports.

Some parties are promoting carbon capture and storage (CCS) for Alberta, which may reduce the GHG footprint. It may even encourage gasification of coke or heavy oil for hydrogen and syngas because CCS would concentrate the carbon dioxide by-product.

Alberta 2007 Bitumen Production

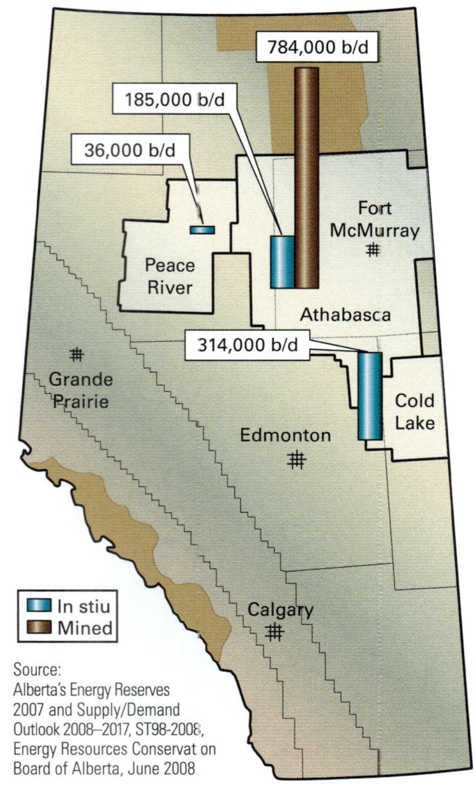

Source: Alberta's Energy Reserves 2007 and Supply/Demand Outlook 2008–2017, ST98-2008, Energy Resources Conservat on Board of Alberta, June 2008

AB: ERCB production view

Bitumen production will more than double by 2017 to 3.23 million b/d, from 1.32 million b/d in 2007, according to a mid-2008 update on Alberta oil sancs by the Energy Resources Conservation Board.

Projects are in various stages of development in Alberta's Athabasca, Cold Lake, and Peace River regions, using both mining and in situ processes. Athabasca produces the most bitumen.

The Canadian Association of Petroleum Producers (CAPP) and the Regional Issues Working Group (RIWG) have estimated tha: industry spending for 2007–11 on oil sands projects likely will be $77 billion (Can.).

Production Forecast

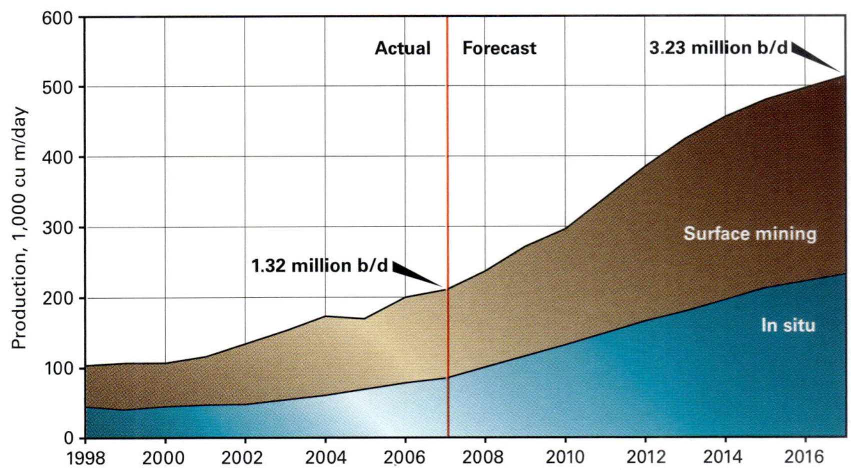

Source: Albert's Energy Reserves 2007 and Supply/Demand Outlook 2008–2017, ST98-2008, Energy Resources Conservation Board of Alberta, June 2008.

ERCB gave this assessment of bitumen resources in Alberta at yearend 2007:

- Initial in place: 1.712 billion bbl
- Initial established: 179 billion bbl
- Cumulative production: 5.9 billion bbl
- Remaining established: 173 billion bbl
- 2007 production: 0.482 billion bbl (1.32 million b/d)
- Ultimate potential: 315 billion bbl.

Oil sands expenditure forecast

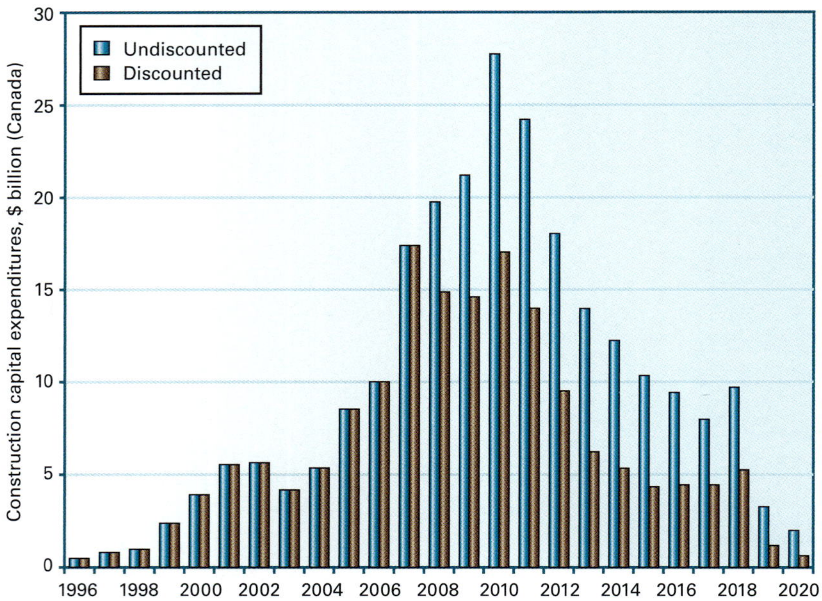

Source: Alberta Oil Sands Industry Update, Alberta Employment, Immigration, and Industry, December 2007.

India interest in oil sands

Aiming to secure oil supplies for its refineries and strengthen India's energy security, state-owned Indian Oil Corp. (IOC) reported in early 2008 that it was seeking equity in oil sands blocks in Canada.

IOC said it had initiated talks with BP PLC (London) and Shell Canada Ltd. (Calgary).

An Indian delegation was in Canada in February 2008 to discuss the plan.

Other Indian firms interested in acquiring oil sands assets in northeastern Alberta include state-run Oil & Natural Gas Corp. and Oil India Ltd.

Initial in-place bitumen estimate

Oil sands area and oil sands deposit	Initial in place, billion bbl	Area, 1,000 acres	Average pay thickness, ft	Average bitumen saturation Mass, %	Average bitumen saturation Pore volume, %	Average porosity, %
Athabasca						
Grand Rapids	54.6	1,703	24	6.3	56	30
Wabiskaw-McMurray (minable)	101.2	633	100	9.7	69	30
Wabiskaw-McMurray (in situ)	841.8	11,841	43	10.2	73	29
Nisku	65.0	1,233	26	5.7	63	21
Grosmont	317.7	10,297	34	4.7	68	16
Subtotal	**1,380.3**	—	—	—	—	—
Cold Lake						
Grand Rapids	108.9	4,223	19	9.5	66	31
Clearwater	59.3	1,070	39	8.9	59	31
Wabiskaw-McMurray	27.0	1,198	18	7.3	59	27
Subtotal	**195.1**	—	—	—	—	—
Peace River						
Bluesky-Gething	69.0	2,511	20	8.1	68	26
Belloy	1.8	64	26	7.8	64	27
Debolt	49.1	746	78	5.1	65	18
Shunda	15.8	353	46	5.3	52	23
Subtotal	**135.6**	—	—	—	—	—
Total	**1,711.0**	—	—	—	—	—

The main deposits covered by the report are the Athabasca Wabiskaw-McMurray (AWM), Cold Lake Clearwater (CLC), and Peace River Bluesky-Gething (PRBG) covering about 139,860 sq km (54,000 sq mi).

Of the 173 billion bbl remaining established reserves, ERCB considers 82% (142 billion bbl) recoverable with in situ methods and the remaining 31 billion bbl recoverable with surface mining methods.

ERCB says bitumen production from in situ projects increased by 7.1% in 2007, compared with 2006, while production from mining projects increased by 4.3%.

In 2007, the minable area produced 286 million bbl, while in situ projects produced 196 million bbl.

Production in 2007 from the three current surface mining projects was:

- Syncrude Canada Ltd., 367,000 b/d.
- Suncor Energy Inc., 267,000 b/d.
- Albian Sands Energy Inc., 150,000 b/d.

Futures sources of production from minable resources include:

- Horizon project of CNRL (production start expected third quarter 2008).
- Jackpine mine of Shell Canada Ltd. (late 2010).
- Fort Hills project of Petro-Canada, UTS Energy, and Teck Cominco (by 2011).
- Multistage Kearl mine of Imperial Oil and ExxonMobil (2011).
- Multistage Joslyn North mine of Total E&P Canada Ltd. (2013).

With these projects, ERCB expects total mined bitumen production to reach 1.76 million b/d in 2017, up from 0.78 million b/d in 2007.

The ERCB outlook does not include the Northern Lights project (Synenco Energy-SinoCanada Petroleum) and the Equinox and Frontier project (UTS), which are not expected to come on stream until after 2017.

Mined Production

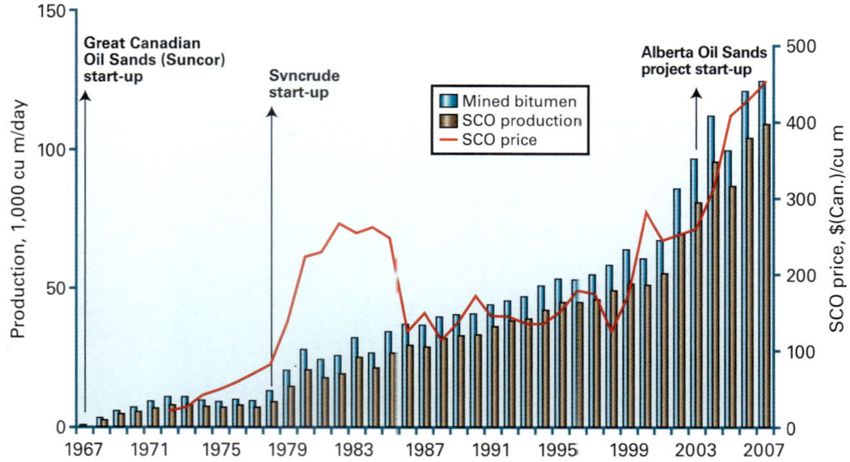

Source: Alberta's Energy Reserves 2007 and Supply/Demand Outlook 2008–2017, ST98-2008, Energy Resources Conservation Board of Alberta, June 2008.

In Situ Production

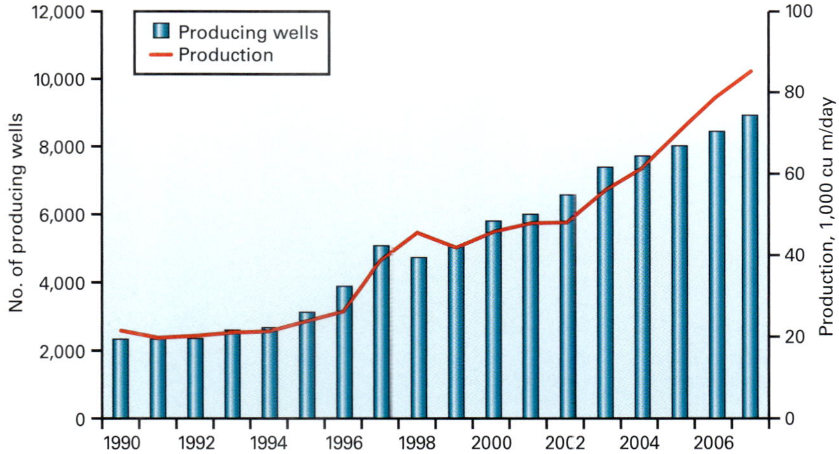

Source: Alberta's Energy Reserves 2007 and Supply/Demand Outlook 2008–2017, ST98-2008, Energy Resources Conservation Board of Alberta, June 2008.

AB: Oil sands pipelines

In mid-2008, Alberta's ERCB reported the following about pipelines in the oil sands region:

- Inter Pipeline Corridor pipeline: plans to increase diluted bitumen capacity to about 465,000 b/d by 2009 and support expansions beyond 2009.
- Horizon pipeline (Pembina Pipeline Corp.; Calgary): completed mid-2008 to provide 250,000 b/d of capacity to the Horizon oil sands project of Canadian Natural Resources Ltd. (CNRL; Calgary).
- Access pipeline: completed early 2008 to deliver 150,000 b/d diluted bitumen from the Christina Lake area to facilities in the Edmonton area.

- Waupisoo pipeline (Enbridge; Edmonton): completed in mid-2008 to move 350,000 b/d of blended bitumen from the Cheecham terminal, south of Fort McMurray, to the Edmonton area.
- Fort Hills pipeline system (Enbridge): planned for 2011 operation to carry blended bitumen (125,000 b/d) from the mine site of Fort McMurray to the upgrader site in Sturgeon County.

Alberta Bitumen Pipelines

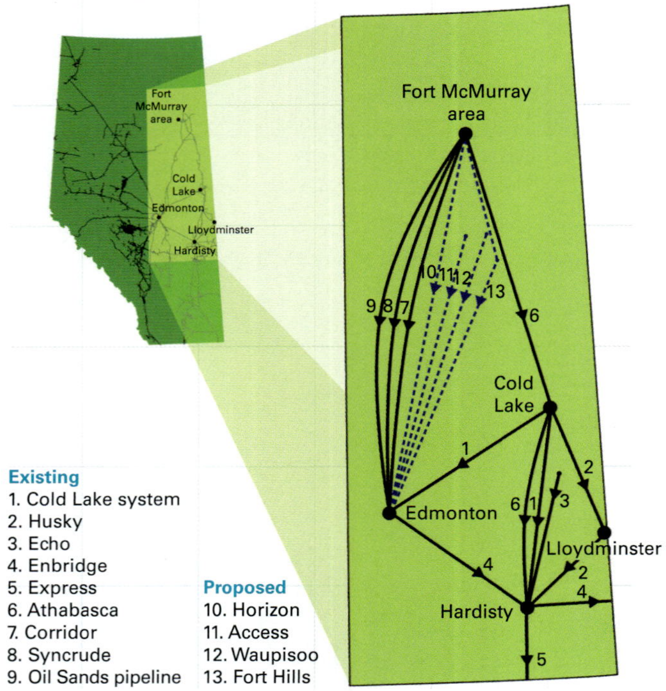

Source: Alberta's Energy Reserves 2007 and Supply/Demand Outlook 2008–2017, ST98-2008, Energy Resources Conservation Board of Alberta, June 2008.

Land Reclamation Certificate

The Alberta government presented its first oil sands reclamation certificate in early 2008 to Syncrude Canada Ltd. (Fort McMurray) for the company's work in converting a 104-hectare (257-acre) area known as Gateway Hill into forest and wetlands, 35 km (22 mi) north of Fort McMurray.

In 2006, the company reported spending $30.5 million on reclamation. It also said it reclaimed more than 80 additional hectares (198 acres) in 2007, and that it planned to reclaim another 100 hectares (247 acres) in 2008.

Pipeline role in exports to U.S.

In a mid-2008 report, Wood Mackenzie Ltd. (Edinburgh) said that two planned pipeline systems "could determine the destination of oils sands products…and influence U.S. import patterns."

Texas Access is being built by Enbridge Inc. (Houston) and ExxonMobil Pipeline Co. (both in Houston) from Patoka, IL, to the Texas Gulf Coast. It will transport crude from the Canadian oil sands region in Alberta and from the upper U.S. Midwest to refiners in Nederland and Houston, TX.

The Gateway Pipeline is to be built by Enbridge to transport 400,000 b/d of oil sands crude from near Edmonton, Alberta, to a tanker terminal in British Columbia for shipment to China, other parts of Asia, and to California.

According to the report, by 2015 only small volumes of Canadian oil sands products will reach the U.S. Gulf (where most heavy oils are now processed) because additional new processing capacity is being added in the U.S. Midwest.

Voyageur expansion plan

Suncor Energy Inc. (Calgary) announced in late 2008 that it would delay by one year its plan to invest $20.6 billion (Can.) to expand oil production capacity at its Voyageur oil sands operation north of Fort McMurray, AB.

The revision means the 200,000-b/d expansion (to 550,000 b/d) would be completed in 2013, not 2012.

The expansion includes constructing four additional stages of in situ bitumen production, a third upgrader to convert the bitumen into higher-value oil, and other facilities and utilities.

Suncor reported that it has already invested $2.5 billion on detailed engineering, site work, and fabrication of major vessels for the expansion.

Pipelines

Oil pipeline capacity

The National Energy Board said in mid-2008 that Canadian oil pipeline systems need additional capacity soon to accommodate growing supply and provide greater market flexibility.

Though there was some spare capacity, NEB noted in their annual report that there were times in 2007 when some pipelines were not able to fully meet shipper demands.

Factors driving pipeline use are growing oil sands production and continued strong U.S. demand.

NEB regulates some 45,000 km (27,960 mi) of oil, natural gas, and product pipelines in Canada.

Keystone oil pipeline

A partnership of subsidiaries of TransCanada Corp. (Calgary) and ConocoPhillips (Houston) got permission from the U.S. State Department to proceed with construction, maintenance, and operation of the proposed 3,457-km (2,148-mi) Keystone oil pipeline.

Construction was slated to start in mid-2008 on the line, which will transport Canadian crude to U.S. markets. An in-service date of fourth quarter 2009 was estimated by TransCanada.

The line is being designed to carry up to 590,000 b/d of crude from Hardisty, AB to U.S. Midwestern markets at Wood River and Patoka, IL, and to Cushing, OK.

First deliveries to Patoka are expected in late 2009.

Mackenzie Delta pipeline

As of mid-2008, operators and the Canadian government were still negotiating an agreement for the proposed $6-billion (Can.), 1,223-km (760-mi) Mackenzie Delta pipeline that would allow for producing the large stranded gas resources in the Northwest Territories. First gas may flow after 2014.

Refining

Placentia Bay refinery

Newfoundland & Labrador Refining Corp. continued work in third quarter 2008 on a new $6.5-billion, 300,000-b/d refinery in the Placentia Bay area of Newfoundland and Labrador.

Early in the year, NLRC selected UOP LLC (Des Plaines, Illinois) to supply technology, basic engineering services, and equipment for the project.

It will be the first refinery built in North America since 1984. Completion is scheduled for 2011.

Eider Rock refinery

BP PLC (London) and Irving Oil Ltd. (St. John, NB) signed a memorandum of understanding in early 2008 to work on initial stages of the proposed 300,000-b/d, $7-billion Eider Rock refinery in St. John.

The companies will consider forming a joint venture, should they decide to proceed.

In December, Irving Oil chose a subsidiary of Fluor Corp. (Irving, Texas), to perform front-end engineering design for the refinery.

The refinery would be built close to Irving Oil's existing 300,000 b/d refinery and the existing Irving Canaport deepwater oil terminal, 104 km (65 mi) from the U.S. border.

This next phase of development could require more than $100 million of investment over the next 12–15 months, Irving Oil said.

If approvals are timely, full-scale construction could begin in 2011. Start-up is expected in 2015.

CCRL refinery expansion

International Alliance Group (IAG; Houston) awarded a detailed design services contract to Mustang Engineering, L.P. (Houston) in mid-2008 for a $1.9-billion expansion of the Consumers' Cooperative Refineries Ltd. (CCRL) refinery in Regina, SK.

IAG is program manager for the grassroots portion of a proposed expansion that would increase capacity to 130,000 b/cd from a current capacity of 100,000 b/cd.

The grassroots expansion also aims to help increase processing of light synthetic crude from Canada's oil sands. The expansion is scheduled for completion in 2012.

Environment

Industry CCS project

Enbridge Inc. (Calgary) announced plans in early 2009 to lead a group of 19 oil companies and electric utilities in the Alberta Saline Aquifer Project (ASAP), a carbon capture and storage (CCS) research and pilot sequestration project.

Others involved (also based in Calgary) include BP Canada Energy Co., Chevron Canada Resources, ConocoPhillips Canada, EnCana Corp., Penn West Energy Trust, and TransCanada Pipelines Ltd.

By yearend 2009, ASAP was aiming to identify deep saline aquifers suitable for long-term storage of carbon dioxide in Canada.

A second phase calls for designing sequestration sites to receive injected carbon dioxide. Later phases could involve large-scale, long-term commercial sequestration.

In February 2008, a joint federal-provincial government panel recommended that Canadian governments invest $2 billion (Can.)—to be matched by energy industry funds—toward CCS research, with the goal of sequestering 5 million tonnes of carbon dioxide by 2015.

Federal GHG emissions program

Canada's federal government has set ambitious targets for reducing emissions of greenhouse gases (GHGs). Its policy statement calls for cuts during 2006–20 of 20% and through 2050 of 60–70%.

The program requires emission reductions at the company level. Firms can lower emissions at their facilities, invest in emission-reducing technology, trade emission allowances, and use the Kyoto Treaty's "clean development mechanism," which credits a company in one country for lowering emissions in another.

The National Round Table on the Environment and the Economy recommended that the government implement an emission tax or a cap-and-trade system, or both. For parts of the economy that don't respond to the price signal, it suggests "complementary regulatory policies."

GHG reduction approach: Alberta

In view of the importance of the oil sands industry to the province's economy, Alberta's government crafted its own path in early 2008 on GHG policy. It calls for a 14% cut in emissions of carbon dioxide against 2005 levels by 2050, with all reductions coming after 2020.

The change reflects expected near-term increases in production of bitumen and heavy oil.

Alberta officials also decided to concentrate on CO_2 capture and storage rather than the federal approach of imposing new emission targets for companies.

The EcoEnergy Carbon Capture and Storage Task Force calls for total cuts in GHG emissions of 5 million tonnes/year through operation of three to five capture-and-storage projects by 2015.

It further recommends that federal and provincial governments supplement industry investments in the initial projects with funding of as much as $2 billion.

Ultimately, it says, such projects might reduce Canadian CO_2 emissions by 600 mtpy.

Company News

EnCana reorganization delay

Uncertainty in commodity and financial markets led independent EnCana Corp. (Calgary) to postpone in late 2008 its plans to split into two publicly traded companies, one focused on natural gas and the other on the Alberta oil sands.

In particular, declining oil prices hurt the economics of a range of oil sands projects.

The company said that the transaction would be reexamined when market conditions improved.

Shell Canada-Duvernay

The Canadian government approved in third quarter 2008 the acquisition by Shell Canada Ltd. of unconventional gas producer Duvernay Oil Corp. for $5.9 billion (Can.). Both are in Calgary.

The deal adds to Shell's acreage in the WCSB. Duvernay has 25,000 boe/d of production, predominantly gas. It had plans to increase production to 70,000 boe/d by 2012.

Total SA-Synenco Energy

Total SA (Paris) acquired Synenco Energy Inc. (Calgary) in mid-2008, including that firm's main asset—a 60% stake in and operatorship of the Northern Lights heavy oil project in Alberta.

Northern Lights, 100 km (62 mi) northeast of Fort McMurray, is estimated to hold 1.08 billion bbl of bitumen.

Total would work with China's Sinopec (Beijing), which holds the remaining 40% stake in Northern Lights.

The group was considering upgrading the bitumen recovered from its three fields in Canada—Surmont, Joslyn, and now, Northern Lights—using a single upgrader. It currently has an upgrader project planned for Joslyn.

Latin America

ARGENTINA

Capital: Buenos Aires
Monetary unit: Peso
Refining capacity: 626,075 b/cd
Oil production: 610,000 b/d
Oil reserves: 2.616 billion bbl
Gas reserves: 15.600 tcf

Exploration and Development

Surubi block

Gran Tierra Energy Inc. (Calgary) reported a stabilized flow of 2,324 b/d of 43.8° gravity oil on a production test of Cretaceous Lower Palmar Largo volcanics and volcaniclastics at the Proa.x-1 exploration well.

The well, on the 90,688-acre (142-sq mi) Surubi block in northern Argentina's Noroeste basin, was drilled to TD 3,745 m (12,920 ft).

The company said that the results improve the prospectivity of identified leads in the Surubi block and in the adjoining Chivil and Palmar Largo blocks where it has working interests.

Cuyana basin

Oromin Explorations Ltd. (Vancouver, BC) won approval to exploit oil and gas on the 7,694-sq km (2,971-sq mi) Santa Rosa Block in the Cuyana basin. The basin has produced more than 940 million bbl of oil.

The area of greatest exploration interest covers a large, untested dome less than 1,200 m (3,933 ft) deep that is defined by four regional seismic lines.

Neuquen basin

Argenta Oil & Gas Inc. (Toronto) said the first well on its Loma El Divisadero (LEDO) Block in the Neuquen basin found 19° gravity oil in the Chorreado member of the Cretaceous Huitrin formation.

ARUBA
CURACAO
GRENADA
Bridgetown
BARBADOS
TRINIDAD and TOBAGO
Port of Spain
COSTA RICA
Pto. Limon
Colón
Panama
PANAMA
Santa Marta
Barranquilla
Cartagena
Maracaibo
Caracas
Lechoso
José
Puerto la Cruz
Penal-Barrackpore-Wilson
Teak
Balata-Bouvallius
Southeast Galeota
Radix
E. Queens Beach
Quarry-Coora-Quinam
Punzan
Grico
Tibu
Hato
Sinco
Palmita
Silvestre
S. Silvestre
Ciudad Bolivar
Barrancabermeja
S. Cristobol
Georgetown
Tambaredjo/Jossikreek
Cantimplora
VENEZUELA
GUYANA
Volcanera
Cupiagua
Cusiana
Apiay Complex
Bogota
Chichimene
Tulua
Cali
Castilla
Andalucia
Dina
Tello
San Francisco
Rubiales
SURINAME
Karanambo
Roraima
Tumaco
Esmeraldas
COLOMBIA
Quito
ECUADOR
Para
Guayaquil
Amistad
Manaus
Amazonas
Tumbes
Zorritos
Piedra Redonda
Talara
Iquitos
Concordia
Jurua
Southwest
Rio Urucu
Coari
Leste do Urucu
Nova Olinda
Paripi
Bayovar
Maquia
B R A Z I L
Aguaytia
Pucallpa
Aguas Calientes
Acre
Porto Velho
PERU
Rondônia
Camisea Fields
Lima
La Pampilla
Conchan
Pisco
Paracas
Mato Grosso
BOLIVIA
Trinidad
Lago de Titicaca
Pirin
La Paz
Surubi
Pto. Villarroel
Cochabamba
Sicasica
GCC
Oruro
Santa Cruz
Colpa
Tita
Rio Grande
Corumba
Arica
Sucre
Tatarenda
Camiri
Boyuibe
Monteagudo
Vuelta Grande
Villemontes
Caigua
Buena Vista
Sanandita
Tigre
Madre Jones
PARAGUAY
0 100 200 300 400 Miles
0 200 400 600 Km
Tocopilla
Toro
Bermejo
Campo Duran
Ramos
Lomitas Tranquitas
Villa Elisa
SOUTH AMERICA
Mejillones
CHILE
Rio Pescado
Caimancito
Jujuy
Salta
ARGENTINA
Asuncion
Antofagasta

SOUTH AMERICA
0 100 200 300 400 Miles
0 200 400 600 Km
Atlantic Ocean
Atlantic Ocean
Georgetown
Tambaredjo/ Jossikreek
Paramaribo
7
Cayenne
GUYANA
SURINAME
FRENCH GUYANA
Amazon River
Belem
Sao Luis
44
Manaus
Rio Tapajos
Nova Olinda
Atum
Curima
1-CES-3
Xareu
Espada
Pescada
Fortaleza
6
Fazenda Belem
Ubarana
Agulha
Canto do Amaro
Serraria Lorena
Guamare
Faz. Pocinhos
Estreito
BRAZIL
Rio Teles Pires
Rio Araguaia
Rio Tocantins
Rio Sao Francisco
Recife
Maceio
Aracaju
269
Salvador
Catu
Cuiaba
Brasilia
Senador Canedo
Ethanol
Sao Mateus
Corumba
145
Betim
Belo Horizonte
Vitoria
Jubarte
PARAGUAY
Aquidauana
Campo Grande
Rio Grande
Parana
Duque de Caxias
232
Campos
RJS-150
Frade
Roncador
10°
5°
0°
5°
10°
15°
20°
55°
45°
35°

SOUTH AMERICA
BRAZIL
PARAGUAY
ARGENTINA
CHILE
URUGUAY
Atlantic Ocean
Falkland Islands
(Islas Malvinas)
Golfo San Matias
Ethanol
Oruro
Arica
Sucre
Santa Cruz
Tita
Rio Grande
Tatarenda
Camiri
Boyuibe
Monteagudo
Vuelta Grande
Buena Vista
Caigua
Sanandita
Madre Jones
Tigre
Toro
Campo Duran
Bermejo
Ramos
Lomitas Tranquitas
Rio Pescado
Caimancito
Jujuy
Salta
Metan
Tocopilla
Mejillones
Antofagasta
Corumba
Aquidauana
Campo Grande
Villa Elisa
Asuncion
Formosa
Resistencia
Corrientes
Paso de los Libres
Uruguaiana
Sao Gabriel
Apucarana
Belo Horizonte
Betim
Campos
Duque de Caxias
Rio de Janeiro
Tremembe
Maua
Sao Paulo
Cubatao
Santos
Sao Sebastiao
Irati
Curitiba
Merluza
Sao Francisco do Sul
Perimbo
Florianopolis
Laguna
Canoas
Porto Alegre
Tramandai
Rio Grande do Sul
Cordoba
San Lorenzo
Mendoza
San Luis
Lumunta
Tupungato
Cacheuta
Concon
Valparaiso
Santiago
Refugio
Colorados
Barrencas
El Sosneado
Campana
Buenos Aires
La Plata
Montevideo
Dock Sud
Las Flores
Cnl. Pringles
Bahia Blanca
Punta Alta
Puerto Rosales
Concepcion
Colonia Catriel
El Medanito
Guanaco
Plaza Huincul
Challaco
Cerro Bandera
Barda Gonzales
San Antonio
Rivadavia
Anticlinal Grande
Cerro Dragon
Canadon Grande
Canadon Seco
Comodoro Rivadavia
Cerro Tortuga
Comodoro
Caleta Olivia
Pico Truncado
Santa Cruz
Rio Gallegos
Punta Arenas
Tierra Del Fuego
Rio Grande
Ushuaia
0 100 200 300 400 Miles
0 200 400 600 Km

Fuel export ban

In early 2008, the Argentinean government suspended all exports of liquid fuels, including oil, gasoline, diesel, fuel oil, and naphtha in the wake of rising domestic prices.

The Ministry of Trade also ordered that state-regulated prices return to levels in place on October 31, 2007, based on a 1974 law requiring suppliers of energy and other products to first meet domestic demand before exporting.

Officials said they would lift the ban once oil companies reduced gas oil and petrol prices to the October 31 levels.

The government of Paraguay expressed concern that supply problems could arise in that country because more than half of the fuel consumed is imported from Argentina.

The well flowed at undisclosed rates without treatment.

The LEDO x-2 exploration well went to TD 1,614 m (5,294 ft) and cut 60 m (197 ft) of gross thickness in three packages of sandstone with resistivities that suggest high oil saturation.

The company was reviewing recent 3D seismic data in early 2008 and examining options to improve production rates.

Tierra del Fuego well

Apache Corp. (Houston) reported that its Seccion Banos-2004 well on the Argentine side of Tierra del Fuego Island was producing 1,635 b/d of oil and 1.3 MMcfd of gas in early 2008 from the Lower Cretaceous Springhill sandstone.

The well is the first large producer drilled following a 1,800 sq-km (700 sq-mi) 3D seismic survey in the area.

Apache said SB-2004, drilled to 1,690 m (5,544 ft), had the highest initial production rate of any well drilled in 40 years of oil and gas activity on the concession, confirming the productivity of an area between three 40-year-old fields—Canadon Piedras, Cabo Nombre, and Bajo Grande.

The firm said that two wells it previously developed at an offshore Cabo Nombre Sur discovery were producing a total of 11.5 MMcfd of gas and 53 b/d of oil.

Offshore seismic program

GX Technology (GXT; Houston) awarded SCAN Geophysical ASA (Oslo) a contract to acquire up to 8,000 km (4,971 mi) of long offset 2D data, a part of which will be used to complete GXT's ArgentineSPAN™ program.

GXT and SCAN agreed to mobilize SCAN's high-capacity 2D vessel to begin data acquisition in late 2008.

Drilling and Production

Los Flamencos field

The LF-1010 well drilled by partners Roch SA (Buenos Aires) and Antrim Energy Inc. (Calgary) in Los Flamencos field in Tierra del Fuego stabilized at more than 10 MMcfd of gas from Cretaceous Springhill in October 2008.

The well extended Los Flamencos west of the LF-1005 well that tested previously at 8 MMcfd.

Tests were planned at LF-1011 and LF-1012 wells to confirm further westward extension of the field.

Gas delivery from eastern Los Flamencos field to the mainland through the San Martin pipeline started in September, and gathering lines are being laid to the four wells in the western part of the field. Deliveries are expected to rise to 40 MMcfd in mid-2009 from 20 MMcfd in mid-2008.

Refining

Saenz Pena refinery

Officials in Chaco province signed a letter of intent with Empresa Nacional de Petroleo SA (Napesa) for construction of an oil refinery in Saenz Pena.

The provincial government said the refinery will be a public/private investment and is slated to begin operating by April 2009 to produce gasoline, fuel oil, asphalt, and diesel.

The provincial government will create a fund to guarantee the supply of crude oil to the plant.

Alternative Fuels

Biodiesel plant near Rosario

Recently established Rosario Bio Energy SA (Rosario, Argentina) was constructing the Complejo Industrial RBE in late 2008, its first biodiesel production plant.

The facility is in Roldan, about 20 km (12 mi) from the port city of Rosario.

The likely major feedstock for the 19-million gal/yr plant will be soy oilseed because the port is the site of major soy oilseed export operations, and there are numerous oil-extraction plants in the region.

BELIZE

Capital: Belmopan
Monetary unit: Belizean dollar
Refining capacity: 0
Oil production: 0
Oil reserves: 0
Gas reserves: 0

Access granted to large two-part block

Providence Energy Belize Ltd. (Houston) acquired exploration and production rights to 1,375 sq km (531 sq mi) in and offshore southern Belize.

The award consists of a mostly-offshore block in the Gulf of Honduras from Punta Gorda to New Haven and another block over and southeast of Glover Reef off Dangriga.

Nearly all of the 57 wells drilled in Belize had oil shows or are producing oil, according to Providence Energy Group. The firm believes that a major oil migration path extends from Southern Mexico and through Guatemala into Belize.

Two independent Denver companies began producing oil in Belize in 2006 near Spanish Lookout, about 88 km (55 mi) west-southwest of Belize City.

That year, a consultant to the Belize government estimated that the next discoveries in Belize would be in the southern offshore basin and also in the southern basin on land near Crique Sarco, Monkey River, and the Punta Gorda Belt.

BOLIVIA

Capital: La Paz
Monetary unit: Boliviano
Refining capacity: 41,200 b/cd
Oil production: 40,000 b/d
Oil reserves: 465.0 million bbl
Gas reserves: 26.5 tcf

Bolivia's proven crude oil reserves have risen in recent years as a result of increased exploration activity and investment. The country has the second-largest natural gas reserves in South America (after Venezuela) but remains the continent's poorest country.

Pres. Evo Morales nationalized the country's oil and gas sector in May 2006, and state-owned Yacimientos Petroliferos Fiscales Bolivianos (YPFB) now runs all oil and gas pipelines and domestic sales and exports.

The nationalization action also hiked taxes to as much as 82% from 18% on some multinationals and ordered YPFB to hold a majority stake in several foreign-owned refineries.

However, Bolivia's Hydrocarbon Chamber said investment in the country's oil and gas industry fell to $120 million in 2006, from $650 million in 2002.

GCC pipeline construction start

After earlier delays, YPFB and nationalized gas transporter Transredes began construction in mid-2008 of the 130-MMcfd Gasoducto Carrasco-Cochabamba (GCC) pipeline.

The $170 million line will extend 250 km (155 mi) from the gas-producing region of Carrasco to the city of Cochabamba.

GCC is intended to help resolve gas supply restrictions for the industry in Cochabamba, La Paz, Oruro, and Potosi arising from the insufficient capacity of the Gasoducto al Altiplano (GAA).

Officials with Transredes said that the firm would prioritize the GCC construction project as well as the extensions of the existing GAA and Gasoducto Villamontes-Tarija (GTV).

In May, before it was nationalized, Transredes had already completed Phase 2 work on the GVT, raising its capacity to 13.8 MMcfd.

The expanded GAA will connect to the GCC, increasing its transport capacity to 32 MMcfd from 27 MMcfd.

BRAZIL

Capital: Brasilia
Monetary unit: Real
Refining capacity: 1,908,278 b/cd
Oil production: 1.815 million b/d
Oil reserves: 12.624 billion bbl
Gas reserves: 12.890 tcf

Overview

According to the U.S. Energy Information Administration (EIA), Brazil is the third-largest energy consumer in the Western Hemisphere, behind the United States and Canada.

Offshore exploration and production contributes 81% of Brazil's oil and gas production.

The Brazilian Petroleum and Gas Institute (IBP) has predicted that oil and gas investments in Brazil through 2012 will total $100 billion and sustain oil self-sufficiency (which was attained in May 2006).

Oil

State-owned Petroleo Brasileiro SA (Petrobras) controls more than 95% of crude oil production in Brazil. Rio de Janeiro state contains about 80% of Brazil's total production.

About 90% of Brazil's proved oil reserves are offshore, with 64% of the concession areas in deep and ultradeep waters, according to Petrobras. These reserves consist of mostly heavy grades.

OPEC: Membership offer declined

Iran invited Brazil in mid-2008 to join OPEC, according to a Brazilian senior government official, but Brazil's Pres. Luiz Inácio Lula da Silva demurred in September, saying that the country has been investing in its refining capacity in order to become an exporter of fuel, not crude oil.

In June, Brazil was asked to attend a meeting of oil producer and consumer nations in Jeddah

Petrobras has aggressively expanded production in recent years and is expected to continue to do so.

Natural gas

The Campos and Santos basins hold the majority of Brazil's gas reserves. Natural gas production has grown slowly in recent years, however, mainly due to a lack of domestic transportation capacity and partly due to low domestic prices.

Brazil's largest natural gas production is in the Campos Basin in Rio de Janeiro state from offshore fields. Most onshore production occurs in Amazonas and Bahia states.

In 2006, Petrobras announced an ambitious plan to increase its natural gas production to 14.7 bcm (520 bcf) by the end of 2008. The project will focus on increasing production from the Marlim and Merluza fields and bringing two new fields onstream in Espirito Santo state.

Petrobras has predicted that gas use will reach 134MMcmd (4.73 bcfd) in 2012, almost triple the level of 2006.

Gas is mostly used as a substitute for fuel oil in industrial and power-generating applications. Further, the introduction of natural gas imports has helped fuel the rapid growth in domestic consumption.

Alternative fuels

Brazil is the world's second-largest ethanol producer after the United States, and is the world's leading exporter of ethanol made from sugar cane.

New growth program based on oil, gas

Energy projects in the strategic plan of Petrobras underpin Brazil's new economic expansion drive, called the Program for Accelerated Growth. The PAC was outlined in mid-2007.

About 65% of PAC energy projects involve oil and gas, mostly in the southeastern states. Of these, 183 are already in

the Petrobras strategic plan, representing $86 billion in Brazil-based investment through 2010.

Policy

Oil law update pondered

The Ministry of Mines and Energy at midyear was exploring possible updates to the country's oil law.

One proposal would create a new state-run firm to manage recent oil discoveries made in the subsalt layer of the Santos basin.

News of the proposal prompted Petrobras to say it would present a new business plan later in 2008, revising 25 projects that would include developments in the presalt offshore area.

Under the current system, oil firms buy a concession from the government to explore within a defined block, often in partnership with Petrobras. The state then gives control of any oil discovered to the operating companies and is paid royalties instead of oil.

Many in the Brazilian government apparently support a change that would establish production-sharing agreements, under which reserves remain state property and international oil companies are given a share of output.

Exploration and Development

Oil development bonanza seen

Several firms issued positive reports and projections in late 2008 on Brazilian initiatives:

- Petrobras reported record exports of 574,000 b/d of oil in October (more than 17.8 million bbl).
- BG Group PLC (Reading, UK) said output from the Tupi, Iara, and Guara deepwater fields in the Santos basin should hit 300,000 boe/d by 2012. First commercial production is expected from its 100,000 b/d Tupi pilot by fourth quarter 2010.
- The Ministry of Mines and Energy estimated that oil and gas production is viable from Brazil's subsalt reserves even with oil at $60/bbl.

SANTOS BASIN SUBSALT EXPLORATION FOCUS AREA, WELLS, AND OTHER FIELDS

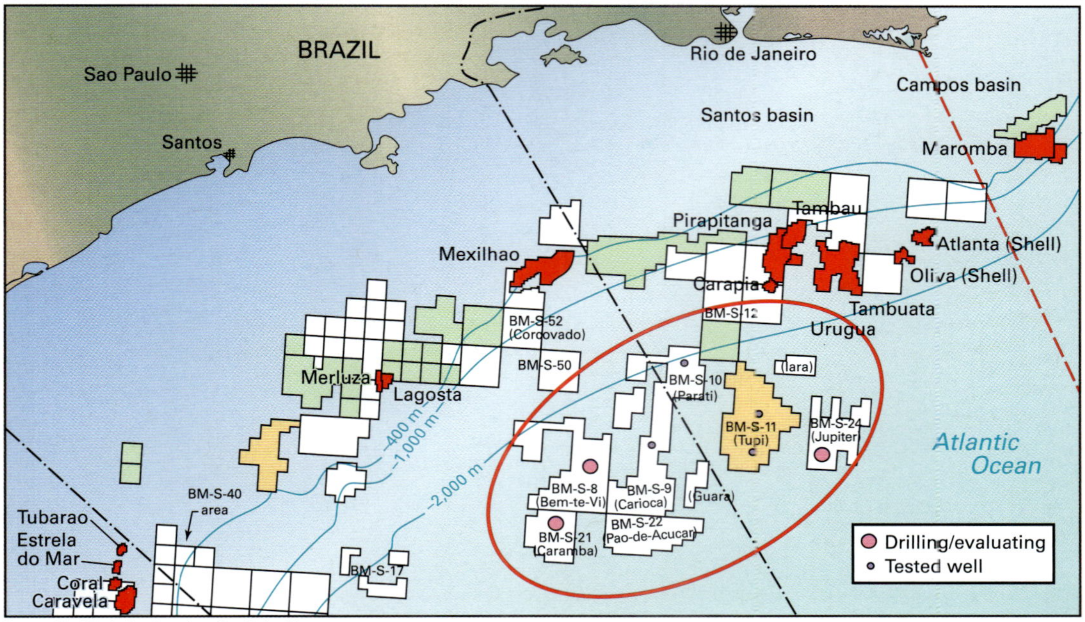

Source: Modified after Petroleos Brasileiro SA

- ExxonMobil Corp. (Irving, TX) said it started drilling its Azulao well (Block BM-S-22) in the presalt layer of the Santos basin in late 2008. First results are expected by early 2009.
- Chevron Corp. (San Ramon, CA) said it expects first oil from its Frade field in the Campos basin in second quarter 2009 and that production could hit 85,000 b/d in 2011.
- Inpex Corp. (Tokyo) and Royal Dutch Shell (The Hague, Netherlands) said an exploration well will likely be drilled in 2009 in offshore Block BM-C-31, about 150 km (93 mi) off the state of Rio de Janeiro.
- Officials in west-central Mato Grosso state said oil and gas exploration may begin there in 2010.
- Brazil's ANP planned a 10th exploration and production rights auction for mid-December 2008, offering 130 onshore blocks in seven basins.

Views on subsalt, presalt search

At a mid-September 2008 conference in Rio de Janeiro, Petrobras and several IOC executives shared perspectives on Brazil's presalt play and on subsalt exploration.

Petrobras

During 2006–08, Petrobras discovered light oil in presalt reservoirs, including eight discoveries in the deepwater Campos basin with 28–30° oil.

Drilling in the Santos basin's Litoral Norte led to discoveries of light oil in complex carbonate reservoirs, typified by Tupi and Iaga. The salt is 2,000 m (6,562 ft) thick, and the high-pressure, acid environment requires special production equipment.

The Tupi production pilot in the Santos basin should begin producing in 2010, and Petrobras approved two additional presalt pilots for 2013–14.

Petrobras said its new presalt fields in the Santos basin should begin commercial production in 2010. By 2017, the presalt cluster in Santos is expected to produce 1.126 million boe/d.

Chevron

Chevron reported that it has made 25 discoveries in deep water and was running 18 deepwater projects, with eight already producing. The company is a major investor with Petrobras on Campos basin Block BC-20, which includes the Frade discovery, and in other areas.

Shell

Shell, involved in Brazil's BC-10 development, said about 10% of its annual production comes from subsalt wells in Gabon, Oman, Kashagan, and the Gulf of Mexico.

The firm has interests in 10 offshore blocks, including five in the Santos basin, and produces about 26,000 boed from Bijupira-Salema. It expects to drill at least 10 new subsalt wells in the coming years in water deeper than 1,500 m (4,921 ft).

Block BM-S-11 in ultradeep Santos Basin waters

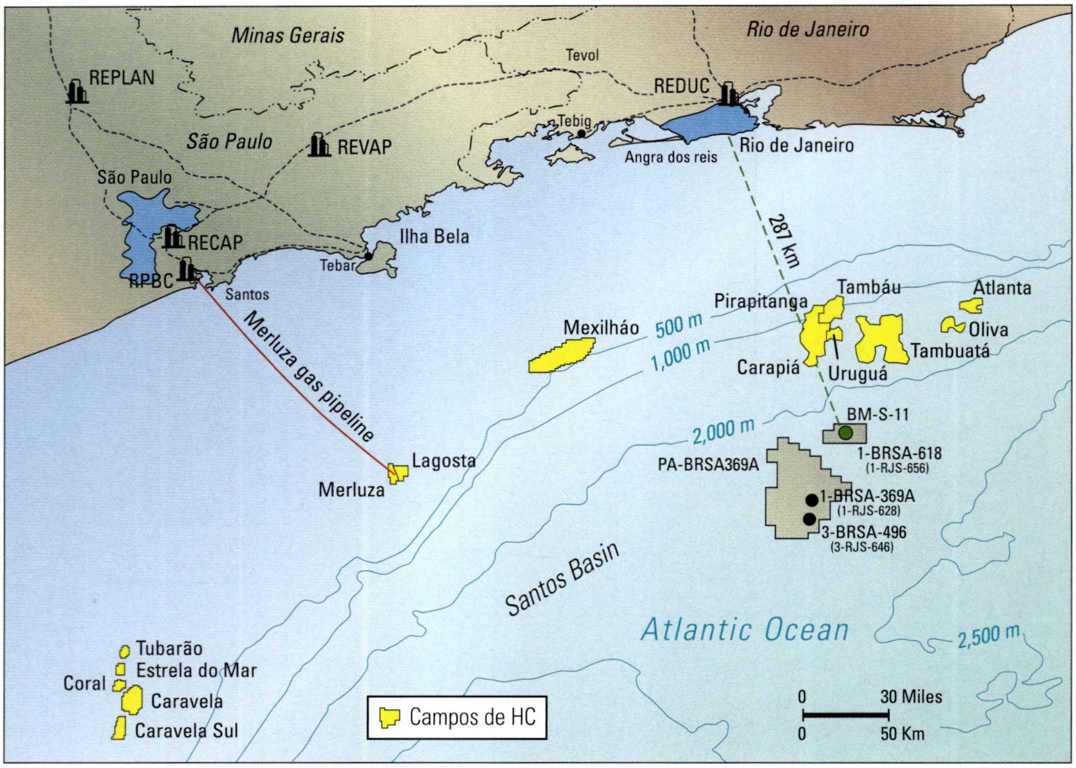

Pre-salt potential: 100 billion bbl?

Brazil's ANP said in November 2008 that newly discovered pre-salt oilfields may contain more than 100 billion bbl.

The resource is trapped beneath a layer of salt, under more than 7,000 m (22,000 ft) of water and rock.

The formation holding the oil is about 200 km (125 mi) off the southern coast from the Santos Basin. The deposits were discovered in 2007. Several drilling concessions were granted before the government suspended annual concession auctions to plan development in more detail.

ANP has approved five new deepwater fields in the Santos basin: Parati (BM-S-10), Tupi (BM-S-11), Caramba (BM-S-21), and Carioca and Guará (both BM-S-9).

Presalt approach pondered

Oil discoveries in the presalt layer off Brazil, triggered by the 2007 Tupi find, are likely to be contiguous, said an exploration expert with Petrobras.

Such contiguity is a factor that will trigger formal cooperation among the seven companies holding concessions in the region: Petrobras, BG, Galp Energia, Repsol-YPF, Royal Dutch Shell, ExxonMobil, and Hess Corp.

Under Brazilian law, according to Petrobras, contiguous concession areas become the object of joint exploration by companies holding the various concessions to avoid drilling by one company in another's area.

Brazilian officials have been seeking ways to maximize government control over fields in the presalt layer, viewing them as a low-risk opportunity for IOCs. One proposal would create a separate company to administer production in the presalt area, considering the oil as a "sovereign asset" to be developed for the benefit of the whole population.

Petrobras reported the Tupi discovery in 2007, estimating that the site could contain 5–8 billion bbl of crude oil. A formal statement on reserves is expected in second-half 2009.

Heavy subsalt, presalt action

Petrobras and others had numerous offshore subsalt and presalt discoveries in 2008.

Petrobras: off Espirito Santo

- Wells: 6-BFR-1-ESS and 6-BAZ-1DB-ESS
- Location: 80 km (50 mi) offshore; just north and south, respectively, of 1-ESS-103A in the producing Jubarte field
- Depth data: Drilled to 1,348 and 1,426 m (4,423 and 6,678 ft) from the water line; salt layer up to 700 m (2,297 ft) thick
- Other data: Recoverable volumes estimated to be at least 1.5 billion boe.

Anadarko (The Woodlands, Texas): Campos basin: off Espirito Santo

- Blocks/well: Block BM-C-30; Wahoo prospect; 1-APL-1-ESS
- Location: 40 km (25 mi) southeast of previous Petrobras presalt discoveries at Jubarte field
- Depth data: Drilled to 5,608 m (18,400 ft); water depth 1,417 m (4,650 ft)
- Other data: Play characteristics similar to those of nearby Jubarte 1-ESS-103A.
- Plans: At least four more wells in deepwater Campos, Espirito Santo basins through mid-2009.

Petrobras: South end of Santos basin

- Blocks/well: Block S-M-1289, part of BM-S-40; 1-BRSA-658-SPS (1-SPS-57)
- Location: 200 km (124 mi) offshore the state of São Paulo
- Depth data: Drilled to approximately 2,060 m (6,759 ft)
- Other data: Sandstone reservoirs like finds by 1-BRSA-607-SPS (1-SPS-56).

Petrobras: Espirito Santo basin

- Well: 4-GLF-23-ESS
- Location: Near infrastructure already installed in Golfinho field
- Depth data: Drilled in 1,374 m (4,508 ft) of water
- Other data: First estimates indicate 150 million bbl recoverable; 27° API.

Petrobras and partners; Santos basin

- Block/well: Ultradeepwater Block BM-S-8 (aka Bem-te-Vi); well 1-BRSA-532A-SPS (1-SPS-52A)
- Location: 250 km (155 mi) off state of Sao Paulo, south of Rio de Janeiro
- Depth: TD 6,773 m (22,220 ft); water depth 2,139 m (7,016 ft)
- Other data: Partners Royal Dutch Shell (The Hague, Netherlands), Galp Energia (Lisbon); 25–28° API.

Petrobras: Santos basin

- Blocks/well: Group of blocks designated BM-S-40; 1-BRSA-607-SPS (1-SPS-56)
- Location: Far southern part of basin
- Depth data: Reservoir at 2,080 m (6,824 ft); water depth 235 m (771 ft)
- Other data: 36° API found above the salt; production rate estimated at 12,000 b/d

Petrobras and partners; Santos basin

- Blocks(s)/well(s): Block BM-S-11; 1-BRSA-618-RJS informally known as Iara
- Location: 230 km (143 mi) off Rio de Janeiro, north of Tupi field
- Depth data: TD 6,080 m (19,948 ft), in 2,230 m (7,316 ft) of water
- Other data: Partners BG Group (Reading, UK), Galp Energia (Lisbon); gross recoverable volumes estimated at 3–4 billion boe.

Sugar Loaf: A giant?

The new Sugar Loaf field in block BM-S-9 may hold 33 billion boe, said ANP in April 2008.

It is on a large structure that extends into the Carioca field in an adjacent block and may also stretch into BM-S-21 and BM-S-22 blocks in the Santos basin.

Petrobras operates BM-S-9 with BG Group (Reading, UK) and Repsol YPF (Buenos Aires).

$12 billion for Santos work

Petrobras said in early 2008 that it planned to invest $12 billion on activity in the Santos basin by 2012.

Production from Mexilhao field alone was expected to reach 15 MMcmd of gas—a volume equal to about 30% of Brazil's current gas demand, the firm said.

Other Santos basin projects include the development of Urugua, Tambau, Caravela, and Cavalo Marinho fields, and construction of a gas treatment unit in Caraguatatuba.

Lease Sales

Fewer blocks for December auction

ANP cut from 162 to 130 the number of exploration and production concession blocks it planned to offer (in eight sectors and seven sedimentary basins) at the 10th auction round, set for December 18, 2008.

The reduction reflected debate about potential legal changes to restrict access to the presalt region because of the high success rate of recent exploration activity. The area could hold massive oil reserves. The Tupi area alone (Santos basin) is believed to contain 5–8 billion boe.

Domestic opponents of the proposed changes are concerned that royalties will most likely go to private investors.

Five deepwater licenses awarded

Perenco (Paris) acquired five exploration licenses in deepwater Espirito Santo basin. Water depths are around 1,500 m (4,921 ft).

Each of blocks ES-M-416, ES-M-418, ES-M-472, ES-M-529, and ES-MM-531 is 720 sq km (278 sq mi). The agreement calls for acquisition and processing of 2D and 3D seismic plus drilling of at least two wells.

Drilling and Production

First oil from Jubarte pre-salt

Production of first oil began in third quarter 2008 from well 1-ESS-103A, tapping the pre-salt layer in the Jubarte field in the Campos basin, said Petrobras.

The first well's production potential was reported to be 18,000 b/d. Development will help Petrobras improve its knowledge about the pre-salt reserves located in Espírito Santo and in other places offshore.

Total investment leading to first production was estimated at $50 million.

A 6- to 12-month test is planned to evaluate how pre-salt oil behaves, both in the reservoir and in the process plant at the JK P-34 platform.

New platform, terminal

In October 2008, Petrobras inaugurated the P-51 semisubmersible platform—the first to be built completely in Brazil.

The P-51 was to be installed in the Campos basin and begin operation by yearend in the Marlim Sul field, 150 km (93 mi) offshore in 1,255 m (4,118 ft) of water.

Interconnected to 19 wells, it can process 180,000 b/d of 22° API oil, plus 6 MMcmd of natural gas. Full operation is expected in 2010.

P-51 start-up coincided with reports that Petrobras also began operating its new $837 million PRA-1 tanker loading terminal in the Campos basin.

The loading terminal is vital to company plan to export large quantities of oil from Roncador and Marlim fields over the next 20 years.

The PRA-1 is designed to receive crude from four deepwater floating production systems and transfer it to tankers through a floating storage and offloading vessel, at a rate of 810,000 b/d.

FPSOs for Santos basin

Petrobras contracted for 10 new FPSO platforms in third quarter 2008 for service in the Santos basin pre-salt areas.

The first two, chartered from third parties, will be used for use in pilot development projects, processing 100,000 bbl/d of oil and 5 MMcmd of natural gas. They will be installed between 2013 and 2014 in areas yet to be defined.

The eight others will belong to Petrobras. To be installed during 2015–2016, they will process 120,000 bbl/d of oil and 5 MMcmd of natural gas.

All of the FPSOs will operate in ultradeep waters, ranging from 2,400 to 3,000 m (7,874 to 9,843 ft).

BM-S-48 and BM-S-55

Repsol (Madrid) started drilling in late 2008 on the Panoramix well in block BM-S-48 in the Santos basin while also making plans to continue drilling the Vampira well, 180 km (112 mi) from the coast in 140 m (459 ft) of water.

Work in the Santos basin will last two years in blocks BM-S-48 and BM-S-55, where two or three wells are to be drilled.

Tupi pilot production: 2009

Petrobras said in August 2008 that it plans to start oil production from its giant Tupi field in the Santos basin off Rio de Janeiro state in March 2009.

Initial output in the pilot project is projected at 20,000–30,000 b/d, rising to 100,000 b/d in 2010.

Tupi holds an estimated 5–8 billion bbl of oil reserves. It is in a frontier presalt reservoir, deeper (2,100 m or 6,890 ft) than discoveries in the Campos basin and also farther offshore—300 km vs. 150 km (186 mi vs. 93 mi). Its development could boost Brazil's oil and gas reserves by more than 50%.

Brazil oil projects

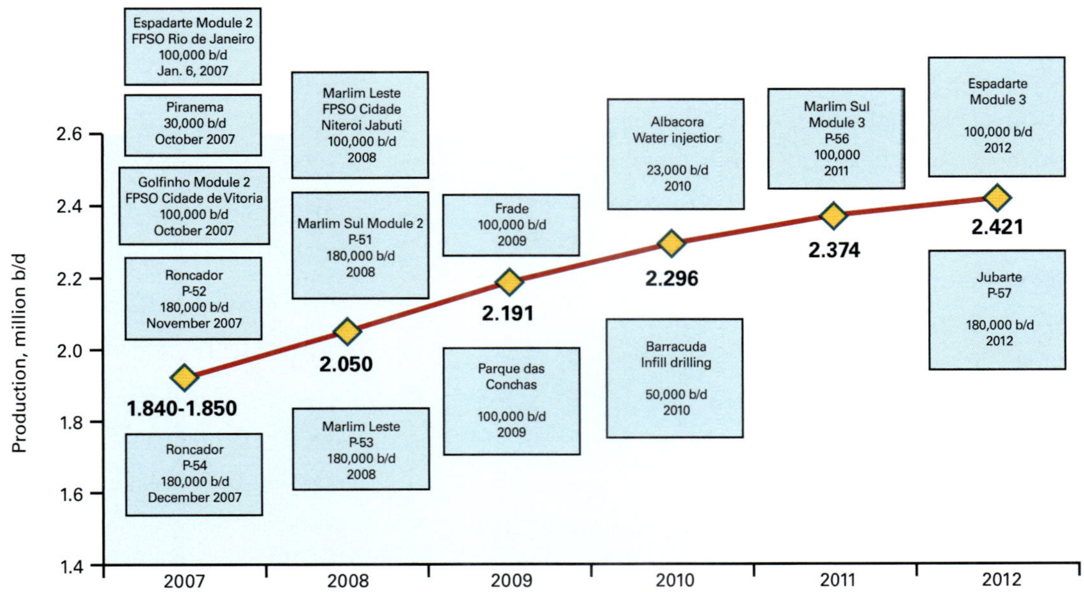

Source: Based on Barbassa, A., and Castro, C.H.D., Petrobras overview, 9th Foro Latibex, Nov. 22, 2007.

Brazil gas projects

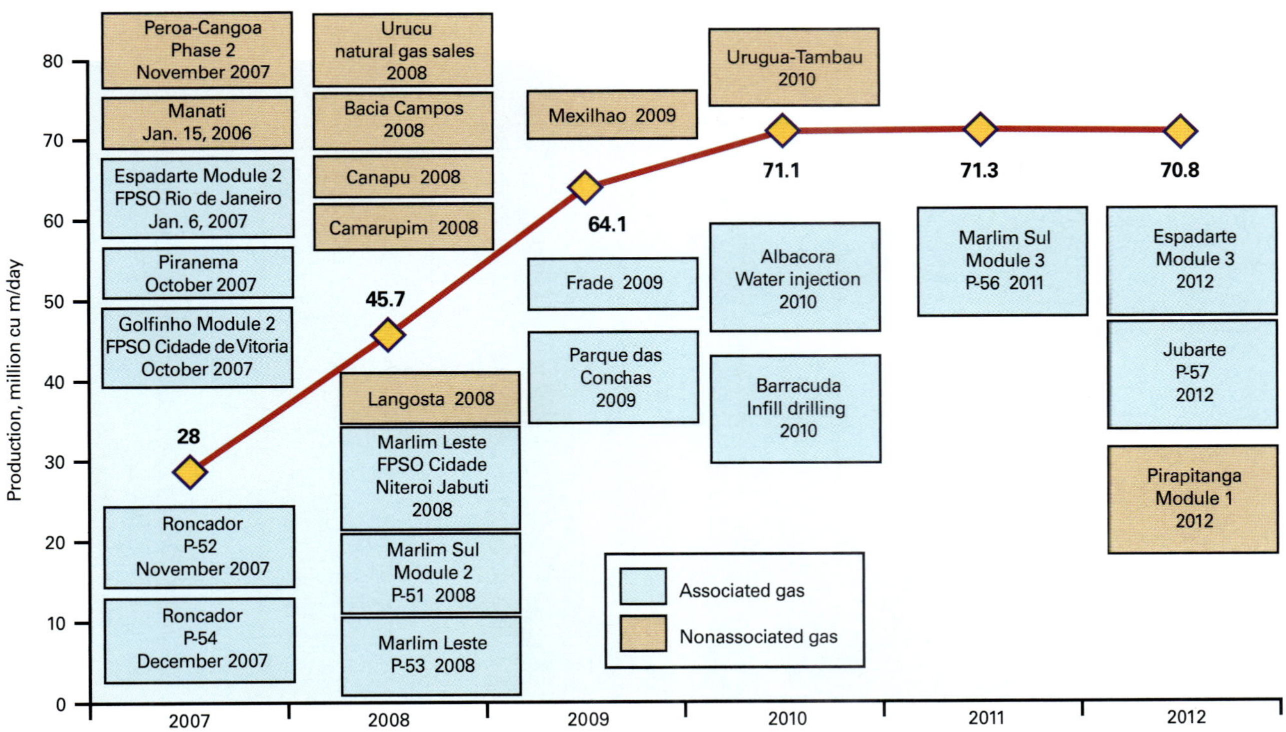

Source: Based on Barbassa, A., and Castro, C.H.D., Petrobras overview, 9th Foro Latibex, Nov. 22, 2007.

Petrobras plans to install two production units every two years to produce pre-salt reserves in the Santos basin, beginning in 2010, and has chosen Tupi to be the first of its pre-salt resources to be developed.

There will be 11 platforms eventually producing in Tupi; future production is estimated at 1 million b/d of oil from its pre-salt strata. The field also will provide around 220 cu m (7,769 cu ft) of gas for each cu m (6.3 bbl) of oil extracted.

The sub-salt reserves extend over an 800-km (500-mi) stretch and may contain enough oil to propel Brazil to the status of a major world oil producer.

Pres. Lula da Silva said in mid-2008 that his goal was to direct revenues from Tupi operation to healthcare and education.

Ambitious facilities plans

Petrobras continued in 2008 its ambitious program of facilities development aimed at expanding the production of oil and natural gas.

The company expects oil production in Brazil to reach 2.421 million b/d in 2012 and 2,812 million b/d in 2015.

Gas production is expected to increase to 70.8 MMcmd in 2012, up from the 28 MMcmd average in 2007.

Petrobras was working to deploy three new oil platforms and one gas platform off Brazil in 2008:

- P-51 in Marlim Sul field, Campos basin (180,000 b/d capacity)
- P-53 in Marlim Leste field, Campos basin (180,000 b/d capacity
- FPSO *Cidade de Niteroi* in Marlim Leste field, Campos basin (100,000 b/d capacity)
- FPSO *Cidade de Sao Mateus* in Camarupim field, Espirito Santo basin (processing capacity 10 MMcmd of gas).

BC-10: Heavy oil, deep water

Shell E&P (Houston) has interests in 10 offshore exploration blocks in the offshore Santos, Campos, and Espirito Santo basins.

In its BC-10 integrated field development, Shell was using a range of new subsea technologies—from caisson separation with electric submersible pumps, to surface BOP with slim risers, to high-power umbilicals.

The heavy oil field will be developed with deepwater, horizontal wells and openhole completions and produced to an FPSO vessel.

Block BC-10 Development, Northern Campos Basin

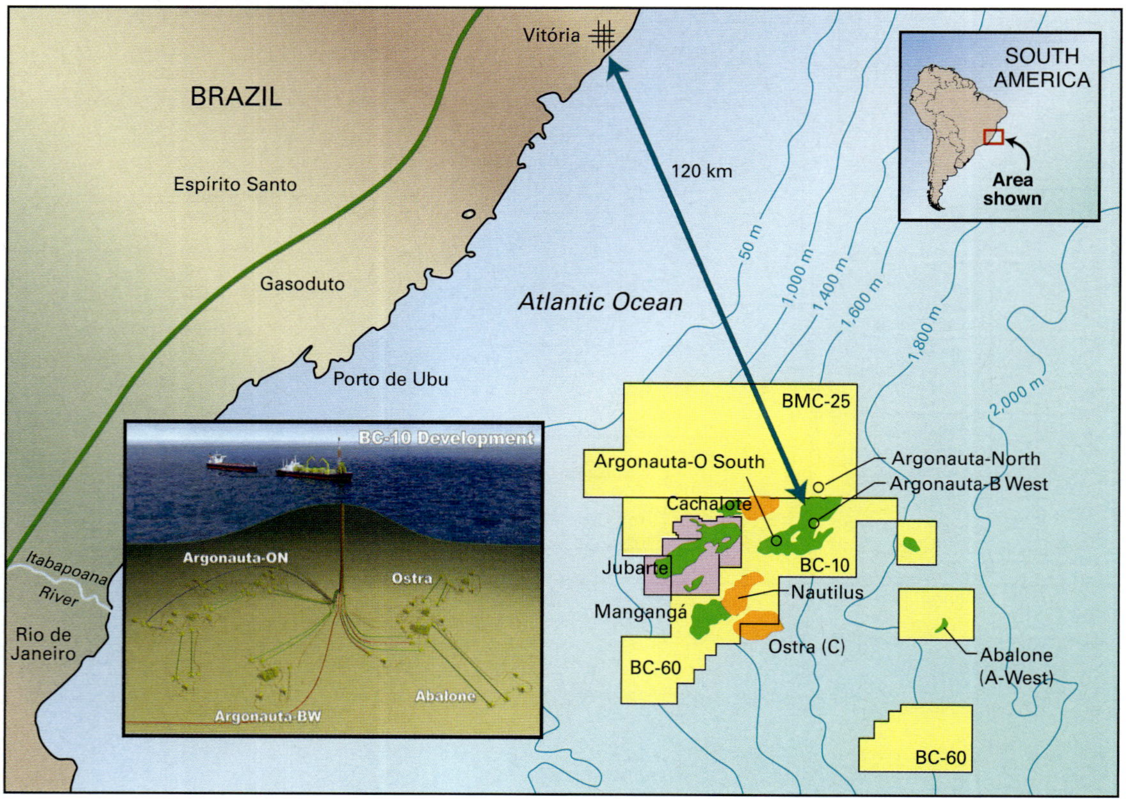

Campos basin

Campos is one of five major basins off Brazil, located between the Espirito Santo basin to the north and the Santos basin to the south. Campos stretches from the State of Espirito Santo south to Cabo Frio, off the northern coast of Rio de Janeiro State.

Campos encompasses several major fields: Albacora, Albacora Leste, Barracuda Caratinga, Bijupira-Salema, Marlim, Marlim Sul, Marimba, Marimba Leste, Roncador, and Jubarte.

Industry believed the basin held the largest oil reserves on the Brazilian shelf, until Petrobras recently discovered the giant Tupi oil field in Santos basin.

Shell has made technical discoveries in four areas of BC-10: Abalone, Argonauta, Ostra, and Nautilus.

All four fields are low pressure and require boosting. Abalone and Ostra contain gas and require subsea separation. Argonauta has less gas and no need for the caisson separation.

First oil from Siri field

The FPSO vessel *Petrojarl Cidade de Rio das Ostras* received first oil in early 2008 from the Siri field, operated by Petrobras offshore Brazil.

The vessel also serves as a test facility for heavy oil production, providing data to help optimize production techniques. It can store up to 214,000 bbl of crude oil and can produce 15,000 b/d of stabilized crude oil.

LNG

First-ever imports

To help cut dependence on imports from Bolivia, Petrobras announced midyear 2008 the signing of two contracts to buy LNG from BG Group PLC (Reading, UK).

The agreements mark Brazil's conversion to an importer of LNG but will give the country more flexibility to meet growing gas demand for electric power.

BG LNG is directed to the import terminals Petrobras is building at Pecem Ceara state) and Guanabara Bay (Rio de Janeiro state) with 75,000–130,000 cu m/day initially.

The first contract provided the first LNG cargo ever delivered into Brazil; it was used to commission the Pecem terminal in July.

The second will supply LNG to either the Pecem or Guanabara terminals on a schedule established by Petrobras.

Petrobras was forced out of Bolivia in 2006 after Pres. Evo Morales nationalized his country's gas industry. Petrobras still depends on Bolivia for 60% of the gas it uses.

In March, Bolivia was producing less than 42 MMcmd and struggling to both supply domestic needs and meet its contracts with Brazil (30 MMcmd) and Argentina (7 MMcmd).

Pipelines

Gascav, Gasene pipelines

The Gascav natural gas pipeline, owned by Petrobras and operated by its subsidiary, Transpetro, began operation in February 2008, after completion by China's Sinopec (Beijing).

The $238-million pipeline runs 303 km (188 mi) from the Cabiunas terminal in Rio de Janeiro to Vitoria in the Espirito Santo state.

This section is the first of the three legs in what Petrobras calls the Gasoducto de Interconexion Sudeste-Nordeste (Gasene) Pipeline, a 1,371-km (852-mi) line to connect Brazil's southeastern and northeastern areas and increase gas transportation capacity.

In May 2008, Petrobras and Brazilian officials inaugurated construction on the third (and longest) leg of the Gasene Pipeline. To be completed by mid-2010, it will be 953 km (592 mi) long and is estimated to cost about $2.5 billion. The new section will connect Cacimbas (Espirito Santo state) to Catu (Bahia state).

Pipe-in-pipe gas line

Construction was under way on a 21-km (13-mi) subsea gas pipeline by Technip (Paris) for Petrobras.

It will connect the Canapu field—in 1,700 m (5,577 ft) of water—to the Cidade de Vitoria floating production facility, which is in 1,400 m (4,593 ft) of water.

It will be the first pipeline in Brazil to use a pipe-in-pipe design (e.g., production pipe surrounded by carrier pipe with insulation in between). It was to be completed by early 2009.

Ethanol pipelines

Petrobras moved ahead with plans in mid-2008 to build two ethanol pipelines. One could be operational by yearend 2009. The lines support the firm's program to export ethanol, mainly to Japan.

One line would run from Senador Canedo in Goias state to Sao Sebastiao, the port in Sao Paulo state, carrying 12 million cu m/year of ethanol made from sugar cane.

The second, for both ethanol and biodiesel, would connect Cuiaba in Mato Grosso state, with Paranagua in Parana state.

The more advanced is the $1 billion, 1,150-km (715-mi) line from Goias to the Paulinia Terminal in Sao Paulo state. A JV agreement between Petrobras, Mitsui & Co. (Tokyo), and Brazilian construction company Camargo Correa (Sao Paulo) has created PMCC Ethanol Transport Projects SA.

For the second planned line, viability and environmental impact studies were under way in late 2008.

In April 2008, Brazil Renewable Energy Co. (Brenco; Sao Paulo) said it plans to invest $1 billion to build a 1,100-km (684-mi), 4-million l./year ethanol pipeline from Alto Taquari in Mato Grosso state to Santos—the nation's largest port—in Sao Paulo state on the country's south Atlantic seaboard.

Refining

Premium refineries planned

In mid-2008, the Ministry of Mines and Energy announced that a "premium" refinery, with a capacity of 600,000 b/d, is to be built at Porto de Itaqui, in Maranhão state.

Petrobras also was exploring with the government of Ceará state the possibility of installing a second premium refinery there, with a processing capacity of 300,000 b/d.

The two plants together would cost $20 billion and would focus on the export of oil products.

Refinery upgrade: Sao Paulo

Petrobras awarded a $125 million contract to Skanska AB (Solna, Sweden) to build a sulfur recovery unit (SRU) and a tail gas treatment unit (TGTU) at its 47,000 b/d Capuava refinery in Sao Paulo, Brazil.

The project is part of a nationwide program to reduce sulfur contaminants in petroleum byproducts.

The SRU facility will recover 20 tpd of sulfur; the TGTU will clean 40 tpd of tail gas.

Work was slated to begin in October 2008 and to be completed in 24 months.

Alternate Fuels

Ethanol joint venture

Petrobras and Mitsui & Co. (Tokyo) announced plans in early 2008 to form a 50:50 joint venture for making investments in renewable energy.

The new JV, called Participacoes Nippo Brasileira em Complexos Bioenergeticos SA, will provide ethanol for the Japanese market.

Company News

Ultrapar-Chevron deal

Two subsidiaries of Chevron Corp. (San Ramon, CA) agreed to sell Chevron's fuels marketing business in Brazil to a subsidiary of Ultrapar Participacoes SA (Sao Paulo).

Ultrapar will acquire 2,000 service stations operating under the Texaco brand, an equity interest in associated terminal operations, and Chevron's commercial and industrial fuels business.

Under the Texaco brand, Chevron sells an average of 120,000 b/d of fuel in Brazil.

Eni, Petrobras cooperation

Eni SPA (Rome) and Petrobras renewed their commitment in mid-2008 to work closely together on upstream and downstream operations and to conduct feasibility studies on renewable fuels in Brazil and elsewhere.

Eni will offer its exclusive slurry technology to convert residues and heavy oils, typical of those produced in Brazil.

Both companies will study the valuation of the natural gas reserves recently discovered by Eni off Brazil, particularly in the Santos basin.

Eni has four offshore exploration blocks in Brazil and a 30-year concession granted in 1999 to distribute natural gas in the northwest area of São Paulo state.

CHILE

Capital: Santiago
Monetary unit: Peso
Refining capacity: 226,800 b/cd
Oil production: 2,300 b/d
Oil reserves: 150.00 million bbl
Gas reserves: 3.46 tcf

Overview

Chile has one of South America's most robust and open economies. In 2007, its real GDP grew by an estimated 5.2%, continuing several years of strong growth, according to the *World Factbook* of the U.S. Central Intelligence Agency.

However, the country has limited domestic energy resources and must import the bulk of the energy it uses.

Exploration and Development

Fell block wildcat

GeoPark Holdings Ltd. (Hamilton, Bermuda) discovered two new gas fields and an oil field in the Fell block as of mid-2008.

In production tests, the Aonikenk 1 oil exploration well tapping the Springhill formation at 2,270 m (7,448 ft) flowed 1,201 b/d of oil and 1.2 MMcfd of gas.

Analyses suggested opportunity for further drilling in the field and GeoPark said it planned to do so in fourth quarter 2008.

The firm earlier discovered two gas fields on the Fell Block with wells Nika Oeste-3 and Bump Hill-1.

Nika Oeste-3—drilled to 3,023 m (9,918 ft) TD—flowed 133,089 cu m/d of gas and 70 b/d of condensate from an interval at 2,960 m (9,712 ft) under test.

The Bump Hill-1 well was drilled to 2,979 m (9,774 ft) TD. It produced test flows of 25,485 cu m/day of gas and small amounts of condensate.

Tranquilo Block, Magellan basin

The government signed a special operating contract in April 2008 on the 6,700-sq km (2,587-sq mi) Tranquilo Block in the Magellan basin with Improved Petroleum Recovery Traquilo Chile (Dallas) and Manas Petroleum Corp. (Baar, Switzerland).

IPR and Manas committed to a three-phase, $33.24 million exploration program. $14.36 million is to be spent in the first three years for 2D and 3D seismic and to drill six wells.

Tranquilo is the largest of 10 blocks awarded in Chile's November 2007 licensing round. IPR, operator, and Manas hold the block 50-50.

The block's petroleum system is proved with Tranquilo gas field and with the Esperanza and Manzano wells.

A 417-km (259-mi) network of gas pipelines connect the main fields to consumption centers, such as Methanex (the world's largest methanol plant). Also, ENAP has four gas processing plants in the basin.

Tamarugal basin

March Resources Corp. (Calgary) had only very modest noncommercial gas flows at its first well in northern Chile's Tamarugal basin.

The company perforated 16 intervals in four zones to evaluate gas shows and drilling breaks in the Cerro Empexa and Cretaceous Guatacondo formations at the Pica-1 well.

The well's main targets were at 2,800 m (9,187 ft) to TD 3,110 m (10,204 ft), but two perforated intervals between 2,000 m (6,562 ft) and TD yielded no pressure increase or gas to surface.

For seven intervals perforated at 1,500–1,800 m (4,922–5,906 ft), some gas surfaced but wellhead pressure was too small to measure.

The Pica-2 well was being drilled at 1,450 m (4,758 ft) in Cerro Empexa.

Lease Sales

Magallanes blocks awarded

Chile signed contracts with several IOCs in May 2008 for exploration of blocks in the southern Magallanes region.

- Apache Corp. (Houston): Russfin, Lenga blocks
- Pan American Energy LLC (Buenos Aires): Coiron Block
- Greymouth Petroleum Holding Ltd. (Auckland, New Zealand): Porvenir, Brotula, Isla Magdalena, Caupolican blocks
- IPR-Manas (Baar, Switzerland): Tranquilo block

Collectively, these firms will invest some $220 million for seismic surveys and exploratory drilling over 3–7 years. Work was expected to begin by yearend.

Brotula and Isla Magdalena are offshore; the remaining blocks are onshore.

Chile's state-run ENAP holds a 50% stake in the Coiron, Caupolican, and Lenga blocks. The other blocks will be 100%-held by the winning companies.

A consortium of Wintershall (Kassel, Germany), GeoPark Holdings Ltd. (Hamilton, Bermuda), and Methanex Corp. (Vancouver, BC) later in the year was awarded the onshore/offshore Otway block. That block had initially been won by Total SA (Paris) but was not pursued by that firm.

Drilling and Production

Otway Block

In July 2008, the Ministry of Mining gave the go-ahead to a consortium to drill in the Otway Block in the Magallanes region.

The consortium of Wintershall (Kassel, Germany), GeoPark Holdings Ltd. (Hamilton, Bermuda), and Methanex Corp. (Vancouver, BC)—collectively "WGM"—had won exploration rights earlier in the year.

WGM said it plans to invest about $30 million on the Otway Block to acquire 2D and 3D seismic data and to drill seven exploratory wells. Chile's state-run ENAP said it plans to invest $300 million.

LNG

Mejillones terminal

GNL Mejillones (GNLM)—a 50-50 JV of GDF Suez Energy International (Brussels) and Chile's national copper company Corporación Nacional del Cobre de Chile (Codelco; Santiago)—obtained a needed environmental permit in early 2008 for a planned LNG regasification terminal in Mejillones in northern Chile.

The terminal will have a planned annual send-out capacity of 5.5 MMcm of gas, sufficient to produce 1,100 MW of electricity.

GNLM planned to start construction by mid-year, with delivery of gas expected to start by early 2010.

COLOMBIA

Capital: Bogota
Monetary unit: Peso
Refining capacity: 285,850 b/cd
Oil production: 578,000 b/d
Oil reserves: 1.355 billion bbl
Gas reserves: 3.739 tcf

Overview

Colombia is an important petroleum producer, although political unrest and stagnant reserves have led to decreased exports in recent years. According to the U.S. Energy Information Administration (EIA), the country aims to boost hydrocarbon exploration to preserve its status as a net oil exporter in the longer term.

Foreign investment in Colombia's petroleum sector is expected to top $5 billion in 2008 compared with $3.5 billion in 2007.

The country's Agencia Nacional de Hidrocarburos (ANH) said it hoped to sign 70–80 exploration and production contracts by yearend compared with 56 last year.

ANH said operators drilled more than 70 wildcat wells in 2007 and that it wanted to boost that figure to 90 in 2008.

Longstanding civil conflict has taken a toll on the country's energy sector. The frequency of attacks against energy infrastructure has declined by 80% since 2002, said government officials, but pipelines and power lines still suffer damage from insurgent groups.

Oil

Colombia's oil production has declined steadily since 1999, when it peaked at 830,000 b/d, due to a lack of sizable new reserve discoveries. Average production was 525,000 b/d in 2007.

Much of the country's crude is lighter and sweeter than that of other major Latin American producers.

All oil production is overseen by state-owned Ecopetrol in contracts of association with foreign companies. Colombia's oil licensing agency ANH has said Ecopetrol aims to raise output to 700,000 b/d by 2015 and to 1 million b/d by 2020.

To attract oil companies, Ecopetrol has cut its mandatory stake to 50–55% in any exploration contract with a private-sector company—down from the 70% required in previous decades.

Natural gas

Colombia's gas reserves are spread across 18 basins, seven of which have active production. Most reserves are in the Llanos basin, although the Guajira basin accounts for most of current production.

The government's Natural Gas Mass Consumption Plan aims to increase domestic natural gas use and establish Colombia as the "gas hub" for the Andean region.

Exploration and Development

Eni, Ecopetrol agreements

Eni (Milan) signed two agreements with Ecopetrol in late 2008. They relate to creating opportunities (a) for Ecopetrol to farm in to Eni's current and future exploration portfolio and (b) for Eni to gain access to potential hydrocarbon projects in Colombia and other South American countries.

Guachiria Block

Solana Resources Ltd. (Calgary) reported an average flow of 5,645 b/d of 41° API oil from the Los Aceites-1 discovery well on the Guachiria Block. The well was drilled in late 2008 in the Tertiary Carbonera formation in the Llanos basin.

Carbonera was drilled to TD of 2,166 m (7,108 ft) and perforated at 2,095–98 m (6,874–84 ft).

After an extended test, Solana expects to build facilities at nearby Primavera-1, site of the company's other 2008 discovery on the block.

Rio Magdalena Block

The Popa-2 exploration well in the Middle Magdalena basin found gas-condensate in the Upper Cretaceous Monserrate formation, said Gran Tierra Energy Inc. (Calgary).

The well went to TD 2,635 m (8,646 ft) in basement on the Rio Magdalena Block.

Drill stem tests at several depths yielded gas flow rates of 1.40 MMcfd to 3.95 MMcfd and 74 to 85 b/d of condensate. The block has numerous other leads and prospects.

Chaza Block

Gran Tierra Energy Inc. (Calgary) finished drilling the Costayaco-4 well in Costayaco field on the Chaza Block in mid-2008. The well reached 2,708 m (8,884 ft) TMD.

The firm said log data indicate potential hydrocarbon pay in the Kg sand unit of the Rumiyaco formation, the U sandstone unit of the Villeta formation, the T sandstone unit of the Villeta formation, and the Caballos formation.

Upper Magdalena Valley basin

Ecopetrol made plans to evaluate an oil and gas discovery in Huila province.

The Tempranillo-1 well was drilled in the Upper Magdalena Valley basin and is part of Ecopetrol's wholly owned 10,184-hectare Brisas-Lomalarga-Dina-Potrerillo Block.

On test, crude flowed at 1,600–2,400 b/d and gas at 2.0–2.75 MMcfd.

Tempranillo-1 is one of 20 wells that Ecopetrol planned to drill in 2008.

Lease Sales

Onshore northern Colombia

The so-called Colombian licensing round, launched in February, offered 43 blocks for exploration on three onshore areas in northern Colombia along the border with Venezuela north of Bogotá.

The acreage is in the natural gas-prone areas of Sinu North and Cesar Rancheria area, as well as the frontier Cordillera area and Crudos Pesados West, a heavy oil region.

Contracts with successful applicants were expected to be signed by October.

Second Ordinary Round

In mid-2008, the government announced an auction for more than 100 blocks. Dubbed the Second Ordinary Round 2008, it offered mostly heavy-oil blocks in the Upper Magdalena basin, the Middle Magdalena basin, the Llanos basin, and the Putumayo basin.

East Llanos basin: heavy oil

Colombia's ANH invited 20 companies to submit formal proposals in May 2008 to develop eight greenfield heavy-oil blocks in the East Llanos basin.

The companies include ExxonMobil Corp. (Irving, TX), Royal Dutch Shell PLC (The Hague, Netherlands), Nexen Inc. (Calgary), and OAO Lukoil (Moscow).

ANH expects to select four to eight companies and sign agreements with them by July.

The winners will have three years to evaluate a large area and then choose a more limited area in which to seek exploration contracts.

Colombia separately offered 17 heavy-oil blocks under its international bidding round to potential operators to sign exploration contracts by October 2008.

Drilling and Production

Vigia field

Emerald Energy PLC (London) announced plans in September 2008 to drill another development well in the Vigia field on the Campo Rico Association Contract.

Vigia-4ST will be drilled directionally to the Une and Gacheta reservoirs south of the Vigia-5 well.

The company said its Vigia-6 development well found oil in both reservoirs and initially flowed 850 b/d of oil under test from clean sandstones in Gacheta.

Refining and Products

Barrancabermeja upgrade

Ecopetrol SA awarded contracts in late 2008 for FEED and project management consultancy for the Barrancabermeja refinery upgrade project in Colombia.

The project will increase refining capacity to 300,000 b/sd from 250,000 b/sd, add heavy crude processing capability, and provide a processing configuration to meet the projected 2013 Colombian clean fuels product specifications.

Pipelines

Rubiales field link

Ecopetrol SA and Pacific Rubiales Energy Corp. (Toronto) awarded a contract in third quarter 2008 for construction of a 235-km (146-mi) pipeline to connect the Rubiales heavy oil field in east-central Colombia to the OCENSA pipeline.

The Oleoducto de los Llanos Orientales will extend from Rubiales field on the Rubiales and Piriri association contract areas in Meta Province to the OCENSA Monterrey station of the Oleoducto Central SA (OCENSA) pipeline in Casanare Province. Completion is set for third quarter 2009.

The line's initial 170,000-b/d capacity could be expanded to 260,000 b/d to accommodate further developments and exploration. The line is seen as supporting future development of the Los Llanos heavy oil basin.

International Activities

India-Colombia agreement

India signed a memorandum of understanding with Colombia to enable bilateral cooperation in oil and gas exploration, including the training of technicians and professionals.

The pact is expected to facilitate participation of Indian firms in oil exploration in Colombia, improve oil recovery from existing fields, train Colombian technologists in India, and establish joint biofuel projects. It will also explore potential investment by Indian firms in Colombia's refining sector.

India's ONGC Videsh Ltd., a subsidiary of Oil & Natural Gas Corporation Ltd. (New Delhi) already has invested in three offshore deepwater blocks on Colombia's Caribbean coast.

Reliance Industries Ltd. (RIL; Mumbai) also has two offshore drilling blocks in Colombia.

A third firm, GAIL (India) Ltd., has expressed interest in a possible major project with RIL for a petrochemical plant and a gas pipeline network in Colombia.

Neighborhood friction

Colombia's hopes to raise its oil production to 1 million b/d by 2020 could be hindered by the actions of neighbors Ecuador, Venezuela, and Nicaragua, said *Oil & Gas Journal*.

One irritant is the left-wing Revolutionary Armed Forces of Colombia, or FARC, suspects in attacks on Colombia's 60,000-b/d trans-Andean pipeline (operated by Ecopetrol SA) in southern provinces near the border with Ecuador.

The FARC forces, though Colombian, are based in Ecuador, but Ecuador says it has no responsibility for their actions.

Colombia conducted a dramatic raid in March 2008 on a FARC base in Ecuador that freed hostages and killed a key rebel leader. It seized documents that were said to show financial support for FARC from Venezuela and coordination with Ecuadorian officials.

The raid angered the Ecuadoran government as well as its left-leaning allies in Venezuela and Nicaragua. The three made hostile gestures toward Colombia and criticized the country for its relationship with (and oil exports to) the United States.

CUBA

Capital: Havana
Monetary unit: Peso
Refining capacity: 301,400 b/cd
Oil production: 50,000 b/d
Oil reserves: 124.0 million bbl
Gas reserves: 2.5 tcf

Overview

Cuba uses more than 150,000 b/d of oil, and buys approximately 100,000 b/d of it from Venezuela at highly discounted prices.

By some estimates, Cuba has allocated $1.7 billion to its energy industry since 2004, working with a variety of international partners.

Two Canadian firms—Sherritt International Corp. (Toronto) and Pebercan Inc. (Montreal) —together have been producing about 60% of all Cuban output, in collaboration with the country's national oil company, Unión Cubapetróleo (Cupet).

In 2006, Sherritt reported average oil production of 30,000 b/d from its 10 production sharing contracts. Pebercan (Montreal) produces another 19,500 b/d of oil from Block 7.

Repsol-YPF SA (Madrid), India's ONGC Videsh, Petrobras (Rio de Janeiro), and StatoilHydro AS (Stavanger, Norway) all were making plans to explore for oil beginning in 2008.

Vietnam's state-owned Petrovietnam also was expected to become more active in the near future, following a cooperation agreement signed in late 2006 with Cupet.

Cuban officials continued to seek additional foreign investment, aiming to raise oil and gas production to 100,000 boe/d by 2009.

On 10 of Cuba's 59 Gulf of Mexico oil and gas exploration blocks, more than 25 oil fields have been discovered and prepared for drilling. The largest, Varadero, has reserves estimated at 3 billion bbl.

There is also an additional projected reserve of 2.8 billion bbl in the southeast section of the Gulf of Mexico, just within Cuba's Exclusive Economic Zone.

Exploration and Development

Block 37

Brazil's Petrobras (Rio de Janeiro) and Companhia Cubana de Petróleo (Cupet) signed a joint oil exploration and production agreement in late 2008 for Block 37 offshore northern Cuba.

Boundary action urged

U.S. Sen. Bill Nelson (D-Florida) asked Pres. George W. Bush in early 2008 to recall a diplomatic note to Cuba that delineates Gulf of Mexico boundaries.

Cuba-U.S. boundaries are delineated by the 1977 Maritime Agreement between the two nations and are typically reiterated every two years through exchange of diplomatic notes.

He requested the action based on an agreement made in January agreement between the national oil companies of Brazil and Cuba.

Petrobras and Cuba's Cupet foresee "assessment of the offshore blocks in the Cuban sector" of the gulf, Petrobras said, and envision cooperation on refining, maintenance, research and development.

Sen. Nelson expressed alarm at increasing activity by several nations in gulf waters between Cuba and Florida, pointing to recent initiatives and expressions of interest by firms in Spain, Canada, India, and China.

He warned of potential damage from oil development to Florida's marine sanctuaries and coastal tourism industry.

The block, in the Florida Straits, is in 500–1,600 m (1,640–5,249 ft) of water between Havana and Matanzas, about 145 km (90 mi) south of Key West, Florida.

Located near Cuban onshore oil reserves, the block covers 1,600 sq km (618 sq mi) and extends 3–12 km (1.8–7.5 mi) off the coast.

Each of four exploration phases will span 18 to 24 months.

In early 2008, Petrobras signed several cooperation agreements with Cupet. This marked Brazil's first involvement in the Cuban oil and gas sector, which Venezuela had previously dominated.

Seismic on Petrovietnam blocks

Petrovietnam Exploration Production Corp. (Ho Chi Minh City) let a contract to SCAN Geophysical ASA (Oslo) to shoot 4,000 line-km of 2D seismic in the Gulf of Mexico off Cuba.

The work was slated for midyear on Petrovietnam's Blocks N31, N32, N42, and N43 west of the island.

Brazil: Inroads into energy sector

Brazil's Petrobras signed agreements with Cuba's Cupet in January 2008 for cooperation in oil and gas exploration and production and other areas.

Petrobras, which has expertise in deepwater exploration and production, said the agreement "foresees the assessment of the offshore blocks in the Cuban sector of the Gulf of Mexico…"

Cuba hopes discoveries in the deepwater gulf will enable the country to become self-sufficient in oil production. The USGS says Cuba's gulf areas could contain 4.6–9.3 billion bbl of crude oil and 9.8–21.8 tcf of gas.

Brazil's official news agency said the agreements coincide with political transition in Cuba following transfer of power in mid-2006 by Cuban Pres. Fidel Castro to his brother Raul, due to health problems.

Agencia Estado said Brazil wants to be in a positive position with Cuba "when trade opens up as part of Fidel's succession process."

Venezuela: Major trading partner

Venezuela—Cuba's biggest trading partner in South America—has until recently blocked Brazil's attempts (see above) to gain a foothold in the island nation.

Since 2003, Pres. Hugo Chavez has been selling petroleum to Cuba at subsidized prices in exchange for physicians, nurses, and hospital equipment.

In December 2007, at the PetroCaribe summit in Havana, Pres. Chavez and Cuba's Raul Castro signed 14 agreements in the energy, mining, and oil sectors. These included a $122 million loan for Cuba to buy tanker ships to transport crude oil and oil products from Venezuela.

Plans also call for Venezuela and Cuba jointly to more than double the capacity of Cuba's Cienfuegos refinery. Cuba also plans to reopen an oil pipeline from the refinery to Matanzas.

Drilling and Production

Sherritt activities

Sherritt International Corp. (Toronto) finalized a production-sharing contract for a $15-million carbon dioxide enhanced oil recovery pilot project in Varadero field, east of Havana.

The firm, which also signed a production sharing contract on Block 8 in south-central Cuba, said it expected its 2008 gross working interest oil production in Cuba to be consistent with the 2007 average of 30,637 b/d.

Refining

Cienfuegos refinery on line

In January 2008, processing of crude oil began at Cuba's revitalized 65,000 b/d Cienfuegos refinery. Gasoline output began in March.

PDV-Cupet SA—a joint venture of Venezuela's PDVSA and CubaPetroleo—owns the refinery, located in the port city of Cienfuegos, 250 km (155 mi) southeast of Havana.

The facility is configured to produce fuel oil, diesel, gasoline, and jet fuel, as well as LNG for the domestic market and for export to other countries in the Caribbean region.

Plans call for the refinery eventually to process 150,000 b/d of crude and supply feedstock for a new petrochemical facility in the area. The $1.3 billion expansion is slated to start in 2009 and be completed in 2013.

ECUADOR

Capital: Quito
Monetary unit: U.S. dollar
Refining capacity: 176,000 b/cd
Oil production: 500,000 b/d
Oil reserves: 4.660 billion bbl
Gas reserves: 315.0 bcf

Overview

Ecuador has oil reserves of about 4.5 billion bbl, the third largest in South America. The country's oil production was estimated at 500,000 b/d in 2007, making Ecuador the fifth-largest producer in South America.

The state-owned petroleum firm is Petroleos de Ecuador (Petroecuador), and its production arm is Petroproduccion.

About a dozen foreign oil companies had contracts and assets in Ecuador in 2006, including Brazil's Petrobras, Repsol-YPF SA (Madrid), EnCana Corp. (Calgary), Agip SFA (Milan, Italy), Perenco SA (Paris), and Occidental Petroleum Corp. (Los Angeles).

The largest refinery in Ecuador is the 110,000-b/d Esmeraldas facility on the Pacific coast.

Exploration and Development

Offshore seismic

Empresa Estatal de Exploratión y Producción de Petróleo del Ecuador was working in late 2008 with SCAN Geophysical ASA (Oslo) to arrange a 2D seismic program offshore.

Ecuador's supergiant heavy oil field

The program will cover most of the Ecuadorian coast, but will focus on Esmeraldas, Manabí, Guayas, and Santa Elena provinces.

Pungarayacu heavy oil appraisal

Ecuador's state-owned Petroecuador and Petroproduccion signed a 30-year contract in October 2008 with a subsidiary of Ivanhoe Latin America to explore and develop Block 20 on the western edge of Ecuador's Amazon basin, 200 km (125 mi) southeast of Quito.

Block 20 contains the 648-sq km (250-sq mi) Pungarayacu heavy oil accumulation. The contract covers project appraisal and the development, including producing and upgrading the heavy oil.

Evaluation of Pungarayacu during the 1980s by Petroproduccion, ARCO and others estimated that the field contains 4.5–7 billion bbl of oil in place.

Ivanhoe said confirmation of these resources "would make the Pungarayacu field the largest accumulation of heavy oil in Ecuador and one of the largest in Latin America."

Ultimate development cost is expected to be several billion dollars.

Petroecuador officials estimated that Ivanhoe could start producing about 108,000 b/d of heavy oil from Pungarayacu in 2009, but this was seen as optimistic.

Guayaquil gulf

Petroecuador announced in July 2008 the creation of a joint company with Chile's ENAP to explore for natural gas in the Guayaquil gulf, off the southwestern coast of Ecuador. Surveys have indicated the possibility of significant natural gas deposits.

Petroecuador is also involved in a similar undertaking with Venezuela's PDVSA, with the purpose of searching for gas in an adjacent area in the gulf.

ENAP and Petroecuador will explore in Block 40 and with PDVSA in Block 5.

In both cases, Petroecuador is expected to hold a majority stake, although with ENAP, discussions were ongoing and the percentages were still to be agreed upon.

Policy

New contract terms for IOCs

Ecuador's Pres. Rafael Correa in January 2008 gave IOCs operating in his country—City Oriente (Panama City-based but U.S. owned), Petrobras (Rio de Janeiro), Perenco (Paris), Repsol-YPF SA (Madrid), and Andes Petroleum (owned by China's CPNC)—45 days to accept new contract agreements.

The government was seeking to change current joint ventures--which allow IOCs to include Ecuadorean reserves on their balance sheets—to subcontracting deals, which do not.

In 2007, the government unilaterally altered the allocation of the windfall produced by the rise in international oil prices. The government now takes 99% of the windfall and the companies get 1%—a decision the companies are fighting in the courts.

Companies could accept the 99% decree, accept the new contracts, or leave the country. If they leave, Correa said, the government will reimburse them for their investment and turn the fields over to Petroecuador.

As of October, Ecuadorian officials reported that Petrobras, Perenco and Andes had signed the new contracts, that Repsol-YPF was close to doing so, and that City Oriente had left the country.

Drilling and Production

Cooperation with Venezuela

In what officials called a sign of regional oil and gas "integration," Ecuador and Venezuela made plans in 2008 for several cooperative projects.

- Petroecuador said that it will use two Venezuelan rigs to drill the Guanto Dureno oil well in the province of Lago Agrio. Venezuelan Pres. Hugo Chavez offered the two rigs and related technicians at a discount, covering all expenses for maintenance, operations, and administration.
- Petroecuador and PDVSA formed the JV company, Rio Napo, to boost output from the Sacha oil field in Ecuador's Amazonian jungle from 40,000 to 70,000 b/d. Reserves are estimated at 480 million bbl.
- The two national oil companies also announced plans to build a new refinery in the province of Manabi (see details below).

Refining

Iran, Ecuador refining pact

Iran and Ecuador signed an MOU under which Iran will help refurbish and upgrade Ecuador's three refineries. The agreement also envisions construction of an oil refinery to be co-financed by Iran, Ecuador and Venezuela.

Iran and Ecuador additionally agreed to construct a petrochemical production unit in Ecuador, and the two countries will likely establish an oil company to implement the joint energy projects.

The agreement also covers training of Ecuador's oil industry employees by Iranian experts and cooperation in maintenance on Ecuadorean oil facilities.

New refinery in Manabi

Petroecuador and PDVSA made plans for a joint venture company to build and operate a 300,000-b/d refinery on Ecuador's Pacific Coast in Manabi province.

Petroecuador said midyear that it will ask foreign companies to bid on constructing the Pacific Refinery-CEM, requiring them to finance 70% of its construction. No start date was specified.

Pres. Rafael Correa previously estimated that the proposed refinery would cost $5 billion, increasing to $10 billion if the partners decided to add a petrochemical plant on the site.

Refinery construction would take about four years. Initial crude supplies would come from Ecuador.

Pipelines

Subsea CP survey

A division of Cathelco Ltd. (Derbyshire, UK) completed in early 2008 a major cathodic protection (CP) survey on the subsea section of the Oleoducto de Crudos Pesados (OCP) off Ecuador.

The subsea section consists of four pipelines from storage tanks at the Esmeraldas marine terminal to two coupling buoys 7 km (4 mi) offshore in 35 m (115 ft) of water. Each line can transport up to 60,000 bbl/hr.

The work was undertaken to satisfy insurance requirements for OCP Ecuador SA, the pipeline's operator.

The 450,000 b/d mainline portion of the OCP runs 500 km (311 mi) from the Amazonas terminal (Nueva Loja, Sucumbios), across the Andes, to Esmeraldas on the coast.

GUATEMALA

Capital: Guatemala City
Monetary unit: Quetzal
Refining capacity: 0
Oil production: 14,000 b/d
Oil reserves: 83.07 million bbl
Gas reserves: 0

Guatemala and Belize are the only oil-producing countries in Central America.

Most of Guatemala's oil is produced from fields in its northern jungles, adjacent to the Mexican border. The government had planned to hold a licensing round in 2008 for acreage in the offshore Pacific basin, but no action had been taken as of the fourth quarter.

Perenco (Paris) produces the majority of the oil in the country. Domestic refining capacity is meager and almost all of their production is exported to the United States. The country must import petroleum products.

In 2006, the United States imported an average of 15,000 bbl/d of crude oil from Guatemala.

Also that year, the Mexican government outlined potential plans for a large new Central American refinery, in either Guatemala or Panama. Discussions continued into 2008.

PetroCaribe membership

In the face of sharply rising energy prices, the Guatemalan government considered exploring new oil fields and negotiating deals with PetroCaribe and PDVSA, said the hydrocarbon ministry.

Guatemala was admitted as the 18th member of the PetroCaribe alliance in mid-2008.

Led by Venezuela, PetroCaribe was created in 2005 as a regional energy security alliance. It sells oil to Caribbean nations under flexible credit terms.

As of mid-2008, members had received almost 60 million bbl of oil, said Venezuelan officials.

In addition, eight facilities worth $552 million were being built to improve refinery, storage and supply infrastructure.

Alliances with Brazil explored

In early 2008, Pres. Alvaro Colom visited the Petrobras Research & Development Centre (Cenpes) in Rio de Janeiro and with Brazil's Pres. Luiz Inacio Lula da Silva in Brasilia to discuss possible alliances in oil and gas exploration and production as well as in downstream operations and renewable fuels.

JAMAICA

Capital: Kingston
Monetary unit: Dollar
Refining capacity: 36,000 b/cd
Oil production: 0
Oil reserves: N/A
Gas reserves: N/A

Jamaica consumed about 27.8 million bbl of petroleum and petroleum products in 2007, according to the Petroleum Corporation of Jamaica (PCJ), the state energy company. The country depends on imported oil to meet over 90% of its energy needs.

Jamaica is among the Caribbean countries that signed the Petrocaribe Initiative in 2005. Under that program, Jamaica can buy crude oil and refined products from Venezuela under favorable financing terms.

The similar San Jose Accord enables Jamaica to receive oil and refined products from Mexico.

LNG option fades

Jamaica's inability to source LNG forced the government to abandon it as the primary energy source for the country's development, said the Energy Ministry in October 2008.

The government expected to get 1.1 mtpy of LNG from Trinidad and Tobago, starting in 2009. But that deal fell through, as did plans for LNG delivery from Venezuela and purchase of compressed natural gas from Colombia.

The country instead will look to coal as its key fuel for power generation, contingent on agreement from the island's alumina refineries, which use one-third of Jamaica's electricity.

The bauxite companies and power producers have tentatively approved a move to coal.

Among the projects predicated on LNG was a US$1-billion expansion of the Jamalco alumina refinery. The plant was to have a new 80-MW gas-fired power plant that would sell excess electricity to the national grid.

The cost of switching to coal was estimated at US$300 million (J$21.6 billion). The government said it would facilitate construction of coal-fueled power plants by private investors.

Jamaica's national domestic energy grid—not including the alumina sector—now generates about 800 MW, but the government estimates that an additional 210 MW will be needed by 2012.

Minor oil spill

In early November 2008 a small oil spill occurred at Jamaica's only refinery, outside Kingston.

Officials with the state-run refinery, Petrojam, were investigating the incident, apparently triggered when a safety valve exploded.

Built in the 1960s, the refinery processes most of the oil consumed in Jamaica. It refines about 36,000 b/d of crude.

MEXICO

Capital: Mexico City
Monetary unit: Peso
Refining capacity: 1,540,000 b/cd
Oil production: 2.800 million b/d
Oil reserves: 10.501 billion bbl
Gas reserves: 13.162 tcf

Overview

The oil sector generates approximately 40% of government revenues. Most of Mexico's oil production is from the Cantarell field in the Gulf of Campeche, off the southeastern coast.

Cantarell is one of the largest fields in the world, producing 2.0 million b/d of crude oil in 2005.

However, state-owned Petroleos Mexicanos (Pemex) has warned that Cantarell has entered a stage of long-term decline. Production is expected to drop by an estimated 14% per year.

The two other major oil production centers in the Gulf of Campeche are Ku-Maloob-Zaap (KMZ) and Abkatun-Pol-Chuc.

Production from the KMZ complex, adjacent to Cantarell, has risen by 50% since 1996. But for Abkatun-Pol-Chuc, production also has declined steadily, down more than 50% from 1996.

Pemex has predicted that the declines in Cantarell production can be offset by new wells coming online in deepwater Gulf of Mexico and in Chicontepec field in Veracruz state.

But many experts say both areas would require at least a decade of exploration and development before significant production begins.

Important onshore production centers in the southern part of the country include Bellota-Jujo and Samaria-Luna.

Oil

Outlook: Energy ministry

Mexico's energy ministry released the country's oil outlook for 2006–16 in late 2007.

Under the best-case scenario, production levels for Pemex will average some 3.255 million b/d over the period, reaching 3.4 million b/d in 2016.

The best-case scenario will depend on annual investments of some 157 billion pesos as well as new production from Chicontepec and from Gulf of Mexico deep water.

US
Texas
Nogales
Nogales
Cananea
Magdalena
El Paso
Ciudad
Juarez
Rio Grande
30°
Sonora
Hermosillo
Celulosa de
Chuhuahua
Chihuahua
Chihuahua
Delicias
Guaymas
Gulf of California
Baja California
Navojoa
Cuidad
Camango
Coahuila
Ulua
Buena Suerte
Jimenez
25°
Topolobanpo
Sinaloa
Durango
Gomez Palacio
Culiacan
La Paz
MEXICO
Durango
Zacatecas
Mazatlan
San Luis
Potosi
Zacatecas
Chicura
Canuch
Jayabo
15°
Pacific
Ocean
Nayarit
Aguas.
Aguascalientes
San Blas
Tepic
Lagos
20°
Guanajuato
Guadalajara
Salamanca
Puerto Vallarta
Lago de
Chapala
245
Jalisco
Morelia
Colima
Manzanillo
Colima
Michoacan
Texoman
100
0
100
200 Miles
100
0
100
200
300 Km
Lázaro Cárdenas
MEXICO-WEST
15°
110°
115°

Texas
Coahuila
US
Gulf of Mexico
Corpus Christi
Ulua
Buena Suerte
Nuevo Laredo
Nuevo/Laredo
Laredo
Lampazos
Monclova
La Presa
Arcos
Sta. Rosalia
Cuitla
Culebra
Canor
Carretas
McAllen
Reynosa
Mission-Trevino
275
Cadereyta
Reynosa
Brasil
Brownsville
Fortuna
Matamoros
Primavera
Cabeza
Matamoros
Monterrey
Ramirez
S. Bernardo
18 de Marzo
Benavides
Chapul
Huizache
Pobladores
San Fernando
Nuevo Leon
Tamaulipas
Zacatecas
Lerma
MEXICO
Ciudad Victoria
San Luis Potosi
190
Aguas
San Luis Potosi
Barcodon
Tamaulipas-Constituciones
Ciudad Madero
Ebano Panuco Fields
Arenque
Tampico
Cabo Nuevo
Arrecife Medio
Tuxpan
Isla de Lobos
Golden Lane Fields
Tiburon
Esturon
Marsopa
Guanajuato
Guanajuato
Queretaro
Hidalgo
Tuxpan
Salamanca
Queretaro
Bagre
Escuala
315
Poza Rica
Tula
Pachuca
Pienche
Morelia
Tlax.
Veracruz
Mexico City
Tlaxcala
Jalapa
Michoacan
Toluca
D.F.
Veracruz
Remudauera
Cuernavaca
Puebla
Miralejos
Capite
M. Pliancho
Mecayucan
Angastura
Morelos
Puebla
Mirador
Coapa
Cacuite
Macuile
Novillero
Azcapotzelco
Veinte
Mata Verde
Guerrero
Chilpancingo
Oaxaca
Oaxaca
Acapulco
Minatitlan
Cardenas
Tabasco
185
Bay of Campeche
Dos Bocas
Sinán
Citam
Bolontikú
Kix
Yum
May
Ciudad del Carmen
Frontera
Ciudad Pemex
Villahermosa
Tehuantepec
Juchitan
Salina Cruz
330
Gulf of Tehuantepec
Tuxtla
Chiapas
CANTARELL COMPLEX
Tunich
Tabay
Zazil-Ha
Maloob
Zaap
Bacob
Lum
Kutz
Ku
Bacab
Cantarell
Tartunich
Ixtoc
Luch
Batab
Nohoch
Och
Abkatun
Alux
Uech
Pic
Chac
Caan
Kax
Kay
Takin
Chem
Merida
Yucatan
Campeche
Quintana Roo
Campeche
Belmopan
Xan
BELIZE
Caribe
W. Chinaja
1 San Diego
Rubelsanto
Yalpemech
GUATEMALA
Guatemala
HONDURAS
22
San Salvador
Sonsanate
EL SALVADOR
25°
20°
15°
100°
95°
90°
100
0
100
200 Miles
100
0
100
200
300 Km
MEXICO-EAST

Straits of Florida
Great Bahama Bank
Gulf of Mexico
Canasi
Santa Cruz
Puerto Escondido
Yumuri
Boca de Jaruco
Seboruco
Varadero
Guanabo
La Luz
Tarara
Havana
122
Matanzas
Matanzas
Bacuranao
Motembo
2
Cabaiguan
76
Cienfuegos
CUBA
Cristales
Jatibonico
102
Santiago
30°
Yucatan
Cozumel
MEXICO
Quintana Roo
Caribbean Sea
JAMAICA
36
Kingston
Chetumal
BELIZE
Belmopan
Gulf of Honduras
Puerto Cortes
15°
HONDURAS
Tegucigalpa
22
San Salvador
La Union
NICARAGUA
20
Managua
Bluefields
Lake Nicaragua
10°
COSTA RICA
24
Limon
San Jose
Punta Cahuita
Panama City
100 0 100 200 Miles
100 0 100 200 300 KM
Golfito
David
PANAMA
La Palma
CENTRAL AMERICA
85°
80°
COLOMBIA

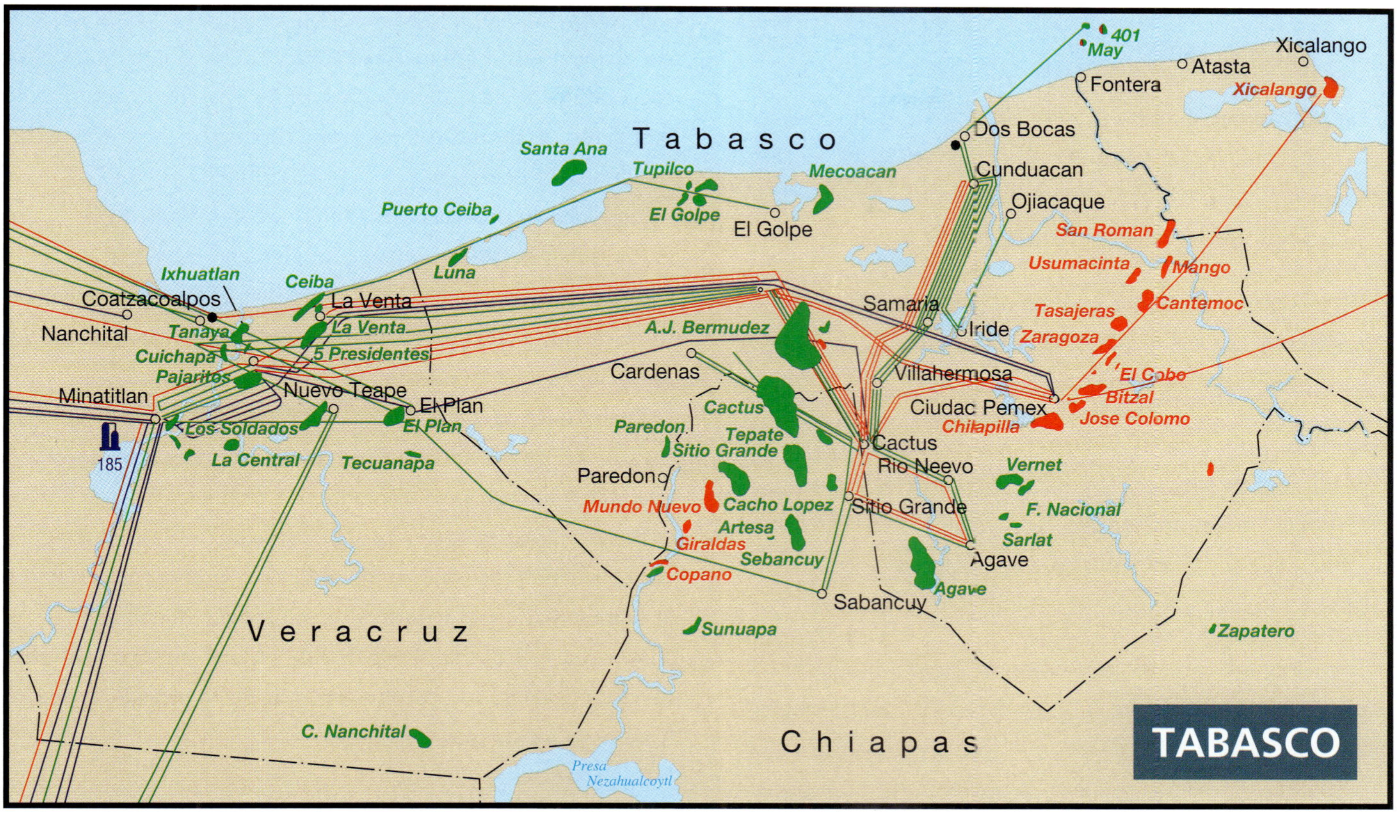

Strengthening of Pemex capabilities and operating capacity was seen as essential to developing deepwater reserves.

Outlook: Pemex

Looking out to 2012, Pemex estimated in December 2007 that the following investment levels would be needed to develop specific reserves:

- Cantarell: $11–14.5 billion
- Chicotepec: $10.5–14.5 billion
- Burgos: $7.5–10.5 billion
- Ku-Maloob-Zaap: $6.5–9 billion.

Much of the new drilling planned by Pemex will be at the Tertiary Gulf Oil project in the northern states of Veracruz and Puebla, where output rose to 41,000 b/d by yearend 2007.

By 2015, Pemex hopes production in the region will rise to nearly 600,000 b/d following investments now pegged at $1.7 billion.

Natural gas

Mexico's natural gas production sites are dispersed. Onshore fields in the north represented 38% of 2005 production, with onshore fields in the south contributing 29%, and offshore fields in the Bay of Campeche contributing the remainder.

The single largest field is Cantarell, in the Gulf of Campeche. Other major fields include Caan, Culebra, and Muspac.

Pemex says that the Southern Region of the country contains the largest share of proven reserves, but that the Northern Region contains almost ten times more probable and possible reserves

Gas production has grown in recent years, but at the same time, consumption also has grown steadily, particularly for power generation.

In 2007, gas consumption was more than 90% greater than it was a decade ago. Mexico imported 480 MMcf/d in 2005.

Pipelines

The extensive Pemex oil pipeline network connects major production centers with domestic refineries and export terminals. More than 450 pipelines span 4,665 km (2,900 miles), concentrated chiefly in the south.

Pemex also operates more than 9,170 km (5,700 miles) of natural gas pipelines in Mexico. The company has 11 natural gas processing centers that produce NGLs, condensates, and LPG.

Oil industry critics

Mexico agreed in mid-2008 to talks with Ejercito Popular Revolucionario, a rebel group that claimed responsibility for major 2007 attacks on oil and gas installations, including the Cactus-San Fernando gas pipeline.

Pemex said the attacks caused no casualties, but some 2,000 businesses in 10 states went without power at a cost of $100 million/day.

Another attack came in 2008 from thousands of demonstrators in Mexico City protesting a government bill to partly privatize Pemex. The reform aims to offer incentive-based oil service contracts and open the refining industry to private capital.

The protest leader (and recent presidential candidate) Andres Manuel Lopez Obrador said the bill would "turn over to private and foreign individuals the exploration, drilling, refining...pipelines and storage of (Mexico's) oil."

The senate energy committee promised a series of national debates on the proposed reforms.

Pemex also operates most of the country's natural gas distribution network, which includes 12 active connections with the United States.

A subsidiary of TransCanada Pipelines Ltd. (Calgary) operates the 130-km (80-mile) Tamazunchale Pipeline, which connects the Pemex gas processing facility in Naranjos to a gas-fired power plant near Tamazunchale.

Tidelands Oil & Gas Corp. (San Antonio, TX) continued to study construction of the 28.3-MMcmd (1-bcfd) Terranova Oriente bi-direction gas pipeline, which would connect a proposed storage facility to the U.S. and Mexican grids.

Exploration and Development

Transboundary gulf reserves

The United States and Mexico began talks in mid-2008 about how to conduct oil exploration operations in the Gulf of Mexico beginning in January 2011, when a 10-year moratorium expires.

A treaty from June 2000 focused on the continental shelf in the western gulf beyond 200 nautical miles. It cited a need for cooperation because of "the possibility that there could exist petroleum or natural gas reservoirs that extend across that continental shelf boundary."

Experts note that there very likely are significant fields in the Perdido Belt between the Western Gap and the coast.

Some Mexican experts suggested that U.S. development on the U.S. side of the gulf boundary could reduce reserves on the Mexican side due to oil migration.

There could be a rush to lease tracts on the U.S. side in 2011 because of significantly higher crude oil prices and improved deepwater exploration and production technology.

Some U.S. experts at an April 2008 forum in Washington questioned whether the issue can be addressed without major reforms at Pemex that could require a constitutional amendment.

Mexican courts may have to rule on any reforms that pass Mexico's congress and become law in order to provide a possible legal basis for contracts.

Drilling and Production

Drilling can expand: Pemex

Pemex said in late 2008 that it could boost exploratory drilling to an average of 1,800 wells/year from the current 600 as the result of new contracting arrangements approved under reform legislation.

More specifically, officials said the changes would enable Pemex to replenish 100% of its reserves by 2012 and begin developing deepwater deposits in 2014–15. For the medium term, a return to output of 3 million b/d was predicted.

The new legislation will enable Pemex to offer incentives to contractors that it said will improve services.

One billionth barrel offloaded

The floating storage and offloading unit Ta'Kuntah—permanently moored in 80 m (262 ft) of water in Cantarell field in the Bay of Campeche—reached an industry milestone in late 2008, offloading a total of 1 billion bbl of oil to more than 1,700 tankers.

Owned and operated by Modec and leased to Pemex, the vessel also marked 10 years of continuous operation on September 14.

The first FSO unit in the Gulf of Mexico, Ta'Kuntah can store more than 2 million bbl of oil.

Oil output drop

Mexico's total oil production fell by 174,000 b/d or 5.3% in 2007 to an average of 3.08 million b/d, said Pemex, due to inclement weather (notably Hurricane Dean) and an expected decline in the Cantarell complex in the Gulf of Mexico.

In the first seven months of 2008, Pemex produced an average of 2.84 million b/d of oil, down 10% from the same period in 2007. This reduced the country's overall oil production and created a sharp decline in exports.

Pemex said the decline was due mainly to the drop in production from Cantarell—down 36% over the 12 months ended July 31. The giant field produced 1.12 million b/d, a figure 472,000 b/d less than during the same period a year before.

Pemex said investment in Cantarell could reach $5 billion in 2008 in efforts to boost production.

Oil decline at Cantarell was partially offset by a 40% increase in production from the Ku-Maloob-Zaap (KMZ) complex. Pemex expects KMZ to reach its highest production level in 2010—averaging 800,000 b/d—then start to decline, along with Cantarell.

Production from the onshore Chicontepec field could compensate for part of the decline in both fields, but observers point out that the terrain will make production from Chicontepec very difficult.

Pemex said overall oil production also was hurt by the need to shut in some wells that have started to produce large amounts of water.

Onshore southern Mexico

Pemex let a three-year, $683 million contract in early 2008 to Halliburton Co. (Houston) to manage the drilling and completion of 58 land wells in the southern region of Mexico.

The contract spans a number of well conditions, including depressurized and high-pressure/high-temperature formations, combined with complex geologies and extreme depths of 3,500–6,500 m (11,483–21,326 ft).

Lower oil exports to U.S. in 2008

Citing in mid-2008 the Pemex affiliate in charge of marketing, *El Universal* newspaper said Mexico will reduce crude oil exports to the United States by an average of 184,000 b/d throughout 2008.

The report said the reduction might continue until 2010 due to Mexico's falling oil output.

The U.S. EIA had earlier predicted a 13% shortfall of imports from Mexico for FY 2008.

The decline in Mexican output is due chiefly to falling production at supergiant Cantarell field, according to EIA.

LNG

Energía Costa Azul terminal

In August 2008, Mexico's Pres. Felipe Calderón and officials of Sempra Energy (San Diego) dedicated Sempra's new Energía Costa Azul LNG terminal in Baja California, Mexico, the first LNG receipt facility on the West Coast of North America.

In operation since May 2008, the $975 million terminal can process up to 28.3 MMcmd (1 bcfd) of natural gas.

The first cargo of imported LNG arrived in April.

LNG will be procured from Shell International Gas Ltd. and from a new liquefaction facility BP and its Tangguh LNG Partners are completing in Indonesia. Shipments from BP are scheduled to begin arriving in second quarter 2009.

Gas processed at Energía Costa Azul will be used in Baja California and the U.S. Southwest.

Manzanillo terminal

In March 2008, Mexico's Federal Electricity Commission (CFE) awarded a $900 million contract to build and operate an LNG receiving terminal on Mexico's Pacific coast.

The award went to Terminal KMS de GNL, a consortium of Mitsui & Co. (Tokyo), Samsung Corp. (Seoul), and Korea Gas Corp. (Seoul).

The terminal will be built at the Port of Manzanillo and is scheduled for startup in mid-2011.

After starting at 90 Mcfd in 2011, supply will be scaled up to 500 Mcfd by 2015. The gas from Manzanillo will feed the 3 GW of power generation capacity CFE intends to install in the region.

CFE awarded Repsol Comercializadora de Gas a 15-year, $15 billion contract in September 2007 to supply the Manzanillo terminal with Peruvian gas.

Pipelines

More Costa Azul LNG gas to U.S.

Sempra Energy (San Diego) completed a $250-million expansion of its 225-km (140-mi) Baja North pipeline, enabling it to carry natural gas from the Energía Costa Azul LNG regas terminal in Baja California, to U.S. markets in California and Arizona.

The expansion was expected to be in operation by mid-2008.

Gasoducto Bajanorte SRL operates the pipeline, which originates at an interconnection with North Baja Pipeline LLC west of Algodones and continues west through Mexicali and Tecate.

The existing 500-MMcfd mainline serves new and existing power plants and industrial customers in northern Baja California and Southern California.

Sempra said it expected to have access by mid-2009 to the under-construction Yuma lateral that will move gas from the terminal to power plants in the Yuma, AZ, area.

Refining

More refineries, pipelines urged

Pemex told Mexico's Congress in mid-2008 that the country needed either additional refining resources or additional pipelines.

The company said construction of a 300,000-b/d, $8.17-billion refinery in Tula (Hidalgo state) would provide best return investment.

The proposed refinery also would require construction of a new oil pipeline from Nuevo Teapa to the Chicontepec region, plus extension of a new multipurpose pipeline to eliminate bottlenecks.

The second priority was said to be a 300,000-b/d facility in Tuxpan (Veracruz state).

The Pemex study also looked at seven other possible sites for a new refinery.

Absent a new refinery, the report said, Mexico will need to invest some $1.61 billion in central and western Mexico for nine new pipeline segments to complement existing distribution systems.

Gasoline imports

Gasoline imports rose to 360,700 b/d in March 2008, the highest level since November 2007. This coincided with a 7.8% decline in oil production in first quarter 2008, largely due to declining output from traditional oil fields.

Pemex said there are increasing numbers of cars in the country and that vehicles consume 41% of total supply.

Gasoline imports are projected to increase by 58% to 489,000 b/d by 2015 unless new refining capacity comes online, according to a recent report by the energy ministry. It noted that Pemex has not built a refinery since 1979.

To eliminate imports, officials say Mexico will need to build a refinery every 3–4 years for the next two decades.

Alternative Fuels

Bioethanol plant planned

BioFields S.A.P.I. de C.V. (Mexico City) said it will start constructing an algae-based bioethanol production plant in 2009 in Puerto Libertad, Sonora, in northeastern Mexico.

The company established a license agreement with Algenol Biofuels Incorporated (Naples, Florida), developer and owner of proprietary technologies for cultivating specialized blue-green algae in bioreactors to produce bioethanol.

NICARAGUA

Capital: Managua
Monetary unit: Córdoba
Refining capacity: 20,000 b/cd
Oil production: 0 b/d
Oil reserves: 0
Gas reserves: 0

Perlas and Tyra blocks

Infinity Energy Resources Inc. (Denver) reported that officials in two Autonomous Regions of Nicaragua ratified the company's contracts in mid-2008 for exploration and development of offshore concessions.

The company holds the Perlas and Tyra blocks that cover on 566,600 hectares (1.4 million acres) in 30–91 m (100–300 ft) of water in the Caribbean.

Infinity was contacting potential partners while awaiting final contract review by Nicaragua's president.

Sandino basin

Norwood Resources Ltd. (Vancouver, BC) began evaluating its discovery wells in western Nicaragua in third quarter 2008.

Analysis of whole core from the Maderas Negras-1 well led to recalculation of the net hydrocarbon thickness in the Brito formation to 87 m (286 ft) rather than 42 m (138 ft).

Testing showed oil and gas from Brito perforations at 1,802–1,813 m (5,915–5,950 ft) for 10 hr.

No flow rates were measured, but the volumes of oil and gas recovered strongly suggested that the earlier calculation of net hydrocarbons was too conservative.

In the firm's San Bartolo well, 4.8 km (3 mi) southwest of Maderas Negras-1, Norwood earlier had identified 118 m (387 ft) of net hydrocarbon pay in the overlying Paleocene Masachapa formation.

Norwood was planning to re-enter the well to run cased-hole logs and determine the need for fracturing.

Both wells are on the margin of the Sandino basin on the Indoklanicsa Concession.

PERU

Capital: Lima

Monetary unit: New sol

Refining capacity: 192,950 b/cd

Oil production: 75,000 b/d

Oil reserves: 415.769 million bbl

Gas reserves: 11.842 tcf

Overview

Oil is the dominant fuel source in Peru, with hydroelectricity second, according to the U.S. EIA. However, the start of production in 2004 from the Camisea natural gas project has led to increased use of natural gas.

An LNG liquefaction plant being built at Pampas Malchorita in 2008 will allow Peru to become an important regional exporter of natural gas.

Oil

Crude oil production in Peru has been in decline for the last decade. The country produced 113,000 b/d of total oil liquids in 2007, of which 68% was crude, said EIA.

In 2007, Peru consumed about 170,000 b/d of oil, importing about 57,000 b/d mostly from Ecuador and other South American countries.

Pluspetrol (Buenos Aires) controls over half of Peru's crude oil production. Other major producers include Occidental Petroleum (Los Angeles), Petrobras (Rio de Janeiro), and Petro-Tech Peruana.

Peru mostly privatized the former state-owned oil company, Petroperu, in 1993. The company still controls the country's only crude oil pipeline and most refineries.

Petroperu said that it wants to become an oil producer again, aiming to produce enough oil to meet at least half of domestic fuel demand by 2015.

In 2006, Petroperu partnered with Petrobras and Ecopetrol (Bogota) to explore six blocks, and in late 2008, reached agreements with Discover Petroleum (Tromsø, Norway), Reliance Industries Ltd. (Mumbai), CNPC (Beijing), and Pluspetrol.

Production of crude oil (mostly a heavy, sour type) is concentrated in the north. The largest oil blocks are Block 1-AB along the border with Ecuador, Block 8 in the northeastern Amazon, Block X in the northwest, and Block Z-2B off the northwest coast. The first two account for almost two-thirds of total production.

Natural gas

Peru had proven natural gas reserves of 11.9 tcf in 2008, the fifth-largest amount in South America. Officials want to increase gas use and reduce reliance on oil and hydroelectricity.

In 2006, the country produced and consumed 1.8 bcm (63 bcf) of natural gas, about 13% higher than in 2005, chiefly due to the start of gas production at the Camisea project. Continued ramp-up of production there will likely boost natural gas production further in coming years.

The government was planning to invest $35 million in 2008 to expand and upgrade the country's gas distribution networks.

The Peru LNG consortium broke ground in January 2006 on an LNG export terminal 169 km (105 mi) south of Lima that is to begin exporting up to 4.4 mtpy in 2010.

Exploration and Development

Block Z-6

Petrotech Peruana reported an oil discovery in the San Pedro field on offshore Block Z-6, offshore from Piura and Lambayeque. It estimated proven probable reserves in early 2008 at 1.13 billion bbl.

Further drilling to the south, off Lambayeque, was planned, and oil production from the block was estimated to begin in third quarter 2009.

Offshore seismic planned

Under a its license agreement with Perupetro, Gold Oil PLC (Surrey, UK) as operator expected to acquire 2,000 km (1,243 mi) of 2D seismic data before the end of 2008 offshore Peru in concession Z34.

Talara-Sechura basin

A group led by Upland Oil & Gas LLC was planning to explore Block XXIV. It is on the coastal plain in northern Peru, with about 32,400 ha (80,000 acres) offshore and 79,300 ha (196,000 acres) onshore, with sections situated in both the Talara and Sechura basins.

Upland and Perupetro signed a contract in mid-2007, and EnDevCo Inc. (Houston) took a 20% participation in March 2008. Several other firms were reported to be working with some success in nearby Blocks XII and Z2B.

EnDevCo said initial mapping of existing offshore seismic in Block XXIV revealed at least four prospects similar to a discovery made in shallow water in Block Z2B.

Preliminary internal estimates of reserves on the offshore prospects on Block XXIV range from 26–169 million bbl, said EnDevCo.

The company was planning to acquire 2D seismic for the onshore portions of the block.

Gas find near Camisea

Operator Repsol-YPF SA (Buenos Aires) made a gas discovery in early 2008 with its Kinteroni X1 exploration well on Block 57 near the Camisea project.

Initial production tests showed gas flows of 1 MMcmd and 198 cu m/d of associated liquids. The field's reserves are estimated at 2 tcf of gas.

Ucayali basin condensate

A group led by Repsol-YPF SA started production tests in early 2008 at a gas-condensate discovery on Block 57 in the Ucayali basin (Cuzco Province).

Early test rates were 35.3 MMcfd and 1,245 b/d from 115 m (377 ft) of pay in an undisclosed formation. Partner Petrobras (Rio de Janeiro) said the Kinteroni structure—some 22-km (14 mi) long—indicated a capacity of up to 2 tcf of gas.

Petrobras said that the discovery well is near Blocks 58 and 110, which it is exploring alone.

Lease Sales

2008 bid round

As of early August, Perupetro said 41 companies had formally expressed interest in participation in Peru's 2008 bid round.

Bids were to be taken until 8 September 8 for 22 blocks on offer.

The 22 included five new blocks added in June: four in the Ucayali Basin and one in the Maranon Basin. The new blocks are primarily comprised of acreage recently relinquished by Hess oil (New York City)—Blocks 118, 119 and 120.

Evaluation criteria for the 2008 round were designed to favor larger companies over those that took part in the successful 2007 bid round, said Perupetro.

Perupetro was aiming to sign 25 exploration contracts in 2008, roughly the same number as in 2007.

2009 bid round

Perupetro announced in October 2008 that it was preparing a new licensing round for 2009.

Specific blocks had not yet been identified, but the agency said they will include areas with both oil and gas potential.

Drilling and Production

Pagoreni gas-condensate startup

The Peruvian government reported that the Consorcio Camisea began production of natural gas at Pagoreni field on Block 56. Located in the Cuzco region, it is part of the Camisea gas-condensate project.

The consortium—comprised of Pluspetrol Peru Corp., Hunt Oil (Dallas), SK Energy (Jongro GU, South Korea), Tecpetrol SA (Buenos Aires), Sonatrach (Algiers), and Repsol-YPF SA (Madrid)—expects to initially produce gas at six wells.

The Pagoreni output could boost liquids production by as much as 45,000 b/d and gas production from 14.2 MMcmd to 35.7 MMcmd (500 MMcfd to 1.26 bcfd).

Production from Block 56 was said to make available a new reserve of 3.4 tcf of gas and 227 million bbl of NGL, increasing the Camisea reserves to 14.8 tcf.

Development of Block 56 is also being referred to as the second phase of the Camisea project because it is adjacent to Block 88, which went into production in mid-2004.

Maranon basin

In early 2008, Perenco (Paris) completed its acquisition of Barrett Resources (Peru) LLC, which owns three heavy oil fields set for development in the Maranon basin, at a cost of $1.5 billion.

The company estimated that the Paiche, Dorado, and Pirana fields on Block 67 have the capacity to produce as much as 100,000 b/d from more than 300 million bbl of proved and probable reserves.

A development plan calls for oil production to start in January 2011.

Perenco Peru Ltd., which operates a similar project in Ecuador's Oriente, will participate in a continuing exploration program in the Peruvian region, including a seismic survey on Block 121.

Refining and Products

Talara refinery upgrade

Petroperu announced plans to modernize Talara refinery, the country's largest, in 2008.

The upgrade would be part of a wider effort to double Peru's production.

In addition to the refinery work, Petroperu also would have to start production from 12 blocks where petroleum exploration was taking place at mid-year.

After delays due to alleged irregularities at the agency in the awarding of exploration licenses, the bid deadline for FEED and EPC work to upgrade the refinery was extended to mid-October.

The plant is in Piura, 1,200 km (746 mi) north of Lima.

The $1 billion upgrade will increase oil production capacity to 90,000 b/d from 62,000 b/d.

Workers will revamp existing units and add facilities for desulfurization of diesel and gasoline, production of sulfuric acid, and electricity generation.

LNG

Pampa Malchorita plant

Work continued in 2008 on the $3.8-billion Pampa Malchorita LNG liquefaction plant under the Peru LNG consortium. The facility is located about 170 km (106 mi) south of Lima on the coast.

In addition to the liquefaction plant, The plant will include marine export facilities, a breakwater, and a 408-km (253-mi) pipeline to bring 17.7 MMcmd (625 MMcfd) of feed gas to the site.

The project is scheduled to operational in 2012, producing 4.4 mtpy of LNG for export.

Consortium members include Hunt Oil Co. (Dallas), SK Corp. (Seoul), and Repsol-YPF (Madrid).

Pipelines

Transporting Camisea output

The Peruvian government reported that gas demand is growing so quickly that the pipeline transporting gas from the Camisea project to the capital region has reached capacity.

The pipeline is operated by a consortium known as Transportadora de Gas del Peru SA, or TGP. According to the Ministry of Energy and Mines, TGP has a contract to increase capacity in the westbound pipeline to 450 MMcfd from the current 290 MMcfd.

In October 2008, the government awarded Kuntur Transportadora de Gas, SAC (Kuntur)—a subsidiary of Conduit Capital Partners LLC—a 30-year concession contract to build a southbound 700-km (435-mi) pipeline to carry gas from Camisea to the port of Ilo in southern Peru.

The cost for the pipeline—the Gasoducto Andino del Sur (GAS)—is estimated at $1.4 billion. Construction is due to begin in 2010, with operations expected to begin by yearend 2012.

Norperuano pipeline

In mid-2008, Petroperu announced plans to invest some $50 million for an extension to the 852-km (529-mi) Norperuano crude oil pipeline that runs to the Pacific coast from fields in the northeastern jungle region.

TRINIDAD and TOBAGO

Capital: Port of Spain
Monetary unit: Dollar
Refining capacity: 168,000 b/cd
Oil production: 113,000 b/d
Oil reserves: 728.30 million bbl
Gas reserves: 18.77 tcf

Exploration and Development

Intrepid Block 5(c)

Canadian Superior reported a major natural gas discovery in mid-2008 at its Bounty exploration well on Intrepid Block 5(c), just east of Dolphin gas field, 97 km (60 mi) off the east coast.

The company said the discovery has the potential for as much as 73.6 bcm (2.6 tcf) of natural gas in place.

The firm said initial test results indicate that Bounty could produce 200 MMcfd from this high-pressure zone.

Earlier in 2008 on the same block, a discovery by Canadian Superior's by the Victory well was estimated to contain 17–34 bcm (600 bcf–1.2 tcf) of gas.

A third prospect, Endeavour, was slated for later evaluation.

Deepwater gas find on Block 22

Petro-Canada (Calgary) reported in early 2008 that its first well on deepwater Block 22 is a large gas discovery.

The Cassra-1 wildcat is 120 km (75 mi) east-northeast of the North Coast Marine Area, north of Tobago Island, in 430 m (1,411 ft) of water.

Petro-Canada estimated that the discovery could contain 17–37 bcm (0.6–1.3 tcf) of recoverable contingent resources.

Plans called for moving about 4.5 km (2.8 mi) south to drill Cassra-2, plus two or three other prospects on Block 22 in an expected five-well 2008 drilling program.

Central Range onshore

Two Calgary independents—Petro Andina Resources Inc. and Voyager Energy Ltd.—were making plans to prospect on 334 ha (214,000 acres) in the underexplored Central Range in central Trinidad.

Petro Andina was to become operator of the onshore Central Range Shallow and Central Range Deep blocks, with drilling likely to start in 2009.

The blocks are subject to production-sharing contracts, and the deal is subject to approval by the Trinidad and Tobago government.

Drilling and Production

South Erin wells

Jasmin Oil & Gas Ltd. (private Trinidad operator) and Maxim Resources Inc. (Vancouver, BC) announced plans in mid-2008 to drill four wells to develop their 2006 oil discovery on the South Erin Property, 64 km (40 mi) southwest of Port of Spain.

The discovery well, on test from mid-2007 to mid-2008, was flowing 85 b/d of oil at 500 psig.

Maxim called the South Erin find Trinidad's largest onshore oil discovery in several decades.

Reserves

Audit shows 30.8 tcf

An audit of the country's natural gas reserves for 2007 showed 872 bcm (30.8 tcf) of gas.

Results of the audit, by Ryder Scott Co., were presented to the nation's standing committee on energy in mid-2008.

Proved reserves totaled 481.365 bcm (16.997 tcf), probable reserves bcm 223.251 bcm (7.883 tcf), and possible reserves 166.752 bcm (5.888 tcf), with a rate of reserves to production ratio of 13 years.

Analysts noted that there had been a virtual 100% replacement of reserves for 2007.

Trinidad and Tobago's annual gas use is approximately 36.8 bcm (1.3 tcf).

The audit also places the Caribbean island's exploration potential at an additional 885.104 bcm (31.253 tcf).

Lease Sales

Five blocks offered

The government sought bids in the third quarter for five new exploration blocks in its 2008 bid round. They include:

- Block 4(b) in 400–800 m (1,312–2,625 ft) of water off the east coast, just east of BG Group's prolific Dolphin field.
- Block 5(d), also off the east coast, north of the Manikin gas discovery in 450–800 m (1,476–2,625 ft) of water.
- NCMA (3), (4), and (5)—all off the northern coast—in water 100–550 m (328–1,804 ft) deep.

Plans called for awarding these blocks in early 2009, and for seeking rebids on three 2006 ultradeep water parcels later in 2009.

Two awards

In mid-2008, the government granted permission for a consortium to explore and develop Block NCMA 2 in the shallow marine area off northern Trinidad.

The consortium members are ONCG Mittal Energy Ltd. (OMEL; London) and Petroleum Co. of Trinidad & Tobago (Petrotrin),

The group plans to drill five exploratory wells to depths of 2,743–4,115 m (9,000–13,500 ft) and to acquire 986 sq km (381 sq mi) of 3D seismic.

The government also granted approval for a consortium comprising Petrotrin and Trinidad Exploration & Development Co. Ltd. (TED) to explore and develop the Southwest Peninsula Shallow and Deep Horizon blocks onshore in Trinidad's Southern basin.

Drilling and Production

Greater Angostura: Phase 2

BHP Billiton (Melbourne) contracted with the National Gas Co. of Trinidad and Tobago (NGC) to buy gas from the second phase of its Greater Angostura field.

During the Phase 2 development, BHP Billiton and partners will install a new gas export platform beside the firm's existing facilities in the Greater Angostura field.

The facility is to be online by mid-2011 with a design capacity of 7.9 MMcmd (280 MMcfd).

The development also will modify the existing Angostura facilities and install new flowlines.

NGC will take delivery of the gas at the new gas export platform and will transport it into their proposed Northeastern Offshore pipeline and Tobago pipeline.

The Greater Angostura field includes oil and gas discoveries at Aripo, Kairi, and Canteen.

LNG

Contract renegotiation sought

Prime Minister Patrick Manning said in late 2008 that Trinidad and Tobago wants to renegotiate the contracts for Trains 1, 2, and 3 of Atlantic LNG Co. (ALNG). The goal is to increase the revenues collected from the gas producers that supply ALNG.

Plans were being made for Manning to travel to Madrid and London for discussions with executives of Repsol-YPF SA and BP PLC.

With regard to Trains 1–3, ALNG buys gas from suppliers and sells freight on board to customers from its Point Fortin port. But ALNG Train 4 operates as a processor of gas.

Refining and Products

Second refinery planned

Trinidad and Tobago's Ministry of Energy and Energy Industries announced in early 2008 that the country will build a $3–4 billion refinery next to its existing 168,000 b/d refinery at Pointe-a-Pierre.

The facility, whose capacity is expected to be in the range of 200,000 b/d, will be export-oriented, said the ministry, producing salable products for the retail fuel market. Bottom-of-the-barrel products from state-owned Petrotrin may be used as feedstock.

Policy

Petrocaribe membership declined

The government of Trinidad and Tobago in early 2008 declined to join Petrocaribe, the trade initiative established by Venezuela to help Caribbean community countries meet their energy requirements in the face of price increases on the global oil market.

The administration said it remains committed to the rival Free Trade Area of the Americas.

Apart from Trinidad and Tobago, Barbados is the only other Caricom state that has not signed the PetroCaribe accord, which several other nations have embraced with enthusiasm.

VENEZUELA

Capital: Caracas
Monetary unit: Bolivar
Refining capacity: 1,282,100 b/cd
Oil production: 2.350 million b/d
Oil reserves: 99.377 billion bbl
Gas reserves: 170.920 tcf

Overview

Venezuela is the fourth-largest producer among the Organization of Petroleum Exporting Countries (OPEC). A report by Business Monitor International in early 2008 concluded that Venezuela will provide about 28% of Latin America's oil supply by 2011.

Oil production has fallen 25% since Pres. Hugo Chávez took office in 1999, largely as a result of the layoffs of more than 20,000 experienced engineers and executives with the state oil company, Petroleos de Venezuela SA (PDVSA), who had joined a 2002–03 strike urging that Pres. Chávez be removed from office.

BMI's *Venezuela Oil & Gas Report* projected that export capability would be about 2.6 million b/d by 2011, with gas production of 48 bcm that year. Gas export potential exists beyond 2011.

The report pointed out the recent changes in licensing, production-sharing, and taxation have reduced foreign involvement.

At the same time, BMI said the country has considerable untapped hydrocarbons potential, with a high reserves-to-production ratio and significant output growth potential.

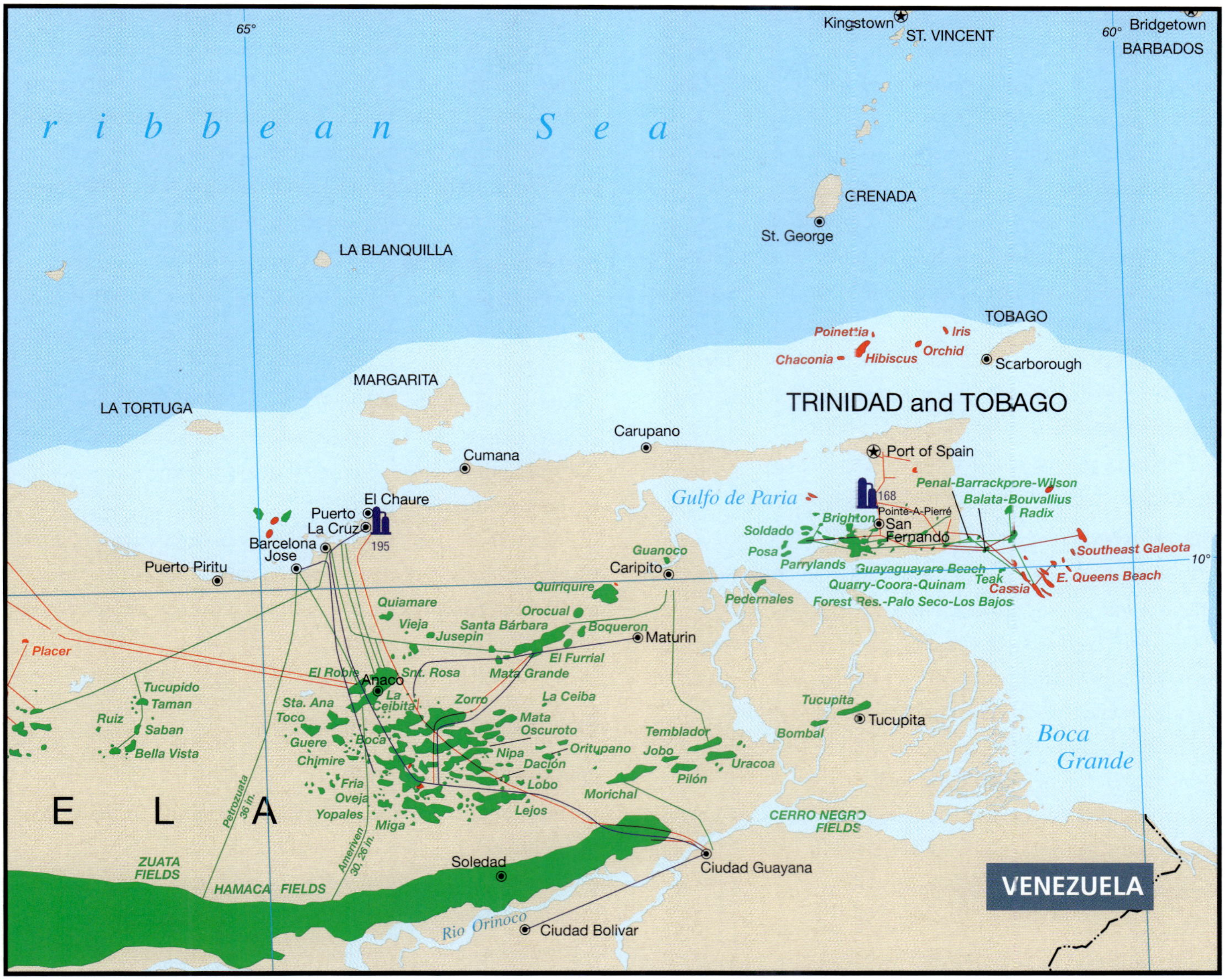

Assessments by reputable sources indicate likely substantial reserves in the Orinoco extra-heavy crude belt, which the government was working to verify in 2008.

In the Orinoco Magna Reserve project, the government expects to certify more than 200 billion bbl of crude oil reserves. If correct, that would give Venezuela the largest hydrocarbons reserves in the world.

Policy

PDVSA shakeups

In a move that could have implications for Venezuelan oil production, Pres. Chávez continued to realign leadership within the country's oil and gas industry and PDVSA.

In late 2008, he made several executive appointments and approved six new members to the PDVSA board of directors and retained the oil minister to be president of PDVSA.

The changes were said to indicate Chávez's desire to keep his most trusted allies atop PDVSA.

On the other hand, he did appoint one of the ousted oil vice-ministers as the country's new governor at OPEC—a man considered a key architect of the Chávez oil policy.

As part of a wider set of reforms, Pres. Chávez issued a decree in July that broadened the role of PDVSA, giving it the responsibility of "participating in ventures intended for the sustained, organic, and integral development of the country, including agricultural [and] industrial activities."

Tech transfer rules changed

Pres. Chávez said in mid-2008 that his country would stop doing business with IOCs that fail to transfer technology as part of their contracts.

The issue of technology transfer is also a sore spot with IOCs themselves.

For example, when ExxonMobil Corp. (Irving, TX) and ConocoPhillips (Houston) withdrew under government pressure in October 2007, they left behind a variety of technology and infrastructure.

That technology could fall into the hands of companies, such as PDVSA, that lack the technical expertise to extract reserves with the same efficiency as the two U.S. firms.

The status and use of the left-behind technology may become the subject of legal challenges from the two companies.

New tax on exported oil

Venezuela increased taxes in April 2008 on crude oil and oil products moving out of the country.

On their exported oil volumes net of imports, companies must pay a per-barrel tax of 50% of the amount by which the monthly average price of Brent crude exceeds $70/bbl.

The move came as the government acted to nationalize key industries, including cement and steel.

The tax increase was the latest in a series of related actions that began when Pres. Chávez was elected in 1998.

Partial compensation for Orinoco

Venezuela said in early 2008 that it will pay $1.1 billion in compensation to StatoilHydro AS (Stavanger, Norway) and Total SA (Paris) for its nationalization of a heavy-oil oil project that they previously controlled.

PDVSA agreed to pay the partners a mix of crude and cash for their reduced stakes in the Sincor heavy oil venture. However, Total and StatoilHydro must pay $130 million of their compensation as a "bonus" to fund the new Sincor joint venture company.

The government in May 2007 took over majority stakes in heavy oil upgrading projects along the Orinoco River basin operated by four international oil companies.

StatoilHydro and Total remained as minority partners under new arrangements with the government, but ConocoPhillips (Houston) and ExxonMobil Corp. chose to leave the country.

PDVSA was said to be still considering compensation for the two U.S. firms that were also former Sincor partners.

Pres. Chávez wants to certify Orinoco belt reserves by 2010—an action that could boost the country's reserves to more than 300 billion bbl from the current 100 billion bbl.

Venezuela could ramp up production and sales of Orinoco heavy crude without OPEC authorization.

Duel with ExxonMobil

In February 2008, ExxonMobil Corp. won U.S. and British court injunctions that froze as much as $12 billion in assets owned by PDVSA.

ExxonMobil had launched lawsuits in U.S., British and Dutch courts, all challenging Chávez's nationalization of the multibillion-dollar Sincor oil project formerly owned by ExxonMobil in the Orinoco belt.

The court decisions prompted threats and accusations by Pres. Chávez against the company and the United States that rattled oil markets. Some traders were said to be worried about a potential cutoff of Venezuelan oil.

Exploration and Development

Cumarebo field

The PetroCumarebo mixed company working in the Falcon basin announced plans in mid-2008 to drill the San Patricio exploration well by yearend in a prospective area east of Cumarebo field, said PetroFalcon Corp. (Caracas).

PetroCumarebo also plans to drill the La Cruz South prospect.

It was also reevaluating six wells—LV-12, 20, and 21, and CU-170, 171, and 173—in Cumarebo and La Vela fields for reperforation. Doing so on the first well, CU-170, boosted production by 4.6 MMcfd of gas and 190 b/d of condensate.

Colon Block

The Baripetrol SA joint venture said its 2008 plans for the Colon Block in western Venezuela called for drilling four development wells in La Palma field, a 165-sq km (64-sq mi) 3D seismic shoot in Socuavo field, and 20 workovers in Los Manueles and Rosario fields.

Baripetrol is a joint venture between PDVSA, Tecpetrol (Buenos Aires) and Perenco (Paris).

International Activities

Russian visitors

Officials from the Russian Energy Ministry and other oil and gas organizations visited Venezuela in mid-2008.

Other Russian oil and gas officials in Shmatko's party noted their country's interest in "possible participation of Russia in the development of not only oil, but also gas fields" in Venezuela and elsewhere in Latin America.

One official observed that "there will be no need as before to link Caracas only with the American market, regarding both oil and natural gas."

Another said Russia and Venezuela were considering "establishment of a joint venture with [PDVSA] on the recovery of heavy and extra-heavy crude in the Orinoco River belt."

Of particular note were Russian comments about OPEC.

One senior official said his country hopes to sign an MOU with OPEC, as a first step in cementing a closer relationship with the group. Potential aims are better interaction between Russian companies and OPEC members and greater cooperation regarding overall oil market analysis and information.

Gazprom spuds well

OAO Gazprom (Moscow) confirmed in November that drilling had begun for the first offshore natural gas well in Venezuelan waters, on the Urumaco I and II Blocks.

Drilling began from the Scorpion Vigilant platform in the presence of Pres. Chávez of Venezuela and senior officials of the Russian Federation Government.

Gazprom's operation is part of an agreement between Russia and Venezuela to cooperate on oil and gas projects. PDVSA retains the exclusive right to produce natural gas in the country.

Potential natural gas reserves of Urumaco I and Urumaco II average 100 bcm, the company said.

Delta Caribe Orientale Project

PDVSA signed agreements September 2009 with eight transnational companies— from the United States, Italy, Portugal, Malaysia., Qatar, Russia, Japan and Argentina—for participation in a major project to exploit Venezuela's natural gas resources.

The Delta Caribe Orientale Project aims to boost Venezuela's gas production to 11.5 billion cfd day by 2012.

A multi-partner company is planned—including ChevronTexaco (San Ramon, CA); Qatar Petroleum (Doha); Mitsubishi Corp., Mitsui & Co., and Itochu Corp. (all based in Tokyo); OAO Gazprom (Moscow); Eni SPA (Rome) and Petronas (Kuala Lumpur)—to produce LNG.

The participants also signed an MOU to develop the offshore gas fields Blanquilla and Tortuga.

Overall investment in the project, planned to start in 2009, is estimated at almost $6 billion.

Orinoco: Junin Block 5

Eni SPA (Rome) announced in early 2008 that it will assess and develop Junin Block 5 in Venezuela's Orinoco oil belt under a strategic agreement with Venezuela's energy ministry and PDVSA.

The joint venture aims to initially produce 30,000 b/d and reach a long-term production plateau of 300,000 b/d.

The 670-sq km (259-sq mi) block is in Anzoategui state, 550 km (342 mi) southeast of Caracas, with a resource estimated at over 2.5 billion boe.

Eni brings to the venture proprietary slurry technology to convert heavy oils, bitumens, and asphaltenes into high-quality light products.

Russian leader visits

Several Russian warships docked in La Guaira, near Caracas, in late November 2008. Russia's first such deployment in the Caribbean since the Cold War, their arrival was timed to coincide with Pres. Dmitry Medvedev's visit, the first ever to Venezuela by a Russian president

Pres. Chávez has sought Russian help to build a nuclear reactor, invest in oil and natural gas projects and bolster his opposition to U.S. influence in the region.

He has proposed a major gas pipeline across South America and has explored Russian investment for the project.

The level of potential mutual assistance, however, was threatened by falling oil prices and the pressures of the world's financial crisis. Pres. Medvedev's tour—to also include Peru, Brazil, and Cuba—was planned before the economic downturn.

China agreements

Chinese Pres. Hu Jintao met with Venezuelan Pres. Chávez in September 2008 in Beijing and agreed to further increase the countries' cooperation in various fields.

Hu said China is willing to increase its participation in Venezuela's upstream and downstream operations in the petroleum sector by continuing to support Chinese investments in Venezuela.

He also said China will establish an economic and trade cooperation zone and help build rail, communication, housing, and water projects in Venezuela.

The presidents signed 12 agreements in four sectors: economy, petroleum, education and justice assistance. Petroleum initiatives included:

- A fuel-oil supply contract between PDVSA and PetroChina Co. Ltd. (Beijing) to increase oil exports to China from the current 500,000 b/d to 1 million b/d 2012.
- A joint study by PDVSA and Sinopec (Beijing) on the feasibility of a new refinery in oil block Junin-8 at the large Orinoco Belt to process up to 300,000 b/d.
- Three additional plants to process Venezuelan heavy oil in southern China's Guangdong province.

Another agreement proposes to double the size of the countries' joint strategic fund from the initial $6 billion agreed on in 2007 to $12 billion.

Refining and Products

Refinery modernization

As part of its on-going Siembra Petrolera plan, PDVSA signed an agreement in October 2008 with JGC Corp. (Yokohama) to form a strategic alliance aimed at improving the country's refining capacity.

The accord will focus on projects involving the conversion of heavy oil, said PDVSA.

JGC helped build the Cerro Negro heavy crude upgrader in the Orinoco belt and is completing basic engineering for an expansion at the Puerto La Cruz refinery.

In September, Venezuela said it signed a $1.2 billion loan with the Japanese Bank for International Cooperation to fund modernization of refineries at El Palito and Puerto La Cruz.

In July, it was reported that PDVSA now plans to invest $78.116 billion in 2007–12, about 40% more than the $56 billion proposed for the same period in 2005.

New Santa Ines refinery

PDVSA started construction of the 100,000-b/d Santa Ines refinery in late 2008, aiming to process more of the country's heavy oil at home.

The refinery, in Barinas state, will cost $1.2 billion. Phase 1 development, to be completed in 2011, will develop an initial capacity of 30,000 b/d. Phase 2 will bring the plant to full capacity in 2014.

Crude oil will come from Barinas and Apure states, as well as from the Orinoco heavy crude belt.

The product slate will include regular and high-octane gasoline, LPG, diesel, kerosene, fuel oil, and asphalt, chiefly for the domestic market.

Three other refineries are either under way or planned: the 50,000-b/d Caripito unit; a 200,000-b/d plant in Zulia state; and a 400,000-b/d facility at Cabruta.

Alternative Fuels

New fuel ethanol plants

In July 2008, Dedini S.A. Industrias de Base (Piracicaba, Brazil) announced its sale of four ethanol distilleries to PDVSA.

Each plant will have the capacity to produce 700,000 liters/d of cane-based ethanol. The fuel will be used to replace a lead derivative added to Venezuelan gasoline.

Equipment is to be delivered by yearend 2009 to set up distilleries in the states of Barinas, Portuguesa, Cojedes, and Trujillo.

The transaction followed a change in government thinking about ethanol.

In March 2008, reversing a 2007 objection to ethanol development, PDVSA made a distinction between sugarcane- and corn-based ethanol, and Pres. Chávez introduced an E7 national ethanol target.

Venezuela said it plans to build more than 300 million gal in domestic ethanol capacity, with a total construction of 14 ethanol plants.

Other action in Latin America

Several other nations in Latin America saw activity in the oil and gas sector in 2008.

Costa Rica

Capital: San Jose

Refinery joint venture

CNPC (Beijing) and Refinadora Costarricense de Petroleo S.A. (Recope) (San Jose) agreed to set up a joint-venture refinery in Costa Rica. This will be the first cooperative oil and gas project in that country for CNPC, China's largest petroleum company.

The companies will upgrade and expand the 25,000-b/d Moin refinery to 64,000 b/d while also improving product quality and the plant's ability to process heavy crude.

The companies will also study building a possible new 214,000-b/d refinery.

Guyana

Capital: Georgetown

Seismic studies

Several firms let contracts to acquire 3D seismic in 2008 in areas of the Atlantic off Guyana.

YPF Guyana Ltd.—a subsidiary of Repsol YPF SA (Madrid)—and partner CGX Energy Inc. (Toronto) want a marine 3D survey on the Georgetown petroleum prospecting license (PPL).

CGX established its own contract for a 3D program on the adjacent Corentyne PPL. That program was to cover four prospects for which an independent firm has calculated a combined best estimate (P50) prospective resource of 2.7 billion bbl oil.

Takutu basin

Groundstar Resources Ltd. (Vancouver, BC) and an unnamed company signed a farmout agreement in March 2008 that sets the stage for possible drilling of at least two exploratory wells by early 2009 in the Takutu basin. Groundstar has three large seismically defined prospects on the license.

Tests on the Karanambo-1 exploration well (drilled by Home Oil Ltd. in 1982) proved that the basin has the essential elements of a petroleum system.

Panama

Capital: Panama City

Pipeline flow reversal

BP Products North America (Houston) signed a seven-year agreement in early 2008 with Petroterminal de Panama SA (PTP) to ship oil through the Trans-Panama Pipeline (TPP).

The 130-km (81-mi) TPP originally carried crude through Panama from the Pacific to the Atlantic.

PTP will modernize the pipeline and reverse its flow.

Instead of sailing around the southern tip of South America, VLCCs will unload their cargoes at Chiriqui Grande, Bocas del Toro, on the Caribbean for pipeline transport across Panama.

The crude will be piped to Charco Azul on the Pacific coast and then reloaded into tankers for shipment to the U.S. West Coast. Construction is expected to take about two years.

Santa Lucia

Capital: Castries

No refinery

Hess Corp. (New York City) denied in late 2008 that it planned to build a $5 billion refinery on the Caribbean island of St. Lucia, though it said it has an option to do so.

The statement came just days after a member of the St. Lucia government said Hess was likely to build the refinery.

Europe

ALBANIA

Capital: Tirana
Monetary unit: Lek
Refining capacity: 26,300 b/cd
Oil production: 9,900 b/d
Oil reserves: 199.140 million bbl
Gas reserves: 30.00 bcf

The oil and gas sector in Albania is administered by the Ministry of Industry and Energy, under which the National Petroleum Agency is responsible for oil and gas licensing activities.

The Albanian Petroleum Corporation (APC) is the parent to Albpetrol, responsible for exploration and production.

Armo, responsible for refineries at Ballsh and Fier and Servcom, also supports Albpetrol in development and drilling.

Despite generally poor results from exploration and exploitation programs, limited drilling is still proceeding.

Blocks A, B, D, and E

A subsidiary of Manas Petroleum Corp. (Baar, Switzerland) launched a 600 line-km 2D seismic survey in mid-2008 in the fold and thrust belt in northwestern Albania, to highgrade identified oil prospects.

The plan called for acquisition of 183 line-km on blocks A, B, and E in 2008, to be followed by additional surveys on blocks B, D, and E in mid-2009.

Kucova field

Bankers Petroleum Ltd. (Calgary) in mid-2008 acquired Sherwood International Petroleum Ltd., thereby gaining the right to redevelop Kucova heavy oil field in south-central Albania.

Original oil in place at Kucova was 490 million bbl, of which about 6% has been produced.

Reservoir depths are estimated to vary from 150–1,400 m (492–4,593 ft).

Kucova, 30 km (19 mi) northeast of Patos-Marinza, is geologically similar to that field.

EUROPE
0 50 100 150 Miles
0 100 200 Kilometers
North Sea
Irish Sea
Celtic Sea
Bristol Channel
Lyme Bay
Bay of Biscay
Mediterranean Sea
Gulf of Cadiz
IRELAND
N. IRELAND
WALES
ENGLAND
NETHERLANDS
BELGIUM
LUX.
GERMANY
DENMARK
FRANCE
SWITZERLAND
SPAIN
PORTUGAL
CORSICA
SARDINIA
MINORCA
MAJORCA
IBIZA
MOROCCO
ALGERIA
TUNISIA
Corrib
Londonderry
Belfast
Sligo
Ballina
Galway
Dublin
Foynes
Limerick
Arklow
Tralee
Wexford
Cork
Waterford
Bantry
Whitegate
Waterford
Kinsale Head
Moffat
Carlisle
Newcastle
Darlington
Teesside
Middlesbrough
Barrow
Lancaster
Heysham
Morecambe
Ballough
Bradford
Formby
Eastham
Manchester
Liverpool
Sheffield
Killingholme
Grimsby
Theddlethorpe
Lincoln
Bacton
Ellesmere Port
Aberystwyth
Birmingham
Milford Haven
Pembroke
Neath
Gloucester
Cambridge
Ipswich
Coryton
Oxford
London
Bath
Exeter
Plymouth
Fawley
Southampton
Portsmouth
Weymouth
Dunkerque
Antwerp
Amsterdam
Rotterdam
Groningen
Emden
Lingen
Gelsenkirchen
Duisburg
Essen
Dusseldorf
Cologne
Rheinland
Bonn
Brussels
Verviers
Maubeuge
Charlesville
Grandpuits
Luxembourg
Metz
Fredericia
Kaergard
Kalundborg
Skaelskor
Heide
Kiel
Cuxhaven
Brunsbuttel
Wilhemlshaven
Schwerin
Harburg
Hamburg
Leer
Bremen
Loningen
Verden
Beckedorf
Salzbergen
Meerdf.
Buckeburg
Herringhausen
Salzgitter
Hamm
Dortmund
Eisenach
Jena
Erfurt
Suhl
Frankfurt-am-Main
Karlsruhe
Vohburg
Ingolstadt
Munich
Notre Dame de Gravenchon
Port Jerome
Cherbourg
Le Havre
Gargenville
Gonfreville L'Orcher
Petit Couronne
Plestin Les Greves
Brest
St. Malo
Paris
Quimper
Lorient
Vannes
Laval
Donges
Nantes
Angers
Orleans
Blois
Cholet
Les Sables
Poiters
Chauvigny
Vierzon
La Rochelle
Niort
Roussiness
Royan
Le Verdon
Limoges
Chazelles
Pauillac
Ambes
Bordeaux
Bergerac
Tulle
Les Ancizes
Issoire
St. Etienne
Le Puy
Villefranche
Toulouse
Castres
Carcassonne
Pamiers
Lourdes
Narbonne
Perpignan
La Junquera
Fos-sur-Mer
Berre l'Etang
Marseille
Toulon
Nice
Monaco
Manosque
L'Isle
Valence
Grenoble
Chamberry
Lyon
Vienne
Feyzin
Bourg
Macon
Annemasse
Geneva
Dijon
Dole
Besancon
Bourbonne
Epinal
Wallbach
Briare
Vert-La-Gravelle
Coulommes-Vaucourtois
Velye-Germinon
Bussey-Lettree
Villeperdue
Dommartin-Lettre
Coudemagnes
Chaunoy
Marolles
Valence en Brie
Villemar
Donnemarie
Flacy
Grandville
Forcelles
St. Martin-de-Bossenay
Gisy-les-Nobles
Joigny
Aufferville
Chateaurenard
Trois Fontaines
Revigny
Gallician
Lavergne
Cazaux
Lugos
Cabeil
Mothes
Parentis
Ychoux
N. Mimizan
Lucats
Lacq sup
Lacq
St. Marcet
Proupiary
Pont d'As
Rousse
Meilon
St. Faust
El Serrablo
Gaviota
Bayonne
Vizcaya
Bilbao
Santander
Ayolvengo
Otomin
Zurich
Zug
Hochst
Bern
Chur
Martigny
Aosta
Lugano
Mandello
Brescia
Padua
Biella
Milan
Chivasso
Cremona
Rovigo
Turin
Asti
Aquata
Busalla
Parma
Savona
Genoa
Pistola
Leghorn
Bastia
Muro
Aleria
Zonza
Porto Torres
Maddalena
Alghero
Olbia
Nuoro
Arbatax
Iglesias
Porto Scuso
Cagliari
Ferrol
La Coruna
Aviles
Logrono
Burgos
Vitoria
Huesca
Barbastro
Monzon
Balaguer
Tarrega
Gerona
Palamos
Zamora
Vallodolid
Zargoza
Porto
Fayon
Mataro
Barcelona
Tarragona
Chipirón NE
Dorado
Angula
Salmonete
Montanaza
Tarraco
Casablanca
Castellon
Amposta
Madrid
Vinaroz
Castellon de la Plana
Vallencia
Cabo Ruivo
Lisbon
Setubal
Sines
Puertollano
Almendralejo
Cordoba
Huelva
Sevilla
Murcia
Cartegena
Granada
Rota
Cadiz
Malaga
Almeria
Algeciras
Algiers
Bejaia
Skikda
Tunis
Oran
Arzew

Discovered in 1928, Kucova has more than 1,700 wells. Production averaged 400 b/d at yearend 2006.

Trans-Adriatic Pipeline

Energy trading company Elektrizitats-Gesellschaft Laufenburg (EGL; Laufenburg, Switzerland) and StatoilHydro (Stavanger, Norway) agreed in early 2008 to a JV to build and operate the Trans-Adriatic Pipeline (TAP).

A section of the line would run through Albania, transporting Caspian and Middle Eastern natural gas to the heart of Europe.

Details are given in the Greece section of this chapter.

Serbia, Kosovo impasse

Kosovo declared its independence from Serbia on February 17, 2008, with the backing of the United States and several European Union nations.

The action led to the fall of Serbia's government March 8, precipitated by an impasse about how Kosovo's action affects Serbia's pursuit of EU membership.

Plans for new elections in Serbia were uncertain.

Serbia considers the territory of Kosovo its historic and religious heartland.

AUSTRIA

Capital: Vienna
Monetary unit: Euro
Refining capacity: 208,600 b/cd
Oil production: 16,000 b/d
Oil reserves: 50.0 million bbl
Gas reserves: 570 bcf

Strasshof, Ebenthal fields

Ost- und Mitteleuropa Verein AG (OMV; Vienna) started natural gas production in August 2008 from Strasshof and Ebenthal fields in the Vienna basin.

Strasshof, which holds an estimated 4 bcm (141 bcf) of reserves. The field was discovered in 2005; since then, OMV has drilled four more wells.

Maximum initial production from Strasshof will be 4,000 boe/d.

Ebenthal holds gas reserves of 1.5 bcm (53 bcf). Its output was expected to reach 3,000 boe/d in September.

OMV said it plans to invest €250 million in 2008 and again in 2009 to optimize production from its mature oil fields as well as exploring for new reserves.

BELGIUM

Capital: Brussels
Monetary unit: Euro
Refining capacity: 797,629 b/cd
Oil production: 0
Oil reserves: 0
Gas reserves: 0

BRC refinery

Petroplus Holdings (Zug, Switzerland) said in late 2008 that its Belgium Refining Co. (BRC) refinery in Antwerp will decrease production by 5–10% in 2009 (from 110,000 b/d to 95,000 b/d).

The company said that a fire-related incident in November was responsible for the decision.

Petroplus also operates the 115,000-b/d APF refinery in Antwerp as well as seven other refineries across Europe.

NORTHERN EUROPE
0 50 100 150 Miles
0 100 200 Kilometers
15°
10°
5°
0°
5°
10°
60°
55°
50°
Norwegian Sea
North Sea
Irish Sea
Celtic Sea
Donegal Bay
Moray Firth
Firth of Forth
Firth of Clyde
Galway Bay
FAROE I.
SHETLAND I.
ORKNEY I.
SCOTLAND
N. IRELAND
IRELAND
WALES
ENGLAND
NORWAY
DENMARK
NETHERLANDS
BELGIUM
LUX.
GERMANY
SWITZERLAND
Midgard
Åsgard
Tyrihans
Draugen
Ormen Lange
Njord
Åsgard Transport
Agat
Troll
Boa
Heimdal
Gekko
Kneler
Corrib
Morecambe
Formby
Galley Head
Seven Heads
Helvick
Waterford
Ardmore
Kinsale Head
Weymouth
Meerdf
Roxforrde
Tjeldbergodden
Trondheim
Nyhamna
Mongstad
Kollsnes
Sotra
Bergen
Oslo
Slagen
Vall oy Tonsberg
Karsto
Kopervik
Stavanger
Sola
Arendal
Mandal
Goteborg
Skagen
Sullom Voe
Lerwick
Kirkwall
Flotta
Wick
Stornoway
Nigg Bay
Inverness
Mallaig
St. Fergus
Cruden Bay
Aberdeen
Dundee
Finnart
Ardyne Point
Grangemouth
Edinburgh
Glasgow
Ardrossan
Moffat
Londonderry
Belfast
Sligo
Ballina
Galway
Ballough
Dublin
Arklow
Foynes
Limerick
Tralee
Wexford
Waterford
Cork
Whitegate
Bantry
Carlisle
Newcastle
Port Clarence
Teesside
Darlington
Middlesbrough
Scarborough
Barrow
Heysham
Lancaster
Bradford
Leeds
Hull
Manchester
Eastham
Liverpool
Grimsby
Theddlethorpe
Lincoln
Ellesmere Port
Nottingham
Bacton
Aberystwyth
Birmingham
Leicester
Coventry
Milford Haven
Pembroke
Llandarcy
Neath
Northampton
Cambridge
Gloucester
Oxford
Ipswich
London
Bath
Reading
Fawley
Bournemouth
Southampton
Portsmouth
Exeter
Plymouth
Dunkerque
Calais
Lille
Brugge
Antwerp
Brussels
Liege
Verviers
Maubeuge
Amsterdam
Rotterdam
Groningen
Emden
Wilhelmshaven
Cuxhaven
Heide
Brunsbuttel
Kiel
Schwerin
Hamburg
Harburg
Salzbergen
Gelsenkirchen
Duisburg
Essen
Dortmund
Dusseldorf
Cologne
Rheinland
Bonn
Eisenach
Suhl
Frankfurt-am-Main
Karlsruhe
Luxembourg
Metz
Fredericia
Kaergard
Esbjerg
Arhus
Kalundborg
Skaelskor
Notre Dame de Gravenchon
Cherbourg
Port Jerome
Le Havre
Gonfreville L'Orcher
Rouen
Gargenville
Petit Couronne
Grandpuits
Reims
Plestin Les Greves
Brest
St. Malo
Quimper
Rennes
Laval
Lorient
Vannes
Donges
Nantes
Orleans
Blois
Paris
Vert-La-Gravelle
Coulommes-Vaucourtois
Velye-Germinon
Bussey-Lettree
Chaunoy
Villeperdue
Soudron
Dommartin-Lettre
Courdemagnes
Marolles
Valence en Brie
Donnemarie
Grandville
Trois Fontaines
Villemar
Aufferville
Flacy
St. Martin-de-Bossenay
Forcelles
Gisy-les-Nobles
Joigny
Chateaurenard
Bourbonne
Reichstett
Vendenheim
Zurich
Munich
Ingolstadt
Vohburg
Baden-Wurttemberg
199
116
206
12
70
106
100
83
268
27
267
221
221
78
93
104
71
210
10
89
266
191
81
373
406
305
357
115
21
227
346
326
163
159
141
223
101
338
154
310
229
85
262
110

BULGARIA

Capital: Sofia
Monetary unit: New lev
Refining capacity: 115,240 b/cd
Oil production: 1,000 b/d
Oil reserves: 15.00 million bbl
Gas reserves: 200.00 bcf

Kavarna, Kaliakra discoveries

In October 2008, Melrose Resources (Edinburgh) reported a gas discovery with the Kavarna No.1 exploration well in the Black Sea.

The well encountered a Paleocene reservoir formation at 801 m (2,628 ft), with reserves initially estimated at 24 bcf.

Replacement well Kavarna No. 2 encountered 38 m (125 ft) of net gas pay, confirming the initial reserves estimate.

Tie back to the Galata field facilities was being carried out, with first production expected in late 2009.

Melrose also reported a discovery earlier in 2008 at its GE3-Kaliakra No. 1 well.

That discovery, with estimated reserves of 47 bcf, also will be tied back to the Galata facilities, aiming for first production in 2010.

A-Lovech Block

Private independent Direct Petroleum Exploration Inc. (Denver) reported in early 2008 that a subsidiary had drilled Deventci-R1, the deepest well in Bulgaria in the last 30 years, to TD m (19,313 ft).

The well, on the A-Lovech Block in the northwest, tapped the Lower Triassic Alexandrovo formation and encountered gas-saturated reservoirs in the Dolni Dabnik member of the Middle Triassic Doirentsi.

The company said other potential reservoirs are in the Upper Triassic Rusinovdel and the Lower Jurassic Ozirovo.

Windfarm near Kavarna

A consortium of banks led by the European Bank for Reconstruction and Development (EBRD; London) approved loans amounting to $276 million toward the construction of a 156-MW windfarm in Bulgaria.

The project—with a total cost estimated at $377 million—would be the largest windfarm not only in Bulgaria but also in Eastern Europe.

The Sveti Nikola (St. Nicholas) project, to be built near Kavarna on the Black Sea coast, would have 52 wind turbines, each capable of generating 3MW. First operation is scheduled for October 2009.

A subsidiary of AES Corp. (Arlington, VA) will own and run the facility.

CROATIA

Capital: Zagreb
Monetary unit: Kuna
Refining capacity: 250,317 b/cd
Oil production: 15,700 b/d
Oil reserves: 79.300 million bbl
Gas reserves: 1.080 tcf

Refinery modernization

State-owned oil and gas company Industrija nafte (INA; Zagreb) said in mid-2008 that it plans to invest $ 1.1 billion to modernize its two oil refineries by 2011.

The upgrades aim to achieve compliance with the quality standards set by the European Union.

INA shares in play

Croatia's government is obliged to reduce its 44 % stake in INA to 25 % before the country joins the European Union, expected around 2011.

In mid-2008, OMV AG told the government that it is interested in buying a 26 % stake in INA.

Hungarian rival MOL Group (Budapest), which holds 25% of INA, also was in talks with the Croatian government to raise its stake.

Analyst said OMV is seeking to buy into INA as a way to thwart the ambitions of MOL.

Should MOL acquire more than 25 %, it would have to launch a public offer for all of INA stock.

Croatia-Iran cooperation?

In mid-2008, Croatia's ambassador to Iran said that his country was fully prepared to expand its relations with Iran in several fields, particularly oil exploration.

Croatia's INA is already active in oil exploration in the Moghan oil block in the northwestern Ardebil province of Iran.

European Energy Issues

European gas balance

Demand for natural gas was falling in Europe because of high prices in 2008 and a deteriorating economic outlook.

But the region also faces a potential of oversupply of gas, said Wood Mackenzie (Edinburgh).

The firm noted that by 2011, as much as 110 bcm (3,885 bcf) of LNG will enter the market, though some planned projects may not be completed.

European-African energy partnership

In September 2008, the African Union Commission and the European Commission established an Africa-European Union energy partnership that could attract more than €1 billion in EU funds for energy investments in Africa.

Details are presented in the Africa chapter of this book.

EU, Egypt energy agreement

In December 2008, the European Commission and Egypt's Foreign Affairs Ministry signed an MOU to enhance energy cooperation between Egypt and the European Union and reinforce energy security for both.

Among the top priorities are establishment of a program to converge Egypt's energy markets with the EU's; development of energy networks such as the Arab Gas Pipeline; and promotion of renewable energy and energy efficiency.

Funding for Nabucco

In late January 2009, two European organizations pledged funding for the proposed Nabucco natural gas pipeline.

The pipeline line would run 3,300 km (2,050 mi) from eastern Turkey to Austria, carrying Caspian gas to Western Europe.

The European Commission said that it plans to invest $330 million in the pipeline.

The EC announcement was viewed as an important initial sign of action to diversify European energy supplies following the January 2009 Russia-Ukraine gas dispute. EU nations depend on Russian gas, 80% of which is piped across Ukraine.

Also, the European Investment Bank (finance arm of the European Union) said that it was prepared to finance up to 25% of the cost of Nabucco—a potential commitment of more than $2.6 billion.

Bank officials stressed that political agreement was needed before it could raise any money.

Turkey has been pushing Austria, Hungary, Bulgaria and Romania to assure secure supplies of low-cost gas for itself as part of the Nabucco deal.

Gas supply from Russia

Russia's Gazprom cut off all gas supplies to Ukraine on January 1, 2009, after the two sides failed to reach agreement on debt payments and on how much Ukraine will pay for gas in 2009.

About 20 European nations suffered significant reduction in gas supply.

On January 28, 2009, gas supplies were restored after both sides signed a contract with new tariffs.

More details are given in the Ukraine section of the Former Soviet Union chapter of this book.

European gas shale study

Three European institutes were making plans to set up by late 2008 a consortium of universities and energy centers to map possible gas shale sites in several European countries.

The three institutes are the GFZ German Research Center for Geosciences (Potsdam); the Institut Francais du Petrole (IFP; Rueil-Malmaison, near Paris); and the Netherlands Organization for Applied Scientific Research (TNO; Eindhoven).

There are shale formations in France, England, Sweden, Ukraine, Poland, and elsewhere in Europe.

The €6-million program is to be financed by BP PLC (London), Royal Dutch Shell (The Hague, Netherlands), ExxonMobil Corp. (Irving, TX), Devon Energy Corp. (Oklahoma City), and others.

Fuel-price protests

In mid-2008, soaring prices for diesel oil led to protests in France, Spain, Portugal, and Italy by fishermen, truckers and farmers.

Climate change plan

European leaders in a December 2008 summit meeting agreed to support an ambitious plan to reduce greenhouse gas emissions and use more renewable energy resources.

It includes a commitment to use of carbon capture and storage technology at about a dozen power stations.

The agreement means member states must fulfill the "three 20s" target. By 2020 each country must:

- bring down its greenhouse gas emissions by 20% from its 1990 level
- bring the share of renewables in the energy mix to 20% from the current 8%
- achieve energy savings of 20%.

In an expanded carbon emissions trading scheme, power generators would pay for 100% of their emissions allowances, beginning in 2013. They currently get most free.

Iran has said that an exploration campaign on Moghan block will be completed by 2012 and that the government has allocated about $140 million to develop the block.

Adria LNG terminal

In early 2008, OMV AG (Vienna) reported that Croatia's government confirmed the Island of Krk as the location for the planned Adria LNG terminal and that construction would begin at yearend 2009.

The terminal is to be built by an OMV-led consortium that includes for other companies in Germany, France, and Slovenia.

The facility will be built for LNG tankers carrying up to 265,000 cu m.

Energy-sector investment

In late 2008, Croatian officials said the country was formulating a new energy development plan that could include investing about €10 billion in its energy sector.

The aim would be to reduce dependence on the international energy market.

The country imports about 80% of its oil, 40% of its natural gas, and 23% of its electric power.

Several major energy projects were under discussion at yearend 2008, including:

- Joint construction with Albania of a nuclear power plant in that country
- A pipeline linking the gas networks of Hungary and Croatia.

CZECH REPUBLIC

Capital: Prague
Monetary unit: Koruna
Refining capacity: 183,000 b/cd
Oil production: 4,500 b/d
Oil reserves: 15.0 million bbl
Gas reserves: 140.0 bcf

Russian oil deliveries

In mid-2008, Russia reduced its oil deliveries to the Czech Republic. At the end of July, the cutback totaled 50%, said Czech officials.

The reductions began in early July, one day after the United States formalized a deal with the Czech Republic to deploy part of an antiballistic missile shield on Czech territory. Russia denied the oil reduction was in retaliation for that agreement.

Russia's actions again called into question its reliability as an energy supplier to Europe.

State-owned Transneft (Moscow) blamed the reduction on technical and commercial reasons but declined to say when full operations would resume through the Druzhba Pipeline, said the Czech officials.

The nations of Eastern Europe are concerned that Russia may gradually shift oil exports to the Baltic Pipeline system, which feeds Northern Europe.

That would mean higher oil-shipment fees for most East European countries.

Carpathian operations

Carpathian Resources Ltd. (Sydney) summarized its 2008 operations in the Czech Republic.

- Average production at Janovice gas field well Ja3a dropped in 2008 to 20,000 cu m/d (706,200 cfd) to stabilize a small amount of water overflow.
- At the Mosnov Mo-1 well on the Skotnice prospect, plans were being made mid-year to hydraulically fracture and acid-treat a Miocene sandstone reservoir of interest.
- On the Skalice and Raskovice-Moravka licenses, Carpathian was considering additional seismic surveys. Earlier seismic did not clearly identify drillable targets.
- At mid-year, the company was working to include in a land-use permit its work on a deep Vazany gas prospect, on the Morava permit in the northern part of the Vienna basin.
- In the Roznov project area, 25 km (16 mi) southwest of the Janovice field, the company was assessing a potentially large gas deposit and working to arrange a drilling location.

WEST GREENLAND EXPLORATION BLOCKS

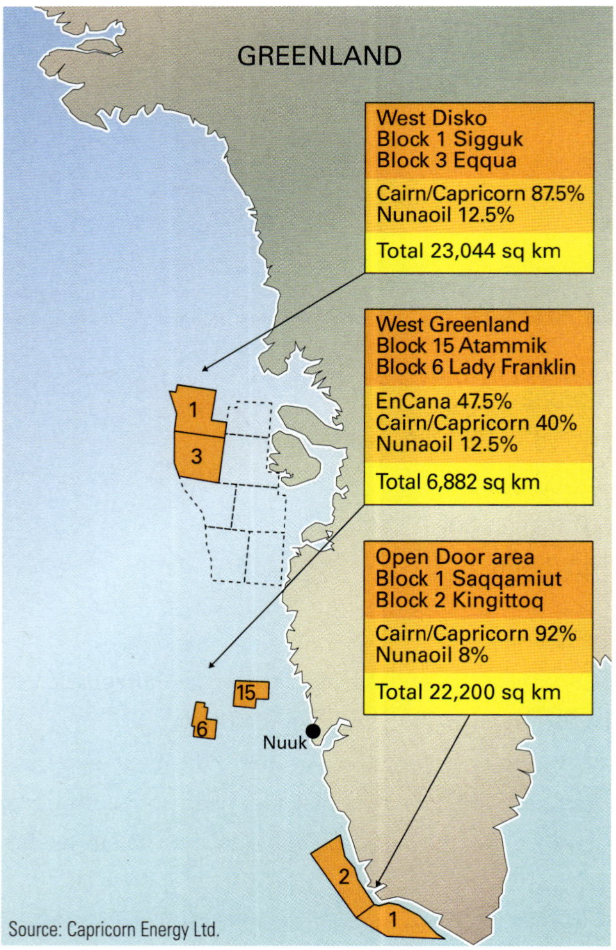

Source: Capricorn Energy Ltd.

GREENLAND ACTION

Separation from Denmark

Since 1979, Greenland has been an "autonomous country" within the Kingdom of Denmark.

In November 2008, a non-binding Greenland referendum on autonomy passed with a 75% approval vote. The Danish parliament said it would honor the results.

The actions described in the referendum are to take effect in June 2009.

Self-rule would be expanded in 30 areas, including language, police and courts, and foreign policy,

Also, changes would be made regarding ownership of oil, gas, and other mineral resources.

One change would give Greenlanders a greater share of annual revenues from oil sales.

Cape Farewell blocks

In October 2008, Greenland awarded Capricorn Energy Ltd., a unit of Cairn Energy Ltd. (Edinburgh), a 92% operated interest in blocks Cape Farewell 1 and 2, off the southern coast.

DENMARK

Capital: Copenhagen
Monetary unit: Krone
Refining capacity: 174,400 b/cd
Oil production: 286,000 b/d
Oil reserves: 1.060 billion bbl
Gas reserves: 2.165 tcf

Exploration and Development

South Tor Pod well

Noreco (Stavanger, Norway) said in mid-2008 that it had discovered oil with its South Tor Pod exploration well in Denmark.

The discovery is 5 km (3 mi) southeast of the producing South Arne field in license 7/89.

Noreco estimated the size of the prospect before drilling at 32 million boe.

The well, seeking hydrocarbons in the Upper Cretaceous Chalk, was drilled to a vertical depth of 2,968 m (9,738 ft) below the sea level.

Blocks 5410–5412

In early 2008, Danica Resources AS (Copenhagen) won a six-year exploration and production license for Block 5410 and parts of 5411 and 5412 in the Baltic Sea offshore Denmark.

The license area covers 6,429 sq km (2,482 sq mi) and includes Lolland and Falster islands.

Preliminary evaluation revealed several structures similar to hydrocarbon-producing structures in northern Poland and Germany.

The remaining interest in each block is held by Nunaoil AS, the national oil company of Greenland.

In total, the two blocks span 20,072 sq km (7,750 sq mi) with water depths from 50–2,200 m (164–7,218 ft).

Capricorn was making plans to acquire 1,780 km (1,106 mi) of 2D seismic over the blocks.

Cairn has said it expects to drill 10 exploration wells over five years off Greenland, starting in 2011,

The USGS estimates there could be 50 billion bbl of oil and gas equivalent offshore Greenland.

Six southwest, west blocks

In early 2008, Capricorn Energy Ltd. (see above) took interests in six blocks off west and southwest Greenland.

The six licenses cover more than 51,800 sq km (20,000 sq mi), in water depths from 300–1,400 m (984–4,593 ft).

Capricorn received about 90% interest in the Sigguk and Eqqua blocks (awarded in the West Disko licensing round) and in the Kingittoq and Saqqamuit blocks (southern Open Door area).

Nunaoil AS holds the remaining interest in those four blocks.

Capricorn also acquired 40% interest in the Atammik and Lady Franklin blocks off West Greenland. Operator EnCana Corp. (Calgary) and Nunaoil hold the remaining interests.

Work programs for al six blocks call for seismic or electromagnetic surveys in 2008–09.

In related action, Iceland was making plans to offer exploration and production licenses in early 2009 in the Dreki area on Jan Mayen Ridge northeast of Iceland.

Survey work

TGS-NOPEC Geophysical Co. (Asker, Norway) launched more than 140,000 km (87,000 mi) of geophysical surveys in two areas off Greenland in early 2008.

The work included 80,000 km (50,000 mi) of aeromagnetic and aerogravity surveys in the Labrador Sea/Ungava basin and 58,000 km (36,000 mi) off northeast Greenland. Data were to be available in fourth quarter 2008.

An 8,300-km (5,160-mi) 2D seismic shoot also was planned in Baffin Bay for third quarter 2008.

Lease Sales

Kúlubokan, Rannva Extension

In late 2008, the Faroe Islands—an autonomous province of Denmark since 1948—awarded two new prospective licenses to Faroe Petroleum (Aberdeen, UK) in the province's third licensing round.

In the Kúlubokan award (33 full and partial blocks in quads 6201, 6202, and 6203), Faroe will share interests with three other parties. The license covers 5,300 sq km (2,046 sq mi).

Rannva Extension award includes nine full and partial blocks in quads 6007 and 6008. Faroe will have 100% interest. The license covers 1,200 sq km (463 sq mi).

The extension is adjacent to Faroe's existing Rannva license, which contains the substantial Rannva "A" lead on the crest of the Wyville-Thomson Ridge.

Drilling and Production

Halfdan field

The Danish Energy Agency gave the go-ahead in mid-2008 to Maersk Oil and Gas AS (Copenhagen) for phase-four development of the Halfdan field.

The firm will expand the well pattern with parallel oil producers and water injectors to the northeast, and build a new process platform with three-phase separation, water disposal, and gas compression.

The new platform, HBD, will be bridge linked to the existing Halfdan HBA platform.

Maersk estimates the work will yield significant additional reserves of both oil and gas.

Nini East field

Dong Energy (Copenhagen) initiated work on an unmanned satellite production platform for its Nini East oil field in the North Sea. The field holds an estimated 17 million bbl of oil.

The site is 7 km (4 mi) northeast of the existing Nini platform.

Dong signed a contract with Bladt Industries (Aalborg Ost, Denmark) for the work.

Production from the new platform is scheduled to start by yearend 2009.

FINLAND

Capital: Helsinki
Monetary unit: Euro
Refining capacity: 255,600 b/cd
Oil production: 0
Oil reserves: 0
Gas reserves: 0

New energy and climate strategy

The government presented its new energy and climate strategy in late 2008, which aims to meet 38% of energy needs with renewable resources by 2020.

That goal is to be achieved with a boost in domestic generation, partly from more use of nuclear power.

State funding of the climate and energy sector will increase from € 350 million in 2007 to € 440 million in 2008 and to € 550 million in 2009.

Priority will be given to power plants that emit little if any greenhouse gases. These could include combined power and heat plants using renewable fuels, as well as water- and wind-power plants.

In addition to continued investments in nuclear energy, the use of wood-based energy, waste fuels, heat pumps, biogas and wind energy will be increased substantially.

Gas, timber deal posed

In late 2008, Russian officials offered to postpone a dramatic increase in export tariffs on round timber (up several hundred percent) if Finland will help the controversial Nord Stream pipeline project advance.

Finland has been pushing hard for a change in the tariffs, which Russia had said it needed in order to strengthen its own timber processing industry.

At the same time, Finland has been reluctant to give the Russian-German Nord Stream consortium access to its continental shelf.

Part of the pipeline will traverse Finnish and Swedish continental waters.

The consortium is aiming to start building the 1,200-km (746-mi) underwater pipeline in 2010.

FRANCE

Capital: Paris
Monetary unit: Euro
Refining capacity: 1,986,197 b/cd
Oil production: 19,500 b/d
Oil reserves: 103.300 million bbl
Gas reserves: 245.000 bcf

Markets

Strike at Marseille

A rolling strike halted oil shipping for several weeks in mid-2008 through the Port of Marseille.

It is the world's third-largest oil port after Rotterdam and Houston, and receives about 40% of France's crude oil.

At one point in late June, 22 tankers were waiting to offload their cargoes at Fos-Lavera terminals, the country's largest oil port.

Port workers protested government plans to privatize the loading activities of seven out of the nine public ports.

LNG

First LNG from Snøhvit

The first shipment to France of LNG from Snøhvit in Norway was delivered in March 2008 to the Montoir-de-Bretagne terminal of Gaz de France (GDF) in Brittany.

The delivery marked the opening of a new LNG supply route capable of providing 700 million cu m/yr of gas for the European market.

LNG accounted for about 31% of GDF's overall gas supplies in 2007. Deliveries from Snøhvit will increase its supply flexibility.

Pipelines

Gazprom plans in Europe

Gazprom officials said in December 2008 that the company wants to acquire gas transport and distribution assets in France and other areas of Europe to bolster the company's gas storage capacities.

Details are presented in the Russia section of the Former Soviet Union chapter of this book.

Nuclear power shift?

French Pres. Nicolas Sarkozy said in mid-2008 that his country was willing to work with Italy on nuclear power as part of a common energy strategy in response to high oil prices.

Details are given in the Italy section of this chapter.

Refining and Products

Provence refinery upgrades

Total SA (Paris) carried out a €100 million turnaround of sections of the 158,000-b/d Provence refinery at La Mede on the French Riviera in October and November 2008.

About 60 projects were carried out to improve safety, cut environmental impact, and optimize production.

The upgrade addressed only the facility's western installations. Eastern units continued to operate.

Fos terminal: Seventh berth

The Marseille Port Board approved in mid-2008 construction of a seventh mooring berth for refined oil products at the Fos petroleum terminal to keep the facility competitive.

The investment will increase the receiving capacity to 21.5 million tons in 2011 from 6 million tons in 2006.

The seventh berth will accommodate 45,000-ton vessels up to 200 m (656 ft) long and provide a 12-m (39-ft) water draft. It will be operational 2011.

Refinery acquisitions

The European Commission approved in early 2008 the $875-million proposal of Petroplus Holdings AG (Zug, Switzerland) to buy two French refineries from Shell International Petroleum Co. Ltd.

The deal was to close by mid-year.

The refineries of interest were the 164,000-b/cd Petit Couronne facility and the 80,000-b/cd Reichstett Vendenheim unit.

Petit Couronne is 130 km (81 mi) northwest of Paris on the Seine River. Reichstett Vendenheim is in Alsace.

Erika tanker verdict

On January 16, 2008, the Paris Criminal Court found Total SA (Paris) liable for damage from the 1999 wreck and sinking of an oil tanker that it had chartered.

When the Erika foundered, 20,000 tonnes of heavy fuel oil polluted the coastline of Brittany.

Observers said that the ruling calls into question long-established procedures in maritime law.

Total said it has already spent €200 million in cleanup efforts and did not dispute the €375,000 fine to compensate victims for "ecological damage."

But it will appeal the court's verdict, disputing the directive requiring it and other parties to pay a total of €192 million in damages for causing the ship to sink.

The court said these damages were caused "through lack of care during the tanker-selection process." Total said it was misled by the certification company, Rina, about the tanker's condition.

Alternative Fuels/Energy

New renewables program

The Environment and Energy Ministry offered in November 2008 a "national plan for renewable energies."

It sets aside, for the time being, any assessment of the European Union's biofuels draft directive to bring the share of biofuels in transport to 10% by 2020.

The plan, involving 50 measures, aims to bring to 23% the share of renewables in France's energy mix by 2020. This corresponds to 20 million tonnes of oil equivalent.

It outlines action to develop biomass, wind, geothermal, hydro, ocean energy, and especially solar energy.

Use of solar energy is targeted to jump to 5,400 MW in 2020 from 13 Mw in 2007.

A €1-billion fund is to be set up during 2009–11, with tax credits extended to 2012.

Futurol project

Total SA (Paris) said in third quarter 2008 that it will participate in Futurol, a major ethanol research and development project.

The project seeks to develop a viable industrial process for producing ethanol by fermenting nonfood lignocellulosic biomass.

Futurol is expected to involve French research organizations, manufacturers, and financial institutions.

First offshore windfarm

Enertrag France SARL (Cergy), subsidiary of Enertrag AG (Dauerthal, Germany), said in late 2008 that it expects to begin construction of the first offshore windfarm in France in second quarter 2009.

The Cote d'Albatre Wind Farm is to be located in the Veulettes-sur-Mer Commune, in the Haut-Normandie region.

Multibrid GmbH (Bremerhaven, Germany) is to deliver 21 model M5000 wind turbines, each with a capacity of 5 MW. Cote d'Albatre total output will be 105 MW.

Turbine installation is to start in July 2009, with construction work to be finished in third quarter 2010.

Company News

GDF-Suez merger

In July 2008, Gaz de France merged with Suez, creating GDF Suez SA, one of the world's largest utility companies. The merged firm is based in Paris.

GDF was a producer, transporter and seller of natural gas in Europe and around the world.

Suez was a French-based multinational corporation with operations in water, electricity, natural gas, and waste management.

Total considers bid for Nexen

In late 2008, Total SA was reported to be pondering a bid of perhaps US $16.1 billion for Calgary-based Nexen, a Canadian oil company.

Nexen has assets in the North Sea, the United States, Canada and Yemen, and with about 250,000 boe/d of production.

GERMANY

Capital: Berlin
Monetary unit: Euro
Refining capacity: 2,417,523 b/cd
Oil production: 60,000 b/d
Oil reserves: 276.000 million bbl
Gas reserves: 6.200 tcf

Stake in Yuzhno Russkoye project

In December 2008, Russia's Gazprom (Moscow) signed an agreement granting German utility E.On AG (Dusseldorf) a stake in the Yuzhno-Russkoye natural gas project in southern Russia.

Severneftegazprom (SNGP) owns and operates the Yuzhno-Russkoye project.

At the bottom line, Gazprom will hold just over 50% in SNGP, while E.On Ruhrgas (a subsidiary of E.On AG) and BASF SA subsidiary Wintershall AG (Kassel) will each have a stake of 25% minus one share.

E.On AG aims to obtain about 6 bcm/yr (212 bcf/yr) from the Russian field, some 60% of its output.

Yuzhno Russkoye is considered one of the world's largest gas fields, with reserves estimated at more than 600 bcm (21.2 tcf).

For 2009, gas production from Yuzhno-Russkoye is expected to reach 25 bcm (883 b cf), about two years ahead of the original timetable.

Renewable energy sector

The *Financial Times* reported in December 2008 that German renewable energy companies were meeting with government and financial officials to discuss ways of preventing the global financial crisis from disrupting the industry.

Government subsidies have helped make Germany a leading manufacturer of solar and wind technology, but those subsidies are likely to be scaled back.

Sales in the renewables sector grew by 10% to US$33 billion in 2007, spurred by rising oil prices. The sector supported about 250,000 jobs.

Projections of growth for such companies were being overshadowed, however, by concerns about project finance, falling prices and delayed orders.

The wind-power industry was seeing the delay of some projects.

Analysts also expect the price of solar components to fall in 2009 amid a supply glut, due in part to greater competition from China.

Germany aims to produce 30% of its energy from renewables sources by 2020.

GREECE

Capital: Athens
Monetary unit: Euro
Refining capacity: 423,000 b/cd
Oil production: 1,300 b/d
Oil reserves: 10.000 million bbl
Gas reserves: 70.000 bcf

Refining

Thessaloniki refinery upgrade

Hellenic Petroleum SA (Athens) awarded EPC contracts to Foster Wheeler Ltd. (Clinton, NJ) in mid-2008 for work at the company's refinery at Thessaloniki in northern Greece.

The upgrade will boost production of low-sulfur fuels and increase the refinery's processing capacity to 26,000 b/sd for light products.

The project is expected to be completed by yearend 2010.

Corinth refinery expansion

Motor Oil (Hellas) Corinth SA (Maroussi) was working in early 2008 to expand its 110,000-b/d refinery in Corinth, about 70 km (43 mi) from Athens.

The company let a contract to Technip (Paris) for EPC work on a 60,000-b/d crude oil distillation unit to be operational early in 2010.

The expansion follows a 2005 upgrade that added a 37,000-b/d mild hydrocracker, a 32,000-b/d gas oil desulfurization unit, and a 65,000-cu m/hr hydrogen production unit.

Pipelines

High-pressure gas network

DESFA SA (Athens), the Hellenic Transmission System Operator, was established in April 2007, after the unbundling

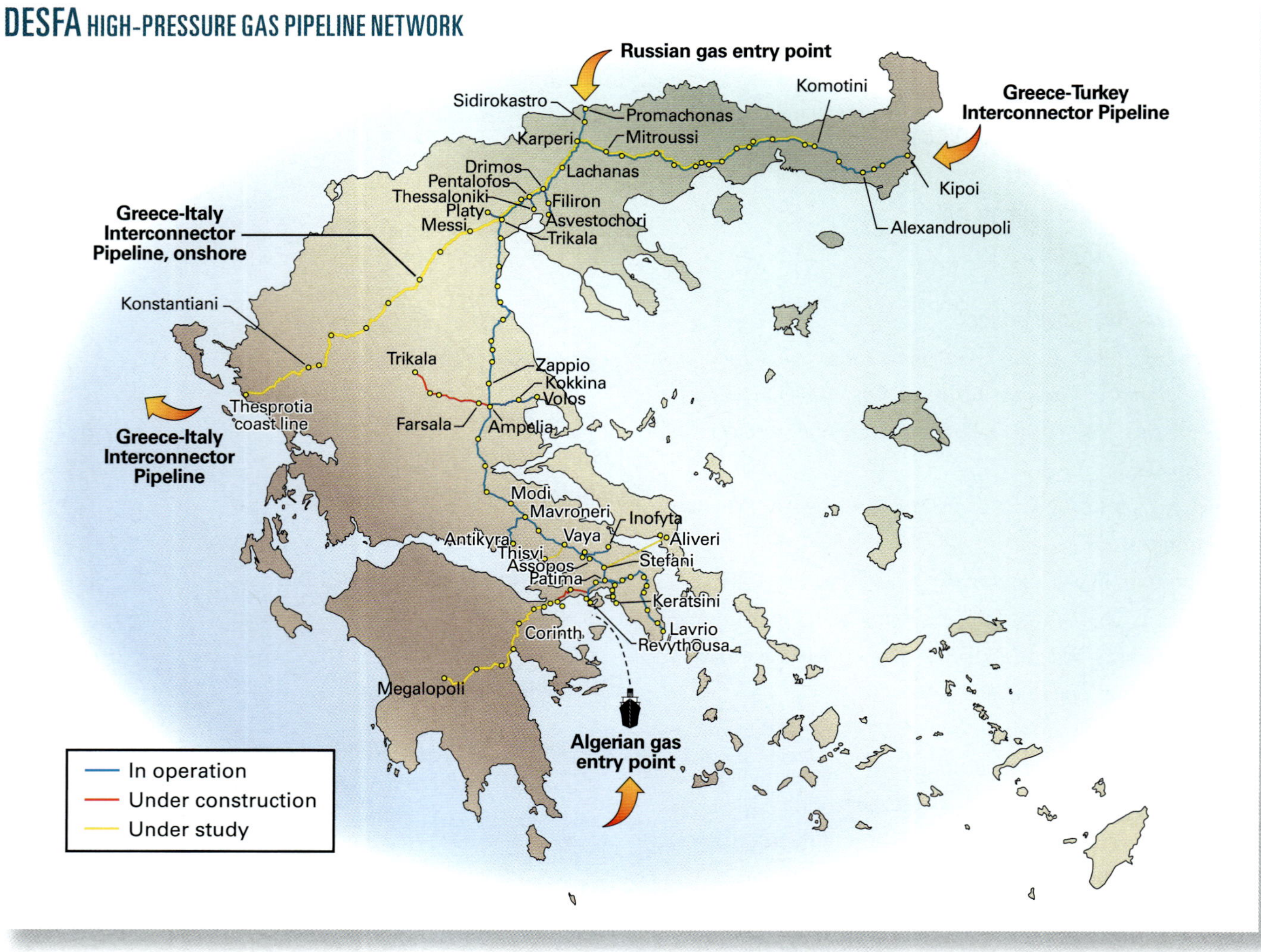

of the transmission and trading activities of the Public Gas Corp. of Greece (DEPA SA; Athens) to liberalize the natural gas market.

DESFA SA owns, operates, and is expanding the Greek natural gas transmission system.

That system comprises more than 1,070 km (665 mi) of high-pressure pipeline, two border metering stations, and one of the largest LNG terminals in Europe (Revythoussa).

Trans-Adriatic Pipeline

Energy trading company Elektrizitats-Gesellschaft Laufenburg (EGL; Laufenburg, Switzerland) and StatoilHydro (Stavanger, Norway) agreed in early 2008 to a joint venture to build and operate the Trans-Adriatic Pipeline (TAP).

The line would transport Caspian and Middle Eastern natural gas to the heart of Europe, providing an alternative to Russian gas.

The $1.5-billion gas pipeline would run 520 km (323 mi) from Thessaloniki through Albanian territory and under the Adriatic Sea to connect with the Italian pipeline grid near Brindisi.

Caspian and Iranian gas would reach Greece by way of an existing network of pipelines that run through Turkey. Greece and Turkey built a gas interconnector that came online in 2007.

A final investment decision on TAP is expected in 2009.

It would become operational no sooner than 2011, with an initial capacity of 10 bcm/yr (353 bcf/yr).

This could rise in the longer term to 20 bcm/yr (706 bcf/yr).

HUNGARY

Capital: Budapest
Monetary unit: Forint
Refining capacity: 161,000 b/cd
Oil production: 14,400 b/d
Oil reserves: 20.180 million bbl
Gas reserves: 286.000 bcf

Tompa block

Toreador Resources Corp. (Dallas) and a JV partner were preparing to drill a deep, high-pressure high-temperature well in late 2008 in the Kiskunhalas trough.

The trough is on the Tompa block in the southeast, near the border with Serbia.

The well was expected to test an 1,800-m (5,905-ft) section of overpressured sands, shales, and conglomerates below a 1,400-m (4,593-ft) shale cap.

The unconventional play in Carpathian gas sands was discovered in the 1980s.

Nyirseg production

Ascent Resources PLC (London) started natural gas production in August 2008 from its PEN-104 well in the Peneszlek area of the Nyirseg permits in eastern Hungary.

Initial production of 48,000 cmd (1.69 MMcfd) was increased over a few days to a target of 85,000 cmd (3.0 MMcfd).

Gas production is transported some 50 km (31 mi) by pipeline to the MOL gas processing facility at Hajduszoboszlo.

The company also planned to shoot 3D seismic over the permits to delineate other gas reservoirs in the vicinity.

Ascent also was considering redevelopment of the Peneszlek field, which produced gas during 1983–91.

Hungary, Romania pipeline connection

Hungary and Romania—citing supply security issues—agreed in mid-2008 to connect their natural gas systems.

FGSZ Natural Gas Transmission Co. (Siófok) is owner/operator of Hungary's gas transmission system.

Transgaz (Arad) is the operator of Romania's gas transmission system.

The interconnecting pipeline will follow the Arad-Szeged route.

FGSZ will build the 47-km (29-mi) Hungarian section of the line, and Transgaz will lay the 67-km (42-mi) Romanian section, of which 36.7 km (23 mi) are already built.

The interconnection should become operational in mid-2010.

Initially, gas will be transported to Romania from Hungary. Later, gas may be transported in the other direction.

REPUBLIC OF IRELAND

Capital: Dublin
Monetary unit: Euro
Refining capacity: 71,000 b/cd
Oil production: 0
Oil reserves: 0
Gas reserves: 350.000 bcf

Offshore blocks

Venture Production PLC (Aberdeen, UK) expected to drill the first two exploration wells on acreage in the southern North Sea offshore Ireland in 2008–09.

Venture was making plans to drill the Carna exploration prospect Block 43/21 during late 2008, and the another well on the Hypnos exploration prospect in block 48/3 in late 2008 or early 2009.

Later in 2009, plans call for drilling an exploration well on the Andrea prospect (block 48/15b in the A fields area), and two appraisal wells on the Cygnus discovery (blocks 44/11 and 44/12).

Venture estimates that the accumulation contains 1 tcf of gas in place.

Permits 1/08, 2/08

Providence Resources (Dublin) was making plans for a mid-2008 2D seismic reflection survey over permits 1/08 and 2/08 offshore.

The survey will acquire up to 3,000 line km (1,864 mi) of data.

The blocks are in 2,000–3,000 m (6,562–9,843 ft) water depths off the southwest coast.

Porcupine basin acreage

Providence Resources and two partners were awarded new license acreage in early 2008, offshore in the Porcupine basin.

The license area covers four blocks and is adjacent to the Providence-operated Spanish Point License, which holds the Spanish Point gas condensate accumulation and the Burren oil discovery.

Providence plans

Providence Resources in mid-2008 was proceeding with plans for a drilling program in the North Celtic Sea basin, offshore southern Ireland

The work is to be guided by data collected from the firm's 50/11-3 oil discovery on the Hook Head structure.

Analysis suggested resources of up to 60 MMbbl of oil recoverable based on a STOIIP of 390 MMbbl of oil. The northwest sector of the Hook Head structure was targeted for first evaluation.

Providence also said it estimates significant oil potential—perhaps 18 million bbl—in an adjacent discovery known as Dunmore, in blocks 50/6 and 50/7.

Corrib project

In December 2008, Royal Dutch Shell PLC (The Hague, Netherlands) withdrew its planning application for a modified onshore pipeline for the Corrib natural gas project.

The project includes a 93-km (58-mi) pipeline—9 km (6 mi) of it onshore—to carry gas from a field 3,500 m (11,483 ft) below sea level.

Part of the line would cross under Broadhaven Bay, an area of conservation under EU law.

As of mid-2008, Shell E&P Ireland said it had spent more than €200 million on the project, and that one-third of its gas processing terminal was already built.

The national planning board had been seeking more information about the new pipeline route, chiefly regarding environmental impacts.

The proposed high-pressure onshore line would run from landfall at Glengad under Dooncarton Mountain, site of several landslides in 2003.

Other factors said to have influenced the Shell decision were a delay until 2009 in laying the offshore pipeline, and new rules about disposal of peat unearthed during pipe-laying.

A revised application under fast-tracking legislation likely will be lodged in 2009. Shell said it would seek minor onshore route realignments to avoid sensitive habitats.

It was not clear that the changes would satisfy local opponents.

They want Shell to process the gas at a shallow-water platform, fearing that onshore processing at Bellanaboy would bring a pipeline too close to homes.

LNG terminal project

Shannon LNG Ltd. (Dublin) has been working since 2006 to build an LNG receiving and re-gasification terminal on a site between Tarbert and Ballylongford.

The firm said in January 2009 that construction was expected to start in late 2009, despite some speculation from a U.S. source about possible delay.

The terminal will consist of up to four LNG storage tanks, several jetties for unloading, and a regasification train capable of supplying 3 mtpy (400 MMcfd) of LNG.

A 30-km (19-mi) pipeline will connect the facility to the Irish national gas grid.

Planners said the facility should be ready to accept its first shipments of gas by 2012.

Carbon sequestration, gas storage

In mid-2008, Providence Resources started the Ulysses project to evaluate the potential for sequestering carbon or storing natural gas in the Kish Bank basin, offshore Ireland.

The study is a 50/50 JV with Star Energy Group PLC (London).

The firms won a three-year licensing option from the Irish government for eight blocks. Work will include oil and gas exploration as well as the Ulysses study.

ITALY

Capital: Rome
Monetary unit: Euro
Refining capacity: 2,337,229 b/cd
Oil production: 99,000 b/d
Oil reserves: 406.500 million bbl
Gas reserves: 3.325 tcf

Italy is Europe's third largest producer of oil and gas. Most new activity in the past decade has focused onshore at fields such as the prolific Val d'Agri in the south.

Offshore, the dominance of Eni SPA (Rome) has been virtually unchallenged since Royal Dutch Shell (The Hague, Netherlands) and BG Group (London) exited their licenses earlier this decade.

Eni's offshore oil production comes mainly from the Vega field off southern Sicily, and Aquila off the coast of Puglia. It also produces gas from numerous fields in the northern Adriatic, although expansion has been constrained for environmental reasons.

In recent years, it has made further small gas discoveries such as Argo offshore Sicily, and Benedetta in the Adriatic.

However, there are numerous other "stranded" fields in Italian waters that the government is eager to monetize.

Exploration and Development

Castello gas field

Italy granted to Po Valley Energy Ltd. a 20-yr production concession for Castello gas field near Milan in late 2008.

The firm, based in Perth, Australia, has an office in Rome.

It was the first concession grant in the Po Valley region of northern Italy since the country's gas sector was deregulated in 1998.

Po Valley Energy had completed construction of a surface plant for Castello, and connection of the pipeline to the Italian national gas grid was to be completed early in 2009.

Gas production is scheduled to begin in second-quarter 2009.

The field will be produced from a single well at an initial production rate of about 2.7 MMcfd of gas. The field's remaining proved gas reserves are put at 4.6 bcf of gas.

Sicilian Strait

Eni SPA (Rome) announced plans in late 2008 to fast-track the development of its offshore discoveries in the Sicilian Strait.

The announcement followed tests that flowed 170,000 cmd (6 MMcmd) of natural gas from the Argo 2 well, located 20 km (12 mi) offshore Agrigento in 500 m (1,640 ft) of water.

Eni also found gas in deeper water in Argo 2.

Panda and Cassiopeia fields are nearby, and Eni plans to link them under its development plan.

In August, Eni had tested 190,000 cmd (6.7 MMcfd) of gas from the Cassiopeia 1 well in 560 m (1,837 ft) of water 22 km (14 mi) off Agrigento. Cassiopeia's reserves are estimated at 16 bcm (565 bcf).

Eni estimated the Sicilian Strait area potential at 18 bcm recoverable.

Ombrina Mare oil field

Mediterranean Oil & Gas PLC (London) was working in mid-2008 to assess contingent and prospective oil and gas resources on the BR 269 GC permit in the Adriatic Sea.

Consulting engineers certified the proved and probable recoverable oil volumes at the Ombrina Mare oil and gas field at 20 million bbl, four times the previously estimated level.

Ombrina Mare is 7 km (4 mi) offshore Italy in the Adriatic Sea, in approximately 20 m (66 ft) of water.

These volumes relate only to that part of the field appraised by the horizontal OM-2 well in May 2008.

The permit is, west of giant Rospo Mare heavy oil field. A number of other oil prospects and gas leads were identified.

OM-2 was said to be the first offshore well operated by a junior oil and gas company in Italian waters in more than 25 years.

Nine Adriatic permits

Petroceltic Elsa Srl—subsidiary of Petroceltic International PLC (Dublin)—was granted nine exclusive exploration permits in mid-2008 in the central Adriatic Sea off Italy: permits d492 and d493.

These interests together with the others it holds give Petroceltic access to a near-contiguous proven hydrocarbon play fairway of over 2,660 sq km (1,027 sq mi) along the Apulian Carbonate platform margin offshore Italy.

The latest permits, in water depth of 30–150 m (98–492 ft), are adjacent to existing oil and gas fields that have demonstrated three working hydrocarbon plays in the region.

Seismic data revealed a number of prospects and leads in the permits.

Petroceltic has 100% working interest in the nine permit awards, which are d492 through d500.

Independents active off southern Italy

Foreign operators were making a comeback offshore Italy in 2008.

A new group of companies was applying for blocks in the Adriatic and Ionian seas, with the encouragement of the country's licensing bureau.

OFFSHORE LICENSES, WEST AND SOUTH OF SICILY

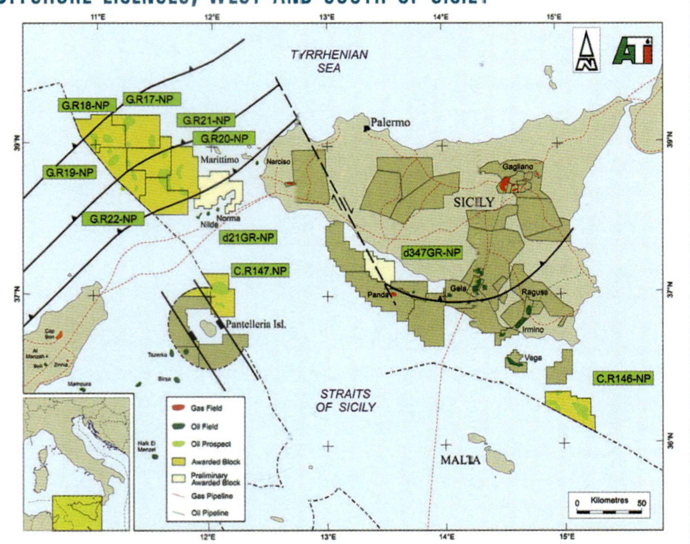

There are numerous "stranded" fields in Italian waters. Some have been made available in 2008 concessions, such as Ombrina Mare off the central eastern coast.

There, Mediterranean Oil & Gas Ltd. (London) was preparing to spud an appraisal well early in 2008 seeking to prove reserves of up to 90 MMbbl.

Farther south, Northern Petroleum PLC (London) was re-evaluating the Rovesti oil discovery of Eni SPA (Rome) and the Giove oil discovery of Enterprise Oil PLC (London) offshore Brindisi.

The company also has proven gas accumulations in a permit offshore Calabria, close to Sicily.

Of the newcomers, said editors at PennWell's *Offshore* magazine, Northern was by far the most committed to offshore exploration.

It acquired operatorship of 11 licenses in the Sicily Channel, the southern Adriatic, offshore Calabria, and the Tiber Delta in the Tyrrhenian Sea since 2004. It also shares three onshore exploration licenses in the Po Valley, and another on the Adriatic coast.

Oil discoveries close to Aquila field

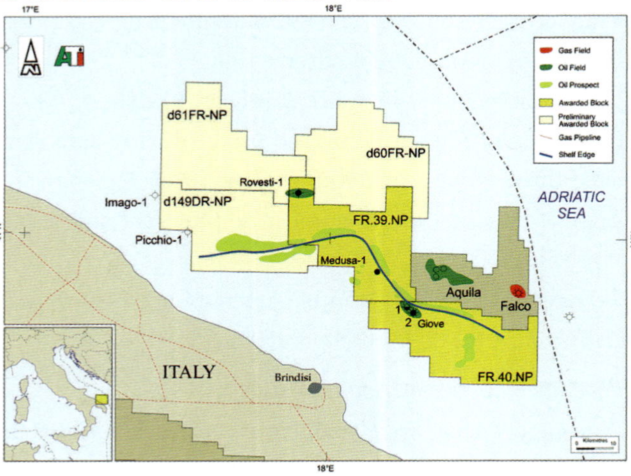

LNG

First offshore LNG terminal

The Adriatic terminal—called the world's first offshore LNG terminal—was being installed off the Italian coast in September 2008.

The gravity-based terminal is expected to be able to supply about 10% of Italy's natural gas requirements.

It is designed to store and re-gasify LNG to deliver 775 MMcfd (8 bcm/yr) of natural gas when it reaches full operational capacity in 2009.

Built in Algeciras, Spain, the terminal is positioned about 15 km (10 mi) off Porto Levante in the Adriatic Sea in about 28 m (95 ft) of water.

Swiss gas deal with Iran

In March 2008, Switzerland's Elektrizitaets-Gesellschaft Laufenburg (EGL; Laufenburg) signed a 25-year gas procurement contract with National Iranian Gas Export Co. (NIGEC), a subsidiary of the National Iranian Oil Company (NIOC; Tehran).

EGL said the deal is intended to secure gas supplies for Switzerland, diversify the supply of natural gas to Europe, and reduce dependence on Russian supplies.

The contracted gas quantities will allow EGL to cover its 50% share of transportation capacity of the planned Trans Adriatic Pipeline (TAP).

EGL will build the TAP with StatoilHydro ASA (Stavanger, Norway).

A first delivery of natural gas by NIGEC to EGL is planned for 2009.

Larger volumes are expected to follow in 2012, when the TAP is to become operational.

Deliveries could reach as much as 5.5 bcm/yr (194 bcf/yr) in a second phase of the contract.

U.S. officials criticized the agreement, saying it "violates the spirit of the sanctions" imposed on Iran by the UN Security Council.

Those sanctions are part of an effort to persuade Iran to give up its program of uranium enrichment and to cooperate further with the International Atomic Energy Agency.

A pipeline will connect the terminal to Italy's natural-gas grid.

Partners in the Adriatic LNG project are ExxonMobil Italiana Gas Srl (Rome), Qatar Terminal (subsidiary of Qatar Petroleum; Doha), and Edison SPA (Milan).

Gas will come to the Adriatic LNG terminal from Qatar's North field.

Five terminals in the works

In October 2007, Italy approved plans to simplify and accelerate the permitting process for new LNG terminals.

As a result, several companies were pursuing development of LNG facilities in mid-2008.

Anonima Petroli Italiana SPA (Rome) applied for approval from the country's Environment Ministry for an offshore regasification terminal with a capacity of about 4 bcm/yr (141 bcf/yr).

The facility will be located in the Adriatic Sea near the company's Falconara refinery.

Gas from the facility would be injected into the grid of Snam Rete Gas SPA (Milan).

Construction was expected to take about six months.

Gas Natural SDG SA (Barcelona) was seeking environmental clearance for an 8-bcm regasification terminal to be developed at Trieste in northern Italy.

Iride SPA (Turin) and Sorgenia SPA—a unit of Compagnie Industriali Riunite (CIR SPA; Milan)—said their planned 12-bcm regasification terminal at Gioia Tauro won environmental approval and was granted access to European Union funding. The facility is to be in service in 2013.

Royal Dutch Shell PLC (The Hague, Netherlands) and ERG Power & Gas SPA (Rome) received local approval to build the Ionio LNG import terminal in Sicily. Construction is to start in 2010, with operation to begin in 2013.

GDF Suez (Paris) and Hoegh LNG (Oslo), after a feasibility study, signed a joint agreement to develop the Triton LNG project, 30 km (18.6 mi) off Italy's Marches region in the Adriatic Sea.

The facility would include a floating storage and regasification unit vessel.

Pipelines

IGI pipeline

The Italy-Greece Interconnector (IGI) is part of a larger project that will deliver 9 bcm/yr (318 bcf/yr) of gas from the Caspian to Italy and Western Europe by way of Turkey and Greece.

Being designed by Edison SPA and Greece's national hydrocarbon company, Depa, the line will help Italy and other nations diversify their natural gas sources.

A 600-km (373-mi) segment is to be built in Greece, and a 200-km (124-mi) marine section will run between the Greek coast and Italy's Apulia region.

IGI Poseidon SA was created in mid-2008 to design, develop and manage the marine section.

To meet a 2012 in-service target date, gas supply and transit agreements must be finalized by the end of 2009 to support final investment decisions.

During the next phase, the IGI consortium will ask interested shippers to submit binding offers to reserve transportation capacity, which will be followed by an allocation stage.

The EU has given the highest priority to the IGI project, calling it a Project of European Interest.

Galsi project

In July 2008, Galsi SPA (Milan) filed an application seeking a permit to build and operate the Italian segment of the Galsi natural gas pipeline and a ruling on the project's environmental compatibility.

Galsi SPA was established in 2003 to develop, build and operate a new 840-km (522-mi) natural gas pipeline to carry 8 bcm/yr from Algeria to Tuscany by way of Sardinia.

The company also asked for the start of the project's environmental impact assessment.

Construction can begin after the regulatory approval process is completed and a comprehensive permit is issued. Galsi SPA signed in late 2007 an agreement with Snam Rete Gas regarding construction.

As of mid-2008, the company said that the pipeline's final route had been identified and an environmental impact study completed.

In July, Snam Rete Gas reportedly found some problems with stability in seismic areas of the original route, but it was not clear if this would delay the project.

The Italian section of the Galsi pipeline will cover a total of 596 km (370 mi).

A first submarine segment of 45 km (28 mi), making landfall at Porto Botte, will be followed by an overland segment of 272 km (169 mi) crossing Sardinia.

EASTERN EUROPE
Copenhagen
Malmo
BORNHOLM
RUGEN
Reinkenhagen
Kolberg
Rostock
GERMANY
Stettin
Schwedt
Berlin
Vestrup
Rudersdorf
Roxforrde
Frankfut-an-der-Oder
Magdeburg
Staakow
Grunberg
Cottbus
Guben
Leipzig
Dresden
Jena
Zeitz
Litvinov
Kralupy
Pardubice
Kolin
Prague
Waidhaus
CZECH REP.
Brno
Trebio
Znojmo
Linz
Vienna
Bratislava
Munich
Salzburg
AUSTRIA
Graz
Maribor
Paluzza
SLOVENIA
Ljubljana
Zagreb
Belluno
Trieste
Koper
Rijeka
Porto Marghera
Ferrara
Bologna
Ravenna
Florence
San Marino
Falconara
Ancona
ITALY
Marta
Rome
Gaeta
Naples
Terracina
Petroliard-Farnesina
Bari
Taranto
Brindisi
Giovinazzo
Crotone
Luna
C. Cimiti
San Leonardo
Vibo Valentia
Rosarno
Lavinia
Bagnara
Messina
Milazzo
Palermo
SICILY
Mazara del Vallo
Marsala
Porta Enpedocle
Augusta
Catania
Gela
Perla
Ragusa
Vega
MALTA
Tyrrhenian Sea
Adriatic Sea
Ionian Sea
Mediterranean Sea
Gulf of Venice
Gdansk
Kaliningrad
Krasnoborskoye
Ushakovskoye
Kaunas
Vilnius
Grodno
Plock
Warsaw
Poznan
Lodz
POLAND
Radom
Pulawy
Kielce
Sandomierce
Rozwadow
Trzebina
Grobla
Partinia
Lubaczow
Jaroslaw
Jaslo
Rudki
Lvov
Czechowice
Strazske
Gorlice
Uzhgorod
Stryj
Dolina
Dashava
Nadvornaya
Kosmachskoye
Krasnoilsk
Kosov
Chernortsy
SLOVAKIA
Miskolo
Budapest
HUNGARY
Szeged
Arad
Timisoara
Resita
Pecs
Baranovich
BELARUS
Minsk
Davydovskoye
Vishanskoye
Mozyr
Brest
Kobrin
Pinsk
Ovruch
Kovel
Lutsk
Rovno
Dubno
Brody
Ternopol
Unecha
Kursk
Kharkov
Lisichansk
Kiev
Berdichev
Staro-Konstantinov
Vinnitsa
UKRAINE
Kremenchug
MOLDOVA
Ugeny
Kishinev
Odessa
Kherson
ROMANIA
Bacau
Dradanesti
Ploiesti
Pitesti
Bucharest
Cimpina
Turnu-Severin
Giurgiu
Constanta
Navodari
Midia
Black Sea
Varna
Kavarna
Tollouknin
BULGARIA
Sofia
Vraca
Stara Zagora
Plovdiv
Burgas
Otmanli
Belgrade
Pancevo
SERBIA AND MONTENEGRO
Aleksinac
Nis
Pristina
BOSNIA & HERZEGOVINA
Sarajevo
Banja Luka
Bosanski Brod
Sisak
CROATIA
Split
Dubrovnik
Gospic
Pula
Shkoder
ALBANIA
Tirana
Durres
Vlore
MACEDONIA
Skopje
Thessaloniki
Kavala
Komotini
Alexandroupolis
GREECE
Larissa
Trikala
Arta
Delfi
Aspropyrgos
Athens
Elefsina
Argos
Tripoli
Agean Sea
Gulf of Corinth
CRETE
Khania
Rethimnon
Iraklion
Rodhos
TURKEY
Istanbul
Izmit
Bursa
Bandirma
Burhaniye
Izmir
Soke
Eregli
S.Eregli
Cst.Eregli
Eregli Deep
Aliphi
S. Akcakoca
Ayazli
Cst. Akcakoca
3 Lines
2 Lines
0 50 100 150 Miles
0 50 100 150 200 Kilometers

A second submarine segment will then reach the coast of Tuscany near Piombino.

In some places, the line will be under 2,800 m (9,186 ft) of water, making it one of the world's deepest offshore pipelines ever laid.

At Piombino, the line will link up with the national transmission network.

Other partners in the Galsi project are Sonatrach (Algeria) Edison SPA (Milan), Enel (Rome), Hera Trading Srl (Bologna), and the Regional Administration of Sardinia through its Sfirs subsidiary.

Trans-Adriatic Pipeline

Energy trading company EGL (Laufenburg, Switzerland) and StatoilHydro (Stavanger, Norway) agreed in early 2008 to a joint venture to build and operate the Trans-Adriatic Pipeline (TAP).

The line would transport Caspian and Middle Eastern natural gas to the heart of Europe, providing an alternative to Russian gas.

One section of the $1.5-billion gas pipeline would run under the Adriatic Sea to connect with the Italian pipeline grid near Brindisi.

Details about the project are shown in the Greece section of this chapter.

Refining and Products

Isab refinery venture

OAO Lukoil (Moscow) reported in mid-2008 that it will pay €1.3 billion to acquire a 49% stake in a new JV with ERG SPA (Milan). ERG will have a 51% share.

The venture will jointly operate ERG's 320,000-b/d Isab refining complex on the coast at Priolo, Sicily.

Both parties require approval from the European Commission and other antitrust bodies.

The refinery will produce mainly kerosene and diesel fuel for European markets.

Refinery relocation to Pakistan

TransAsia Refinery Ltd. (Karachi) said in mid-2008 that it plans to relocate a 100,000-b/d refinery from Naples to Port Qasim, Karachi.

The company was awaiting approval from environmental and government officials in Pakistan.

The project is scheduled for completion by April 2010.

Alternative Energy

Enel Green Power

Enel SPA launched a renewable energy division called Enel Green Power in late 2008.

Enel Green will focus on expanding Enel's renewable energy business and generating power from clean sources of energy in Italy and abroad.

The new company will take over Enel's existing operations in the wind, solar, geothermal, hydroelectric and biomass renewable energy sectors across Europe, North America and Central and South America.

Around the world, Enel Green has an installed capacity 4,300 MW and a generation output of over 17 billion kW-hrs.

Windfarm in Sicily

Theolia SA (Aix-en-Provence, France) said in late 2008 that it was proceeding with plans to build a 30-MW windfarm in the Giunchetto area of Sicily.

Theolia will work with its Italian subsidiary, Maestrale Green Energy Srl (Milan).

In February 2008, Sicily approved the environmental-impact evaluation for the $63.5-million project.

Some 35 wind turbines will be sited between Leonforte, Nicosia and Nissoria.

The 85-kW turbines will be delivered and installed by Vestas Wind Systems AS (Randers, Denmark).

Construction is expected to begin in early 2009, with project completion in early 2010.

The windfarm will provide energy for almost 27,000 households.

Environment

First CCS project

Eni SPA and utility Enel signed a strategic cooperation agreement in late 2008 under which they will integrate their carbon capture and storage (CCS) projects to build a pilot plant in Brindisi.

According to a separate MOU, the two will also work with the Environment Ministry on developing CO_2-capture and renewable-energy technologies.

At the Brindisi thermal power station, Enel was working on a pilot plant that will be able to remove 2.5 tonnes/hr of gas and transport it to the Cortemaggiore site in fall 2009.

Eni has started an injection project that could offload 8,000 tpy of CO_2 into depleted, exhausted Stogit field at Cortemaggiore (Piacenza) in fall 2010.

Both firms have committed to build a pilot dense-phase CO_2 transport line at the Brindisi site to bolster their experience in transporting CO_2.

Company News

Exploration deal with PNG

Eni SPA (Rome) signed an agreement in late 2008 with Papua New Guinea (PNG) for a long-term partnership to pursue sustainable development of PNG's hydrocarbon resources.

Details are given in the Papua New Guinea section of the Asia-Pacific chapter of this book.

Eni, Petrobras cooperation

Eni SPA (Rome) and Petrobras (Rio de Janeiro) renewed their commitment in mid-2008 to work closely together on upstream and downstream operations and to conduct feasibility studies on renewable fuels in Brazil and elsewhere.

Details are given in the Brazil section of the Latin America chapter of this book.

MALTA

Capital: Valletta
Monetary unit: Lira
Refining capacity: 0
Oil production: 0
Oil reserves: 0
Gas reserves: 0

Area 4 PSC

In October 2008, Leni Gas & Oil (London) announced that a production sharing contract with its JV partner Mediterranean Oil & Gas PLC (MOG; London) was formally approved.

The license covers Area 4 offshore Malta, consisting of Blocks 4, 5, 6, and 7.

Area 4 lies on the Pelagian Shelf south of Malta, abutting the boundary with Libyan offshore Block 44.

Malta Oil Pty Ltd.—a subsidiary of MOG—will be contractor and operator.

The three most mature prospects are in Block 7 along the ramp setting of the Melita-Medina Graben.

The work program will acquire electromagnetic and gravity data, execute depth re-processing on the acquired 3D seismic, and acquire and interpret non-seismic data.

The prospects and leads on the 5,700-sq km (2,200-sq mi) Area 4 have been delineated, with its most likely prospective recoverable oil resources estimated at 1.47 million bbl of oil.

A 2010 drilling campaign is planned.

Areas 2 and 7 exploration

The Oil Exploration Department said in November 2008 that it expects a unit of Heritage Oil Ltd. (St. Helier, Jersey, Channel Islands, UK) to drill an offshore oil exploration well by 2010.

Heritage Oil International Malta Ltd.—a wholly owned subsidiary of Heritage Oil Ltd.—holds a PSC with the government that requires the drilling of at least one exploration well by 2010.

Heritage Oil holds Areas 2 and 7, south-southeast of Malta, and has committed to invest at least US$22 million in exploration activities from 2008–2010.

The company has identified several prospects with potential reserves of over 1.3 billion bbl of oil in water depth ranging from 80–350 m (262– 1,148 ft).

Libya opposes award to Heritage

In mid-2008, National Oil Corporation (NOC; Tripoli) protested the December 2007 award to Heritage Oil Ltd. of Areas 2 and 7 offshore Malta.

The two concessions total almost 18,000 sq km (6,950 sq mi).

Malta and Libya have been negotiating joint exploration of the region between them for some time but have not concluded an agreement.

Libya contends that the Heritage concessions lie on the Libyan continental shelf and that part of the area was licensed by NOC to Sirte Oil Co. (Mersa al Brega, Libya).

A determination of sovereignty likely will be needed before any exploration operations can proceed.

NETHERLANDS

Capital: Amsterdam
Monetary unit: Euro
Refining capacity: 1,207,714 b/cd
Oil production: 35,000 b/d
Oil reserves: 100.000 million bbl
Gas reserves: 50.000 tcf

M7-A project

Cirrus Energy Corp. (Calgary) was anticipating first gas from its M7-A project in the Dutch North Sea by mid-2009.

The firm expected that a pipeline connecting M7A with the host L09-FF process platform would be in place by September 2008.

North Sea platforms

Total E&P Nederland BV (Groningen, Netherlands) established a five-year agreement in early 2008 for construction of 21 production platforms in the Dutch sector of the North Sea.

The agreement is with Fabricom Oil & Gas BV (Ijmuiden, Netherlands).

Bergermeer gas storage project

Gazprom Exports (Moscow) and TAQA Energy BV, a wholly owned subsidiary of Abu Dhabi National Energy Co. (Abu Dhabi, UAE), signed an MOU in late 2008 to partner on the Bergermeer project.

The project would convert the depleted Bergermeer gas reservoir north of Amsterdam into Europe's largest new seasonal gas storage facility.

Cost is estimated at $1.5 billion for the reservoir, which will have a capacity of 4.1 bcm (145 bcf).

Also in the Bergermeer consortium are Energie Beheer Nederland (Heerlen, Netherlands), Dyas BV (Utrecht, Netherlands), and Petro-Canada Ltd. (Calgary).

If a final investment decision is made as planned in first quarter 2009, construction would begin by midyear, aiming for commercial operation in mid-2013.

The Bergermeer reservoir project is of great interest to Gazprom because it will be along the route of the planned Nord Stream pipeline that will carry Russian gas to Europe..

Gazprom has said that it likely will spend $2.8 billion in 2009 to support construction of the overland portion of Nord Stream.

Energy ports

Dutch port operator Havenbedrijf Rotterdam NV and state-owned Qatar Petroleum (QP; Doha) signed an agreement in mid-2008 regarding port management.

Details are given in the Qatar section of the Middle East chapter of this book.

Gate LNG terminal

In mid-2008, Dutch officials approved construction of the Gate LNG terminal, the first LNG import terminal to be built in the Netherlands. It is slated to come on line by 2012 in Rotterdam.

Also, E.On Ruhrgas AG became the fourth firm to formalize a supply agreement with Gate LNG, agreeing to send 3 bcm/yr (106 bcf/yr) of regasified LNG through the facility.

The other supply agreements are with Dong Energy (Copenhagen) EconGas GmbH (Vienna), and Essent NV (Arnhem, Netherlands).

Gate project partners are Neder Nederlandse Gasunie NV (Groningen) and Royal Vopak (Rotterdam).

Planned future expansion will raise total throughput capacity to 16 bcm/yr (565 bcf/yr) from the initial 9 bcm/yr (318 bcf/yr).

LyondellBasell

LyondellBasell Industries (Rotterdam)—the world's third-largest petrochemicals company—filed in early January 2009 to place 79 of its global affiliates under bankruptcy protection.

The move followed an unsuccessful attempt to restructure its $26 billion debt.

The company was created in 2007 when Basell Holdings BV (Hoofddorp, Netherlands) bought Lyondell Chemical Co. (Houston).

Several hundred affiliates in Europe, Asia and other regions were not included in the bankruptcy filing, according to the company.

LyondellBasell had been struggling to restructure the debt it took on as part of its 2007 merger.

NORWAY

Capital: Oslo
Monetary unit: Kroner
Refining capacity: 315,000 b/cd
Oil production: 2.125 million b/d
Oil reserves: 6.680 billion bbl
Gas reserves: 81.680 tcf

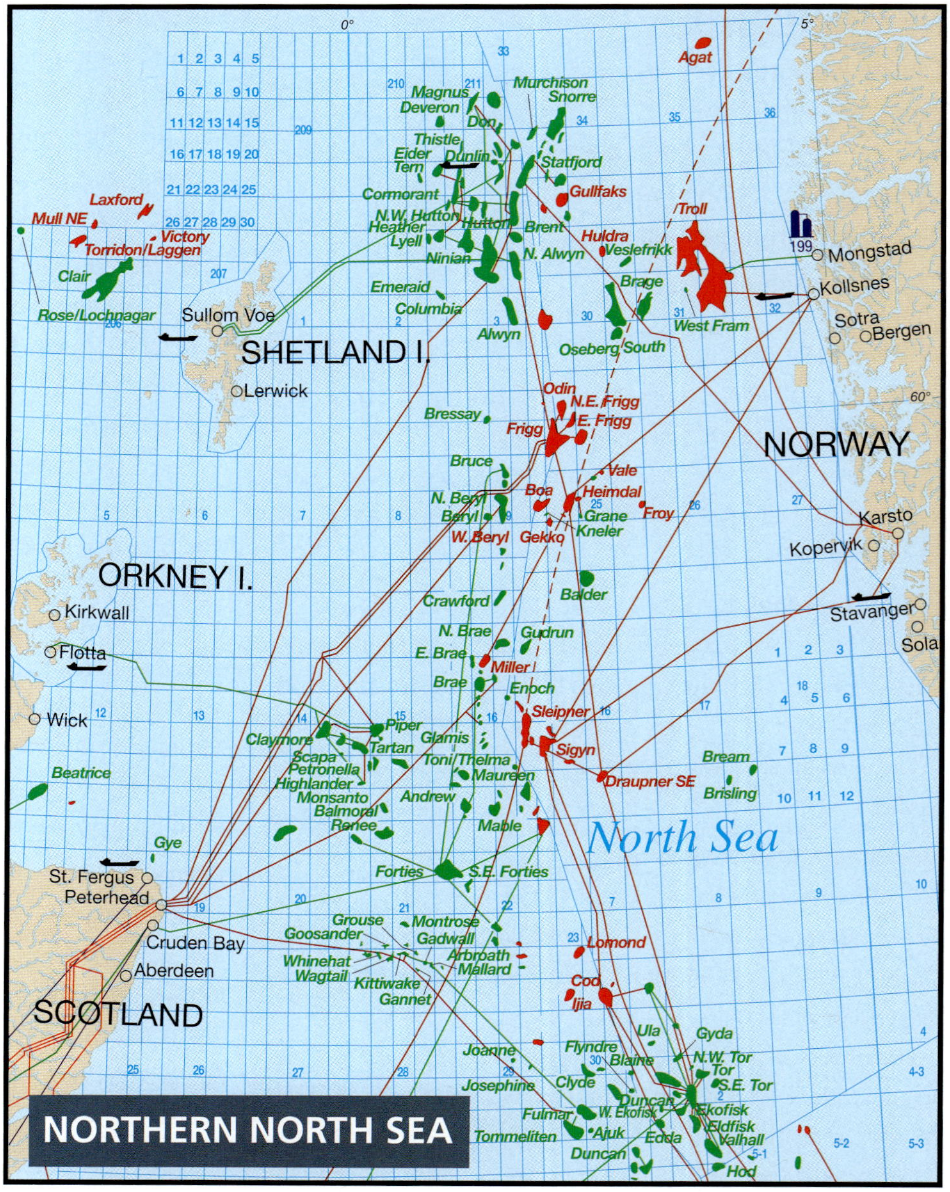

Overview

According to the Norwegian Petroleum Directorate (NPD), Norwegian oil production, including NGLs, was about 2.25 million b/d in 2007, slightly lower than in 2006.

However, production was expected to gradually trail off in coming years because of maturing Norwegian continental shelf fields.

The agency also projected that gas sales would reach 100 bcm in 2008—about 10% higher than in 2007. Sales are expected to reach 110 bcm (3.88 tcf) in 2009.

Norway is the world's third largest gas exporter, delivering 85 bcm/yr (3 tcf/yr).

When the giant Ormen Lange field reaches plateau production in 2010, total gas exports from Norway are expected to increase to120 bcm/yr (4.23 tcm/yr).

Operators were planning 125 billion kroner of investment in 2008, including exploration, about 10% higher than in 2007.

Norwegian Continental Shelf (NCS)

Norway's petroleum ministry said operators spent about $17.6 billion on the NCS in 2007—about $2.2 billion more than the agency's previous estimate.

Operators were expected to drill 35–40 exploration wells on the NCS in 2008, said NPD early in the year.

NPD reported 32 wells were spudded in 2007, resulting in 12 small discoveries.

Of the 32 total, 20 were wildcat wells and 12 were appraisal wells.

NPD said that the shelf produced 1.5 billion boe in 2007, some 164 million boe less than the record set in 2004.

Around ten plans for development and operation (PDOs) were expected to be submitted to NPD in 2008.

Production started from four oil and gas fields in 2007: Blane, Enoch, Ormen Lange, and Snøhvit.

Three more fields were expected to start producing in 2008: Alvheim, Vilje, and Volve.

Exploration and Development

StatoilHydro: Plans for 2009

StatoilHydro ASA (Stavanger) said in January 2009 that it planned to cut its exploration spending and is likely would have to borrow to fund its investment program because of low oil prices.

(Statoil merged with the oil and gas business of Norsk Hydro in 2007 to create StatoilHydro.)

Nonetheless, the company told the *Financial Times* that its strategy remained unchanged, and stuck with its target for production growth by 2012.

StatoilHydro said it plans to invest about $13.5 billion in capital spending for 2009, down from $16 billion in 2008, although that included about $3.5 billion for acquisitions.

Spending on exploration for new sources of oil and gas is being cut by about 13%, to $2.7 billion.

The company plans to drill 65–70 exploration wells in 2009, down from 79 in 2008, focusing on its best prospects.

StatoilHydro: Long view of NCS

Officials with StatoilHydro said in August 2008 that they expect new field projects and actions to increase oil and gas production to add 600,000 b/d to the firm's production from the Norwegian continental shelf (NCS) by 2015.

New field development projects are designed to add 300,000 boe/d in 2015 alone. StatoilHydro has field development plans for the Gudrun, Trestakk, Valemon, Astero, Peon, Luva, Lavrans, and Dagny fields.

Enhanced recovery from fields already on stream will account for 40–50% of the company's total new production on the NCS in the next 5–10 years.

Targets for enhanced recovery and for extension of field life and facilities already include Gullfaks, Oseberg, Grane, Sleipner West and Tune.

The company said it plans to drill some 80 new production wells and some 35 exploration wells on the NCS in 2008.

As of mid-2008, StatoilHydro:

- submitted PDOs for three development projects: Yttergryta, Morvin and further development of the Troll field.
- put six new fields onstream on the NCS: Gulltopp, Oseberg Gamma Main Statfjord, Vigdis East, Theta Cook, Oseberg Delta, and Vilje.

StatoilHydro, Chesapeake pact

StatoilHydro will pursue unconventional gas opportunities and gas shale development under an agreement expected to be finalized by yearend 2008 with Chesapeake Energy Corp. (Oklahoma City).

The companies have committed to jointly look for gas in China, Romania, and Ukraine, said Statoil officials.

StatoilHydro expects to spend $3.38 billion for a 32.5% share in Chesapeake's gas assets in the Marcellus shale region in Pennsylvania, West Virginia, and New York.

StatoilHydro already has capacity rights for the Cove Point (Maryland) LNG terminal, a gas trading and marketing organization, and interests in gas-producing assets in the U.S. Gulf of Mexico.

StatoilHydro, Ecopetrol in GoM

StatoilHydro signed an agreement in late 2008 with the U.S. affiliate of Ecopetrol SA (Bogota, Colombia) for joint exploration in the Gulf of Mexico.

Terms call for the companies to drill three or more wells in the next several years and to move mature prospects along to the point of potential drilling.

Drilling is expected to start in 2009.

Barents Sea: Block 7222/11

In October 2008, StatoilHydro confirmed the presence of hydrocarbons in an exploration well in the Caurus prospect in Block 7222/11 in the Barents Sea.

The hydrocarbons were found in Middle and Late Triassic sandstones.

Exploration well 7222/11-1 S was drilled to TD of 2,825 m (9,268 ft) below sea level, at a water depth of 356 m (1,168 ft).

The firm estimated that the Late Triassic find could hold 2–14 bcm (70–494 bcf) of recoverable gas.

A delineation well for the Obesum discovery was to be drilled in late 2008.

Hyme, South East Tor

Faroe Petroleum Ltd. (Aberdeen, UK) began drilling in early 2009 on Norwegian License PL006C well 2/5-14 S, which holds the Hyme oil prospect.

That prospect is on the western flank of the South East Tor discovery.

The company planned to drill the South East Tor appraisal well 2/5-14 A soon thereafter.

South East Tor oil discovery well (2/5-3)—drilled in 1972—lies in 67 m (220 ft) of water near the producing reservoirs Ekofisk, Valhall, and Tor fields.

Both South East Tor and Hyme are within tie-back distance of several potential production hosts.

Pan/Pandora prospect

A wildcat well targeting the Pan/Pandora prospect, south of the Visund field, proved gas and oil in the upper sections of the

Iceland: Exploration offshore

At the September 2008 Iceland Exploration Conference, industry and government officials concluded that unexplored areas offshore may hold significant reserves of oil and gas.

The Dreki area of the Jan Mayen Ridge, northeast of Iceland, was cited as most promising.

The government was making plans for a first round of offshore licensing in the Dreki area, between Iceland and Jan Mayen Island.

Licenses were expected to be offered in early 2009.

Major challenges to exploration and production offshore Iceland will be the distance to shore and the lack of infrastructure. Also, areas of most interest are in water 1,000–2,000 m (3,281–6,562 ft) deep.

The National Energy Authority (Orkustofnun) would license any prospecting, exploration and production of hydrocarbons.

InSeis (now Wavefield-InSeis; Oslo) won a prospecting license on the southern Jan Mayen Ridge in 2001.

And TGS-NOPEC Geophysical Company (Asker, Norway) was awarded a prospecting license in 2002, in a region further south on the ridge.

Iceland Northern Dreki Licensing Round

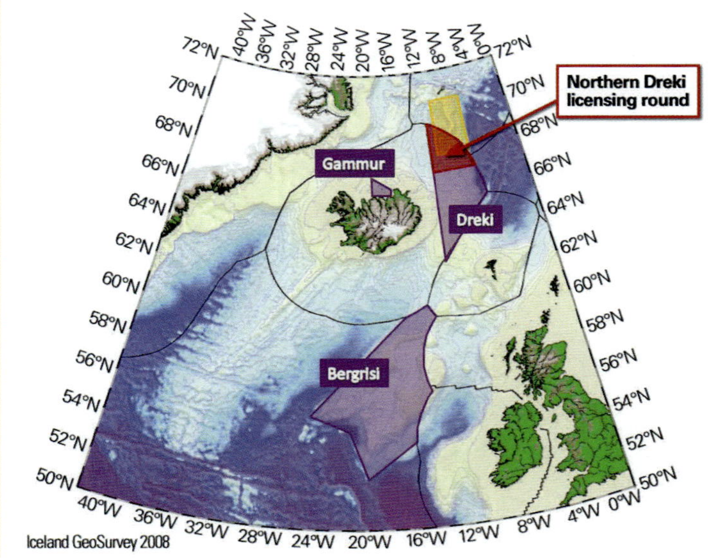

Both companies acquired seismic data and have put that data up for sale.

Several other firms also have shown interest in prospecting on the Icelandic continental shelf.

Iceland and Norway agreed in 1981 on an area of cooperation straddling the delimitation line between the economic zones of Iceland and Jan Mayen.

In this area, each nation has a 25% stake in any hydrocarbon discoveries made in the other country's part of the area.

Brent group in October 2008, said the Norwegian Petroleum Directorate (NPD).

Vertical well depth was 3,077 m (10,095 ft) below the sea surface, tapping the Drake formation in the Lower Jurassic. Water depth was 292 m (958 ft).

A sidetrack was planned to delineate the discovery towards the north.

Knockandoo-Talisker well

Oil was discovered in October 2008 in the Knockandoo-Talisker exploration well near the producing Brage field, said Norwegian Energy Co. ASA (Noreco; Stavanger).

The company estimates 0.3–1 MMcm (11–35 MMcf) of recoverable oil.

The well was drilled from the Brage platform to a vertical depth of 2,513 m (8, 245 ft) below the sea surface, and was concluded in the Statfjord group in the Lower Jurassic.

Production of up to 10,000 boe/d gross was anticipated after planned completion by yearend 2008.

Frøy field

Det norske oljeselskap ASA (DNO; Trondheim) and its partners submitted plans in late 2008 for development and operation of the Frøy field in production license 364.

Recoverable reserves from the field are estimated at 56 MMbbl, with initial production of 28,000 b/d.

An eight-well production unit is expected to begin operation on the Frøy field in late 2012.

Noatun prospect

StatoilHydro said in September 2008 that an exploration well in the Noatun prospect, north of the Njord field in the Norwegian Sea, struck gas and condensate.

Well 6407/7-8, drilled to a vertical depth of 5,110 m (16,765 ft), terminated at the Åre formation in the lower Jurassic.

The size of the find was not confirmed.

Arenaria prospect

StatoilHydro said in third quarter 2008 that it discovered natural gas within the Arenaria prospect on Block 7224/6 in the Barents Sea.

The prospect is some 160 km (99 mi) north of Nordkapp, Norway.

Commercial viability could not be confirmed because of poor quality Triassic reservoir rocks.

The discovery well—the first on Exploration License 394—reached a vertical depth of 2,315 m (1,439 ft) below sea level. Water depth was 265 m (869 ft).

Galtvort prospect

StatoilHydro was completing the drilling of the second of two exploration wells in mid-2008 on the Galtvort prospect.

The prospect is in the Norwegian Sea, some 9 km (6 mi) from the producing Draugen field.

StatoilHydro estimated preliminary recoverable gas volume in the Galtvort discovery at 1–3 bcm (35–106 bcf).

Development planning is under way.

Alve field

Exploration extension drilling in Alve field in the Norwegian Sea confirmed natural gas and condensate in a reservoir deeper in mid-Jurassic sandstone than was previously known.

The field is in the Norwegian Sea, some 16 km (10 mi) southeast of the Norne field.

Operator StatoilHydro said that well 6507/3-5S was drilled to 3,834 m (12,579 ft) subsea in 368 m (1,027 ft) of water. Preliminary estimates indicate 3–5 bcm (106–177 bcf) of gas, StatoilHydro said.

Production was expected to start in early 2009.

Ververis prospect

In mid-2008, StatoilHydro confirmed the existence of gas in the Ververis prospect in the Barents Sea, on the NCS.

Exploration well 7226/2-1 was drilled to a vertical depth of 2,992 m (9,816 ft) below the sea surface and completed in the lower Triassic Havert formation.

Ververis is the first well in production license 395.

StatoilHydro said that more analyses and evaluations were needed to determine resource potential.

Oseberg area

StatoilHydro discovered commercial quantities of oil and gas in mid-2008 in exploration well 30/9-21 S in the Norwegian North Sea.

Oil and gas were found in the upper parts of the Brent Group of production license PL104.

Drilling took place 7 km (4 mi) west of the Oseberg Sor platform and was concluded in the Middle Jurassic.

StatoilHydro estimated that the discovery held 16 million recoverable boe.

StatoilHydro also found oil in the Theta structure through its long-reach exploration well from the Oseberg C platform, targeting the Cook formation.

Natalia prospect

StatoilHydro announced in first quarter 2008 a large gas discovery in the Natalia prospect in the Norwegian Sea.

Preliminary estimates said the discovery could contain 1.5 bcm (53 bcf) of recoverable gas.

Natalia is northwest of the Midgard structure on the Åsgard field on the Halten Bank.

No formation testing had been done, but major data collection and sampling had been carried out, including core drilling and fluid sampling.

Yttergryta development plan

Gas production from Yttergryta field, 33 km (21 mi) east of Asgard B platform in the Norwegian Sea, was slated to start first quarter 2009, under plans submitted by StatoilHydro.

Yttergryta is expected to produce 1.75 bcm (62 bcf) of gas over 3–5 years.

A subsea template tied back to the Asgard B platform will send 3.5 MMcmd (124 MMcfd) of gas through the Asgard gas pipeline to the processing plant at Kårstø.

Ipswich prospect

Revus Energy ASA (Stavanger) and Dong E&P Norge (Copenhagen) announced an oil discovery on the Ipswich prospect in mid-2008, in the southern part of the Norwegian North Sea.

The firms said that further investigation was needed to determine commercial feasibility.

Exploration well 1/3-11 and sidetrack 1/3-11 T2 demonstrated a 60 m (197 ft) oil column with reservoir properties comparable to Oselvar.

The main well was drilled to 3,232 m (10,603 ft) and the sidetrack to 3,465 m (11,368 ft) subsea.

Both were terminated in rocks of early Paleocene age.

Jordbaer, Torphins prospects

Revus Energy ASA (Stavanger) confirmed two gas/oil discoveries in mid-2008 on two prospects.

The Jordbaer prospect is in the Norwegian North Sea, in PL 373S.

The Torphins prospect is in the central UK North Sea sector.

Operations were ongoing to gather further data for continued evaluation.

PL 373S

BG Norge AS (Stavanger) reported finding oil in mid-2008 in exploration well 34/3-1 S in the Norwegian North Sea.

The NPD said the well was drilled in 410 m (1,345 ft) of water on production license 373 S.

It reached a total vertical depth of 4,057 m (13,311 ft) below sea level, discovering oil in a Jurassic reservoir.

The well is northeast of StatoilHydro's Snorre field.

The resource potential was to be evaluated with a sidetrack.

North Sea, PL 292

BG Norge AS found oil and gas in mid-2008 with exploration well 15/12-19 on Production License 292 (PL 292) in the Norwegian North Sea.

Combined with the nearby 6/3-1 discovery, the find could hold up to 3–5 million cu m of oil equivalent (19–31 boe).

Drilled south of Varg field, the well tested 750 cmd (26,483 cfd) from Jurassic and Triassic reservoirs.

The well was drilled in 86 m (282 ft) of water to 3,169 m (10,397 ft) TVD subsea.

Afrodite prospect

Eni Norge (Oslo) confirmed a "promising" gas and condensate discovery on the Afrodite prospect in the Norwegian North Sea in first quarter 2008. No specific reserves estimate was issued.

Well 34/12-1 was drilled in 373 m (1,224 ft) of water in the Vikinggraben in production license 293.

The location is around 30 km (19 mi) northeast of Kvitebjorn field.

The NPD said that the well proved gas and condensate in Mid-Jurassic strata before being terminated at a total depth of 4,688 m (15,380 ft).

Eni said it needs more extensive evaluation before establishing the resource potential.

Draupne prospect

In early 2008, Det norske oljeselskap ASA (DNO; Trondheim) discovered oil and gas on Draupne prospect (production license 001B), near the Grane field in the North Sea.

The discovery well was drilled to a vertical depth of 2,511 m (8,238 ft), encountering 44-m (1,732-ft) thick sandstone from the Middle Jurassic period.

The well has proven recoverable resources between 4–5 MMcmoe (141–177 MMcfoe) which, together with two other discoveries located in the area (Hanz and West Cable), may contain 90 MMbbl of recoverable oil and gas.

DNO expects to drill an appraisal well in 2009.

Drilling and Production

Volve production starts

StatoilHydro began oil and gas production from Volve field in the Norwegian North Sea in February 2008.

Volve is 200 km (124 mi) west of Stavanger in the southern section of the NCS.

The company had drilled eight wells to develop reserves of 78.6 million bbl of oil and 1.5 bcm (53 bcf) of gas. Five more wells are planned.

Production is expected to plateau at 50,000 b/d by mid-2009.

The gas will be sent to the Sleipner A platform for final processing and export. Production is expected to last 4–5 yrs.

Oil to FPSO from Vilje field

StatoilHydro delivered first oil in August 2008 from Vilje field to an FPSO vessel.

The field—in the northern part of the Norwegian North Sea, north of the Heimdal field—holds an estimated 52 million bbl of recoverable oil.

Production from Vilje was expected to reach a plateau of 35,000 b/d by mid-2008.

A 19-km (12-mi) pipeline connects the FPSO to the field.

Gulltopp completion: world record

Its Gulltopp well on the Gullfaks field, completed in April 2008, set a world record for the longest producing well from an offshore platform, said StatoilHydro (Stavanger).

Drilling of the 9,910-m-long (32,513-ft) horizontal Gulltopp well from the Gullfaks-A platform, in 134 m (439 ft) of water, actually began in early 2005.

Hydrocarbons flowed from the shallow Brent reservoir at 2,430 m (7,972 ft) subsea.

The 10-km (6-mi) drill pipe was controlled from the drilling rig at the sea surface. It ran 150 m (492 ft) down to the seabed, and then through various rock strata.

Contractor Seawell Ltd. (Hamilton, Bermuda) drilled most of the wellpath at an inclination of 7 degrees. To counteract friction, workers filled the 8-km (5-mi) casing with air instead of drilling mud, "floating" the casing in the nearly horizontal well.

The Gullfaks oil field is in the Norwegian sector of the North Sea, some 175 km (109 mi) northwest of Bergen.

Gulltopp is in the Tampen area, 5 km (3 mi) north of Gullfaks satellite fields Gullveig and Rimfaks.

Reservoir units are sandstones of early- and middle-Jurassic sandstones, 2,000 m (6,562 ft) subsea and several hundred meters thick.

Gulltopp has estimated recoverable reserves of 4 million scm of oil and 500 million scm (17.7 billion scf) of gas.

The company expects a production plateau of 16,000 b/d of oil.

GULLTOPP FIELD, PL 050/B, NORWEGIAN SECTOR, NORTH SEA

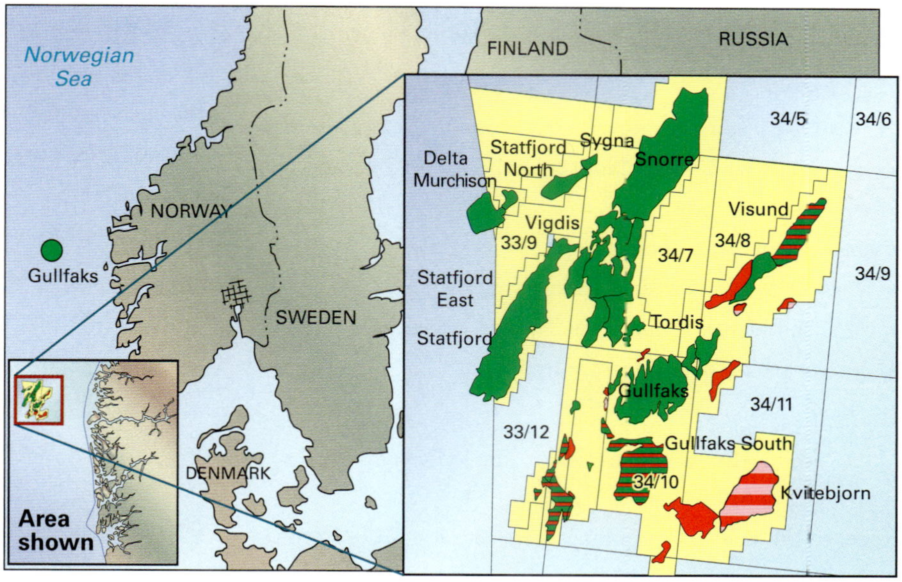

CHALLENGING DRILLING REGIME, GULLTOPP WELL

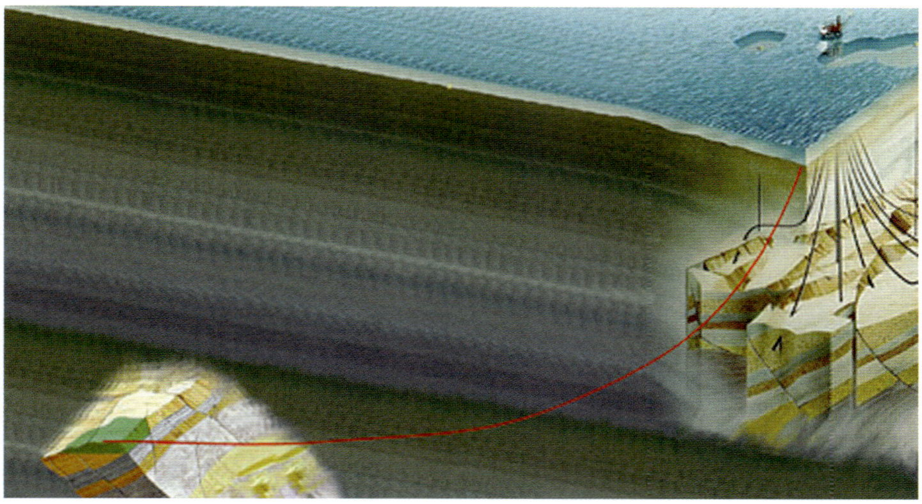

Greater Ekofisk platform

ConocoPhillips Co. (Houston) awarded a front-end engineering and design contract in late 2008 for its 2/4 A wellhead platform in the Greater Ekofisk area.

The platform, in the Norwegian North Sea, will be installed in 70–75 m (230–246 ft) of water.

The FEED project is expected to be complete by yearend 2009.

Decommissioning Ekofisk units

ConocoPhillips Co. (Houston) initiated plans in early 2008 to decommission its nine Ekofisk platforms in the Norwegian and UK sectors of the North Sea by 2013.

Heerema Marine Contractors (HMC; Leiden, Netherlands) won the contract for the engineering, offshore preparation, removal, and onshore recycling of the platforms

The Ekofisk topsides and jackets will be removed and taken to Vats, Norway, for recycling and disposal. An estimated 96–98% of the material recovered will be recycled by 2014.

Marsvin prospect

Faroe Petroleum Ltd. (Aberdeen, UK) began drilling in third quarter 2008 on well 3/7-7 on Norwegian license PL289, which holds the Marsvin oil prospect.

The license is in water about 65 m (213 ft) deep.

The Marsvin target is a substantial Upper Jurassic oil prospect, on the south west flank of the Søgne Graben, straddling the border between Norway and Denmark.

Alvheim field

In June 2008, the Alvheim field came onstream in the Norwegian North Sea.

Partners in the field's development were Marathon Oil Corp., ConocoPhillips Co. (both in Houston), and Lundin Petroleum AB (Stockholm).

Plateau production of about 120,000 b/d was expected by yearend.

The NPD estimates reserves of 154 MMbbl of oil and 5.7 bcm (201 bcf) of gas for Alvheim.

Lease Sales

Awards under APA 2008

In mid-December 2008, the NPD issued offers for 34 new production licenses on the NCS.

The agency had invited bids on all or part of 18 blocks in May 2008.

The December offers were issued after the NPD considered applications from 47 companies in the Awards in Predefined Areas (APA) 2008 program.

The deadline for the companies to respond is January 9, 2009.

Of 34 licenses, 21 are in the North Sea, 11 in the Norwegian Sea, and two in the Barents Sea.

Eight of the licenses are additional acreage in existing production license areas. Three of these border known discoveries or fields.

Among the companies who received offers were:

Lundin Petroleum AB (Stockholm): 11 blocks

Dana Petroleum Norway AS: 8 blocks

Faroe Petroleum Ltd. (Aberdeen, UK):
3 prospective licenses

20th Licensing Round

In its 20th licensing round, Norway sought bids from operators in mid-2008 on 79 blocks or parts of blocks on the NCS.

Operators were invited to apply for 51 blocks in the Norwegian Sea and 28 blocks in the Barents Sea.

Several blocks are in the deep waters of the Voring basin, which is a relatively underexplored area of the Norwegian Shelf.

Winners will be announced by mid-2009.

Awards under APA 2007

In February 2008, Norway offered 52 production licenses to 19 operators that had applied to develop blocks in the Norwegian North Sea, the Norwegian Sea, and the Barents Sea.

The Awards in Predefined Areas 2007 program drew 113 applications.

The APA initiative aims to attract new and smaller companies to the NCS.

In most of the APA blocks, petroleum activities have previously taken place, said the Norwegian Petroleum Ministry said. Relatively small discoveries are likely.

Successful bidders included BG Norge AS, Centrica Resources Norge AS, Dana Petroleum Norway AS, E.On Ruhrgas Norge AS, Endeavour Energy Norge AS, Eni Norge AS, Esso Norge AS, Genesis Petroleum, and Faroe Petroleum Norge AS.

LNG

Snøhvit LNG costs

In October 2008, StatoilHydro increased its estimated investment costs for the first phase of the Snøhvit LNG project in the Barents Sea by 3 billion kroner. A 2005 estimate had put project cost at 48.1 billion kroner (nominal).

The adjustment reflected problems associated with delivering the full 4.3 mtpy of LNG capacity at the onshore liquefaction plant.

Snovhit started production in September 2007, but struggled to regularly deliver at full capacity. There were two scheduled shutdowns in mid- and late 2008.

First Snøhvit LNG to France

The first shipment to France of LNG from Snøhvit in Norway was delivered to the Montoir-de-Bretagne terminal of Gaz de France (GDF) in Brittany in March 2008.

Details are given in the France section of this chapter.

Snøhvit LNG to Cove Point

The first export cargo of European LNG to the United States was delivered in February 2008 from the Hammerfest liquefaction plant of StatoilHydro.

The cargo arrived at the terminal of Dominion Cove Point LNG LP in Maryland.

SNØHVIT FIELD

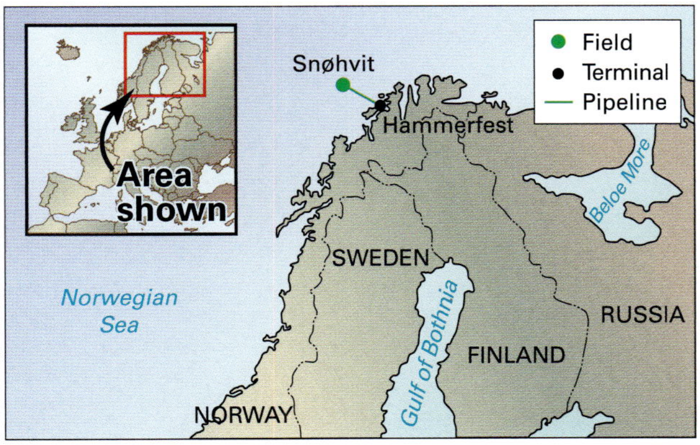

Refining

Mongstad refinery upgrade

StatoilHydro let an engineering and procurement contract in mid-2008 for the upgrade of the coker units at its 186,000-b/d Mongstad refinery near Bergen.

The work will be done by to M.W. Kellogg Ltd. (West London, UK).

The upgrade, coordinated with a scheduled refinery turnaround, is to be completed by mid-2010.

Pipelines

Statfjord gas piped directly to UK

As of October 2008, all gas from the Statfjord field in the North Sea was piped directly to the UK.

An innovative remote-controlled, subsea procedure tied the Tampen Link pipeline into the Statfjord interfield pipeline system (Sips).

Before the procedure, gas from the Statfjord A and C platforms was piped through the Statpipe system to continental Europe. All gas from the field now flows directly to St. Fergus in the UK.

Environment

Troll subsurface seismic

StatoilHydro launched a seismic subsurface survey in late 2008 of the Troll field in the North Sea.

The project will determine whether the site is suitable for storage of large volumes of carbon dioxide (CO2) from carbon-capture facilities at industrial plants at Mongstad and Kårstø.

The Norwegian Petroleum Directorate has studied three possible storage sites: the Johansen formation and two potential storage sites connected to the Utsira formation in the Sleipner area.

The Johansen formation is south of the Troll area in the North Sea at a depth of 2,500 m (8,202 ft) below the Troll oil and gas reservoirs.

Kårstø CO_2 capture unit

Aker Clean Carbon and the Norwegian government announced plans in late 2008 to establish a 725 million kroner CO_2 capture and sequestration unit at Kårstø on the west coast of Norway.

Operations are to begin in 2009.

The plant will have the capacity to remove 100,000 tpy of CO_2 from exhaust gases.

POLAND

Capital: Warsaw
Monetary unit: Zloty
Refining capacity: 492,950 b/cd
Oil production: 16,500 b/d
Oil reserves: 96.375 million bbl
Gas reserves: 5.820 tcf

Mazowieckie, Lubelskie provinces

In late 2008, Poland's environment ministry granted permission to ExxonMobil Poland to explore for natural gas deposits in eastern and southeastern parts of the country.

ExxonMobil Poland, which holds two five-year licenses for conducting related operations, was planning seismic surveys and drilling in the Mazowieckie and Lubelskie provinces.

The first concession covers 1,200 sq km (463 sq mi) near Wolomin, northeast of Warsaw.

The second covers 1,000 sq km (386 sq mi) near Zamosc, in the southeast.

ExxonMobil Poland's exploration activities are expected to last into 2011.

Plock expansion

Polski Koncern Naftowy (PKN Orlen; Plock) announced plans in early 2009 to double the capacity of its 800,000-tpy petrochemical plant in Plock by 2013.

The expansion will increase production of both teraphthalic acid and paraxylene by 600,000 tpy.

The firm aims to be one of the primary petrochemical suppliers in Poland. It operates a total of eight refineries in the Czech Republic, Lithuania and Poland.

PKN Orlen also operates 2,700 retail outlets for petroleum distribution in the same three countries, plus Germany and Hungary.

PKN Orlen's investment comes despite predictions of slower growth for the economies of Central Eastern Europe.

Coal gasification project

In late 2008, the largest manufacturer of fertilizer in Poland said it would join with a state-owned coal mining company to build a $1.2-billion coal gasification plant.

Zaklady Azotowe Pulawy SA (ZAP; Pulawy) will pursue the project with state-owned coal mining firm Lubelski Wegiel Bogdanka SA (LWB; Puchaczow).

ZAP currently uses about 1 MM cm/yr (35 MMcf/yr) of natural gas. The new plant, using up to 1.3 billion tpy of coal, is expected to yield enough gas to satisfy half of ZAP's needs.

Construction is expected to be completed in 2013.

CCS for power stations

Alstom SA (Levallois-Perret, France) signed an MOU in December 2008 with power plant operator PGE Elektrownia Belchatow SA (Belchatow, Poland) for a project to reduce CO_2 emissions.

The two will develop technology for carbon capture and storage (CCS).

PGE operates the 4,400-MW Belchatow station. It is the largest conventional power plant—and the largest single emitter of CO_2—in Europe.

The first phase of the project will install a pilot CCS unit at Unit 12 of the Belchatow station by 2011.

That unit will be designed to capture 100,000 tpy of CO_2 using advanced amines technology developed jointly by Alstom and Dow Chemical Co. (Midland, MI).

Alstom also began building a 858-MW coal-fired unit for PGE in late 2008. That plant, to be completed by 2015, will incorporate a full-scale CCS system developed by Alstom.

The European Union launched in 2007 a program to cut CO_2 emissions by 60% by 2050 to combat climate change. One element of that program aims to establish 10–12 large CCS pilot plants across Europe by 2015.

The plants would demonstrate the commercial viability of various CCS technologies for use by 2020.

PORTUGAL

Capital: Lisbon
Monetary unit: Euro
Refining capacity: 304,172 b/cd
Oil production: 0
Oil reserves: 0
Gas reserves: 0

Seismic survey: Offshore Atlantic

A group led by a Dutch subsidiary of Petróleo Brasileiro SA (Petrobras; Rio de Janeiro) let a contract in mid-2008 to SCAN Geophysical ASA (Oslo) to shoot more than 8,500 line-km of 2D seismic in the Atlantic off Portugal.

The three group members hold four blocks that cover 12,159 sq km (4,695 sq mi) in 200–3,000 m (656–9,843 ft) of water in the Lusitanian basin.

46-MW PV power plant

In December 2008, Amper Central Solar SA (Moura, Portugal)—a subsidiary of Spanish power company Acciona Energy (Sarriguren, Spain)—inaugurated what it called the world's largest photovoltaic (PV) solar power plant.

The 46-MW Central Solar de Amareleja power plant covers 2.5 sq km (1 sq mi) in the Alentejo region.

Yingli Green Energy Holding Co. (Baoding, China) supplied the 262,000 polycrystalline PV modules that are mounted on more than 2,500 solar tracking devices.

Connected to the national grid, the plant will provide power for up to 30,000 homes.

240-MW windfarm

Also in December 2008, Empreendimento Eolico do Vale do Minho SA (EEVM; Braga, Portugal) began commercial operation of what it called Europe's largest onshore windfarm.

The 240-MW Ventominho Windfarm is close to the Spanish border in northwestern Portugal.

The facility uses 120 wind turbines, each of 2-MW capacity, from Enercon GmbH (Aurich, Germany).

EEVM was created as a joint venture to build and operate windfarms in the Portuguese region of Vale do Minho.

Parties with interests in EEVM are EDF EN Portugal [a subsidiary of EDF Energies Nouvelles (Paris)] and Eolverde [a subsidiary of Endesa SA (Madrid)].

Evaluation of various data confirmed a 39-m (128-ft) gas column with a net pay of 23 m (75 ft) in high-quality reservoir sands with good permeability, said the company.

The well also encountered a thin gas-bearing reservoir in a shallower horizon that was being evaluated.

MAJOR EUROPEAN GAS PIPELINE PROJECTS

ROMANIA

Capital: Bucharest
Monetary unit: New leu
Refining capacity: 516,557 b/cd
Oil production: 92,500 b/d
Oil reserves: 600.0 million bbl
Gas reserves: 2.225 tcf

Offshore Neptun block

In November 2008, the Romanian unit of ExxonMobil Exploration and Production (Irving, TX) signed an agreement with Petrom SA (Bucharest) to help explore deepwater portions of the offshore Neptun block.

The two firms will cooperate on 3D seismic acquisition and evaluation in 2009.

Ana-2 appraisal

Sterling Resources Ltd. (Calgary) said in third quarter 2008 that its Ana-2 appraisal well in the Romanian Black Sea confirmed gas pay consistent with the Ana-1 discovery well.

The well encountered gas within the Doina Main sand formation at 1,121 m (3,678 ft) TVD subsea.

SERBIAN OIL, GAS FIELDS

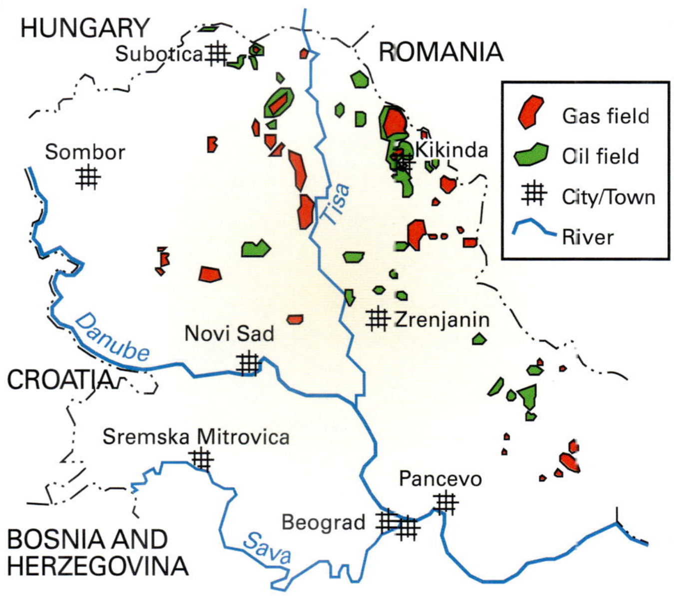

Pipeline connection with Hungary

Hungary and Romania—citing supply security issues—agreed in mid-2008 to connect their natural gas systems.

Details are presented in the Hungary section of this chapter.

SLOVAKIA

Capital: Bratislava
Monetary unit: Koruna
Refining capacity: 115,000 b/cd
Oil production: 300 b/d
Oil reserves: 9.0 million bbl
Gas reserves: 500.0 bcf

Slovakia is neither a large producer nor a large consumer of oil and gas. According to the International Energy Agency (IEA; Paris), more than 80% of the country's net primary energy supply is imported, chiefly from Russia.

The IEA has applauded the rapid progress Slovakia has made toward developing an open market economy over the past decade, including relevant changes in the energy sector to meet increasing energy demand.

Continuing challenges include energy-resource diversification, opening energy markets and integrating them into the EU, and increasing energy efficiency to offset rising energy prices and to help control carbon emissions.

Carpathian Fold Belt

JKX Oil & Gas (London) entered into an agreement in early 2008 to acquire an interest in three exploration licenses in the Carpathian Fold Belt in Slovakia.

The licenses cover a total area of 2,278 sq km (880 sq mi).

A JKX subsidiary will acquire a 25% interest in the Svidnik, Medzilaborce, and Snina Blocks.

Those blocks are currently held and operated by three subsidiary companies of Aurelian Oil & Gas PLC (London).

JKX said that advances in seismic technology have led to recent discoveries in Poland in structures analogous to the Carpathian Fold Belt.

A 2D seismic program was being planned for 2008.

Nuclear plant

Slovakian officials announced in January 2009 a plan to address the country's gas shortages by reopening a power generator at the aging Jaslovski Bohunice nuclear plant.

The plant was shut down in 2008 as one condition for the country's entry into the European Union.

Officials in Austria objected to the plan, on the basis of both the EU accession agreement and concerns about the safety of the nuclear reactor at the site.

EU officials were reviewing the situation.

SPAIN

Capital: Madrid
Monetary unit: Euro
Refining capacity: 1,271,500 b/cd
Oil production: 2,600 b/d
Oil reserves: 150.000 million bbl
Gas reserves: 90.000 bcf

Ayoluengo field

Leni Gas & Oil PLC (London) started a waterflood pilot in June at Ayoluengo field in the northern Burgos region.

The field was judged capable of recovering another 4–6 million bbl of oil, beyond the 17 million bbl produced since discovery in 1964.

Producing oil from the Lower Cretaceous Sargentes sandstone at 1,250 m (4,100 ft), Ayoluengo, had averaged 115 b/d in early 2008.

Lukoil bid for Repsol

A controversy swirled in 2008 regarding possible purchase of a significant stake of oil group Repsol YPF SA (Madrid) by Russia's Lukoil (Moscow).

Repsol's main shareholder—financially struggling Spanish construction group Sacyr Vallehermoso SA (Madrid)—was seeking to sell its 20% stake (to Lukoil or another company).

And Lukoil also was said to be talking with Repsol's second-largest stakeholder, Criteria CaixaCorp SA (Barcelona).

Criteria held a 9.1% direct stake and another 5% through an investment vehicle.

MEDGAZ ROUTE

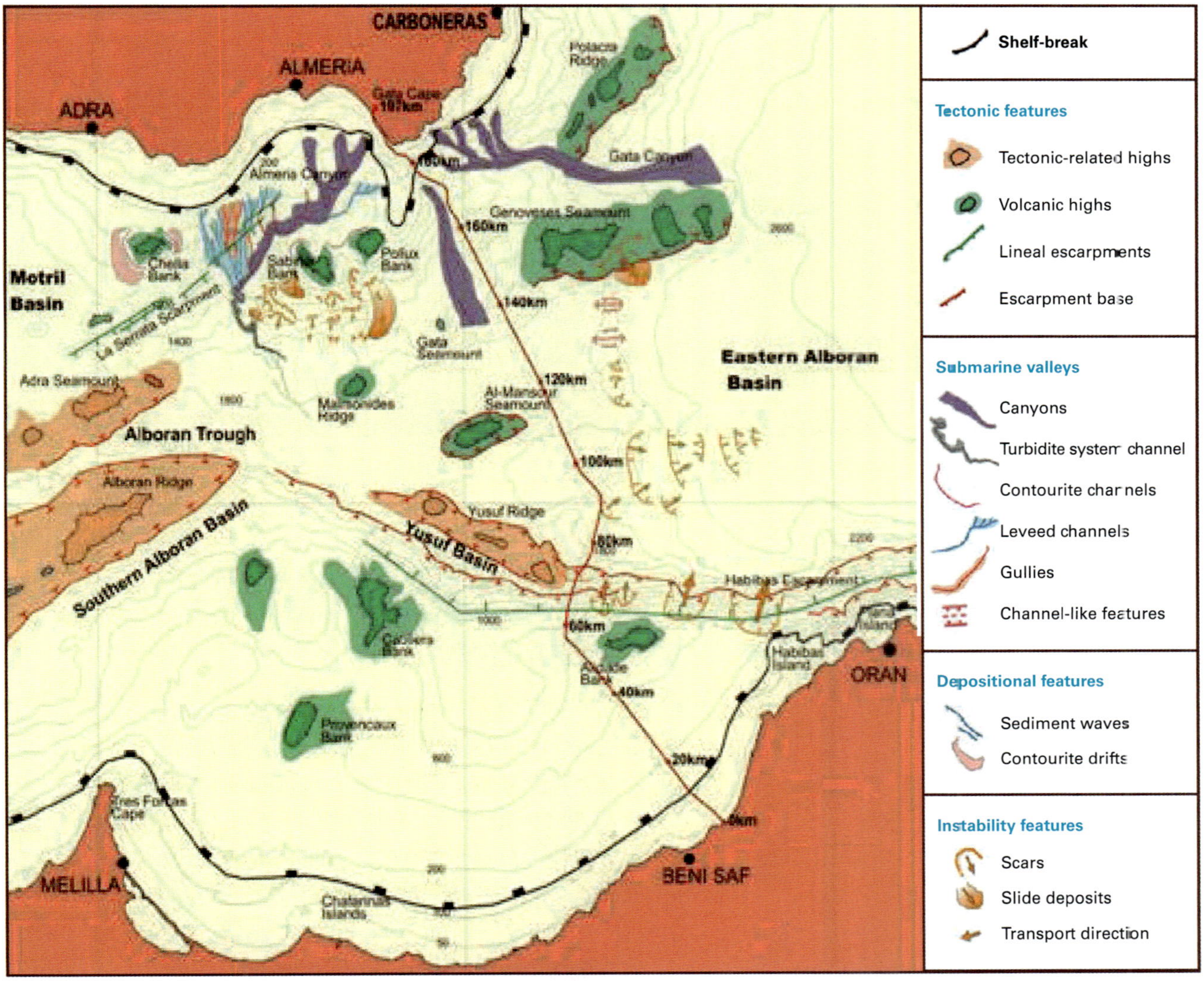

In summary, it was possible that Lukoil could buy as much as 34.1% of Repsol—a scenario described by some in Spain as a threat to national security.

Additional concerns arose when it was reported that, under Spanish takeover law, a shareholder must launch an offer for the entire company once it passes the 30% threshold.

Spain's Industry Ministry said in late 2008 that it would do everything it could to keep Repsol "independent and in Spanish hands."

Medgaz pipelay begins

Medgaz is a consortium of five international companies: Sonatrach (Algeria), CEPSA (Madrid), Iberdrola (Bilbao), Endesa (Madrid) and GDF Suez (Paris).

In May 2008, the group began the offshore pipelaying phase for the Medgaz pipeline, to run 210 km (130 mi) from Beni Saf, Algeria, to Almeria, Spain, along the bottom of the Mediterranean Sea.

The "S-lay" method was used for the shallow-water portions of the line [down to 550 m (1,804 ft)], first from Almeria and then from Algeria.

The deepwater pipelay [down to 2,160 m (7,086 ft)], using the "J-lay" method, was slated for completion in third quarter 2008.

Medgaz also investigated the limits of using S-lay vessels for deepwater, large-diameter pipelay.

Cartagena refinery

As noted in *Oil & Gas Journal*'s semiannual *Worldwide Construction Update* in early 2009, Repsol-YPF SA was planning to double the capacity of its 100,000-b/d Cartagena refinery in southern Spain.

It had awarded major contracts to Fluor Corp. (Irving, TX) and Foster Wheeler Ltd. (Clinton, NJ) for work on the project.

Fluor called it the largest refinery upgrade and expansion effort in Spain in the past 30 years.

The $4.8-billion project includes design and construction of new units as well as refurbishment and expansion of existing units.

One key component will be a new delayed coker complex, including of a 53,000-b/sd delayed coker unit, a gas concentration unit, and a 90,000 b/sd vacuum distillation unit.

Commissioning is scheduled for 2011.

OFFSHORE PIPELAY METHODS

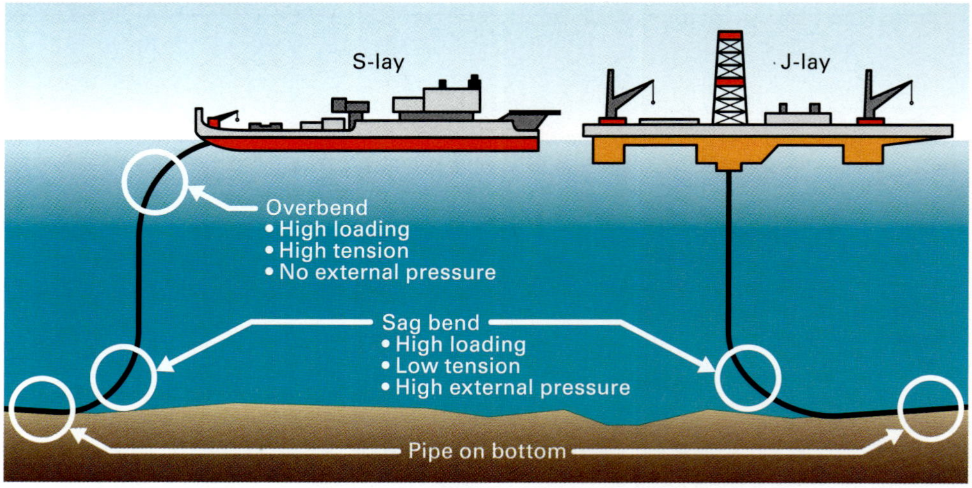

SWEDEN

Capital: Stockholm
Monetary unit: Krona
Refining capacity: 437,000 b/cd
Oil production: 0
Oil reserves: 0
Gas reserves: 0

Coker for Lysekil refinery

Refiner Preem Petroleum AB (Stockholm) was seeking environmental permits in early 2008 to build a new 4-mtpy coker unit near its 220,000-b/d Lysekil refinery on the west coast.

The expansion supports the firm's strategic move from fuel oil into transportation fuels, according to Preem officials.

The company was planning to make a final investment decision on the project by yearend 2008.

Use of a coker unit would produce more carbon emissions, so Preem was also investigating methods of carbon capture and sequestration with several research institutions.

TURKEY

Capital: Ankara
Monetary unit: Lira
Refining capacity: 714,275 b/cd
Oil production: 41,200 b/d
Oil reserves: 300.000 million bbl
Gas reserves: 300.000 bcf

Exploration and Development

ExxonMobil: Black Sea exploration

An affiliate of ExxonMobil Corp. (Irving, TX) signed an agreement in late 2008 with Turkey's state-owned Turkiye Petrolleri Anonim Ortakligi (TPAO; Ankara) to explore two large deepwater blocks.

The deal marked ExxonMobil's entry into Black Sea exploration.

The ExxonMobil affiliate will be operator in the initial exploration phase in the 8,500-sq km (3,282-sq-mi) Samsun Block and the 21,000-sq km (8,108-sq-mi) eastern portion of the 3921 Block.

The blocks, in 2,000 m (6,562 ft) of water in the Black Sea basin, are separated by a lease held by TPAO and Brazil's Petrobras (Rio de Janeiro).

The development and production phases are scheduled for completion in 2009.

Sea of Marmara seismic

Toreador Resources Corp. (Dallas) completed shooting in late 2008 and was processing a 1,000 line-km 2D seismic survey performed in shallow waters of the Sea of Marmara in the Thrace basin.

Toreador is operator of the permits with 50% working interest. Thrace Basin Natural Gas Turkiye Corp. and Pinnacle Turkey Inc. each have 25% working interest in the joint venture.

Offshore seismic

In late 2008, TPAO awarded Wavefield Inseis AS (Oslo) a 3,000-sq km (1,158-sq mi) 3D seismic project offshore in the Black Sea.

Pipelines/Transportation

Yalova storage terminal

Royal Vopak NV (Rotterdam) was making plans in late 2008 to build an oil products storage terminal on the coast of the Sea of Marmara near Yalova.

Marketing and technical plans were being prepared to support a final investment decision in 2009.

The terminal would serve the Istanbul area, storing liquid bulk chemical and vegetable-oil products as well as crude oil.

Drilling and Production

Thrace basin

A group led by Incremental Petroleum Ltd. (Perth) said it plans to start gas production in mid-2009 from the Edirne license in the Thrace basin, northwest of Istanbul.

The group awarded contracts for the design and engineering of a gas processing plant and pipeline.

The JV will be Turkey's first to produce and sell onshore gas into the Botas gas distribution network.

Incremental Petroleum was also involved in an oil and gas exploration project on 30,000 acres (47 sq mi) in the Tuz Golu basin, south of Ankara.

Ayazli field

Operator TPAO began producing natural gas in early 2008 from Ayazli gas field in the Black Sea, according to partner Stratic Energy Corp. (Calgary).

Ayazli is the third gas field in which production has begun under Phase 1 of the South Akcakoca sub-basin development program.

Production began in 2007 from Akkaya and East Ayazli fields.

Officials estimate that production from the Ayazli field will average more than 283,000 std cu m/d (10 MMscfd). Production from Akkaya and East Ayazli was about 481,000 std cu m/d (17 MMscfd) gross.

Pipelines

Nabucco draws shipper interest

Potential shippers made (nonbinding) commitments by late 2008 for more than 100% of the capacity of the proposed Nabucco gas pipeline.

Construction of the 3,300-km (2,050-mi) line, from the Caspian Sea to Austria via Turkey and the Balkan states, will occur in two phases.

The first phase, with an initial capacity of 8 bcm/yr (282 bcf/yr), is to start in 2010 and be completed in 2013. The second phase will start in 2013 and be finished at yearend 2014.

The line will carry 31 bcm/yr (1.1 tcf/yr) of gas to the European Union from the Middle East and Central Asia.

The EU supports the line as a way to reduce its reliance on Russian gas supplies.

Cyprus seeks LNG, storage

In mid-2008, Cyprus said it would launch a tendering process for LNG supplies and gas storage infrastructure.

A government spokesman said that the Cypriot cabinet had approved both a floating LNG regasification terminal and a gas storage facility.

The government was seeking tenders for a temporary offshore storage facility as a stop gap measure as it also solicited interest in building a permanent onshore site.

Construction of either type of facility would likely take at least five years.

Some critics argued that the technology for a floating LNG storage unit was untested.

The island relies mainly upon heavy fuel oil for electric power generation.

Gas pipeline project

Turkey and Syria signed an agreement in early 2008 aimed at enhancing cooperation between their two countries in several fields, including development of a major pipeline.

The 100-km (62-mi) pipeline would carry Egyptian natural gas from Aleppo, Syria, to Turkey's southeastern town of Kilis.

Details are presented in the Syria section of the Middle East chapter of this book.

Med Stream pipeline

Initial discussions were under way in late 2008 on the proposed Med Stream pipeline system.

The line would run from Turkey to Israel, across the Mediterranean.

Construction of the $12-billion pipeline is expected to start in early 2010.

In addition to crude oil and natural gas, the line would transport fresh water as well as electrical and fiber-optic cables.

Turkey would source crude oil and natural gas for the line from the Caspian region.

Currently, Israel sources oil from the Black Sea and transports it by ship to its ports.

Indian officials also have expressed interest in extending the line to their country.

Tensions with Iraq

In early 2008, Turkey launched air strikes and artillery attacks, and then sent its troops across the border into the Khakurk region of the Kurdish-held autonomous north of Iraq.

Turkey said it aimed to destroy elements of the Kurdistan Workers'Party (PKK).

The PKK, described by U.S. officials as terrorists, had previously conducted cross-border raids into Turkey.

In late February 2008, Turkey withdrew its troops.

In early 2009, the Iraqi Foreign Ministry said that Turkey, Iraq and the United States were making plans to set up a center in the Iraqi city of Irbil to coordinate further actions against the PKK.

Pipeline security

Iraq's northern crude export system runs from the northern Kirkuk oil fields to the Turkish Mediterranean export terminal at Ceyhan. Part of the line runs through southeast Turkey.

The two parallel pipelines have a nominal combined capacity of 1.5 million b/d.

Since 2003, however, capacity has been reduced due to erratic oil production in northern Iraq as well as attacks by insurgents.

In November 2008, two bomb attacks damaged the Turkish section of the pipeline.

The PKK claimed responsibility for one of the attacks.

Analysts at BMI (London) said that the PKK attacks are likely to continue unless there is a wider agreement allowing the Kurdish Regional Government in northern Iraq to export through the line.

UNITED KINGDOM

Capital: London
Monetary unit: Pound
Refining capacity: 1,857,668 b/cd
Oil production: 1.400 million b/d
Oil reserves: 3.410 billion bbl
Gas reserves: 12.110 tcf

Overview

In 2007, the UK met 75% of its natural gas needs through domestic production and was self-sufficient in oil, according to the Oil & Gas UK trade association (London).

And a late-2007 government report gave a generally optimistic view of the UK energy situation.

It was prepared by the Department for Business, Enterprise, and Regulatory Reform (DBERR) and the Office of Gas & Electricity Markets.

Natural gas

The UK will have sufficient capacity in pipelines and LNG terminals to receive natural gas imports in the near and medium term. Prices will determine how much gas is sent to the country.

LNG will play a crucial role in meeting the UK's gas needs along with pipeline gas as the country's indigenous gas supplies fall.

Gas import capacity quadrupled from 2004–2007.

In 2008, the UK was expected to import 40% of its gas demand, with that figure rising to 75% by 2015.

Further investment will be needed to avoid market tightness in the next 10–20 years.

LNG is expected to make up 28% of the UK's total supply by 2017, according to the 2008 forecast of National Grid PLC (London).

Energy efficiency

Energy efficiency will be important to managing energy demand. However, planning delays could threaten the development of major energy projects, as could the growing shortage of skills in the engineering and construction sectors.

Oil and refined products

Imports are likely to rise as domestic production wanes. However, increased complexity of international routes and growing resource nationalism could disrupt availability.

Transportation is expected to be the biggest consumer of oil products. Oil use by industrial and domestic consumers is expected to fall.

Refiners face growth in demand for diesel and jet fuel and falling demand for gasoline. Increasing investment in units to process lower-quality feedstocks appears to be prudent.

Renewables

The government said it would scale down its goal to source 20% of Britain's energy supply from renewables by 2020. The new target is likely to be 15%.

Financial crisis: Impacts on energy industry

In December 2008, a committee of Parliament warned that Britain was threatened by an "energy crunch" with severe social and economic consequences because of the impact of the unfolding global financial crisis on the energy industry's investment plans.

The Business and Enterprise Committee of the House of Commons said that the government should ensure that energy companies invest in new power stations and gas storage facilities, in the same way that it intervened to support the banks earlier in the year.

The MPs noted that generating capacity equivalent to nearly one-third of current electricity demand will need to be replaced by 2020.

There was also concern about investment in new nuclear power stations.

The committee noted the competing demands on the energy industry, which is under pressure to cut bills for consumers, but also to invest more to ensure adequate energy supplies.

New energy bill

The new UK energy bill unveiled in January 2008 calls for a simpler regime for offshore gas supply infrastructure.

The aim, said government officials, is to enable energy companies to invest with greater clarity and with reduced costs and risks.

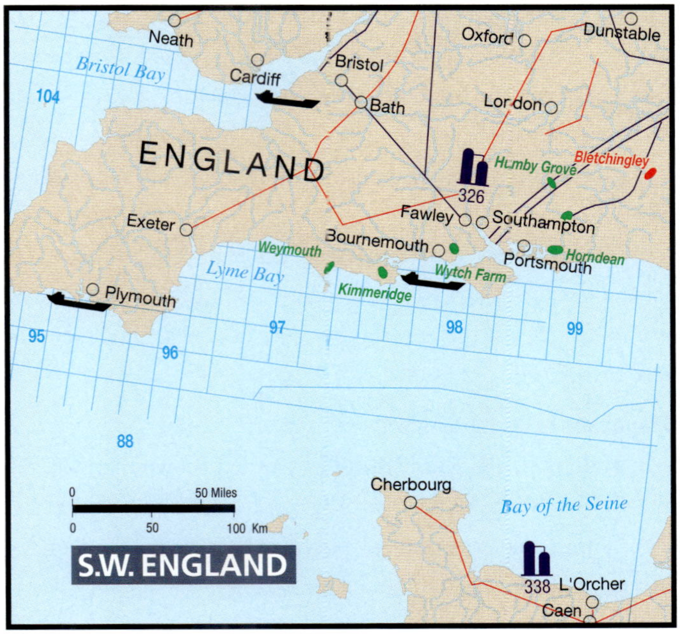

Skills shortage

Offshore operators and contractors in the UK agreed in mid-2008 to a partnership to share experiences and best practices on new learning products and services.

The aim is to help address the skills shortage in the nation's oil and gas industry.

OPITO-The Oil & Gas Academy (Aberdeen and Norfolk, UK) and the Engineering Construction Industry Training Board (ECITB; Hertfordshire, UK) will work with Oil & Gas UK (London) and the Offshore Contractors Association (Aberdeen, UK).

The bill also aims to push smaller players in the North Sea into offshore oil and gas decommissioning by ensuring adequate environmental and taxpayer protections.

Decommissioning costs are estimated at £15–19 billion, with 500 oil and gas installations in the UK North Sea and more than 6,000 miles of pipeline.

Energy companies also will be encouraged to invest in carbon capture and storage (CCS) technology to help address climate change. The government will create a licensing regime for storing CO_2 offshore.

The UK also has proposed building a fleet of new nuclear power stations, with some possibly in operation by 2020.

The nuclear plants would be paid for entirely by energy companies, with each costing an estimated £1.5 billion. Companies also would pay the full costs of decommissioning and their full share of waste management costs.

High oil prices have made nuclear power a much more attractive option.

Ernst & Young (London) projected that UK reliance on natural gas for electric power generation could soar to as much as 65% by 2025 if the government does not fully implement its new energy bill.

Responses to rising oil prices

In mid-2008, UK Prime Minister Gordon Brown met with oil industry leaders to discuss ways to control soaring fuel prices.

Parliament and the general public called for policy changes on road and fuel taxes to help families.

Also of concern were possible fuel shortages caused by several disruptive strikes at mid-year by fuel-truck drivers upset about low wage scales.

Officials with DBERR, which grants licenses for the UK North Sea, said the agency had made changes in its licensing regime to increase production there.

Also under discussion were tax incentives to promote production from the remote and technically challenging West of Shetlands region.

Exploration and Development

UKCS production challenges

Operators on the UK Continental Shelf (UKCS) spent £4.9 billion to develop new reserves in 2007, compared with £5.5 billion spent in 2006, said a mid-2008 economic report from trade association Oil & Gas UK (OGUK; London).

The decline, coupled with rising exploration and production costs, means that capital is only one-third as efficient as it was five years ago, said the group, which threatens new oil and gas production efforts.

OGUK said the industry required significant investment beyond the £21 billion planned by companies to produce 9.8 billion boe from 2008–2013.

The association believes that as much as 25 billion bbl could be recovered.

OGUK also called for tax incentives to attract investment in the mature province that produced 2.8 million boe/d in 2007.

West Breagh wells

Faroe Petroleum PLC (Aberdeen) hit pay in late 2008 with the 42/13-5 and 42/13-5z wells on the West Breagh gas field in the UK southern North Sea gas basin.

The 42/13-3 vertical appraisal well drilled in the same location in 2007 tested at rates up to 17.6 MMcf/d.

The 42/13-5z well was being prepared for flow testing.

East Breagh

EnCore Oil PLC (London) proved gas with its East Breagh well 42/13-4 in the southern North Sea in late 2008.

The well tested two zones at combined controlled rates of up to 10.2 MMcf/d.

The company said that the well results suggest that "this accumulation has the potential to become one of the larger discoveries of recent years in the southern North Sea."

Cairngorm well

Stratic Energy PLC (London) reported the start of drilling operations in November 2008 the Cairngorm discovery area in the UKCS.

Well 16/2b-5, the firm's first operated well in the UKCS, was to appraise multiple fracture zones in the granite reservoir that are thought to have been damaged in the 16/3a-11z discovery well.

That well flowed at rates in excess of 2,000 b/d of oil.

Bowstring East prospect

In November 2008, Sterling Resources Ltd. (Calgary) announced a wildcat oil discovery on the Bowstring East prospect in the North Sea.

The Cladhan well encountered an oil column of more than 110 ft in an Upper Jurassic sequence below m (9,450 ft).

The well, on Block 210/29a in license P1064, was drilled in 163 m (535 ft) of water and reached total measured depth of 2,967 m (9,735 ft).

Jacky field

The UK subsidiary of Ithaca Energy Ltd. (Calgary) said in late 2008 that it expects to produce 7,500 b/d of oil from one well on Jacky field in the UK North Sea.

The field is scheduled to come on stream in early 2009.

Holding 5.2 million bbl of proved and probable reserves, Jacky lies 10 km northeast of Beatrice field in the Inner Moray Firth area of the UKCS.

Produced oil will be processed at Ithaca's nearby Beatrice facilities.

Pink, Ettrick, Blackbird

Nexen Petroleum Inc. (Calgary) hit 17 m (57 ft) of net oil pay with the Pink discovery in the UK sector of the North Sea in late 2008. A sidetrack well encountered 41 m (134 ft) of net oil pay.

The firm also reported that first production from the Ettrick field has been pushed back to early 2009 following commissioning delays with the company's leased FPSO.

Production volumes are expected to average 15,000–20,000 boe/d in 2009.

Also, Nexen suspended its Blackbird well on Block 20/2a, 6 km (4 mi) south of Ettrick field.

The well struck 34 m (111 ft) of net pay in multiple zones, and drillstem tests indicated an average of 3,800 b/d of oil production potential.

Torphins prospect

Lundin Petroleum AB (Stockholm) discovered oil and gas in mid-2008 with exploration well 21/8-4Z on the Torphins prospect in block 21/8 on the UKCS.

Well 21/8-4 and sidetrack 21/8-4Z reached TD of 2,492 m (8,177 ft) and 2,210 m (7,250 ft), respectively. Both wells tapped Paleocene sandstones.

Moth discovery

In mid-2008, Oilexco North Sea [a subsidiary of Oilexco Ltd. (Calgary)] reported that its 23/21-6z Moth discovery well flow tested gas-condensate at significant rates.

The well is in the UK sector of the North Sea.

Flow averaged 20.3 MMcf/d of natural gas with 2,110 bbl/d of condensate (5,493 boe/d).

Maximum flow rates were 24.4 MMcf/d and 2,460 b/d of condensate (6,526 boe/d).

The company said its calculations suggest that the Moth well could be capable of 44 MMcf/d and 4,400 b/d of condensate (11,800 boe/d).

West and East Rinnes fields

Dana Petroleum PLC (Aberdeen) discovered oil in the West Rinnes structure in early 2008, close to its producing Hudson and Melville fields in the UK Northern North Sea.

Well 210/24a-11, drilled to a total measured depth of 1,972 m (6,470 ft), encountered a full Brent reservoir sequence with high quality sands.

A drill-stem test flowed 100% clean oil at up to 7,800 b/d, similar in quality to output from Hudson, 5 km (3 mi) away.

Shortly thereafter, a sidetrack well discovered the East Rinnes oil field in block 210/24a.

Falkland Islands prospects

Prospects identified from 850 sq km (328 sq mi) of 2007 3D seismic conducted in 2007 on PL032 and PL033 in the North Falkland basin could contain a mean 1.9 billion bbl recoverable.

That was the conclusion of Rockhopper Exploration PLC (Salisbury, UK).

Other leads are under investigation, and prospects identified on PL023 and PL024, could have an estimated 2.5 billion bbl recoverable.

The new mapping identified five hydrocarbon plays and multiple drilling targets.

Rockhopper was planning additional analysis and modeling.

East Rinnes, the fourth field under Dana's operatorship in the area, was drilled to a total measured depth of 6,935 ft (2,114 m) and encountered a full Brent reservoir sequence.

Initial analysis indicated that the East Rinnes oil is similar to the high-quality crude discovered at West Rinnes and to that being produced by Dana at the nearby Hudson oil field.

Jasmine field

ConocoPhillips Co. (Houston) made a discovery in early 2008 in a structure adjacent to the Jasmine gas condensate field in the UK central North Sea.

The Jasmine field development straddles blocks 30/6 (P.011) and 30/7a (P.032) and is west of the Judy production platform.

The exploration well has confirmed the presence of hydrocarbons in a structure northeast of the Jasmine discovery well (30/6-6), which was drilled in 2006.

Bentley field

Xcite Energy Ltd. (Aberdeen) said it expects that commercially viable production rates can be achieved from its early-2008 oil find on the Bentley field, block 9/3b, license P.1078.

The firm said that horizontal well technology and artificial lift will likely be needed.

The 9/3b-5 appraisal well was drilled to 1,224 m (4,105 ft) TD and encountered heavy-oil-bearing Paleocene Upper Dornoch formation sands.

Lease Sales

25th Licensing Round

DBEER said it received 193 applications covering 277 blocks in its 25th Offshore Licensing Round, the highest number since 1974.

The following are some of the firms that were awarded licenses in late 2008.

Dana Petroleum PLC (Aberdeen) won 11 new licenses, comprising interests in 29 full or partial offshore blocks.

The licenses concerned a range of blocks in: the Rinnes oil area (northern North Sea); the prospective West of Shetland region (Quads 205, 208 and 209); the Greater Kittiwake oil area and near the Enoch oil field (central North Sea); and near the producing Johnston gas field (southern North Sea).

Stratic Energy Corp. (Calgary) was awarded three blocks in the Southern Gas basin adjacent to the Breagh gas discovery, and part of a block in the central North Sea.

Lundin Petroleum AB (Stockholm) was awarded six exploration licenses 14 blocks or part-blocks in the central North Sea.

13th Onshore Licensing Round

In mid-2008, the government awarded 54 companies onshore licenses in its 13th Licensing Round.

A total of 97 petroleum exploration and development licenses went to a variety of independent companies, including Island Gas Resources PLC (IGas), Northern Petroleum (both in London); and Providence Resources PLC (Dublin).

Island Gas Resources PLC (IGas; London) a CBM developer, will explore seven blocks in northwestern England.

Northern Petroleum PLC (London) will operate two blocks on the Isle of Wight.

Providence Resources PLC will examine an oil discovery and the Burton Down exploration prospect on Block SU91.

Others offered licenses were Egdon Resources PLC (Hampshire, UK) and Europa Oil & Gas PLC (Abingdon, UK).

Drilling and Production

Grouse field

The Grouse field came onstream in December 2008, said Venture Production PLC (Aberdeen).

On Block 21/19 in the UK central North Sea, the field is expected to produce 10,000 b/d of oil and 92,042 std cu m/d (3.25 MMscfd) of natural gas in 2009.

A single subsea production well ties the field back to the Kittiwake platform, the production hub for the company's Greater Kittiwake Area (GKA) operations.

Initial well productivity was in line with previous estimates: about 5,000 boe/d net to Venture.

The firm said that Grouse was the third GKA satellite brought into production since Venture assumed operatorship in 2003.

Curlew C development

Production began in September 2008 from the Curlew C development in the North Sea, said Shell UK Ltd. (London).

The development marks the first time that the company has produced hydrocarbons from a chalk formation in the North Sea.

Curlew C lies in block 29/7 of the UK central North Sea, 210 km (130 mi) east-southeast of Aberdeen.

Dumbarton field extension

Production began in mid-2008 from the first phase of a major extension of the Dumbarton field project, said Maersk Oil North Sea UK (Aberdeen).

The field is in blocks 15/20a and 15/20b of the UK sector of the North Sea.

The extension work involves phased drilling of up to eight new horizontal production wells and up to a further two produced water disposal wells in the Dumbarton and nearby Lochranza fields.

Bringing the first two wells onstream raised total field production to more than 40,000 b/d of oil.

The second phase drilling will continue into 2009.

Don fields

Petrofac Energy Developments Ltd. (London) spudded the first well in mid-2008 on the West Don field in the UK North Sea, on Blocks 211/13b and 18a.

West Don is expected to start production in second quarter 2009, along with Don Southwest field.

The fields, 150 km (93.2 mi) northeast of the Shetland Islands, are in 170 m (558 ft) of water.

Petrofac was planning to drill seven wells in West Don in 2008, then move to Don Southwest early in 2009.

Don Southwest's oil production is expected to hit 12,000 b/d at peak rate.

At West Don, Petrofac hopes to produce 20,000 b/d of oil at peak rate.

Wissey gas field

Faroe Petroleum PLC (Aberdeen) reported in mid-2008 the start of production from the Wissey gas field in block 53/4D in the UK Southern Gas basin.

Initial flow rates are approximately 70 MMscf/d, or 11,600 boe.

The Wissey gas field is tied back to the Horne and Wren platform.

Faroe estimates its net reserves from the field are at least 5.9 bcf.

Heavy oil: UK sector north

PennWell Corporation's *Offshore* magazine reported in mid-2008 that heavy oil in the UK North Sea was drawing increased interest from producers.

Getting most attention are Quadrant 9 blocks in the far north, where a hub could soon emerge serving four long-neglected fields.

According to DBERR, there are 19 heavy oil discoveries throughout the UK shelf, with in-place reserves totaling 9 billion bbl. None were developed until Gryphon field came onstream in 1993.

Six active heavy oil fields collectively were generating 150,000 b/d in mid-2008. Most prolific are the Captain and Alba fields in the central North Sea.

Offshore Activity, Far North UK

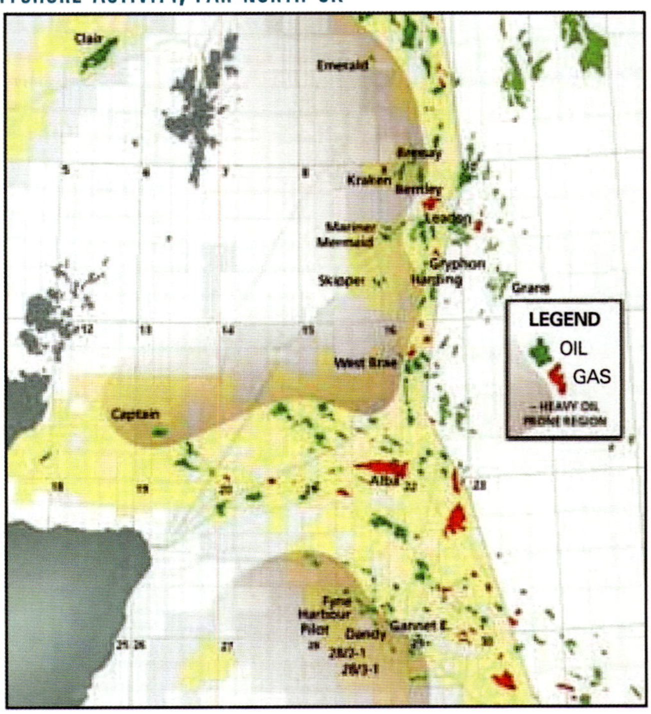

Mariner in block 9/11c and Bressay in two blocks adjoining Quadrant 9's northern border together could hold combined in-place oil of more than 2 billion bbl.

Horizontal wells and use of more-durable electric and hydraulic submersible pumps are transforming the economics of these far-north fields, said Offshore staff.

Starling, Caravel and Shamrock fields

Shell UK Ltd. (London) reported the start of production in mid-2008 from Shamrock field in the UK sector of the North Sea.

Shamrock was developed using Shell's proprietary Monotower platform, powered by wind and solar energy, which the firm says costs much less to build than a traditional offshore platform.

Shamrock was brought on stream shortly after Shell's Starling and Caravel fields.

Production capacity through the Shamrock monotower will average 120 MMscfd.

Chiswick field

Venture Production PLC (Aberdeen) brought a second production well in the Chiswick gas field onstream in early 2008.

The field is in block 49/4a, about 18 km (11 mi) northwest of Markham and 150 km (93 mi) northeast of Bacton in the southern North Sea.

The second well brought overall production from the Chiswick field to more than 1.9 MMcmd (70 MMcf/d). That rate is expected to increase to 2.3–2.5 MMcmd (80–90 MMcf/d).

Work was ongoing to prepare for up to three more wells targeting incremental reserves in the Gamma fault block and the undrilled Beta fault block.

Jura field

Total SA (Paris) said it expected that production from Jura gas-condensate field in the UK North Sea would start in second quarter 2008.

The company will produce 45,000 boe/d at plateau and will connect the field via a 3-km (1.86-mi) pipeline to the Forvie North subsea wellhead.

Jura holds more than 170 million boe of proved and probable reserves in the Alwyn area, 160 km (99 mi) east of the Shetland Islands and 440 km (273 mi) northeast of Aberdeen.

The Alwyn area holds the Alwyn North, Dunbar, Ellon, Grant, and Nuggets fields.

Decommissioning

North Sea decommissioning

North Sea production platforms and other offshore structures eventually will have to be removed at the end of their economic lives.

Of the two main producing sectors, the need is more imminent on the mature UK side, said *Oil & Gas Journal* in early 2008.

In that sector, equipment already is retiring and companies and trade groups grapple with associated financial, regulatory, and operational issues.

According to trade association Oil & Gas UK (London), of 470 offshore installations on the UKCS, an estimated 10% are floating structures, 30% are subsea, 50% are small steel, and 10% are large steel or concrete. Each type of structure requires its own decommissioning method.

The group estimates decommissioning costs of the UKCS to be £15–20 billion.

In addition, the UKCS has 10,000 km (6,214 mi) of pipelines, 15 onshore terminals, and 5,000 wells that eventually must be decommissioned.

Of 400–500 fields on the UKCS, fewer than 100 have financial provisions in place to cover decommissioning, according to Oil & Gas UK.

In early 2008, North Sea operators welcomed tax changes proposed by the UK Chancellor of the Exchequer aimed at extending the life of marginal oil fields.

The changes would reform the petroleum revenue tax and help operators cover the treatment of decommissioning costs of UK North Sea facilities.

Decommissioning Projects Under Way

Three major decommissioning projects were under way or being planned in early 2008 in the North Sea. They are:

- The North West Hutton platform of BP PLC (London) in the UK sector
- The Frigg platforms of Total SA (Paris) that straddle the UK-Norwegian boundary
- The Ekofisk-1 platform of ConocoPhillips Co. (Houston) in Norwegian waters.

LNG

LNG cargo from Qatar

Centrica PLC (Windsor, UK) was making plans to import at yearend 2008 its first cargo of LNG from Qatar to commission the second expansion phase of the Isle of Grain LNG terminal in Kent.

The cargo from RasGas Co. Ltd. was expected to be the UK's largest-ever LNG delivery, said Centrica.

National Grid PLC (London) owns the terminal. Centrica has the rights to import 3.4 bcm/yr of gas for the next 20 years.

Centrica was expecting about 5 bcf (50 million therms) of gas to be delivered in a high-capacity Q-Flex supertanker.

Pipelines

Statfjord gas: Direct to UK

As of October 2008, all gas from the Statfjord field in the North Sea was piped directly to the UK.

Details are presented in the Norway section of this chapter.

Environmental Issues

First carbon permits auction

The first auction of carbon permits in Great Britain took place in November 2008, netting the government £54 million.

Almost 4 million permits were sold, each equivalent to a tonne of carbon dioxide, under the European Union's greenhouse gas emissions trading scheme.

Bidders were not identified, but electricity producers were expected to be the main buyers.

Alternative Fuels

Biodiesel plans shelved

The *Financial Times* reported in late 2008 that Ineos Group Ltd. (Hampshire) had put on hold its plans to build four biodiesel plants in Europe.

The company said that it was cutting back on "all non-essential growth capital expenditure" in the face of the "continued and prolonged global economic downturn."

The four plants—at Antwerp, Lavera (France), an unidentified site in Germany, and Grangemouth, near Edinburgh—were intended to come on stream during 2010–2012.

Their combined capacity was to have been about 2 mtpy (roughly 40,000 b/d).

2.5% biofuels rule

A new UK policy took effect in April 2008, requiring that all fuels, including gasoline and diesel, must contain 2.5% biofuels.

Critics claimed that the measure was pushing up food prices and damaging the environment through land conversion and increased use of chemical fertilizers.

The new policy is part of the UK Renewable Transport Fuels Obligation (RTFO), developed to ensure that it meets European Union regulations.

By 2010, the requirement will rise to 5% and is expected to cut CO_2 emissions by 2.5 million tonnes.

The Department for Transport will review biofuels use by June 2009, after which ministers may postpone or drop plans to increase the biofuel percentage to 10% by 2020.

Company News

ONGC Videsh, Imperial Energy

ONGC Videsh Ltd. (OVL; New Delhi), foreign investment arm of the Indian state oil company, was poised to buy Imperial Energy PLC (London) at year-end 2008.

A minimum of 90% acceptance by shareholders was required to finalize the £1.3-billion deal. OVL had received acceptances from 96%.

The transaction for Imperial, which has most of its assets in Russia, was expected to close in mid-January 2009.

OVL said the deal would give it a presence in western Siberia, one of the world's largest oil and gas producing regions.

Imperial has probable and possible reserves of 3.4 billion boe and was planning to lift production to 80,000 b/d by 2011.

EDF, British Energy

In December 2008, the European Commission (Brussels) gave permission (with conditions) to Électricité de France (EDF; Paris) to buy British Energy (BE; East Kilbride).

BE owns most of Britain's nuclear power stations.

Negotiation of final details was under way.

EDF, which owns France's 58 nuclear reactors, has said it plans to build four of its European Pressurized Reactors in the UK.

The firm also was making plans to secure a strategic U.S. partnership with Constellation Energy (Baltimore, MD).

Following the EDF-BE decision, Centrica PLC (Windsor, UK) said it was moving toward a deal in January 2009 to take a 25% stake in British Energy (East Kilbride) for about £3.1billion.

Centrica seeks more non-fossil fuel power plants to reduce its vulnerability to volatility in gas prices.

Activity in Scotland

Amos prospect

In late 2008, OMV AG (Vienna) made two discoveries in the North Sea.

Exploration well 204/25a-8 on the Amos prospect, West of Shetlands, was completed in April 2008 as an oil discovery, said OMV.

The firm believes that the Amos prospect discovery could be commercial because of its proximity to the producing Schiehallion field.

Moray Firth oil

A group led by Petro-Canada Ltd. (Calgary) reported in early 2008 a potentially commercial discovery 110 km (68 mi) north of Aberdeen, in the UK North Sea's Inner Moray Firth area.

The 13/21b-7 well went to TD 2,398 m (7,867 ft) in 89 m (292 ft) of water and found oil in two Lower Cretaceous sandstone reservoirs.

Petro-Canada said that the reservoirs have the potential to yield commercial flow rates through the use of horizontal wells and appropriate completion technology.

TNK-BP saga

TNK-BP Ltd.—under pressure from its billionaire Russian backers, Alfa-Access-Renova group (AAR)—said that the firm's executive vice-president for downstream, Tony Considine, would leave in September 2008 to "pursue other opportunities."

That announcement came just a month after TNK-BP chief executive, Robert Dudley, left Russia in the face of what he called "sustained harassment" by a Moscow court and labor officials.

He continued to run TNK-BP from the UK.

Dudley's leadership is said to threaten to break up the five-year partnership between BP PLC (London) and AAR. The Russian backers have called for Dudley's resignation.

People close to the company have suggested that Denis Morozov, former chief executive of Norilsk Nickel, likely will become the new chief executive in 2009.

Middle East

BAHRAIN

Capital: Manama
Monetary unit: Dinar
Refining capacity: 262,000 b/cd
Oil production: 170,000 b/d
Oil reserves: 124.560 million bbl
Gas reserves: 3.250 tcf

Bahrain has become a major regional financial center, but its economy still depends heavily on oil. Petroleum revenues make up about two-thirds of government revenue and export earnings, according to the U.S. Energy Information Administration (EIA).

Bahrain's two major industries are refining and aluminum smelting, so oil prices dramatically affect its economy.

Relatively high oil prices of recent years helped Bahrain's real GDP to grow at over 5% for 2004 through 2006, said EIA. A rate of more than 6% is projected for 2007.

Gas exploration bid round

Bahrain launched a bidding round in late 2008 for exploration of its onshore natural gas fields, said the nation's oil ministry. Officials said that 25 international companies expressed interest in bidding and that the open bidding process will end in second quarter 2009.

The country aims to develop deep subsea gas fields to meet rising domestic demand.

In 2007, Bahrain consumed 1.3 bcm (45.9 bcf) of gas and its consumption is expected to rise to more than 2 bcm/yr (70.6 bcf/yr) in the long term.

PERSIAN GULF
IRAN
IRAQ
KUWAIT
SAUDI ARABIA
BAHRAIN
QATAR
UNITED ARAB EMIRATES
OMAN
TURKMENISTAN
AFGHANISTAN
DIVIDED ZONE
Caspian Sea
Persian Gulf
Strait of Hormuz
Gulf of Oman
0 50 100 150 Miles
0 100 200 Km
Tehran
Shahr-e Rey
220
Karaj
Qazvin
Rasht
Lulebur
Bandar-e Pahlavi
Hashtpar
Astara
Ardebil
Saveh
Qum
Hamadan
Arak
170
Isfahan
284
Ardakan
Yazd
Bafq
Shahreza
Yazd-e Khvast
Kerman
Mashiz
Baft
Shiraz
40
Marv Dasht
Kazerun
Firuzabad
Ganaveh
Bushehr
Deyyer
Asaluyeh
Bandar-e Maqam
Bandar-e Lengeh
Bandar 'Abbas
232
Qotbabad
Mashhad
Sabzevar
Shahrud
Damghan
Shahi
Neka
Behshahr
Gorgan
Shah Pasand
Dasht
Bojnurd
Quchan
Ashkoad
Sarakhs
Khangiran
Dauletabad Donmez
Torbat-e Haydari
Husan Kui
Okrem
Korpedzhe
S. Bugdaili
Kamyshldzha
N. Okarem
Okarem
Keimir
S. Okarem
Akpatlaukh
Chikishlyar
Veyzenhar/ Malah kuh
Sarkan
Samand
Jelow Gir
Qaleh Nar
Abu Ghirab
Lab-i-Safid
Andimeshk
Lali
Karun
Palangan
Zeloi
Masjid-I-Suleiman
Shushtar
Par-E-Siah
Naft Safid
Haft Kel
Lordegan
Doudrou
Ramin (Mulla Sani)
Kuh-I-Rig
Ramhormuz
Shurom
Nazarabad
Kuh-E-Bangestan
Parsi
Karanj
Ramshir
Bandar-e Mah Shahi
Gachsaran
Ardakan
Ahwaz
Marun
Jufeyr
Ab Teymour
Nahr Umr
Basra
150
Rumaila
Zubair
Khorramshahr
Abadan
350
Bandar-e Khomeine
Rag-E-Safid
Hendijan
Binak
Bahrgansar
Radak-Milaatun
Sarvestan
Nowrooz
Abuzar
Nargesi
Doroud (Darius) Kharg
Bushgan
Kuh-i-Kaki
Dalan
Aghar
Kuh-i-Mand
Bandubast
Kangan
Nar
Varavi
Pars
Fa
Assaluyeh
Lamard
Tabnak
G
F
South Pars
North
LAVAN I.
30
BH
Alpha
Reshadat
Resalat
Sirri A
Sirri E
Salman
Abu Al Bukhoosh
Maydan Mahzan
Bul Hanine
Nasr
Mandoos
Belbazem
Zakum
Umm Al Dalkh
Jebel Ali
Margham
TFP
Sharjah
71
Dubai
120
Mo'Ayyid
S. Juwaiza
Khubai
Sajaa
Al Fujairah
90
Fujayrah
Umm al Qaiwain
Salakh
Hanquan
Saleh
Bukha
W. Bukha
Baih
Limah
Gavarzin
S. Gashu
Suru
HD
Sarkhun
W. Namak
Tuba
Sabriyah
Ar Rawdatayn
Bahrah
Raudhatain
190
Al Kuwait
Riquah
Abdullyah
443
Mina al Ahmadi
Burgan
Mina Abdullah
257
Mina Su'ud
Minagish
Umm Gudair
Wafra
Fawaris
S. Wafra
Dorra
Soroush
Hout
Lulu/Esfandiar (Iran)
Marjan/Foroozan
BB
Bushire
Hamur
Khafjii
Safaniya
Salsal
Manifa
Kurayn
Karan
Hasbab
Sharar
Watban
Bakr
Dhib
Jaladi
Fadhili
Ras al Ju'aymah
Jana
Fadhi
Abu-Safah
290
Al Jubail
Berri
Samin
Ras Tanura
550
Qatif
Al Rayyan
Ad Dammam
Manama
Awali
262
Abqaiq
Al Uqayr
Al Mubarraz
Dukhan
Umm Bab
200
Doha
Idd Al Shargi
Umm Said
Jarnain
Bu Jufair
Dalama
Hair Dhalma
Ghasha
Bu Tini
Hail
Taweelah
150
Abu Dhabi
Fateh
Falah
Zol
Rumaitha
Zubaya
Arjan
Ruwais
350
Bida Al Qimzan
Tarif
Bab
Habshan
Sahil
Bu Hasa
Hawaila
Asab
Qusahwira
Shah
Zayqat
Mender
Lekhweir
Deleel
Nati
Al' Ayn
Mahdah
Suhar
Al Khaburah
Muscat
Mina al Fahal
85
Shams
Safah
Nadir
Al Sahwa
Hafar
Shibka
M. Huraymah
Habiba
Abeer
Baqlah
Khurais
Ghawar
Qirdi
Abu Jifan
Farhah
Mazalij
Harmaliyah
Haradh
Dilam
As Salamiyah
Ad Dilam
Sahba
Midrikah
Tinat
Lughfah
Jawb
Raghib
Naim
35°
30°
25°
50°
55°

IRAN

Capital: Tehran
Monetary unit: Rial
Refining capacity: 1,451,000 b/cd
Oil production: 3.900 million b/d
Oil reserves: 136.150 billion bbl
Gas reserves: 991.600 tcf

Overview

Despite steep production declines (8%/yr onshore and up to 13%/yr offshore) and problems with oil field development in recent years, FACTS Global Energy (FGE; Honolulu) expects Iran's crude and condensate production capacity to reach 4.7 million b/d by 2012, almost all from South Pars field.

Iran has very large gas reserves, but FGE sees a ceiling of 20–30 mtpy on Iranian gas exports. Domestic demand is kept high by subsidy pricing, oil field needs for gas injection of about 283 MMcmd (10 bcfd), large gas-based petrochemical projects, the use of compressed natural gas to supplement gasoline supplies, and political opposition to gas exports.

In late 2008, Iran was importing 20–23 MMcmd (700–810 MMcfd) of gas from Turkmenistan, or about 5% of its total consumption.

Iran's only gas exports were to Turkey, in the range of 20 MMcmd (700 MMcfd).

A researcher at Harvard University observed in mid-2008 that Iran's clerical leadership was pondering three gas-sector strategies:

- Reinjection of gas into oil fields to boost oil production (assuming oil prices recover)
- Supply of more gas to domestic markets to encourage industrial development and reduce gas prices for residential and commercial users
- Greater focus on gas development and LNG technology to boost gas exports.

Because of underinvestment in its refining sector, Iran must import about one-third of its domestic liquid fuel, and the government started a system of rationing to curb demand.

The oil ministry said that fuel price subsidies also might be removed to further reduce demand, possibly by mid-2009. A presidential election is slated for June 2009.

Iran has been inviting outside participation in its oil sector, but lack of confidence in the country's political environment and concerns about its economy remain a deterrent.

Close the Straits of Hormuz?

The standoff between Iran and the UN Security Council continued in 2008 over concerns that the country's nuclear development program has a military component.

In early August, after unproductive talks, Iranian military officials threatened to use anti-ship missiles to close the Straits of Hormuz, if necessary.

According to the U.S. EIA, some 17 million b/d of crude oil are shipped through the straits. The United States has pledged to keep the route open.

The UN Security Council has imposed three sets of limited penalties on Iran since 2006, but the Iranian government has continued efforts to secure the technology it claims to need.

In early June, Israel staged a large military exercise in the eastern Mediterranean that some analysts viewed as practice for a potential bombing attack on Iran's nuclear plants.

Oil industry development plans

Oil ministry officials said in late 2009 that the nation's oil industry likely will need about $145 billion in investments over the next decade to boost its role as a major producer of petrochemicals and natural gas.

Iran has invited greater participation and investment from the private sector as well as from other nations to boost the country's oil and gas sector.

In late 2008, officials in Turkey were considering an investment of $12 billion in Iran for offshore gas development in the southern region. A new 1,850-km (1,150-mi) gas pipeline also was discussed, to run from the Assalouyeh gas field to the Turkish border.

Iranian firms are developing the country's largest onshore oil field at Azadegan, which has reserves of 42 million bbl.

Russia's Gazprom (Moscow) has signed agreements to help develop Iran's oil and gas fields. The firm also is likely to be involved in other ongoing and proposed projects such as construction of refineries and the $12 billion MedStream offshore pipeline project.

Exploration and Development

Azadegan, South Pars interest

In early 2008, ONGC Videsh Ltd. (OVL; New Delhi) agreed to an MOU with Hinduja Group (London) to jointly explore acquisition of overseas acreages, with a focus on Iran.

OVL is the overseas arm of India's state-run Oil & Natural Gas Corp. (ONGC; Dehradun).

The two firms were said to be exploring joint opportunities in Iran's onshore Azadegan oil field and offshore South Pars Phase 12 gas project.

Azadegan, with reserves estimated at 33 billion bbl, is one of the largest oil fields discovered during the past 30 years.

South Pars gas field, together with its extension, Qatar's North field, is considered the largest offshore gas field in the world.

Dayyer gulf block

Iran signed a contract in early 2008 with Edison S.p.A. (Milan) for exploration rights to the 2.1-million acre (3,281-sq mi) Dayyer offshore block.

The block lies generally between Iran's coastal towns of Kangan and Bushehr, north across the Persian Gulf from the Qatar peninsula.

New gas field in Gulf

A gas field with an estimated 311 bcm (11 tcf) in reserves was discovered in early 2008 by an unnamed Indian company in the Persian Gulf offshore Iran.

Drilling and Production

South Pars: More output needed

Production from South Pars gas field Phases 6, 7, 8, and 9 was targeted for completion in late 2008—output seen as critical to preventing a severe energy crisis in Iran in the winter of 2008–09, according to FACTS Global Energy (FGE; Honolulu).

In the winter of 2007–08, a gas shortage resulted in massive reduction of gas reinjection into Iranian oil fields and a gas-supply cut to electric power plants, industrial projects, and 200,000 residential and commercial consumers. Iran also halted gas exports to Turkey.

To control growing domestic gas demand and decrease subsidies for high-volume end users, Iran raised domestic prices substantially in 2008 (by nearly 800% for some customers), but this was not successful in cutting demand.

Gas demand growth of 7–8% is expected from 2008–20, according to FGE.

IRAN'S SOUTH PARS, QATAR'S NORTH FIELD

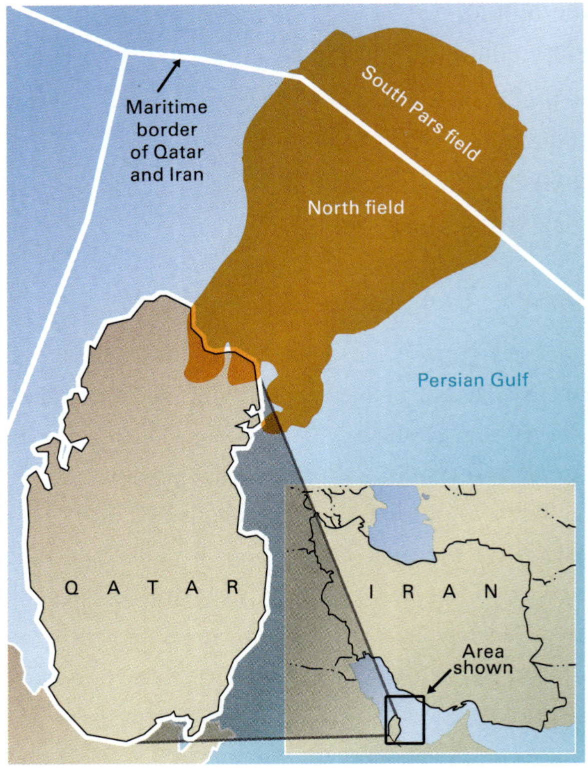

Source: Energy Information Administration, adapted from IHS Energy

South Pars: Near-term plans

In 2008, gas production was expected to increase by 0.9 bscfd by developing Tange Bijar gas field and drilling more wells in Nar, Kangan, Homa, Shanul, Varavi, or Tabnak gas fields.

But output from South Pars Phases 9 and 10 would be more substantial and would be more urgently needed, said FGE.

Phases 9 and 10—nearing completion in September 2008—would release 2 bscfd of gas to the domestic market by yearend.

Phases 6–8 are expected to produce another 3.6 bscfd of gas (plus 152,000 b/d of condensate and 1.6 mtpy of LPG). However, the gas is scheduled to be piped to Khuzestan Province for oil-field reinjection.

Phase 6 was scheduled for completion in mid-2008, and gas was expected to flow from Phases 7 and 8 in late 2008 and early 2009, respectively.

South Pars: Later phases

Several other phases of South Pars development have longer time horizons, said FGE.

- Phase 12: Completion in early 2012, yielding 3 bscfd of gas and 110,000 b/d of condensate

- Phases 17 and 18: Completion in 2013–14, yielding 2 bscfd of gas and 80,000 b/d of condensate from 27 wells
- Phases 15 and 16: To yield 2 bscfd of gas and 80,000 b/d of condensate. Completion date uncertain.

Unlikely to be completed before 2015 are Phases 11 and 19–24.

South Pars: Phase 13

Royal Dutch Shell PLC (The Hague, Netherlands) and Repsol-YPF SA (Madrid) suspended plans in early 2008 to develop the proposed $10-billion LNG export plant that is part of Phase 13 for the South Pars gas field.

The firms had planned to develop Persian LNG on Tombak Island, building a $10-billion LNG plant in partnership with National Iranian Oil Co. to export 16 mtpy of LNG to Europe, Asia, and the Far East.

Factors cited for the decision were deteriorating relations between the United States and Iran and increasing development costs.

Pipelines

Iranian gas to Italy

In March 2008, the NIGC contracted with Elektrizitaets-Gesellschaft Laufenburg (Dietikon, Switzerland; EGL) to supply gas valued at €18–27 billion ($28–42 billion) over the 25-yr contract life.

Iran will deliver 5.5 bcm/yr (194.2 bcf/yr) of gas to Europe, providing a fourth avenue for gas to enter Western Europe apart from Russia, the North Sea, and North Africa. The anticipated route is from Turkey through Greece and Albania.

For the gas to reach Italy, a pipeline may be constructed under the Adriatic Sea, connecting Greece and Italy. Construction plans were being discussed in late 2008; completion in 2012 is possible.

Analysts said the deal could provide leverage for the EU with Gazprom because many Russian gas supply contracts are to be renegotiated in 2010.

Gas from Iran is seen as a more viable and secure option than gas from Azerbaijan or Iraq. However, critics wonder whether Iran can be a stable supplier, pointing to curtailed gas shipments to Turkey in 2006 and 2008. The latter action grew from price disputes that led Turkmenistan to stop its gas shipments to Iran.

Nabucco

Gas-price disputes in 2008 between Iran and Turkmenistan (and their implications for reliable gas delivery) raised questions about the viability of the proposed $7.5 billion, 3,000-km (1,864-mi) Nabucco pipeline.

Construction of the line, to carry 30 bcm/yr (1.06 tcf/yr), is expected to start in 2010.

It would allow Iranian gas to flow through pipelines from eastern Turkey to Austria's Baumgarten gas hub, by way of Bulgaria, Romania, and Hungary, and transport some 30 billion cu m/yr of gas.

However, there are fears that an export halt anywhere along the line could affect the whole network and prompt participating countries to hoard gas for their domestic markets.

The Nabucco discussions were said to have encouraged Russian support for the competing South Stream pipeline, to run from Russia to Europe along the bottom of the Black Sea.

IPI Pipeline

In October 2008, Iran and Pakistan decided to go ahead with the $7.4-billion Iran-Pakistan-India (IPI) gas pipeline project bilaterally, saying India could join the project later if it chose to.

Proposed in 2005, the project stalled due to difference on pricing and transit fees and concerns about security. India said it wants these issues addressed before the project could be implemented.

Iran's Products Production vs. Consumption

Product	1980	1985	1990	1995	2000	2005	2006
				1,000 b/d			
NIOC output of refined products							
Gasoline	74.9	74.9	122.4	149.1	182.6	208	274
Kerosine	101.3	70.5	86.2	124.5	156.9	178	133
Distillates	126	225.7	229.2	287.2	357.3	403	506
Residuals	243.1	201.2	230.9	255.8	397.6	436	483
Others	37	75	82	125.5	222	215	52
Total	**582.3**	**647.3**	**750.7**	**942.1**	**1,316.4**	**1,440**	**1,448**
				1,000 b/d			
Products consumption in Iran							
Gasoline	82.6	105.7	129	182.2	225	421	463.1
Kerosine	124.8	163.8	167.3	169.2	193.9	133.4	122.5
Distillates	169.4	281.7	393.1	398.8	404	473.8	533.3
Residuals	141.3	200.7	240.9	263.6	232	213.1	226.3
Others	44.4	60.4	71.8	127	134	217.8	233.4
Total	**562.5**	**812.3**	**1,002.1**	**1,140.8**	**1,189.1**	**1,459.2**	**1,578.6**

Source: OPEC Annual Statistical Bulletin, 2007

The total length of the IPI line would be 2,670 km (1,660 miles): about 1,115 km (690 miles) in Iran, 705 km (440 miles) in Pakistan and 850 km (530 miles) in India.

Construction likely will take 4–5 years.

Refining and Products

Refinery projects: 2008–2012

The National Iranian Oil Refining and Distribution Company (Tehran) was working in late 2008 to set up seven new refineries.

Aggregate investment cost is projected to be $21 billion, mostly from the private sector, according to news reports.

All are to begin operation by 2012–13, designed to increase Iran's refining capacity of crude oil and gas condensate by 1.57 million b/d, to a level of 3.1 million b/d.

Gasoline production is to be raised by 110 million l./d (29 million gal/day; 691,000 b/d).

- Persian Gulf Star refinery, Assalouyeh (southern Iran). Produce 360,000 b/d of gas condensate plus gasoline, jet fuel and other products. On stream in 2011.
- Caspian refinery, Gorgan (northeastern Golestan province). Refine 300,000 b/d of Caspian crude. Produce up to 20 million l./d (5.3 million gal/d; 126,000 b/d) of gasoline and 11 million l./d (2.9 million gal/d; 69,000 b/d) of gas oil.
- Hormoz refinery, Bandar Abbas. Cost: $4 billion. Refine 300,000 b/d of heavy and extra heavy crude oil. Produce gasoline and middle distillates compliant with European standards. Construction about 5% complete at yearend 2008.
- Khuzestan refinery, near Abadan City. Cost: about $4 billion. Refine 180,000 b/d of heavy crude from Azadegan and Yadavaran fields. Produce diesel oil, Euro IV-compliant super gasoline, liquefied gas, jet fuel, and sulfur. To be completed 2011.
- Shahriar refinery, Tabriz. Cost: $2.8 billion. Refine 150,000 b/d of crude; produce gasoline. Construction about 3% complete at yearend 2008.
- Anahita refinery, Kermanshah province. Cost: $3.5 billion. Produce 150,000 b/d.
- Pars refinery in Shiraz. Cost: $1.1 billion. Refine 120,000 b/d of crude. Construction about 3% complete at yearend 2008.

The country also was planning to invest $7 billion to enhance its nine existing refineries by 2012.

Refinery projects: 2010–2015

Also in late 2008, the oil ministry said that the government had formulated a $25-billion plan designed to boost Iran's gas and oil refining capacity to about 5 million b/d by 2015, utilizing both domestic and foreign investment.

Development projects in five categories will have a combined production capacity of 1.75 million b/d and all are to be implemented during 2010–15.

Gasoline consumption

Although it is one of the world's largest oil producers, Iran imports roughly 40% of its gasoline, due to limited refining capacity as well as rapidly rising domestic demand.

Fuel rationing and modest use of CNG as a vehicle fuel have had some impact on controlling demand growth, said FACTS Global Energy (FGE; Honolulu). The government also said in late 2008 that it intends to reduce (and eventually eliminate) gasoline subsidies.

The longer-term strategy, however, is to expand or upgrade refineries in Abadan, Arak, Isfahan, Bandar Abbas, Tabriz and Tehran. FGE forecasts that these actions should yield a surplus of supply in 2012–2013.

NIOC REFINERY CAPACITIES

Location	1980	1981	1982	1985	1989	1991	1994	1997	2000	2003	2006
	1,000 b/d										
Abadan	635	635	0	0	117	297	297	297	400	400	460
Arak	0	0	0	0	0	0	135	135	150	150	170
Tehran	220	220	220	220	220	220	220	220	225	225	230
Esfahan	240	240	240	240	254	254	254	254	265	265	270
Tabriz	80	80	80	80	99	99	99	99	112	112	112
Shiraz	40	40	40	40	40	40	40	40	40	40	42
Kermanshah	15	15	NA	15	27	27	27	27	30	30	30
Lavan	20	20	20	20	20	20	20	20	20	20	20
Masjid-i-Suleiman	70	70	70	0	0	0	0	0	0	0	0
Bandar Abbas								116	232	232	232
Total	**1,320**	**1,320**	**670**	**615**	**777**	**957**	**1,092**	**1,208**	**1,474**	**1,474**	**1,566**

Source: OPEC Annual Statistical Bulletin, 2007

Iran's planned refining capacity increases

Refinery	Existing capacity (2006), 1,000 b/cd	2011 capacity, 1,000 b/cd	Cost, €	Expected completion	Status
Modifications, upgrades planned at existing refineries					
Esfahan (stabilize capacity)	375	370	2.5 billion		Basic design. Two of five construction contracts awarded
Arak	180	250	2.5 billion	45 months	EPC, detailed design under way
Bandar Abbas capacity expansion	265	320	38 million	3 years	Under construction IGC and Oil Design & Construction Co.
Bandar Abbas Gasoline production increase			400 million	3 years fast track	Basic engineering, site prep., civil works under way. PIDEC is EPC contractor.
Bandar Abbas-Esfahan pipeline	540 km of 20-in. pipeline, 3 pump stations, expand terminal capacity		140 million	27 months	Rafsanjan-Esfahan section complete
Tabriz	110	110			
Tehran	250	250	230 million	3 years	Feasibility study complete. Feed change. Refining from heavy products to light.
Shiraz	55	55			
Kermanshah	25	25			
Levan	30	45			
Grassroots refineries planned					
Persian Gulf Star gas condensate	0	360	1.7 billion	3 years	Basic design complete by ODCC and Indian Oil Co. EPC contractor selected.
Abadan	385	360	1.5 billion	3 years	New 180,000 b/sd train. Modify processes.
Hormoz extra-heavy crude	0	300	3 billion	42 months	Feasibility study complete. Basic design and EPC tenders issued to contractors.
Khoozestan extra-heavy crude	0	180	1.5 billion	3 years	Snamprogetti completed feasibility study. Preparations to start basic design.
Shiraz	0	120			
Kermanshah	—	150	1.3 billion	3 years	Feasibility study complete. Preparations to start basic design. Engineering bid out.
Tabriz Shahriar	0	110	1.2 billion	42 months	Combine new refinery with older one.
Total	**1,675**	**3,005**			

Source: National Iranian Oil Engineering & Construction Co.

Petrochemicals

JV investments: $4.5 billion

Iran's National Petrochemical Company (NPC) announced in October 2008 that its joint-venture investments with foreign firms had reached $4.5 billion.

Partnership countries include the Philippines, Indonesia, Oman, Singapore, India, Venezuela and the UAE.

The oil ministry reported that NPC plans to double its petrochemical output to 68 mtpy by 2024 and to 100 mtpy by 2026.

Completing delayed projects

In third quarter 2008, FACTS Global Energy (FGE; Honolulu) concluded that Iran was likely to finish by yearend 2009 many delayed petrochemical projects now being built.

Key projects still being developed are expected to bring more than 29 mtpy of new capacity online for the production of ethylene, propylene, methanol, and other products by 2015. These include plants in the Pars and Bandar Iman Petrochemical Zones.

Iran petrochemicals

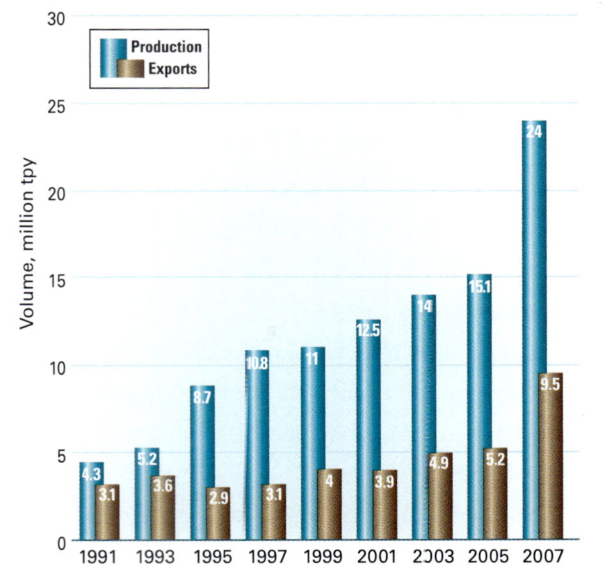

By mid-2009, Iran's petrochemical exports are expected to increase to 12–13 million tpy, according to the report. By 2015, petrochemical production will nearly double.

Iran is also building an ethylene pipeline, currently under construction from Assaluyeh in the south of Iran to Mahabad in the northwest of the country.

More-reliable feedstock for petrochem projects in western and northwestern Iran could materialize after completion of a 2,163-km (1,344-mi) ethylene pipeline (the West Ethylene Pipeline) from the south to the west and northwest of the country.

The pipeline was about 50% complete in third quarter 2008.

Also planned is a 3,500-km (2,175-mi) ethylene pipeline to the east and northeast (East Ethylene Pipeline), and a 600-km (373-mi) pipeline to central Iran.

Iran Petrochemical Projects

Project	Capacity, 1,000 tpy	Status
9th Olefin (Arya Petrochemical)	1,000	Commissioning
10th Olefin (Jam Petrochemical)	2,100	Commissioning
Ethylbenzene (Pars Petrochemical)	631	97% complete, start-up in 2008-09
6th Urea & ammonia (Ghadir Petrochemical)	1,150	97% complete, start-up in 2008-09
6th Methanol (Zagross 2nd Methanol)	1,650	95% complete, start-up in 2008-09
HDPE (Mehr Petrochemical)	300	88% complete, start-up in 2009-10
PVC (Arvand Petrochemical)	1,507	Completed in 2009-10
5th Olefin (Morvarid Petrochemical)	540	86% complete, start-up in 2009-10
LDPE (Amir Kabir Petrochemical)	300	Start-up in 2009-10
HDPE (Kermanshah Petrochemical)	300	Start-up in 2009-10
13th Olefin (Ilam Petrochemical)	738	Start-up in 2010-11
11th Olefin (Kavian Petrochemical)	2,180	Under construction, start-up in 2011-12
Gachsaran (Olefin)	1,090	Under construction, start-up in 2012-13
Shiraz Urea & ammonia	1,050	Under construction, start-up in 2012-13
LLDPE (Mahabad Petrochemical)	307	Under construction, start-up in 2012-13
LLDPE (Lorestan Petrochemical)	300	Under construction, start-up in 2012-13
LLDPE (Kordestan Petrochemical)	300	Under construction, start-up in 2012-13
12th Olefin (Persian Gulf Petrochemical)	5,214	Completed in 2013-14

Propane dehydrogenation plant

Uhde GmbH (Dortmund, Germany) signed an agreement in late 2008 with the National Iranian Petrochemical Company to build a 350,000-tpy propane dehydrogenation plant at the Imam Khomeini port in Khuzestan province.

The $32-million project is expected to be complete in 2012.

Feedstock will come from the Imam Khomeini petrochemical complex.

International Activities

China oil deal

In January 2009, Iran signed a $1.7-billion oil contract with China, bypassing international sanctions on its nuclear program, according to a Financial Times report.

China's CNPC and the National Iranian Oil Company will work together to develop an oil field.

Iran-Russia-Qatar gas cartel?

According to news reports, Iran, Russia and Qatar took the first serious steps in October 2008 toward forming an OPEC-style cartel on natural gas—an idea first proposed in early 2007 by Iranian leaders.

The action raises concerns among western nations that closer strategic ties between Russia and Iran could hinder efforts to rein in the Iranian nuclear programs.

The European Union expressed strong opposition to any natural gas cartel. Because Russia provides nearly half of the natural gas imported into Europe, a cartel is seen as a threat to supply security and price stability.

Moscow in recent years has cut gas exports to Georgia and Ukraine, causing disruptions in European nations further downstream.

No timeframe was given regarding finalization of cartel details.

Sri Lanka refinery upgrade

In May 2008, Iranian officials launched a $1.2-billion project to upgrade Sri Lanka's sole refinery at Sapugaskande, outside Colombo.

Details about the project are presented in the India section of the Asia-Pacific chapter of this book.

Refinery help for Ecuador

Iran and Ecuador signed an MOU in mid-2008 under which Iran will help refurbish and upgrade Ecuador's refineries. Venezuela also will participate.

Details are presented in the Ecuador segment of the Latin America chapter of this book.

Iran, Malaysia agreements

Iran and Malaysia made several agreements in late 2008 regarding development of LNG facilities and the supply of crude oil.

The agreements involve:

- Development of the Golshan and Ferdousi gas fields in Iran
- Development of an LNG facility near the port of Deir
- Export of heavy and ultra-heavy crude from Iran to the Kedah refinery in Malaysia
- Sale of gas condensates from Iran to Malaysia

Oil-trade transformation

The Middle East will emerge as a new export force in the world oil market, according to a July 2008 report from FACTS Global Energy (FGE; Honolulu).

The firm said that in the Middle East, bumper oil revenues, strong domestic demand growth, an influx of foreign investment, and the launch of a Middle East sour crude futures contract are driving a rapid evolution of the oil market.

The region is a major crude supplier with spare capacity and is exporting increasing volumes of oil products as its refining capacity grows. As a result, said FGE, product-trade patterns in the Middle East will see significant changes.

Some key points from the study:

- For the period 2007–12, Middle East crude oil production will rise by 2.8 million b/d, but spare crude production capacity also will grow, from 2.5 million b/d to 3.2 million b/d.
- Production of NGLs and condensates will grow by about 2 million b/d, 2007–12.
- Demand for oil products in the Middle East will grow 7.5% from 2006–10, and 4.3%/yr from 2010–20.

Key trends driving petroleum-product demand in the region include growing petrochemical activity, strong transportation fuel demand, and rising fuel-oil consumption for power generation.

Middle East oil demand is forecast to grow 2.9 million b/d during 2007–15.

In parallel, Middle East refining capacity is projected to grow by 3.3 million b/d in 2008–15, with more than 70% of capacity additions being export-oriented.

Total petroleum-product exports from the region are projected to reach nearly 4.4 million b/d by 2015, up from 2.8 million b/d in 2007.

- Development of refineries in Indonesia, Syria, and at Shiraz (in Iran)
- Development of a petrochemical refinery complex in Malaysia.

More details are given in the Malaysia section of the Asia-Pacific chapter of this book.

IRAQ

Capital: Baghdad
Monetary unit: Dinar
Refining capacity: 597,500 b/cd
Oil production: 2.365 million b/d
Oil reserves: 115.00 billion bbl
Gas reserves: 111.94 tcf

Overview

Rebuilding

Prime Minister Nouri al-Maliki and Iraq's Council of Representatives (national parliament) continued efforts to establish an effective central government, ensure security, rebuild infrastructure, and reestablish social and commercial activities in the country in 2008.

Sectarian violence and attacks against U.S. and coalition forces continued in Iraq in 2008, but at a much reduced level. March marked the fifth anniversary of the U.S.-led invasion to depose then-president Saddam Hussein.

An important milestone was reached in late 2008 regarding a schedule for withdrawal of U.S. troops from the country (see box).

Oil and gas situation

Iraq holds the world's third largest proven reserves of crude (some 115 billion bbl) and an estimated 3.17 tcm (112 tcf) of natural gas. It currently produces about 2.4 million barrels of oil per day and seeks to boost this to 6 million b/d by 2018.

Decades of wars, UN sanctions, violence and sabotage have battered the oil industry and impaired the country's ability to ramp up production.

Work continued in 2008 on formulation of a federal law regarding the development of the country's oil sources, the sharing of revenues from such development, and the appropriate role of foreign companies in oil exploration, production and sale.

A long-standing dispute between the federal government and the Kurdistan Regional Government (KRG) in the north has hindered formulation of a national petroleum law for upstream projects. The federal government wants centralized control of the oil sector; the KRG wants decentralized control.

As of November 2008, no federal law had been passed.

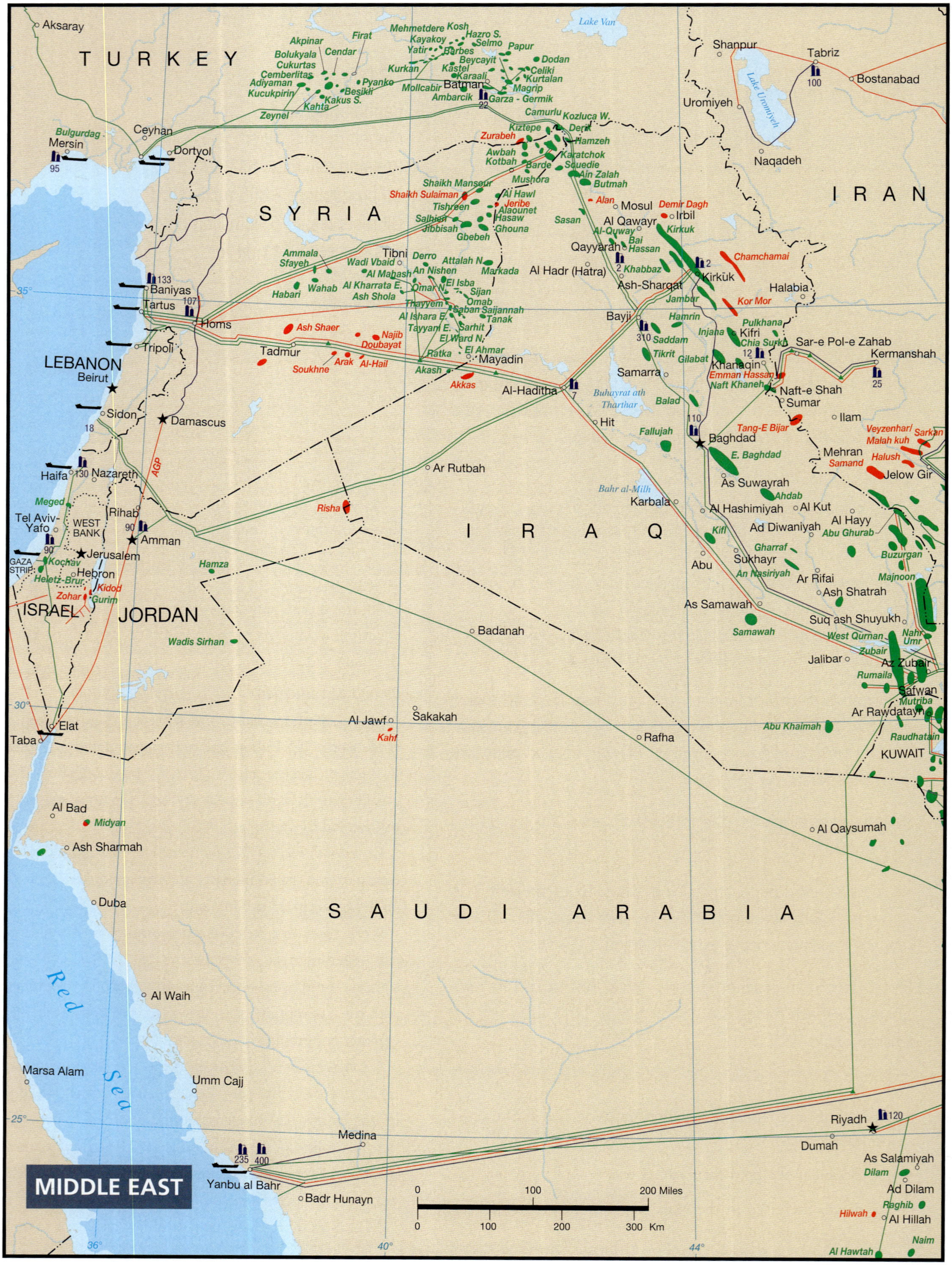
MIDDLE EAST
TURKEY
SYRIA
IRAQ
IRAN
LEBANON
ISRAEL
JORDAN
WEST BANK
GAZA STRIP
KUWAIT
SAUDI ARABIA
Red Sea
Lake Van
Lake Uromiyeh
Buhayrat ath Tharthar
Bahr al-Milh
Aksaray
Ceyhan
Dortyol
Mersin
Bulgurdag
Batman
Tabriz
Bostanabad
Shanpur
Uromiyeh
Naqadeh
Mosul
Irbil
Al Qawayr
Qayyarah
Al Hadr (Hatra)
Kirkuk
Ash-Sharqat
Halabja
Bayji
Samarra
Kifri
Khanaqin
Sar-e Pol-e Zahab
Kermanshah
Naft-e Shah
Sumar
Ilam
Mehran
Hit
Baghdad
As Suwayrah
Karbala
Al Hashimiyah
Al Kut
Al Hayy
Ad Diwaniyah
Sukhayr
Abu
Ar Rifai
Ash Shatrah
As Samawah
Suq ash Shuyukh
Jalibar
Az Zubair
Safwan
Ar Rawdatayn
Al-Haditha
Ar Rutbah
Mayadin
Tibni
Homs
Tartus
Baniyas
Tripoli
Tadmur
Beirut
Sidon
Damascus
Haifa
Nazareth
Tel Aviv-Yafo
Rihab
Amman
Jerusalem
Hebron
Elat
Taba
AGP
Badanah
Al Jawf
Sakakah
Rafha
Al Qaysumah
Al Bad
Ash Sharmah
Duba
Al Waih
Marsa Alam
Umm Cajj
Medina
Yanbu al Bahr
Badr Hunayn
Riyadh
Dumah
As Salamiyah
Ad Dilam
Al Hillah
Akpinar
Firat
Mehmetdere
Kosh
Hazro S.
Selmo
Papur
Kayakoy
Yatir
Barbes
Beycayit
Dodan
Celiki
Kurtalan
Bolukyala
Cendar
Cukurtas
Cemberlitas
Adiyaman
Kucukpirin
Pyanko
Besikli
Kakus S.
Kahta
Kurkan
Kastel
Karaali
Magrip
Mollcabir
Ambarcik
Garza - Germik
Zeynel
Camurlu
Kozluca W.
Kiztepe
Derik
Zurabeh
Hamzeh
Awbah
Kotbah
Karatchok
Barde
Sauedie
Mushora
Ain Zalah
Butmah
Shaikh Mansour
Shaikh Sulaiman
Al Hawl
Jeribe
Tishreen
Alaounet
Salhieh
Hasaw
Jibbisah
Ghouna
Gbebeh
Alan
Demir Dagh
Sasan
Al-Quway
Bai Hassan
Kirkuk
Chamchamai
Khabbaz
Ammala
Sfayeh
Wadi Vbaid
Derro
Attalah N.
An Nishen
Al Mahash
Markada
Wahab
Habari
Al Kharrata E.
Omar N.
Ash Shola
El Isba
Sijan
Omab
Thayyem
Saban
Saijannah
Tanak
Al Ishara E.
Tayyani E.
Sarhit
El Ward N.
El Ahmar
Ratka
Ash Shaer
Najib
Doubayat
Arak
Al-Hail
Soukhne
Akash
Akkas
Jambur
Kor Mor
Hamrin
Pulkhana
Injana
Chia Surkh
Saddam
Tikrit
Gilabat
Emman Hassan
Naft Khaneh
Balad
Tang-E Bijar
Fallujah
E. Baghdad
Veyzenhar/ Malah kuh
Sarkan
Halush
Samand
Jelow Gir
Ahdab
Kifl
Abu Ghurab
Gharraf
An Nasiriyah
Buzurgan
Majnoon
Samawah
West Qurnan
Nahr Umr
Zubair
Rumaila
Mutriba
Raudhatain
Abu Khaimah
Risha
Meged
Kochav
Heletz-Brur
Kidod
Zohar
Gurim
Hamza
Wadis Sirhan
Kahf
Midyan
Dilam
Raghib
Hilwah
Naim
Al Hawtah
35°
30°
25°
36°
40°
44°
0
100
200 Miles
0
100
200
300 Km

The Kurdistan Regional Government (KRG) in northern Iraq separately has signed exploration and production contracts with several foreign firms, incurring the wrath of Baghdad and stirring controversy in the international industry.

Oil and gas outlook

Business Monitor International Ltd. (BMI; London) noted that Iraq has ambitious targets for its oil sector, aiming for a long-term goal of 6 million b/d.

However, the firm sees major risks ahead, including attacks on oil installations, soundness of infrastructure, Iraq's OPEC entitlement, and the degree of success of new energy policy to stimulate investment by international oil companies.

IOCs are expected to enter production-sharing agreements with the state in 2008–2009, which should help accelerate the growth in oil output.

In late 2000, BMI estimated average oil production of 2.3 million b/d for 2008, growing to 3.15 million b/d in 2012. Production in May 2008 was 2.5 million b/d.

BMI expects an increase of 86.5% in Iraqi oil production during 2007–18, with crude volumes rising steadily to 4 million b/d by the end of 2018.

Iraq's crude oil exports in October 2008 were reported to be 1.703 million b/d: 1.385 million b/d from the Basra oil terminal, 309,000 b/d by pipeline to Turkey's Mediterranean port of Ceyhan, and 9,000 b/d trucked to Jordan.

The oil ministry said exports averaged 1.76 million b/d in November and 1.85 million b/d in December.

Also in late 2008, Exclusive Analysis (EA; London) concluded that crude exports from Iraq in 2009 were likely to be disrupted by crumbling production and pipeline systems than by security issues.

Iraqi natural gas production is expected to climb to 35 bcm (1.23 tcf) by 2018.

Exploration and Development

Ahdab oil field

The Iraqi oil ministry renegotiated an agreement in 2008 that will allow state-owned China National Petroleum Co. (CNPC; Beijing) to develop the Ahdab oil field, in Wasit province.

The $3-billion, 22-yr contract requires approval of the Iraqi and Chinese governments.

As part of the agreement, some of the oil will be delivered to the al-Zubaidiya power station in Wasit province.

Troop withdrawal agreement

In mid-November 2008, the leadership of the Iraqi government and the U.S. government agreed on a "Status of Forces Agreement" (SOFA), to replace a UN mandate for the presence of foreign troops that was set to expire December 31.

The SOFA establishes that U.S. combat forces will withdraw from Iraqi cities by June 30, 2009, and that all U.S. forces will be out of Iraq by December 31, 2011.

Iraq also will gain strict oversight over American troops on the ground.

The Iraqi parliament and the presidential council approved the SOFA with some modification soon after it was signed.

A public referendum on the agreement is to be held by July 2009 in Iraq.

The deal represents the first upstream energy contract with Iraq since the fall of Saddam Hussein.

CNPC will provide technical advisers, oil workers, and equipment to increase production at the field, about 160 km (100 mi) southeast of Baghdad.

First output from Ahdab, at a rate of 110,000 b/d, is expected in 2011.

The field is estimated to contain about 4.54 billion bbl of original oil in place, of which 955 million bbl is believed recoverable.

IOC deals rejected

Iraqi oil ministry officials had hoped to sign contracts with several international oil companies by mid-2008, including Royal Dutch Shell PLC, BP PLC, and ExxonMobil Corp.

The no-bid, short-term technical support contracts—worth about $500 million each—would give the IOCs the first major foreign involvement in Iraq's oil sector in nearly 40 years.

The short-term contracts were to be followed by longer-term development contracts, for which dozens of foreign energy firms would likely bid.

However, Iraqi officials in September criticized IOCs for trying to overcharge the war-torn nation and for ignoring what officials called their "humanitarian" duty to help develop Iraq's battered oil industry.

At mid-year, protests arose in several countries over what was seen as IOC attempts to "gain control" of Iraq's oil resources.

Taq Taq field

Addax Petroleum Corp. (Calgary) tested 470 b/d of oil in November 2008 at the No. 11 appraisal well in Taq Taq field. Oil flowed from a 52-m (170-ft) gross oil column in the Eocene Pila Spi formation at 1,000 m (3,280 ft).

Oil from this part of the field would supply demand in Kurdistan.

Kurdistan: Kalar-Bawanoor Block

WesternZagros Resources Ltd. (Calgary) spud its Sarqala-1 wildcat in early May 2008 on the Kalar-Bawanoor Block in Iraqi Kurdistan.

The well, first of a multiwell program scheduled for the next three years, will evaluate four potential reservoirs.

Lease Sales

First licensing round: Mid-2008

In June 2008, Iraq opened its first postwar round of bidding for contracts to develop six major oil fields and two gas fields. Of the 120 oil companies that applied to participate, 34 were selected.

Among those chosen were Royal Dutch Shell PLC (The Hague, Netherlands), BP PLC (London), ExxonMobil Corp. (Irving, TX), Chevron Corp. (San Ramon, CA), and Total SA (Paris).

The ministry plans to sign the contracts from the first round in mid-2009.

IRAQ'S CURRENTLY ENVISAGED LICENSE SCHEME

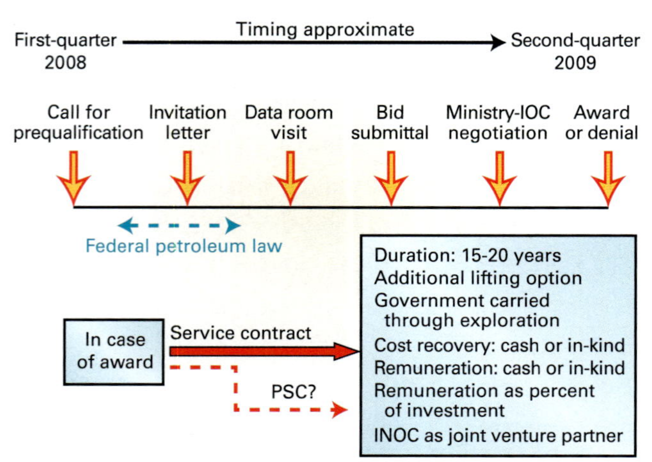

Second licensing round: Early 2009

In late 2008, the oil ministry announced plans to offer 10 undeveloped or underdeveloped gas and oil fields in early January 2009, in Iraq's second postwar licensing round.

Officials expect to sign contracts from the second round at the end of 2009.

Fields offered will include Majnoon and West Qurna Phase 2 in the south, which hold reserves of some 12 billion bbl of crude each. They produce far below their individual output potential of 600,000 b/d.

The ministry said the focus will be on fields shared with Iran and Kuwait, or that are located near borders. Some Iraqi officials are concerned that neighboring countries are extracting oil from the shared fields.

Contracts from the mid-2008 and early-2009 licensing rounds could boost production by as much as 4.5 million b/d within four to six years, said the oil ministry.

Oil and Gas Exports

Oil export policy

In late 2008, Iraq's central government and regional authorities in Kurdistan were making progress on mechanisms for possible export of oil from the northern region's fields in 2009.

In a key development, the Iraqi oil ministry and Kurdish officials agreed in principle to connect the Tawke and Taq Taq oil fields in Kurdistan to Iraq's main northern export pipeline, which runs to the Turkish port of Ceyhan.

More-difficult issues, particularly revenue sharing, would still have to be resolved.

Linking Tawke to the northern pipeline could help sustain oil exports because the field is far from the area where the pipeline is frequently attacked by the Kurdish PKK rebel group.

Analysts estimate that the combined output from Tawke and Taq Taq could reach 250,000b/d by the end of 2009.

Iraq gas to Europe?

In early 2008, European Commission officials said that Iraq had offered to increase its supply of natural gas to the European market over the next three years. The offer came during talks between the Iraqi oil ministry and the EU energy commissioner.

Total exports were said to be on the order of 5 bcm (177 bcf) by 2011.

Drilling and Production

Miran license

Heritage Oil Ltd. (Channel Islands) began drilling the Miran West-1 well in the Kurdistan Region of Iraq in early 2009.

It is the first exploration well to be drilled on the Miran license, which is west of the city of Suleimaniah.

Planned drilling depth is 2,500–3,000 m (8,202–9,842), targeting the Shiranish, Kometan, and Qamchuqua reservoirs.

Heritage has acquired seismic for the two large structures in the license area, Miran West and Miran East.

The firm also has an agreement with the KRG to establish a 50:50 JV company to build, own, and operate a 20,000-b/d oil refinery near the license. Start-up is planned for 2011.

Oil infrastructure in KRG territory

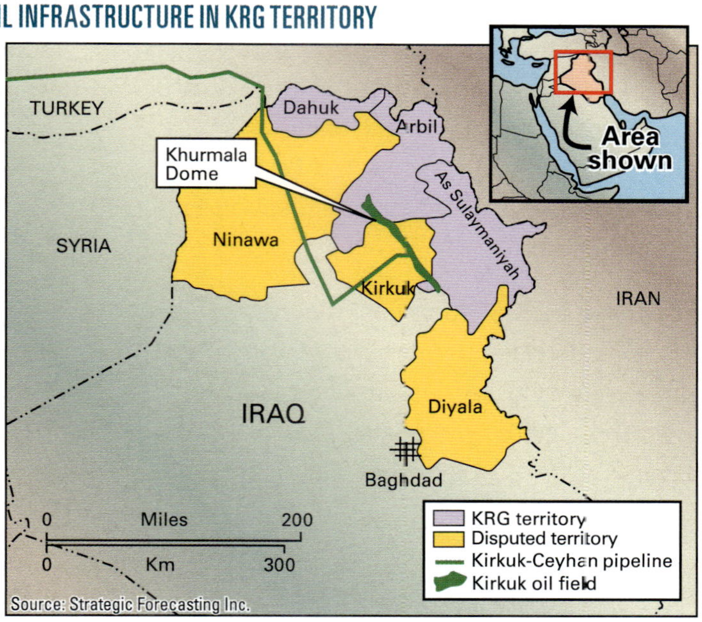

Source: Strategic Forecasting Inc.

Syria-Jordan border fields

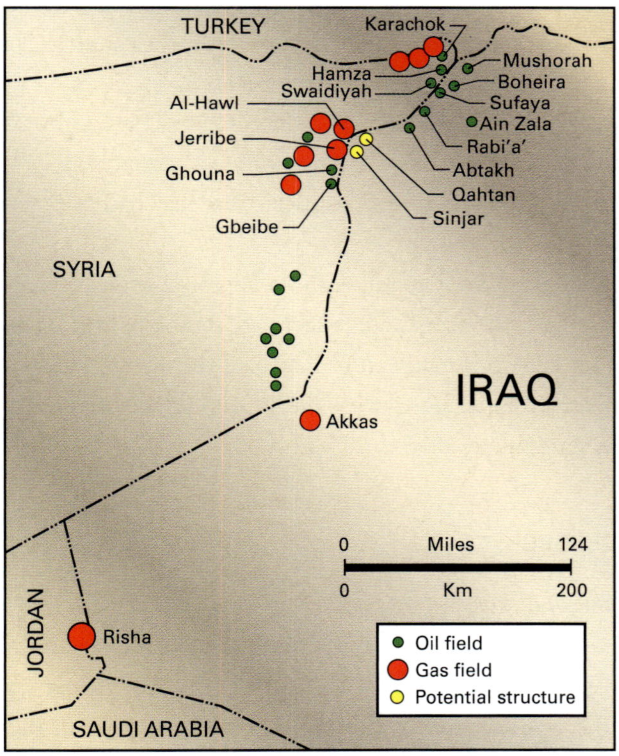

Note: The location of the borders and fields are approximate and are not to scale.

Countries adjacent to Iraq

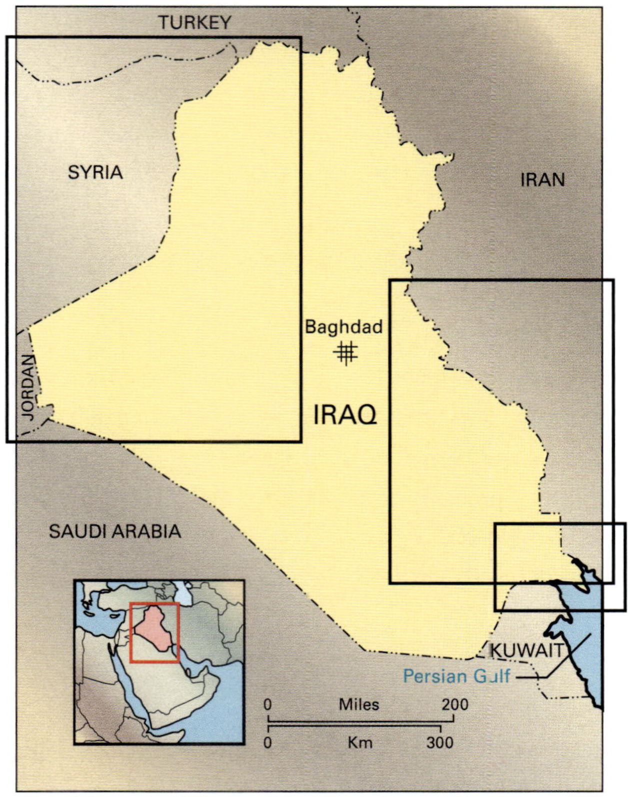

Infrastructure issues

Although Iraq has an ambitious plan to boost oil output over the coming decade, industry analysts have expressed concern about the fitness of the underlying infrastructure to support greater oil production and exports.

For example, a 2008 report from the U.S. EIA said that the underwater pipelines that carry nearly 2 million b/d into tankers at Iraq's Gulf coast export point are so corroded that they could fail at any time.

A more specific 2008 report by Foster Wheeler Corp. (Clinton, New Jersey) said that the country's southern port requires a complete overhaul, including:

- dredging deeper shipping channels
- building new pipelines
- erecting storage tanks able to hold 18 million bbl of oil

It also said that the scale of some needed new facilities would require that Iraq first resolve its long-running maritime border dispute with Iran and Kuwait.

Khor Mor field

Dana Gas and partner Crescent Petroleum (both in the UAE) started natural gas production in late 2008 from Khor Mor field in Iraq's northern Kurdistan region.

Output is expected to rise from the initial 75 MMcfd to 300 MMcfd by mid-2009.

A 180-km pipeline will carry the produced gas to two electric power plants.

Khor Mor field has estimated gas reserves of 1.4 tcf.

Developing border fields

An early-2008 study by the Centre for Global Energy Studies (CGES; London) advocated joint development and operation of the many fields that lie along Iraq's borders with adjacent countries.

CGES conducted the study in reaction to claims in the Iraqi press that neighboring countries were stealing Iraq's oil.

The study suggested that joint analysis could verify whether fields are connected or are in separate reservoir systems.

Even for two separate fields, a joint production system sharing some surface facilities could offer economies and other benefits.

Associated gas recovery

In third quarter 2008, the Iraqi government approved a plan to recover and use associated gas that has typically been flared.

The deal with Royal Dutch Shell (The Hague, Netherlands) paves the way for the firm to gather 500–600 MMcfd of associated natural gas in the Basra area.

The gathered gas will be used mostly to generate electric power, but some could be exported as LNG.

Refining

Micro-refinery program

Iraq was poised to inaugurate by yearend 2008 the 20,000 b/d al-Diwaniya refinery–one of several so-called micro-refining projects under development.

The country has launched several such projects since 2006, due to the amount of time needed to repair the country's larger refineries and ready them for full operation.

As of November 2008, the country's three main refineries—Dora, Shuaiba, and Beiji—were running at no more than 50% of the 597,500-b/d installed capacity that they had prior to the 2003 U.S.-led invasion.

In August, the government also approved a major contract to upgrade the Samawah refinery in southern Iraq. Located 370 km (230 mi) southeast of Baghdad, the 30,000-b/d facility suffered 90% damage during the 1991 Gulf War and has only been partially repaired.

Pipelines

Repair of Kirkuk-Banias line

Iraq's state-owned North Oil Co. signed a preliminary agreement in early 2008 with Russia's Stroytransgaz for repair of the Iraqi section of the 880-km (547-mi) Kirkuk-Banias oil pipeline.

The Kirkuk-Banias pipeline had a carrying capacity of 1.4 million b/d when built in the 1950s. Just before the U.S.-led invasion in 2003, its capacity was 300,000 b/d.

International Activities

Japanese aid to oil, gas industry

Japanese and Iraqi officials met in mid-2008 to discuss cooperation in rebuilding Iraq's oil and natural gas industry.

Topics included reconstruction of existing energy-related facilities as well oil and gas development.

In January, the Japan Bank for International Cooperation said it had signed an agreement with Baghdad for as much as ¥182.7 billion to restore war-damaged facilities, including the oil export facility at Basra; the Al-Mussaib thermal power plant; the Basra water supply system; and Kurdistan's power grid.

Japanese officials also offered to train 100 Iraqi engineers per year during 2009–13.

Nippon Oil plans

In late 2008, Nippon Oil Exploration (NOE; Tokyo) was talking with Iraq about $10–20 billion worth of investments that would include exploration as well as construction of a 300,000-b/d refinery.

Iraq in 2008 approved construction of four new refineries with total combined capacity of 750,000 b/d.

The oil ministry said Iraq is seeking help from foreign companies to stem the decline in oil production, develop new fields, and boost refinery capacity.

ISRAEL

Capital: Jerusalem
Monetary unit: Shekel
Refining capacity: 220,000 b/cd
Oil production: 0
Oil reserves: 1.940 million bbl
Gas reserves: 1.075 tcf

Overview

In the Eastern Mediterranean region, Israel has the largest and most developed economy. According to the U.S. CIA World Fact Book, growth of real GDP for 2008 was estimated to be 4.2%.

Oil

Israel produces only a minimal amount of oil. The country imported approximately 334,000 b/d of crude oil and products in 2005 to meet domestic needs, primarily from Egypt, the North Sea, West Africa, and Mexico.

However, in recent years, Israel has increased imports from Russia and the Caspian, including Azeri oil transported through the new Baku-Tbilisi-Ceyhan (BTC) pipeline.

More than 20 small independent firms are active in hydrocarbon exploration throughout the country. Israel's Petroleum Commission has estimated that the country could contain 5 billion bbl of oil reserves.

Natural Gas

Gas production in 2006 was estimated at 2.3 bcm (81.2 bcf). However, Israel plans to dramatically increase the share of natural gas in its fuel mix, especially for electricity generation; demand is expected to reach 7.9 bcm (282 bcf) by 2010.

Supply options being developed include imports from Egypt's Nile Delta and production from domestic offshore reserves.

Israel also is considering building an LNG regasification facility or an underwater pipeline to bring gas from Turkey.

Several energy companies have discovered small but commercially viable quantities of natural gas off the coast of Israel and the Gaza Strip. Israel's and the Palestinian Authority's offshore gas reserves are under concession to two main groups:

- Yam Thetis consortium, comprising U.S.-based Nobel Energy, Israel's Avner Drilling (Petach Tikva), and Delek Exploration (Netanya)
- A partnership of British Gas (BG; London) with Isramco (Houston) and others.

Yam Thetis operates Israel's two largest gas fields, Mari and Noa (Border), with estimated combined reserves of 1.5–1.7 tcf. Yam Thetis supplies natural gas to Israel Electric Corp. and the Ashdod refinery.

BG first discovered gas in 1999 in the Palestinian Authority's territorial waters with its Gaza Marine-1 well. The company's 25-yr contract includes gas exploration and set-up of a distribution network.

Israeli, British, and Palestinian Authority officials as well as private-sector parties have talked about a BG plan to import natural gas from fields in Gaza.

BG also is considering the export of gas to LNG plants in Egypt for sale to Europe and North America.

Israeli-Hamas violence in Gaza

Hostilities between Israel and militant Hamas members in the Gaza Strip erupted again in late 2008, raising geopolitical tensions and roiling oil markets.

A ceasefire agreement signed in June 2008 had led to a marked drop in the number of rockets fired into Israeli territory from Gaza through early November.

But tensions increased, and when the ceasefire expired on December 19, the rocket and mortar fire into Israel increased sharply.

In response, Israel launched "Operation Cast Lead" with an air attack on December 27 and then a ground offensive on January 3, 2009.

UN Secretary General Ban Ki-moon and several world leaders appealed for a ceasefire, and a range of proposals were made to try to reduce arms smuggling into Gaza.

In mid-January, medical officials said that more than 1,000 Palestinians had been killed and about 4,000 wounded in the Israeli assault.

The Israeli death toll was reported to be 13, ten of whom were soldiers killed in the campaign.

Israel declared a unilateral cease fire on January 17. Two days later, the king of Saudi Arabia said that his country would donate $1 billion to help rebuild the Gaza Strip.

Exploration and Development

Ma'anit-Rehoboth-2 well

Zion Oil & Gas Inc. (Dallas) said in planned to spud the Ma'anit-Rehoboth-2 well in the onshore Pleshet basin in early 2009, expecting to drill directionally to at least 5,500 m (18,000 ft).

Dependent on funding, Zion also intends to drill (a) a second well on the Joseph license to the Triassic at 4,695 m (15,400 ft) and/or the Permian at 5,500 m (18,000 ft), and (b) another well to similar depths on the Asher-Menashe license.

International Activities

MedStream pipeline project

The proposed $12-billion MedStream project is a multipurpose subsea pipeline project intended to transport crude oil across the Mediterranean Sea from Turkey to Israel.

A 10-month technical feasibility study was to begin by early 2009, with construction expected to start by late 2009 or early 2010.

Indian officials expressed interest in September 2008 in possibly extending the line to India.

Participants in Med Stream are expected to be Gazprom (Moscow); Indian Oil Corp. Ltd. (Mumbai); Eni SPA (Rome); Calik Holding A.S. (Istanbul); and Royal Dutch Shell (The Hague, Netherlands).

The Turkey-to-Israel line also might carry natural gas, fresh water, and electrical and fiber-optic cables.

The MedStream project would consist of several stages, extending from Russia to India.

The completed project could provide 90% of Israel's energy needs, providing an alternative to importing oil from the Black Sea region by ship.

JORDAN

Capital: Amman
Monetary unit: Jordanian dinar
Refining capacity: 90,400 b/d
Oil production: 0
Oil reserves: 1.000 million bbl
Gas reserves: 213.0 bcf

Oil shale exploitation

Jordan's Ministry of Energy and Mineral Resources signed a three-year MOU with Brazil's Petrobras and Total SA (Paris) in September 2008 to study oil shale development.

The Jordanian government estimates that the country has more than 50 billion tonnes of oil shale resources.

Also, in December, Jordanian officials were poised to approve a production sharing agreement with Royal Dutch Shell PLC (The Hague, Netherlands) to exploit Jordan's oil shale resources.

The company would use its patented thermal *in situ* conversion process that heats the subsurface for an extended period to extract oil and gas.

The $20-billion project is expected to be completed in 15–20 years.

Surveys and development would be conducted over 22,000 sq km (8,495 sq mi) in the southern and central Jordan.

Official estimates indicate 40 billion tons of oil shale present in more than 20 locations, including the districts of Beit Ras, Buweida, Karak, Maan and Madaba.

Jordan currently pays more than $1 billion/yr for the imports needed to meet 95% of its energy needs.

Nuclear power pondered

Complicating the oil shale development picture somewhat in Jordan is evolving interest in nuclear power in the country.

In November 2008, Jordanian officials halted all oil shale activity in central Jordan for 18 months to enable the government to explore for uranium.

In 2007, King Abdullah II unveiled a civil nuclear energy program as an alternate way to generate electricity and desalinate water. Under the program, Jordan would set up a nuclear power plant by 2015, and by 2030, generate about 30% of its power from nuclear sources.

In late 2008, Jordan signed a cooperation agreement under which South Korea would take part is a range of Jordanian infrastructure projects, including construction of nuclear power plants.

Also, in August, Jordan agreed to cooperate with China on peaceful uses of nuclear energy, and also took initial steps to establish a joint venture with Areva SA (Paris) to explore for uranium in central Jordan and to build a nuclear power plant.

Jordanian officials estimate that the country has uranium reserve of 140,000 tons.

Russia is actively encouraging nuclear energy cooperation with Middle Eastern countries such as Egypt, Saudi Arabia and Turkey, among others.

Nuclear power companies in South Korea, Canada, France and the United States also are promoting their skills and products in the region.

KUWAIT

Capital: Kuwait City
Monetary unit: Dinar
Refining capacity: 889,200 b/cd
Oil production: 2.320 million b/d
Oil reserves: 101.500 billion bbl
Gas reserves: 62.860 tcf

Neutral Zone production

State-owned Kuwait Gulf Oil Company (Kuwait City) announced in late 2008 that it will invest $3.6 billion through 2012 to boost oil production.

The investment is part of the 2020 plan formulated earlier to increase production capacity in the neutral zone common to both Kuwait and Saudi Arabia. The two countries share the zone's oil output of 550,000 b/d.

Al-Zour refinery

Kuwait announced plans in mid-2008 to increase the country's refining capacity to 1.415 million b/d (from the current 889,000 b/d) by building a new 615,000-b/d refinery at Ras al-Zour and upgrading two existing refineries.

Al-Zour is scheduled to start operating in 2012, two years later than initially planned.

Members of parliament had expressed concern about the bidding and contract-award process for the al-Zour plant, but oil ministry officials were confident that legislative support would be forthcoming.

Collapse of K-Dow deal

After months of negotiation in a deteriorating global economic environment, the government of Kuwait in December 2008 cancelled a $17.4-billion petrochemicals joint venture with Dow Chemical (Midland, Michigan).

The K-Dow Petrochemicals project between Dow and Kuwait's state-owned Petrochemicals Industries Company (PIC; Safat, Kuwait), was part of Dow's strategy to shift from a reliance on commodity chemicals (produced from oil and gas) to high-profit specialty chemicals.

K-Dow was to manufacture and distribute polypropylene, ethyleneamines, polyethylene, polycarobonates and various catalyst and licensing technologies worldwide.

PIC was to pay Dow $9.5 billion to contribute five Dow units to the 50-50 joint venture.

Dow expected to use proceeds from the deal to fund a planned \$15-billion takeover of specialty chemicals group Rohm and Haas Co. (Philadelphia).

As of early 2009, legal actions were under way by various parties unhappy with the cancellation. There were also calls from Kuwait parliamentarians for an investigation.

OMAN

Capital: Muscat
Monetary unit: Riyal
Refining capacity: 85,000 b/cd
Oil production: 715,000 b/d
Oil reserves: 5.500 billion bbl
Gas reserves: 30.00 tcf

Stake in CPc sold

In late 2008, Oman sold to Russia its 7% stake in the Caspian Pipeline Consortium (CPC) led by Chevron Corp. (San Ramon, California).

The consortium owns and operates the 1,580-km (982-mi) Tengiz-Novorossiisk oil pipeline.

The Omani government gave no reason for the sale, but industry sources said the government had grown frustrated over protracted disagreements among consortium members about funding the expansion of the pipeline.

Russia had long opposed doubling the line's capacity, but reversed its stand when (a) most of the partners agreed to raise the shipping tariff to \$38/tonne from last year's \$30.24/tonne and (b) investors agreed to cut interest rates by 50% on a \$5-billion loan to CPC.

BP PLC (London) still opposes expansion. The firm said it might sell its 6.6% stake—held through Lukarco and Kazakhstan Pipeline Ventures—if the CPC partners do not reach a compromise.

Shareholders are Russia, the Kazakh government, Chevron, Lukarco, a Rosneft-Shell joint venture, ExxonMobil, Eni S.p.A., BG, Kazakhstan Pipeline Ventures, and Oryx Caspian Pipeline.

LNG SECTOR OPERATION

Qatar operates its LNG sector as two companies.

Qatargas is a set of joint ventures between Qatar Petroleum and major oil companies, including Shell, ConocoPhillips, Total and ExxonMobil. In July 2008, Qatargas started up the last of a series of four mega-trains, each of which can produce more than 7 mtpy of LNG.

Rasgas is a series of joint ventures in which ExxonMobil holds 25–30% and state-owned Qatar Petroleum holds 65–70% along with other (mainly Asian) investors.

The company plans to complete its sixth and seventh trains in 2009.

The government is aiming to raise total LNG production capacity from the current 31 mtpy to 77 mtpy.

QATAR

Capital: Doha
Monetary unit: Riyal
Refining capacity: 200,000 b/cd
Oil production: 855,000 b/d
Oil reserves: 15.210 billion bbl
Gas reserves: 891.945 tcf

Exploration and Development

Offshore Block 4N

Qatar granted Wintershall AG (Kassel, Germany) an offshore exploration license for Block 4N (Khuff), close to North Field. The exploration and production-sharing agreement has a 25-yr term.

The 544-sq km (210-sq mi) area, which has gas potential, lies in 70 m (230 ft) of water.

Operator Wintershall plans to reprocess seismic data and shoot additional seismic and drill two exploration wells by 2010.

The company also plans to shoot seismic and drill two exploration wells by 2010 on Block 3.

Drilling and Production

Record extended-reach well

Transocean Inc. (Houston) said in mid-2008 that its jackup GSF Rig 127 set a world record for the longest extended-reach well ever drilled.

Drilled offshore Qatar in 36 days, it measured 40,320 ft (12,289 m) deep with a 35,770-ft (10,902-m) horizontal section.

With the new record of 7.6 mi (12 km), it is also the first well in the history of offshore drilling that exceeds 40,000 ft (12,191 m).

It beats by about 2,000 ft (610 m) the prior extended-reach record of 38,322 ft (11,680 m) MD set by Parker Drilling Co. (working for ExxonMobil Corp.) with a land rig drilling to a target offshore Sakhalin Island in early 2008.

LNG

Production outlook

Wood Mackenzie (Edinburgh) told the *Financial Times* in late 2008 that it expects no further large-scale LNG developments in Qatar, based on estimates of North Field gas reserves [generally put at 25.5 bcm (900 tcf)].

Qatar, Russia, Iran: Gas cartel?

Russia, Iran and Qatar mace the first serious moves in October 2008 toward forming an OPEC-style cartel for natural gas.

The action raised concerns that Russia could further boost its influence ove- energy markets from Europe to South Asia.

Details are given in the Iran section of the Middle East chapter of this book.

If all projects currently planned are completed, Qatar will be producing about 650 MMcmd (23 bcfd).

Refining and Products

Al Shaheen refinery

Because of uncertain market conditions, Qatar Petroleum (Doha) delayed for about six months the solicitation of bids for design and construction of the proposed $5-billion Al Shaheen refinery at the Mesaieed Industrial City.

The 250,000-b/d refinery is now slated for commissioning in early 2012.

A 90-km (56-mi) pipeline will carry oil from the offshore Al Shaheen oil and gas field; a second 110-km (68-mi) pipeline segment will run overland to Mesaieed.

The delay of the Al Shaheen project follows a similar decision by Qatar Petroleum to postpone a $2.6-billion petrochemical plant in the Mesaieed Industrial City by one year.

The unit, with a combined production capacity of 900,000 tpy of propylene and polypropylene, is now slated to go on stream in 2012.

Pipelines and Transportation

Dolphin gas to Oman

Abu Dhabi's Dolphin Energy began supplying Qatari natural gas to Oman by pipeline in late 2008, under a long-term gas sales agreement signed in 2005.

Qatar, the UAE, and Oman have supported development of the Dolphin Gas Project.

The gas is produced from Dolphin's production wells off Qatar, processed at the Ras Laffan gas plant, and piped to Abu Dhabi.

The gas is then piped along Dolphin's Eastern Gas Distribution System to Al Ain in Fujairah, where the line connects to a new Omani pipeline at the Oman border.

Oman will be receiving 5.66 MMcmd (200 MMscfd) of gas under its agreement with Dolphin.

Several other customers also have 25-yr gas supply agreements for Dolphin gas from Qatar.

Mubadala Development Co. (run by the government of Abu Dhabi), Total SA (Paris) and Occidental Petroleum Corp. (Los Angeles) have interests in the Dolphin project.

Energy ports: International collaboration

State-owned Qatar Petroleum (QP; Doha) and Dutch port operator Havenbedrijf Rotterdam NV signed an agreement in mid-2008 regarding port management.

QP said it wants to develop its port at Ras Laffan to the highest international standards, while Havenbedrijf wants to strengthen its position as Europe's main energy port, especially for LNG and associated hydrocarbons.

Algerian officials also have expressed interest in developing an LNG terminal at Maasvlakte, near Rotterdam, plus LNG storage facilities.

The Netherlands appears to be focusing on increased imports as a way to retain its declining reserves.

Former Saudi oil minister views

In July 2008, as oil prices topped $145/bbbl, a former Saudi oil minister said speculation was contributing to high world oil prices, not imbalances between supply and demand.

Sheik Ahmed Zaki Yamani offered his comments in a published interview.

"So much money is flowing into the market," he said, "it's almost like gambling."

Yamani felt that the price would fall to $70/bbl by 2010, but noted that a spike in the other direction—perhaps to $200/bbl—could occur if there were military attacks on Iran.

Gas Processing

LPG recovery projects

Oil & Gas Journal's semiannual *Worldwide Construction Update* noted that cryogenic LPG recovery projects were under construction in early 2008 for QatarGas III and QatarGas IV in Ras Laffan.

The projects, each with capacity of 41 bcmd (1.45 bcfd), are scheduled to be completed in 2009 and 2010, respectively.

SAUDI ARABIA

Capital: Riyadh
Monetary unit: Riyal
Refining capacity: 2,080,000 b/cd
Oil production: 8.900 million b/d
Oil reserves: 264.210 billion bbl
Gas reserves: 257.970 tcf

Exploration and Development

Eight oil, gas finds

Saudi Arabia's ministry of petroleum and mineral resources reported in January 2009 that state-owned Saudi Aramco had discovered five more oil fields and three gas fields in the country's Eastern Province.

Four of the oil fields are on land and one in the Persian Gulf. They include:

- Jaouf-11
- Ramthan-9: 400 km (249 mi) northwest of Dhahran
- Nayashin-1
- Jareed-101: 130 km (81 mi) north of Dhahran
- and Khorsaniya-114: 138 km (86 mi) northwest of Dhahran

The three gas fields–all offshore some 125–200 km (78–124 mi) northeast of Dhahran–are Arabiya-1, Rabib-1, and Hisbah-16.

Major projects delayed

By late 2008, in light of declining oil prices and the deepening global economic crisis, Saudi Aramco decided to cancel four major development projects. They included:

- The restart of production from the Dammam oil field
- The 400,000-b/d Yanbu refinery
- The 900,000-b/d Manifa heavy oil field
- The 400,000-b/d refinery project at Jubail.

The company said that its short-term projects were on track and that the kingdom would reach its target of increasing production capacity to 12.5 million b/d by the end of 2009.

Drilling and Production

Saudi production cut: January 2009

The Organization of Petroleum Exporting Countries (OPEC) announced in December 2008 a record 2.2-million b/d cut in oil output.

In early January 2009, Saudi Arabia said it had unilaterally cut oil production by 300,000 b/d more than its agreed OPEC limit. This would make the country's output 7.7 million b/d.

The January reduction means that Saudi Arabia had cut its output by 20% since July.

Analysts said the Saudi move indicated how severely the cartel—which supplies about 40% of the world's oil—is struggling to reverse the slide in oil prices.

Saudi leadership said in late 2008 that it views $75/bbl as the appropriate price for oil.

On January 28, 2009, light, sweet crude for March delivery settled at $42.16/bbl on the NYMEX.

Aramco production capacity

Saudi Aramco was taking steps to boost production by yearend 2008 by a combined 850,000 b/d from three fields.

The firm planned to bring Khursaniyah and Nuayyim on stream (500,000 b/d and 100,000 b/d, respectively); and boost output at Shaybah by 250,000 b/d (from 500,000 b/d).

In the longer term, the firm said output should grow by another 900,000 b/d in fourth quarter 2009 with the start of production from its large Manifa heavy oil field.

Saudi Arabia's overall plan is to raise its daily production capacity—including output from the neutral zone shared with Kuwait—to 12.5 million b/d of crude oil by yearend 2009.

> **$75/BBL OIL**
>
> The Saudi oil minister said at a mid-December 2008 conference that Saudi Arabia supports development of alternative energy to complement fossil fuels and that an oil price of $75/bbl would be high enough to stimulate investment in both.

Continued production dominance

In a late-2008 report, Business Monitor International Ltd. (BMI; London) concluded that Saudi Arabia will account for 20.9% of Middle Eastern regional oil demand by 2012 while providing a dominant 40.7% of supply.

BMI's *Saudi Arabia Oil & Gas Report* also said that regional oil use—which stood at 8.24 million b/d in 2001 and rose to 10.61 million b/d in 2007—should average 10.86 million b/d in 2008 and rise to 11.81 million b/d by 2012.

Regional oil production—22.87 million b/d in 2001 and 25.24 million b/d in 2007—is set to rise to 28.94 million b/d by 2012.

Regarding natural gas, the region in 2007 consumed 370 bcm (13.1 tcf), with demand of 541 bcm (19.1 tcf) targeted for 2012.

Production of 363 bcm (12.8 tcf) in 2007 should rise by 58% across the region and reach 575 bcm (20.3 tcf) in 2012, which implies net exports rising to 34 bcm (1.2 tcf) by the end of the period.

Prices: OPEC market basket

In second quarter 2008, Business Monitor International Ltd. (BMI; London) projected an OPEC basket price average of $106/bbl for 2008, compared with the $81 estimate it developed in the first quarter.

Based on recent price differentials, the firm said this implies Brent at $109.71, WTI averaging $110.64/bbl, and Urals at $106.88/bbl for 2008.

BMI projections

In second quarter 2008, BMI estimated that Saudi Aramco would increase output to 11.78 million b/d by 2012, from 10.41 million b/d in 2007.

The firm said Saudi gas production should reach 96.2 bcm (3.40 tcf) by 2012, up from 75.9 bcm (2.68 tcf) in 2007.

For 2007–18, BMI projected an increase in Saudi oil production of 29.7%, rising steadily to 13.5 million b/d by the end of the 10-yr forecast period.

Gas production is expected to rise to 117 bcm (4.13 tcf) by the end of the period, from 76 bcm (2.68 tcf).

Khursaniyah field

Saudi Aramco reported the start of production in September 2008 from its Khursaniyah oil field.

Production volumes could eventually ramp up to as much as 500,000 b/d.

Refining and Processing

PetroRabigh refinery upgrades

As of mid-2008, Rabigh Refining & Petrochemical (PetroRabigh)—a joint venture of Saudi Aramco and Sumitomo Chemicals (Tokyo)—said it planned to complete the $10-billion upgrades to its 400,000-b/d refinery and petrochemical complex in Rabigh by yearend 2008.

The upgraded complex will produce additional volumes of transportation fuels, plus 2.4 mtpy of petrochemical solids and liquids.

Jizan refinery

In mid-2008, Saudi Arabia delayed for the third time the solicitation of bids for a projected 250,000- to 400,000-b/d refinery to be built in the southwestern province of Jizan.

Several major foreign companies had created consortia with Saudi firms to bid for the project.

Various reasons were cited for the earlier delays, including possible adverse environmental effects in the original location, concerns about funding, and uncertain participation by IOCs.

The Kingdom also will have to build a pipeline to supply the plant.

In January 2009, however, Saudi officials said they aim to award the contract by yearend 2009.

Petrochemicals

"World's largest" PP train

Dow Technology Licensing (Midland, MI) and Saudi European Petrochemical Co. (Ibn Zahr)—a Saudi Basic Industries Corp. (Sabic) JV—started up in late 2008 what they called the world's largest single polypropylene (PP) train.

The unit, at Al Jubail industrial city on the Persian Gulf coast, has a nameplate capacity is 500,000 tpy of PP resins.

International Activities

Saudi Aramco, Sinopec MOU

Saudi Aramco and China Petrochemical Corp. (Sinopec Group) signed an MOU in mid-2008 to increase cooperation between the two companies and to strengthen their relations and business development opportunities.

Analysts said the agreement reflected China's view of the importance of Saudi oil in world energy markets as well as the Kingdom's view of China's crucial role in the global economy.

In 2007, Sinopec imported nearly 460,000 b/d of crude from Saudi Aramco. That was expected to grow to 650,000 b/d in 2008, and to as much as 1.0 million b/d in 2010.

SYRIA

Capital: Damascus
Monetary unit: Syrian pound
Refining capacity: 239,865 b/cd
Oil production: 385,000 b/d
Oil reserves: 2.500 billion bbl
Gas reserves: 8.500 tcf

Exploration agreements signed

In September 2008, Total SA (Paris) signed three oil and gas agreements with Syria.

The first renews the Deir Ezzor oil license for 10 years, to 2021. It is wholly owned by Total and jointly operated by Total and state-owned Syrian Petroleum Co (SPC) through the Deir Ezzor Petroleum Co. JV.

The earlier agreement, signed in 1988, allowed Total to produce 30,000 b/d at Deir Ezzor.

Total said the extension will allow it to prolong and optimize oil production from the Jafra, Qahar, and Atalla fields.

The second agreement aims to enhance output from the Tabiyeh gas and condensate field to increase domestic gas deliveries.

The third agreement (an MOU) establishes a strategic partnership for developing common projects among Total, SPC and state-owned Syrian Gas Co.

In related news, the oil ministry also said plans were proceeding for a second refinery at Deir Ezzor—capacity 140,000b/d—to be built jointly with Kuwait's Noor Investment Group.

CNPC (Beijing) previously built a 100,000-b/d refinery near the Total concession.

Block 26 oil discovery

The Yousefieh-1 exploration well on Block 26 in northeastern Syria—near Khurbet East oil field—discovered an accumulation of oil in late 2008.

Partners Gulfsands Petroleum PLC and Emerald Energy PLC (both in London) were moving to appraise the discovery with a well at the north end of Khurbet East field.

Yousefieh-1 cut 63 m (207 ft) of net oil pay in the target Cretaceous formations and flowed 900 b/d from the shallowest 19-m interval. TD is 2,139 m (7,018 ft). Test flow averaged 1,450 b/d for 3 hours.

The companies were planning to shoot more 3D seismic to identify other prospects.

Cumulative production has exceeded 1 million bbl from Khurbet East field.

Gas treatment projects

Petrofac (London) won two major contracts in early 2008 for Syrian gas-field development projects.

The first contract is for construction of a 4-MMcfd plant to treat gas from Jihar field near Palmyra, which is operated by Hayan Petroleum (a Croatian-Syrian JV).

The second, awarded by Petro-Canada (Calgary), is for processing gas and liquids from the Al-Shaer and Cherrife fields in central Syria.

The fields are expected to produce 2.5 MMcmd (88.3 MMcfd) of sales gas and 150 tpd of LPG, starting in 2010.

Gas from Egypt

In mid-2008, Egypt began sending natural gas to Syria via a pipeline that extends through Jordan.

The Syrian oil ministry said the new line would provide the country with 2.5 MMcmd (88.3 MMcfd) of gas, rising to 6.0 MMcmd (212 MMcfd) over the next nine years.

The line is part of a larger project to export Egyptian gas to several countries in the Middle East and, eventually, to Europe.

Egypt has agreed to supply Jordan, Lebanon, and Syria with gas for 30 years.

Nearly 99 bcf/yr of Egyptian gas has exported to Jordan under a 15-yr agreement.

The 1,200-km (746-mi) Arab Gas Pipeline—a project initiated in 2001—currently runs from Egypt to Jordan's Red Sea port of Aqaba; then under the Gulf of Aqaba (to avoid crossing Israeli territory); then to Rihab, Jordan; then to Syria's Deir Ali power station south of Damascus.

Further extensions are planned to Lebanon and Turkey (see below).

Gas pipeline project

Turkey and Syria signed an agreement in early 2008 aimed at enhancing cooperation between their two countries in several fields, including development of a major pipeline.

The 100-km (62-mi) pipeline would carry Egyptian natural gas from Aleppo, Syria, to Kilis in southeastern Turkey.

Gas could flow from Kilis to European markets if a link with the planned Nabucco pipeline can be established.

Turkey will have the right to as 1.5 bcm (53 bcf) of gas for its domestic needs.

The two countries also discussed joint exploration for oil and gas, given their similar geologies.

UNITED ARAB EMIRATES

Capital: Abu Dhabi
Monetary unit: Dirham
Refining capacity: 781,250 b/cd
Oil production: 2.616 million b/d
Oil reserves: 97.800 billion bbl
Gas reserves: 214.400 tcf

Exploration and Development

MOU for gas exploration, Abu Dhabi

Royal Dutch Shell PLC (The Hague, Netherlands) signed an MOU with Abu Dhabi National Oil Co. (ADNOC) to carry out possible deep offshore gas exploration in the country.

In 2007, Shell's production in Abu Dhabi averaged 146,000 boe/d.

Shell already holds interests in a joint venture with Abu Dhabi Onshore Co. and in Abu Dhabi Gas Industries.

Abu Dhabi fields and discoveries

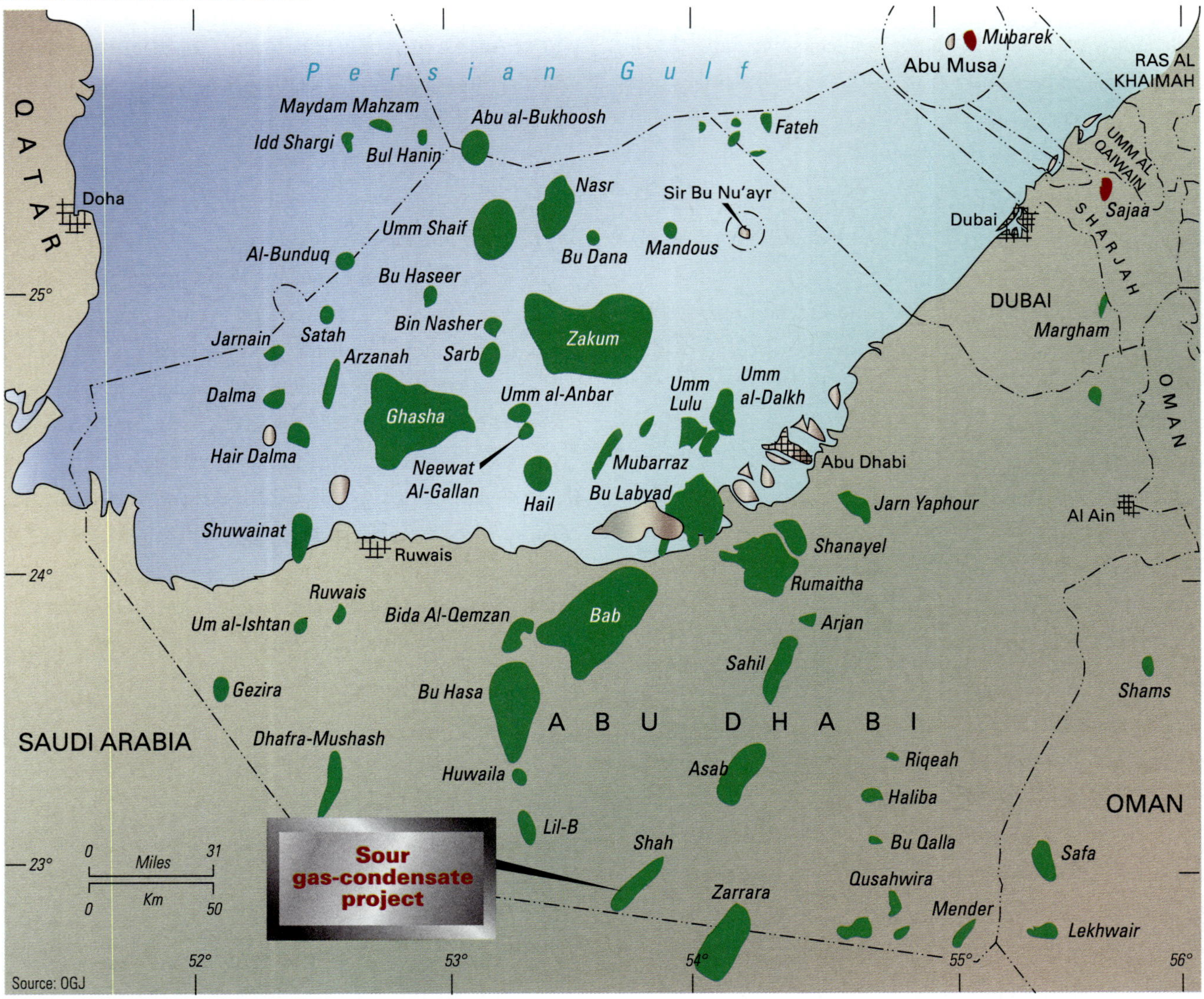

Jarn Yaphour, Ramhan fields, Abu Dhabi

Occidental Petroleum Corp. (Oxy; Los Angeles) signed a preliminary agreement in late 2008 with ADNOC to appraise and develop the Jarn Yaphour and Ramhan oil and gas fields in Abu Dhabi.

Oxy will operate both fields.

Jarn Yaphour field is onshore near the capital city of Abu Dhabi. Development activities were to start by early 2009. First production, expected in 2009, will see 10,000 boe/d of gross production from initial development.

The Ramhan discovery is near the Abu Dhabi refinery. Appraisal also was slated to start by early 2009, with first production as early as 2011 in the range of 10,000 boe/d.

Shah field, Abu Dhabi

ConocoPhillips (Houston) and ADNOC signed an interim agreement in mid-2008 to develop the Shah sour gas-condensate field onshore in southern Abu Dhabi.

The $10-billion project involves production of sour gas-condensate, stripping the liquids and sulfur for pipeline shipment to the Persian Gulf, and reinjecting the gas for pressure maintenance.

Final agreements were to be completed by yearend.

Shah field is 180 km (119 mi) southwest of Abu Dhabi City. Discovered in 1966, it had original reserves of 500–600 million bbl of oil in Cretaceous carbonates at 2,438 m (8,000 ft). The gas-condensate is in the deeper Jurassic Arab formation.

OGJ estimated Abu Dhabi's proved gas reserves at 198.5 tcf.

Western Offshore, Sharjah

Dana Gas (UAE) was moving ahead in late 2008 with plans for exploration and development of the Western Offshore concession in Sharjah, as part of a 25-yr agreement with that emirate.

It is Dana's first offshore venture.

The concession covers over 1,000 sq km (386 sq mi) offshore and will develop the offshore Zora gas field in Sharjah by mid-2009.

Processed gas will move through a 25-km (15-mi) offshore pipeline to be installed by Dana.

Drilling and Production

Boosting oil production

In mid-2008, *Oil & Gas Journal* noted that Abu Dhabi Company for Onshore Oil Operations (ADCO) was working to expand production in various Abu Dhabi fields by 560,000 b/d.

Also, ExxonMobil was involved in the Upper Zakum redevelopment project to increase production by about 250,000 b/d from that field.

Refining

Ruwais expansion

Abu Dhabi Oil Refining Co. (Takreer), an arm of state-owned Abu Dhabi National Oil Co. (ADNOC), was working to double the capacity of the Ruwais refinery to 817,000 b/d by 2014.

The refinery—240 km (149 mi) west of Abu Dhabi City, will produce propylene, unleaded gasoline, naphtha, LPG, aviation fuel, and other products. Basic engineering design was under way in third quarter 2008.

After expansion, Ruwais is to be integrated with a petrochemicals complex and an oil lubricants plant, due online in 2012 and currently under construction.

Pipelines

New Abu Dhabi oil line

CNPC (Beijing) and the International Petroleum Investment Company (Abu Dhabi) signed an EPC contract in late 2008 for a $3.29-billion crude oil pipeline project in Abu Dhabi.

CNPC said it was the largest EPC contract CNPC has signed marketing an overseas project.

The export-oriented, 400-km (248-mi) pipeline will link the Habshan oil field in western UAE to the Fujairah Port in the east.

Designed transmission capacity will be 1.5 million b/d.

Commissioning is slated for third quarter 2011.

TFP gas line

The UAE's Dolphin Energy awarded a contract to Russia's Stroytransgaz in mid-2008 for construction of the Taweelah-Fujairah natural gas pipeline (TFP).

Construction of the 240-km (149-mi) line was to start by yearend 2009, with completion expected in 2010. The line will link Dolphin's receiving facilities at Taweelah in Abu Dhabi with Fujairah on the eastern coast of the UAE.

Dolphin already supplies the UAE with gas from Qatar through a 364-km (226-mi) subsea export pipeline that connects the company's Ras Laffan gas processing plant in Qatar with the receiving facilities at Taweelah.

That pipeline can carry as much as 56.6 MMcmd (2.0 bcfd) of gas or 20.7 bcm/yr (731 bcf/yr).

Habshan-Fujairah pipeline

International Petroleum Investment Co. (Abu Dhabi) was proceeding in mid-2008 with work on the 1.5-million b/d Habshan-Fujairah oil pipeline, which is to begin operating in early 2010, about one year later than first planned.

No explanation was cited for the delay of the 320-km (199-mi) pipeline, which will join the Habshan fields of state-owned ADNOC with the Port of Fujairah.

The project also includes oil storage and terminal facilities for crude exports at Fujairah.

The pipeline will provide an alternative to shipment through the Strait of Hormuz, generally seen as a strategic oil-transport choke point.

Environmental Issues

CCS project

Hydrogen Energy (Surrey, UK) said it was developing the world's first commercial carbon capture and sequestration/ storage (CCS) project, in Abu Dhabi.

The project envisions an eventual network to capture carbon dioxide from factories and manufacturing plants across the UAE and use the gas to enhance recovery from oilfields.

In late 2008, the $3-billion CCS project was in the FEED stage. It is expected to come on stream by early 2013.

Injected into aging oilfields, carbon dioxide maintains pressure to enhance oil recovery. As oil comes out of the fields, the carbon dioxide remains trapped in the earth.

It has been estimated that up to 3 bbl of oil can be recovered per ton of carbon dioxide injected.

The Abu Dhabi project is being designed to capture 50 million tons of carbon dioxide over a 25-yr life and to recover about 2 billion bbl of oil.

The project initially will capture around 2 mtpy of carbon dioxide.

Terrorist Attacks

Al Qaeda and its affiliates attacked several elements of Yemen's oil infrastructure in 2008, including:

- a pipeline belonging to Total SA in the Sah Valley (March)
- a facility of an unidentified Chinese oil company operating in Al-Khish'a (March)
- a housing and office complex of Safer E&P Operations Co. in Sana'a (April).

YEMEN

Capital: Sana'a
Monetary unit: Rial
Refining capacity: 140,000 b/cd
Oil production: 305,000 b/d
Oil reserves: 3.000 billion bbl
Gas reserves: 16.900 tcf

Raising oil production

The oil ministry said in early 2008 that the government plans to increase oil production to 500,000 b/d by 2010. (*Oil & Gas Journal* estimates 2008 production at 305,000 b/d.)

The country will both sell new licenses and improve output at existing sites.

According to the U.S. EIA, total oil production was 360,000 b/d in 2007, down from 400,000 b/d in 2005 and 438,000 b/d in 2002.

Yemen's Petroleum Exploration and Production Authority (PEPA) said the drop is due to declining production in Masila and Marib, the country's two largest basins.

Block S2

OMV AG (Vienna) said in early 2008 that it plans to drill new 40 oil wells by 2010 in Yemen's S2 block (Al Uqlah) in the Armah district of Shabwa governorate. The firm aims to increase production to more than 30,000 b/d.

In 2007, OMV announced the start of oil production at 1,000 b/d from its Kharwah-1 well on Block S2 and that it planned to deliver 11,000 b/d of oil by late 2008.

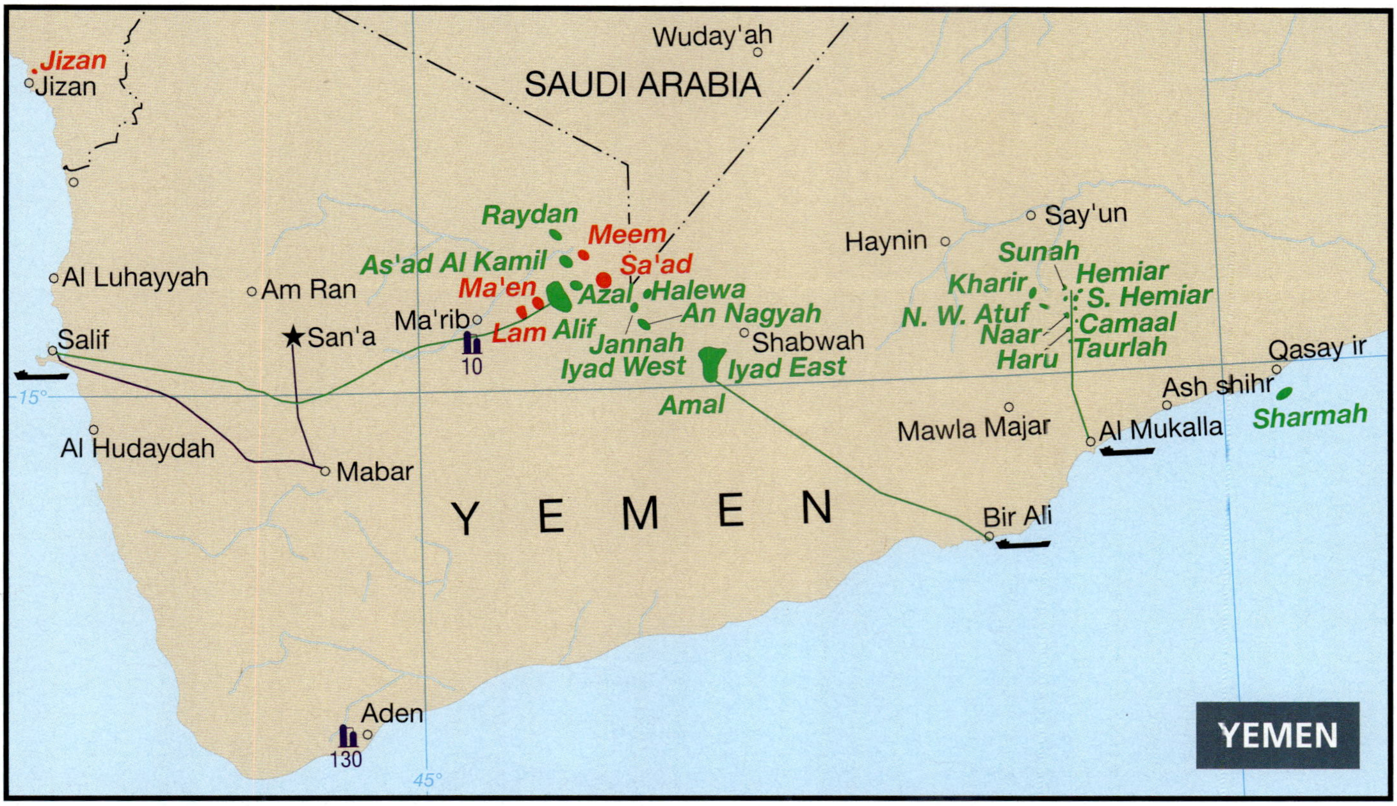

LNG financing

In mid-2008, Yemen LNG Co. Ltd. (Sana'a) announced that it had obtained $2.8 billion in financing to build its LNG liquefaction plant on the Gulf of Aden at Belhaf.

The funds include $1.1 billion from Total SA (Paris) and $1.7 billion from export agencies in France and South Korea.

A two-train liquefaction plant—capacity 6.7 mtpy—will be built, as well a 320-km (199-mi) pipeline and two 144,000-bbl storage containers. Gas will come from Block 18 in Marib.

Yemen LNG has long-term contracts to supply South Korea's Kogas (2 mtpy), Suez (2.55 mtpy), and Total Gas & Power (2.15 mtpy).

Africa

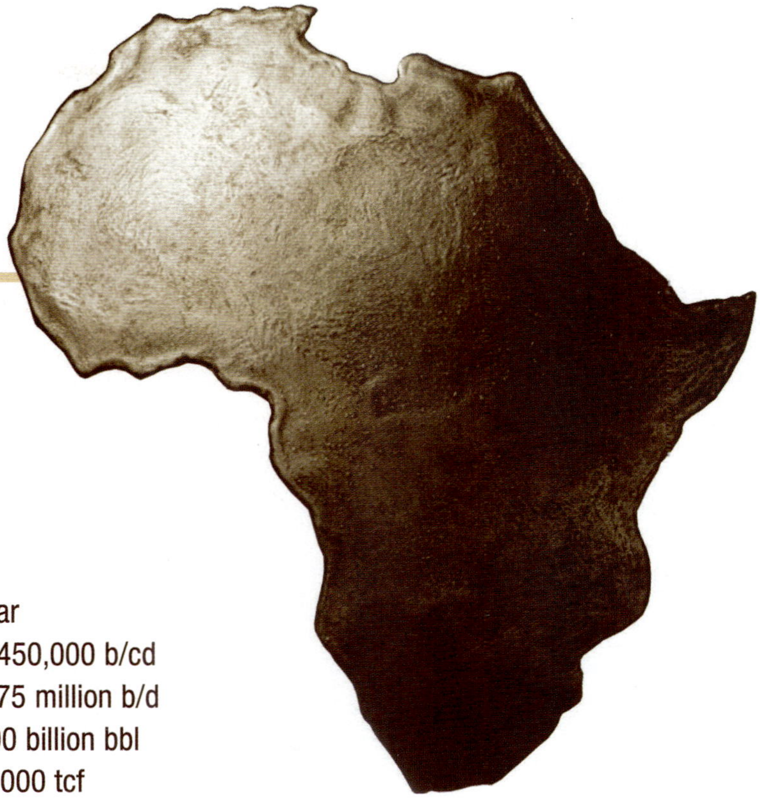

ALGERIA

Capital: Algiers
Monetary unit: Dinar
Refining capacity: 450,000 b/cd
Oil production: 1.375 million b/d
Oil reserves: 12.200 billion bbl
Gas reserves: 159.000 tcf

Exploration and Production

In Salah fields

BP Exploration (El Djazair) Ltd. let a FEED contract in late 2008 to Foster Wheeler Ltd. (Clinton, NJ) for development of the southern fields of the In Salah project in Algeria.

The work supports the second phase of the In Salah gas (ISG) development. The aim is to develop the four fields south of the current development: Garet el Befinat, Hassi Moumene, In Salah, and Gour Mahmoud.

Under the second phase, gas output will be maintained at plateau levels when production from the three northern fields starts to drop.

ISG currently produces 9 bcm/yr (318 bcf/yr) from three northern fields. Seven proved gas fields are under development in the southern Sahara, 1,200 km (745 mi) south of Algiers.

ISG is also the world's first full-scale carbon dioxide capture project at an onshore gas field.

The partners in the ISG venture are Sonatrach, BP PLC (London), and StatoilHydro (Stavanger, Norway).

Menzel Ledjmet East

First Calgary Petroleums Ltd. (Calgary) continued its work in 2008 on the Menzel Ledjmet East (MLE) development in the Berkine basin.

One rig was drilling delineation wells at midyear in the MLE area, and the first section of a housing complex Hassi Messaoud was to be ready for occupancy by December 2008.

First hydrocarbons production from the project is scheduled for second quarter 2011.

Plans calls for combined production first from MLE and then from the Central Area Field Complex.

WEST AFRICA
Atlantic Ocean
Mediterranean Sea
Ionian Sea
Gulf of Guinea
Bight of Biafra
Cape of Good Hope
Lake Chad
MOROCCO
ALGERIA
TUNISIA
LIBYA
SICILY
GREECE
CRETE
WESTERN SAHARA
MAURITANIA
MALI
NIGER
CHAD
SENEGAL
GAMBIA
GUINEA-BISSAU
GUINEA
SIERRA LEONE
LIBERIA
IVORY COAST
BURKINA FASO
GHANA
TOGO
BENIN
NIGERIA
CAMEROON
CENTRAL AFRICAN REPUBLIC
EQUATORIAL GUINEA
GABON
CONGO (FORMER ZAIRE)
CABINDA
ANGOLA
NAMIBIA
BOTSWANA
REPUBLIC OF SOUTH AFRICA
Rabat
Mohammedia
Casablanca
Kechoula
Essaouira
Djebel-Jeer
Safi
Sidi Rhalem
Sidi-Kacem
Las Palmas
Sidi Ifni
Tarfaya
Oran
Arzew
Algiers
Skikda
Bejaia
Bizerte
Tunis
Gela
Athens
Tripoli
Sider
Tobruk
El Golea
Hassi Messaoud
2 Lines
3 Lines
Bel Rhazi
Touat
Ilatou
SBAA
Azzene
Reggane
In-Salah
In Amenas
Wafa
Murzuk
Amasralad
Le Camp
Farrud
Umm Farud
Amal
Nafoora
Augila
Defa
Messla
Sarir
Chinguetti
Nouakchott
Bamako
Niamey
Ouagadougou
Bissau
Conakry
Freetown
Monrovia
Kano
Kaduna
N'Djamena
Abuja
Makurdi
Komé
Bolobo
Miandoum
Doba
Abidjan
Espoir
Panthere
Bellier
Foxtrot
Saltpond
Accra
Lome
Porto Novo
Lagos
Yaounde
Bangui
Ceiba
Kribi
Sao Tome
Libreville
Cape Lopez
Port Gentil
Tschengue
Rembo-Kotto
Asswe
Sette-Cama
Gamba-Ivinga
Mossaka
Brazzaville
Kinshasa
Pointe Noire
N'Kossa
Emeraude
GCO
Essungo
Mavanga
Sulele
Guntala
Maleva North
Benfica
Tobias
Luanda
Lobito
Windhoek
Walvis Bay
Gaborone
Luderitz
Johannesburg
Kudu
Ibhubesi
Cape Town
Mossel Bay
Sable
F-A-
Oribi
0 300 600 Mi.
0 200 400 600 Km.

EAST AFRICA
SICILY
Ionian Sea
Aegean Sea
Athens
GREECE
CRETE
TURKEY
SYRIA
CYPRUS
Nicosia
LEBANON
Beirut
Damascus
ISRAEL
Jerusalem
Amman
JORDAN
IRAQ
Baghdad
IRAN
PAKISTAN
Kuwait
Persian Gulf
Manama
Doha
Abu Dhabi
Ghawar
Riyadh
Hilwah
SAUDI ARABIA
OMAN
YEMEN REPUBLIC
Sanaa
Jizan
Ma'bar
Aden
Gulf of Aden
Mediterranean Sea
Bizerte
El Kala
Annaba
Tunis
Gela
TUNISIA
Tripoli
Mellitah
Hassi Messaoud
Wafa
In Amenas
Murzuk
Le Camp
Sider
Farrud
Umm Farud
Tobruk
Alamein
Amal
Nafoora
Augila
Defa
Sarir
Messla
LIBYA
Abu Gharadig
Abu-Qir
Alexandria
Cairo
Suez
Ramadan
Ras Gharib
Midyan
Mango
Asyût
Geisum
EGYPT
Port Sudan
Red Sea
Khartoum
Kassala
SUDAN
Kosti
Sheraf
Muglad
Heglig
Tabaldi
Munga
Unity
Adar
Melut
Abu Gabra
Bentiu
ERITREA
Asmera
Assab
DJIBOUTI
Djibouti
Lake Tana
Addis Ababa
ETHIOPIA
Calub
SOMALIA
Mogadishu
(Idle)
NIGER
CHAD
Lake Chad
N'Djamena
Kano
NIGERIA
Makurdi
Bolobo
Miandoum
Doba
Kome
CAMEROON
Yaounde
Kribi
CENTRAL AFRICAN REPUBLIC
Bangui
CONGO (FORMER ZAIRE)
UGANDA
Lake Albert
Kampala
KENYA
Lake Victoria
Nairobi
Mombasa
Libreville
GABON
CONGO
Tschengue
Rembo-Kotto
Sette-Came
Asswe
Gamba-Ivinga
Mossaka
Brazzaville
Kinshasa
N'Kossa
CABINDA
Cabinda
Emeraude
GCO
Essungo
Mavanga
Guntala
Maleva North
Sulele
Luanda
Benfica
Tobias
RWANDA
Kigali
Bujumbura
BURUNDI
Lake Tanganyika
Dodoma
Zanzibar
Dar es Salaam
TANZANIA
Indian Ocean
Kilwa
Songo-Songo
Lobito
ANGOLA
Ndola
ZAMBIA
MALAWI
Lake Malawi
Lilongwe
Lusaka
MOZAMBIQUE
Lake Kariba
Harare
ZIMBABWE
Mutare
Beira
Mozambique Channel
MADAGASCAR
Antananarivo
NAMIBIA
Windhoek
Walvis Bay
BOTSWANA
Gaborone
Thohoyandou
Pande
Temane
Pretoria
Johannesburg
Maputo
Mbabane
SWAZILAND
Sasolburg
Luderitz
Kudu
Ibhubesi
Maseru
LESOTHO
Durban
REPUBLIC OF SOUTH AFRICA
Umtata
Cape Town
Mossel Bay
Sable
F-A
Oribi
Cape of Good Hope
0 300 600 Miles
0 200 400 600 Km

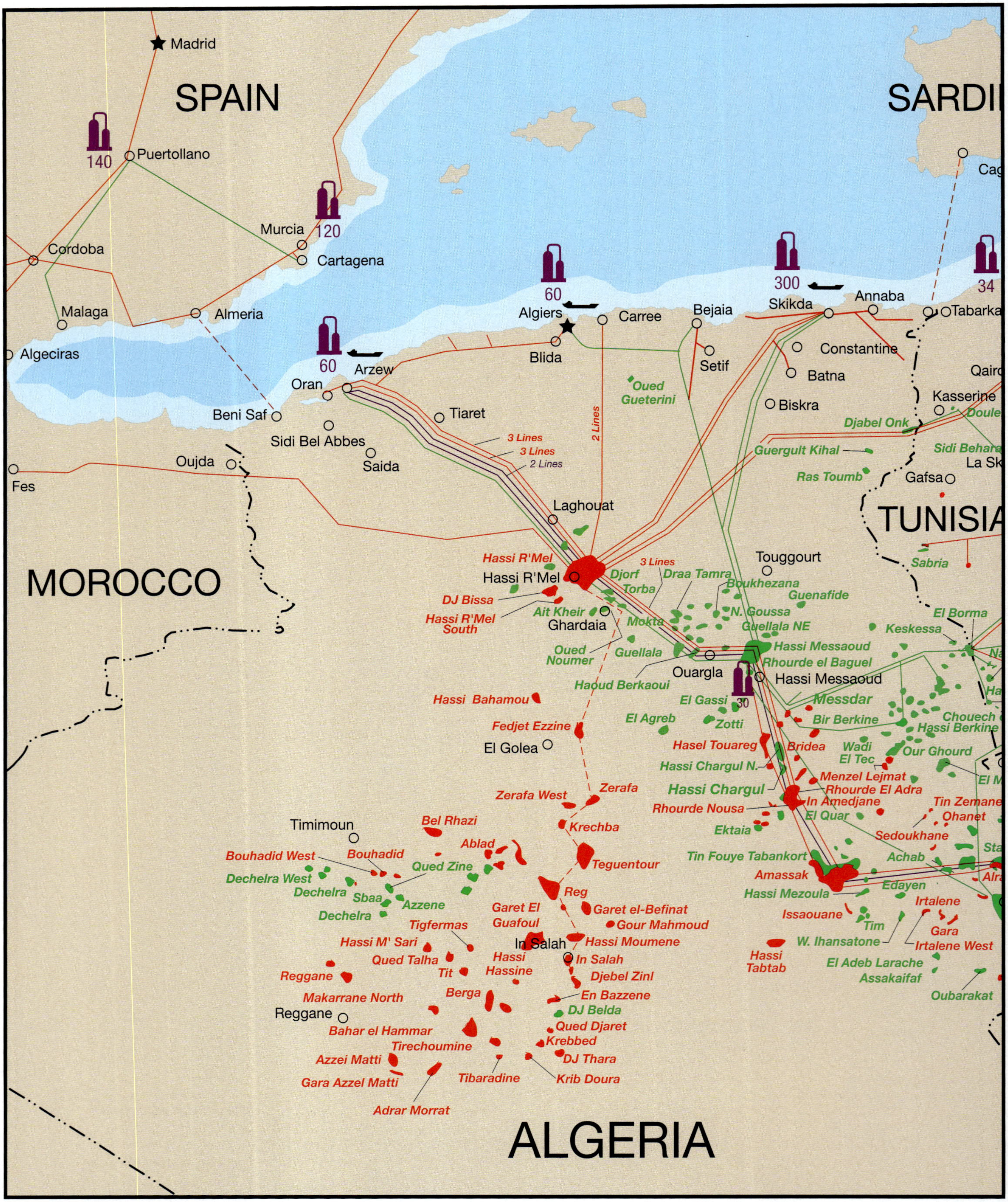
SPAIN
SARDI
MOROCCO
TUNISIA
ALGERIA
Madrid
Puertollano
140
120
Murcia
Cartagena
Cordoba
Malaga
Algeciras
Almeria
60
Oran
Arzew
Beni Saf
Sidi Bel Abbes
Saida
Tiaret
Oujda
Fes
60
Algiers
Blida
Carree
Bejaia
Setif
300
Skikda
Annaba
Constantine
Batna
Biskra
34
Tabarka
Cag
Qair
Kasserine
Gafsa
La Sk
Oued Gueterini
3 Lines
3 Lines
2 Lines
2 Lines
3 Lines
Laghouat
Hassi R'Mel
Hassi R'Mel
DJ Bissa
Hassi R'Mel South
Djorf
Torba
Draa Tamra
Boukhezana
Guenafide
Touggourt
Ait Kheir
Ghardaia
Mokta
N. Goussa
Guellala NE
Oued Noumer
Guellala
Ouargla
Haoud Berkaoui
Hassi Messaoud
Rhourde el Baguel
Hassi Messaoud
30
Messdar
El Gassi
Zotti
El Agreb
Hassi Bahamou
Fedjet Ezzine
El Golea
Hasel Touareg
Bridea
Bir Berkine
Hassi Chargul N.
Hassi Chargul
Rhourde Nousa
Ektaia
Wadi El Tec
Our Ghourd
Menzel Lejmat
Rhourde El Adra
In Amedjane
El Quar
Tin Zemane
Ohanet
Sedoukhane
Achab
Tin Fouye Tabankort
Amassak
Hassi Mezoula
Edayen
Issaouane
Tim
Irtalene
Gara
Irtalene West
W. Ihansatone
El Adeb Larache
Assakaifaf
Oubarakat
Hassi Tabtab
Djabel Onk
Guergult Kihal
Ras Toumb
Sidi Behara
Doule
Sabria
El Borma
Keskessa
Chouech
Hassi Berkine
Zerafa West
Zerafa
Krechba
Teguentour
Reg
Garet el-Befinat
Gour Mahmoud
Hassi Moumene
In Salah
In Salah
Djebel Zinl
En Bazzene
DJ Belda
Qued Djaret
Krebbed
DJ Thara
Krib Doura
Timimoun
Bel Rhazi
Ablad
Bouhadid West
Bouhadid
Qued Zine
Dechelra West
Dechelra
Sbaa
Azzene
Dechelra
Garet El Guafoul
Tigfermas
Hassi M' Sari
Qued Talha
Tit
Hassi Hassine
Reggane
Makarrane North
Berga
Reggane
Bahar el Hammar
Tirechoumine
Azzei Matti
Gara Azzel Matti
Tibaradine
Adrar Morrat

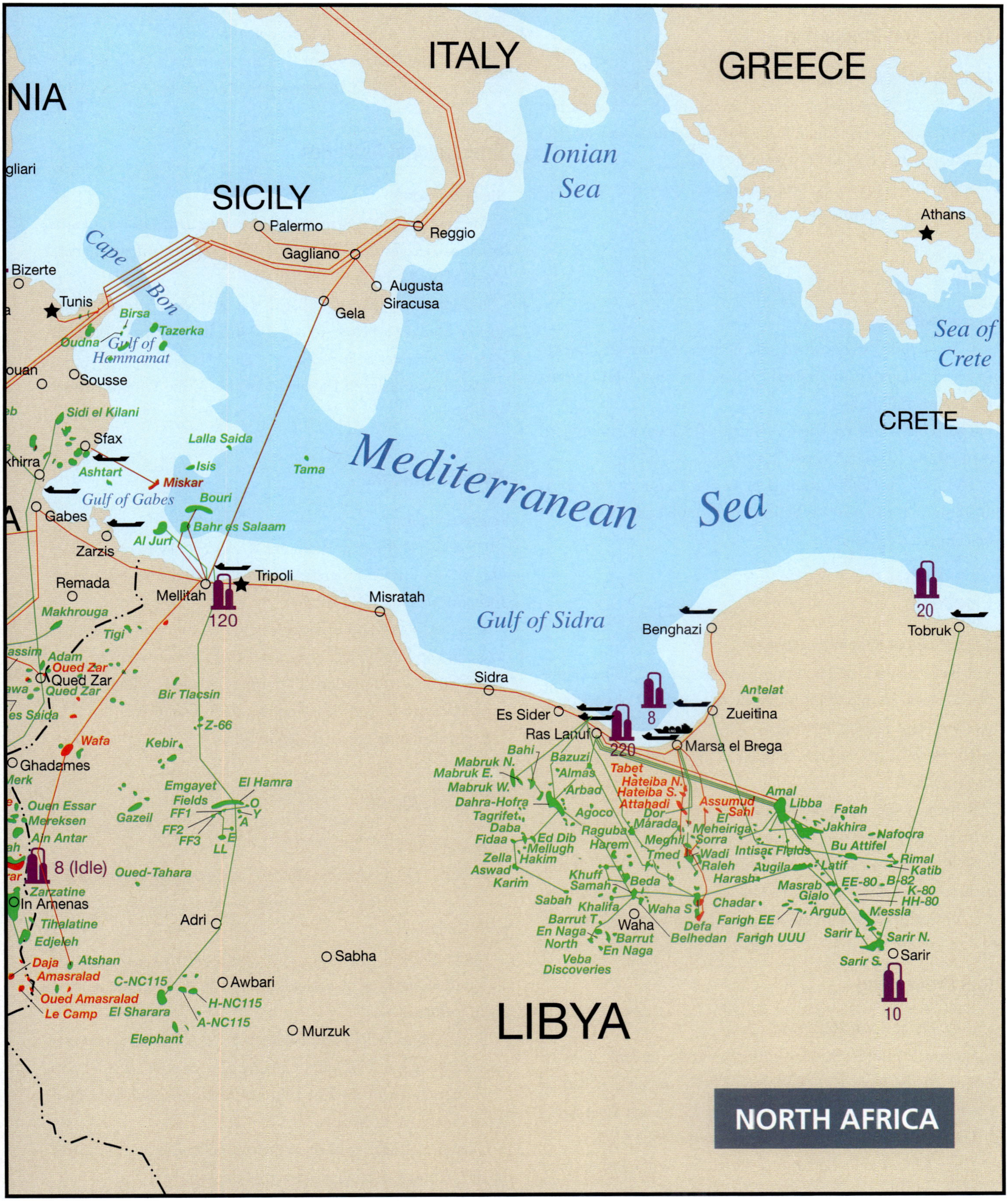
ITALY
GREECE
NIA
gliari
SICILY
Ionian Sea
Palermo
Reggio
Gagliano
Augusta
Siracusa
Gela
Athans
Sea of Crete
CRETE
Cape Bon
Bizerte
Tunis
Birsa
Tazerka
Oudna
Gulf of Hammamat
ouan
Sousse
Sidi el Kilani
Sfax
Lalla Saida
khirra
Ashtart
Isis
Tama
Miskar
Mediterranean Sea
Gulf of Gabes
Bouri
Gabes
Bahr es Salaam
Al Jurf
Zarzis
Tripoli
Remada
Mellitah
120
Misratah
Gulf of Sidra
Benghazi
20
Tobruk
Makhrouga
Tigi
assim
Adam
Oued Zar
Qued Zar
awa
Qued Zar
Bir Tlacsin
es Saida
Z-66
Sidra
Antelat
Es Sider
8
Zueitina
Ras Lanuf
220
Marsa el Brega
Wafa
Kebir
Ghadames
Bahi
Bazuzi
Mabruk N.
Mabruk E.
Mabruk W.
Almas
Tabet
Hateiba N.
Hateiba S.
Attahadi
Arbad
Emgayet Fields
El Hamra
Amal
Libba
Fatah
Ouen Essar
Mereksen
Gazeil
FF1
FF2
FF3
O
Y
A
E
LL
Dahra-Hofra
Tagrifet
Daba
Fidaa
Agoco
Assumud
Sahl
El Meheiriga
Dor
Marada
Jakhira
Nafoora
Ain Antar
Raguba
Ed Dib
Mellugh
Hakim
Harem
Meghil
Sorra
Intisar Fields
Bu Attifel
8 (Idle)
Oued-Tahara
Zella
Aswad
Karim
Tmed
Wadi
Raleh
Augila
Latif
Rimal
Katib
Khuff
Samah
Harash
Beda
Masrab
Gialo
EE-80
B-82
K-80
HH-80
Zarzatine
In Amenas
Sabah
Khalifa
Waha S
Chadar
Argub
Messla
Tihalatine
Adri
Barrut T.
Waha
Defa
Farigh EE
Edjeleh
En Naga North
Barrut
Belhedan
Farigh UUU
Sarir L.
Sarir N.
En Naga
Sabha
Sarir S.
Sarir
Daja
Atshan
Veba Discoveries
Amasralad
C-NC115
Awbari
10
Oued Amasralad
H-NC115
Le Camp
El Sharara
A-NC115
LIBYA
Murzuk
Elephant
NORTH AFRICA

Drilling and Production

Rhounde El Baguel field

In mid-2008, redevelopment of several fields in Algeria continued, including Rhounde El Baguel field, the largest oil field in Algeria, which went on stream in 1962.

Sonatrach (Tripoli) expects production from this redevelopment to peak at 100,000 b/d in 2009.

LNG

Arzew LNG train

In mid-2008, state-owned oil company Sonatrach (Tripoli) awarded Saipem/Snamprogetti (Milan) a €2.8-billion contract for a new LNG train.

The project will be a joint venture of Saipem/Snamprogetti (Milan) and Chiyoda Corp. (Yokohama).

The 4.7-mtpy GL3Z train is to be built near Arzew, about 400 km (248 mi) west of Algiers, with completion. Work is to be completed by yearend 2012.

Pipelines

Assessment of Sonatrach pipelines

Sonatrach contracted Solutions International BV (The Hague, Netherlands) in mid-2008 to review the integrity of part of its pipeline transmission network.

The aim is to ensure compliance with stringent legislative standards.

Sonatrach operates more than 16,000 km (9,940 mi) of transmission pipelines.

Shell will evaluate management processes as well as the technical integrity of part of the pipeline system and then recommend ways to improve system reliability and availability.

Gas Processing

LPG facility at Hassi Messaoud

Sonatrach also let a €1.3-billion contract in late 2008 to Saipem/Snamprogetti to develop an LPG processing facility at the Hassi Messaoud oil and gas complex in central Algeria.

The site is about 900 km (560 mi) southeast of Algiers.

Three LPG production trains are to be built, with a total capacity of 8 million cu m/day.

The work is to be completed by first-half 2012.

Supergiant Hassi Messaoud is the country's largest oil field, with estimated ultimate recovery of 9 billion bbl.

International Activities

Medgaz pipelay

Medgaz is a consortium of five international companies: Sonatrach (Algeria), CEPSA (Madrid), Iberdrola (Bilbao), Endesa (Madrid) and GDF Suez (Paris).

In May 2008, the group began the offshore pipelaying phase for the Medgaz pipeline, to run 210 km (130 mi) from Beni Saf, Algeria, to Almeria, Spain.

Details are given in the Spain section of the Europe chapter of this book.

Petrochemicals

Arzew complex

Oil & Gas Journal's semiannual *Worldwide Construction Update* (April 2008) noted progress on plans to build a $3 billion petrochemical complex in Arzew.

Total SA (Paris) and Sonatrach signed a framework agreement to build the facility, including a 1.4-mtpy ethane cracker.

The facility will produce 1.1 mtpy of ethylene, which will be processed into polyethylene and monoethylene glycol.

Commissioning is scheduled for 2012.

Company News

Gazprom office

In mid-2008, Gazprom (Moscow) opened an office in Algeria, aimed at developing opportunities in Africa and in particular with Sonatrach.

Under an MOU signed in 2006, both parties are committed to jointly developing oil and gas projects in Algeria, Russia, and third countries.

The deal also includes probable swaps of assets and operations, with exchanges of LNG and pipeline gas, for efficiency.

The action marks the first time that Gazprom has established a representative in Africa.

Gazprom officials said that cooperation would cover all parts of the oil and gas value chain across the continent.

ANGOLA

Capital: Luanda
Monetary unit: Kwanza
Refining capacity: 39,000 b/cd
Oil production: 1.870 million b/d
Oil reserves: 9.040 billion bbl
Gas reserves: 9.530 tcf

Angola has overtaken Libya as the region's second most prolific oil producer.

The country's production is predominantly from deep water overlaying thick salt layers.

In mid-2008, there were 21 deepwater fields on production. Start-ups in 2007–2008 included the Greater Plutonio complex on Block 18 and Kizomba C on Block 15.

In 2009, Tombua and Landana fields on Block 14 are scheduled to start production.

And in 2011, the Pazflor field project on Block 17 is expected to produce first oil.

Africa, EU energy cooperation

In September 2008, the African Union Commission and the European Commission jointly established an Africa-European Union energy partnership that could attract more than €1 billion in EU funds for energy investments in Africa.

The action represented the first 2008–10 action plan adopted at the partnership's December 2007 Lisbon summit.

One high priority is developing energy interconnections between Africa and Europe. This included revival of the 4,300-km (2,672-mi) Trans-Saharan Gas Pipeline project.

It would link Nigeria to Algeria, carrying 20 bcm/yr (706 bcf/yr) of Nigerian gas to the EU in 2015 via either Spain or Italy.

Exploration and Development

Dione prospect

State oil company Sociedade Nacional de Combustiveis de Angola (Sonangol) and BP Exploration (Angola) Ltd. made an oil discovery on the Dione prospect in ultra deepwater block 31 in late 2008

The discovery is about 9 km (6 mi) southwest of the Juno-1 discovery.

Dione was drilled in a water depth of 1,696 m (5,564 ft), some 390 km (242 mi) northwest of Luanda. The well reached a total depth of 3,272 m (10,735 ft) below sea level.

Tests confirmed the capacity of the reservoir to flow in excess of 5,000 b/d under production, said BP.

Block 15/06

Eni SPA (Rome) and Sonangol EP reported an oil discovery in late 2008 on the western area of Block 15/06 off Angola.

The Ngoma-1 well, 350 km (217 mi) from Luanda, reached a TVD of 3,383 m (11,099 ft).

Drilled in 1,421 m (4,662 ft) of water, it struck an oil column in high-permeability Miocene sands.

At midyear, Eni and other partners had success with their first well, Sangos-1, drilled on the same block.

That well went to TVD of 3,343 m (10,968 ft) in water 1,349 m (4,426 ft) deep, also finding oil in Miocene sands.

Block 15/06 lies within the Lower Congo basin and covers 2,984 sq km (1,152 sq mi).

Onshore Cabinda South Block

In mid-2008, Roc Oil Co. Ltd. (Perth) spud the seventh and final well in an exploration program on the 1,073-sq km (414-sq mi) Cabinda South onshore block in the Lower Congo basin.

Also, a group led by Roc Oil spudded the Arroz-1 exploration well in the eastern part of the Cabinda South block. The aim was to test a subsalt prospect, with drilling to be completed in September.

The group's first well, Massambala-1, discovered a shallow heavy oil accumulation on the western side, and plans called for drilling as many as six shallow appraisal wells in the second half of 2008.

Block 31

Several new offshore deepwater discoveries on Block 31 were expected to come on stream by 2011.

Sonangol gave approval to operator BP Exploration (Angola) Ltd. and its partners in mid-2008 to develop Pluto, Saturn, Venus, and Mars (PSVM) fields, in the northeast part of the block.

Development was to start in 2008, with oil production expected to peak at 150,000 b/d by 2012.

The fields lie in 2,000 m (6,562 ft) of water and are 400 km (248 mi) northwest of Luanda.

As of mid-2008, the consortium had discovered 15 fields on the acreage and said it will develop the others in a similar manner to the PSVM group.

The partners have begun to plan the second development in the southeast part of the block.

Also, in early 2008, Sonangol EP and BP reported the Portia oil discovery on Block 31. BP said it was the company's 15th discovery on the block.

Tests were said to confirm that the reservoir, about 10 km (6 mi) southwest of Pluto field, has the capacity to flow more than 5,000 b/d under production conditions.

Drilling and Production

Block 17: Pazflor development

Operator Total SA (Paris) awarded principal contracts in early 2008 for the giant Pazflor oil field development.

It will be the third development center on offshore Block 17, following Girassol and Dalia, both on production.

Drilling operations are planned to start in 2009, with production slated to begin in 2011.

Pazflor lies in 600–1,200 m (1968–3937 ft) of water about 150 km (93 mi) offshore and 40 km (25 mi) northeast of Dalia field. It covers 600 sq km (232 sq mi).

The project aims to bring four fields into production: Perpetua; Hortensia and Zinia (Upper Miocene); and Acacia (Oligocene).

Kizomba C

Esso Exploration Angola (Block 15) Ltd.—a subsidiary of ExxonMobil Corp. (Irving, TX)—started production in 2008 from the Mondo, Saxi and Batuque fields as part of the Kizomba C project offshore Angola.

The entire Kizomba C development was expected to produce 200,000 b/d of oil before yearend 2008, and an eventual total of 600 million bbl.

Kizomba C is 90 mi (145 km) offshore Angola in about 2,400 ft (800 m) water depth. The development includes two FPSOs and 36 subsea wells.

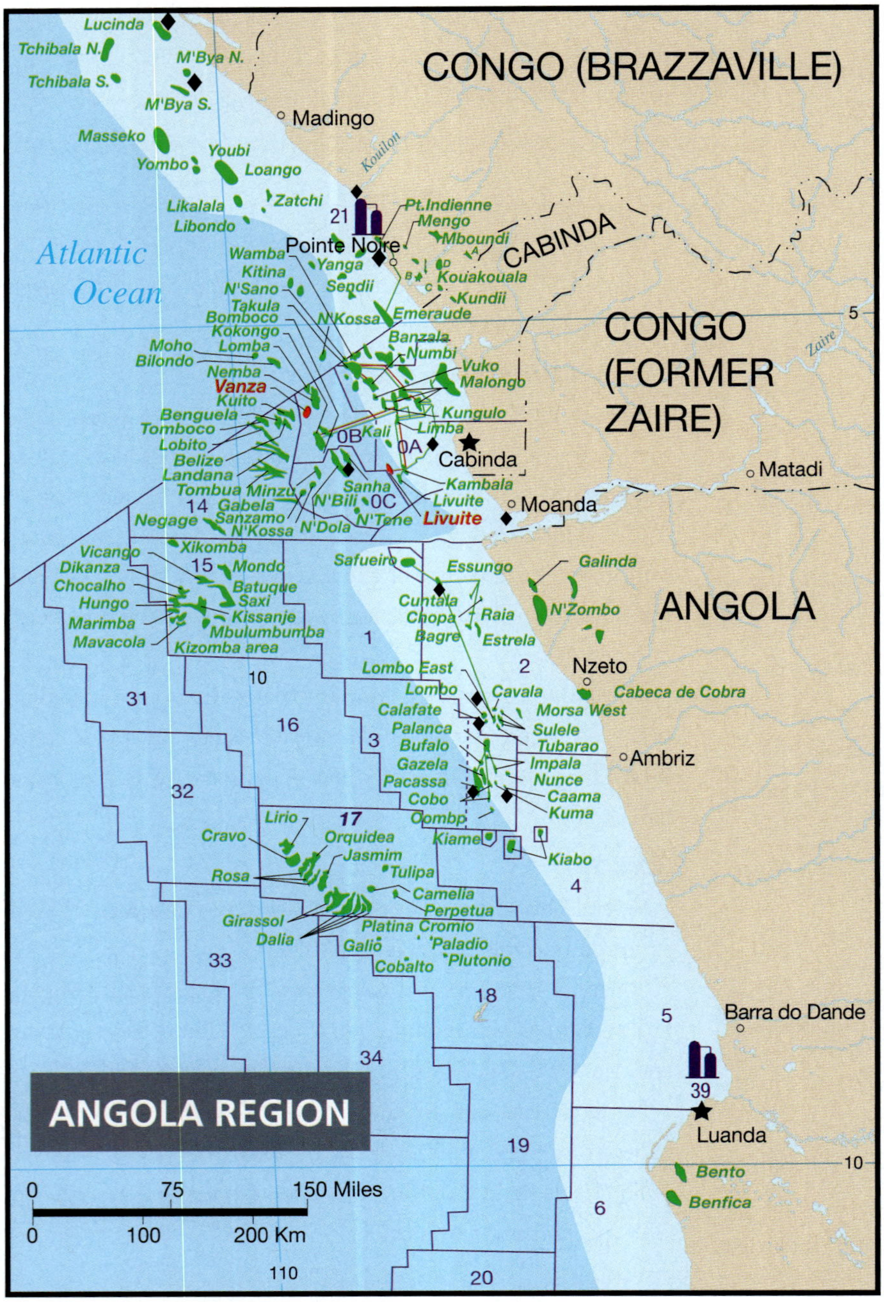

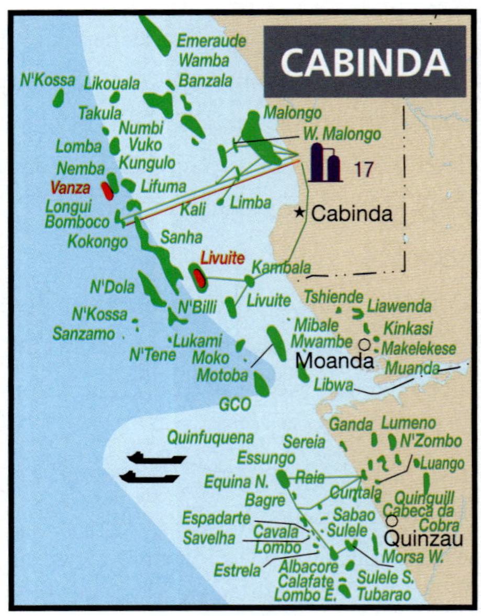

Takula processing platform

Cabinda Gulf Oil Co. Ltd.—a subsidiary of Chevron Corp. (San Ramon, CA)—installed the Takula gas processing platform on Block 0 off Angola, aiming for start-up by yearend 2008.

The platform is in 57 m (187 ft) of water with the capacity to process 100 MMscfd of sour gas for injection and delivery to the new onshore Cabinda gas plant.

The platform is seen as key to Angola's determination to reduce onshore and offshore gas-flaring.

Refining

Lobito refinery

At yearend 2008, Sonangol awarded an EPC and site-development contract to KBR Inc. (Houston) for the 200,000-b/d Lobito refinery about 600 km (373 mi) south of Luanda.

The grassroots refinery will process heavy crudes and reduce Angola's product imports.

LNG

First LNG plant

Angola LNG Ltd. was making plans in early 2008 to build the country's first gas liquefaction facility near Soyo in Zaire Province. The facility will have a nominal capacity of 5.2 mtpy of LNG.

It will include storage for LNG, LPG, and condensate, and a loading jetty for tankers as large as 210,000 cu m. A subsidiary of Bechtel Corp. (San Francisco) will build the facility.

First LNG from the project is expected by early 2012.

Angola LNG Ltd. shareholders are affiliates of Sonangol; Chevron Corp. (San Ramon, CA); BP PLC (London); Total SA (Paris); and Eni SPA (Rome).

CAMEROON

Capital: Yaounde
Monetary unit: CFA franc
Refining capacity: 37,000 b/cd
Oil production: 85,000 b/d
Oil reserves: 200.000 million bbl
Gas reserves: 4.770 tcf

Nigeria relinquishes Bakassi

In August 2008, Nigeria ceded control over the potentially oil-rich Bakassi Peninsula to Cameroon, amid threats from armed groups protesting the transfer.

The handover ended a territorial dispute that has almost triggered war in the past.

Bakassi is a 1,600-km (995-mi) border area that juts into the Gulf of Guinea.

In 2002, the International Court of Justice ruled that Nigeria should relinquish control following the border dispute between the countries.

However, legal and political disputes and violence have delayed the process. Opponents in Nigeria said that parliament never ratified the agreement to hand over the territory.

About 50 people died in 2007 in clashes related to this issue, said Nigerian reports.

In November 2008, even after the handover, kidnappers held members of an oil service vessel for 11 days as a protest to the change of status.

Analysts said that Cameroon must stabilize security before it can begin oil exploration.

CHAD

Capital: N'Djamena
Monetary unit: CFA franc
Refining capacity: 0
Oil production: 145,000 b/d
Oil reserves: 1.500 billion bbl
Gas reserves: 0

First oil refinery

Construction began in late 2008 on Chad's first oil refinery. Start-up is scheduled for 2011.

It is a joint-venture project between China National Petroleum Corporation (CNPC; Beijing) and state oil firm Societe des Hydrocarbures du Tchad (SHT; N'Djamena).

The refinery will process about 20,000 b/d of crude oil to produce gasoline, diesel, kerosene and liquefied petroleum gas, said a CNPC official.

The refinery will supply products for both domestic and export markets.

The project also includes a 40-MW power plant, and about 20 MW will be supplied to N'Djamena. The project is CNPC's third oil-refinery investment in Africa.

Chad-Cameroon pipeline

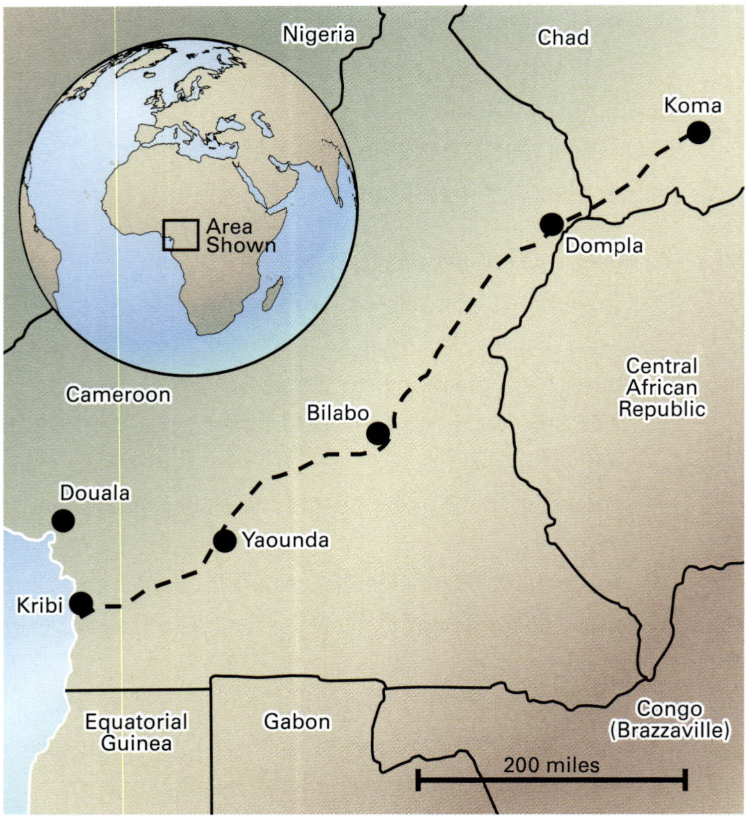

Doba basin fields

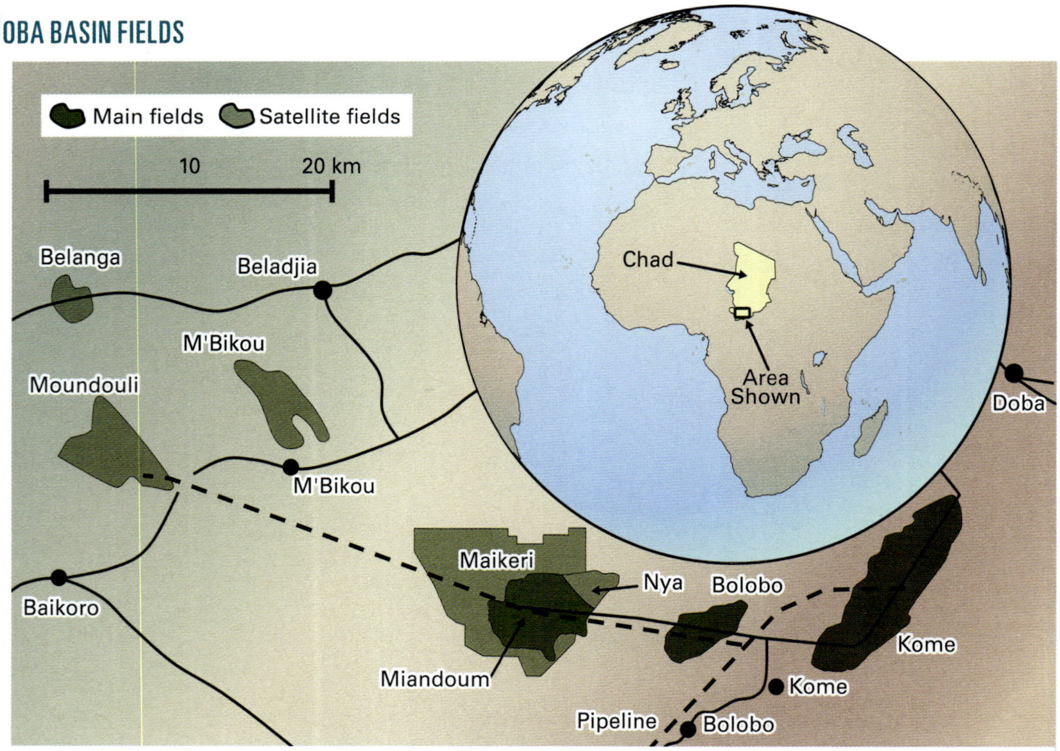

CONGO (BRAZZAVILLE)

Capital: Brazzaville
Monetary unit: CFA franc
Refining capacity: 21,000 b/cd
Oil production: 240,000 b/d
Oil reserves: 1.600 billion bbl
Gas reserves: 3.200 tcf

Note: Congo (Brazzaville) is also referred to as the Republic of the Congo.

It should not be confused with Congo (Former Zaire), which is also known as the Democratic Republic of the Congo (DR Congo).

Moho Nord Marine-3 well results

Total SA (Paris) reported positive appraisal results in late 2008 for its Moho Nord Marine-3 well, 80 km (50 mi) offshore, in 1,030 m (3,379 ft) of water.

The well was drilled to TD 2,300 m (7,546 ft) northwest of Moho Nord Marine-2 and north of Moho Nord Marine-1. It encountered a 60-m (197-ft) column of oil-bearing Upper Miocene sands.

Phase 1 development of the southern part of Moho-Bilondo is under way, comprising 14 subsea wells tied back to a floating production unit (FPU).

Phase 1 came onstream in April 2008, and is expected to climb to plateau production of 90,000 b/d.

Moho-Bilondo production start

In mid-2008, Chevron Corp. (San Ramon, CA) reported first oil production from the deepwater Moho-Bilondo project in the Republic of the Congo.

Moho-Bilondo consists of several subsea well clusters that flow into an FPU. Maximum total production of 90,000 b/d of crude oil is expected in 2010.

The Moho field lies 9 mi (15 km) off the Congolese coast.

Two exploration wells, Moho-Nord Marine 1 and 2, were drilled in the permit area in 2007.

Chevron's partners are Société Nationale des Pétroles du Congo (SNPC; Pointe-noire) and Total E&P Congo.

Marine XIV and XI Blocks

In late 2008, the Congolese subsidiary of SOCO International PLC (London) entered into a farm-in agreement to acquire a partial working interest in the offshore Marine XIV Block.

Other interests in the block are PA Resources Congo; SNPC; Raffia Oil SARL; and a subsidiary of Lundin Petroleum AB (Stockholm).

The block is in the Lower Congo Basin in shallow water, adjacent to SOCO's Marine XI Block.

The concession covers approximately 265 sq km (102 sq mi).

Earlier in the year, the SOCO subsidiary agreed to farm-out part of its interest in the offshore Marine XI Block to Petrovietnam Exploration Production (Ho Chi Minh City).

Other interests in the block are SNPC, Raffia Oil SARL, Africa Oil & Gas, and a subsidiary of Lundin Petroleum.

The block covers approximately 1,400 sq km (541 sq mi). A multi-well drilling program was planned for 2008.

Oil sands development

Eni SPA (Paris) said in mid-2008 that it had discovered and will develop a major oil sands deposit that holds an estimated 0.5–2.5 billion bbl of recoverable oil.

The sands are in the Tchikatanga and Tchikatanga-Makola permit areas of Kouilou Department, in two blocks covering a total of 1,790 sq km (691 sq mi).

Eni said that from 2008 to 2011, it will invest US$3 billion in the Congo, generating an expected equity production of 150 million boe/d.

M'Boundi field

Tullow Oil PLC (London) announced in January 2008 that it had sold its interest in the onshore M'Boundi Concession to Korea National Oil Company (KNOC; Seoul).

The M'Boundi Field, discovered in 2001, is one of the largest oilfields found onshore Africa in recent years. Operator Eni SPA (Rome) was producing approximately 37,000 b/d from the field.

CONGO (FORMER ZAIRE)

Capital: Kinshasa
Monetary unit: Zaire
Refining capacity: 0
Oil production: 25,000 b/d
Oil reserves: 180.000 million bbl
Gas reserves: 35.000 bcf

Nganzi Block

SOCO International PLC (London) got government approval in early 2008 to proceed with planned work on the Nganzi Block.

Government-owned Congolaise des Hydrocarbures (Cohydro) is the other interest-holder in the block.

The block covers approximately 800 sq km (309 sq mi) on the eastern flank of the prolific coastal basin.

The company had already acquired or conducted aeromagnetic, gravity and geochemical surveys. Seismic work was slated to begin in mid-2008.

Block V

In early 2008, the government granted to Dominion Petroleum Ltd. (Hamilton, Bermuda) a production sharing contract for the new exploration concession, Block V.

Located in the eastern part of the country, adjacent to the border with Uganda, the block covers 7,105 sq km (2,743 sq mi) of land and lake areas.

It includes part of Lake Edward and adjoins Block 4B in Uganda, where Dominion has been active.

Both blocks are part of the Albertine Rift system where major oil discoveries were made in 2006–07.

Dominion and partners first will carry out geological and geophysical work, acquire seismic data, and drill two exploration wells.

Besides Dominion, other interests are SOCO International PLC (London), and Cohydro.

Block III

South Africa Mineral Resources Corp. (Samroc; Cape Town) said in early 2008 that it was purchasing South Africa Congo Oil Company (SacOil; Pretoria), including its oil concessions in Congo.

Samroc said it adopt the SacOil name and focus on pan-African oil and gas exploration rather than on minerals development.

The company expected to begin exploration by mid-2008 in Congo's Albertine Graben, on Block III, south of Lake Albert in Ituri Province.

EGYPT

Capital: Cairo
Monetary unit: Pound
Refining capacity: 726,250 b/cd
Oil production: 680,000 b/d
Oil reserves: 3.700 billion bbl
Gas reserves: 58.500 tcf

Overview

The Egyptian government moved ahead with plans to invest $7 billion into oil and natural gas exploration and oil field reconstruction in the 2008–09 fiscal year, which began July 1, 2008.

In addition, officials said the government plans to invest $850 million by yearend 2009 to speed up the construction of natural gas pipelines leading to Upper Egypt to provide energy needed for industrial development and urban needs.

Oil production

The Gulf of Suez accounts for about half of Egypt's oil production; the rest is from its three other key areas: The Western Desert, the Eastern Desert, and the Sinai Peninsula. The Gulf of Suez is likely to continue to be the main oil producing region, according to the Ministry of Petroleum.

But aggressive exploration in the deep water of the Mediterranean, and onshore in the Nile Delta and Western Desert, has resulted in several recent oil finds.

Gas production

Due to major recent discoveries, natural gas is likely to be the primary growth engine of Egypt's energy sector for the foreseeable future, according to the U.S. Dept. of Energy's Energy Information Administration (EIA).

The government wants to add 30 tcf to its proven gas reserves by 2010 and to double gas exports by 2011. Most of this increase would come from new natural gas discoveries offshore from the Nile Delta, and in the Western Desert.

In the Nile Delta, recent offshore developments include Port Fuad, South Temsah, and Wakah fields.

Two areas in the Western Desert that have shown strong potential are Obeiyed field and the Khalda concession.

Gas exports

Through a variety of projects involving both Egyptian and foreign companies, Egypt was exporting to other countries (by pipeline and as LNG) about 29% of the gas it produced in 2008, said *Oil & Gas Journal.*

Egyptian gas exports have risen from 226 MMcm (8 bcf) in 2003 to 1.93 bcm (68 bcf) in 2006, according to EIA. But growing domestic demand has caused the government to consider how best it should use its resources and raised debate about how much gas the country should export.

Changing policies

In a late-2008 evaluation, Wood Mackenzie (Edinburgh) noted that Egyptian officials had begun taking action to draw more interest from IOCs in bidding for new oil and gas blocks.

The government proposed raising (in stages) the price cap on gas sold in the domestic market.

Also, state-owned Egyptian Natural Gas Holding Co. (EGAS) told companies such as BP PLC (London), Eni SPA (Rome), and BG Group (London) that it would allow negotiation of gas price once a discovery is made.

As a result, experts expect there to be greater interest among IOCs in Egyptian projects.

Exploration and Development

West Mediterranean: Block 1

Hess Corp. (New York City) announced at yearend 2008 that the Dekhila-1x well in the West Mediterranean (Block 1) deepwater area cut a gross hydrocarbon section of 148 ft (45 m) at multiple intervals.

The well, 72 km (45 mi) offshore within the Nile Delta basin, was drilled to 7,707 m (8,881 ft) in 1,184 m (3,883 ft) of water.

Results were being incorporated into engineering studies for the wider West Med development.

The 2,371 sq km (916 sq mi; 586,000-acre) West Med block contained the Abu Sir, El King, El Max, and Al Bahig gas discoveries.

North Ramadan concession

ONGC Videsh (Mumbai) and IPR Red Sea Inc. (Cairo) made a second oil discovery in the North Ramadan Concession in the Gulf of Suez in late 2008.

Discovery well, NR-2 is north of IPR's first oil discovery, NR-1A ST1, which produced 3,000 b/d of oil and 42,480 cu m/d (1.5 MMcfd) during tests.

NR-2 was drilled to TD 3,566 m (11,700 ft) in the Lower Miocene Mheiherrat formation. A total of 34 m (113 ft) was perforated over a gross interval of 87 m (286 ft).

Tests produced up to 800 b/d of oil and 14,160 cu m/d (500,000 cfd) on natural flow.

ONGC Videsh and IPR were aiming to drill one remaining exploratory well (NR-3) by yearend, and then to pursue other prospects already identified.

West Kalabsha concession

Apache Corp. (Houston) said its WKAL-C-1X discovery in late 2008 in the West Kalabsha concession in the Western Desert flow-tested at 4,746 b/d of oil and 124,510 cu m/d (4.4 MMcfd) of gas.

The firm called the discovery the westernmost oil ever discovered in Egypt.

West Kalabsha involves the Jurassic Safa formation in Faghur basin, 80 km (50 mi) east of the Libyan border.

Drilled to 4,450 m (14,600 ft) TD, the well logged 14 m (46 ft) of Safa oil pay and 11 m (35 ft) of pay in the Alem El Bueib 3C formation.

Eastern Nile Delta

Dana Gas (Sharjah, UAE) estimated that its Al Tawil-1 discovery in late 2008 in the eastern Nile Delta has located 2.6 bcm (90 bcf) of gas and 4 million bbl of condensate recoverable.

The discovery well, in the West Manzala onshore concession 15 km (9 mi) south of the company's El Wastani gas processing plant, flowed 0.67 MMcmd (23.5 MMcfd) and 1,027 b/d on test.

It encountered 34 m (116 ft) of net pay in the Miocene Qawasim formation. TD was 3,163 m (10,377 ft) measured depth.

The company plans to drill more exploratory wells in the West Manzala and West Qantara concessions in 2008–09.

North Shadwan block

Gulf of Suez Petroleum Co.—the JV of BP PLC (London) and Egyptian General Petroleum Corp. (Cairo)—was drilling an appraisal well in late 2008 at the Burtocal oil discovery on the North Shadwan block in the southern Gulf of Suez.

Two shallow, nearshore discoveries each found 8 million bbl proved and probable recoverable, while Burtocal identified 24 million bbl in the Cretaceous Nubia sandstone.

Drilling venture launched

In late 2008, Egypt has awarded a drilling contract to the recently formed Egyptian Offshore Drilling Co. (EODC), a JV of trading firm Toyota Tsusho, Egyptian Natural Gas Holding Co., and Ganoub El-Wadi Petroleum Holding Co.

EODC in late 2008 awarded to PPL Shipyard (Singapore) a $425 million contract to build two jackup rigs. Drilling could begin as early as 2011.

Toyota Tsusho, a subsidiary of Toyota Motor Co., will manage the project. Egyptian Natural Gas will operate them and carry out test drilling.

The JV has signed a $500-million loan agreement with a syndicate of Japanese and Egyptian banks. In September, Japan Bank of International Cooperation (JBIC) agreed to participate in that syndicate.

Disouq concession

RWE Dea AG (Hamburg) discovered gas in mid-2008 for the second time within the South Sidi-Ghazy 1x-well on the Disouq concession in the Egyptian Nile Delta.

The well is 8 km (5 mi) south/south-west of the North Sidi Ghazy-1x discovery.

South Sidi-Ghazy 1x reached TD 3,188 m (10,459 ft) within the Late Miocene and found gas within the Messinian formation.

Tests confirmed gas in the intervals of 2,747–2,771 m (9,012–9,091ft).

The company was planning next to drill nearby exploration well North West Khilala-1x, looking for additional hydrocarbons in the Messinian structure.

Apache plans for 2008

Apache Corp. (Houston) said it planned to drill 282 wells in Egypt in 2008, including 40 exploration wells that will target the following:

- Jurassic formations with gas-condensate potential near the recent Hydra, Qasr, and Jade discoveries
- oil potential near Apache's Neith, Heqet, and Buchis development leases
- new Bahariya and Abu Roash oil fields in the Abu Gharadig basin
- a 10.5-million-acre (16,406-sq mi) farmout from Tharwa Petroleum.

Faghur basin

Apache Corp. (Houston) said it has proved a Jurassic Safa formation oil play in mid-2008 that could cover as much as 320 sq km (830 sq mi) in the Western Desert.

The Heqet-2 well was drilled in the Greater Khalda area of the Faghur basin, 106 km (66 mi) southwest of Qasr field and the same distance east of the border with Libya.

The well was drilled to TD 4,481 m (14,700 ft), producing 2,100 b/d of oil.

West El Burullus area

In mid-2008, Dana Petroleum PLC (Aberdeen, UK) reported a gas discovery with its first exploration well in Egypt.

The WEB-1X well was the first drilled by the Dana Petroleum and GDF-Suez (Paris) partnership at the West El Burullus prospect within the Nile Delta region, off northeast Alexandria.

Drilled to 2,403 m (7,884 ft) TVD, the well targeted a Pliocene prospect, encountering good quality gas bearing sands.

Under constrained test conditions, the well flowed gas at rates of up to 0.76 MMcmd (27 MMcfd).

Nile Delta gas

RWE Dea AG (Hamburg) signed a long-term agreement in early 2008 with Egyptian General Petroleum Corp. (Cairo) and EGAS (Cairo).

The deal aims to pursue the first phase of development of gas fields discovered in the offshore concession North Idku in the Nile Delta.

Production is scheduled to begin in 2010.

North Idku concession, Nile Delta region

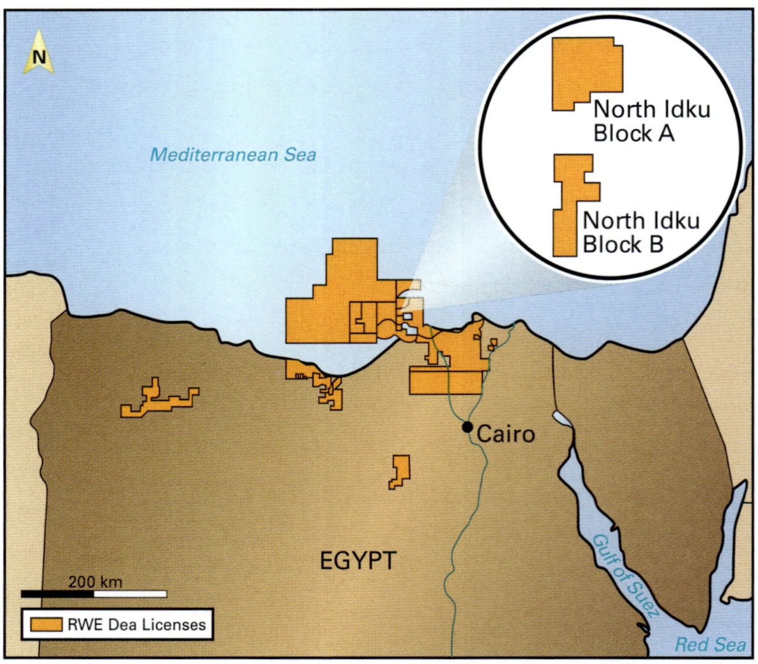

North El Burg concession

In early 2008, BP PLC (London) was evaluating its first discovery of deep Oligocene nonassociated gas offshore in the Nile Delta.

BP said that the Satis-1 well—in 90 m (295 ft) of water, 50 km (31 mi) north of Damietta—was drilled to a Nile Delta record TD of more than 6,500 m (21,325 ft).

Nile Delta Oligocene rocks include the Ghazalat, Qatrani, and Abu Zaabal formations, but BP did not identify the producing zone.

Satis-1 is on the North El Burg concession, which lies between the Ras El Barr development concession and the Offshore Baltim development concession.

Western Desert: Shushan C concession

In early 2008, Apache Corp. (Houston) reported a hydrocarbon discovery in the Western Desert with its Hydra-1X exploration well.

The well, in the Shushan "C" concession, logged 54 m (178 ft) of net pay in the Lower Safa—one of the thickest pay zones in that formation.

On test, Hydra-1X flowed 1.18 MMcmd (41.6 MMcfd) of natural gas and 1,313 b/d of condensate.

The firm said that Hydra-1X penetrated the same Lower Safa sandstone found 40 km (25 mi) away at Qasr—the largest field Apache ever discovered, with estimated proved reserves of 65 bcm (2.3 tcf) of gas and 80.4 million bbl of condensate.

The latest discovery extends the known Lower Safa production trend an additional 14 km (9 mi) north of the Kahraman B-22 Lower Safa discovery, which flowed 0.45 MMcmd (16 MMcfd) of gas and 480 b/d of condensate after fracture stimulation in 2006.

The well also encountered 14 m (45 ft) of probable condensate pay in the Jurassic Alam El Bueib (AEB) Unit 6 sand and 9 m (30 ft) of probable oil pay in the Lower Cretaceous AEB Unit 3.

Offshore permits offered

In September 2008, Egypt offered seven offshore exploration permits in Mediterranean areas potentially rich in natural gas. Companies were told to submit bids by February 9, 2009.

Drilling and Production

Taurt gas field

Operator Gulf of Suez Petroleum Co. (Gupco) started natural gas production from Taurt field on the offshore Ras El Bar concession in mid-2008.

The field lies 70 km (43 mi) northeast of Port Said in the West Nile Delta. Gas production will supply the Damietta LNG plant.

Gupco is a joint venture of BP PLC (London), Egyptian General Petroleum Corp. (Cairo), and Eni SPA (Rome) affiliate International Egyptian Oil Co. (IEOC).

Taurt is BP's first subsea development in Egypt.

Saqqara field

In mid-2008, Gulf of Suez Petroleum Co. (Gupco) began delivering more than 30,000 b/d of oil from offshore Saqqara field in the Gulf of Suez. Saqqara was discovered in 2003.

A 13-km pipeline carries output to a separation and gas processing plant at Ras Shukeir, Gulf of Suez.

Gupco is a joint venture of BP PLC (London) and Egyptian General Petroleum Corp. (Cairo).

Ras Budran: 25 years of production

Operator RWE Dea AG (Hamburg) was celebrating in early 2008 the 25th year of production from the crude oil reservoir Ras Budran in the Gulf of Suez, 4 km (2 mi) off the coast of Sinai.

A JV between RWE Dea as operator and the Egyptian General Petroleum Corp. developed the field. As of early 2008, approximately 267 MMbbl of crude had been produced.

The concession was renewed in 2007 for another ten years.

Komombo and Nile Delta

The 2008 exploration and development program for the Egypt assets of Dana Gas (Sharjah, UAE) called for one exploration and one development well in the Komombo concession in Upper Egypt

Also, in the Nile Delta, 14 exploration wells are planned. Five will target the Sidi Salem formation at a depth of approximately 4,000 m (13,123 ft).

Pipelines

Arab Gas Pipeline

The AGP began transporting gas in 2003 from El Arish, Egypt, to Aqaba, Jordan, and was extended from Aqaba to El Rehab, Jordan, in 2005.

The line is part of a larger project to export Egyptian natural gas to the Middle East and, eventually, to Europe. Ultimate capacity is expected to be 27 MMcmd (950 MMcfd).

Egypt began sending gas through Jordan to Syria in mid-2008. The Syrian officials said that the new line would likely provide that country with 6 bcmd (212 MMcfd) by 2017.

Also, in late 2008, the AGP was being extended to Homs and Kilis on the Turkish border

EU, Egypt energy agreement

In December 2008, the European Commission and Egypt's Foreign Affairs Ministry signed an MOU to enhance energy cooperation between Egypt and the European Union and reinforce energy security for both. Details are given in the Europe chapter of this book.

The EU energy commission said it expects pipeline-gas imports to the EU of about 2 bcm/yr (71 bcf/yr) from Egypt in early 2010 through the AGP.

Egypt announced a delay in AGP gas exports to Lebanon, from July 2008 until at least January 2009.

Officials in Lebanon said in late 2008 that it appeared Egypt would only be able to supply half of the agreed volumes—1.7 MMcmd (60 MMcfd)—when supplies are launched.

Gas to Israel

Gas began flowing in February 2008 through a 100-km (62-mi) subsea pipeline from El Arish in eastern Egypt to Ashkelon, Israel. Capacity is 4.67 MMcmd (165 MMcfd).

Gas Processing

Salam gas plant

Khalda Petroleum Co. (Cairo)—a JV of Apache Corp. (Houston) and Egyptian General Petroleum Corp. (Cairo)—let a contract in mid-2008 to Petrofac Ltd. (London) for engineering and procurement services related to building an additional gas train at the Salam natural gas plant.

The facility is on Apache's Khalda Concession in the Western Desert.

Expected to come online in 2010, it will be Khalda's fifth gas processing facility at Salam.

Gas trains 3 and 4 were already under construction; each has the capacity to process 2.8 MMcmd (100 MMcfd) of gas and 14,000 b/d of condensate.

Trains 1–4 will provide total processing capacity of 20.1MMcmd (710 MMcfd) of gas and 66,000 b/d of condensate. The facility processes gas from Apache's Jurassic formation gas reserves.

Petrochemicals

Ain Sokhna complex

Egypt Hydrocarbon Corp. (EHC; Cairo) was proceeding in late 2008 to set up a greenfield olefins complex at Ain Sokhna on the Suez coast.

The facility will consist of a 2.5-mtpy naphtha cracker and two polyethylene units. The company plans to convert 900,000 tons of ethylene into polyethylene. The plant will be capable of producing 900,000 tpy of ethylene and 400,000 tpy of propylene.

EQUATORIAL GUINEA

Capital: Malabo
Monetary unit: CFA franc
Refining capacity: 0
Oil production: 320,000 b/d
Oil reserves: 1.100 million bbl
Gas reserves: 1.300 tcf

Block I: Benita

Noble Energy Inc. (Houston) announced in mid-2008 that its Benita oil appraisal well I-5 encountered 13 m (42 ft) of net oil pay on offshore Block I, east of Bioko Island.

On test, the "high-quality" Miocene reservoir yielded 6,250 b/d of oil and 152,930 cu m/d (5.4 MMcfd) of natural gas.

Development wells are expected to produce approximately 10,000 b/d. First production is expected in 2012.

In January, test flows on the Benita I-4 well on Block I measured 818,465 cu m/d (28.9 MMcfd) of gas and 1,634 b/d of condensate from a high-quality Miocene reservoir.

Block I: Diega

Earlier in 2008, Noble also had discovered multiple reservoirs containing oil and/or gas condensate on the Diega prospect, north of Benita, on Block I.

Subsequent drilling in August yielded test flows of 2,700 b/d of oil and 76,465 cu m/d (2.7 MMcfd) of natural gas.

Noble's modeling indicated potential production rates of approximately 10,000 b/d of oil plus associated natural gas.

The company estimates gross resource potential at 20–80 MMboe.

Gas processing plans

Officials with state-owned Sonagas announced plans in late 2008 to establish a company to gather gas in the Gulf of Guinea and direct it to key projects in the nation.

The country wants to become a gas processing hub, gathering gas from Nigeria, Cameroon, and from other sources in Equatorial Guinea that could be fed into the country's LNG expansion, petrochemicals, LPG plants, and methanol systems.

The facility could also help to eliminate gas flaring.

Devon assets sold

Devon Energy Corp. (Oklahoma City) completed in mid-2008 the $2.2-billion sale of its oil and gas business in Equatorial Guinea to the national oil company, GEPetrol.

GEPetrol acquired a 23.75% participating interest in offshore Zafiro oil field on Block B, where proved reserves were estimated at 55 million bbl of oil at yearend 2007.

The other assets are Devon's interests in undeveloped offshore Blocks C and P.

Devon said it sold its interests in order to concentrate on its North American operations.

GABON

Capital: Libreville
Monetary unit: CFA franc
Refining capacity: 24,000 b/cd
Oil production: 235,000 b/d
Oil reserves: 2.000 billion bbl
Gas reserves: 1.000 tcf

Major Anguille redevelopment

Total Gabon [subsidiary of Total SA (Paris)] announced plans in early 2008 for a two-phase redevelopment by 2011 of the Anguille field.

The goal is to boost oil recovery to 23% from 13%.

The $2-billion project is expected to yield additional proved and probable reserves of about 150 million bbl, with production peaking at more than 30,000 b/d in 2014.

Phase 1 will include work to debottleneck surface facilities, increase the number of drainage points, and conduct hydraulic fracturing and massive waterfloods.

In Phase 2 (2009–2011), the company will: install new offshore infrastructure; decommission obsolete process units; building an onshore plant for fluid treatment, and gas compression; and drill 30 more wells.

The redevelopment also will help to reduce gas flaring and coastal discharges of production water.

Anguille lies 20 km (122 mi) off Port-Gentil in the Grand Anguille Marine concession.

Ebouri field: Vaalco

In November 2008, Vaalco Energy Inc. (Houston) began drilling a development well in the Ebouri field and an appraisal well (North Ebouri) in the Etame block, offshore.

First oil from the Ebouri field well is expected in January 2009. Vaalco expects the well to raise the company's total production to approximately 25,000 b/d.

In the Etame block, in addition to the North Ebouri appraisal well, the company also was planning two wells (North Etame and South East Etame) on newly mapped structures.

Ebouri field: Addax

Addax Petroleum Ltd. (Calgary) reported successful drilling in late 2008 of two pilot holes to delineate the Gamba sandstone formation above oil-water contact.

The holes were drilled in the main area of the offshore Ebouri field. The first was south of the original Ebouri discovery. The second was to the northwest, about 100 m (328 ft) from the lease line of the Gryphon Marin license area.

A horizontal development well was being completed, aiming for initial production in early 2009.

Omoueyi permit

In mid-2008, Maurel & Prom SA (M&P; Paris) tested 5,510 b/d of oil from the Omko exploration well.

The well, drilled to 1,800 m (5,905 ft) on exploration permit Omoueyi, is 6 km (4 mi) east of Onal field. M&P was working to appraise the discovery by yearend.

Oil was produced from two separate intervals, the Kissenda and the base sandstone.

The company was planning to follow this well with the Alonha B, Alonha A, Onal East, and Maroc exploration wells.

Also, M&P said it will drill exploration well N'Zamo to study the southwestern area of the field.

GHANA

Capital: Accra
Monetary unit: Cedi
Refining capacity: 0
Oil production: 6,000 b/d
Oil reserves: 15.000 million bbl
Gas reserves: 800.000 bcf

Exploration and Development

Jubilee field

Substantial work continued in 2008 by Tullow Oil PLC (London), operator Kosmos Energy LLC (Dallas), and their partners to appraise more completely and to further develop the large offshore Jubilee oil field.

> **MANAGING OIL DEVELOPMENT**
>
> Ghana's leaders have stressed their determination not to suffer an economy-distorting "resources curse" as their country becomes an exporter of oil.
>
> The government launched a Gas Oil Forum in 2008 to gather ideas on how it can regulate and manage oil discovery to promote economic growth and sustainable development.
>
> It has received advice on these issues from the British and Norwegian governments.

One of the largest oil fields discovered off West Africa in the past 10 years, Jubilee is estimated with 90% probability to hold 170 million bbl of recoverable oil. Ultimate upside potential is estimated to be 1.8 billion bbl.

The field, in the offshore Tano Basin, on the West Cape Three Points Block, lies in 900–1,700 m (2,953–5,577 ft) of water and 2,100 m (6,890 ft) below the sea bed.

Jubilee was discovered in 2007, with the drilling of the Mahogany-1 exploration well.

Several appraisal wells have since validated the significance of the field:

- Mahogany-2 (mid-2008) and Mahogany-3 (late 2008) on the West Cape Three Points Block
- Hyedua-2 (late 2008) on the adjacent Deepwater Tano Block.

Drillstem tests on Mahogany-2 and Hyedua-2 indicated that individual wells could produce more than 20,000 b/d of oil.

Mahogany-2 encountered a gross reservoir interval of 243 m (797 ft). It contained 63 m (206 ft) of high-quality stacked reservoir sandstones and net hydrocarbon-bearing pay of 50 m (164 ft).

Log analysis and pressure testing indicated that the combined hydrocarbon columns exceed 600 m (1,969 ft).

Aiming for first production in 2010, the field will be developed using a subsea production and control system tied back to a turret-moored FPSO vessel.

Mahogany Deep

In addition to validating the Jubilee field, the Mahogany-3 well in late 2008 also discovered the Mahogany Deep, a significant new find of good-quality oil pay in an untested, stratigraphically deeper and separately trapped reservoir beneath Jubilee.

The well was drilled to a depth of 4,028 m (13,215 ft) in a water depth of 1,236 m (4,055 ft).

It encountered 17 m (56 ft) of good-quality, oil-bearing reservoir sandstones at levels significantly deeper than the oil water contacts previously intersected on Jubilee.

Ebony-1 well

Tullow Oil said that its late-2008, shallow-water Ebony discovery could be an indication of further prospectivity in deeper water.

The Ebony-1 exploration well was drilled to TD 2,640 m (8,662 ft) in the southwest part of the 983-sq km (380-sq mi) Shallow Water Tano license.

Discoveries, prospects off western Ghana

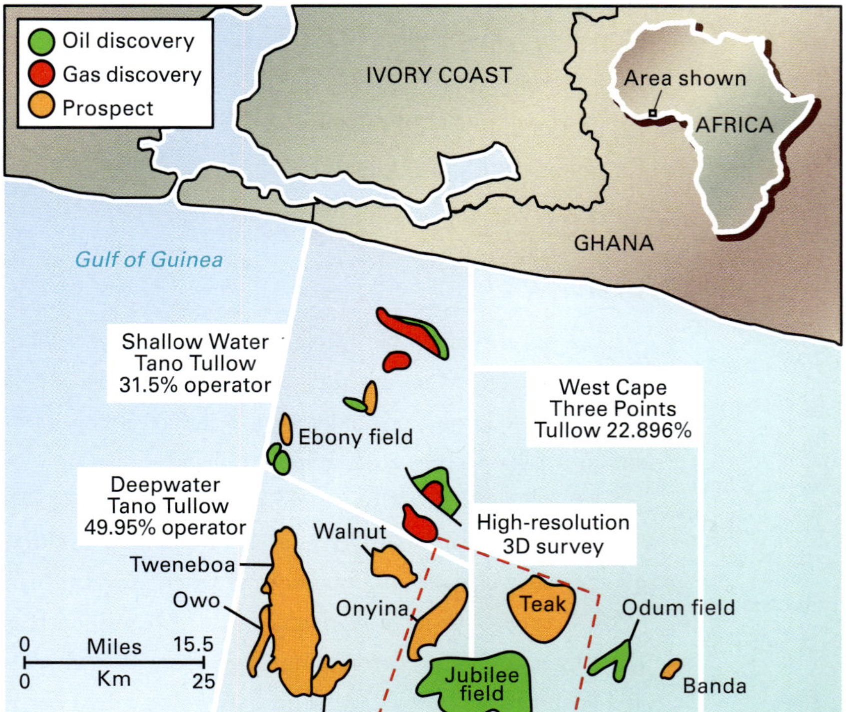

Source: Tullow Oil PLC

It intersected two hydrocarbon bearing intervals in Late Cretaceous turbidite sands: a 4-m (13-ft) oil bearing interval at 2,053 m (6,736 ft) and a 2-m (7-ft) high-pressure gas-condensate interval at 2,570 m (8,432 ft).

Analysis suggests possible connection between these sands and the material downdip Tweneboa oil prospect, which lies mainly in the Deepwater Tano license. Tweneboa is set for drilling in the first quarter of 2009.

Cuda-1X well

In the Keta basin off eastern Ghana, Afren PLC (London) said it spudded the Cuda-1X exploration well in November 2008.

It targets a prospect estimated large enough to hold up to 325–642 million bbl of oil. The play type is similar to the Jubilee and Odum discoveries.

The well targets the Upper Cretaceous deepwater sandstones and was aiming to reach a TVD of 4,800 m (15,750 ft).

Cuda-1X will be drilled in 1,700 m (5,577 ft) of water.

Ten wells already have been drilled in the Keta basin.

Odum-1 discovery

In February 2008, Tullow Oil announced an oil discovery with its exploratory Odum-1 well.

Located on the West Cape Three Points License, the well opens a second new play fairway in the Tano basin.

Odum-1 was drilled to 3,387 m (11,112 ft) in 955 m (3,133 ft) of water, some 13 km (8 mi) from Jubilee field. It encountered 22 m (72 ft) of net pay in a 60-m (197-ft) gross oil column.

The discovery is considered commercial because of its proximity to Jubilee.

South Deepwater Tano

In late 2008, Ghana National Petroleum Corp. (GNPC) and government authorities awarded Aker ASA (Oslo) a major oil and gas deepwater production license in the Gulf of Guinea.

Aker also will operate the offshore exploration area, South Deepwater Tano.

The Tano exploration area covers 3,500 sq km (1,351 sq mi). Water depths in the area range from 2,000–3,000 m (6,561–9,842 ft).

South Deepwater Tano is located further offshore and at greater water depths than the Jubilee field, but shares much of the same geology.

The seismic survey program is to start in first quarter 2009, with drilling of the first exploration well at South Deepwater Tano scheduled for 2011.

Refining

Takoradi refinery

In late 2008, Arabian Gulf Oil Ltd. (Tokyo) announced that it will invest $2 billion to build an export-oriented, 200,000-b/d oil refinery in the southwestern port town of Takoradi.

Ghana's National Petroleum Authority has approved the project.

The facility will process crude oil from Oil Field No. 3 off of Ghana's West Cape into gasoline and natural gas products, mainly for European and American markets.

Some oil products also may be exported to nearby Burkina Faso, Mali and Niger.

Future production from the field has been estimated at more than 60,000 b/d.

IVORY COAST

Capital: Abidjan
Monetary unit: CFA franc
Refining capacity: 63,990 b/cd
Oil production: 30,000 b/d
Oil reserves: 100.0 million bbl
Gas reserves: 1.000 tcf

New offshore play

Tullow PLC (London) said in early 2008 that its exploration success off Ghana at the Jubilee discovery (formerly Mahogany and Hyedua) had opened a new play off Ivory Coast in Upper Cretaceous turbidite sands.

The new play—and a previously discovered play in tilted Albian fault blocks—both extend throughout the West Africa Transform Margin.

Tullow holds nine licenses in these fairways (six off Ivory Coast and three off Ghana), and was making plans to drill in both plays in 2008.

Refinery at Abidjan

In its early-2008 *Worldwide Construction Update*, the staff of *Oil & Gas Journal* noted that Ivory Coast had started construction on its second refinery, a 60,000-b/d facility in the Vridi district of Abidjan.

The project is being funded by Energy Allied International Corp., WCW International (both in Houston), and Ivory Coast's state-owned Petroci Holdings (Abidjan).

Devon sells assets

Devon Energy Corp. (Oklahoma City) completed the sale of its operations in Cote d'Ivoire to Afren PLC (London) for $205 million in late 2008.

The assets include Devon's interests in offshore producing block CI-11, offshore undeveloped block CI-01, and the onshore Lion gas processing plant.

With completion of the deal, Devon had divested all of its oil and gas producing properties in Africa.

Devon had decided to exit Africa in early 2007.

LIBYA

Capital: Tripoli
Monetary unit: Dinar
Refining capacity: 378,000 b/cd
Oil production: 1.720 million b/d
Oil reserves: 43.660 billion bbl
Gas reserves: 54.380 tcf

Overview

Libya has the largest oil reserves in Africa (41 billion bbl) and is the eighth-largest oil producer among the members of OPEC. Libya wants to increase oil production to 2 million b/d by 2008–10 and to 3 million b/d by 2015, from 1.7 million b/d in 2007.

State-owned National Oil Company (NOC) also has said it wants to boost Libya's oil reserves to 20 billion by 2015 through an aggressive offshore and frontier exploration program.

NOC expects to increase production to 3.5 million b/d by 2020 by encouraging the drilling of at least 50 wildcats/year and acquiring at least 4,000 sq km (1,544 sq miles) per year of 3D seismic data and 10,000 km (6,214 miles) per year of 2D seismic data.

Libya is the second-largest oil supplier to Europe.

NOC established subsidiaries with offices in Europe: Umm Al-Jawaby Oil Service Co. Ltd. (London) in 1983, and Mediterranean Oil Services GmbH (Düsseldorf) in 1988.

Since the United States lifted economic sanctions against Libya in 2004, NOC has reinstituted lease sales, which are attracting the attention of international players and national oil companies.

Exploration and Production

Arous Al-Bahar prospect

The A1-54/01 well in the offshore Arous Al-Bahar prospect hit hydrocarbon pay, according to a yearend 2008 report from Hess Corp. (New York City).

The well was drilled to 11,077 ft (3,376 m) in 2,807 ft (856 m) of water in the offshore extension of the Sirte basin.

It encountered a gross hydrocarbon section of approximately 500 ft (152 m) at various intervals.

Sirte basin

State-owned National Oil Co. (NOC) said in October 2008 that RWE Dea AG (Hamburg, Germany) made its eighth oil discovery in the NC-193 concession area in the Sirte basin with exploration well G1-NC 193.

MEDITERRANEAN'S 7TH OF NOVEMBER BLOCK

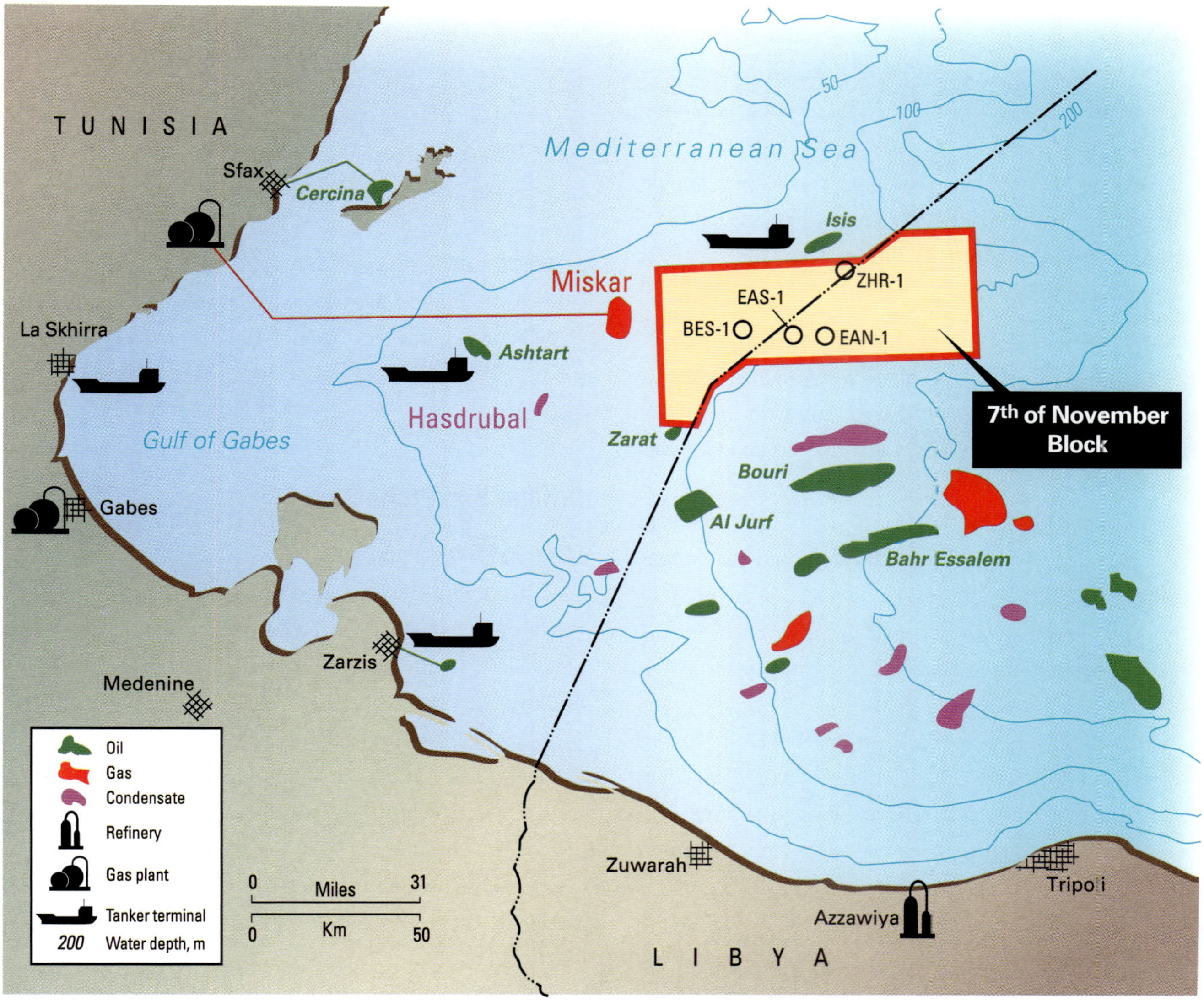

Well G1-NC 193 hit oil at 1,412–1,421m (4,631–61 ft) and tested flow at a net rate of 426 b/d.

7th of November Block

In late 2008, Joint Oil awarded the 7th of November Block in the southern Mediterranean to Canadian Superior Energy Inc. (Calgary).

Joint Oil is owned jointly by the Tunisian and Libyan governments.

The initial work commitment calls for drilling three exploration wells and one appraisal well and shooting 777 sq km (300 sq mi) of 3D seismic on the 3,035-sq km (1,172-sq mi; 750,000-acre) block.

Canadian Superior will operate the block, 75 miles offshore in 250–375 ft of water.

Parts of the block lie along the productive trends of giant Bouri oil field off Libya and giant Ashtart field off Tunisia, where cumulative production is 750 million bbl and 250 million bbl of oil, respectively.

The first exploration well was being planned for first quarter 2009, to be drilled on the Fisal prospect, about 10 km (6 mi) northeast of the Zarat appraisal wellsite.

In an associated swap, Joint Oil gets an overriding royalty interest and optional participating interest in Canadian Superior's Mariner Block in the Atlantic off Nova Scotia.

NOC deals with Oxy, OMV

NOC signed exploration and production agreements in mid-2008 with Occidental Petroleum Corp. (Oxy; Los Angeles) and OMV AG (Vienna) to upgrade their existing petroleum agreements.

NOC said the companies would train Libyans and carry out enhanced oil recovery projects.

Oxy—Libya's largest land holder—has a contract that covers fields with 2.5 billion bbl of oil reserves. The firm owns a net working interest in nine exploration blocks covering 130,000 sq km (50,193 sq mi).

It will carry out major field redevelopment and exploration programs in the Sirte basin, where it operates Blocks 106 and 124.

Oxy said it plans to spend $5 billion from 2008–2013 to triple its gross production to 300,000 b/d.

Ghadames basin

Verenex Energy Inc. (Calgary) made a seventh oil discovery in October 2008 on production-sharing contract Area 47 in the Ghadames basin; it was also the firm's second discovery on Block 4.

Exploration well C1-47/04, about 150 km (93 mi) southwest of Tripoli, was drilled to 3,095 m (10,155 ft) and encountered two productive sandstone intervals.

On test, oil flowed at 900 b/d and gas at 118,946 cu m/d (4.2 MMcfd) of gas from the Memouniat Formation at 3,036 m (9,962 ft).

Also, in the Lower Acacus formation at 2,383 m (7,819 ft), flow tests saw 150 b/d of oil and 127,443 cu m/d (4.5 MMcfd) of gas.

Verenex said that eight new field wildcat exploration wells and two appraisal wells in the area have tested at a maximum aggregate rate of 93,800 b/d.

NOC deal with ExxonMobil

NOC signed an Exploration and Production Sharing Agreement in early 2008 with ExxonMobil Libya. The agreement covers Area 21 in the offshore basin.

The minimum work program calls for extensive 2D and 3D seismic and an exploration well.

ExxonMobil also will fund training or scholarships and improve schools in Libya.

Area 21 encompasses four blocks covering a total of 2.5 million acres in the Sirte basin, 177 km (110 mi) offshore.

BP target: Ghadames basin

BP PLC (London) announced plans in early 2008 to drill its first onshore well in the Ghadames basin in 2010. The $2-billion work program will include the drilling of 17 exploratory wells.

Seismic acquisition was to begin in third quarter 2008.

In 2007, BP and partner Libya Investment Corp. (Tripoli) signed an agreement with NOC to explore 54,000 sq km (20,850 sq mi) of the Ghadames and offshore frontier Sirt basins.

Drilling and Production

Libya: Regional energy role

Analysts with Business Monitor International (BMI; London) projected in late 2008 that Libya will supply 17.37% of Africa's regional oil supply by 2012, while accounting for about 7.45% of African oil demand.

BMI said African regional oil production—10.13 million b/d in 2007—is set to rise to 12.66 million b/d by 2012. Regional oil use was 3.58 million b/d in 2007, and should increase to around 4.05 million b/d by 2012.

Regarding natural gas, the region in 2007 consumed 100 bcm (3,529 bcf). Demand in 2012 is expected to reach 171 bcm (6,039 bcf).

Gas production was 193 bcm (6,816 bcf) in 2007; it should reach 331 bcm (11,687 bcf) in 2012, implying an increase in net exports from 94 bcm (3,320 bcf) in 2007 to 161 bcm (5,686 bcf) in 2012.

Verenex Libyan Ghadames Basin Well History

Well name	Type	Total depth, ft	Tested oil rate, gross bo/d*	Perforated interval, ft	Status
A1-47/02	NFW	11,550	12,500	174	Discovery
B1-47/02	NFW	1,030	23,800	312	Discovery
C1-47/02	NFW	9,900	23,570	188	Discovery
D1-47/02	NFW	9,720	7,742	157	Discovery
E1-47/02	NFW	9,639	1,216	11	Discovery
F1-47/02	NFW	10,300	7,215	18	Discovery
A2-47/02	Appraisal	10,400	7,352	48	Interested water oil contact
D2-47/02	Appraisal	9,850	Trace	6	At water oil contact
A3-47/02	Appraisal	10,500	Trace	22	At water oil contact
A1-47/04	NFW	10,400	—	—	Preparing to test
A4-47/02	Appraisal	—	—	—	Drilling
B1-47/04	NFW	—	—	—	Drilling

*Test rates are not necessarily indicative of ultimate production rates.
Source: Verenex Energy Inc.

Refining

Mellita refinery

Zwara Oil Refinery Co. Ltd. (Zorco; Zwara) was proceeding with plans for a new 200,000-b/sd refinery at Mellita, near Zwara.

Foster Wheeler Italiana SPA (Milan) will carry out consultancy and project management services for the $4 billion refinery, which is to be completed in 2014.

The state-of-the-art facility will produce premium quality gasoline, jet fuel, and diesel with minimal fuel oil production.

Ras Lanuf refinery

NOC signed a 50:50 JV agreement in mid-2008 with the Star Consortium of TransAsia Gas International LLC and Star Petro Energy LLC (both in Dubai) covering improvement of the 220,000-b/d Ras Lanuf export refinery.

Stage 1, to start by yearend, will refurbish the existing plant to increase capacity and facilitate product marketing.

In stage 2, the companies will expand the refinery and add technology to improve efficiency and to convert fuel oil into high-value products that meet international quality standards.

Completion of both stages is expected by 2013.

The Ras Lanuf complex produces naphtha, kerosene, and light and heavy gas oil, as well as ethylene and polyethylene.

Company News

Libya's stake in Eni

Italian officials approved in December 2008 Libya's decision to buy a stake of up to 10% in Eni SPA (Rome).

As part of the deal, Rome agreed to pay compensation totaling a reported $5 billion over 25 years for its period of colonial rule in Libya, official sources said.

The Italian government owns about 30% of Eni.

Eni is the largest and longest established foreign oil company in Libya. It extracts more than half of Libya's production of some 1.7 MMb/d. Italy also is the largest buyer of Libyan oil.

MADAGASCAR

Capital: Antananarivo
Monetary unit: Malagasy franc
Refining capacity: 0
Oil production: 0
Oil reserves: 0
Gas reserves: 0

Tsimiroro field

Madagascar Oil SA (a subsidiary of Madagascar Oil Ltd., Bermuda) in late 2008 increased its estimate of reserves contained in its heavy oil Tsimiroro field.

The field is in the onshore Monrondava Basin on Block 3104.

The company raised its estimate by 30%, to 1.7 billion bbl of oil.

Tsimiroro is one of two heavy oil projects that the company is developing in the country.

The company said that 12 of 17 exploration wells drilled within the Tsimiroro Block have produced discoveries, indicating oil-in-place of 900 million to 1.4 billion bbl.

First oil (just 45 b/d) flowed in early 2008 from a steam-assisted gravity drainage pilot project. The firm called it the first oil production in the country in more than 60 years.

Although not a large volume, Madagascar Oil said the initial production proves that the field is technically exploitable.

Bemolanga heavy oil license

Total SA (Paris) signed a farm-in agreement in mid-2008 with Madagascar Oil SA to operate the Bemolanga heavy oil license with a 60% interest.

The partners will use mining technology to develop the estimated 10 billion bbl of heavy oil in place.

The onshore license area covers 16,835 sq km (6,500 sq mi).

Total has previously developed heavy oil projects in Venezuela and Canada.

Block 1101

Candax Energy Inc. (Calgary) reported in mid-2008 the start of its 2D seismic acquisition program over onshore Block 1101.

The program will acquire up to 228 km (142 mi) of data across the Ankaramy and Ambilobe areas.

The results will guide Candax and its partners in planning for a first exploration well in 2009.

Blocks 3103, 3110, 3115

Essar Exploration & Production (EEPL; Mumbai) reported in mid-2008 that it had started airborne survey work on onshore exploration blocks 3103, 3110, and 3115.

The firm will acquire 12,000 km (7,456 mi) of airborne gravity and magnetic data to aid interpretation of the subsurface geology.

EEPL called it the largest airborne survey ever conducted in Madagascar to facilitate oil and gas exploration by any operator.

MALI

Capital: Bamako
Monetary unit: CFA franc
Refining capacity: 0
Oil production: 0
Oil reserves: N/A
Gas reserves: N/A

Blocks 7, 11

Heritage Oil Corp. (Channel Islands, Jersey, UK) completed in early 2008 its farm-in into two onshore exploration blocks and became their operator.

Heritage obtained its stakes in Blocks 7 and 11 from Mali Oil Development, a subsidiary of Centric Energy Corp. (London).

The two blocks cover more than 72,000 sq km (27,800 sq mi) in the Gao Graben, which is considered geologically similar to the Albert Basin in Uganda and the Muglad Basin in Sudan..

Near-term work programs on both blocks include acquisition of 2D seismic and drilling of one exploration well.

Victoria O&G and Falcon Petroleum

Victoria Oil & Gas PLC (London) agreed in early 2008 to a 12-month option to acquire Falcon Petroleum Ltd. (Lagos, Nigeria) and its exploration prospects in Mali.

Falcon has a production sharing agreement for 50% of the 20,000-sq km (7,722-sq mi) Block 17 in the Nara Trough in western Mali. The project is in a relatively unexplored but highly prospective region.

MAURITANIA

Capital: Nouakchott
Monetary unit: Ouguiya
Refining capacity: 0
Oil production: 0
Oil reserves: 100.000 million bbl
Gas reserves: 1.000 tcf

Banda NW well

Premier Oil PLC (London) said in early 2008 that the Banda NW appraisal well in offshore PSC B encountered promising gas sands.

The well was sidetracked into PSC A, reaching 2,727 m (8,947 ft) TD.

The oil and gas intervals found were similar to those in the discovery well Banda-1, nearby to the east.

Logging data indicated that the sidetrack well intersected an 85-m (279-ft) gross gas column and a 15-m (49-ft) gross oil column.

Banda East-1B well

Affiliates of state-owned Petronas (Kuala Lumpur) cut gas and oil columns in late 2008 at the Banda East-1B appraisal well, offshore in the Atlantic. Banda oil and gas field spans PSC Areas A and B.

The well intersected 22 m (72 ft) of net gas pay and 7 m (23 ft) of net oil pay. TD was 2,716 m (8,911 ft).

The well, in 300 m (984 ft) of water, was to be cased and tested for gas deliverability.

In July 2008, Petronas said that a well drilled 2 km (1 mi) away from its original Banda-1 discovery well confirmed the existence of "significant" quantities of oil and gas.

Further exploratory work was being planned to determine the overall size of the reservoir, but Petronas said gas resources could exceed 28.3 bcm (1 tcf).

Coup threatens oil, gas industry

A coup by Army commanders in mid-2008 created additional uncertainty for Mauritania's nascent oil and gas industry, said *Oil & Gas Journal.*

The industry already was facing threats from the al-Qaeda terrorist organization.

Four top military leaders, after being fired by Mauritania's president for their criticism of him, seized the president and held him at the presidential palace.

They then announced formation of a new "state council," led by one of the fired generals.

In July, al-Qaeda's North Africa network said it planned to attack interests held by the United States, which it said was establishing military bases and seeking control of the region's energy sources.

MOROCCO

Capital: Rabat
Monetary unit: Dirham
Refining capacity: 154,901 b/cd
Oil production: 200 b/d
Oil reserves: 0.752 million bbl
Gas reserves: 53.000 bcf

Ouled N'Zala gas well

Circle Oil PLC (Wokingham, UK) tested 94,025 cu m/d (3.32 MMscfd) of natural gas in late 2008 from exploration well ONZ-6 on the Ouled N'Zala permit.

The permit lies northeast of Rabat in the Rharb basin in Morocco.

ONZ6 was the first successful well for Circle in the country.

The earlier ONZ-4 well was already on continuous production at 1.5 MMcfd.

The company later drilled another exploration well on the Sebou permit, in the same general area as the Ouled N'Zala permit.

That CGD-9 exploration well sustained 250,920 cu m/d (8.86 MMcfd) of gas from the Lower Guebbas formation and was being readied for an extended production test.

Essaouira Shallow, Deep Offshore areas

Canamens Energy Ltd. (London) was embarking in early 2009 on a round of exploration in two areas offshore Morocco.

The action followed the award of contracts from and agreements with Morocco's Office National Hydrocarbures et des Mines (ONHYM).

Canamens plans first to reprocess and acquire new 2D seismic data in the Essaouira Shallow Offshore, then to determine whether to convert the license into an exploration permit.

In the Essaouira Deep Offshore area—with water deeper than 500 m (1,640 ft)—Canamens plans to reprocess and acquire new 2D data, with an option to extend into a 3D and drilling commitment.

The acreages totals more than 11,000 sq km (4,247 sq mi).

Jorf Lasfar refinery

In mid-2008, Abu-Dhabi's state-owned energy investment firm International Petroleum Investment Co. (IPIC) approved a plan to build a refinery at Jorf Lasfar in Morocco.

A company to administer the project was being set up with Moroccan partners, not named.

The Moroccan energy ministry had announced plans in mid-2007 for a 10-million-tonne, $3-billion refinery at Jorf Lasfar, which would be the country's second facility. Completion by 2015–16 was proposed.

Negotiations were said to be under way with Spanish, UAE, and Kuwaiti companies and other firms.

MOZAMBIQUE

Capital: Maputo
Monetary unit: Metical
Refining capacity: 0
Oil production: 0
Oil reserves: 0
Gas reserves: 4.500 tcf

Mozambique has three gas discoveries—Temane, Pande, and Buzi—all near the coast in the southern part of the country.

Gas deliveries began in early 2004 from the Temane fields. This is Mozambique's only hydrocarbon production.

Temane-to-Secunda pipeline

In early 2008, Foster Wheeler South Africa Pty Ltd. (Midrand) won a contract from The Republic of Mozambique Pipeline Investments Co. Pty Ltd. (Rompco) for a pipeline expansion project.

Foster Wheeler will increase the capacity of Rompco's pipeline from Temane gas field in Mozambique to the coal-to-liquids complex of Sasol (Johannesburg) in Secunda, South Africa.

A new compressor station with two new compressors will be added to the line, which extends for 865 km (537 mi), including 340 km (211 mi) in South Africa.

Mozambique's Instituto Nacional de Petroleos estimates Temane's reserves at 1 tcf, with a planned 25-year production life from initial start-up in early 2004.

NIGERIA

Capital: Abuja
Monetary unit: Naira
Refining capacity: 505,000 b/cd
Oil production: 1.940 million b/d
Oil reserves: 36.220 billion bbl
Gas reserves: 184.160 tcf

Overview

Nigeria is the largest oil exporter in Africa. At the end of 2008, the country was exporting an estimated 1.88 million b/d.

The country has several LNG projects on tap for monetizing primarily associated gas. Development of deepwater resources also continues and is expected to add more than 1 million b/d of oil production in the next few years.

Legislation also has been introduced to eliminate gas flaring by yearend 2008, though security issues have slowed progress. Flaring has amounted to an estimated 3.5 billion boe from 1982–2008.

The administration is aiming to achieve better balance between domestic gas and export projects.

The country will expand gas-fired domestic electric power generation to nearly 15 GW by 2012, from none in 2007. Gas use for power would expand to more than 170 MMcmd (6 bcfd) by 2011.

Gas demand for domestic industrial use also would grow to more than 170 MMcmd (6 bcfd) by 2013.

For export, by both pipeline and LNG, gas demand is expected to reach more than 311.5 MMcmd (11 bcfd) by 2014 from only 113.3 MMcmd (4 bcfd) in 2007.

To meet this level, NLNG Train 7 will come on line in 2012, and both Brass LNG Train 1 and OKLNG Trains 1–4 are to start up by 2013. Combined peak production is expected to reach 311.5 MMcmd (11 bcfd) in 2014.

Combined natural gas domestic and export demand are expected to peak at more than 453 MMcmd (16 bcfd) by 2013.

Nigeria relinquishes Bakassi

In August 2008, Nigeria ceded control of the potentially oil-rich Bakassi Peninsula to Cameroon.

Bakassi is a 1,600-km (995-mi) border area that juts into the Gulf of Guinea.

Details are given in the Cameroon section of this chapter.

Attacks at Addax facility

Addax Petroleum Corp. (Calgary) reported in early June 2008 that militants in southern Nigeria ambushed a naval vessel guarding an offshore concession operated by the firm.

One person was killed and several wounded.

The attack was on Oil Mining License (OML) 126, located 90 km (56 mi) south of Port Harcourt, which produced about 41,250 b/d in 2007.

Shut-in production

Industry reports said that more than 600,000 b/d of Nigerian crude production remained shut in in early June 2008 due to militant assaults, amounting to some 20–25% of the country's typical oil production.

Streamlined energy ministry

Nigerian Pres. Umaru Yar'Adua re-shuffled his cabinet in November 2008 in conjunction with other restructuring related to the petroleum sector.

The Ministry of Petroleum Resources had been split earlier in the year into power, gas, and petroleum sections, each with its own minister. At the same time it was renamed the Energy Ministry.

However, Pres. Yar'Adua chose to again consolidate the three divisions under the banner Ministry of Petroleum Resources.

Other action created a new ministry for the Niger Delta to coordinate a response to attacks by militants in that area on oil company facilities. It also will create environmental and youth empowerment policy initiatives.

Exploration and Development

Ofon field

A subsidiary of Total SA (Paris) discovered oil in late 2008 near its producing Ofon oil field in southeastern OML 102 offshore southeastern Nigeria.

Violence in Niger Delta

Fighting continued in 2008 between the government and a rebel group called the Movement for the Emancipation of the Niger Delta (MEND). The group seeks to localize control of Nigeria's oil and to secure reparations from the federal government for pollution caused by the oil industry.

Military crackdown

A Nigerian general pledged in September 2008 to "break the will" of armed groups responsible for sabotage against the oil industry in the Niger Delta by launching more raids on their hideouts.

The delta saw the most intense series of attacks against the oil industry since early 2006 that month. Militants blew up pipelines in retaliation for a military raid on one of their camps.

Observers feared that the military escalation could lead to wider fighting in the region or unite disparate militant factions.

Attack at Alakiri

Also in September, Royal Dutch Shell PLC (The Hague, Netherlands) reported a major attack on its Alakiri flow station, gas plant, and field logistics base southwest of Port Harcourt.

The militant MEND group threatened to burn the facilities to the ground. They called the attacks the first stage of an "oil war" of sabotage on oil and gas installations.

Pipeline attacks

In May, MEND claimed credit for an attack on a Shell facility near Awoba flow station. A pipeline was blown up and, in an ensuing gunfight, several Nigerian soldiers were reported killed.

In August, Shell declared *force majeure* on its export commitments at its Bonny Island terminal after militants damaged its Nembe oil trunk pipeline in the Niger Delta.

The Etisong-1 well went to TD 2,207 m (7,240 ft) in 70 m (230 ft) of water and tested at more than 6,000 b/d.

Total said it was exploring the feasibility of a new development pole on OML 102 that would combine production from the Etisong main discovery and as-yet-undrilled surrounding structures.

Total E&P Nigeria said it plans to start production from Akpo deepwater oil field on OML 130 in early 2009. The company also was studying development of the Egina oil field on the same block.

Total's offshore operated production in Nigeria comes from OML 99, 100, and 102 in joint venture with the Nigerian National Petroleum Corp.

The main fields are Amenam-Kpono, Ofon, and Odudu area fields.

First 4D seismic

Nigeria's first 4D ("time-lapse") seismic acquisition in deepwater oil and gas operations was completed in late 2008 at Bonga field, 120 km (75 mi) southwest of the Niger Delta, according to operator Shell Nigeria Exploration and Production Co. Ltd. (SNEPCo).

The results will guide efforts to maximize the field's lifecycle production.

Pirate attacks

Nigeria has become the world's leader in piracy attacks, primarily targeting the oil industry in that country, reported the International Maritime Bureau (London) in mid-2008.

Nigeria accounted for 10 of the 49 pirate attacks registered worldwide in first quarter 2008. It marked the first time in 16 years of reporting that another country has surpassed Indonesia for that position.

The most dangerous zones are around Lagos and in the southern Niger Delta.

OML 13

In early 2008, Frontier Oil Ltd. (Lagos) spud the first of two planned appraisal-development wells on OML 13 at Uquo, an oil and gas discovery offshore in the eastern Niger Delta.

Uquo-5, projected to 2,591 m (8,500 ft), is to be followed by Uquo-6.

The company aims to start oil production in 2009 through the Qua Iboe terminal, 8 km (5 mi) south.

The discovery well tested both oil and gas in more than nine reservoirs.

Anambra basin

A joint venture of Afren PLC (London) and Global Energy Co. Ltd. (Lagos) announced plans for drilling in 2009 to appraise decades-old gas-condensate and gas and oil discoveries in the nonproducing Anambra basin.

Blocks of interest are OPL 907 and 917, totaling 3,500 sq km (1,351 sq mi).

The company refers to Anambra as Nigeria's second most prospective basin. It has a gas resource estimated at more than 5 tcf, 30 wells and six discoveries.

Kita Marine

Addax Petroleum Corp. (Calgary) appraised the Kita Marine oil discovery off southeastern Nigeria, raising the possibility of the first development and production in the northeastern part of OML 123.

The KTM-6 well, just southwest of the 2005 discovery well, cut 53 m (173 ft) of gross oil column in four zones. Tests indicated medium-gravity oil, consistent with the Antan blend produced from OML123.

Kita Marine is in 10 m (33 ft) of water near the Cameroon marine border and 10 km (6 mi) northeast of the North Oron/Oron West producing facility.

Usan field

Partners in Usan oil field off Nigeria were moving in early 2008 to develop the 2002 discovery, which lies in 750–850 m (2,460–2,789 ft) of water on OML 138.

The block is about 100 km (62 mi) off the eastern Niger Delta coast.

Development plans include an FPSO vessel with a storage capacity of 2 million bbl of oil.

First production is expected in 2011, with peak production of 180,000 b/d.

Co-venturers in the development are operator Elf Petroleum Nigeria Ltd.; Chevron Nigeria Deepwater Ltd.; Esso Exploration & Production Nigeria (Offshore East) Ltd.; and Nexen Petroleum Nigeria Ltd.

The government had approved conversion of Oil Prospecting License (OPL) 222 into OML 138 and 139, each covering half the original area of OPL 222.

Domestic gas market

Producers of natural gas in Nigeria will be expected to sell gas at an affordable price to the domestic market under a new pricing policy approved in early 2008 by Pres. Yar'Adua.

Gas producers also will be expected to allocate a share of their production and resources to the domestic market rather than export it.

Gas markets of particular interest are electric power generation and fertilizer and methanol production,

Drilling and Production

Agbami oil field

Star Deep Water Petroleum Ltd., a Nigerian affiliate of Chevron Corp. (San Ramon, CA), started oil production from offshore Agbami field in mid-2008.

First flow was achieved from an FPSO vessel. Initial production was expected to reach more than 100,000 b/d of oil.

Discoveries in Nigeria's Anambra Basin

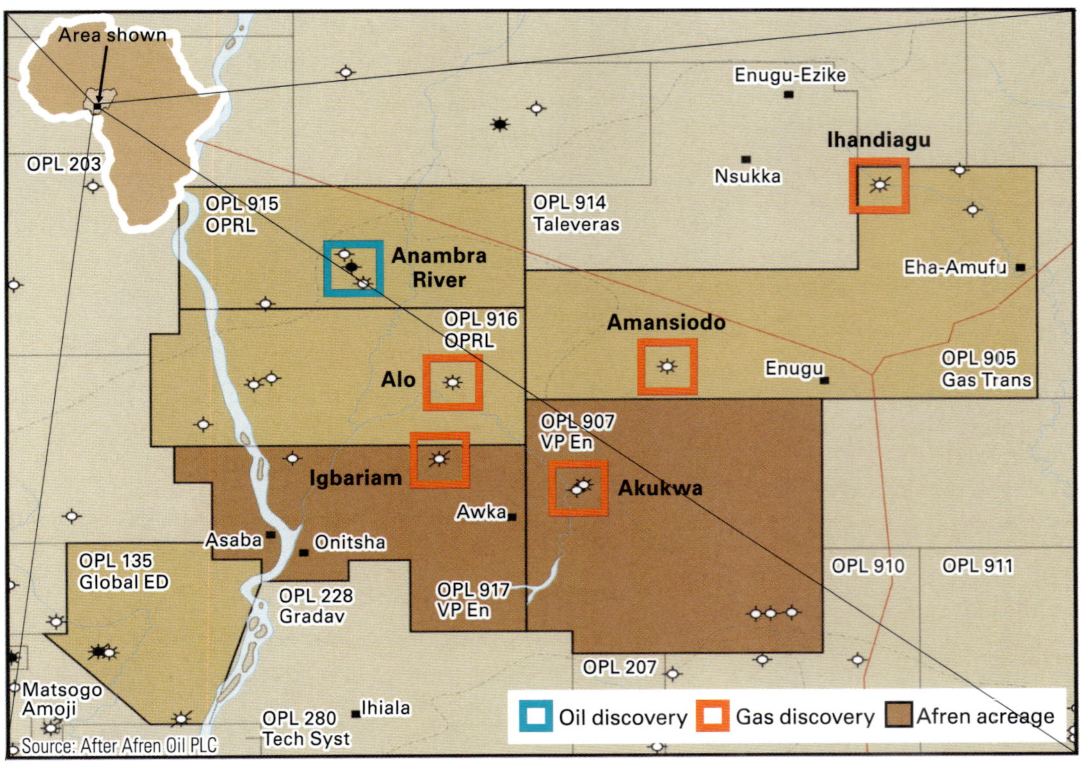

Source: After Afren Oil PLC

International interest in gas development

Portugal, Russia, China, and India all have expressed interest in helping Nigeria to develop its natural gas potential.

Nigeria wants to ramp up domestic gas use and develop export projects. It also has committed to eliminating gas flaring by the end of 2008.

The potential involvement of Gazprom (Moscow) raised fears in Europe and the West about that company's potential domination of gas supplies. Gazprom officials have identified Africa as one priority in plans to seek energy assets outside Russia.

Nigeria has been seeking foreign investment for projects such as power plants, liquefaction facilities, and the Trans Sahara gas pipeline, which would transport gas through Algeria to Europe.

Chevron expects production at the field—the largest deepwater discovery in Nigeria—to increase to 250,000 b/d of oil and natural gas liquids by yearend 2009.

The firm estimates that the field holds potentially recoverable volumes of 900 million bbl.

Shell in Ogoniland

As of mid-2008, Shell Petroleum Development Co. (SPDC) was still operating its facilities in Ogoniland in Rivers State, despite an earlier comment by Pres. Yar'Adua that another operator would replace Shell by yearend.

One issue driving the situation is the government's directive to end gas flaring in the Niger Delta by yearend 2008. Shell officials have said it would cost the company $5 billion to meet this target.

SPDC said it missed previous gas-flaring deadlines because JV partner, Nigeria National Petroleum Corp. (NNPC), failed to contribute its share of funding to set up gas-gathering infrastructure.

SPDC said it has spent $3 billion in "commissioning facilities to gather and process gas."

Okoro, Setu fields

Afren PLC (London) achieved first oil from the Okoro-Setu project in offshore OML 112 in mid-2008.

Initial production from the first of two wells drilled exceeded 3,000 b/d of oil, in line with expectations.

The remaining wells were to be batch drilled, with peak production of 15,000 b/d of oil expected by yearend 2008.

The Okoro and Setu fields have combined 2P reserves of 42 million bbl (management case) and 2P Independent Certified reserves of 32 million bbl (NSAI).

Setu is 7 km (4 mi) north of Okoro.

Milestone on OML 123

In July 2008, Addax Petroleum Corp. (Calgary) achieved 200 MMbbl of oil production from offshore OML 123. The company said it expects that the block will continue to be a major production hub for a several years.

Shell restructuring plans

In early 2008, Shell Petroleum Development Co. (SPDC) suspended plans to restructure its JV in Nigeria following a request from NNPC to resolve its oil production problems and improve efficiency.

A major driver behind the suspension was concern about continuing attacks on Shell's facilities in Nigeria by militants and vandals.

The development left in question the jobs of 5,000 employees, most of whom are Nigerians.

Shell had announced in late 2007 its plans to reduce costs by cutting jobs and to boost efficiency and productivity in its JV. Shell estimated that its actions could save $200 million/year.

JV partners include NNPC, Total SA (Paris), and Agip [part of Eni SPA (Rome)].

A Shell official said that the company used to produce 1 million b/d, but due to the Niger Delta crisis, was struggling to produce half that amount.

Ofrima North

Addax Petroleum Corp. (Calgary) said in early 2008 that a combination of discoveries at Ofrima North, 50 km (31 mi) south of Brass, justifies standalone development with an FPSO vessel and subsea tiebacks.

The company also sees potential for future gas exports from OML 137.

Production could start by late 2009.

In 2007, the Ofrima-2 discovery well cut 52 m (170 ft) of gross oil-bearing interval in the H42 reservoir and several gas-bearing intervals.

The later Ofrima-3 well, in the field's west fault block, cut a gross 22 m (72 ft) comprising deposits of light oil, rich gas, liquids-rich gas, and light oil.

In the main fault block, Ofrima-3A confirmed the western continuity of the H42 oil reservoir.

AJE field

Providence Resources PLC (Dublin) reported in mid-2008 that its offshore AJE-4 well on OML 113 evaluated additional deeper exploration targets, including the Albian, confirming the field's commercial potential.

AJE-4 was drilled in water 1,000 ft (305 m) deep to evaluate the lateral extent of the oil and gas bearing reservoirs that had been successfully tested in the AJE-1 and -2 wells.

It encountered Cretaceous gas condensate and an oil-bearing Turonian reservoir.

LOCATION OF AJE DISCOVERY

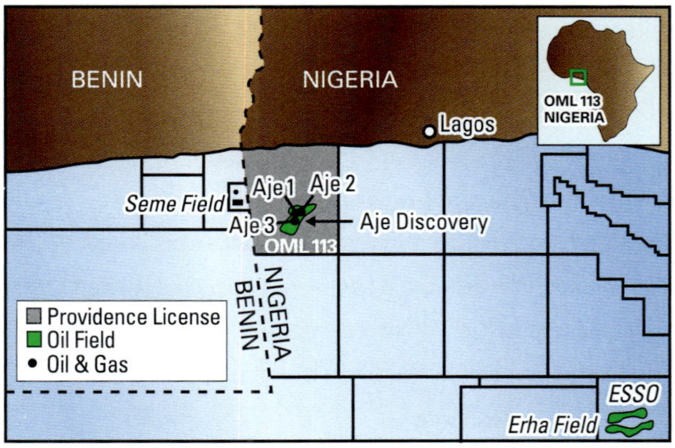

Gas Processing

NGL project

ExxonMobil Corp. unit Mobil Producing Nigeria began operating its $1.3-billion East Area Natural Gas Liquids II project on Bonny Island in mid-2008.

Located about 27 km (17 mi) off Nigeria, the project is expected to recover 275 million bbl of NGL from associated natural gas produced in East Area reservoirs on Blocks OML 67, 68, and 70.

In addition, the East Area NGL II project will produce at its peak about 50,000 b/d of NGL and ultimately recover 275 million bbl of NGL from about 950 MMscfd.

The NGL project is part of an integrated approach to reduce flaring in conjunction with the existing East Area Additional Oil Recovery project, which started up in mid-2006.

Pipelines

Gas pipeline plan

A Nigerian "gas master plan" under development for six years includes proposals to build three gas pipeline transmission systems and associated compressor stations.

North-south pipeline

A 1,135-km (705-mi) pipeline between Akwa-Ibom in Calabar state and Kaduna via Abuja would carry 85 MM cu m/d (3 bcfd) to eastern, central, and northern Nigeria. It also would carry about 57 MM cu m/d (2 bcfd) of gas for the proposed Trans-Sahara Gas Pipeline.

South Korean investors have committed $5 billion to building the 1,200-km (746-mi) Ajaokuta-Kano link. Pipeline operations are expected to start in 2013.

The government is seeking investors to construct the Calabar-Ajakuta leg.

Western pipeline

A 700-km (435-mi) pipeline network would link the existing Escravos-Lagos pipeline system to Shagamu with a 200-km (124-mi) offshore extension. The proposed capacity is 36.8 MMcmd (1.3 bcfd). The system would extend to Jebba with a spur serving the OK LNG plant.

Interconnector

A 200-km (124-mi) system would link the Obiafu central gas processing facility and Ajaokuta to deliver gas from the east to the western and north-south transmission systems.

It would have a capacity of 48.1 MMcmd (1.7 bcfd).

Refining

Warri refinery

In early 2008, NNPC was preparing to restart its 125,000-b/d refinery in Warri. Thee refinery had been idle since early 2007.

Vandalism had halted feedstock transport, and $100 million in repair work was required.

Refinery legislation

The government was proceeding with plans in early 2008 for legislation to require international oil companies working in the country to refine a percentage of their crude oil production there.

The move is seen as a way to reduce dependence on imported fuel.

Nigeria can produce only 445,000 b/d of oil products, while demand is currently around 600,000 b/d, said officials with the NNPC.

The country has four state-owned refineries, and NNPC hopes to build two more, each with a capacity of 200,000–300,000 b/d. Incentives will be offered to encourage IOCs to take a share in the facilities.

In August 2007, the Department for Petroleum Resources said it had received 26 applications from private companies wishing to build refineries in Nigeria.

New refinery planned

ONGC Mittal Energy Limited (OMEL) (Nicosia, Cyprus) is a 51:49 joint venture between ONGC Videsh Limited (OVL) (New Delhi) and Mittal Investments Sarl (MIS) (Luxembourg).

The JV was moving forward in late 2008 with plans to construct a 192,000-b/d refinery in Nigeria.

Nexant Inc. (San Francisco) was performing a feasibility study, to be completed by yearend, for the proposed $4-billion refinery.

LNG

Exports cut

In late 2008, Nigeria was forced to cut exports of LNG after Royal Dutch Shell (The Hague, Netherlands) closed the Soku gas plant in the Niger Delta to repair damage caused by thieves.

Shell Petroleum Development Company (SPDC) said it had been forced to temporarily shut down the facility to repair damage caused by vessels illegally siphoning fuel from connecting pipelines.

Illegal connections raised safety risks that required urgent repairs.

Soku supplies about 40% of the gas used by NLNG, Nigeria's sole LNG export facility.

Much of the gas from NLNG, on Bonny Island, goes to terminals in the United States and Europe, but some also goes to India and Japan.

Train 6 at the facility was completed in January 2008. It has a design capacity of 4 mtpy. Train 7 is scheduled to come on line in 2013.

New plant studied

A consortium including Afren PLC (London), E.On Ruhrgas AG (Dusseldorf), and African LNG Holdings Ltd. announced in early 2008 that it will assess the feasibility of building an LNG plant and related gas facilities in the Anambra basin in southeastern Nigeria.

Goals are to reduce gas flaring and ensure that domestic demand can be satisfied and export commitments met.

LNG shipping firm

NNPC established a new shipping company, Nikorma Shipping Services Ltd., in early 2008 to handle the additional volume of Nigerian LNG exports expected in the coming years.

NNPC holds a 51% share, but was planning to reduce this to 30% and to offer remaining shares to private Nigerian investors.

Nigeria's LNG exports are projected to rise to 60 mtpy by 2013 from the 2008 level of 17.9 mtpy.

Alternative Fuels

Fuel ethanol factories

China National Heavy Machinery Corporation (CNHMC; Beijing) started construction in the third quarter 2008 of a $70-million fuel ethanol factory in Iyemero, in southern Nigeria.

The factory is designed to produce 180,000 l./d (47,520 gal/d) of ethanol from locally grown cassava.

Nigeria's Crownek Green Energy Ltd. is a partner with CNHMC.

Also, Global Biofuels Ltd. (Lagos) has opened a 49-acre farm for commercial production of sweet sorghum in Arigidi, Ondo State.

The company plans to expand the farm to 7,400 acres in 2009 and to 24,700 acres later.

The ethanol factory in Ondo has a 90,000-l./d (23,760 gal/d) production capacity.

Environment

Flaring reduction threatened

News reports in late 2008 indicated that lower oil and gas prices and the global credit crisis were threatening Nigeria's December 2008 deadline for developing its natural gas anti-flaring program.

Delays would cause the country to continue losing money due to lost gas sales.

Shell estimates that about half of the 2 bcfd of associated gas produced in Nigeria is flared.

The new industry strategy is to collect the associated gas and process it into LNG for export and for domestic industrial use and into natural gas liquids (NGL) for cooking.

SOUTH AFRICA

Capital: Cape Town
Monetary unit: Rand
Refining capacity: 485,297 b/cd
Oil production: 14,000 b/d
Oil reserves: 15.000 million bbl
Gas reserves: 0 tcf

Outside West Africa

South Africa's outlook for 2008 and beyond is promising, with a doubling of onstream fields. The area saw six fields for 2003 through 2007 and estimates show 12 fields from 2008–2013. Libya has seen a four-fold increase in fields, from two to eight. The count for countries outside West Africa from 2003 through 2007 was 33, while the estimate for 2008 through 2013 calls for 49 new fields.

Country	2003–07	2008–13
Egypt (Mediterranean)	13	36
Republic of South Africa	6	12
Tunisia	3	14
Egypt	7	5
Libya	2	8
Mauritania	1	4
Tanzania	1	1
Others	0	2
Grand Total	33	82

Exploration prospects

In 2008, the Petroleum Agency SA (PASA; Cape Town) was evaluating projects covering about 1.1 million sq km (425,000 sq mi) offshore over Orange, Outeniqua, and Durban Zululand basins.

PASA provides prospecting maps and data in anticipation of exploration opportunities.

Project Mthombo refinery

The Petroleum, Oil and Gas Corporation of South Africa Pty Ltd. (PetroSA; Cape Town) let a feasibility and FEED contract in December 2008 to KBR, Inc.(Houston) for its 400,000-b/d Project Mthombo refinery in the Coega industrial development zone outside Port Elizabeth.

A final investment decision is slated for late 2010. If the project proceeds, operation of the $11-billion refinery, which will be Africa's largest, is expected to start in 2014.

The refinery will help meet the nation's fast-growing demand for refined fuels. Diesel use is forecast to grow at 6% and gasoline at 2% per year during 2009–20.

PetroSA said that without the refinery investment, South Africa would have to import 10 billion l. (2.64 billion gal) of fuel/year by 2015, about 20% of the national requirement.

Project Mafutha CTL plant

Sasol Technology (Pty.) Ltd. hired Foster Wheeler South Africa (Pty.) Ltd. to conduct a prefeasibility study for the Project Mafutha coal-to-liquids (CTL) plant in South Africa.

The greenfield CTL facility would produce 80,000 b/d of synthetic fuel using Sasol's proprietary low- and high-temperature Fischer-Tropsch processes as well as advanced refinery technology to convert the Fischer-Tropsch products into fuel products.

A location for the plant was still under study in April 2008. Some reports said the northernmost province of Limpopo was under consideration.

SUDAN

Capital: Khartoum
Monetary unit: Pound
Refining capacity: 121,700 b/cd
Oil production: 490,000 b/d
Oil reserves: 5.000 billion bbl
Gas reserves: 3.000 tcf

New Block: Ea

Sudan's National Petroleum Commission (NPC) mapped a new oil concession in the southern region of the country in early 2008, naming it Block Ea. The action followed an application for the acreage by privately owned H Oil Group (Limassol, Cyprus).

Ea Block adjoins Blocks B and C as well as Block 5A of Total SA (Paris). Those three blocks are the site of Thar Jath oil field, 900 km (559 mi) south of Khartoum in the Muglad basin.

Sudan signed an agreement in 2005 with White Nile Petroleum Operating Co. for development of Thar Jath and Mala fields on Block 5A.

White Nile: Change of direction

Oil/gas explorer White Nile Ltd. (Guernsey, Channel Islands, UK) decided in late 2008 to abandon hydrocarbon development efforts in Sudan.

Sudan unrest continues

Political and military unrest continued to plague the Darfur region of Sudan in 2008.

Darfur. The UN Security Council asked the International Criminal Court (ICC; The Hague, Netherlands) in March 2005 to investigate the violence in the western region of Darfur. The United States government has called the violence genocide; Khartoum denies the charge.

Some experts estimate that 200,000 people have been killed and 2.5 million driven from their homes in Darfur since 2003, when rebels took up arms against the government, charging it with neglect.

Khartoum disputes those figures, has argued that the ICC has no jurisdiction over its nationals, and has said that it would not allow any of its citizens, including rebels, to be tried outside Sudan.

In May 2007, the ICC decided to issue arrest warrants against a Sudanese minister and a Janjaweed leader, both charged with war crimes and crimes against humanity for actions in Darfur in 2003–2004.

In late December 2007, a 9,000-member UN peacekeeping force, including African Union troops, arrived in Darfur. By late 2008, the UN force numbered 10,000.

In July 2008, the chief prosecutor of the ICC asked judges to issue an arrest warrant for Sudanese Pres. Omar Hassan al-Bashir. The prosecutor accused him of orchestrating genocide in Darfur.

In early 2009, the ICC was expected to issue a formal arrest warrant in March for al-Bashir. It would be the first such action by the court against a sitting head of state.

Civil war. January 2005 brought a settlement of Sudan's 20-year civil war between the Arab and Muslim north and the largely black African Christian south.

The peace accord calls for a referendum in 2011 on independence for the south.

Western oil companies pulled out of the country because of human rights abuses during the conflict. After the peace settlement, Chinese, Malaysian and Indian companies moved into Sudan's oil sector and now dominate it.

While its two largest oil fields are in the south, Sudan has begun to sell exploration concessions in the Darfur and northern areas.

In mid-2008, a leader of Sudan's rebel group, Justice and Equality Movement (JEM) said that threats would increase for energy installations and their personnel in Sudan, particularly Chinese oil companies in the southwest.

JEM rebels attacked Khartoum, the capital city, in May 2008.

In late 2008, the Sudanese army reinforced its presence in South Kordofan, an oil-rich region bordering Darfur. Southern Sudan officials said the action violated the January 2005 civil war settlement.

The company said that the economic environment and political situation in southern Sudan were no longer conducive to funding nonproducing, early-stage oil and gas exploration.

The firm said it planned to change its name to Agriterra and to focus on investment in African agriculture.

The firm had been struggling to clarify its exploration rights on Block Ba, and concluded that it was unlikely to do so by January 2011, when the region holds a referendum on total independence.

Port Sudan refinery

Rising costs forced Petronas (Kuala Lumpur) to defer plans in mid-2008 for its 100,000-b/d refinery project at Port Sudan on the Red Sea.

The export-oriented refinery, which would be owned 50:50 by Sudan and Petronas, was due to come on stream in 2009.

The estimated cost for the refinery soared to $5 billion from the originally estimated $2 billion.

TANZANIA

Capital: Dodoma
Monetary unit: Shilling
Refining capacity: 14,900 b/cd
Oil production: 0
Oil reserves: 0
Gas reserves: 230.0 bcf

Exploration and Development

Nyuni Block

The Kiliwani North-1 natural gas discovery of Key Petroleum Ltd. (West Perth, Western Australia) and partners was declared a commercial development option in mid-2008.

Flow tests confirmed a flow of 1.13 MM cu m/day (40 MMcfd).

Key also said that the Nyuni Block that surrounds the discovery has prospects of up to 14.2 bcm (500 bcf) of gas in place. Total undiscovered gas in the block could be more than 56.6 bcm (2 tcf).

A pipeline is planned to connect Kiliwani North-1 with a gas processing plant in neighboring Songo Songo field.

South Block

In early 2008, the government granted exclusive rights to Beach Petroleum Ltd. (Adelaide) to negotiate to explore onshore and the southern half of Lake Tanganyika in western Tanzania.

The South Block covers 5,400 sq km (2,084 sq mi) and forms part of the 6,400-km-long (3,977-mi) East African Rift system as does Lake Albert.

Beach was making plans to negotiate a production-sharing agreement.

Songo-Songo West

Aminex PLC (London) was awarded a Production Sharing Agreement in mid-2008 for an area called Songo-Songo West.

That area is adjacent to the western boundary of the producing Songo-Songo gas field and close to Aminex's recent gas discovery at Kiliwani North.

Songo-Songo West covers 504 sq km (195 sq mi) and has comprehensive, existing seismic coverage. The PSA calls for drilling one firm well in the first three-year period of the 11-year agreement.

TUNISIA

Capital: Tunis
Monetary unit: Dinar
Refining capacity: 34,000 b/cd
Oil production: 85,000 b/d
Oil reserves: 425.00 million bbl
Gas reserves: 2.300 tcf

Exploration and Development

7th of November Block

In late 2008, Joint Oil awarded the 7th of November block in the southern Mediterranean to Canadian Superior Energy Inc. (Calgary). The block overlaps Libyan and Tunisian offshore territory.

Canadian Superior will operate the block, 121 km (75 mi) offshore in 76–114 m (250–375 ft) of water.

Details are given in the Libya section of this chapter.

Pantelleria

Kairiki Energy Ltd. (West Perth, Western Australia) acquired offshore exploration permit G.R15.PU (Pantelleria) in late 2008 in the Mediterranean.

The block is contiguous to the company's recently acquired Kerkouane permit

The firm noted the recent success of Eni SPA (Rome) offshore Sicily and the estimates of 18 bcm (636 bcf) of recoverable gas resources in the Sicily Channel region.

Chouech Essaida, Ech Chouech

In mid-2008, Winstar Resources Ltd. (Calgary) finished shooting seismic covering the Chouech Essaida and Ech Chouech concessions and was proceeding with plans to drill two wells by yearend.

An 80-km (50-mi) pipeline to move 425,000 cu m/day (15 MMscfd) of gas from Chouech Essaida to the El Borma field sales point is to be in service by early 2009.

Didon North prospect

In late 2008, PA Resources AB (Stockholm) made an oil discovery at the Didon North prospect in the offshore Zarat permit, in the Gulf of Gabes.

The DN-1 exploration well was drilled to TD 3,170 m (10,400 ft) and encountered 14 m (46 ft) of oil pay in the El Gueria formation.

The well was drilled to test both the El Gueria formation and the deeper Chouabine formation.

The company was analyzing well data to decide on next steps.

Didon field

PA Resources reported on its high level of activity in 2008 on Didon offshore field.

Didon is 75 km (47 mi) west of the Tunisia-Libya border in 75 m (246 ft) of water. Production from the field averaged 20,000 b/d in early 2008.

Early in the year, the company spudded Didon-8, aiming to test the field's western compartment.

PA also completed the 4,015-m (13,172-ft) Didon-7 horizontal production well in first quarter 2008, which is now producing 12,500 b/d of oil.

In mid-year, PA spudded horizontal oil producer Didon-9, which was also probing the western compartment.

Production from the firm's Didon-5 well was expected to start in third quarter 2008.

Drilling and Production

Chergui field

Petrofac Energy Developments Ltd. (London) began natural gas and condensate production in mid-2008 from Chergui field on Kerkennah Island.

The $100-million development produced first commercial gas from two wells.

Tunisia's state gas and electricity company is buying the output. A 57-km (35-mi) pipeline carries the gas to shore at the utility's plant at Ain Turkia.

Production plateau rates of 556,400 cu m/day (20 MMscfd) are expected for at least four years.

El Bibane field

Candax Energy Inc. (Toronto) and partner PA Resources reported in early 2008 the start of production from the EBB-4 well on offshore El Bibane field.

Initial production was 1,400 b/d of oil, but Candax expected that level to rise considerably when production from EBB-4 is optimized and the EBB-3 well is brought on stream.

EBB-3, the second production well in the El Bibane redevelopment project, was completed shortly before EBB-4 production started.

UGANDA

Capital: Kampala
Monetary unit: Ugandan shilling
Refining capacity: 0
Oil production: 0
Oil reserves: 0
Gas reserves: 0

Block 1: Giraffe well

In January 2009, Heritage Oil Corp. (Jersey, Channel Islands, UK) announced its largest strike to date at the Giraffe well in Block 1.

The Giraffe well appears to be an extension of the large Buffalo oil well struck in December 2008, indicating a connected reservoir with reserves of more than 400 million bbl.

It is also possible, said Heritage, that the Buffalo-Giraffe reservoir could extend further to the north and east, which could push total reserves in Block 1 alone to more than 1 billion bbl.

Oil also has been struck in the other Albert Basin fields: Block 2, owned and operated by Tullow; and Block 3A, operated by Heritage with ownership split between Heritage and Tullow Oil PLC (London).

Kingfisher field

Tullow Oil PLC said in late 2008 that drilling at its Kingfisher field in Block 3A showed the field is substantially larger than previously thought.

The firm said it considered Kingfisher, around Lake Albert, the largest light oil discovery in east Africa.

The areal extent of the field's structure is estimated at 45 sq km (17 sq mi), based on a 3D seismic survey in mid-2007.

Butiaba: Kigogole-1 well

In late 2008, Tullow Oil PLC said its Kigogole-1 exploration well in the Butiaba region along Lake Victoria found light, sweet oil in reservoirs just below 400 m (1,312 ft).

The company said this is the shallowest section in which oil has been encountered in Uganda.

The well, on Block 2, went to TD 616 m (2,021 ft) and encountered two oil zones with a net pay of 10 m (33 ft).

Kigogle-1, about 10 km (6 mi) northeast of the Kasamene-1 discovery, is the third test on the Victoria Nile Delta play in the Lake Albert Rift basin.

Several adjacent prospects will be tested during Tullow's 2009 Butiaba drilling campaign.

Butiaba: Taitai-1 well

Tullow Oil PLC reported in mid-2008 that the Taitai-1 oil discovery in the Butiaba area of Block 2 cut 5 m (16 ft) of net gas pay and at least 8 m (26 ft) of net oil pay.

The well was drilled to a TD 1,006 m (3,300 ft). Tests confirmed the presence of a potential oil column as thick as 80 m (262 ft) in sands above the basement play.

Tullow was making plans to sidetrack the well and core the oil-bearing zones to collect further data.

Block 2

Tullow Oil PLC discovered oil and gas in mid-2008 within the Ngege-1 exploration well on the northern part of Block 2 in the Lake Albert Rift basin.

Ngege-1, located 41 km (25 mi) northeast of the Taitai-1 discovery well, reached TD of 640 m (2,100 ft). It found more than 5 m (16 ft) of net oil pay and dry gas in 9 m (30 ft) of net pay through separate sands.

The Victoria Nile Delta play is in a new fairway along the axis of the Lake Albert Rift basin.

UGANDA'S LAKE ALBERT RIFT BASIN

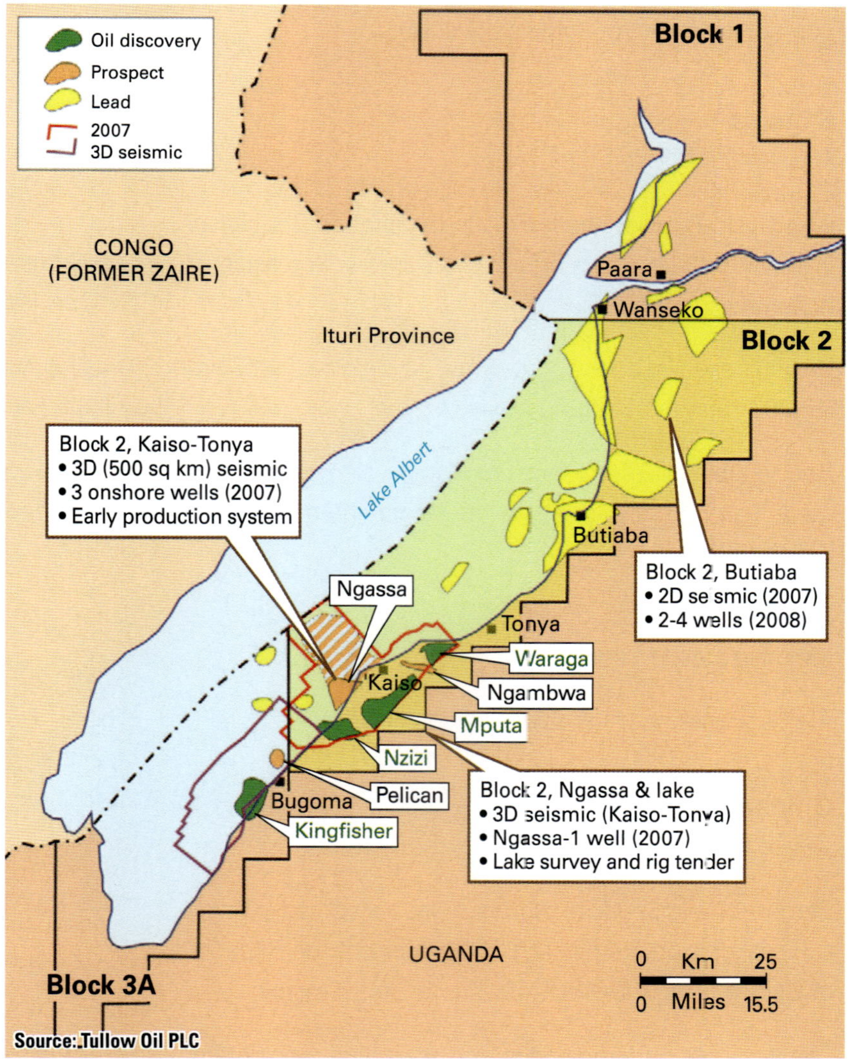

Source: Tullow Oil PLC

New oil pipeline?

Asian and Middle East backers have offered to fund and build a 1,300-km (808-mi) pipeline needed by Tullow Oil PLC (London) and Heritage Oil Corp. (Jersey, Channel Islands, UK) to exploit a large Ugandan oil find, said Heritage in early 2009.

Heritage calls the Albert basin the largest onshore oil discovery in sub-Saharan Africa in 20 years.

The pipeline from Uganda to Kenya's port of Mombasa could be attractive to Asian governments eager to put their surplus steel, idle engineers and uninvested cash to work.

Firms or agencies in China and Japan are the most probable backers, said Heritage.

Analysts say that it will take at least 400 million barrels of oil from remote Ugandan fields to justify the expensive export project.

Several other African nations saw exploration and development action in 2008 and early 2009.

Ethiopia

Capital: Addis Ababa

Geophysical survey work by Afar Exploration Co. (Tulsa) in early 2008 covered parts of the Danakil depression of nonproducing Ethiopia.

Results revealed an evaporite basin to depths of 4,572 m (15,000 ft) in the Danakil depression.

Recoverable oil—from two sandstone formations both more than 305 m (1,000 ft) thick—is expected to exceed 6 billion bbl.

The large Dallol salt dome, discovered in early 2007, is the main area of interest.

Afar was making plans to drill by yearend 2008.

Guinea

Capital: Conakry

Bergen Oilfield Services (Norway) was scheduled to begin a 3D seismic program in late 2008, offshore the Republic of Guinea along the West African transform margin.

The surveys, in preparation for drilling of exploration wells, will cover up to 5,000 sq km (1,931 sq mi).

The contract area already has at least 40 discoveries and an estimated 8 billion boe of reserves.

Namibia

Capital: Windhoek

Tower Resources PLC (London) reported in early 2008 that seismic evaluation for License 0010 in northern offshore Namibia confirmed the presence of giant structures with significant oil-bearing probability.

The company estimated reserve potential at 10 billion bbl in three prospects, with large volumes of associated gas also present.

Tower was planning more extensive seismic evaluation of the three prospects in 2008–2009, aiming to get data to guide drilling of a first well by 2010.

Niger

Capital: Niamey

CNPC International (Tenere) Ltd. is a unit of the China National Petroleum Corp. (Beijing).

The firm was planning early 2008 to shoot seismic and drill the Facai-1 exploratory well by mid-year.

The goal was to test a Cretaceous play in the north-eastern Tenere rift basin.

The location of interest is 100 km (62 mi) north of the Saha-1 and Fachi West-1 wells drilled in 2007.

In late 2008, CNPC also launched construction of Niger's first refinery, about 50 km (31 mi) north of Zinder City. CNPC will hold a 60% stake in the 20,000-b/d facility.

It is expected to begin operation by 2011, processing crude from the Agadem oil block.

The Agadem Integrated Project also includes include exploration of the Agadem block (Niger's largest oil-exploration zone, with reserves of 324 million bbl); and construction of a 2,000-km (1,242-mi) pipeline.

Somaliland

Capital: Hargeisa

TGS-NOPEC Geophysical (Asker, Norway) acquired geophysical data in early 2008 in Somaliland (an autonomous region within Somalia).

The 5,100 km (3,169 mi) of 2D seismic, gravity, and magnetic data characterized both shallow and deep-water offshore areas.

The survey area is geologically analogous to Yemen, where several oil fields have been discovered.

WHO HOLDS ETHIOPIAN EXPLORATION BLOCKS

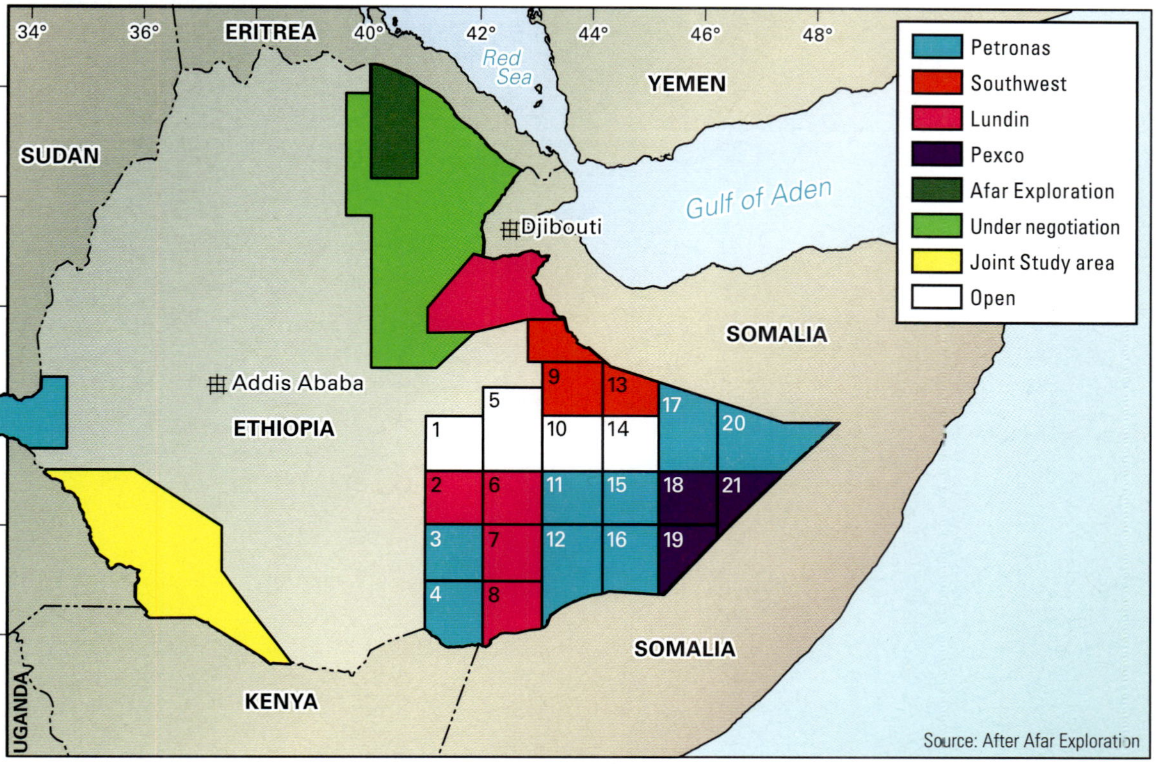

The government said it intended to conduct an open bid round after the data were made available to TGS clients in third quarter 2008.

Western Sahara (Saharawi Arab Democratic Republic: SADR)

Largest city: El-Aaiún

SADR extended the bidding date of its second licensing round to March 31, 2009.

The republic is offering three onshore and six Atlantic blocks in the underexplored Aaiun and Tindouf basins that total 48 million acres (194,249 sq km; 75,000 sq mi), in water up to 3,600 m (11,811 ft) deep.

Eight companies that acquired nine licenses in the first bid round in early 2006 are evaluating prospects.

Zambia

Capital: Lusaka

Zambia announced plans in early 2009 to invite bids to explore for oil in the country.

A mining ministry official told the *Financial Times* that parliament has approved a legal framework for exploration and that the government has held preliminary talks with oil companies "mainly in Europe and some in Africa."

State media quoted a senior mining ministry official in December 2008 as saying that 29 blocks had been demarcated for oil and gas prospecting.

Even if commercially viable oil were found, production would be at least five years away, analysts said.

Former Soviet Union

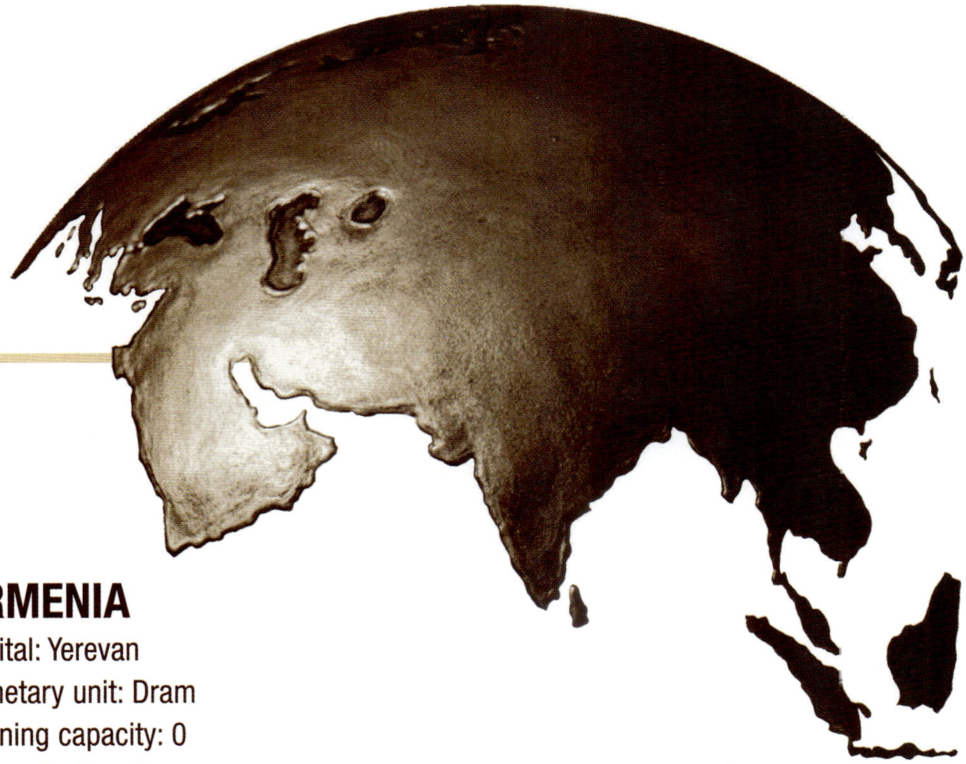

ARMENIA

Capital: Yerevan
Monetary unit: Dram
Refining capacity: 0
Oil production: 0
Oil reserves: 0
Gas reserves: 0

Nizhnekamsk refinery upgrade

In Tatarstan (northeast of Stalingrad), an upgrade program was under way in mid-2008 for the 120,500-b/d Nizhnekamsk refinery.

The facility, 250 km (155 mi) from the city of Kazan, will have a refining capacity of 7 mtpy.

After completion, expected in 2012, the unit will be renamed the OJSC Taneco refinery.

Central Depression evaluation

Blackstairs Energy PLC and Vangold Resources Ltd. (both in Vancouver, BC) made plans in early 2009 to explore blocks 4, 5, and 6 covering 3.4 million acres (5,312 sq mi) in Armenia's Central Depression.

The basin called the Central Depression covers the middle of the country and extends east to the border with Azerbaijan and south towards the border with Iran.

Hydrocarbon exploration of the area prior to 2007 has been described as patchy and unsystematic.

The joint venture aims to develop a new geological model for the area. Initial work will include geological studies, gravity and geochemical studies, and seismic evaluation.

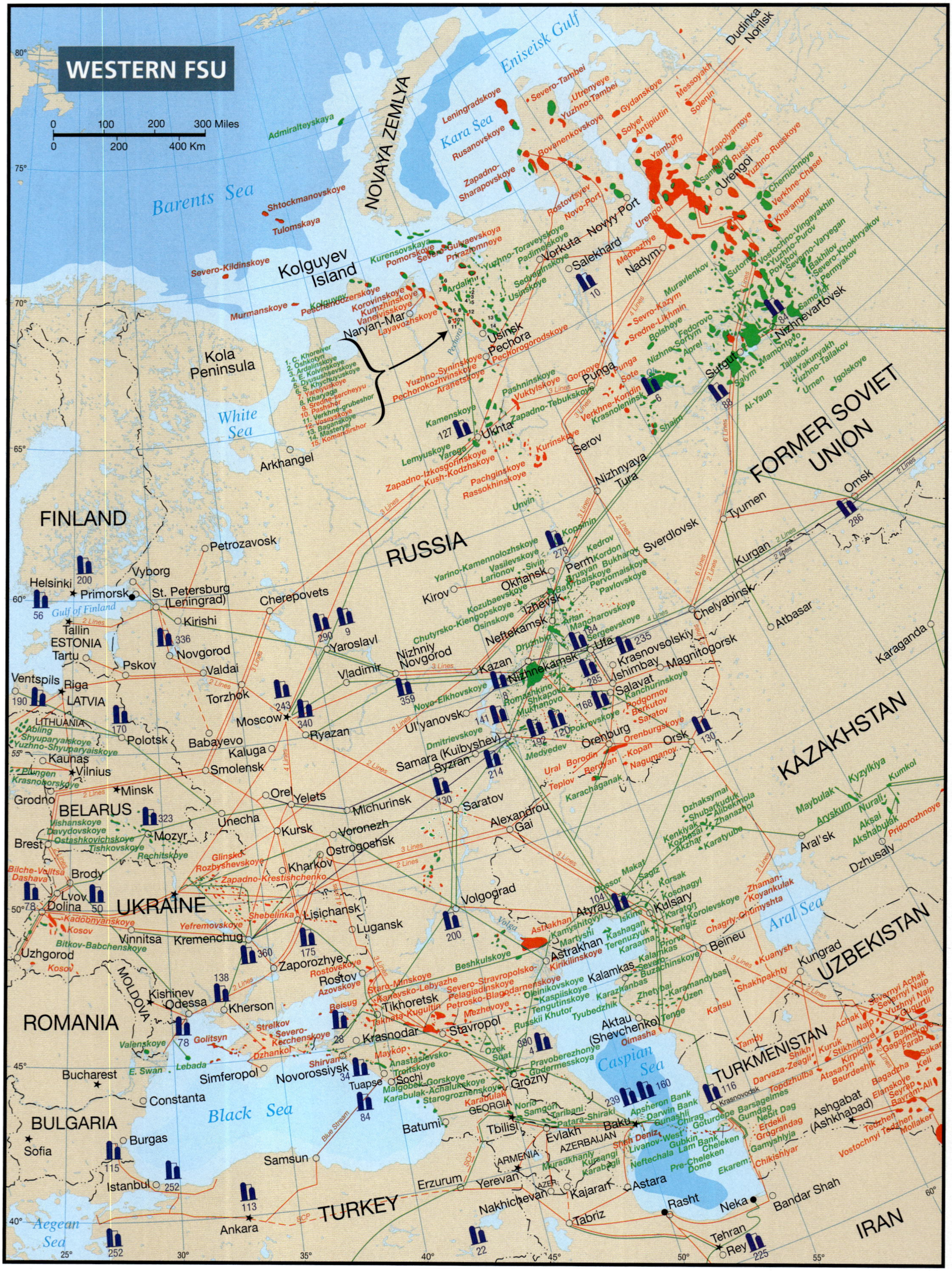
WESTERN FSU
0 100 200 300 Miles
0 200 400 Km
Barents Sea
Kara Sea
Eniseisk Gulf
NOVAYA ZEMLYA
Kolguyev Island
Kola Peninsula
White Sea
FINLAND
RUSSIA
FORMER SOVIET UNION
KAZAKHSTAN
UZBEKISTAN
TURKMENISTAN
UKRAINE
BELARUS
ESTONIA
LATVIA
LITHUANIA
ROMANIA
BULGARIA
MOLDOVA
GEORGIA
ARMENIA
AZERBAIJAN
TURKEY
IRAN
Black Sea
Caspian Sea
Aral Sea
Aegean Sea
Gulf of Finland
Arkhangel
Petrozavosk
Helsinki
Primorsk
Vyborg
St. Petersburg (Leningrad)
Tallin
Tartu
Pskov
Novgorod
Kirishi
Valdai
Cherepovets
Yaroslavl
Ventspils
Riga
Kaunas
Vilnius
Minsk
Grodno
Brest
Polotsk
Smolensk
Mozyr
Unecha
Torzhok
Moscow
Babayevo
Kaluga
Ryazan
Vladimir
Nizhniy Novgorod
Kirov
Kazan
Nizhnekamsk
Neftekamsk
Izhevsk
Okhansk
Perm
Sverdlovsk
Chelyabinsk
Kurgan
Tyumen
Omsk
Atbasar
Karaganda
Ufa
Krasnovsolskiy
Ishimbay
Magnitogorsk
Salavat
Orenburg
Orsk
Ul'yanovsk
Samara (Kuybyshev)
Syzran
Saratov
Orel
Yelets
Michurinsk
Kursk
Voronezh
Ostrogozhsk
Kharkov
Alexandrou Gai
Volgograd
Lisichansk
Lugansk
Brody
Lvov
Dolina
Uzhgorod
Vinnitsa
Kremenchug
Zaporozhye
Rostov
Kishinev
Odessa
Kherson
Tikhoretsk
Krasnodar
Stavropol
Novorossiysk
Tuapse
Sochi
Simferopol
Bucharest
Constanta
Sofia
Burgas
Istanbul
Ankara
Samsun
Batumi
Tbilisi
Grozny
Erzurum
Yerevan
Nakhichevan
Kajaran
Tabriz
Evlakh
Baku
Astara
Rasht
Tehran
Rey
Neka
Bandar Shah
Ashgabat (Ashkhabad)
Krasnovodsk
Aktau (Shevchenko)
Beineu
Kungrad
Aral'sk
Dzhusaly
Atyrau
Kulsary
Astrakhan
Ukhta
Pechora
Usinsk
Naryan-Mar
Vorkuta
Salekhard
Novyy Port
Nadym
Urengoi
Surgut
Nizhnevartovsk
Punga
Serov
Nizhnyaya Tura
Dudinka
Norilsk
Admiralteyskaya
Shtockmanovskoye
Tulomskaya
Severo-Kildinskoye
Murmanskoye
Leningradskoye
Rusanovskoye
Severo-Tambei
Kharasavei
Bovanenkovskoye
Yamburg
Zapolyarnoye
Medvezhye
Samotlor
Orenburgskoye
Karachaganak
Tengiz
Kashagan
Astrakhanskoye
Korolevskoye
Kalamkas
Zhetybai
Uzen
Karabulak
Nebit Dag
Gubkin
Dome
Kumkol
Shatlyk
Dauletabad
Gazli
Achak
Kurinskoye
Romashkino
Shebelinka
Dashava
Kosov
Blue Stream
SCP
Volga
2 Lines
3 Lines
4 Lines
6 Lines
25°
30°
35°
40°
45°
50°
55°
60°
40°
45°
50°
55°
60°
65°
70°
75°
80°

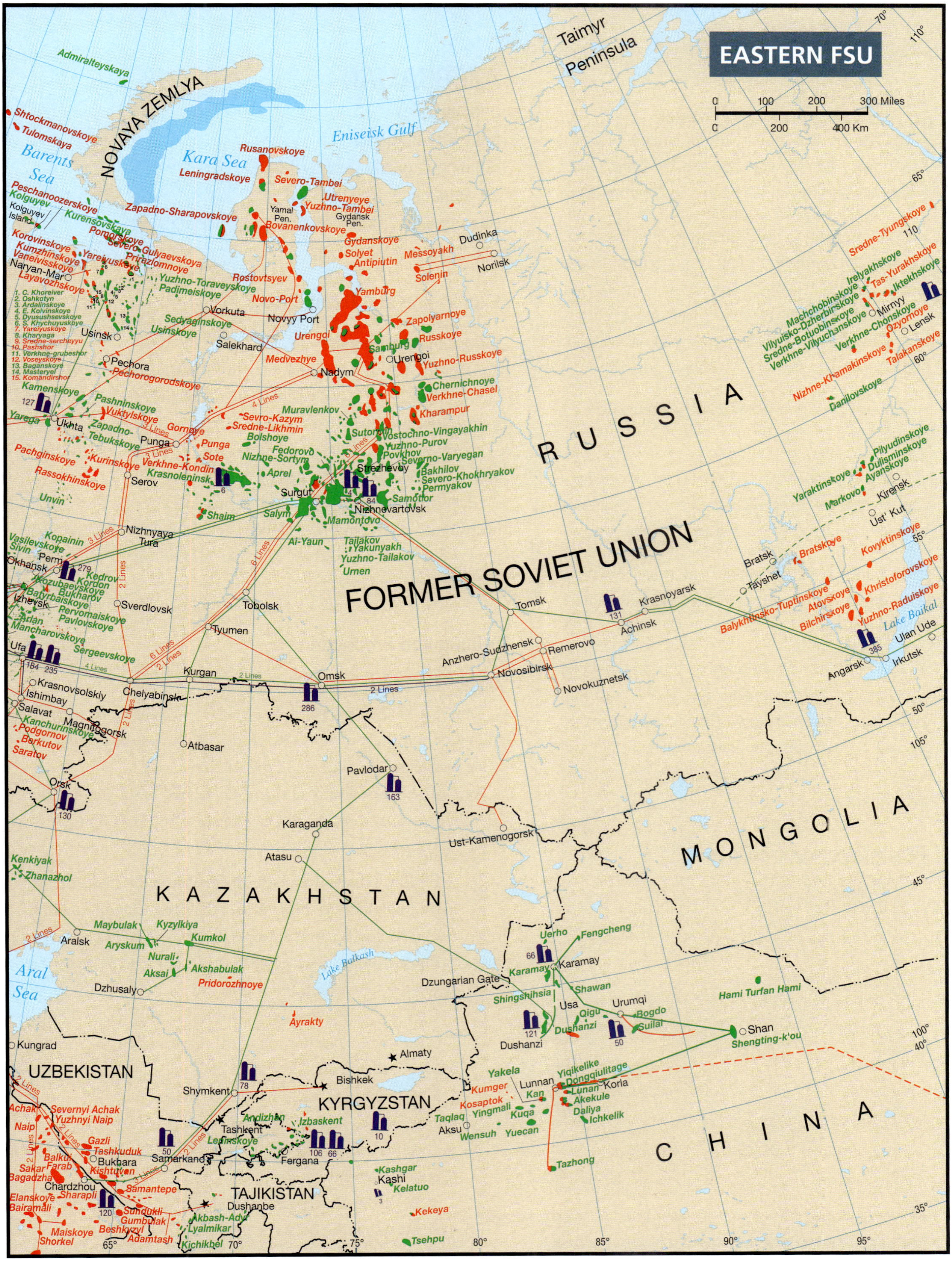
EASTERN FSU
0 100 200 300 Miles
0 200 400 Km
Taimyr Peninsula
NOVAYA ZEMLYA
Kara Sea
Barents Sea
Eniseisk Gulf
Admiralteyskaya
Shtockmanovskoye
Tulomskaya
Rusanovskoye
Leningradskoye
Severo-Tambei
Utrenyeye
Yuzhno-Tambei
Yamal Pen.
Gydansk Pen.
Zapadno-Sharapovskoye
Bovanenkovskoye
Peschanoozerskoye
Kolguyev
Kolguyev Island
Kurensovskaya
Korovinskoye
Kumzhinskoye
Vaneivisskoye
Naryan-Mar
Layavozhskoye
Gulyaevskoya
Prirazlomnoye
Yuzhno-Toraveyskoye
Padimeiskoye
Rostovtsyev
Novo-Port
Gydanskoye
Solyet
Antipiutin
Messoyakh
Solenin
Dudinka
Norilsk
Yamburg
Vorkuta
Novyy Port
Sedyaginskoye
Usinskoye
Usinsk
Salekhard
Urengoi
Zapolyarnoye
Russkoye
Samburg
Medvezhye
Yuzhno-Russkoye
Pechora
Pechorogorodskoye
Nadym
Chernichnoye
Verkhne-Chasel
Kharampur
Kamenskoye
Pashninskoye
Vuktylskoye
Yarega
Ukhta
Zapadno-Tebukskoye
Muravlenkov
Sevro-Kazym
Sredne-Likhmin
Bolshoye
Gornoye
Punga
Sote
Sutormin
Vostochno-Vingayakhin
Yuzhno-Purov
Povkhov
Severno-Varyegan
Bakhilov
Severo-Khokhryakov
Permyakov
Fedorovo
Nizhne-Sortym
Aprel
Pachginskoye
Kurinskoye
Verkhne-Kondin
Krasnoleninsk
Rassokhinskoye
Serov
Strezhevoy
Surgut
Samotlor
Nizhnevartovsk
Shaim
Salym
Mamontovo
Unvin
Kopainin
Nizhnyaya Tura
Ai-Yaun
Tailakov
Yakunyakh
Yuzhno-Tailakov
Urnen
Vasilevskoye
Sivin
Perm
Okhansk
Kedrov
Kordon
Bukharov
Pervomaiskoye
Pavlovskoye
Izhevsk
Sverdlovsk
Tobolsk
Tyumen
Mancharovskoye
Sergeevskoye
Ufa
Kurgan
Omsk
Chelyabinsk
Krasnovolskiy
Ishimbay
Salavat
Magnitogorsk
Kanchurinskoye
Podgornov
Berkutov
Saratov
Atbasar
Orsk
Pavlodar
Karaganda
Atasu
Kenkiyak
Zhanazhol
RUSSIA
FORMER SOVIET UNION
Tomsk
Krasnoyarsk
Achinsk
Anzhero-Sudzhensk
Remerovo
Novosibirsk
Novokuznetsk
Ust-Kamenogorsk
Bratsk
Tayshet
Bratskoye
Kovyktinskoye
Khristoforovskoye
Atovskoye
Bilchirskoye
Balykhtinsko-Tuptinskoye
Yuzhno-Raduiskoye
Lake Baikal
Ulan Ude
Angarsk
Irkutsk
Kirensk
Ust' Kut
Yaraktinskoye
Markovo
Pilyudinskoye
Dulisminskoye
Ayanskoye
Danilovskoye
Nizhne-Khamakinskoye
Talakanskoye
Ozyornoye
Lensk
Mirnyy
Verkhne-Chonskoye
Machchobinskoye
Irelyakhskoye
Tas-Yurakhskoye
Iktekhskoye
Sredne-Tyungskoye
Vilyuisko-Dzherbinskoye
Sredne-Botuobinskoye
Verkhne-Vilyuchanskoye
MONGOLIA
KAZAKHSTAN
Maybulak
Kyzylkiya
Kumkol
Aryskum
Nurali
Aksai
Akshabulak
Pridorozhnoye
Aralsk
Aral Sea
Dzhusaly
Lake Balkash
Ayrakty
Uerho
Fengcheng
Karamay
Dzungarian Gate
Shawan
Shingshihsia
Usa
Urumqi
Bogdo
Qigu
Dushanzi
Suilai
Hami Turfan Hami
Shan
Shengting-k'ou
Kungrad
UZBEKISTAN
Almaty
Shymkent
Bishkek
KYRGYZSTAN
Yakela
Kumger
Lunnan
Yiqikelike
Donggiulitage
Korla
Lunan
Akekule
Daliya
Ichkelik
Kan
Kosaptok
Yingmali
Kuqa
Taqlaq
Aksu
Wensuh
Yuecan
Tazhong
CHINA
Andizhan
Izbaskent
Tashkent
Leninskoye
Fergana
Samarkand
Kashgar
Kashi
Kelatuo
Kekeya
Tsehpu
TAJIKISTAN
Dushanbe
Akbash-Adyr
Lyalmikar
Kichikbel
Achak
Severnyi Achak
Yuzhnyi Naip
Naip
Gazli
Tashkuduk
Bukhara
Kishtuvan
Balkui
Farab
Sakar
Bagadzha
Chardzhou
Samantepe
Sharapli
Elanskoye
Bairamali
Sundukli
Gumbulak
Beshkyzyl
Maiskoye
Shorkel
Adamtash
2 Lines
3 Lines
4 Lines
6 Lines
65°
70°
75°
80°
85°
90°
95°
100°
105°
110°
35°
40°
45°
50°
55°
60°

Caspian Sea conference

The Caspian Sea's five littoral states—Russia, Kazakhstan, Turkmenistan, Iran, and Azerbaijan—held a two-day summit in Tehran in mid-October 2007.

Under 1921 and 1940 treaties, the sea was divided between Iran and the former Soviet Union. More recently, Kazakhstan, Azerbaijan, and Russia agreed to divide the upper portion of the Caspian seabed in a way that allows each some sovereign rights to mineral resources.

However, Turkmenistan and Iran are arguing for a bigger share of the sea.

Iran and Russia—wary of American influence—warned at the summit against outside interference by resource-rich countries in the Caspian area.

Participants also discussed potential oil pipeline routes in the region and related environmental issues.

At the end of the conference, the five nations appeared no closer to resolving the border and other legal issues raised. Conflicting claims to seabed oil and gas deposits remain, as do the arguments of Azerbaijan regarding its borders with Turkmenistan and Iran.

Proposed Caspian Sea division

AZERBAIJAN

Capital: Baku
Monetary unit: Manat
Refining capacity: 398,978 b/cd
Oil production: 925,000 b/d
Oil reserves: 7.000 billion bbl
Gas reserves: 30.000 tcf

Exploration and Development

Yalama offshore prospect

GDF Suez (Paris) acquired a stake in the offshore exploration-production license D-222—owned by Lukoil Overseas and the national Azeri company SOCAR—in the Caspian Sea offshore Azerbaijan.

It contains the Yalama prospect, due to be drilled in early 2009. The exploration period will extend until late 2011.

GDF Suez said that successful development could increase its hydrocarbon reserves by 35 MMboe. The firm aims to increase its proved and probable reserves to 1,500 MMboe from the current 670 MMboe.

Drilling and Production

Absheron block

Total SA (Paris) signed a heads of agreement document in mid-2008 with state-owned SOCAR regarding a production sharing agreement (PSA) for a license in the Absheron area.

Total will be the operator of the block, located in the Caspian Sea, 100 km (62 mi) from Baku, in water depth of 500 m (1,640 ft).

Gunashli field

BP PLC (London) announced the start of oil production in early 2008 from the third phase of the Azeri-Chirag-Gunashli (ACG) field, offshore Azerbaijan in the Caspian Sea.

Plateau production of 320,000 b/d is expected from the deepwater Gunashli (DWG) platform.

Oil will be delivered through the Baku-Tbilisi-Ceyhan (BTC) pipeline from the Caspian Sea to Turkey's Mediterranean coast.

More predrilled wells were expected to come on stream during 2008. Total ACG production is expected to top 1 million b/d, including oil from Chirag, East Azeri, West Azeri, and Central Azeri.

Some of the associated gas produced will be piped to the Azerigas system for domestic use.

Pipelines

Kazakh oil to Azerbaijan

In November 2008, State Oil Co. of the Azerbaijan Republic (SOCAR) and Kazakhstan's state-owned KazMunaiGas agreed on the basic principles of a new project to bring Kazakh oil across the Caspian Sea starting in 2013.

Kazakhstan will build a pipeline network from the Tengiz and Kashagan oil fields to export terminals on its Caspian coast. The oil will then move by tanker to terminals to be constructed on the Azeri coast.

The system's initial shipment capacity will be 500,000 b/d, eventually rising to 0.75–1.2 million b/d.

Once in Baku, oil will be shipped through the 1,700-km (1,056-mi) Baku-Tblisi-Ceyhan (BTC) oil pipeline, or by trans-Caucasus rail to Georgian ports on the Black Sea.

BTC carries Kazakh oil

The first shipments of oil produced in Kazakhstan by the TengizChevrOil (TCO) consortium began to flow through the Baku-Tblisi-Ceyhan (BTC) line in October 2008.

Crude from the Tengiz field is shipped by tanker across the Caspian Sea to the Apsherob Peninsula in Azerbaijan and then loaded into the BTC line.

SOCAR said that the BTC line could eventually transport as much as 100,000 b/d of Kazakh oil.

Baku-Supsa line

BP PLC (London) reported in early November 2008 the reopening of the Baku-Supsa pipeline (also known as the Western Route Export Pipeline). It had been closed as a safety measure in August during hostilities between Russia and Georgia.

Before the closure, the line had been open for only one week, after 18 months of maintenance work.

The Baku-Supsa line can carry 145,000 b/d but was expected to move only 90,000 b/d through yearend 2008.

Attacks on Azeri export lines

During the fighting between the two countries in August 2008, Georgia accused Russia of attempting to bomb the main pipelines through the country:

- the Baku-Tbilisi-Ceyhan (BTC) oil line
- the Baku-Supsa oil line
- the South Caucasus Pipeline that carries gas to Turkey.

Russian officials denied the allegation, but independent press reports suggested that Russian warplanes did target the lines, as well as the main railway line connecting Azerbaijan with Georgia's Black Sea oil export terminals.

Russia blamed shutdown of the BTC line on terrorist attacks along the line's Turkish sector, but did not mention the closure of the other oil and gas export routes.

More details about the Russian-Georgian conflict are presented in the Georgia section of this chapter.

Azeri oil swap with Iran

In response to reports of damage to oil, gas and rail lines during the Russian-Georgian conflict, SOCAR made arrangement to export Azeri Light crude to Iran in a swap arrangement.

Under the deal, SOCAR was authorized to sell up to 300,000 tonnes via the Persian Gulf over two months, shipping oil from the offshore Azeri-Chirag-Guneshli fields across the Caspian Sea to the Iranian port of Neka.

U.S. VP visit

In early September 2008, U.S. Vice President Dick Cheney visited Baku. Analysts said the visit was intended to secure U.S. and European energy interests in the Caspian-Central Asian region following Russian-Georgian hostilities.

Azerbaijan and Georgia are parts of a corridor supplying energy resources to Europe, and the United States wanted assurances that Azerbaijan would continue making oil and gas deliveries through Georgia.

Analysts said that Cheney also wanted to ensure continued Azerbaijani support for the proposed 3,300-km (2,050-mi) Nabucco gas pipeline.

That line would run from the Caspian region through Turkey and the Balkans to Austria.

Azerbaijan is seen as the key potential supplier for the project. Construction is scheduled to start in 2009 and be completed in 2013.

Russia backs the rival South Stream gas pipeline, being built by Gazprom and Eni SPA (Rome). It will include a gas pipeline under the Black Sea from Russia to Bulgaria, with further branches to Austria and Italy.

Petrochemicals

Major petrochemical project

The Asia Pulse news service reported in September 2008 that the Azerbaijan state petrochemical company, Azerkimya, was about to sign a memorandum of cooperation and confidentiality with a half-dozen major companies from Europe, the United States and the UK to build a petrochemical complex in Sumgayit.

Construction could begin in 2009.

Russian-Georgian conflict

On August 7, 2008, after months of growing tension, Georgian military forces attacked the capital city (Tskhinvali) of South Ossetia province, which was controlled by a separatist government backed by Russia.

There were conflicting reports of the actions that triggered the attack.

In response, Russia sent troops into the province in central Georgia. Russia also landed troops August 8 on the coast of a second separatist Georgian province, Abkhazia.

Russian officials said the actions (and later movements deeper into Georgia) were taken to protect its citizens as well as peacekeeping forces that were already deployed in the two regions.

Early in 2008, Russia established semi-official ties with the separatists and sent extra troops to Abkhazia to beef up a peacekeeping contingent there.

Hundreds of civilians were reported dead in the fighting, and witnesses said Tskhinvali was devastated. The hostilities were called the worst since South Ossetia won *de facto* independence in a war against Georgia that ended in 1992.

After five days of fighting, Georgian forces were pushed out of both South Ossetia and Abkhazia.

Of concern to the world energy market were reports of damage done to Georgia's energy infrastructure.

Georgia and Russia signed a ceasefire August 16.

Twelve days later, Russia formally recognized the independence of both provinces, an act that drew criticism from the United States, the UK, France and Germany.

Based on international agreements, Russia withdrew its main forces from Georgia on October 8, but left some troops in both provinces by agreement with Georgian officials.

GEORGIA

Capital: Tbilisi
Monetary unit: Lari
Refining capacity: 0
Oil production: 1,000 b/d
Oil reserves: 35.000 million bbl
Gas reserves: 300.0 bcf

Exploration and Development

Mtsare Khevi and Taribani fields

Frontera Resources Corp. (Houston) was working in late 2008 to establish oil production from Mtsare Khevi field in western Block 12.

Two initial wells were producing from Horizon I of the Upper Pliocene Akchagil formation. Three production horizons total 20–30 m (66–98 ft) of gross pay, 200–315 m (656–1,034 ft) deep.

A 2007 study indicated as much as 5 million bbl recoverable, said Frontera. Further potential exists in Miocene sandstones as deep as 1,100 m (3,609 ft).

In mid-year, the firm also said it established apparently sustainable oil production at its Dino-2 well in the Taribani field unit in Block 12 in eastern Georgia's Kura basin.

Frontera plans to drill at least 40 more wells in 2009.

Lease Sales

Five blocks licensed

In January 2009, the International Oil and Gas Agency of Georgia signed a 25-yr agreement qualifying two Indian investment companies—Jandal Petroleum and Ensearch Petroleum—to conduct exploration operations and produce oil and gas on five blocks: VIIA, IX, X, XIIIA and XIV.

Jandal will develop three blocks, Ensearch two.

Georgia offered the blocks in mid-2008. They are in the regions of Kvemo Kartli, Guriya, Ajaria and Kakheti.

State-owned Saknavtobi estimates potential oil and gas reserves on the blocks at 400 million tonnes.

Saknavtobi says one block has gas reserves of up to 7.7 bcm (272 bcf).

By 2014, the firms are obliged to invest $150 million.

Pipelines and Transportation

Pipelines reopened

On August 25, BP PLC (London) reopened the Baku-Tbilisi-Ceyhan (BTC) oil pipeline that runs through Georgia.

The line provides 1 million b/d of Caspian Sea crude to international markets. It had been closed for more than two weeks after a fire on its Turkish stretch. Kurdish rebels claimed responsibility for the blaze.

The company's ability to export Caspian crude was sharply curtailed by both the fire and by fighting between Russia and Georgia in the provinces of South Ossetia and Abkhazia (see box).

BP shut down its 150,000-b/d Baku-Supsa oil pipeline—which runs Baku in Azerbaijan to Supsa on Georgia's Black Sea coast—on August 12 because of security concerns. The line is also known as the Western Route Export Pipeline (WREP).

It was reopened November 5, but was pumping only about 90,000 b/d at yearend, well below capacity.

In late August, BP also restored exports of natural gas from Azerbaijan into the South Caucasus Pipeline (SCP). Operation of that line also had been suspended August 12, based on reports of possible attacks by Russian warplanes.

During the uncertainty over the resumption of BP's oil pipeline operations, the Russian government sent mixed signals, saying it is prepared to increase oil flows along the Baku-Novorossiisk line, while allegedly preventing shipments of oil from Georgian ports on the Black Sea.

Baku-Tbilisi-Ceyhan oil pipeline

BP's pipelines in the Caucasus

Baku-Tbilisi-Ceyhan (BTC) pipeline

The 1,768-km BTC pipeline began operating in June 2006. It extends 443 km through Azerbaijan, 249 km through Georgia, and 1,076 km through Turkey to the Ceyhan marine terminal. Built at a cost of $4 billion, the BTC pipeline is the first direct transportation link between the Caspian and the Mediterranean seas. At full capacity, it can deliver 1 million b/d of oil.

South Caucasus gas pipeline (SCP)

The 692-km South Caucasus Pipeline (SCP) started delivering natural gas from Shah Deniz field in the Caspian Sea to Azerbaijan and Georgia from December 2006, and in Turkey from July 2007, representing an important new source of gas supply for the region.

Western Route Export Pipeline (WREP)

The 830-km Western Route Export Pipeline (WREP), also known as the Baku-Supsa pipeline, transports oil from the Chirag field in the Caspian Sea, via the Sangachal terminal in Azerbaijan, to the Supsa terminal in western Georgia. WREP, which became fully operational in 1999, can transport as much as 155,000 b/d of oil.

Oil by rail to Georgia disrupted

BP also was forced to suspend rail shipment of 50,000–70,000 b/d of oil from Azerbaijan to Georgia in late August after the destruction of a key bridge on the main east-west train line between the two nations.

Georgian officials accused Russian troops of blowing up the bridge, but Russian officials denied any involvement. Repairs were completed several weeks after the attack.

KAZAKHSTAN

Capital: Astana
Monetary unit: Tenge
Refining capacity: 345,093 b/cd
Oil production: 1.385 million b/d
Oil reserves: 30.000 billion bbl
Gas reserves: 85.000 tcf

Exploration and Development

Kashagan: Second phase

In January 2009, Agip KCO—a fully owned subsidiary of Eni SpA (Milan)—chose an engineering consortium led by WorleyParsons Europe Ltd. (Surrey, UK) to design facilities for the second phase of Kashagan oil field.

The value of FEED services for onshore and offshore facilities is put at $31 million.

Kashagan has 34.5 billion bbl of oil in place but is a challenging field. It has high pressure and high hydrogen sulfide levels and is in an environmentally sensitive area with difficult weather.

An initial output of 150,000 b/d from the field was originally for 2005, but it has been delayed until late 2012 due to the technical challenges, cost increases, and a reconfiguration of the offshore plant.

Located 80 km (50 mi) southeast of Atyrau in the North Caspian Sea, Kashagan will be developed in three phases.

WorleyParsons and Aker Solutions ASA (Oslo) secured the engineering services, fabrication, and hookup contract for the first phase, which is expected to cost $19 billion.

Nursultan Block, Caspian Sea

State-owned KazMunaiGas (KMG; Astana) signed an MOU understanding in late 2008 with ConocoPhillips Corp. (Houston) and Mubadala Development Co. (Abu Dhabi) for joint exploration and production of the Nursultan Block in the Caspian Sea.

Nursultan (also known as N Block) lies 30 km (19 mi) south-southwest off Aktau and covers some 8,100 sq km (3,127 sq mi). KMG describes the block as highly prospective for both oil and gas.

Reserves are estimated at 637 million toe.

The Kazakh government has granted exclusive exploration and production rights to KMG.

Tasaran reservoir

The AKK-15 exploration well in western Kazakhstan flowed gas at a restricted rate of 7.1 MMcfd from a 6-m (20-ft) interval in the Tasaran sandstone reservoir, said Tethys Petroleum Ltd. (Guernsey, Channel Islands, UK) in mid-2008.

The company estimated that a flow rate of 17 MMcfd was possible through larger tubing.

AKK-15 is 9 km (5 mi) east of AKK-14, the area's first Tasaran discovery.

Tethys estimated that AKK15 could go on line by early 2009.

In early 2008, the AKK-14 (on Akkulka Block, North Ust Yurt basin) flowed a combined 13.3 MMcfd of gas from two intervals. It appears to be the first commercial discovery in Cretaceous Tasaran sand.

In November, Tethys Petroleum Ltd. said it gauged gas at more than 10.2 MMcfd at the AKK-16 well in the North Ustyurt basin and named the discovery Southeast Akkulka.

AKK-16 is 36 km (22 mi) southeast of Kyzyloi gas field.

Kul-Bas Block

The KUL01 well on the 2.14-million acre (3,343-sq mi) Kul-Bas exploration block in the Ust-Yurt basin discovered gas in mid-2008, north of the tie-in point of Kyzyloi gas field, said Tethys Petroleum Ltd.

Drilled to 593 m (1,945 ft), the well topped the Kyzyloi sand at 358 m (1,175 ft), about 100 m (328 ft) shallower than in Kyzyloi field.

A 3-m (10-ft) perforated interval with 35% porosity flowed on test at 2 MMcfd. Absolute open flow potential is 3.2 MMcfd.

Tengiz expansion

In September 2008, Chevron Corp. (San Ramon, CA) announced the completion (by affiliate TengizChevrOil; TCO) of its expansion at the Tengiz field, boosting production capacity to 540,000 b/d.

TCO is a JV of Chevron; KazMunaiGas (KMG); ExxonMobil Kazakhstan Ventures Inc.; and LukArco [a JV formed in 1997 by Lukoil (Moscow) and Arco)].

In 2007, the consortium was the largest oil producer in Kazakhstan, with output of 13.93 million tonnes, or 20% of the country's total production.

TCO has been struggling with its JV partners to double the capacity of the Caspian Pipeline Consortium (CPC) export route to 60 mtpy, to accommodate rising output in Kazakhstan.

The CPC line runs from the Tengiz field in Kazakhstan to Russia's Black Sea port of Novorossiisk.

Tsentralnaya structure

TsentrCaspneftegaz—a JV set up by Gazprom and Lukoil (both in Moscow)—reported in mid-2008 a major oil and gas condensate field discovery within the Tsentralnaya structure.

The structure is in the central Middle Caspian Sea on the border between Russia and Kazakhstan.

Well surveying is under way; plans called for 3D seismic operations to further define the reservoir.

Snapshot: Three major fields

In mid-2008, *Oil & Gas Journal* gave this snapshot of the expected development schedule for three major fields in Kazakhstan:

- Tengiz: Phased development was continuing, with a $9.5-billion expansion project that will increase oil production by 260,000 b/d after 2010.
- Karachaganak: An expansion will increase production by 1.6 bcfd in 2012.
- Kashagan (offshore): Production from the first phase of the 13-billion bbl field is now expected no earlier than 2012. With future phases, the field's production may surpass 1 million b/d.

Drilling and Production

Akzhar and Besbolek fields

In October 2008, the Central Committee for Development approved plans by Arawak Energy Ltd. (St. Helier, Jersey, UK) to develop Akzhar and Besbolek oil fields in the Precaspian basin.

A five-rig drilling program was established for the Besbolek fields.

The fields had NI 51–101 proved and probable reserves of 23.4 million and 7.9 million bbl of oil, respectively, as of yearend 2007.

However, in December, Arawak had to cut oil production at Akzhar and Besbolek fields, as well as at Karataikyz and Alimbai fields, in connection with high taxes and export duties and falling oil prices.

Arawak cut output to the minimum needed to maintain processing facilities.

Production from Saigak field, governed by a production sharing agreement, was not affected.

Kashagan field

In October 2008, Kazakhstan ended a 15-month dispute with foreign investors over its Kashagan oil field.

The government and the IOCs developing the field agreed to double the stake of state-owned KazMunaiGas (KMG; Astana) in the project to 16.81% and to allow KMG to receive a larger share of the field's eventual 1.6-million b/d production, when world oil prices are high.

The parties were at odds over production delays and rising costs.

Other consortium members are Royal Dutch Shell PLC (The Hague, Netherlands), ExxonMobil Corp. (Irving, TX), Total SA (Paris), ConocoPhillips Corp. (Houston), and Inpex Holdings Inc. (Tokyo).

The group has reportedly invested $12 billion in the offshore Kashagan field, called the most complex oil project ever undertaken. High sulfur levels, deep water, and foul weather hamper development.

Kashagan development is led by Eni SPA (Rome). Production is expected to start in late 2013.

Shell will oversee offshore construction and Eni onshore construction while Total takes responsibility for coordinating oil transport.

Karachaganak: Delays and changes

In November 2008, BG Group (Reading, UK) deferred indefinitely a planned investment in Phase III development of Karachaganak fields.

The decision was called one of the most significant energy-project delays resulting from the deepening global economic crisis.

Production from Phase III had been planned to begin by the end of 2012, raising output to 16 bcm/yr (565 bcf/yr) and 335,000 bbl of liquids/day.

BG said in June 2007 that Phase III would cost $8 billion, but independent estimates have risen sharply.

The Kazakh finance ministry said in May that the consortium would have to pay a new oil export duty

The Karachaganak consortium is comprised of Eni SPA (Rome); BG Group; Chevron Corp. (San Ramon, CA); and Lukoil (Moscow).

Refining

Ceyhan and Batumi refineries

In early 2009, state-owned KazMunaiGas (KMG; Astana) decided not to proceed with construction of refineries in Ceyhan (Turkey) and Batumi (Georgia).

KMG said the global economic crisis and related economic feasibility issues led to the decision.

Instead, the company will enhance production of three refineries in Kazakhstan.

The proposed 428,000-b/d, $5-billion Ceyhan refinery at the port of Ceyhan, Turkey, was to have begun construction in early 2009 and to have started operation by 2012.

KMG also had planned to invest more than $1 billion in the 100,000-b/d Batumi refinery, on the Georgian Black Sea coast. KMG decided to postpone the construction after a feasibility study.

Pipelines

New lines from Tengiz, Kashagan

In November 2008, KazMunaiGas (KMG) and Azerbaijan's SOCAR agreed on the basic principles of a new project to bring Kazakh oil across the Caspian Sea starting in 2013.

Under the agreement, Kazakhstan will build a pipeline network from its Tengiz and Kashagan oil fields to export terminals on its Caspian coast.

The oil will then move by tanker to terminals to be constructed on the Azeri coast.

Initial shipment capacity will be 500,000 b/d, eventually rising to 0.75–1.2 million b/d.

The estimated cost of the trans-Caspian transport project, including terminals and tankers, is $3 billion.

By 2015, Kazakhstan expects to be exporting 100 million tonnes of crude.

Pipelines for Caspian gas

The Oxford Institute for Energy Studies (Oxford, UK) reviewed in late 2008 the various routes available to Kazakhstan for export of its natural gas resources.

The country is an increasingly important potential supplier to both Europe and Asia, and its geographic location makes it a potential transit nation for gas supplies moving from Uzbekistan and Turkmenistan to Europe.

Projects evaluated include:

- Expansion and upgrade of the Central Asia-Center (CAC) pipeline network
- Construction of the new Caspian Littoral pipeline
- Construction of the Turkmenistan-China pipeline
- Completion of the Trans-Caspian pipeline
- Completion of the Western Kazakhstan-Western China pipeline

The Oxford review also commented on the Nabucco pipeline project, the South Stream pipeline, the Trans-Asian pipeline from Kazakhstan to China, and the Western Kazakhstan-Western China pipeline.

CASPIAN EXPORT PIPELINE PROJECTS

Pipeline	Capacity, billion cu m/year	Length, km	Construction	Export price, $/thousand cu m (2008)	Suppliers	Transit countries
CAC expansion	80-100.2	1,968	2012-15	150-190	Turkmenistan, Uzbekistan, Kazakhstan	Uzbekistan, Kazakhstan
Caspian Littoral	20-30	1,390-1,700	2009-15	150-190	Turkmenistan, Kazakhstan	Kazakhstan
Western Kazakhstan-Western China	10	1,480	2009-12	195	Kazakhstan	None
Turkmenistan-China	30-40	2,051	2008-10	195	Turkmenistan	Uzbekistan, Kazakhstan
Trans-Caspian	28-32	1,592 (to Turkey)	None set	120-130	Turkmenistan, Kazakhstan, Azerbaijan	Kazakhstan, Caspian Sea, Azerbaijan

Source: Gazprom, Kazmunaigaz

CAC, CASPIAN LITTORAL PIPELINES

Trans-Asian, Kazakhstan-China Pipelines

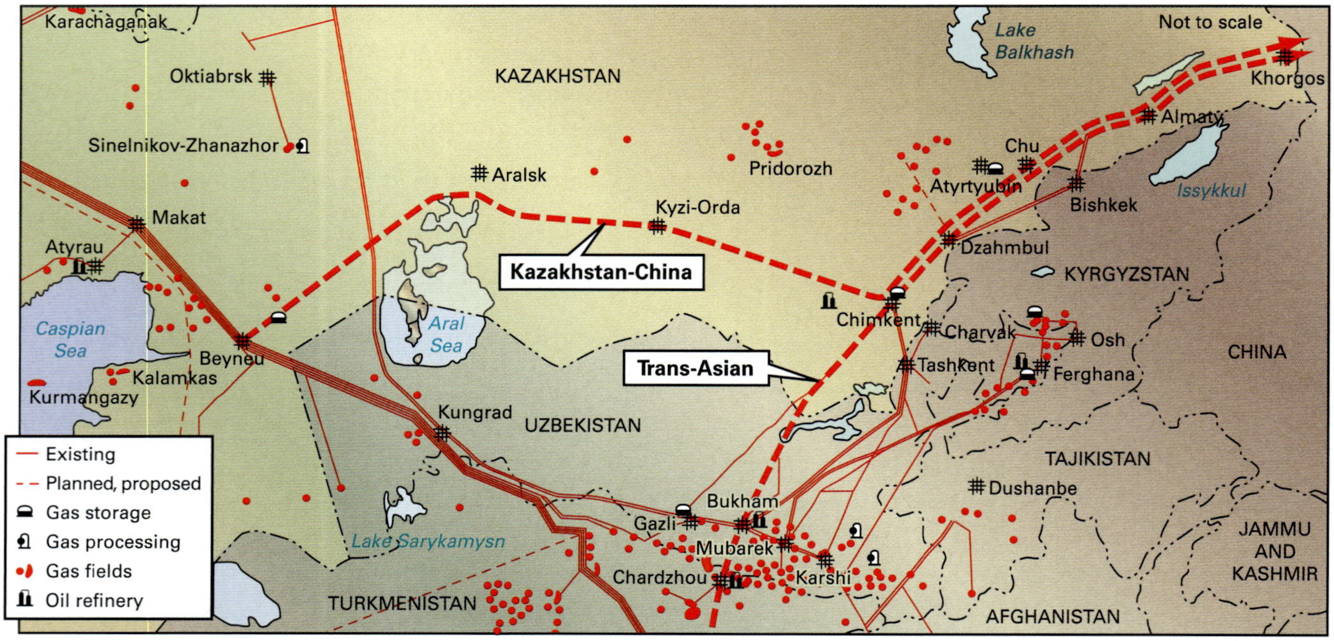

Mitigating against construction of the Trans-Caspian and Nabucco pipelines are:

- competition from other projects
- Russia's opposition to these projects and Central Asia's participation in them
- the still-unclear legal status of the Caspian Sea.

Tengiz oil through BTC pipeline

BP PLC (London) announced plans to start shipping oil from Tengiz oil field through the 1,770-km (1,100-mi) Baku-Tbilisi-Ceyhan (BTC) pipeline beginning in October 2008.

BP said the oil would be transported by barge from Kazakhstan to Azerbaijan and then loaded into the 1 million-b/d BTC pipeline, with exports from Ceyhan expected to begin in mid-November.

The plan is similar to one attributed in September to TengizChevrOil.

BP said that initial rates would likely be relatively low in volume.

Also in October, KazMunaiGaz began talks aimed at buying BP's stake in the 565,000 b/d Caspian Pipeline Consortium (CPC).

The 1,580-km (982-mi) CPC pipeline transports Kazakh oil to the Yuzhnaya Ozereevka terminal at Novorossiisk on Russia's Black Sea coast.

BP was said to be interested in selling its stake due to an ongoing impasse with the Russian government over plans to expand the pipeline's capacity.

Global Insight Inc. (Lexington, MA) said it expected a change in the makeup of the CPC consortium in order to expand the pipeline to its design peak capacity of 1.34 million b/d.

Kazakh use of ESPO pipeline?

Russian officials said in late 2008 that Kazakhstan's state-owned KazTransOil was interested in transporting Kazakh oil through Russia's East Siberia-Pacific Ocean (ESPO) pipeline.

Kazakhstan participated in the launch of the pipeline section from Talakan to Taishet.

The 1,100-km (684-mi) pipeline was expected to supply about 220,000 tonnes of oil to the Angarsk petrochemical company by yearend 2008, from Talakan and Verkhnechonskoye fields.

In a related development, the TNK-BP consortium said it planned to begin supplying oil to the ESPO line in October 2008 from its Verkhnechonskoye field.

The Taishet-Talakan section of the ESPO line was completed in September 2008; the remaining stretch to Skovorodino, near the border with China, is slated for completion by yearend 2009.

Major oil, gas line investment

Fitch Ratings (New York City) said in August 2008 that Kazakhstan's oil and gas pipeline operators were poised to launch ambitious investment programs over the next five years to capitalize on favorable oil and gas industry fundamentals, as well as increased domestic energy demand.

KazTransGas (KTG), the state-owned operator of gas pipelines, was said to be preparing to invest more than $8 billion to build three gas pipelines, including the West-South gas pipeline, the China gas pipeline, and the By-Caspian gas pipeline.

State-run KazTransOil (KTO) was said to be ready to invest more than $2 billion to build construction of two new oil pipelines.

The Kenkiyak-Kumkol route would connect western Kazakhstan to China.

The second line would link Kashagan oil field with a new export terminal on the Caspian Sea.

Fitch also reported that the consortium operating the Caspian Pipeline Consortium (CPC) pipeline was considering a possible $2.5-billion pipeline capacity expansion.

Petrochemicals

Atyrau plant

Work continued in late 2008 on design of a 629,000-tpy grassroots petrochemical plant to be integrated into the existing 104,500-b/cd KazakhOil refinery at Atyrau.

To be completed in 2013, the petrochemical plant will produce 496,000 tonnes/year of paraxylene and 133,000 tonnes/year of benzene from naphtha.

The refinery processes 5 mtpy of oil from various fields in western Kazakhstan.

New plants at Karabatan

Kazakhstan Petrochemical Industries Inc. continued its planning in late 2008 for three new petrochemical plants.

They will have a combined capacity of 1.3 billion tpy of polyolefins. Scheduled for start-up in 2013, all will be built at the petrochemical complex of Karabatan.

Caspian Sea: Legal status

Iran and the Soviet Union signed bilateral treaties regarding the Caspian Sea in 1921 and 1940 but never established seabed boundaries or held consultations regarding oil and natural gas exploration in the area. This complicates resource development and pipeline development in the region.

The primary current problem centers on whether to define the Caspian as an inland lake or a sea. If defined as a sea, the Law of the Sea Convention would apply and both full maritime boundaries and mineral rights of the five states would be established accordingly.

The Law of the Sea, however, does not apply to inland lakes, requiring that the Caspian either be developed jointly or that surrounding states reach agreements regarding its development.

An agreement on the legal status of the Caspian Sea will require not only Russian but also Iranian participation. The latter is highly unlikely given current tensions between Iran and the US.

Russia demands all Caspian countries be consulted before any pipeline project commences.

Azerbaijan and Kazakhstan must also reach a consensus with Turkmenistan on the division of hydrocarbon resources in the area.

KYRGYZSTAN

Capital: Bishkek
Monetary unit: Som
Refining capacity: 10,000 b/cd
Oil production: 1,000 b/d
Oil reserves: 40.000 million bbl
Gas reserves: 200.000 bcf

Tuzluk Block

Santos International Holding Pty. Ltd. (subsidiary of Santos Ltd.; Adelaide, Australia) committed to Phase 2 of its five-block exploration program in the North Ayzar prospect in Tuzluk Block in Kyrgyzstan as a farmee of Manas Petroleum Corp. (Baar, Switzerland).

Santos completed the first phase, spending $10 million for geological studies and reprocessing and shooting seismic surveys.

Several shallow exploration prospects had been identified for drilling beginning in late 2008, and four deep prospects were slated for drilling in 2009.

Santos will drill three exploration and three appraisal wells in phase two.

LITHUANIA

Capital: Vilnius
Monetary unit: Litas
Refining capacity: 190,000 b/cd
Oil production: 3,000 b/d
Oil reserves: 12.0 million bbl
Gas reserves: 0

First refinery in Latvia

Privately held Latvian company Dinaz Group KU SIA announced plans in early 2009 to start construction of the country's first refinery in 2010, aiming to reduce product imports.

Company officials said the $3.7-billion, 6-mtpy (120,500-b/d) refinery would be built near Daugavpils, just north of the Belarus-Lithuania border. The greenfield site is 4 km (2.5 mi) from Druzhba pipeline, which transports Russian oil to Europe.

The company was conducting a feasibility study for the refinery and planned to start the environmental evaluation process in 2009. Dinaz also is seeking partners in developing the refinery.

Separately, Dinaz said it also plans to construct a 10-mtpy (200,820-b/d) oil terminal in Riga.

New energy ministry

Pres. Valdas Adamkus signed into law in January 2009 a bill creating the new energy ministry.

Businessman Arvydas Sekmokas is likely to head the new ministry.

In addition to coordinating energy conservation programs, the new ministry also will oversee energy security projects such as construction of the new nuclear power plant, electric energy "bridges" to Poland and Sweden, a gas pipeline to Poland, and an LNG terminal.

Energy discussions with Poland

In January 2009, Lithuanian officials were scheduled to discuss several key energy issues with their Polish counterparts.

Those issues include a possible Polish-Lithuanian electric "energy bridge," the future of Lithuania's Ignalina nuclear power plant, and the future of the Mazeikiu Nafta refinery, owned by Polish refiner PKN Orlen.

Regarding the nuclear power station, Lithuanian, Latvian, Estonian, and Polish energy companies agreed in 2006 to build a nuclear power plant in Lithuania. The second and last reactor of the exiting Ignalina plant is due to be shut down by December 31, 2009.

Negotiations among the countries have continued since 2006 with little progress on parameters for future cooperation,

Sale of the Lithuanian state-run oil terminal Klaipedos Nafta to PKN Orlen has been suggested, but the Lithuanian government is not is a rush to sell its stakes in the facility.

Power link to Sweden?

An undersea cable linking the energy grids of the Baltic States with Sweden could be a reality within ten years, said the Swedish foreign ministry in early 2009.

Sweden will lead the EU in the second half of 2009 and is expected to pursue ambitious plans for political and economic cooperation within the Baltic Sea area.

Currently both Lithuania and Latvia have competing claims for an underwater cable across the Baltic Sea to Sweden.

The electricity systems bequeathed to the Baltics by 50 years of Soviet occupation which ended in 1991 have hindered attempts at full energy integration.

Currently, a cable from Estonia to Finland called Estlink is the only major connection joining the Baltics to the Nordic grid. A further link, Estlink 2, is in the advanced planning stage in Estonia.

Lithuania has EU backing for a land link to Poland, and both Lithuania and Latvia have competing claims for an underwater cable across the Baltic Sea to Sweden.

IRENA membership

The International Renewable Energy Agency (IRENA), established on the initiative of Germany, is the first international governmental organization focusing on development and use of renewable energy sources.

In January 2009, Lithuania became a member of the agency.

The government aim is to have renewable energy resources provide at least 20% of total energy needs by 2020.

RUSSIA

Capital: Moscow
Monetary unit: Ruble
Refining capacity: 5,428,497 b/cd
Oil production: 9.760 million b/d
Oil reserves: 60.000 billion bbl
Gas reserves: 1,680.000 tcf

Exploration and Development

Lagansky block: Caspian Sea

Lundin Petroleum (Stockholm) announced progress on several projects.

The Laganskaya-1 exploration well was spudded in September. It is in the Lagansky block, in the Russian sector of the northern Caspian Sea.

The well, with a planned depth of 2,000 m (6,562 ft), is targeting Cretaceous and Jurassic Sandstone reservoirs. Gross unrisked prospective resources for the Laganskaya prospect are estimated at 106 MMboe.

The Morskaya-1 exploration well, also drilled on Lagansky block earlier in the year, encountered a major oil accumulation in the Aptian and Neocomian sandstone reservoirs.

Morskaya is one of several giant and supergiant discoveries of varying recoverable volumes in the northern Caspian, none yet placed on production.

Morskaya-1 was drilled to 2,082 m (6,830 ft). The Morskaya structure is on trend with several major oil and gas discoveries made by Lukoil.

Lundin was planning to drill the Petrovskaya-1 well by yearend to target a large anticline to the northwest of and on trend with the Morskaya discovery.

NORTH CASPIAN OIL FIELDS, DISCOVERIES

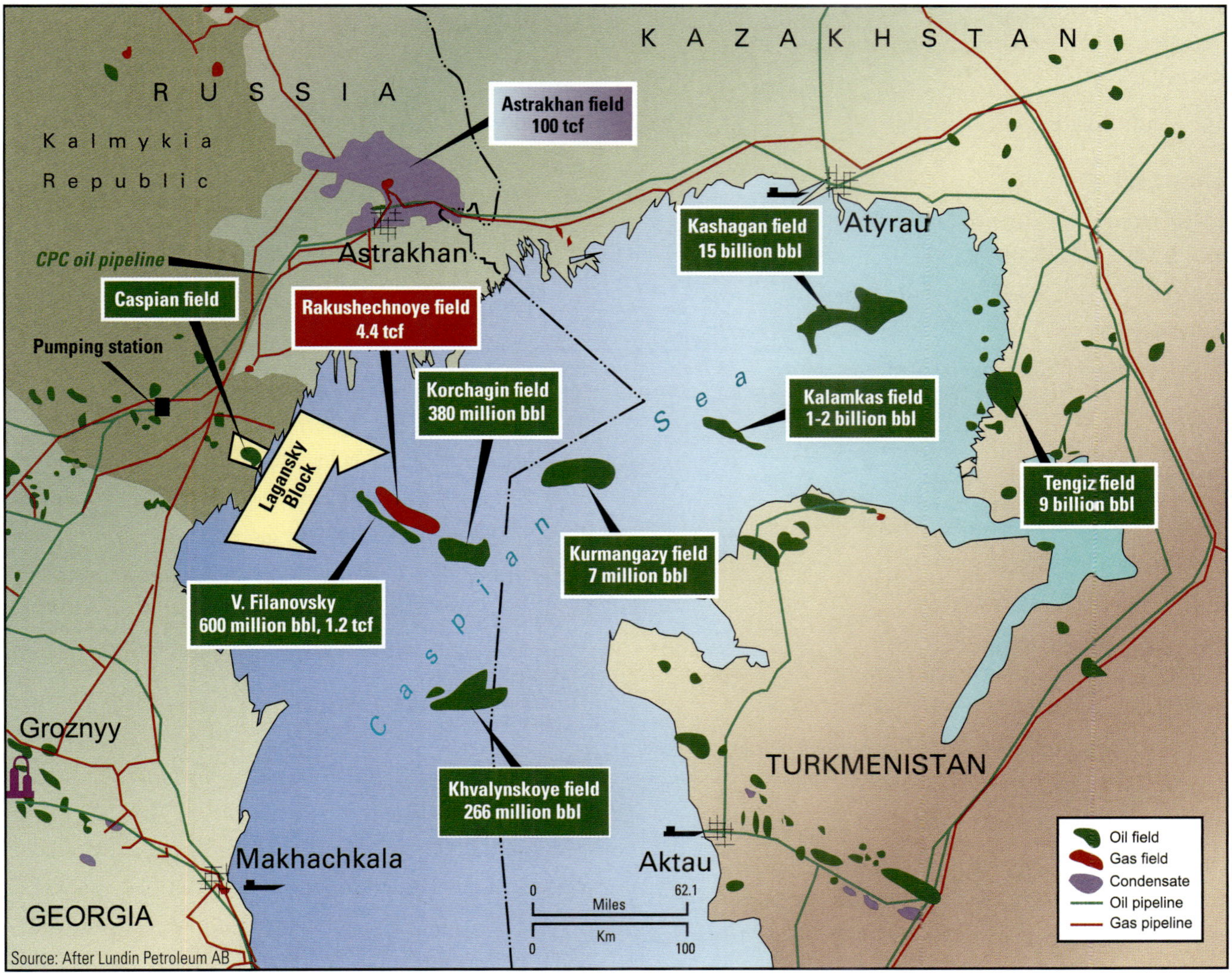

Western Siberian arctic potential

A mid-2008 report by the USGS estimated 3.66 billion bbl of undiscovered oil and 18.4 tcm (651 tcf) of natural gas north of the Arctic Circle in Russia's West Siberian Basin Province.

That province is the world's largest petroleum basin, covering nearly 2.2 million sq km (849,000 sq mi) and with a total discovered volume of more than 360 billion boe.

The report was part of a Circum-Arctic Resource Appraisal program.

Among several giant oil and gas fields discovered to date are Urengoi gas field, with more than 3,500 tcm (100 tcf) of estimated ultimate recovery, and Samotlor oil field, with EUR of more than 28 billion bbl of oil.

The USGS evaluated two assessment units in the province's northern part above the Arctic Circle:

- the Northern West Siberian Onshore Gas AU
- the South Kara Sea Offshore AU.

Kamennoi field

Early planning and engineering work began in early 2008 by TNK-BP for the Kamennoi development project, involving seven Russian oil and gas fields.

The first phase of the project anticipates the recovery of 30 million tons of oil, from 330 wells to be drilled. An export pipeline and additional oil processing facility will be built.

The second phase involves long-term development of the most reliable reserves.

Assessment units in Arctic Western Siberia, Russia

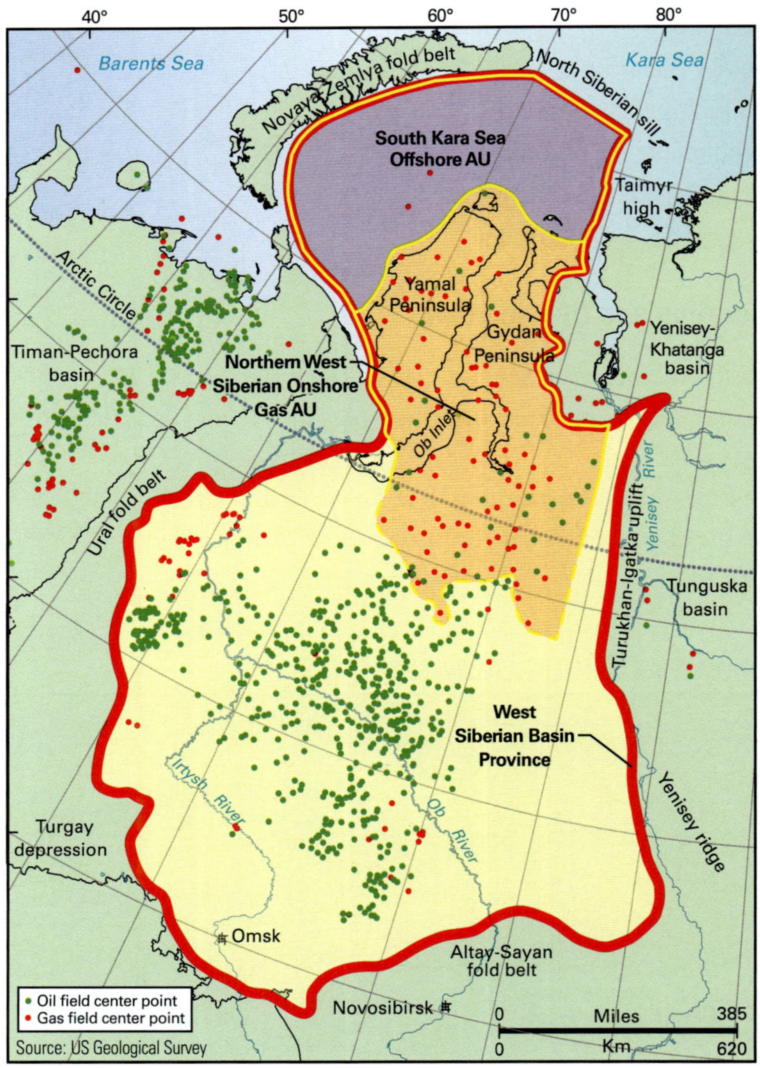

Drilling and Production

Yuzhno Khylchuyu field

A Lukoil-ConocoPhillips Corp. JV began first-stage oil and gas production from Yuzhno Khylchuyu field in the Nenets Autonomous District of northwestern Siberia in mid-2008.

The first 32 development wells were producing oil that moved by pipeline to the Varandey export terminal on Kola Bay on the Barents Sea. The field's oil is of higher quality than the Urals export blend.

Yuzhno Khylchuyu holds more than 500 million bbl of proved oil reserves, and is considered one of the largest fields in Russia's Timan Pechora oil and gas province.

It is the anchor field for $4 billion in developments by the Naryanmarneftegaz joint venture.

A second-stage start-up was planned for December to produce from 32 new wells.

Design capacity for the project is estimated to be 150,000 b/d in 2009.

Major gas projects

In a mid-2008 assessment of pending major gas production projects worldwide, *Oil & Gas Journal* identified two in Russia.

The largest proposed development in Russia is the 3.68-tcm (130-tcf) Shtockman field in the Barents Sea. The first phase at

Shtockman may start producing gas in 2011 at 60 bcmd (2.1 bcfd). Subsequent phases may increase production to 246 bcm (8.7 bcfd) in 2014–19.

A second large development is the 2-tcm (70-tcf) Kovykta field in eastern Russia. Operator Gazprom is now planning the new pipelines that will be needed to move the gas to markets, such as in China.

Gazprom gas production outlook

Forecasts in mid-2008 called for Gazprom gas production to reach 570 bcm (20.2 tcf) for 2010, then 610–615 bcm/yr (21.5–21.7 tcf/yr) by 2015, and 650–670 bcm/yr (23.0–23.7 tcf/yr) by 2020.

By 2020, new fields will account for nearly a half of all gas produced.

Gazprom officials said that in 2007, successful exploration increased discovered reserves to 585 bcm (20.7 bcf), while production was 548.5 bcm (19.4 tcf).

Russia in OPEC?

Speculation about Russia joining OPEC arose in late 2008 after the country sent its most senior delegation in a decade to OPEC's September 9 meeting in Vienna.

Russia is the second-largest oil exporting country after Saudi Arabia.

At the meeting, Russian officials proposed extensive cooperation between Russia and OPEC to meet global energy needs, possibly including joint projects between Russian and OPEC national oil and gas companies, joint investments and the sharing of technology.

But nothing was said about OPEC membership.

Oil & Gas Journal put little stock in the speculation, finding it hard to imagine that Russian officials would accept limits to production imposed by anyone.

At best, said *OGJ*, expect continued uneasy cooperation between OPEC and Russia.

In 2006, exploration increased reserves by 590.9 bcm (20.9 tcf), while production totaled 556 bcm (19.6 tcf).

The 2005 figures were 583.4 bcm (20.6 tcf) and 555 bcm (19.6 tcf), respectively.

Gazprom said it expects three new gas production centers to be operating by 2010: the Arctic shelf, Yamal Peninsula, and East Siberia-Far East.

In the "near future," the company said, Shtokman and Bovanenkovo—in the Arctic shelf and Yamal, respectively—will be put into operation.

International Activities

Russian-Georgian conflict

In late summer 2008, after months of growing tension, outright military hostilities erupted between the Georgian army and Russian forces based in the province of South Ossetia, which was controlled by a separatist group.

Five days of fighting raised concerns about the security and continued operation of key oil and gas pipelines that span Georgia.

Details about the confrontation are presented in the Georgia section of this chapter.

Rosneft, CNPC refinery venture

Initial planning was under way in September 2008 for a JV between CNPC (Beijing) and Russia's Rosneft (Moscow) for building a 320,000-b/d refinery project in Tianjin City.

The venture would be one of the largest refinery projects between China and Russia.

CNPC was working to enhance its market share in downstream products. The company plans to increase its crude oil processing capacity by 2.1 million b/d by 2015.

At yearend 2007, CNPC owned 29 refineries able to process a total of 2.95 million b/d.

Venezuela relationship

In October 2008, a consortium of Russian energy companies—Gazprom, Lukoil Holdings, Rosneft, TNK-BP, and Surgutneftegaz— registered with officials in Venezuela to pursue oil development with state-owned PDVSA.

The Russian group would replace ExxonMobil Corp. (Irving, Texas) and ConocoPhillips Corp. (Houston) in Venezuela.

The venture initially will focus on heavy-oil field development in Venezuela's Orinoco River basin. Future efforts are expected to address other oil and gas projects in Venezuela, then elsewhere in Latin America.

A strategic cooperation agreement was to be signed by yearend 2008. PDVSA will have the controlling stake in the consortium.

Russia, Iran and Qatar: Gas cartel?

In October 2008, Russia, Iran and Qatar made the first serious moves toward forming an OPEC-style cartel on natural gas, raising concerns that Moscow could boost its influence over energy markets spanning from Europe to South Asia.

Details are presented in the Iran section of the Middle East chapter of this book.

Gas dispute with Ukraine

Gazprom cut off all gas supplies to Ukraine on January 1, 2009, after the two sides failed to reach agreement on debt payments and on how much Ukraine will pay for gas in 2009.

About 20 European nations suffered significant reduction in gas supply.

On January 28, 2009, gas supplies were restored after both sides signed a contract with new tariffs.

More details are presented in the Ukraine section of this chapter.

Gas transport, storage in Europe

Gazprom officials said in December 2008 that the company wants to acquire gas transport and distribution assets in France and other areas of Europe to bolster the company's gas storage capacities.

Gazprom Marketing & Trading France, based in Paris, already has joint gas storage projects.

Gazprom agreed in December 2006 to extend Gaz de France's long-term contracts to 2030. At the same time, the partners agreed that Gazprom could directly sell as much as 1.5 bcm/yr (53 bcf/yr) in France.

The company aims to achieve these volumes by 2011, up from 500 MMcm (1.77 bcf) in 2008.

Policy

Deal with Germany

In October 2008, Gazprom dropped demands for stakes in Germany's energy distribution infrastructure.

Analysts saw the action as a sign that Russia wants to improve relations with the west, damaged by the Georgian conflict in August.

News of the change came during announcement of a major gas exploration deal between Gazprom and German utility, E.On Ruhrgas (Dusseldorf).

Sakhalin Island Project

Sakhalin-1

Extended-reach drilling record

ExxonMobil Corp. (Irving, TX) announced in early 2008 that it set a new world record for reaching offshore deposits with land-based extended-reach technology, drilling from an onshore well at its Sakhalin-1 project in eastern Russian to a target area seven miles offshore.

The firm said it drilled the record Z-12 well in half the time needed by conventional technology.

The well was drilled to Chayvo field, which contains 17 of the world's 30 longest extended-reach-drilling wells, and set a record by achieving a measured depth of 11,680 m (38,320 ft).

[Editor's note: This ExxonMobil record was broken later in 2008 by Transocean (Houston) with an offshore well drilled in Qatar. See the Middle East chapter of this book for details.]

Purchase of gas output

Sakhalin-1 officials in February 2008 said they expected that Gazprom would sign an agreement by midyear to buy all natural gas produced at the project [estimated at 11 bcm (388 bcf) per year].

Russian officials have estimated the gas reserves of Sakhalin-1 fields at 500 bcm (17.7 tcf). Gas from the project would supply growing demand in far eastern Russia and other nations in the Far East.

Sakhalin-1 comprises the Chayvo, Odoptu, and Arkutun-Dagi field blocks, which contain potential reserves of about 2.3 billion bbl of oil and 485 bcm (17.1 tcf) of gas.

Partners in Sakhalin-1 are Exxon Neftegas; Japan's Sakhalin Oil & Gas Development Co. Ltd.; RN-Astra and Sakhalinmorneftegas-Shelf (affiliates of Rosneft); and ONGC Videsh Ltd. (New Delhi).

Sakhalin-2

Oil exports from Aniva Bay

Sakhalin-2 is developing oil and gas from Piltun-Astokhskoye field—which includes Piltun and Astokh oil and gas areas—and Lunskoye field.

In December 2008, oil and gas from several platforms in those areas started flowing to a new export terminal at Prigorodnoye port in Aniva Bay, at the south of Sakhalin Island.

The new terminal now includes a subsea pipeline to carry oil to an offshore tanker loading unit. This marked the start of year-round oil-export capability, rather than operation only in ice-free months.

The 800-km (497-mi) TransSakhalin pipeline system connects the offshore Molikpaq, Piltun-Astokhskoye-B, and Lunskoye platforms in the northeast to the new terminal.

An offshore tanker took on the first cargo of 100,000 tonnes of oil in late December for shipment to Yosu, South Korea.

At full operation, Prigorodnoye port will serve 100 Aframax oil tankers and 160 LNG carriers per year. Operation of two-train LNG processing facility is scheduled to start in 2009 (see below).

Sakhalin Island Energy Co. has produced more than 100 million bbl of Vityaz crude since 1999. SEIC partners are Gazprom, Royal Dutch Shell PLC (The Hague, Netherlands), Mitsui & Co. Ltd., and Mitsubishi Corp. (both based in Tokyo).

LNG deliveries: 2009

Shipments of LNG to Japan from Sakhalin-2 are expected to begin in early 2009, said the two Japanese partners in the project, Mitsubishi Corp. and Mitsui & Co.

The LNG project had been delayed by environmental concerns expressed by Russian officials.

In early October 2008, the Japan Bank of International Cooperation (JBIC) joined two other banks to finance the building (now complete) of three LNG tankers for the project.

SEIC has signed long-term supply contracts with Japanese, South Korean, and U.S. buyers for nearly all LNG from the island's 9.6-mtpy plant.

At mid-year 2008, JBIC and four private-sector banks were finalizing plans to provide a $5.3 billion loan for the Sakhalin-2 project, aiming to sign a lending agreement with Gazprom.

Gazprom and its SEIC partners have already committed about $15 billion to the project.

E.On will receive a stake of almost 25% in Yuzhno-Russkoye, a Siberian field and one of the world's largest gas reserves. In return, it will give up almost half (3%) of its 6.5% position in Gazprom.

BASF (Ludwigshafen) already holds a stake in the Yuzhno field through its energy unit, Wintershall.

Some analysts also think that Russia hopes its action can calm opposition to the Nord Stream pipeline, which would bring Russian gas under the Baltic Sea to Germany.

BASF and E.On are developing that line with Gazprom, and Yuzhno would be the main supply for the pipeline.

Cuts on oil export duties

In mid-November 2008, Russia announced plans to cut oil export duties by 33% by yearend.

Exporting oil from Russia became unprofitable for many companies because of falling oil prices and heavy taxation in the second half of the year.

A reduction in exports also was intensifying Russia's growing financial crisis, cutting government revenues. Russian crude exports dropped by 6% in the first 10 months of 2008.

Prices for the Urals Russian crude export blend fell in early November to below $50/bbl for the first time since January 2007 as demand weakened in Europe, Russia's biggest market.

Russian gas to Abkhazia?

Gazprom officials said in mid-2008 that the firm was exploring construction of a gas pipeline to Georgia's breakaway Abkhazia region on the Black Sea coast, according to a Reuters news report.

(Note: This news came before the outbreak of fighting in August between Georgian forces and Russian-backed separatists in Abkhazia and South Ossetia.)

The proposal generated alarm among Western governments who said Russian support for the separatists would stir tensions.

Georgia also viewed the idea dimly, seeing Russia as trying to annex the Black Sea region. It views as illegal any business dealings with the separatist administration.

Georgian officials also have alleged that Gazprom intends to explore for oil in Abkhazia.

Pipelines

Nord Stream

In late 2008, the controversial 1,200-km (746-mi) Nord Stream gas pipeline to link Russia and Germany under the Baltic Sea still appeared to be on track, despite financial and environmental doubts raised about the project.

On the financial front, Gazprom had warned that the credit crisis had weakened its ability to finance debts and obtain new loans, creating doubts about its financing of Nord Stream. Total construction costs have been estimated at $9.5 billion.

Experts have also wondered if Gazprom can afford to develop the giant Arctic Shtokman field that is intended to help supply Nord Stream.

Partners Gazprom, BASF/Wintershall (Kassel, Germany), E.On Ruhrgas (Dusseldorf), and Nederlandse Gasunie (Groningen, The Netherlands) are to pay 30% of Nord Stream costs. A group of banks are to fund the rest.

Political opposition also has created concern. Hungary, Poland, Latvia, Lithuania and Estonia have argued about the line's route, and Scandinavian nations have expressed environmental concerns.

The United States also has said that it opposes Nord Stream. One official described it as "Russia's energy weapon".

Nonetheless, Nord Stream financial officials said they expected completion by 2012.

ESPO pipeline

Snapshot

The ESPO pipeline system is being built in two stages. The first is comprised of the pipeline from Taishet [680 km (423 mi) northwest of Irkutsk] to Skovorodino (near Russia's border with China), plus the oil terminal in Kozmino (on Russia's Pacific Coast).

The second stage consists of a pipeline from Skovorodino to Kozmino.

Under the first stage, as much as 30 million tonnes of oil will be delivered to Skovorodino, then shipped by rail to Kozmino.

In the second stage, an additional 50 million tonnes will be delivered through the line directly to Kozmino.

As of early 2008, Russian officials had not yet stated clearly when ESPO would be fully built and ready to be commissioned.

Export terminal construction

Transneft said it would start construction of the export terminal of the ESPO pipeline in March 2008, at Kozmino, on Russia's Pacific Coast.

Its first tanker berth will accommodate 80,000-dwt ships; a later berth will handle 150,000-dwt ships.

Plans call for an initial 300 visits/yr, rising to 800 when the second berth is completed.

Spur to China

In late 2008, Transneft and China's CNPC (Beijing) signed a major agreement to build a 67-km (41.6-mi), 300,000-b/d pipeline spur from the ESPO pipeline to the Chinese border.

Details are presented in the China chapter of this book.

Kazakh interest

Russian officials said in late 2008 that Kazakhstan's state-owned KazTransOil was interested in transporting Kazakh oil through the ESPO system. For details, see the Kazakhstan section of this chapter.

Siberian oil source

Irkutsk Oil was developing oil fields in mid-2008 in Siberia that could aid extension of the ESPO oil pipeline from Skovorodino to the Sea of Japan.

A JV set up by Irkutsk Oil and Japan Oil, Gas, & Metals National Corp. (JOGMEC; Kanagawa), said it might start test-drilling for oil in the region by yearend 2008.

Irkutsk Oil is to invest more than $500 million by 2015; the European Bank for Reconstruction and Development also will provide partial financing.

Vankor, Talakan oil fields

Rosneft said it would undertake development of Vankor oil fields in connection with the ESPO project. Those fields in the Krasnoyarsk Territory are expected to supply most of the crude for the ESPO line.

Commercial production at Vankor was slated to start in mid-2008, with construction of the 20-mtpy Vankor-Purpe pipeline to be completed in September.

Rosneft expects Vankor output to rise from 9.8 million tonnes in 2009 to 18.7 million tonnes in 2011.

Surgutneftegas—also aiming to partly supply the ESPO line with crude oil—said that it was preparing to begin commercial development of the Talakan oil field.

The field has reserves estimated at 124 million tonnes of oil and 47 bcm (1.66 tcf) of natural gas.

Plans were to start filling the pipeline in mid-2008. This would allow commercial pumping to begin with the launch of the Talakan-Taishet section of the pipeline, planned for fourth quarter 2008.

Production of 1 million tonnes was estimated for 2009.

Imperial Energy pipelines

Imperial Energy Corp. PLC (London) announced in early 2008 that oil from its Snezhnoye field, Block 77, Tomsk, Western Siberia, was now flowing into the Transneft national oil transportation system through Imperial's transfer station at Zavyalovo.

The direct connection unlocks the potential of Imperial's Blocks 74 and 77, northwest of the Ob River and renders the company's North Western pipeline fully operational.

Both of Imperial's west-of-the-Ob pipelines—the Maiskoye-Luginetskoye pipeline and the Snezhnoye-Zavyalovo pipeline are now fully operational and flowing oil.

The firm also said that all construction permissions had been granted for a third pipeline being built to link its Block 80, east of the Ob River, to its Zavyalovo custody transfer station.

First oil through that line was expected in late 2008.

Oman stake in CPc bought

In late 2008, Oman sold to Russia its 7% stake in the Caspian Pipeline Consortium (CPc) led by Chevron Corp. (San Ramon, CA).

The consortium owns and operates the 1,580-km (982-mi) Tengiz-Novorossiisk oil pipeline.

Further details are in the Oman section of the Middle East chapter of this book.

Serbian stretch of South Stream

Russia signed several agreements for oil and gas with Serbia in early 2008, aimed at affirming its political support for the Balkan country.

The main development was an agreement to create a joint company that will build the Serbian stretch of the South Stream gas pipeline.

The agreement followed the decision by the United States and several EU member states to recognize the declaration of independence made by the former Serbian state of Kosovo.

Russia, siding with Serbia, said it will block Kosovo from joining the UN or other international organizations.

Regarding South Stream, the new company would draft details for the pipeline segment by late 2009. Construction could start by late 2011.

Negotiations also are planned with Italy, Bulgaria, Bosnia and Greece as possible partners.

The $14.6-billion South Stream project by Gazprom and Eni SPA (Rome) is designed to carry at least 10 bcm/yr (353 bcf/yr) of Siberian gas to Western Europe.

Serbian NIS acquired

It was reported in early 2008 that Serbia agreed to sell to Gazprom a controlling interest in state oil company NIS. Terms were not disclosed, but Gazprom in December 2007 had offered €400 million for a 51% stake.

Gazprom also promised to invest additional funds and to route the South Stream gas pipeline through Serbia.

South Stream is a €10-billion project, running 898 km (558 mi) under the Black Sea, and then by way of Bulgaria to Italy.

Gazprom and Eni SPA reached commercial terms on the project in December 2007.

Feasibility and marketing studies on the 30-bcm/yr (1.06-tcf/yr) project were to be completed by yearend 2008. Construction could start in 2009, with an in-service date of 2013.

BPS-2

Russian prime minister, Vladimir Putin announced in December 2008 that work would begin in 2009 on the design and construction of the second phase of the Baltic Pipeline System (BPS-2).

The line will carry oil between Unecha in the Bryansk region and Ust-Luga in Leningrad. It is seen as an alternate to the Druzhba pipeline, which supplies oil and natural gas to many EU countries.

Unecha is on the Ukrainian border and Ust-Luga on the Baltic coast. Tankers will carry oil from Ust-Luga to various countries.

The first phase of the BPS, with a capacity of 12 mtpy, has been operational since 2001.

Construction of the $4.5-billion BPS-2 is expected to take two years, with first oil through the pipeline in 2012. Maximum design capacity for the 1,287-km (800-mi) line is 50 mtpy.

Siberian gas to South Korea

The South Korean ministry of energy announced in October 2008 that the country plans to import up to 10 bcm/yr (353 bcf/yr) of gas from Russia for 30 years, beginning in 2015.

The gas would move through a $3-billion pipeline, to run from Vladivostok to South Korea through North Korea.

The line is part of a $90-billion deal under which South Korea would import Siberian gas, providing Russia with another market in addition to Europe.

Details are presented in the South Korea section of the Asia-Pacific chapter of this book.

TAJIKISTAN

Capital: Dushanbe
Monetary unit: Ruble
Refining capacity: 0
Oil production: 0
Oil reserves: 12.000 million bbl
Gas reserves: 200.000 bcf

Gazprom gets licenses

Gazprom reported in mid-2008 an agreement with the government of Tajikistan to explore and develop four oil and gas fields.

The four prospects are estimated to hold 3 tcm (106 tcf) of natural gas and a significant volume of oil.

Gazprom will conduct geological and geophysical studies prior to exploratory drilling. Output from the fields is expected by 2013.

It was thought that the four fields are the Sarikamysh Block, Rengan Block, Shaambary Block, and Sargazon Block in southern Tajikistan.

Tethys operations

Tethys Petroleum Ltd. (Guernsey, UK) in late 2008 updated its work in Tajikistan.

Works was underway on the Komsomolsk gas field, near Dushanbe, on three wells: KOM180, KOM183, and KOM007. Gas previously tested from KOM180 at 10,000 cmd (353,000 cfd).

At the Khoja Sartez gas condensate field, north of Kulob, gas was successfully tested from the Khoja Sartez 22 well at 80,000 cmd (2.8 MMcfd).

A Tethys subsidiary signed an MOU with the ministry of energy to take over drilling operations on the East Olimtoi exploration well (EOL09) in the KPL's Bokhtar Production Sharing Contract area.

At the Beshtentak oil and gas field, north of Kulob, preparatory work was under way for workovers and field enhancements.

TURKMENISTAN

Capital: Ashgabat
Monetary unit: Manat
Refining capacity: 236,970 b/cd
Oil production: 220,000 b/d
Oil reserves: 600.000 million bbl
Gas reserves: 94.000 tcf

Dzheitune (Lam) wells

The Dzheitune (Lam) A/129 development well yielded a combined rate of 2,123 b/d of oil from short and long strings following initial testing in August 2008, said Dragon Oil (Dubai).

The well was drilled to 4,135 m (13,566 ft). Further testing and optimization was planned.

Dzheitune (Lam) 22/130 also was being drilled, from the Dzheitune (Lam) 22 platform, with a target depth of 3,840 m (12,598 ft). Production was expected to begin in October.

The next phase of drilling was slated for the Dzheitune (Lam) 28 platform, to be followed by spudding of Dzheitune (Lam) A/131.

The A/131 well, with a target depth of 4,164 m (13,661 ft), was expected to come online in November.

Seeking energy cooperation

At a London conference in early 2008, Turkmen officials said the country would allow IOCs to develop Caspian Sea offshore deposits under production-sharing agreements (PSAs).

The officials said that 32 licenses for oil and gas exploration in the Turkmen section of the Caspian Sea had already been issued.

Turkmenistan said it will reserve the exclusive right of domestic companies to extract oil and gas from onshore deposits, but that foreign partners may participate by providing services.

A 25-yr trade agreement concluded in 2003 between Russia and Turkmenistan currently delivers 50 bcm/yr (1,766 bcf/yr) of gas to Russia's Gazprom. Deliveries are expected to increase to 70–80 bcm (2,472–2,825 bcf) in the next few years.

Turkmen officials also said a gas pipeline to China would become operational in 2009, moving 30 bcm (1,059 bcf) of gas from eastern Turkmenistan to China by way of Uzbekistan and Kazakhstan.

Other export possibilities under consideration include a gas pipeline to Pakistan via Afghanistan, and a gas pipeline to Europe via the southern Caspian-Black Sea corridor or Nabucco gas pipeline.

Turkmen gas for Europe

During an early-2008 visit to Turkmenistan, EU officials got assurances from Turkmen officials that 10 bcm (353 bcf) of natural gas would be available for EU members every year, beginning in 2009.

This was said to be the first time that a volume of gas supply was specified.

The EU is working to diversify its gas sources away from Russia's dominant position for reasons of both political security and safe supplies.

The EU also has confirmed agreements with Azerbaijan, Kazakhstan, and Ukraine as well as with Egypt, Morocco, and Jordan. Negotiations continued with Libya, Algeria and Iraq.

A deeper EU partnership was also being sought with Gulf Cooperation Council countries (Bahrain, Qatar, Kuwait, Oman, Saudi Arabia, and the UAE).

UKRAINE

Capital: Kiev
Monetary unit: Hryvnia
Refining capacity: 879,759 b/cd
Oil production: 75,000 b/d
Oil reserves: 395.000 million bbl
Gas reserves: 39.000 tcf

Gas supply from Russia

Russia's Gazprom cut off all gas supplies to Ukraine on January 1, 2009, after the two sides failed to reach agreement on debt payments and on how much Ukraine will pay for gas in 2009.

The cutoff replayed a similar crisis that occurred in early 2008. It took until early March 2008 for Russia and Ukraine to resolve a disagreement that began in October 2007 regarding the supply, pricing, and transit of Central Asian natural gas.

The EU gets about one-quarter of its gas from the Russian company. Around 80% of EU gas imports pass through the Sudzha gas station in Ukraine, which pumps more than 76.6 MMcmd (2.7 bcfd).

About 20 European nations suffered significant reduction in gas supply in January 2009, with some suffering their coldest temperatures in decades.

On January 28, 2009, gas flow was restored after both nations signed a contract with new tariffs.

Ukraine has been buying gas at rates much lower than the European market prices. Russia has been demanding that Ukraine pay the same rates as other buying countries.

Under the new agreement, Russia will sell gas to Ukraine at a discount of 20% on the market price.

Relations between the two nations are strained in part by the efforts of Ukrainian Pres. Viktor Yushchenko to win NATO membership and by his support for Georgia in the August 2008 fighting with Russia.

Lelyaki oil field

The Kashtan Petroleum Joint Venture was planning to spud a second development well in Lelyaki oil field in east-central Ukraine in mid-2008.

The directional No. 307 well was to go to 1,960 m (6,430 ft) to further develop the P1 and 2 and K1 reservoirs.

Shelton Canada Ltd. (Calgary) and Ukraine's state Ukrnafta make up the Kashtan venture.

Kashtan also was planning to reenter and sidetrack four other suspended wells by yearend 2008.

The venture's first well has averaged 185 b/d since coming on production in December 2007.

Odessa-Brody oil line

In mid-2008, the Odessa-Brody oil pipeline began carrying light crude oil in its originally-planned direction, from the Caspian Sea toward Europe, according to Ukrainian officials.

The line was originally built to transport light Caspian crude to Europe, but it has never operated as planned. Instead, it has carried Russian oil to Odessa.

The reversal is part of a larger project aimed at extending the pipeline and using it to supply Europe with oil from the Caspian region.

In October 2007, five nations—Azerbaijan, Georgia, Lithuania, Poland, and Ukraine—agreed to set up the so-called Samartia consortium, aiming to enable transportation of oil from the Caspian Sea region—particularly Azerbaijan and possibly Kazakhstan—through Poland to western Europe.

In April 2008, Ukraine and Poland authorized a feasibility study for a new oil network that would extend the Odessa-Brody pipeline from Ukraine to Poland.

That plan calls for the extension of the pipeline to the central Polish city of Plock, site of the country's largest refinery, allowing new supplies to be shipped on to Poland's Baltic Sea port of Gdansk.

UZBEKISTAN

Capital: Tashkent
Monetary unit: Soum
Refining capacity: 222,271 b/cd
Oil production: 105,000 b/d
Oil reserves: 594.000 million bbl
Gas reserves: 65.000 tcf

Chinese-Uzbek oil venture

The CNPC subsidiary of PetroChina Co. Ltd. (Beijing) said in late 2008 that it will form a JV with state-owned Uzbekneftegaz to develop Mingbulak oil field in Namangan province on the northern edge of the Fergana Valley in northeastern Uzbekistan.

CNPC in January 2008 started building the Central Asia-China (CAC) pipeline, which will transport gas from Turkmenistan, through Uzbekistan and southern Kazakhstan, to western China's Xinjiang region.

Scheduled to start operations in 2009, the line will have a capacity of 31.5 bcm/yr (1.1 tcf/yr).

CNPC said production from Mingbulak field, with reserves estimated at more than 30 million tonnes, is expected to reach 2 mtpy.

Russian-Uzbek LNG venture

Uzbekistan's Pres. Islam Karimov approved in early 2009 creation of a $221.5-million joint venture for the production of LNG by Russia's Stroytransgaz and Uzbekistan's Uzbekneftegaz.

The LNG would be produced at the Mubarek natural gas processing plant.

A formal agreement is expected to be signed in March.

The facility will process 12 bcm/yr (424 bcf/yr) of gas and produce 270,000 tonnes of LNG and 70,000 tonnes of gas condensate.

Stroytransgaz, Uzbekneftegaz, and an unnamed Chinese bank will fund the venture.

China

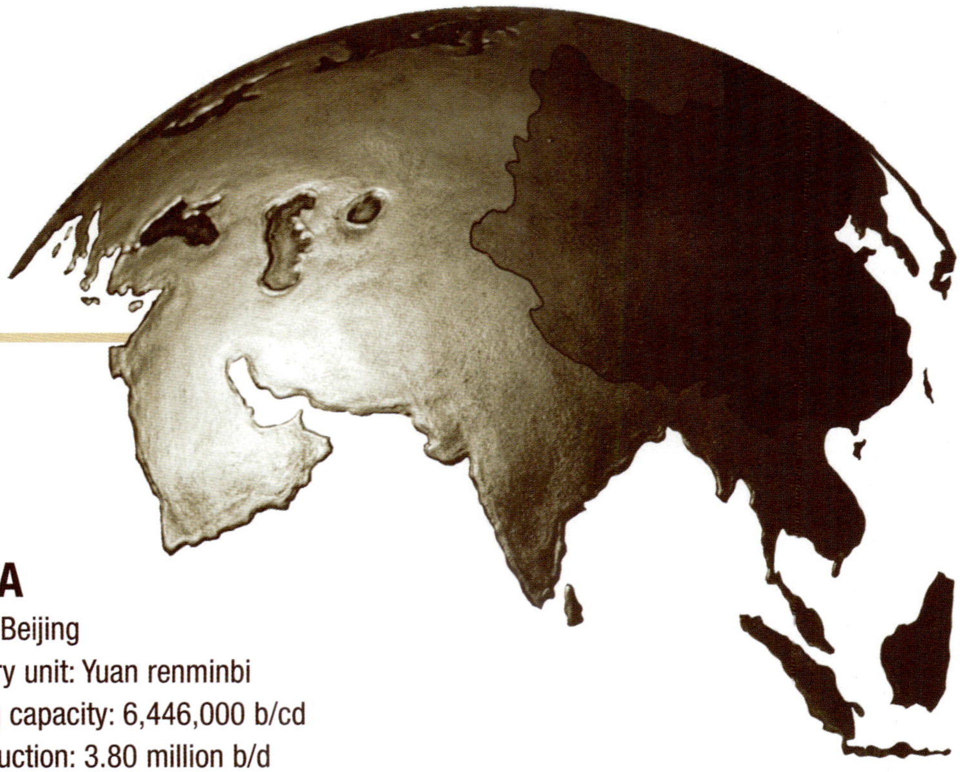

CHINA

Capital: Beijing
Monetary unit: Yuan renminbi
Refining capacity: 6,446,000 b/cd
Oil production: 3.80 million b/d
Oil reserves: 16.00 billion bbl
Gas reserves: 80.00 tcf

Overview

General

China has rather limited oil and gas reserves. However, since the early 1990s, China's demand for energy has been surging. Securing supplies from abroad has become a major component of foreign policy, which is causing some international concern.

Oil and gas resources are controlled by three state firms: China National Petroleum Corp. (CNPC), China Petroleum & Chemical Corp. (Sinopec), and China National Offshore Oil Corp. (CNOOC). The government maintains a majority stake in all of them.

The National Development and Reform Commission (NDRC) is China's main energy planner.

CNPC and Sinopec operate almost all of China's refineries and the domestic pipeline network, while CNOOC has the most expertise with international transactions.

The country's first private sector company, the Great Wall Petroleum Group, was founded in 2005.

In 2006 and 2007, the government also began to take action to build a strategic petroleum reserve.

CHINA'S 2000-20 ENERGY MIX

Fuel	Annual growth rate, 2000-20	Energy mix 2000	Energy mix 2010	Energy mix 2020
	%			
Coal	4.22	69.9	66.7	63.2
Oil	5.10	25.0	25.2	26.7
Gas	9.44	2.8	5.2	6.7
Other	—	2.3	2.9	3.4

Source: Development Research Center of the State Council, Overview of the National Energy Strategy, 2006

Oil production

China has sought assistance from international oil companies to boost oil recovery and extend the life of producing fields.

In the mid-2000s Daqing, Liaohe, Shengli, Xinjiang, Changqing, Sinopec Star, and Zhongyuan fields produced most of China's oil.

WESTERN CHINA
RUSSIA
KAZAKHSTAN
MONGOLIAN REPUBLIC
KYRGYZSTAN
PAKISTAN
INDIA
PEOPLE'S REPUBLIC OF CHINA
NEPAL
BHUTAN
BANGLADESH
MYANMAR (BURMA)
THAILAND
LAOS
VIETNAM
Xinjiang
Qinghai
Xizang
Gansu
Ningxia
Shaanxi
Sichuan
Guizhou
Yunnan
Guangxi
Pavlodar
Karaganda
Atasu
Ust-Kamenogorsk
Lake Balkhash
Dzungarian Gate
Ulan Bator
Almaty
Bishkek
Uerho
Fengcheng
Karamay
Shingshihsia
Shawan
Usa
Dushanzi
Qigu
Urumqi
Bogdo
Suilai
Shanshan
Hami Turfan Hami
Hami
Shan
Shengting-k'ou
Zunnbayan
Tsagaan Els
Yakela
Kumger
Kosaptok
Lunnan
Yiqikelike
Dongqiulitage
Taqlaq
Yingmali
Kan
Aksu
Kuqa
Lunan
Korla
Akekule
Wensuh
Yuecan
Bostan
Daliya
Ichkelik
Wug Chat
Kashgar
Kashi
Kelatuo
Kekeya
Tsehpu
Tazhong
Pal-yang-ho
Shih-you-kou
Ya 'exia
Laojunmiao
Danbei
Baiyanghe
Shiyougou
Yumen
Gaotai
Xianshuiquan
Lenghu
Shizigou
Qigequan
Huatugo
Youshashan
Wunan
Jiandinshan
Nanyishan
Youquanzi
Kaitemilike
Lenghusanhao
Lenghushihao
Mahai
Lenghuwuhao
Yuka
Toufengshan
Sebeiyihao
Sebeiehao
Yanhu
Golmud
Tianjum
Koko Nor
Xining (Sining)
Lanzhou
Tianshui
Yinchuan
Shenglijin
Liujiazhuang
Lizhuangzi
Majiatan
Jingbian
Dashuikeng
Baiyanjin
Yuancheng
Mafang
Huachi
Maling
Chenghao
Zhilou
Qingyang
Yan'an
Yongping
Ganguye
Xiashiwan
Shih-lian-miao
Houma
Wushan
Xi'an (Sian)
Manali
Hoshiarpur
Nagqu
Lhasa
Brahmaputra
Katmandu
Thimphu
Ganges
Jiangyou
Zhongba
Yingshan
Deyang
Wubaiti
Penglaizhen
Gong
Nanchong
Chengdu
Leshan
Wolonghe
Chongqing
Weiyuan
Shiyougou
Yibin
Xishui
Luzhou
Guiyang
Kunming
Tengcong
Liuzhou
Tiandong
Nanning
Beihai
Weizhou 11-1
Weizhou 10-3
Weizhou 6-1
Hanoi
Haiphong
Tien Hai
Gulf of Tonkin
Dongfang
Shillong
Chhatak
Sylhet
Haripur
Kailastila
Patharia
Rashidpur
Kamta
Beanibazar
Fenchugani
Kishorganj
Titas
Habiganj
Bakhrabad
Begamganj
Semutang
Dhaka
Feni
Chittagong
Kutubdia
Patiya
Mandalay
Ayadaw
Yenangyaung
Chauk/Lanywa
Mann
Htaukshabin
Paduakpin
Pyaye
Pyalo
Prome
Myanaung
Shwepyitha
Bay of Bengal
Irrawaddy
Mekong
Mae Fang
Mae Suhn
Louanphrabang
Utiaradit
Mae Sot
Vientiane
0 50 100 150 200 250 Miles
0 100 200 300 400 Kilometers

EASTERN CHINA
RUSSIA
MONGOLIAN REPUBLIC
PEOPLE'S REPUBLIC OF CHINA
NORTH KOREA
SOUTH KOREA
JAPAN
TAIWAN
VIETNAM
LAOS
THAILAND
PHILIPPINES
OKINAWA
Sea of Japan
Yellow Sea
East China Sea
South China Sea
Bohai Gulf
Gulf of Tonkin
Heilongjiang
Jilin
Liaoning
Nei Mongol
Ningxia
Gansu
Shanxi
Hebei
Shandong
Henan
Shaanxi
Jiangsu
Anhui
Hubei
Sichuan
Hunan
Zhejiang
Jiangxi
Fujian
Guizhou
Yunnan
Guangxi
Guangdong
HAINAN DAO
Beijing
Tianjin
Shanghai
Daqing
Harbin
Shenyang
Dalian
Nanjing
Wuhan
Chongqing
Chengdu
Xi'an (Sian)
Lanzhou
Kunming
Guangzhou
Hong Kong
Taipei
Pyongyang
Seoul
Ulan Bator
Khabarovsk
Komsomolsk
Hanoi
Vientiane
0 100 200 300 Miles
0 200 400 Km

CHINA'S OIL CONSUMPTION VS. PRODUCTION

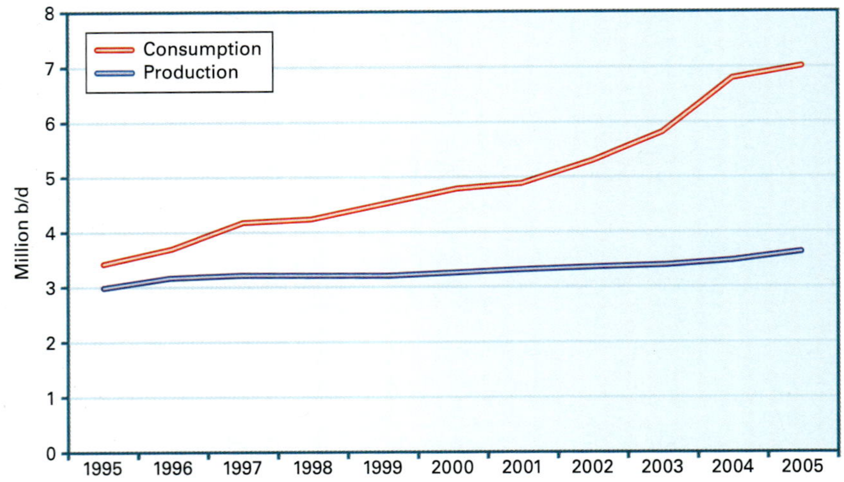

However, production from two major onshore fields, Daqing and Liaohe, is dropping, and production from new finds made in 2004 are not likely to offset this decline.

Only about 15% of China's oil production capacity is from offshore resources, but that production has been growing. The government is aiming for 67 million tonnes (491 million bbl) by 2010.

Interest in offshore exploration has centered on the Bohai Sea, east of Tianjin—believed to hold more than 1.5 billion bbl of reserves—and on the Pearl River Mouth.

CHINA'S TOP OIL IMPORT SOURCES

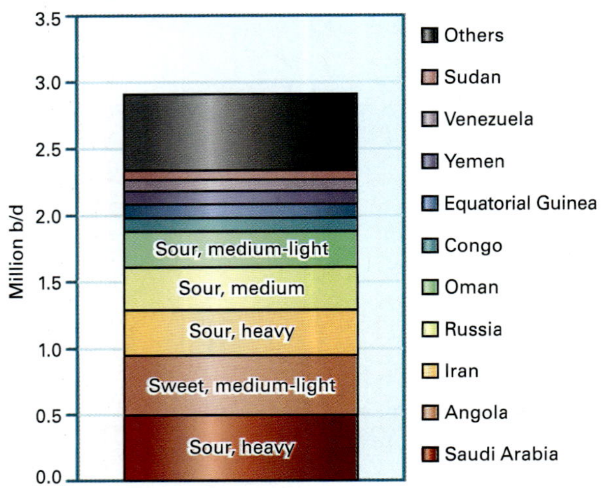

Oil consumption, imports

According to the Energy Information Administration (EIA), part of the U.S. Department of Energy (DOE), China's demand for oil is increasing by some 15% per year. The country consumed 7.4 million b/d in 2006, an increase of nearly a 0.5 million b/d from 2005.

Oil consumption is projected to rise to 15 million b/d by 2030.

Oil consumption—net imports plus output—rose 7.3% to 346 million tonnes (2.54 billion bbl) in 2007.

Imports account for about 46% of China's oil consumption. In 2005, China imported 130 million tonnes (953 million bbl) of crude oil, up 3.3% over 2004.

Nearly 60% of China's oil imports come from the Middle East. Imports from Saudi Arabia more than doubled, from 8.8 million tonnes (64.2 million bbl) in 2001 to more than 20 million tonnes (147 million bbl) in 2005.

The government estimates that it will need to import 70% of its oil (and 50% of its gas) by 2020.

Total oil imports could top 348 million tonnes (2.55 billion bbl) in 2020, driven by rising motor vehicle ownership and the country's energy-intensive basic industries.

CHINA'S 2005 OIL CONSUMPTION

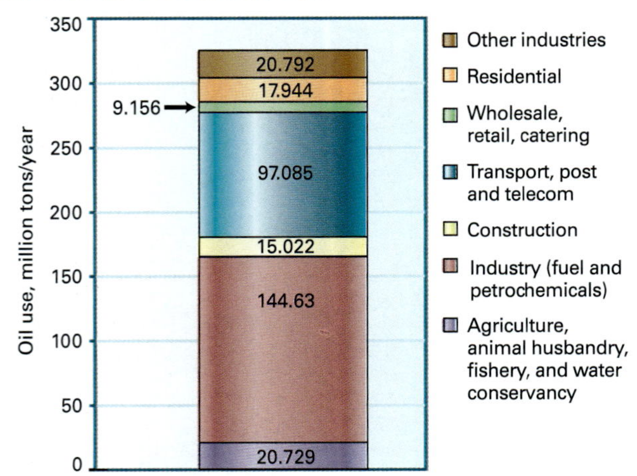

Source: China Strategic Advisory

Natural gas, LNG

China's use of natural gas will grow from increased domestic production in the west, new pipeline supplies from Central Asia, and a growing LNG terminal population.

Natural gas has not been a major fuel source, but EIA projects that natural gas use will jump from 1.2 tcf to 7.0 tcf between 2003 and 2030.

The country has begun a major expansion of its gas infrastructure. Most notable is construction of the West-East pipeline, connecting the Tarim and Ordos basins to the populous and industrial southern coast. China also is considering importing gas by pipeline from Russia and Kazakhstan.

Currently China has one operating LNG terminal, in Guangdong, importing about 3 mtpy from Australia. A second is being commissioned in Fujian, and three more are under construction at Shanghai, Jiangsu, and Dalian. As many as six more are approved or proposed and likely.

Refining and SPR

Strong Chinese oil demand growth will affect the refining and petrochemical sectors in the same way that it is reshaping the global upstream and oil shipping businesses.

CNPC and Sinopec are China's two chief refiners, but the Chinese refining industry has more than 120 individual refining enterprises. The average Chinese refinery processes about 52,000 b/d, about one-half the international average refinery throughput.

China refined 326.79 million tonnes (2.4 billion bbl) of crude in 2007, a growth of 6.4% over 2006.

Plans to boost refining capacity include: a new refinery in Huizhou (Guangdong province) to be operational in 2008; expansion of the Quanzhou refinery in Fujian plus construction of a new one there; and expansion of the Dushanzi refinery in Xinjiang.

China also has begun building a strategic petroleum reserve. It will hold state-controlled crude stocks at Dalian, Huangdao (Qingdao), Zhenhai, and Zhoushan (Ningbo).

Earthquake takes terrible toll

A massive earthquake struck western Sichuan province on May 12, 2008. Official casualty estimates in mid-June put the death toll at more than 69,000 with more than 17,000 people still missing.

The 8.0 magnitude quake toppled eight schools several chemical plants and at least one hospital.

The epicenter was in Wenchuan county, about 100 km (62 mi) from the Sichuan provincial capital Chengdu, Xinhua news agency said. Landslides cut three major rail lines leading to Chengdu, stranding dozens of trains.

Sichuan has long been the leading natural gas producing province in China.

The Jamestown Foundation (Washington, DC) noted in a July 2008 assessment that Sichuan province is a key regional petroleum market, with annual oil product consumption of about 10 Mt in 2007.

The foundation said PetroChina Company Ltd. (Beijing) might reconsider plans to build a $5.7 billion refinery and petrochemical complex in Pengzhou. It had been scheduled for completion by 2010.

Sinopec said that 97% of its Sichuan gas wells had resumed production two weeks after the earthquake.

Jamestown analysts said that China might speed up strategic oil reserve development and the lifting of subsidies on gas prices to speed economic recovery in damaged regions.

Policy

New energy agency

China's NDRC established a new, nine-department government agency, the National Energy Administration (NEA) to secure energy supplies.

Observers say the consolidation signals that China will focus more on energy strategy and planning for sustainable development.

NEA will chart energy strategy and policies as well as manage separate sectors for oil, gas, coal, electricity, nuclear, and renewable energy. It also will manage China's strategic oil reserves.

Oil Supply and Demand Outlook

Peak oil models

Using three peak oil models, researchers with China Petroleum University and Sinopec (both in Beijing) found a widening gap between China's oil demand and production.

Their analyses used a 21.2-billion tonne (155.4-billion bbl) estimate for recoverable reserves in China.

The generalized Weng model predicted peak oil production in China of 196 million tonnes (1.44 billion bbl) in 2026, and the Hubbert model indicates a peak oil demand in 2034 of 633 million tonnes (4.64 billion bbl).

Oil discovery trend

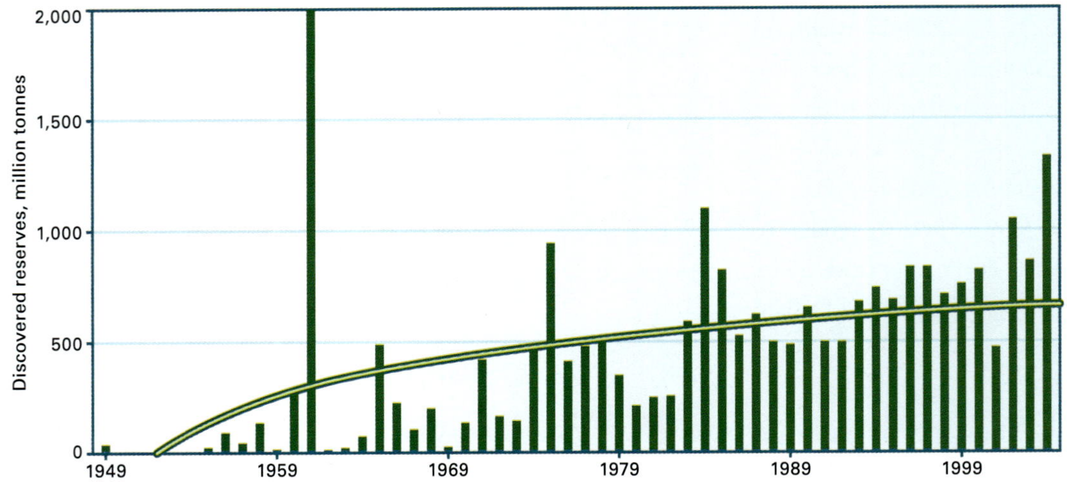

Oil demand, GDP

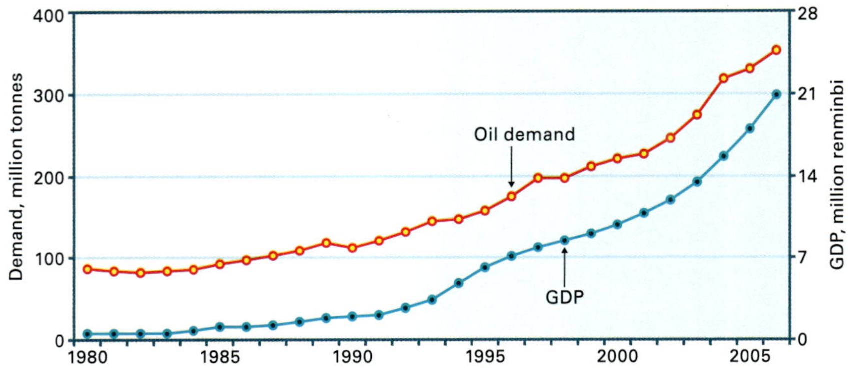

Oil demand, production gap

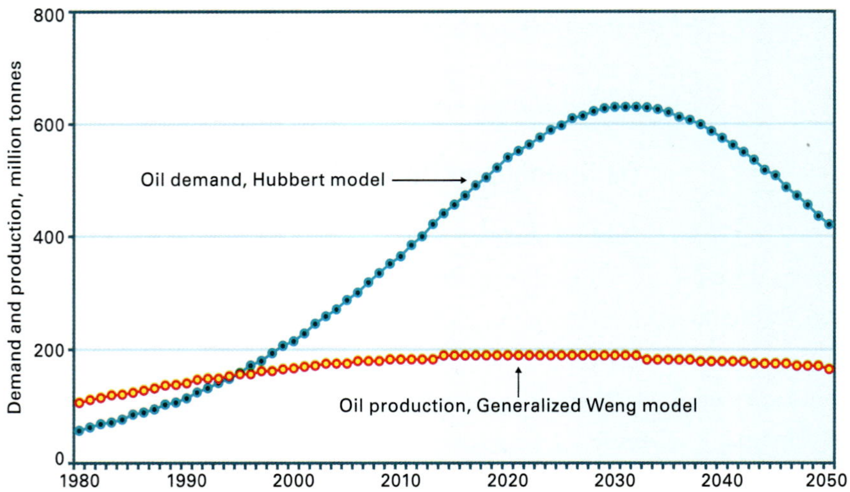

Tibet unrest

The fiercest anti-China protests in Tibet in almost two decades drew international scrutiny in 2008.

Monks led protests in Lhasa, Tibet, that began March 10, the anniversary of a failed 1959 uprising against Chinese rule. But the actions grew increasingly violent, even in areas outside the Tibetan Autonomous Region. Hundreds of Chinese troops were reportedly sent to Tibetan areas in response.

Estimates of fatalities ranged as high as 90.

Peak oil forecast

Model	Peak year	Peak production, million tpy
Hubbert	2005	172
Generalized Weng	2026	194
HCZ	2012	188

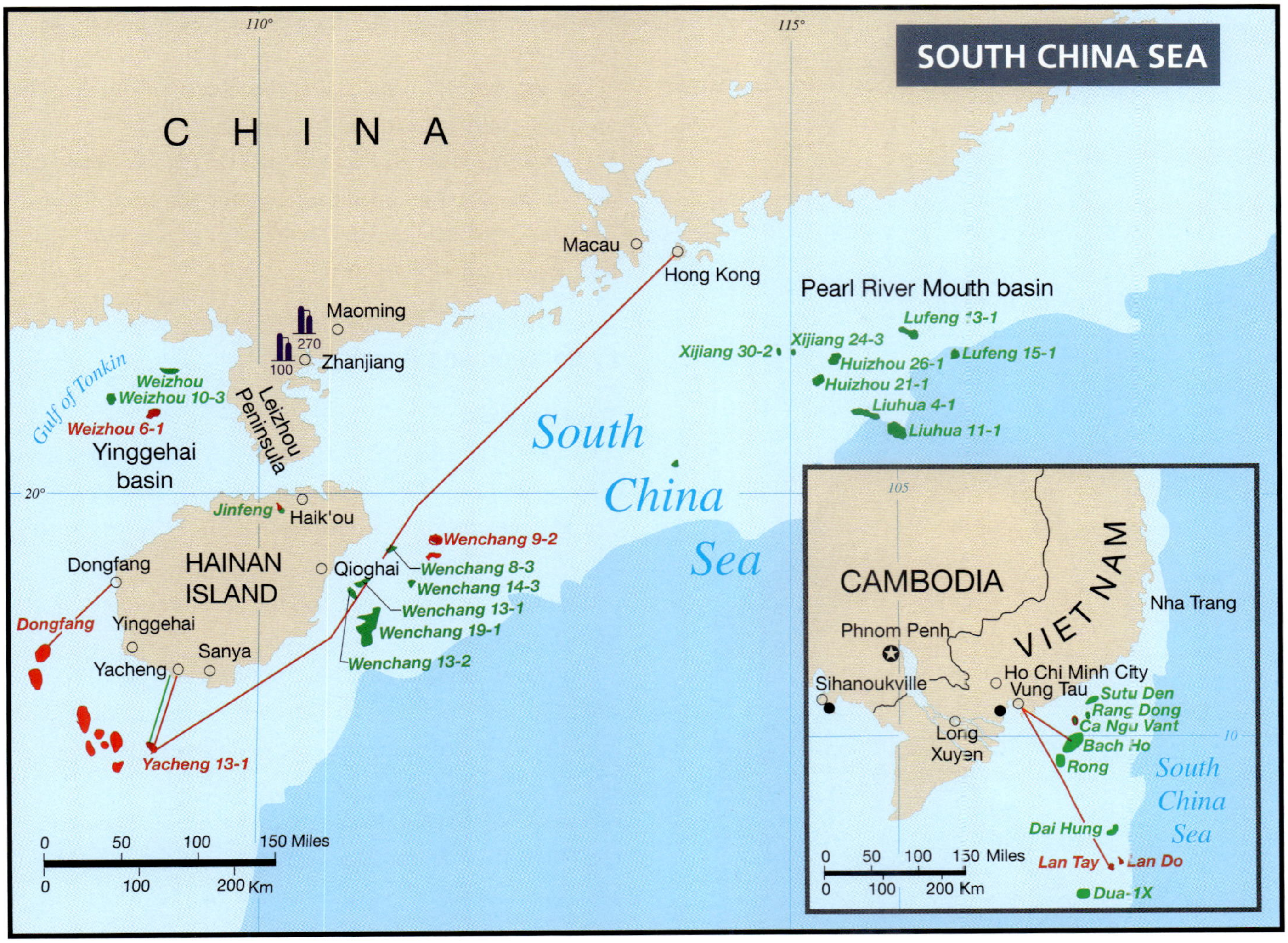

Oil Production

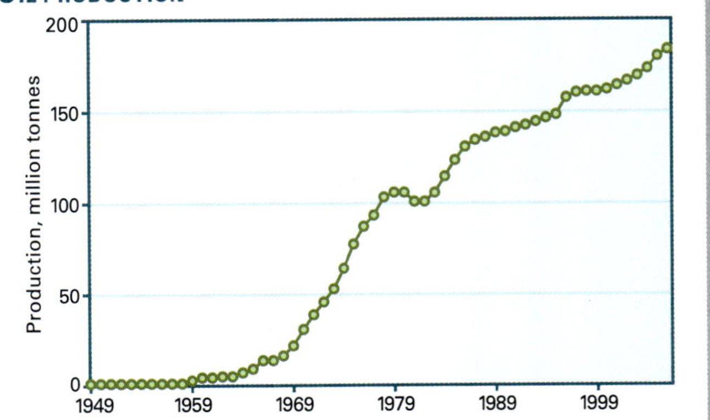

Oil production in eastern China continues to decrease, but total production will remain on a plateau because of increases in the middle, west, and offshore areas.

The middle and western areas have the oldest oil fields in China. With recent successes in the Ordos, Junggar, and Chaidamu basins, these areas have again attracted attention.

Oil Production Distribution

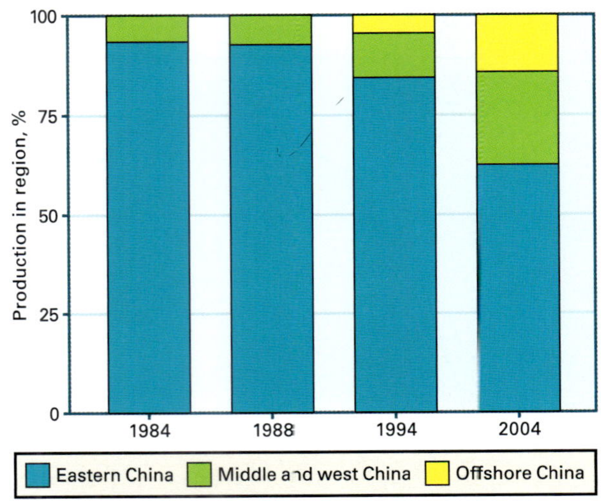

Offshore production in 2004 produced about 14% of China's oil, compared with only 0.27% in 1984. It may peak at 40 mtpy in the future.

Supply, Demand Gap

Year	Oil production	Oil consumption	Gap
		Million tpy	
2005	181	327	146
2010	184	368	184
2020	193	536	343
2030	194	631	437
2050	178	420	242

Oil Production Forecast by Field

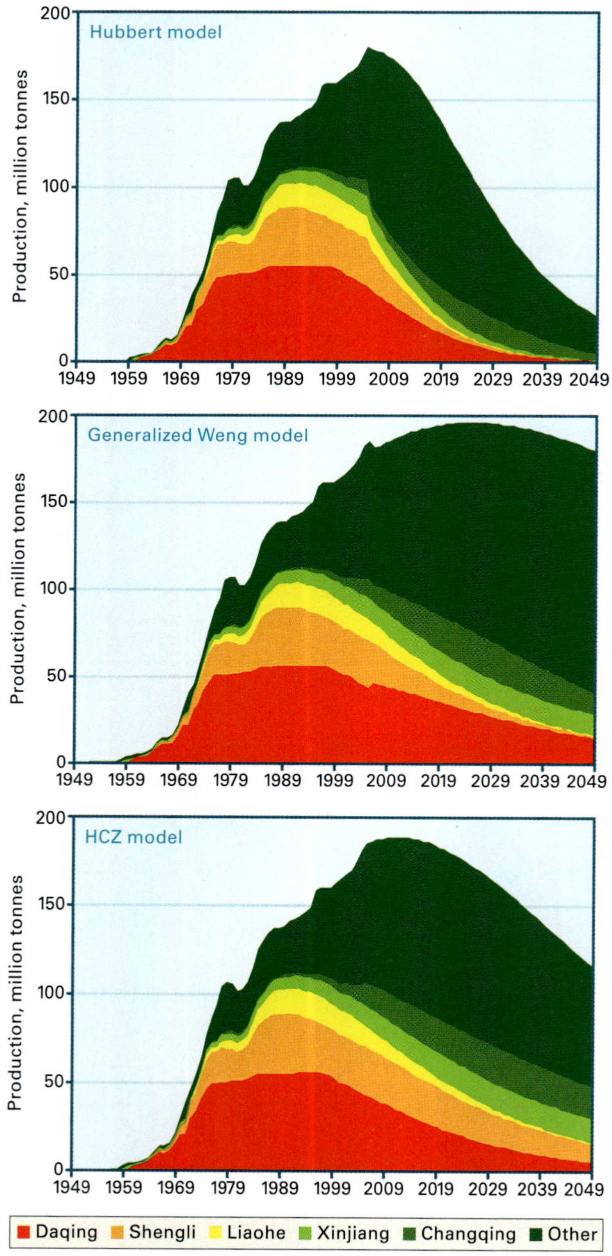

The BP Statistical Review of World Energy 2007 noted that China's oil demand growth increased by 7.4% per year in the recent decade. Demand of 350 mtpy in 2006 was second only to U.S. demand. But oil production grew only by 1.5% in the last decade.

The gap between demand and production now exceeds 160 mtpy and is expected to grow.

The researchers' demand forecast used the Hubbert model combined with oil consumption data from the Statistical Review.

In 2006, China imported 47% of the oil it consumed. That is forecast to reach 69% in 2030.

Exploration and Development

Bohai Bay finds

CNOOC Ltd. (Beijing) made a new discovery in September 2008 at Bozhong (BZ) 35-2 in the Yellow River Mouth Sag. The well was drilled to a total depth of 3,235 m (10,614 ft) in 20 m (66 ft) of water.

Under test, flows averaged 560 b/d of oil and approximately 32,000 cfd of gas.

In August, wildcat BZ29-5-1 on structure BZ 29-5 was drilled to 1,885 m (6,184 ft) TD, with a water depth of about 21 m (69 ft). In drill stem tests, the well flowed 1,500 b/d of oil and 220,000 cfd of gas.

Since 2006, CNOOC has successfully made six discoveries in the Yellow River Mouth Sag.

In July, CNOOC also drilled wildcat Ken Li (KL) 10-1-1 in Bohai Bay, in the Laizhou Bay Sag.

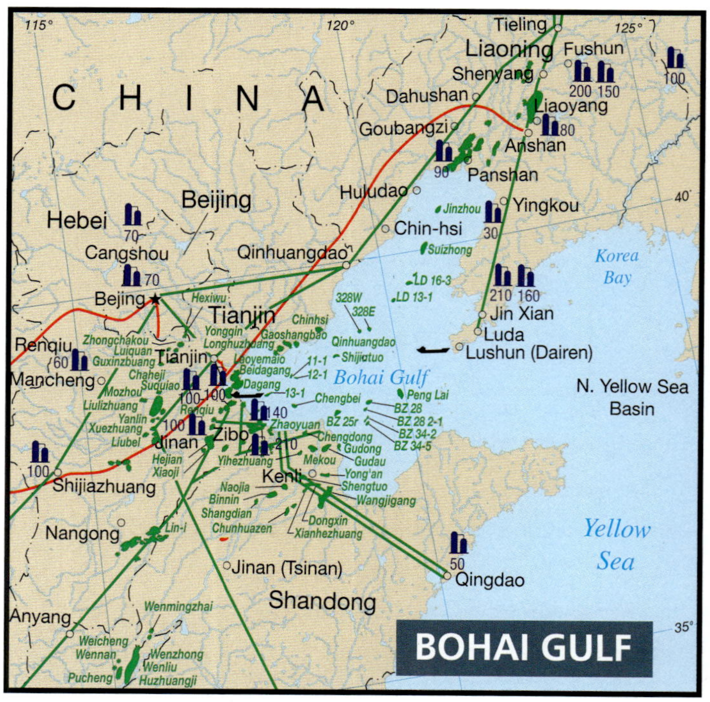

Vietnam deal challenged

The Chinese government warned ExxonMobil Corp. (Irving, TX) to pull out of an exploration deal off Vietnam, describing the project as a breach of Chinese security, said a Hong Kong newspaper.

The protests are said to involve a preliminary cooperation agreement between Petrovietnam and ExxonMobil covering exploration in the South China Sea off Vietnam's south and central coasts.

See the Vietnam section of the Asia Pacific chapter of this book for more details.

Qiongdongnan, Yinggehai basins

Husky Energy Inc. (Calgary) signed a contract with CNOOC for the 1,777-sq km (686-sq mi) Block 63/05 in the Qiongdongnan basin, 100 km (62 mi) southeast of Hainan Island in the South China Sea.

The block is a similar distance southwest of Block 29/26, where Husky Energy planned to start delineation drilling in late 2008 at its Liwan 3-1 deepwater discovery.

Liwan 3-1, in 1,300 m (4,265 ft) of water in the southwestern Pearl River Mouth basin, is the deepest water well drilled off China.

Husky said it has 750 sq km (290 sq mi) of 3D seismic on Block 35/18 in the Yinggehai basin west of Hainan Island and expects to drill exploration wells in 2009–10 on Yinggehai blocks 35/18 and 50/14 and Liwan blocks 29/06 and 29/26.

Drilling and Production

C4 oil field onstream

Production began in October 2008 from seven wells in the C4 oil field and the extended reach area (ERA) of the C and D oil fields in the Zhao Dong block, offshore China, said Roc Oil (Sydney).

The firm also planned to bring four more wells on production by the end of its 2008 drilling program.

A drilling platform at the Zhao Dong field was to start operation by yearend, and an associated processing platform was slated for startup by mid-2009.

Shale-to-oil projects

Mudanjiang City (Heilongjiang province) announced that construction on China's first shale-oil distillation refinery—the Zhongshi Xiandao Pilot Plant—started in August 2008.

The plant—a project of CNPC (Beijing)—will process 1.2 million tons of oil shale and produce more than 700,000 bbl of shale oil.

Initial evaluation found that the area around Mudanjiang holds 46.2 million tons of oil shale, from which more than 26 million bbl of oil could be extracted.

Earlier, Jilin Guangzheng Mineral Co. signed a deal with Royal Dutch Shell (The Hague, Netherlands) to jointly explore and develop oil shale resources in Jilin province.

If the shale deposit is found to be of commercial value, excavation could begin as early as 2010.

Changbei field: innovative drilling

Shell Exploration & Production Co. Beijing and partner PetroChina Company Ltd. (Beijing) used a range of advanced technologies to accelerate development of Changbei field, including the use of dual lateral horizontal wells.

The field area on the Ordos plateau, in the southernmost Gobi Desert, is twice the size of Beijing city.

Over 2.5 years, the firms completed eight wells but were aiming in 2008 to complete at least six more dual laterals. One goal was to drill the longest horizontal wells ever attempted in mainland China.

The Changbei project is Shell's largest onshore foreign investment and the company's first operated development project in China, with a total capital expense exceeding $640 million.

CHANGBEI FIELD, CHINA

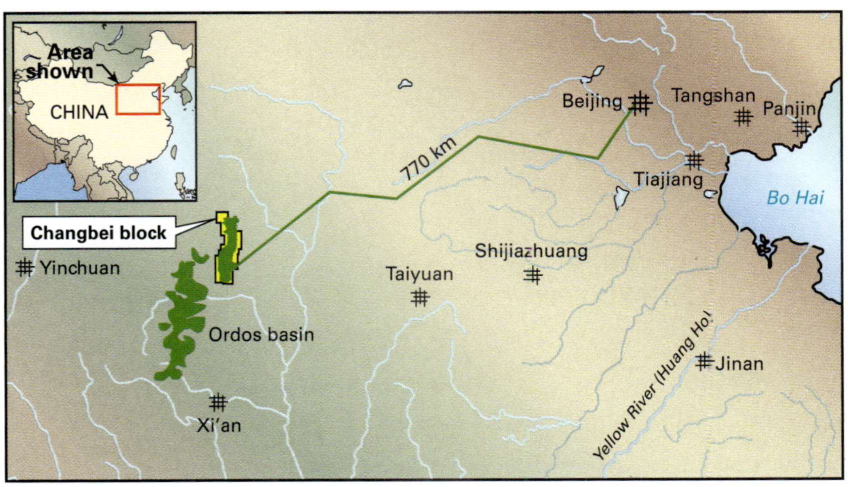

Changbei well design

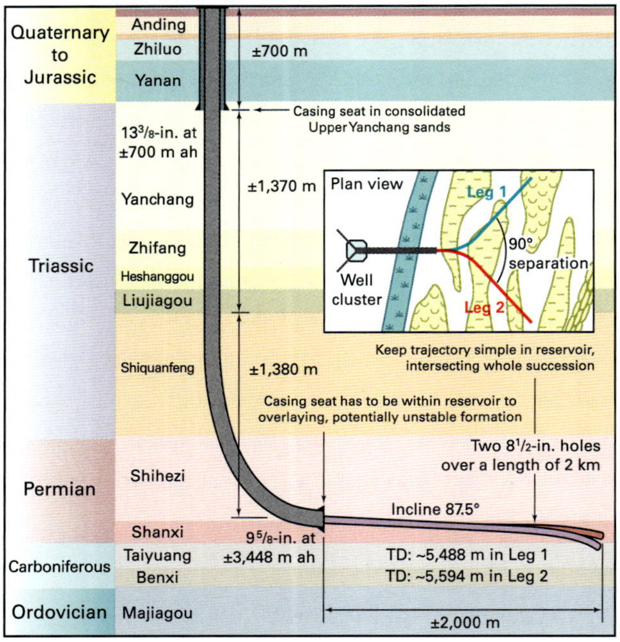

It is also the largest onshore Sino-foreign exploration and production venture to date, said Shell.

Coalbed methane interest

Zhong Yuan Ltd. of Xinjiang province signed an MOU with TrueStar Petroleum Corp. (Vancouver, BC) regarding exploitation of coalbed and coal mine methane there.

Zhong Yuan provides electricity to the province and holds interests in more than 2,800 sq km (1,080 sq mi) of coal acreage in the Zhunggar and Tuha basins.

Gas Processing

Big methanol project: Hainan

CNOOC Ltd. (Beijing) began building a $147 million, 800,000-tpy methanol project in Dongfang Chemical City in Hainan province. The project is funded by CNOOC's Chemical Company Ltd.

The plant will be the downstream auxiliary facility for the development of the Ledong gas field.

Production is expected to begin in late 2010 at an estimated 2,500 tpd.

LPG production up

China's 2007 LPG output maintained its steady increase since 2002, while imports continued their 5-year slide, according to a study by FACTS Global Energy (FGE; Honolulu).

Urban and rural residential use dominated Chinese demand (54% and 11%, respectively), with little impetus for LPG growth in the chemical market. But use of the fuel in fleet vehicles may grow, said FACTS, as Chinese cities work to improve their air quality.

The country produced nearly 615,000 b/d of LPG, up from 2006 by 10.8%.

LNG

Gorgon LNG to China

A subsidiary of PetroChina Company Ltd. (Beijing) signed a long-term agreement in late 2008 to buy LNG with a subsidiary of Royal Dutch Shell (The Hague, Netherlands).

Shell will supply 2 mtpy from the Gorgon Project in Western Australia and the Shell LNG portfolio to PetroChina for 20 years.

In April 2008, CNPC signed a 25-year long-term sales and procurement agreement with Qatar Gas and Shell to procure 3 million mtpy of LNG every year from Phase IV of the Qatar Natural Gas Project.

Petroleum companies in China also reportedly have long-term LNG supply contracts with Iran, Malaysia and Total SA (Paris).

Shanghai: First Petronas LNG

Shanghai Gas Group Co. (SGG) received its first spot cargo of LNG from Malaysia's Petronas in mid-November 2008 at Shanghai's 120,000-cu m capacity Wuhaogou terminal.

Construction of two 50,000-cu m LNG storage tanks at the Wuhaogou terminal was completed in October. The terminal is the smaller of two in Shanghai and receives only spot cargoes.

Petronas subsidiary Malaysia LNG Tiga Sdn. Bhd. is scheduled to supply Shanghai's other regas facility, which has three 165,000-cu m storage tanks.

The Malaysian firm will start LNG deliveries in February 2009 under a 25-year contract, with first-phase receiving capacity of 3 mtpy. The terminal will enter a trial run in April 2009 and will come online in July.

International Activities

Sinopec-KPC oil project

Kuwait Petroleum Corp. (KPC; Safat) and Sinopec (Beijing) launched plans for a massive refinery and petrochemical plant in Guangdong province.

The plant—to process 15 mtpy (300,000 b/d) of Kuwaiti crude oil as well as produce petrochemicals—could start up as early as 2010.

The project could be the biggest international energy joint venture of its kind in China, with a construction cost that could top $8 billion.

KPC has said it wants to be one of China's top five crude suppliers by 2011. In 2008, its target for oil shipments to China is 115,000 b/d.

Russian refinery venture

CNPC (Beijing) reached an understanding with Russia's Rosneft (Moscow) on a JV for a 320,000-b/d refinery project in Tianjin City. An official agreement was to be signed by yearend for one of the largest refinery projects between China and Russia.

In 2006, CNPC and Rosneft established Oriental Energy, a JV mainly engaged in oil and natural gas resource exploration and exploitation in Russia.

The tentative name for the Tianjin refinery joint venture is China-Russia Oriental Petrochemical (Tianjin) Company Ltd.

China-Venezuela: Several agreements

Chinese Pres. Hu Jintao and Venezuelan Pres. Hugo Chavez agreed in September 2008 to increase cooperation and exchange in four sectors, including the petroleum industry.

See the Venezuela section of the Latin America chapter of this book for details.

Sinopec-Saudi Aramco MOU

China Petrochemical Corp. (Sinopec Group) and Saudi Aramco signed an MOU to strengthen their relations and explore business opportunities.

Crude oil trade between the two parties has been steadily increasing. In 2007, Sinopec imported nearly 23 million tons of crude from Saudi Aramco.

With the ongoing expansion of Fujian Refinery, Sinopec wanted to import 32.4 million tons in 2008.

The average import rate is expected to increase to 1.5 million b/d by 2015.

Pipelines/Transportation

ESPO oil line spur

Russia's OAO Transneft and CNPC signed a landmark agreement for construction of a 67-km (42-mi), 300,000 b/d pipeline spur from the East Siberian Pacific Ocean (ESPO) pipeline.

The news about the $800 million spur coincided with an announcement by Russia that the Chinese government will provide Russian oil firms with loans of perhaps $20–25 billion in return for greater supplies of Russian oil.

The deal could give the Chinese access to 15 million tpy over the next 20 years.

The spur will branch from the ESPO pipeline at Skovorodino and extend to the Chinese border to join the Chinese pipeline network to reach the oil hub of Daqing.

The ESPO line is being built in two phases. Phase one will build a 30-mtpy segment from Taishet to Skovorodino. Phase two will then extend the line to the port of Kozmino on Russia's Pacific coast and increase capacity to 80 mtpy.

Until the completion of phase two, oil will be piped from Taishet to Skovorodino and then sent by rail from Skovorodino to Kozmino where an export terminal is under construction.

Russian officials said the spur would not be ready before 2009 at the earliest.

Second WENGP: Eastern section

China approved in late 2008 construction of the eastern section of the country's Second West-East Natural Gas Pipeline Project.

The line would start in Khorgos Port in Xinjiang Autonomous Region and run 9,100 km (5,655 mi) to Shanghai, with branches to Guangzhou and Hong Kong. It would carry 30 bcm (1.06 tcf) per year.

CNPC (Beijing) would operate the line, with Turkmenistan as the major gas source.

Construction of the western section, from Xinjiang to Zhongwei City in Ningxia Autonomous Region, began in February 2008.

The east section, from Zhongwei to the Pearl River Delta in southern China, will span nearly 6,000 km (3,728 mi).

Kazakhstan-China gas pipeline started

Work began mid-year on the 1,300-km (808-mi) Kazakhstan-China natural gas pipeline near Almaty.

The line will be built in five stages, with first shipments expected in 2010. Initially, it will pump 4.5 bcm; full capacity of 40 bcm is expected by 2013.

The new transit route is part of a larger project to build two parallel pipelines connecting China with Central Asia's vast natural gas reserves. The pipelines will stretch more than 7,000 km (4,400 mi) from Turkmenistan, cross Uzbekistan and Kazakhstan, and enter China's northwestern Xinjiang region.

Uzbekistan started building its part in mid-2008; Turkmenistan launched its segment in 2007.

Turkmenistan will be the major supplier of gas for the pipeline. Kazakhstan plans to connect it to its own gas fields near the Caspian as those fields reach production.

The line is the first significant independent gas link connecting the former Soviet region with eastern markets while bypassing Russia.

China, Myanmar pipelines

The Japanese Nikkei newspaper reported in November 2008 that China and Myanmar agreed to build a $2.5 billion oil-and-gas pipeline project connecting the Myanmar port of Sittwe with Kunming, the provincial capital of Yunnan province.

Partners CNPC and Myanmar Oil & Gas Enterprise will build a $1.5 billion oil pipeline, a $1.04 billion gas line, and oil and gas storage tanks.

China will likely import crude oil from the Middle East and natural gas from Myanmar's offshore fields through the pipelines. The lines would be an alternative to shipping through the Straits of Malacca.

Chuanhu gas pipeline

Construction of the Chuanhu gas pipeline, to move gas out of Sichuan province, was slated to conclude in late 2008 or early 2009.

One segment will run 1,674 km (1,040 mi) from Pugang to Shanghai.

Another from Yichang to Poyang would span 842 km (523 mi).

Three branch lines will carry gas to Changshou, Dazhou, and Nanking.

Refining and Products

New refinery: Ningxia

CNPC (Beijing) started building a 100,000-b/d refining project in the Ningxia Hui Autonomous Region in late 2008. The $1.2 billion project is to be completed in 2010.

The facility is designed to annually produce more than 600 million gal of gasoline, more than 745 million gal of diesel, more than 100 million gal of LPG, as well as 100,000 tons of polypropylene, benzene and sulfur.

The gas and diesel will meet Euro-III and Euro-IV standards.

CNPC also was working on three 214,000-b/d refineries. The Qinzhou refinery in the Guangxi Zhuang Autonomous Region is expected to begin production by the end of this year. The two other refineries will be in Liaoning province and Yunnan province.

New refinery: Chongqing

CNPC) (Beijing) was moving in late 2008 to finalize plans for a 200,000-b/d oil refinery in Chongqing City at the Changshou Chemical Industry Park.

Feasibility and environmental impact studies have been approved, and the project now awaits final approval from the State Council.

New refinery: Shanghai

Sinopec (Beijing) signed a framework agreement with the Shanghai Municipal Government in late 2008 aimed at enhancing cooperation on an integrated oil-refining and chemical project in Shanghai Chemical Industry Park.

Asia ethylene capacity seen rising

A report from FACTS Global Energy (Honolulu) said that the Asia-Pacific region will increase its ethylene production capacity to more than 64 mtpy (from 40 mtpy) by 2015.

"Asia's Petrochemical Industry: Implications for Future Naphtha Demand" predicts a wave of additions 2009–11.

Most new capacity will be added in China. The second-largest global producer of ethylene, China will add 14 mtpy by 2015 to its existing 10 mtpy of production capacity.

"Firm and likely additions" will occur mainly in 2009 (4.95 mtpy) and 2011 (3.8 mtpy), said the report.

China will increase its naphtha consumption to more than 1.1 million b/d in 2015 from 600,000 b/d in 2007 because most of the new ethylene production capacity is based on naphtha feedstocks.

China was projected to become the largest naphtha consumer in Asia by 2011.

Sinopec's final partner is still to be determined, and the project is also subject to NDRC approval.

The first project will be a 250,000-b/d oil refinery (Sinopec's third in Shanghai).

In July, Sinopec got approval to relocate and expand its No. 1 ethylene unit from 150,000 tpy to 600,000 tpy at Shanghai Petrochemical.

New refinery: Qinzhou

Construction of a 200,000-b/d, $2.25 billion refinery by PetroChina Co. Ltd. (Beijing) in Qinzhou was on track for completion by yearend 2008. Full operation is slated for late 2009.

The refinery will supply southwest China with gasoline, diesel, LPG and polypropylene.

PetroChina also expects to import about 75 million bbl/yr of crude and export about 6.75 million bbl/yr of refined oil through Qinzhou.

Qinzhou is closer than PetroChina's northern China refineries to export destinations such as Indonesia, Vietnam and Singapore.

The new refinery is China's first large-scale oil refining company in the southwest.

PetroChina said it plans to build six 200,000-b/d refineries by 2010 and eighteen 200,000-b/d refineries by 2020.

Expansion plus ethanol plant: Fujian

Sinopec (Beijing) reported near-completion in late 2008 of a refinery-expansion and ethanol project in Fujian province. Start-up was expected in early 2009.

Other investors include Saudi Aramco (Dhahran) and a subsidiary of ExxonMobil Corp. (Irving, TX).

The project expands refining capacity from 85,000 b/d to about 255,000 b/d for processing sour crude from Saudi Arabia. New facilities include an 800,000-tpy ethanol cracking unit, an 800,000-tpy polyethylene unit, and a 400,000-tpy polypropylene unit.

The factory will be able to provide about 155,000 b/d of Euro-III standard refined oil.

Refinery expansion: Guangdong

CNOOC (Beijing) started expanding a $6.6 billion expansion of its refinery in Huizhou, Guangdong province, to nearly double refining capacity and add 1 million tons of ethylene production capacity.

Refining capacity will grow from about 250,000 b/d to about 470,000 b/d.

The entire project is slated to be finished in 2011.

New refinery: Guangxi

CNPC (Beijing) reported in September 2008 that construction of its refinery in Guangxi was on schedule to be completed by yearend.

By the end of 2009, the Guangxi refinery it is to be fully operational.

This refinery, the first large-scale refinery in southwest China, will process 214,000 b/d of crude to produce 19,000 b/d of refined oil.

Refinery upgrades: Changling, Baling

Sinopec (Beijing) said in early 2008 that it will invest about $2.2 billion to raise the production capacity of two domestic refineries.

The company will spend about $1.4 billion to double the production capacity (to 10 mtpy) of its Changling refinery in Hunan province and $834 million to upgrade its Baling plant, which processes oil from Changling into chemicals.

Dalian refinery gets Saudi crude

The recently expanded Dalian refinery of PetroChina Co. Ltd. (Beijing) got its first cargo (2 million bbl) of Saudi Arabian crude in late August 2008. In June, new units were commissioned that doubled its crude capacity to 410,000 b/d.

Diesel, gasoline shortages

In the first half of 2008, China faced shortages of diesel and gasoline as the gap widened between rising international crude prices and centrally controlled fuel markets.

Shortages first reported in southern and inland China spread to the wealthier areas in the north and east as filling stations struggle to obtain fuel shipments from refiners.

In Beijing and Shanghai, rationing and outright inadequate deliveries were reported.

With inflation at its highest level in a dozen years, Chinese officials resisted pressure from refiners for price increases.

With the world crude price near $100/bbl, state-owned Sinopec and PetroChina reportedly limited sales to independents to ensure supplies for major cities, farming and public transport.

Impact of refining expansions

In a review of refining trends in Asia and China, a researcher at the U.S. Naval War College concluded that refining expansions by China could reshape global oil trade. Key points:

- New refinery projects in Asia will account for about 40% of the total 10-million b/d expansion expected by 2015 in global oil refining capacity. China will dominate Asian activity.
- China's refining-sector growth is driven by rapid oil product demand growth, a push to slash feedstock costs, and a desire to augment China's ability to refine a wider variety of crude oils.
- Chinese oil demand-growth of 5% per year equates to a need to add 300,000 b/d of refinery capacity each year.
- If Beijing continues to control national oil-product prices and restrict exports of refined products, Chinese refiners may strongly consider investing in refining projects abroad.
- China will add 1.8 million b/d of refining capacity by 2010 while shuttering 400,000 b/d of old capacity at small, uncompetitive plants.
- Chinese firms are slated to build at least eight 200,000 b/d refineries, most designed to refine heavy, sour, and high-acid crude (to cut feedstock costs).
- China now takes roughly 30% of its crude imports from Africa, working to diversify away from the Middle East.
- Government price subsidies—30 billion yuan ($3.7 billion) in 2005—could hit 330 billion yuan ($43 billion) by 2012.
- Growing Chinese pull on sour crude supplies may help trim price differentials between sweet and sour feedstocks.
- Expanded Chinese crude acquisition options could shift trade patterns for Middle East and Venezuelan sour crudes.
- After 2010, products from new Chinese and Indian refineries could greatly reduce regional refining margins, which could affect margins even in U.S. West Coast and Gulf of Mexico markets.

Chinese refiners' product mix, expected shifts

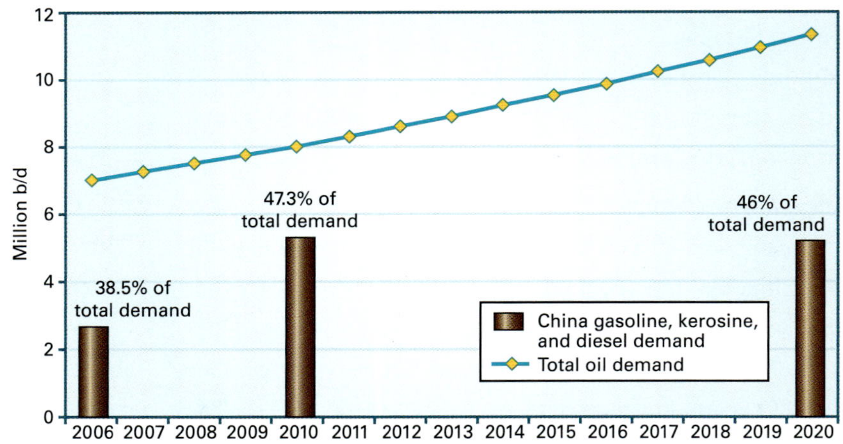

Source: Xinhua, China Oil & Gas

Petrochemicals

SK Energy: Stake in ethylene project

SK Energy Company Ltd. (Seoul) said it plans to invest about $730 million for a 35% stake in an 800,000-tpy ethylene project in Wuhan, Hubei province. The firm signed a framework agreement with Sinopec (Beijing), the project's main developer.

The project is part of a 170,000-b/d oil refining project developed by Wuhan Petrochemical, a subsidiary of Sinopec. The ethylene project is expected to be complete by 2011.

The plant also will produce epoxy ethane; high-density and low-density polyethylene; glycol; and polypropylene.

The venture awaits approval by the central government.

The Wuhan project is SK Energy's first in China's refining industry.

New PP facility: Chandu

PetroChina Sichuan Petrochemical Co. Ltd. (PCSP) began planning for construction of its new 450,000-tpy polypropylene (PP) production facility at Chengdu (Sichuan province).

When it comes on stream in 2010, the PCSP facility will be one of the largest single-train PP facilities in China, part of PCSP's 800,000-tpy ethylene complex at Chengdu.

The PCSP license agreement includes production capability for a range of PP resins.

CHINESE CRUDE OIL CONSUMPTION* BREAKDOWN

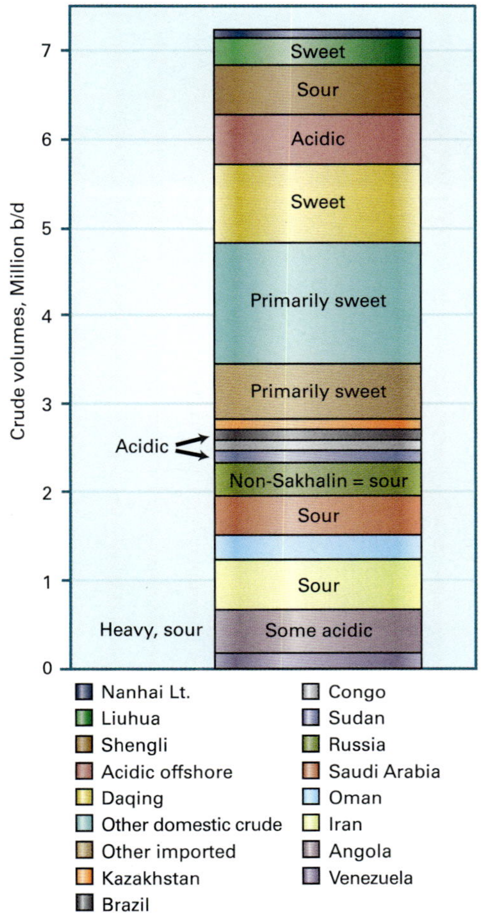

*Imports plus domestic production.
Sources: Energy Intelligence Group, Xinhua, EIA

New petrochem plant: Tianjin

Sinopec (Beijing) and Saudi Basic Industries Corp. (Sabic) took initial steps in early 2008 on a possible 50-50 joint venture to build a $1.7-billion ethylene derivatives complex in Tianjin.

The proposed complex would produce 1 mtyp of ethylene derivatives—600,000 tpy of polyethylene and 400,000 tpy of ethylene glycol. It is proposed to be completed by September 2009. All ethylene feedstock would be supplied by Tianjin Petrochemical Co., a branch of Sinopec.

This would be Sabic's first JV in China.

Alternative Fuels

Biodiesel: Hainan

CNOOC (Beijing) began building a 14.9-million-gal/yr biodiesel project in southern China's Hainan province Hainan in September 23.

The project—the first biodiesel project in Hainan—is expected to start commercial operation by mid-2009. It is located at CNOOC's Dongfang Chemical City.

The project is one of three demonstration biodiesel plants and will use oil from the seeds of the tung tree as feedstock. The output will be used to promote sale of BD5 and BD10 biodiesel in Hainan.

Biodiesel: Chongqing, Hunan

Gushan Environmental Energy Ltd., China's largest producer of biodiesel (measured by annual production capacity) was preparing to bring two new biodiesel production plants at its Chongqing and Hunan sites online by yearend 2008. The plants are designed to produce 30,000 tpy each.

Biodiesel: Pilot projects

NDRC approved three biodiesel pilot projects using jatropha oil feedstock to produce a total of 170,000 tpy of fuel.

PetroChina's Nanchong refinery will build a 60,000-tpy biodiesel plant in Sichuan province; Sinopec will build a 50,000-tpy plant in Guizhou province; and CNOOC will build a 60,000-tpy plant in Hainan province. No construction timetables were given.

Biomass targets

In third quarter 2008, China set major targets for development and use of biomass energy for 2010, according to information released at the Second Biomass Energy Utilization Symposium in Huangshan City (Anhui province). The following production targets were set:

- Biomass: 5,500 MW
- Biofuels: 2 million metric tons
- Marsh gas: 19 bcm (670 bcf)
- Solid fuel: 1 million metric tons

The Ministry of Finance will create policies to support the development of biomass and biochemical industries. The Ministry of Science and Technology will invest $22 million by 2010 to support the agricultural forest biomass program.

Asia Pacific

AFGHANISTAN

Capital: Kabul
Monetary unit: Afghani
Refining capacity: 0
Oil production: 0
Oil reserves: 0
Gas reserves: 1.750 tcf

MOU with Uzbekistan

In early 2008, officials from the Ministry of Energy and Water Resources held talks with counterparts at Uzbekenergo, the state joint stock company of Uzbekistan.

An MOU was signed regarding expanded cooperation in the energy sector and the potential implementation of joint projects.

Gas, oil resources in southeast

The ministry of mines announced in early 2008 that major gas and oil resources had been discovered in Katawaz, in Paktika Province, in the southeast.

In addition, the ministry believed that there are more gas and oil resources in northern provinces than in other parts of the country.

The government said it is interested in having private investors bid for exploration rights regarding both hydrocarbon and mineral resources.

Afghanistan signed a protocol agreement in 2004 with the United States, and Afghan officials said that that the U.S. Geological Service will help with further efforts to identify potential resources.

Fuel imports from Pakistan

The fixed regulatory duty imposed by Pakistan on diesel and kerosene exports to Afghanistan was cut by almost 50% in November 2008.

Details are given in the Pakistan section of this chapter.

AUSTRALIA

Capital: Canberra
Monetary unit: Dollar
Refining capacity: 696,346 b/cd
Oil production: 453,000 b/d
Oil reserves: 1.500 billion bbl
Gas reserves: 30.000 tcf

Exploration and Development

Continental shelf acreage extended

In early 2008, the UN ruled that Australia could extend its continental shelf, after 15 yrs of lobbying.

The UN okayed an extension of 2.5 million sq km (965,000 sq mi)—five times the size of France.

The new area takes in extensions of the Exmouth Plateau and Wallaby Plateau in the west, the Great Australian Bight in the south, and the Lord Howe Rise in the east.

All are areas that the government scientific body Geoscience Australia thinks could hold petroleum reserves and are future potential exploration areas.

A number of the new areas are adjacent to existing producing areas, including the Browse and Carnarvon basins off Western Australia.

Woodside Energy: Going deep

Woodside Energy Ltd. (Perth) said in late 2008 that the search for new "elephants" (large oil and gas fields) in Australia was heading to deepwater.

From 1979 through 2008, the company discovered 16.5 billion boe in shallow waters off Australia. In deepwater, for the same period, the company discovered 9.3 billion boe.

Woodside said it plans to drill at least 39 wells in deepwater Carnarvon basin permits [>500-m (1,640-ft) water depth] through 2011, more than double the number it drilled in deepwater 2000–2008.

In the Browse basin, it plans to spud at least six deepwater wells through 2011, compared to four deepwater spuds from 2000–2008.

Nimblefoot

Hess Corp. (New York City) reported in third quarter 2008 that its Nimblefoot-1 exploration well on the WA-390-P permit on the Northwest Shelf found 28 m (93 ft) of net natural gas pay.

It is the third of four initial exploration wells being drilled on the WA-390-P permit by Hess in 2008. Both of the two previous wells—Glencoe-1 and Briseis-1—were announced as discoveries in mid-year.

Nimblefoot-1 was drilled in 1,115 m (3,658 ft) of water.

Queensland: Oil shale ban

The Queensland government in third quarter 2008 placed a 2-yr moratorium on oil shale projects.

The action halted the proposed $14-billion (Aus.) Queensland Energy Resources Ltd. (QER) project in the McFarlane deposit, about 15 km (9 mi) south of Proserpine on the state's central east coast.

The project would entail exploration of about 400,000 tonnes of oil shale material in the area.

QER says the deposits have the potential to produce 1.6 billion bbl of shale oil over the next 40 yrs.

Community and environmental groups maintain that the regional tourism industry and the Great Barrier Reef would be at risk if the project proceeded.

The government wants to investigate the environmental impacts of oil shale mining.

Montara development

Coogee Resources Ltd. (Perth) slipped the onstream date for its Montara oil development project in the Timor Sea to yearend 2008, citing delays in conversion of an FPSO vessel.

The Montara, Skua and Swift/Swallow oil fields are all approximately 650 km (404 mi) west of Darwin, within Australian territory. Montara is some 82 km (51 mi) southwest of the existing operations at Challis.

Phase 1 includes production from two wells on Skua field and one well each on Swift and Swallow fields. Phase 2, to be completed in 2009, includes two wells at Montara field itself and one more at Swift.

Coogee also completed feasibility studies into the potential commercialization of the smaller gas fields surrounding Montara, Jabiru, and Challis.

Otway basin

An offshore Otway basin JV discovered 22 m (72 ft) of natural gas pay in primary reservoir target sands with its Netherby-1 wildcat drilled on permit Vic/P44 off Victoria.

JV partners are Santos Ltd. (Adelaide), Australian Worldwide Exploration Ltd. (AWE; Sydney) and Mitsui E&P Australia Ltd. (Sydney).

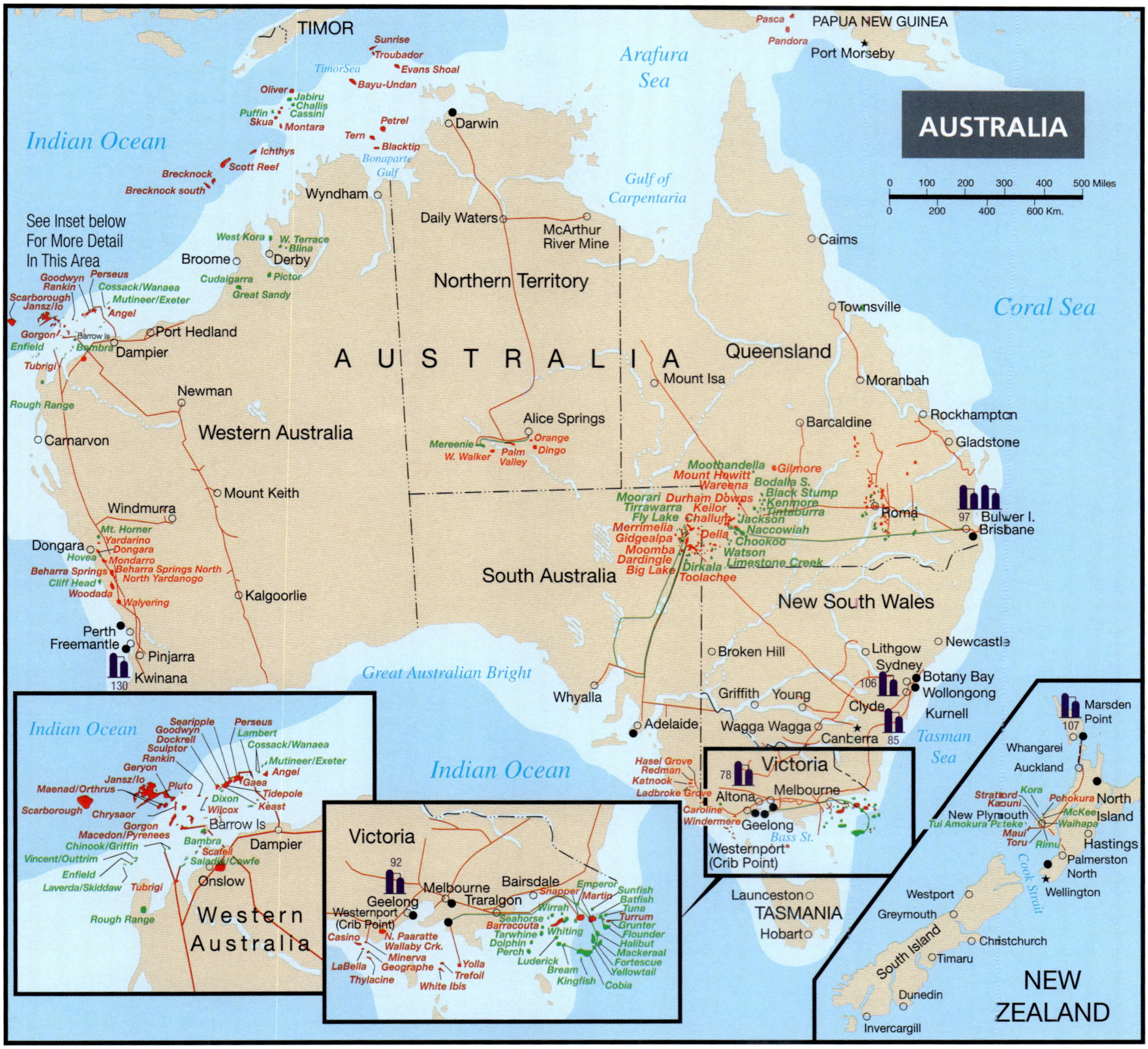

The well, drilled to TD 1,870 m (6,135 ft), was undergoing tests in late 2008.

Netherby is the second of three planned appraisal and development wells in this sector of the permit.

Netherby-1 lies 4 km (2.5 mi) north of Henry gas field.

Browse basin

In mid-2008, Total E&P Australia (Perth) and operator Inpex Browse Ltd. (London), partners in the WA-344-P joint venture, made a new gas discovery from the Mimia-1 well in the Browse basin.

A 72-m (236-ft) gas column was found and a production test confirmed gas and condensate.

Mimia-1 was drilled in 254 m (833 ft) of water approximately 20 km (12 mi) northeast of WA-285P, which contains the Ichthys gas/condensate field.

Inpex and Total are preparing to develop Ichthys field as well as the Mimia discovery. Inpex estimates Ichthys reserves at 362 bcm (12.8 tcf) of gas, and 527 million bbl of condensate.

WA-390-P

Hess Corp. (New York City) discovered natural gas in mid-2008 in Jurassic and Triassic sandstones in the Briseis-1 exploration well on the northwest shelf.

The well encountered 151 ft (46 m) of net gas pay, in line with pre-drill estimates.

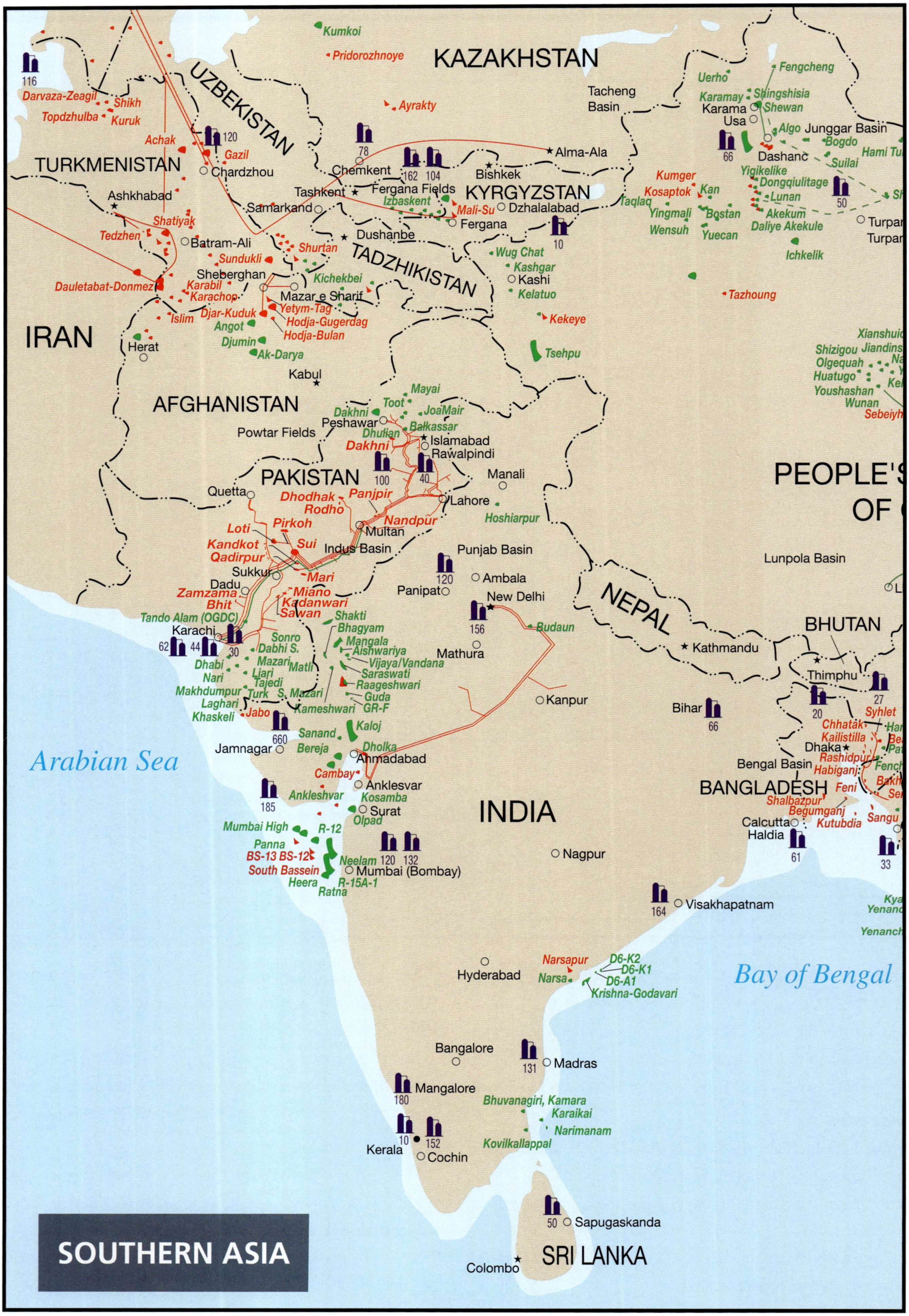
KAZAKHSTAN
UZBEKISTAN
TURKMENISTAN
KYRGYZSTAN
TADZHIKISTAN
IRAN
AFGHANISTAN
PAKISTAN
NEPAL
BHUTAN
BANGLADESH
INDIA
SRI LANKA
Arabian Sea
Bay of Bengal
SOUTHERN ASIA

MONGOLIA
Ulan Bator
N. KOREA
Pyongyang
S. KOREA
Seoul
E'S REPUBLIC
F CHINA
MYANMAR
(BURMA)
LAOS
THAILAND
VIETNAM
CAMBODIA
MALAYSIA
PHILIPPINES
TAIWAN
HAINAN
SABAH
SARAWAK
Yellow Sea
South China Sea
Andaman Sea
Beijing
Tiamin
Qingdao
Shanghai
Hangzhou
Wuhan
Lanzou
Lhasa
Lenghu
Guiyang
K'unming
Guangzhou
Hong Kong
Macao
Maoming
Beihai
Haik'ou
Fuzhow
Santou
Taipei
Kaohsiung
Manila
Limay
Batangas
Vietiane
Fang
Khon Kaen
Bangkok
Sriracha
Phnom Penh
Nha Trang
Ho Chi Minh City
Yangon (Rangoon)
Syriam
Chittagong
Chauk
Digboi
Sibsagar
Jorhat
Nowgong
Myitkyina
Fenchugani
Brunei
Lutong
Erlian Basin
Eerdous Basin
Qinghai Hu Basin
Jiuqan (Juixi) Basin
Lhasa Basin
Bose Basin
Dian-Qian-Gui Basin
Shiwan-Dashan Basin
Nanyang Basin
Jianghan Basin
Sanshui Basin
Zhujiuang Basin
Quingjiang Basin
Subei Basin
Bohai Basin
N. China Basin
E. China Basin
Shoal Basin
Turpan Basin
Unggi
Chinkanshou
Kongchuling
Changchun
Tangluka
Fahniu
Shuguang
Chin-hsi
Huanxing
Yuoy
Jinzhou
GanSuizhong 36-1
Layemalo
Chinshsi
Gaoshangbao
Pohi Bay
Takang
Bozhong 28-1
Bozhong 34-2/4E
Shengil
Yanlin
Dawangzhuang
Lin-4
Zhongyuan
Wecheng
Wenmingzhai
Wennan
Wenzhong
Pucheng
Wenliu
Liuzhuang
Machang
Ahhui
Fumim
Lianmengzhuang
Huanggue
Donghai
Bo Yuan Ting
Ping Hu
Wangil
Sha-Yen
Zhangdian
Xiamen
Welgang
Shuanghe
Dongzhuang
Qianjiang
Guanghau
Wanchang
Zhangang
Xijiayuan
Tuoshi
Huayuan
Gaochang
Xingqou
Shenglinjin
Liujiazhuang
Majiatan
Lizhuangzi
Hongjinzi
Donghongchuang
Dashuikeng
Zichang
Baiyanjin
Wuji
Yongping
Maling
Quinghazhi
Danbei
Gangua
Shengting-k'ou
Pai-yang-ho
Ya'exia
Shih-you-kou
Danbei
Baiyanghe
Laojunmiao
Shiyougou
Lenghusihao
Nanyishan
Youquanzi
Kelfenglike
Lenghuwuhao
Toufengshan
Yanhu
Sebefehao
Zhongba
Jinhuazhen
Leiyinpu
Zhongtaishan
Wubaiti
Guanan
Gihua
Wolongha
Penglaizhen
Wanshunchang
Longnushi
Daxinxi
Weiyuan
Shiyoug
Shilongxnou
Wanglongchang
Taihehums
Tiandong
Lufeng 13-1
Lufeng 15-1
Hizhou 26-1
Xijang 30-2
Weizhou 11-1
Fuizhou 21-1
Liuhua 11-1
Liuhua 4-1
Weizhou 10-3
Weishou 6-1
Tien Hai
Jinfeng
Wenchang 9-2
Wenchang 8-3
Wenchang 14-3
Wenchang 13-1
Wenchang 19-1
Wenchang 13-2
Dongfang
Yacheng 13-1
Chuwei
Paoshan
Yungoshan
Haripur
Beani Bazar
Patharia
Fenchugani
Bakhrabad
Semutang
Patiya
Sangu
Ayadaw
Chauk/Lanywa
Yenangyaung
Mann
Paduakpin
Htaukshabin
Kyaukoyauk
Pyato
Pyaye
Prome
Myanaung
Shwepyitha
Ledang
Mae Fang
Mae Suiti
Yadana
Martaban
Mae Sot
Sirikit
KDN-1
Phu Horm
Nam Phong
Pyaye
Chonnabot
Pyato
Shwepyitha
Bung Ya
Sawng
Neung
Yetagun
Plamuk
Kapong
Surat
Platong
Nang Nuan
Platong S.
Pladang
Tantawan
Chaba
Dara
Trat
Erawan
Sabun
Baanpot
Jakrawan
Pailin
Punan
Arthit
Songkhla
Bongkot
Bua Ban
Pilong
17-B-1
Damar
Jemeh
Bintang
Sepat
Noring
Inas
Bedong
Laho
Sutu Den
Rang Dong
Bach Ho
Rong
Dai Hung
Dua-1X
Bursa
Natuna
Guloc
Malampaya
West Linapacan
Camago
Matinlok
Nido
Cadlao
S.W. Emerald
Barton
Tembungo
S. Furious
Erb W.
St. Joseph
Kinarut
Erb S.
Samarang
Glayze
Jintan
Bijan
B12
B11
E11
M1
M3
M4
F6
F28
E8
Burya Nibung
Tarakan

Australia: Apache 2008 drilling

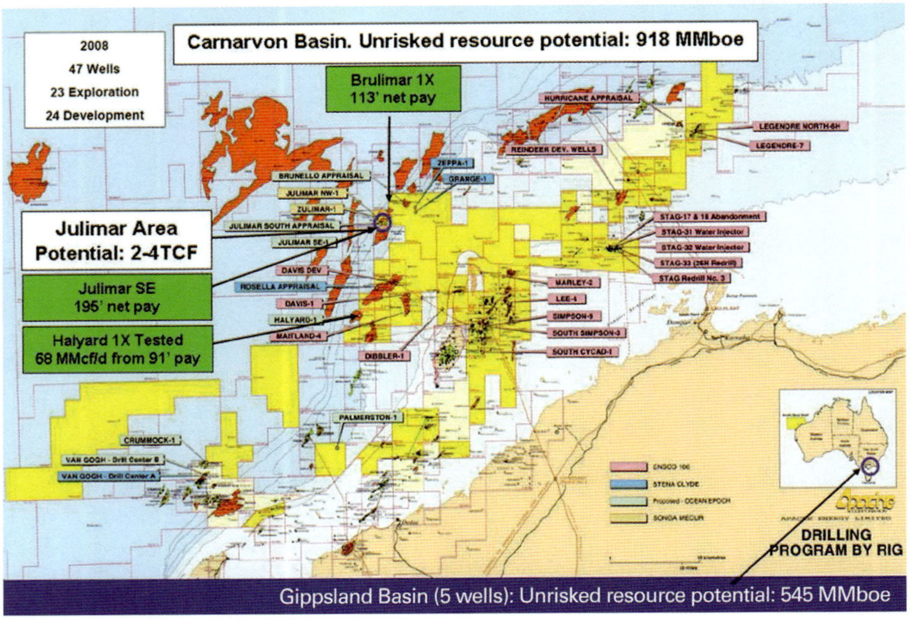

Briseis-1, drilled in 3,668 ft (1,118 m) of water, was the second of four exploration wells being drilled on WA-390-P in 2008. The first, Glencoe-1, was announced as a discovery weeks earlier.

Turrum field

The Gippsland Basin Joint Venture—which includes Esso Australia [a subsidiary of ExxonMobil Corp. (Irving, TX)] and BHP Billiton Ltd. (Melbourne)—announced in mid-2008 that it will invest $1.1 billion to develop more than 270 MMboe from Turrum field in the Bass Strait.

Turrum, 42 km (26 mi) offshore in 60 m (197 ft) of water, holds approximately 28 MMcm (1 tcf) of gas and 110 MMbbl of oil and natural gas liquids.

Construction of one new platform and upgrading of a second are expected to begin in 2009.

Oil production is planned for 2011, with first gas sales in 2015.

Kipper field

The Kipper Unit Joint Venture of Esso Australia, BHP Billiton, and Santos Ltd. (Adelaide) announced in late 2007 it had secured $1 billion to develop natural gas from the Kipper field in the Bass Strait.

Key engineering contracts were awarded in early 2008, and construction was slated to start by 2009.

Workers will install several subsea wells, piped back to existing infrastructure at Longford. First gas is expected in 2011.

The Kipper resource holds approximately 620 bcf (17.6 bcm) of recoverable gas and 30 MMbbl of condensate/LPG.

Arkaringa basin

Sapex Ltd. (Adelaide) was undertaking a six-week seismic survey in the Arkaringa basin in South Australia in early 2008—the first seismic work in the basin since the early 1980s.

Half of the program was to focus on the Boorthanna Trough, where the company was planning to spud two wells by yearend 2008, depending on seismic results.

The second part of the survey was to determine the trough's southward extension.

Apache: Northwest Shelf report

In early 2008, Apache Corp. (Houston) reported on two discoveries on the Northwest Shelf.

The Julimar Southeast-1 discovery logged 195 ft (59 m) of net pay across five intervals of the Triassic Mungaroo Sandstone. It was the fifth gas discovery Apache drilled on license WA-356-P.

Julimar Southeast-1 was drilled in 502 ft (153 m) of water but not tested.

The firm also said that its Halyard-1 discovery test-flowed at a peak rate of 1.9 MMcmd (68 MMcfd) of gas and 936 b/d of condensate from the Cretaceous Halyard sandstone.

Halyard-1 was drilled in 366 ft (112 m) of water in WA-13-L, the same production license as the Apache-operated East Spar field.

The company says the discovery is on trend with its producing John Brookes field.

Policy

Offshore Petroleum Act 2006

As of July 1, 2008, the Petroleum (Submerged Lands) Act 1967 (PSLA)—Australia's primary legislation for the administering its offshore petroleum resources since the 1960s—was replaced by the new Offshore Petroleum Act 2006 (OPA).

The Resources and Energy Ministry said that the new act made no changes to major policies but would make the legislation easier to understand and cut costs for both industry and government.

Condensate excise tax

The Resources and Energy Ministry said in mid-2008 that the government might use the $2.5 billion (Aus.) gained over the next 4 yrs from the imposition of excise taxes on North West Shelf condensate production to push forward struggling multibillion-dollar gas project proposals.

Observers saw the announcement as an indication that the government wants to set a framework to spur progress on the next generation of LNG, GTL and similar projects.

Lease Sales

Offshore E&P permits

In late 2008, Australia awarded 13 offshore exploration permits in its 2007 release scheme.

Winners pledged a total of $500 million (Aus.) in exploration investment over the next 6 years.

Award winners were:

- Woodside Petroleum Ltd. (Perth): permits WA-415-P, WA-416-P, WA-417-P in offshore Canning basin, inshore of the Browse basin region.
- Hunt Oil Co. (Dallas): WA-413-P, WA-414-P
- Murphy Australia Oil Pty Ltd. (subsidiary of Murphy Oil Corp.; El Dorado, AR): WA-429-P
- Nexus Energy Ltd. (Melbourne): WA-424-P
- Finder Exploration Pty Ltd. (Perth): WA-418-P in the Carnarvon basin.
- Essar Exploration & Production Ltd. (part of Essar Global Ltd; Mumbai): NT/P77 and NT/P78 in the Northern Territory area of the Bonaparte basin
- Albers Group (Melbourne): WA-420-P, WA-421-P, and WA-422-P Bonaparte permits in the Western Australian area.

Offshore Western Australia

In May 2008, Australia awarded nine new offshore leases, receiving a total of $425 million (Aus.).

Five permits are in the Bonaparte basin, three in the Browse basin, and one in the Carnarvon basin.

- France's Total SA (Paris): Browse basin permit WA-408-P off northwest Western Australia.
- Sinopec (Beijing): NT/P76 in the Troubadour Terrace area of the Bonaparte.
- Apache Corp.(Houston) and Total: Bonaparte Block AC/P42 and AC/P43
- Finder Exploration Pty Ltd. (Perth): AC/P44 and AC/P45.
- Partnership of Santos Ltd. (Adelaide), Chevron Corp. (San Ramos, CA), and Inpex Corp.(Tokyo): WA-410-P
- Partnership of Santos, Inpex and Beach Petroleum (Adelaide): WA-411-P
- Partnership of Cue Exploration (Melbourne) and Gascorp Australia Ltd.(Melbourne): WA-409-P in the Carnarvon basin

Drilling and Production

North West Shelf: Redevelopment

The North West Shelf joint venture (operated by Woodside Energy Ltd.; Perth) approved in late 2008 a $1.8-billion (Aus.) redevelopment of Cossack, Wanaea, Lambert, and Hermes oil fields to support production beyond 2020.

The aim is to extend the life of the four fields—about 135 km (84 mi) northwest of Karratha off Western Australia in 75–135 m (247–443 ft) of water—and add production from nearby yet-to-be-drilled exploration prospects.

Facilities are to be completed and fully operational by early 2011.

North West Shelf: Angel field

In October 2008, the North West Shelf gas project JV brought the $1.6-billion (Aus.) Angel gas field development on stream.

Angel has the capacity to flow 800 MMcfd of gas plus as much as 50,000 b/d of condensate.

The field lies 115 km (71 mi) off Western Australia, inshore from North Rankin and Goodwyn fields. The three are the original North West Shelf fields found in 1972.

The development comprises a new platform in 80 m (262 ft) of water, a 50-km (31-mi) undersea pipeline tied back to the existing North Rankin A platform, and associated infrastructure.

Angel gas flows to the JV's LNG-domestic gas facility on the Burrup Peninsula near Karratha.

Australian Drilling Activity, 2000-07

	Active onshore wells					Active offshore wells			
	Wildcat	Appraisal	Development	Coal seam methane	Total, onshore	Wildcat	Appraisal	Development	Total, offshore
2000-Q1	5	1	18	2	**26**	16	1	5	**22**
2000-Q2	9	3	20	7	**39**	17	3	9	**29**
2000-Q3	7	5	28	5	**45**	19	3	3	**25**
2000-Q4	10	7	24	9	**50**	14	1	4	**19**
2001-Q1	14	4	26	15	**59**	14	6	6	**26**
2001-Q2	14	4	19	25	**62**	12	1	8	**21**
2001-Q3	15	5	17	31	**68**	11	2	9	**22**
2001-Q4	13	13	9	23	**58**	16	2	5	**23**
2002-Q1	7	7	22	7	**43**	14	3	8	**25**
2002-Q2	2	7	22	23	**54**	7	3	17	**27**
2002-Q3	7	7	10	39	**63**	5	4	15	**24**
2002-Q4	8	9	12	25	**54**	8	7	12	**27**
2003-Q1	8	3	13	9	**33**	19	2	14	**35**
2003-Q2	5	1	16	14	**36**	12	4	7	**23**
2003-Q3	8	3	25	23	**59**	15	4	15	**34**
2003-Q4	22	7	12	32	**73**	9	1	21	**31**
2004-Q1	17	9	7	27	**60**	5	2	16	**23**
2004-Q2	20	7	11	29	**67**	10	11	21	**42**
2004-Q3	19	7	14	28	**68**	7	2	17	**26**
2004-Q4	18	9	12	15	**54**	13	3	13	**29**
2005-Q1	8	5	25	11	**49**	17	3	8	**28**
2005-Q2	11	13	21	10	**55**	11	5	5	**21**
2005-Q3	23	12	14	9	**58**	11	7	2	**20**
2005-Q4	26	10	17	15	**68**	15	5	2	**22**
2006-Q1	14	6	25	14	**59**	12	1	3	**16**
2006-Q2	29	10	31	21	**91**	10	3	3	**16**
2006-Q3	30	22	9	19	**80**	12	2	10	**24**
2006-Q4	36	31	22	14	**103**	16	6	20	**42**
2007-Q1	19	20	21	0	**60**	4	4	9	**17**
2007-Q2	25	23	10	0	**58**	24	13	14	**51**
2007-Q3	27	28	19	0	**74**	13	13	21	**47**
2007-Q4	18	18	21	0	**57**	12	11	20	**43**

Source: Data summarized from Quarterly Drilling Statistics issued by the Australian Petroleum Production & Exploration Association Ltd.

Wells Planned off Australia, 2008

Operator	Total number	Basin	Lease: well names	Rig
Apache Corp.	4 wildcats	Gippsland	VICP 59: Hermit; Elver; Dory/Nemo; Oyster	Diamond's Ocean Epoch semisub
	47 total: 23 exploration 24 development	Carnarvon	WA-155-P: Van Gogh Drill Center A WA-334-P: Dibbler-1	Stena clyde semisub; ENSCO 106 and West Triton jack ups
Beach Petroleum Ltd.		Bass	T/39P: PeeJay-1 T/38P: Spikey Beach-1	West Triton
		Otway	VIC/P46: Fermat-1	
		Gippsland	BMG area wells	Songa Venus semisub
BHP Billiton Petroleum	1 well	Gippsland		
	6 wells	NW Shelf		
Coogee Resources Ltd.	9 wells	Bonaparte	Montara	West Atlas jack up
Esso Australia Pty. Ltd.		Bass Strait		Nabors platform rigs: ISDL 173, Pool 453
Hess Corp.	4 wells	NW Shelf	WA-390P: Glencoe-1, Briseis-1, Nimblefoot-1, Glencoe-1	Jack Bates semisub
MEO Australia Ltd	1 well	NW Shelf	WA-361P: Zeus-1	Songa Venus
Mosaic Oil NL	12 wells	Surat-Bowen		
Nexus Energy Ltd.	1 development	Gippsland	Longtom-4	West Triton
OMV	1 appraisal well	Browse	AC/RLG: Audacious-5	Songa Venus
Peak Well Management				Wilcraft jack up
Santos Ltd.	7 exploratory wells; many development wells and injectors, wildcats	Barrow, Houtman, Sorell, Otway, Bonaparte, Browse	Frigate Deep-1	Ocean Patriot, Songa Mercur, Songa Venus semisubs
Shell Development Australia Ltd.		Browse	AC/P41: Libra	Frontier Discoverer drillship; Ocean Epoch semisub
Stuart Petroleum Ltd.	1 exploration	Gippsland	VIC/P53; Bazzard 1	West Triton
Victoria Petroleum NL	3 wells	Surat		
Woodside Energy Ltd.	2 exploration + 8 in 2009	Browse	Pluto	4 semisubs; 2 platform rigs

Source: Operator reports

Australian basins

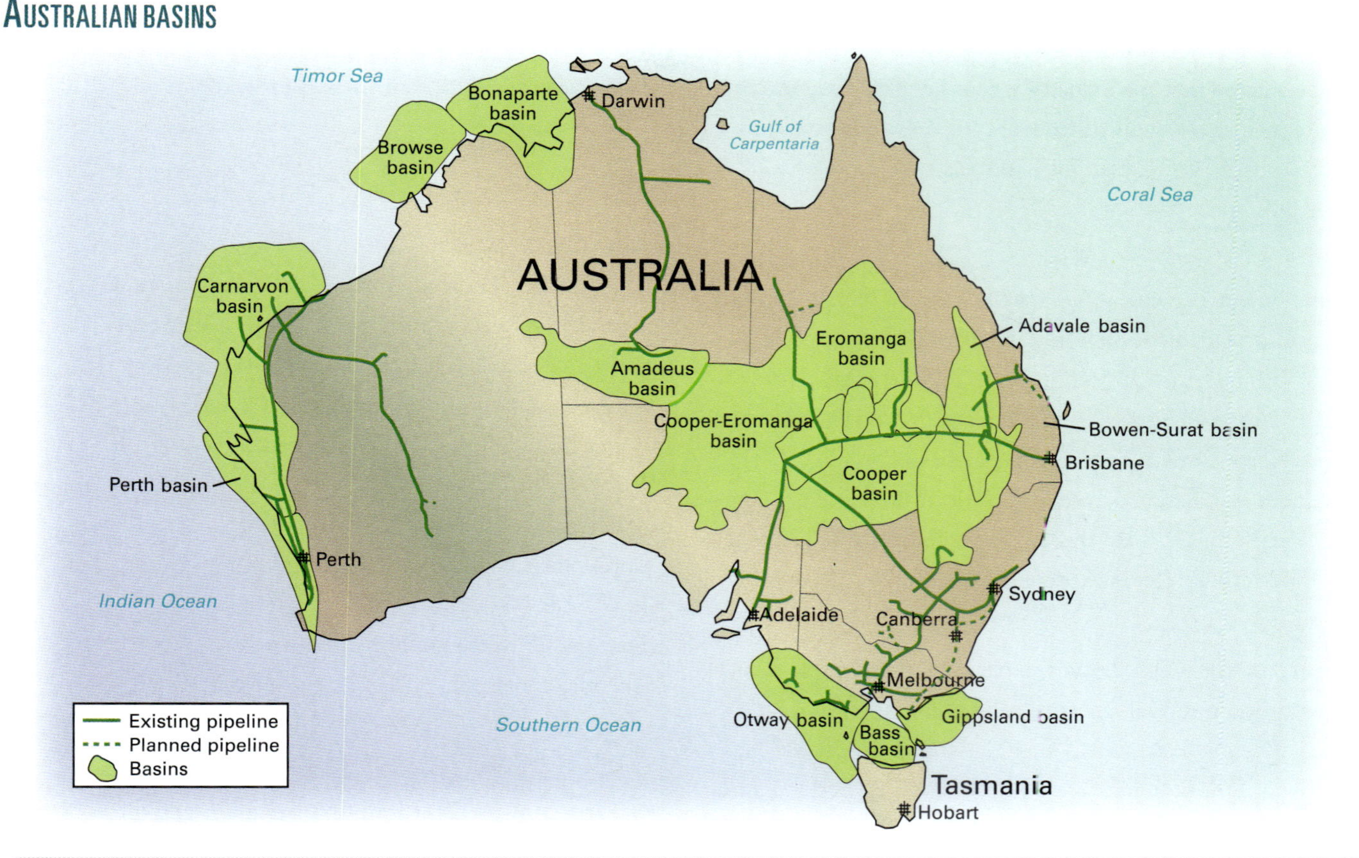

Drilling in Australia: 2008

A summary by *Oil & Gas Journal* showed the following companies pursuing drilling operations in Australia in 2008:

- AED Oil Ltd. (Melbourne)
- Apache Corp. (Houston)
- Beach Petroleum Ltd. (Adelaide)
- BHP Billiton Petroleum Ltd. (Melbourne)
- Central Petroleum Ltd. (Perth)
- Chevron Australia Pty. Ltd. (Perth)
- ConocoPhillips Co. (Houston)
- Coogee Resources Ltd.(West Perth)
- Cooper Energy Ltd. (South Perth)
- Empire Oil & Gas NL (Perth)
- Magellan Petroleum Corp. (Hartford, CT)
- Mosaic Oil NL (Sydney)
- New Standard Energy Ltd. (West Perth)
- Nexus Energy Ltd. (Melbourne)
- Santos Ltd. (Adelaide)
- Stuart Petroleum Ltd. (Adelaide)
- Victoria Petroleum NL (Perth)
- Woodside Energy Ltd. (Perth)

North West Shelf: Train 5

The fifth LNG train was brought on stream in mid-2008 at the North West Shelf gas project JV.

The Train 5 project included the fifth train, a jetty extension, a second LNG loadout berth, and several other ancillary systems.

The new train increased the capacity of the facility by 4.4 mtpy, to a total output of 16.3 mtpy.

Woodside operates site for a six-member consortium.

Gas drilling trends

In a late-2008 analysis, *Oil & Gas Journal* observed that natural gas is driving drilling in Australia, while oil production continues to decline.

LNG trains have opened the market for Australia's abundant natural gas, including associated gas and coal seam gas (CSG).

Energy economics group EnergyQuest (Adelaide) said in late 2008 that Australian CSG production rose to 22.8 million boe for the year ending June 2008, up 38.9% from the previous year.

Domestic gas production rose 2.3%, even though conventional production was down 2.1%.

OGJ also noted that the industry is consolidating, as shown by a series of mergers of small Australian companies with coal seam gas reserves that can be used for LNG gas export projects.

North Rankin B

In early 2008, the North West Shelf gas project JV made its final investment decision to go ahead with the North Rankin B platform development off Western Australia.

The $5-billion (Aus.) project, led by Woodside Petroleum Ltd. (Perth), is aimed at recovering the remaining low-pressure gas from the North Rankin and Perseus gas fields.

It includes installation of a second platform to house gas compression facilities, utilities, and a living-quarters module. The North Rankin B Project is expected to come on stream in 2012.

Crux project

Partners in the Crux liquids project in the Browse basin made progress on several fronts in early 2008.

Nexus Energy Ltd. (Melbourne), Vanguard Oil & Gas International Ltd. (London), and Viking Shipping Services Ltd. (N. Yorks, UK) secured in early 2008 a tanker to be converted into an FPSO vessel.

The group also obtained a bank loan to support drilling of the Crux-3 and Crux-4 wells.

Production start-up from Crux is expected in mid-2010.

LNG

Greater Gorgon project

The Greater Gorgon project is the largest among several potential LNG export projects in Western Australia. The project's estimated resource base exceeds 1.41 tcm (40 tcf).

Gas will come from several fields 200 km (124 mi) offshore in water up to 1,300 m (4,265 ft) deep.

Partners in the $15-billion (Aus.) project are the Australian subsidiaries of Chevron Corp. (San Ramon, CA), ExxonMobil Corp. (Irving, TX), and Royal Dutch Shell PLC (The Hague, Netherlands).

First gas deliveries could be in 2013.

The major elements of the project include:

- Development of the Greater Gorgon gas fields, including subsea pipelines to Barrow Island.
- Three 5-mtpy LNG trains on Barrow Island for liquefying produced gas.

GORGON, OTHER WESTERN AUSTRALIA LNG

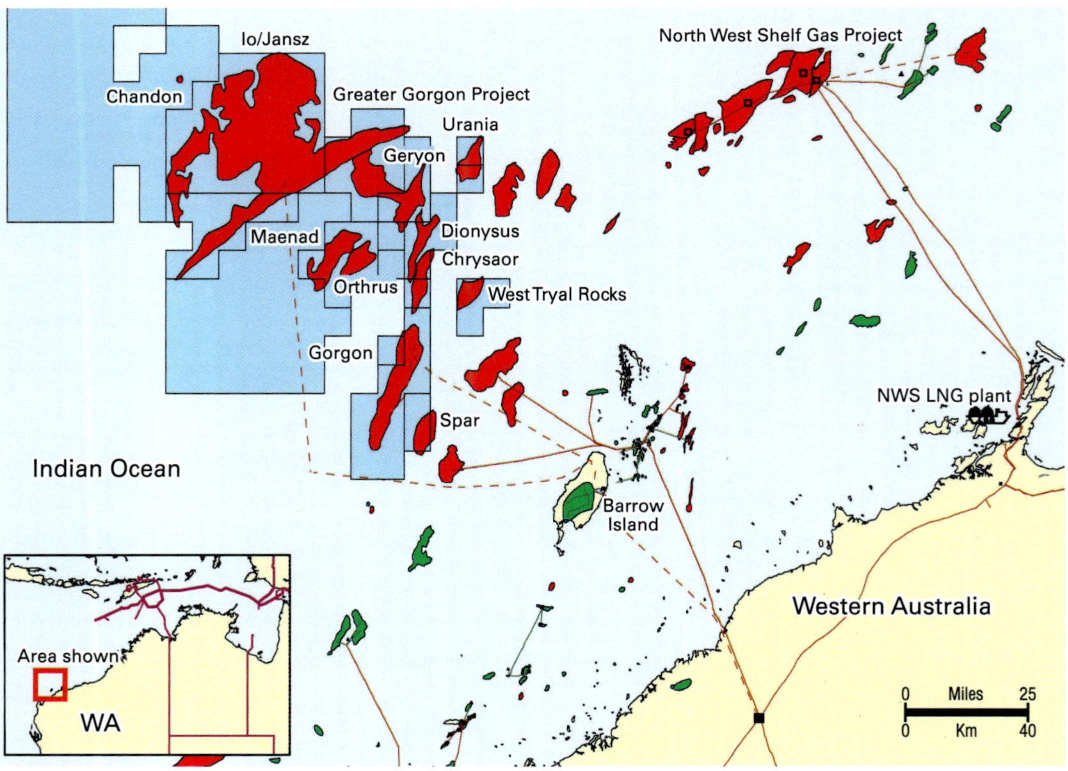

Source: Wood Mackenzie PathFinder.

- An LNG shipping facility.
- Assessment of a domestic gas plant and pipeline to deliver gas to the mainland.
- Greenhouse gas management by injecting/sequestering CO_2 extracted from the natural gas into deep formations beneath Barrow Island.

The Gorgon Joint Venture committed in mid-2008 to including a 300-terajoules/d (279 MMcf/d) domestic gas processing plant alongside the LNG facilities planned for Barrow Island.

Earlier in the year, the JV proposed to increase from two to three the number of 5-mtpy LNG trains on Barrow Island, pending approval from the government and environmental authorities.

Chevron said that FEED work should begin by mid-2009.

Pluto progress

Several major contracts were awarded in 2008 for work on the Pluto LNG project off northwest Karratha, Western Australia. Woodside Energy Ltd. (Perth) is leading the project.

One award concerns construction of a riser production platform.

About 45.3 MMcmd (1.6 bcfd) of gas will be piped from the platform to an onshore single-train liquefaction plant having the capacity to produce 4.3 mtpy of LNG.

The production platform will be moored in more than 89 m (275 ft) of water.

Other contracts concern development options for a proposed second LNG train.

The Pluto system includes 27 km (17 mi) of subsea flowlines to the production platform and 180 km (112 mi) of subsea trunkline to the onshore Burrup plant.

Browse: FLNG plus CCG?

Shell Australia in early 2008 proposed two actions related to the environmental impact of developing its Prelude gas find in the Browse basin, 450 km (280 mi) northeast of Broome.

One proposal is to build a 3.5-mtpy, $2.1-billion (Aus.) floating LNG (FLNG) system to avoid the sensitivities of bringing gas ashore on the Kimberley coast. A pipeline to shore could cost $3 billion.

The company also was considering use of CO_2 capture and geo-sequestration (CCG) technology along with the FLNG facility. CCG would reduce the development's carbon footprint.

Observers said that the FLNG unit could be a catalyst for developing 'stranded' gas elsewhere.

The FLNG unit would be a vessel 480 m (1,575 ft) long and 70–80 m (230–263 ft) wide, anchored to the seabed and designed to survive a once-in-10,000-years cyclone.

The company expected to issue a tender to build a first FLNG vessel in third quarter 2008.

Prelude field has estimated reserves of 2–3 tcf of gas.

Shell estimates an on-stream date of 2012.

Ichthys development

In late 2008, Inpex Australia [a unit of Inpex Holdings Inc. (Tokyo)] and JV partner Total SA (Paris) chose Blaydin Point at Darwin's Middle Arm Peninsula as landfall for development of the 269-bcm (9.5-tcf) Ichthys gas-condensate field in the Browse basin off northwest Western Australia.

Operator Inpex said several benefits offset the costs of the 850-km (528-mi) undersea pipeline that will be needed to deliver the gas from the field to Darwin.

The project partners are aiming to bring first LNG on stream by early 2015.

FEED work was to start in early 2009 to enable a final investment decision by early 2010.

Ichthys field in permit WA-285-P holds an estimated 363 bcm (12.8 tcf) of gas and 527 million bbl of recoverable condensate.

Initial plans call for two trains producing more than 8 mtpy of LNG, 1.6 mtpy of LPG, and more than 100,000 b/d of condensate.

Queensland: Origin-ConocoPhillips project

Origin Energy Ltd. (Sydney) finalized a deal in late 2008 with ConocoPhillips Co. (Houston) to form a 50:50 joint venture called Australia Pacific LNG.

The initiative would develop coal seam gas (CSG) and use that gas as input to a liquefaction facility. The JV is to be called Australia Pacific LNG.

Origin will act as the CSG supplier; ConocoPhillips will be the LNG operator.

The proposal is to develop Origin's CSG resources in the Surat and Bowen basins in Queensland to supply a proposed 14-mtpy LNG plant at Gladstone.

First investment decision for Train 1 is due by the end of 2010, and first LNG production is scheduled for 2014.

Origin turned down an offer from BG Group PLC (London) to form a similar CSG-LNG venture.

Queensland: QGC-BG project

Queensland Gas Co. (QGC; Brisbane) and BG Group (London) proposed construction of a 3- to 4-mtpy gas liquefaction plant on the east coast of Queensland.

The facility would use coal seam gas (CSG) from the Surat basin fields of QGC in the state's southeast.

Production from the new LNG facility—likely to be built near the port city of Gladstone—would begin in 2013. A 380-km (236-mi) pipeline also would be built.

BG Group would buy all of the planned LNG production under a 20-year contract.

Queensland: Santos-Petronas project

In mid-2007, Santos Ltd. (Adelaide) unveiled plans to build a 3- to 4-mtpy LNG plant at a cost of about $6 billion that would use Queensland coal seam gas as feedstock.

In mid-2008, Santos arranged a deal in which Malaysia's Petronas (Kuala Lumpur) will participate in the project for an upfront payment of $2 billion,

The project would also include a 450-km (280-mi) pipeline to Gladstone.

A final investment decision is expected by yearend 2009, with first LNG production in 2014.

At yearend 2008, Bechtel Corp. (San Francisco) was chosen for FEED work for the proposed plant.

Queensland: Arrow, Sunshine projects

In addition to the BG and Santos CSG-to-LNG projects noted above, two other smaller projects with similar goal have been proposed.

Arrow Energy Ltd. (Brisbane) has signed an agreement with LNG International to supply CSG to a 1.3-mtpy LNG plant.

Sunshine Gas Ltd. (Brisbane) has plans to build a 500,000-tpy LNG plant, in conjunction with trading company Sojitz Corp. (Tokyo).

Sunrise, Troubadour projects

Development of the Sunrise and Troubadour projects off East Timor and Australia may again start up after being on hold for a number of years, said *Oil & Gas Journal*.

The projects would develop about 8 tcf of gas for processing at an expanded Bayu Udan LNG plant.

Wheatstone project

In early 2008, Chevron Australia Pty. Ltd. (Perth) reported plans to develop the Wheatstone gas field—discovered in 2004 on the North West Shelf—as a stand-alone LNG project, with facilities on the northwest coast of mainland Western Australia.

A 5-mtpy LNG facility would be built first, with capability for adding more trains later.

FEED work was slated to begin in 2009.

Wheatstone field, with estimated gas reserves of 127 bcm (4.5 tcf), lies on retention lease WA-17R about 145 km (90 mi) off Dampier in 200 m (656 ft) of water.

Many observers thought that the field (a) had more potential as a gas-to-liquids project, or (b) should be an addition to the North West Shelf LNG project (in which Chevron is a participant).

Chevron also made a significant gas discovery in mid-2008 at Iago field that could help expand the nearby Wheatstone LNG project.

Common-use LNG hub?

In early 2008, Western Australia and the new federal government agreed to assess the Kimberley region as a site for a common-user LNG hub.

It might also host associated regional activities to serve proposed gas fields to be developed in the offshore Browse basin.

Government officials want to prevent piecemeal development that would draw fire from environmentalists. The region is regarded as a pristine wilderness.

Observers said that the energy industry would like to see a final decision by yearend 2008.

Environment

Carbon trading scheme

A draft report prepared in mid-2008 at government request recommends that Australia introduce a broad carbon trading scheme as soon as possible, across industries and including transport fuels.

The report declared that coal-powered electricity generation should not be given any compensation for having to pay a carbon tax.

However, the report did urge study to determine as soon as possible whether there can be a low-emissions future for coal.

CO_2 injection in Naylor field

Australian officials worked with the U.S. Department of Energy (DOE) and other partners in 2008 to launch the Otway basin pilot project.

The project will inject carbon dioxide into the Naylor depleted gas field in southeastern Australia, aiming to demonstrate the feasibility of storing CO_2 in the Waarre formation.

The pilot is one of 19 sequestration projects endorsed by the international Carbon Sequestration Leadership Forum (CSLF).

The project is directed by the Cooperative Research Centre for Greenhouse Gas Technologies (CO_2CRC) of Canberra.

Up to 100,000 tonnes of CO_2 will be injected more than a mile beneath the earth over a period of 1–2 years. Researchers will monitor and model the storage reservoir for several years beyond that.

DOE-sponsored research developed instruments to track the CO_2 plume during and after injection.

Company News

BG Group-Queensland Gas Co.

In late 2008, BG Group Ltd. (London) was poised to make a cash takeover offer worth about $5.5 billion (Aus.) for Queensland Gas Co. (Brisbane).

The deal would give BG a leading role in the development of coal seam gas reserves in Australia.

BG had attempted earlier in the year to win Origin Energy Ltd. (Sydney), a larger firm also with extensive CSG assets.

BG wants Australian gas reserves to support its sales of LNG in lucrative Asian markets and to complement its strong position in the Atlantic basin.

Arrow-Pure

In late 2008, Brisbane-based Arrow Energy Ltd. (Brisbane) made a cash and share takeover offer for Pure Energy Resources Ltd. (Brisbane).

With the acquisition, Arrow would control additional reserves in the Walloon Coal Measures in the Surat basin of southeast Queensland—adjacent to its proposed Surat-Gladstone pipeline—as well as further acreage in the Bowen basin adjacent to its proposed Moranbah-Gladstone pipeline.

A successful acquisition would bring Arrow's total uncontracted 2P coal seam gas (CSG) reserves to 56.6 bcm (2 tcf).

AGL-Sydney Gas

AGL Energy Ltd. (Sydney) made a $171-million takeover bid in late 2008 for Sydney Gas Ltd.

The deal would give AGL full control of producing and prospective CSG acreage in the Camden and Hunter Valley regions of the Sydney basin in New South Wales.

Earlier, AGL bought the prospective CGS permit PEL 285 (Gloucester basin, New South Wales) from Molopo Australia Ltd. (Melbourne) and AJ Lucas (Ryde, New South Wales) for $370 million.

Arc-AWE

In third quarter 2008, Australia's Federal Court approved the merger plans of Arc Energy Ltd. (Perth) and Australian Worldwide Exploration Ltd. (AWE; Sydney).

The new company will have a portfolio of interests on and offshore holding proved and probable reserves of more than 78 million boe.

The approval was said to pave the way for a spin-off of Arc's Canning basin assets in Western Australia into an exploration vehicle called Buru Energy Ltd.

The new Arc-AWE entity will own major stakes in the producing Cliff Head oil field in the offshore Perth basin and in the BassGas gas-condensate project in Bass Strait.

BANGLADESH

Capital: Dhaka
Monetary unit: Taka
Refining capacity: 33,000 b/cd
Oil production: 4,000 b/d
Oil reserves: 28.000 million bbl
Gas reserves: 5.000 tcf

Lease sale

Bangladesh launched its first offshore bidding round in February 2008 for 20 deep sea blocks and eight shallow water blocks.

In May, seven IOCs submitted bids for 15 blocks, in what was termed a lukewarm response. Observers said one factor was doubts about the maritime boundary of some blocks close to India and Myanmar (see box).

Petrobangla officials said the oil companies offered to invest $1.6 billion

Santos International Ltd. (Adelaide) was top bidder, proposing $852 million for six blocks in partnership with British Cairn Energy PLC (Edinburgh).

Other bidders were ConocoPhillips (Houston) and a U.S.-China joint venture, Longwoods Resources.

Disputed waters

According to media reports, Bangladesh was seeking formal dialogue in 2008 with both India and Myanmar to resolve disputes about hydrocarbon-related activities in the Bay of Bengal.

Several offshore blocks border the maritime boundaries of the three nations.

Bangladesh proposed a suspension of all exploration and extraction in the Bay of Bengal until decisions are made regarding sovereignty and ownership rights.

Block 9

Petrobangla, the national oil company, reported that operator Tullow Oil PLC (London) started gas production from its Bangora gas field onshore Bangladesh in Block 9 in October 2008.

The new well, 100 km (62 mi) southeast of Dhaka, was producing 850,000 cmd (30 million cfd) of natural gas.

The new production brings total output from the field to approximately 2.83 million cmd (100 million cfd) and increases the country's total gas production to 51.3 million cmd (1.81 bcfd) from some 20 gas fields.

This figure was said to still be some 7.1 million cmd (250 million cfd) short of the nation's requirements.

Cairn reports on efforts

Cairn Energy PLC (Edinburgh) gave an update in early 2008 on its drilling program in Bangladesh.

The company did discover a new gas reservoir on Sangu Field in Block 16, which Petrobangla estimates will contribute 708,000–850,000 cmd (25–30 million cfd) of natural gas to the nation's total production.

However, neither of two earlier wells in Block 16 was successful: Magnama-1 and Hatiya-1.

Cairn was considering whether to conduct a 3D seismic survey by early 2009 to determine if further exploratory drilling would be warranted for either well.

Sangu field

In mid-2008, the government was considering a proposal from Cairn Energy PLC to invest nearly $ 9.5 million to try to find more natural gas at Sangu offshore field.

The firm also asked the government to allow it to sell any resulting new gas at higher prices.

Without further development, said Cairn, the field could be commercially exhausted by yearend 2009.

Cairn said that, as of mid-2008, it had invested nearly $ 600 million in Sangu.

The field began producing gas in 1998 at approximately 5.7 million cmd (200 million cfd), but output has dropped to below 1.4 million cmd (50 million cfd).

Blocks 17 and 18

PTT Exploration & Production (PTTEP; Bangkok) reported in early 2008 signing a farm-in agreement with Total E&P (part of Total SA; Paris) to become a partner in its operated Blocks 17 and 18.

The move is PTTEP's first venture in petroleum exploration in South Asia.

Total also was planning to conduct a new 3D seismic survey over the two blocks by third quarter. They cover 13,724 sq km (5,299 sq mi) and are 150 km (93 mi) from Chittagong.

Gas potential is considered high because of the location of the blocks: just 40 km (25 mi) south of producing Sangu gas field, and 80 km (50 mi) north of the Shwe gas field in Block A1 in Myanmar.

Other partners are Total; Tullow Bangladesh; Okland Bangladesh; and Rexwood Offshore Holdings.

BRUNEI

Capital: Bandar Seri Bagawan
Monetary unit: Dollar
Refining capacity: 8,600 b/cd
Oil production: 180,000 b/d
Oil reserves: 1.100 billion bbl
Gas reserves: 13.800 tcf

Block B

In mid-2008, Total SA (Paris) reported a significant discovery of gas and condensates on Block B in the offshore Maharaja Lela-Jamalulalam (MLJ) field with its MLJ2-06 well.

The company called the well the deepest ever drilled in Brunei in a high-pressure/high-temperature reservoir. Final depth was 5,850 m (19,193 ft).

Total estimated that the well would come onstream by yearend.

In late 2008, Total SA (Paris) struck a new gas-bearing compartment of over 400 m (1,312 ft) thickness on Block B. The company said that the column found by the ML-4 well was equivalent to those already in production in the field.

ML-4 reached a TD of 5,227 m (17,149 ft) in water 62 m (203 ft) deep, about 50 km (31 mi) offshore.

The MLJ field produced 28,500 boe/d in 2007, with its gas going to the Brunei LNG liquefaction plant.

Total was planning further exploration drilling in 2009.

Block L

A 3D seismic survey for Loon Energy Inc. (Calgary) began in October 2008 in eastern onshore Brunei. A total of 350 sq km (135 sq mi) was to be surveyed by yearend.

Block L, covering 550,000 acres (859 sq mi), covers most of eastern Brunei, with some offshore acreage.

Giant Champion and Seria oil fields lie within 25 km (16 mi) of the block, which has had some historic production but no exploration and development for 20 years.

Second refinery

In late 2008, Brunei TRC Energy Bhd (PetroBru) got government approval to proceed with a detailed feasibility study for the country's second oil refinery, to be built in the newly developed Pulau Muara Besar (PMB) port region.

The study will be conducted in association with Foster Wheeler Corp.(Clinton, NJ) and Drewry Shipping Consultants (London). The initial cost estimate for the new refinery is $4 billion.

The complex will refine and store 200,000 b/d of diesel and jet fuel. Plant construction is planned to start in 2010, with start-up in 2013.

Brunei Shell Petroleum (Seria) operates the country's existing refinery as a JV between the government and Royal Dutch Shell PLC (The Hague, Netherlands). It has a refining capacity of 10,000 b/d.

INDIA

Capital: New Delhi
Monetary unit: Rupee
Refining capacity: 2,255,540 b/cd
Oil production: 668,000 b/d
Oil reserves: 5.625 billion bbl
Gas reserves: 37.960 tcf

Overview

With high rates of economic growth and over 15% of the world's population, India has become a significant consumer of energy resources. According to the U.S. Energy Information Administration (EIA), India was the sixth largest oil consumer in the world in 2006.

The global financial crisis and credit crunch of 2007–2008 have slowed India's significant economic growth, particularly in the manufacturing sector. GDP growth rates dropped from 9.3% in 2007 to 5.3% in the fourth quarter of 2008.

But despite a recent slowing economy, India's energy demand continues to increase. Energy use in the transport sector is expected to be particularly high. Ownership of four-wheel vehicles is forecast to increase rapidly once the global economic crisis abates and domestic spending levels resume.

Strong growth in population is also expected to continue, to 1.46 billion by 2030 from 1.09 billion in 2006. That means increasing energy demand as well.

The EIA projects that total energy demand will grow to 29.3 quads in 2025 from 14 quads in 2002.

India was already the fifth largest consumer of oil in the world during 2006, according to EIA.

Major oil and gas companies

Several companies (state-owned unless otherwise stated) dominate India's energy sector.

- Oil and Natural Gas Corp. (ONGC; Dehradun) is the largest and most active operator in India.
- Oil India Ltd. (OIL; Duliajan) is India's second largest domestic exploration and production company.
- Gujarat State Petroleum Corp. (GSPC) operates about a dozen oil and gas fields in India.
- GAIL (India) Ltd. (New Delhi)—formerly Gas Authority of India Ltd.—is the country's principal gas transmission and marketing company. It is also involved in exploration, production, and processing.

- Cairn India Ltd. (Gurgaon) was established as a separate entity in early 2007 by Cairn Energy PLC (Edinburgh, Scotland). Cairn Energy, a major participant in India's hydrocarbon sector since the late 1990s, holds a major interest in Cairn India Ltd.

Oil

Most of India's oil reserves are located in the Mumbai High, Upper Assam, Cambay, Krishna-Godavari, and Cauvery basins. The offshore Mumbai High field is by far India's largest producing field.

There are substantial undeveloped reserves in the offshore Bay of Bengal and in Rajasthan state.

Rising oil consumption and fairly stable production levels make India increasingly dependent on imports. In 2006, the country produced an average of 648,000 b/d, was crude oil, but consumed an estimated 2.63 million bbl/d.

EIA estimated demand growth of 100,000 b/d for 2006 through 2008.

Overseas E&P

Indian NOCs have been actively seeking equity stakes in exploration and production projects overseas.

The most active has been ONGC Videsh Ltd., the overseas investment arm of ONGC. As of January 2007, the firm held interests in 25 oil and natural gas projects in 15 countries.

Terrorist attack: Mumbai

On November 26, 2008, terrorists attacked hotels and other buildings in Mumbai, killing at least 195 people and causing extensive damage.

India blamed Lashkar-e-Taiba (LeT), a Pakistani militant group against Indian control of Kashmir, for the attack.

In February 2009, the Pakistani government admitted that the operation was planned and executed from inside Pakistan and implicated three LeT leaders. It said that ten of the terrorists left from Karachi by boat to conduct the attack on Mumbai.

Pakistan charged eight men with abetting, directing, conspiring and facilitating a terrorist act.

Integrated Energy Policy

In December 2008, the Indian cabinet approved an Integrated Energy Policy (IEP), which aims to help the country achieve energy independence.

The report includes recommendations regarding energy security, energy standards, commercialization of alternate energy resources, market-based pricing strategy for coal, and use of fuel subsidies.

Refining

The Jamnagar refinery owned by Reliance Petroleum is India's largest. Reliance brought online a second facility at that site in early 2009, making it the largest refining complex in the world, with a total capacity of more than 1 million b/d.

IOC, which operates 10 refineries, has announced plans to double its refining capacity by 2012.

Natural gas

EIA estimates that India produced 31.1 bcm (1.1 tcf) of natural gas in 2007, up slightly from 2006. The bulk of the production comes from the western offshore regions, especially the Mumbai High complex. ONGC and OIL account for the bulk of natural gas production.

The onshore fields in Assam, Andhra Pradesh, and Gujarat states are also significant sources of natural gas. The Bay of Bengal has also become an important source of natural gas for the country.

In 2007, India consumed roughly 4.2 bcm (1.5 tcf) of natural gas, some 2.8 bcm (100 bcf) more than in 2006.

Gas demand is expected to grow considerably, largely driven by demand in the power sector. The power and fertilizer sectors account for nearly three-quarters of natural gas consumption in India.

By 2030, EIA expects that gas use in India will be a major component of a doubling of natural gas demand in Asia.

India has been a net importer of natural gas since 2004. Net imports reached 10 bcm (353 bcf) in 2007.

Gas pipelines

Four major pipeline projects to import gas from Bangladesh, Myanmar, Iran, and Turkmenistan have been discussed, but formidable geopolitical and commercial hurdles suggest that pipeline imports in 2015 will be negligible.

In mid-2007, Indian officials also were exploring participation in a proposed Trans-Saharan Gas Pipeline, primarily designed to deliver gas from Nigeria, across Algeria, to Europe.

Exploration and Development

Cambay basin

Five wells drilled in the Cambay basin were reported as discoveries by GAIL Gujarat State Petroleum Corp. in late 2008.

The company has drilled a total of 15 exploration wells and two appraisal wells on the Sanand/Miroli block, according to partner GeoGlobal Resources Inc. (Calgary).

The discovery wells are M-1, M-6, SE-2, SE-4, and SE-8.

In addition, SE-11 well went to TD 2,500 m (8,202 ft) and was awaiting testing, while SE-10 went to 1,900 m (6,234 ft) and was being logged.

The SE-8-A1 appraisal well evaluating the Kalol pay zone went to TD 1,450 m (4,757 ft) and was under test.

Earlier in the year, Oilex Ltd. (Sydney) was seeking to establish oil and gas production in abandoned Cambay gas field.

ONGC reports activity

ONGC struck oil in late 2008 with the G-4-6 well in nomination Block-IG offshore. The well, drilled in water 225 m (738 ft) deep, encountered three hydrocarbon-bearing sands over a gross interval of 3,757–4,185 m (12,326–12,730 ft).

ONGC plans a fast-track hub development program, clustering geographically proximal finds including G-2, GS-29, Padmawati, and Kanakadurga, using an FPSO vessel.

Initial production is expected to be 20,000 b/d by 2012–13, peaking at 150,000 b/d by 2016–17.

The company also made a discovery in mid-2008 with its GS-15-OA well in the Krishna-Godavari basin off the east coast.

ONGC was also exploring possible construction of an FPSO vessel to develop marginal fields and to conduct extended testing.

In mid-2008, the firm reported that it had achieved an oil recovery rate of close to 32% in 18 of its major producing fields.

Officials said that recent discoveries have indicated the potential for huge hydrocarbon reserves in deep and ultra deep reservoirs but that a shortage of rigs was hampering appraisal drilling.

Mahanadi basin

Niko Resources Ltd. (Calgary) discovered gas early in 2008 in the NEC 25 block located in the Mahanadi basin, off the Orissa coast in the Bay of Bengal.

This shallow water block covers 10,775 sq km (4,160 sq mi) in water depths ranging between 20 and 600 m (66 and 1,968 ft). This was the eighth discovery in the block.

The well NEC 25-J1 was drilled on the southern portion of the area of a 2007 3D seismic survey.

It was drilled at a water depth of 478 m (1,568 ft) to the target depth of 2,926 m (9,600 ft).

India-Colombia agreement

India signed a memorandum of understanding with Colombia in late 2008 to enable bilateral cooperation in oil and gas exploration, including the training of technicians and professionals.

Details are given in the Colombia section of the Latin America chapter of this book.

Lease Sales

NELP VII bidding

ONGC dominated bidding on 45 of the 57 blocks offered under the seventh round of India's New Exploration Licensing Policy (NELP VII).

Bids were reviewed in mid-2008; winners were to be named by yearend by the Directorate General of Hydrocarbons (DGH).

Other bidders included BHP Billiton (Melbourne), Gujarat State Petroleum Corp Ltd., and OIL Corp.

BP PLC (London) joined Reliance Industries Ltd. (Mumbai) to bid for one single deepwater block, while Cairn Energy PLC (Edinburgh) made a bid for another deepwater block.

Block KG-DWN-2005/2

Reliance Industries Ltd. (Mumbai) and BP PLC (London) won deepwater block KG-DWN-2005/2 in late 2008. It was offered under India's NELP VII.

RIL has 70% participating interest; operator BP holds the remainder.

The block is 40 km (25 mi) off the east coast of India, covering 1,949 sq km (752 sq mi).

The consortium first will conduct 2D and 3D seismic surveys and reprocess existing seismic data.

Reliance Industries Blocks in India's Basins

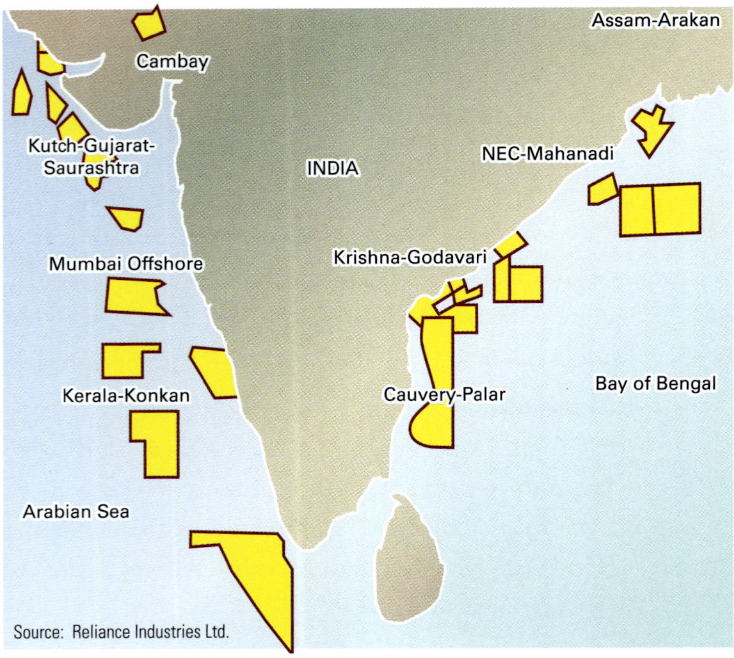

Source: Reliance Industries Ltd.

Drilling and Production

MA oil field

Reliance Industries Ltd. (Mumbai) and Niko Resources Ltd. (Calgary) announced in September 2008 the start of production from deepwater MA oil field in the Krishna-Godavari basin in the Bay of Bengal.

The early production rate was 5,000 b/d.

The field is in 2,438 m (8,000 ft) of water on Block KG-D6, some 50 km (31 mi) off Kakinada.

It is expected to reach target peak production of 40,000 b/d by early 2009.

Niko Resources pointed out that MA oil field:

- is India's first deepwater hydrocarbon producing area.
- produces a grade of oil that can be processed at any refinery in India.
- went on production just 30 months after its discovery.

The FPSO vessel serving the field can operate in water more than 1,200 m (3,937 ft) deep.

Gas Hydrates

Offshore Andaman Islands

An international team led by the USGS and India's DHG released the results in early 2008 of what they described as the world's most comprehensive gas hydrate field venture to date.

Gas hydrates are a naturally occurring, ice-like combination of natural gas and water, formed by high pressure and low temperatures in the world's oceans and polar regions.

In the Krishna-Godavari basin, India's National Gas Hydrate Program (NGHP) Expedition 01 was carried out in 2006.

The report said that one of the thickest and deepest gas hydrate occurrences yet known was discovered off the Andaman Islands, with gas hydrate-bearing layers as deep as 600 m (1,969 ft) below the seafloor.

It also said that a fully developed gas hydrate system was established in the Mahanadi basin.

Development of gas hydrates would pose difficult technical challenges.

Pipelines

Paradip-Haldia oil pipeline

In early 2009, IOC commissioned a 330-km (205-mi) crude oil pipeline in eastern India, connecting the port city of Paradip, Orissa, to Haldia, West Bengal.

The addition of this line brought the firm's total pipeline network to more than 10,000 km (6,214 mi), with a cumulative throughput capacity of 1.5 million b/d of crude oil.

The $2.95-billion project installed a single-point mooring facility at the Paradip port for offshore tanker unloading, built a shore tank farm in Paradip, and laid the Paradip-Haldia pipeline itself to connect with IOC's existing Haldia-Barauni crude oil pipeline.

The 235,000-b/d Paradip-Haldia line traverses about 228 km (142 mi) along the Orissa coast and about 102 km (63 mi) in West Bengal before terminating at IOC's crude oil tank farm at Haldia.

IOC was also working in 2008 on about a dozen other pipeline-related projects with a total investment in the range of $520-$540 million.

Vijaipur-Dadri-Bawana gas pipeline

In late 2008, GAIL India Ltd. (New Delhi) was proceeding with plans for construction of the $1.1-billion Vijaipur-Dadri-Bawana pipeline.

The project will build pipelines to carry regasified LNG from GAIL's dispatch terminal in Vijaipur, Madhya Pradesh, to the firm's receipt terminal in Dadri, Uttar Pradesh, and further to Bawana, Haryana.

Upon completion, the pipeline will have a transmission capacity of 78 MMscmd (2.75 bscfd) of regasified LNG.

Himachal Pradesh gas pipeline

In late 2008, GAIL (India) Ltd. (New Delhi) signed an agreement with the Himachal Pradesh government to link the state with the national gas grid.

The plan is to extend the proposed Dadri-Bawana-Nangal gas pipeline through Uttar Pradesh, Delhi, Haryana and Punjab to the state. Project cost is estimated at $520 million.

Himachal Pradesh will be linked to the national gas grid through a 64 km (40-mi) pipeline from Rajpura town in Punjab to Baddi in the state. This section is planned to be completed by early 2011.

Pipeline investment: 2008–2013

A 2008 study by the Associated Chambers of Commerce and Industry of India and Ernst & Young Global Ltd. (London) estimated that $9.2 billion will be invested from 2008–2012 for laying gas pipelines in India.

The study is part of a paper, "Indian Oil & Gas Sector: Rising Business Opportunities," jointly published by the two organizations.

The rapid growth in the Indian petroleum sector calls for a robust country-wide pipeline infrastructure.

The existing Indian gas pipeline network spans 10,000 km (6,214 mi), of which GAIL (New Delhi) owns 55%. The remainder is held in partnership with the private sector.

MedStream multipurpose pipeline

The proposed $12-billion MedStream project aims to build a multipurpose subsea pipeline to transport crude oil across the Mediterranean Sea from Turkey to Israel.

Indian officials expressed interest in September 2008 in possibly extending the line to India.

Details are given in the Israel section of the Middle East chapter of this book.

IPI Pipeline Project

In October 2008, Iran and Pakistan decided to go ahead with the $7.4-billion Iran-Pakistan-India (IPI) gas pipeline project bilaterally, saying India could join the project later if it chose to.

Details are given in the Iran section of the Middle East chapter of this book.

Refining and Products

Export-oriented Gujarat unit

Reliance Petroleum Ltd. (Mumbai) started up in January 2009 what it called the world's largest single-location refinery complex on December 25, 2008, bringing onstream a 100% export-oriented refinery unit next to the existing unit in the Jamnagar Special Economic Zone in Gujarat.

The two units can process a combined total of 1.24 MMb/d of crude oil and constitute the world's sixth-largest refining complex.

The $6-billion export-oriented refinery unit has a processing capacity of 580,000 b/d of crude and adds 20% to India's total crude oil refining capacity.

The unit will initially produce gasoline and diesel that complies with the Euro-IV grade. Quality will be subsequently upgraded for compliance with Euro V norms after auxiliary units come onstream.

Mangalore expansion

Mangalore Refinery and Petrochemicals Ltd —a subsidiary of ONGC Ltd.—announced plans in late 2008 for a $2.5-billion capacity-expansion project.

The expansion will enable the refinery to process more high-sulfur crude.

The refinery's capacity will increase from about 207,000 b/d to 320,000 b/d by late 2011.

The initial completion date had to be pushed back due to labor and land-acquisition issues.

At the same time, MRPL put on hold its plans for a new 320,000-b/d grassroots refinery and petrochemical plant. The estimated cost for the project has increased by more than $800 million since 2006.

Vadinar expansion

Essar Oil Ltd. (Jamnagar)—part of Essar Group (Mumbai)—said in late 2008 that it will expand the refining capacity of the Vadinar refinery in Jamnagar to about 700,000 b/d from the current 200,000 b/d.

The facility produces diesel and gasoline for use in India and international markets. It can also produce LP gas, naphtha, light diesel oil, kerosene and aviation fuel from a range of sweet, sour and heavy crude.

The first phase of the $6 billion expansion is to be carried out by mid-2010, adding 340,000 b/d.

Another 380,000 b/d of capacity will be added by yearend 2010.

Refinery near the Taj Mahal

IOC Ltd. was awaiting government approval to proceed with an expansion of its refinery at Mathura, about 40 km (25 mi) from the Taj Mahal.

Holding up the project were concerns of environmental groups that increased sulfur dioxide emissions from the facility would further damage the 17th-century monument.

IOC plans to increase the capacity of the Mathura plant from 8 mtpy to 11 mtpy and also build a petrochemical plant. The refinery processes crude from India, the Middle East and Nigeria.

In early 2009, officials were considering planting thousands of tulsi saplings near the Taj Mahal in an attempt to mitigate the effects of airborne pollutants.

Other IOC refinery plans

IOC controls 10 of the 19 refineries in India and has a total refining capacity of 60.2 mtpy, about 40% of the country's total refining capacity.

IOC has planned expansion investments of over $10.8 billion for 2007–12.

- Under way is an upgrade of the Panipat refinery in Haryana, to be finished by early 2009.
- In progress is an expansion of the Haldia refinery in West Bengal, to be completed by yearend 2009
- The company also wants to acquire 100 acres next to the Haldia refinery to set up a paraxylene plant by yearend 2009.
- The refinery and petrochemical project at Paradip, Orissa, is to be completed by 2012.
- An upgrade of the refinery at Gujarat is to be completed by early 2010.
- The 291-km (181-mi) Chennai-Bangalore product pipeline project was awaiting environmental clearance as of late 2008.
- The Paradip-Haldia pipeline to transport crude oil to the Haldia and Barauni refineries from Paradip was to be completed by yearend 2008.

LNG

LNG infrastructure

From April-October 2008, India imported 4.8 million tons of LNG, compared to 4.7 million tons imported for the same period in 2007. India began importing LNG in 2004.

The country has two terminals. The one in Dahej is operated by Petronet LNG Ltd. (New Delhi)—a joint venture of IOC Ltd. (Mumbai), GAIL (India) Ltd. (New Delhi), Bharat Petroleum Corp. Ltd. (Mumbai) and ONGC Ltd. (New Delhi).

Expansion of the Dahej terminal was completed in January 2009, doubling its capacity to 10 mtpy.

Petronet LNG buys about 5 mtpy of LNG from RasGas (Doha, Qatar) under a 25-yr agreement.

RasGas has agreed to supply an additional 2.5 mtpy on a long-term basis under the same contract beginning in September 2009 to for use in the Dabhol and Pragati power plants in Maharashtra.

The second LNG facility is the 3.6-mtpy import terminal in Hazira, run by Royal Dutch Shell Ltd. (The Hague, Netherlands). The company has said it intends to increase capacity by 46%.

Shell imports LNG from Oman, Abu Dhabi, Qatar, Malaysia, Nigeria and Egypt.

India has expressed interest in importing 2 mtpy of LNG from Russia's Sakhalin LNG project. Gazprom (Moscow) was still reviewing the proposal in early 2009.

Also, ONGC Videsh Ltd. (New Delhi) holds a stake in the Sakhalin-1 oil and gas fields in eastern Russia that is operated by ExxonMobil Corp. (Irving, TX).

Ennore facility

IOC Ltd. had proposed to build a third LNG regasification plant in Ennore, near Chennai, but as of early 2008, the company was unable to arrange for long-term supplies from Algeria.

A supply commitment is needed before work can begin on the 2.5-mtpy terminal.

The company was exploring other potential supply arrangements with Qatar and Australia.

Dabhol power project

In late 2008, the national government was evaluating a proposal to allow bids from private- as well as public-sector entities to build a 5-mtpy LNG terminal near the power plant in Dabhol.

The power plant traces its origins back to an initiative by now-bankrupt Enron Corp.

If the proposal is approved, it could open the way for private-sector players to bid for the project, including Reliance Industries Ltd. (Mumbai), Reliance-Anil Dhirubai Ambani Group (Mumbai), BG Group PLC (Reading, UK), Essar Oil Ltd. (Jamnagar), and Adani Enterprises Ltd. (Ahmedabad).

Petrochemicals

Petrochemical plant in Bihar

GAIL and IOC Ltd. (Mumbai) signed an MOU late in 2008 to explore the feasibility of building a petrochemical plant in Barauni, Bihar.

The proposed $2-billion, 300,000-ton complex would use naphtha produced by IOC's Barauni refining plant as well as natural gas that GAIL intends to transport from eastern offshore sources.

In addition, imported LNG would be supplied through the planned Jagdishpur-Haldia pipeline.

If an initial feasibility study is positive, construction would likely take five years.

Pata expansion

GAIL was making plans in late 2008 to increase the production capacity of its gas-based petrochemical plant in Pata in the Auraiya district of Uttar Pradesh.

The capacity of the polymer line would be boosted from the current 410,000 tpy to 500,000 tpy in the first expansion phase.

A second phase would boost production capacity of the whole complex to 800,000 tpy.

The facility produces high-density polyethylene, linear low-density polyethylene, and ethylene.

Visakhapatnam facility

In a joint venture with Total SA (Paris), Hindustan Petroleum Corp. Ltd. (Mumbai), Mittal Energy (Singapore), and OIL Ltd. (New Delhi), GAIL was considering building a major facility in Visakhapatnam, Andhra Pradesh.

It would comprise a 15-mtpy export-oriented refinery and a 1-mtpy petrochemical unit.

The pre-feasibility study of the $6 billion project was under way in late 2008.

Construction time is estimated at five years.

Alternative Fuels

Biofuels from jatropha

A subsidiary of Tata Chemicals Ltd. (Mumbai) signed in late 2008 a JV agreement with the jatropha seedling firm, JOil Singapore Pte Ltd. (Singapore).

The agreement will give Tata Chemicals exclusive marketing rights for the firm's jatropha seedlings in India and East Africa. Oil-seeds from the plant have found growing use in producing biodiesel.

Tata Chemicals was running trials for jatropha cultivation in the semi-arid regions of Andhra Pradesh, Gujarat, Maharashtra, and Tamil Nadu. Trials in Kenya also were being planned.

Sorghum-based ethanol

Tata Chemicals also was making plans in late 2008 to commission India's first ethanol plant based on sweet sorghum feedstock, in Nanded.

After an initial investment of $10 million for a prototype, the firm plans to invest up to $150 million by 2012. Praj Industries Ltd. (Pune) will design and build the facility.

The plant will initially produce 30,000 l./d (7,920 gal/d), with plans to eventually produce 8 million l./d (2.11 million gal/d).

Rajasthan state: Renewable energy

Rajasthan state was poised in late 2008 for private-sector investment of $1.6 billion in its renewable energy sector.

State officials said that the government had signed MOUs with 14 companies to build plants across the state to generate power from wind energy and biomass.

Five wind energy projects with a combined power generation capacity of 1,600 MW are slated for construction in Jaisalmer, Barmer and Jodhpur.

Eight biomass projects are planned for the Tonk, Sirohi, Baran, Alwar, Kota, Sawai Madhopur, Nagaur, Hanumangarh and Jalore districts.

The projects are expected to be operational by 2011–12.

Rajasthan already has 492 MW of wind energy systems and biomass generation capacity of 46.3 MW.

With nearly 7,660 MW of wind power capacity, India ranks fourth globally in grid-connected wind power installations.

Company News

ONGC-Imperial Energy

At yearend 2008, ONGC Videsh, the foreign investment arm of ONGC, was poised to buy Imperial Energy Ltd. (London).

It had received acceptances from more than the minimum 90% of shareholders required to proceed with the £1.3-billion transaction. The deal for Imperial, which has most of its assets in Russia, was expected to close in mid-January 2009.

ONGC Videsh said the purchase would give it a presence in western Siberia, one of the world's largest oil and gas producing regions.

INDONESIA

Capital: Jakarta
Monetary unit: Rupiah
Refining capacity: 992,745 b/cd
Oil production: 855,000 b/d
Oil reserves: 3.990 billion bbl
Gas reserves: 106.000 tcf

Overview

State-run PT Pertamina has had difficulty in recent years producing and delivering sufficient oil and gas to meet domestic needs.

Oil

The government wants to increase oil production to 1.3 million b/d by 2009 from the 0.84 million b/d in 2007, but there is skepticism about its ability to do so.

The economy also has been struggling to keep pace with the cost of increased oil imports. Indonesia consumes 1.35 million b/d of oil and imports about 390,000 b/d.

Upstream oil and gas regulator, BPMigas, said oil production is falling by 1.2% a year, while domestic oil demand is rising by 1.5%.

Natural gas

The government also has set a goal increasing gas output by 30% by 2009.

In 2006, Indonesia produced a total of 8.22 bcf, of which 54% was exported. But declining output that year prevented the government from fully meeting its export commitments.

Officials with the Ministry of Energy and Mineral Resources said early in 2007 that Indonesia expected a shortage in natural gas production of some 300 MMscf for the year.

Officials have promised to push for improved efficiency in domestic gas use and for expanded exploration. They point to exploration already under way in Java, Papua, and the Natuna islands.

Policy

OPEC membership

Indonesia—the only Southeast Asian member of OPEC—decided in mid-2008 to terminate its membership, effective at yearend. It had been a member since 1962.

The Energy and Mineral Resources Ministry said that Indonesia had become a net oil importer, but some observers said the action was also a protest against high global oil prices and a reaction to the country's inability to influence OPEC decisions.

The nation's oil and fuel imports reached $13 billion in May 2008, while exports earned just $7.56 billion. The deficit was projected to reach $15 billion by December.

Purvin & Gertz Inc. (Singapore) saw no major downside for Indonesia's decision to leave OPEC, apart from a loss of prestige. OPEC officials declined to comment on the departure.

Oil, banking linked

The Indonesian government said in late 2008 that within two months, it would order oil and gas companies to keep funds for energy projects in domestic banks.

Oil and gas regulator BPMigas said that the action was to boost liquidity and shore up the falling rupiah.

The mandatory regulation would apply to both national and international oil and gas operators. Their expenses would not be reimbursed unless they used domestic bank services.

BPMigas estimated that oil and gas operators would spend about $11.8 billion in 2009.

Territorial dispute with Indonesia

Late in 2008, the Malaysian government tried to downplay concerns about a possible military conflict with Indonesia regarding rights to and development of the Bukat Block.

Officials stated that the block, operated by Eni SPA (Rome), falls outside the territory disputed with Indonesia.

Details are given in the Malaysia section of this chapter.

Pertamina: Parliamentary scrutiny

In late 2008, state-owned PT Pertamina (Jakarta) was coming under increased scrutiny by parliament.

Investigators alleged irregularities and corruption in the awarding of contracts for importing foreign oil products.

Earlier charges said that the firm imported a sub-standard oil type in 2007.

BPMigas VPs in IOCs

In a bid to exercise greater control over contract holders, the government said in mid-2008 that regulator BPMigas can assign some of its officials to work as executives within IOCs.

In parallel, there were calls for the government to impose a windfall profits tax on IOCs to help ease the impact of high gas prices on Indonesians.

There had been earlier reports of efforts to step up the monitoring of the upstream sector.

The VPs would be assigned to holders of oil and gas cooperation contracts.

E&P cost-recovery process

Indonesian officials met in mid-2008 with top foreign oil and gas company executives.

No official purpose was given for the closed-door meeting, but it coincided with concerns in the country over cost-recovery claims submitted to the government by oil and gas contractors.

A government audit published in late 2007 of the 2005 accounts of nine oil and gas blocks unveiled some $525 million in questionable claims.

Regulatory changes

In April 2008, the new head of BPMigas vowed to streamline the processes that oil and gas companies undergo in order to obtain exploration licenses.

The aim is to help boost oil production, which has declined over the past 5 yrs due to depletion at aging fields and reduced exploration for new ones.

BPMigas also said that more care would be given to assuring that cost-recovery funds would only be given for producing fields.

BPMigas said that in 2007, the government paid $8.33 billion to oil and gas producers for recovery costs, up 6.4% compared to 2006.

International Activities

China, Indonesia agreements

Indonesia and China signed eight energy-related agreements in late 2008, valued at more than $3 billion, during the third Indonesia-China Energy Forum in Jakarta.

The eight agreements and amount of finance include:

- Oil and gas 20-yr contract extension in Madura Strait (BPMigas, CNOOC, and Husky Madura Ltd.; $642 million).
- Biodiesel development in Jambi and South Sumatra (PT Kurnia Selaras and China Development Bank; $255 million).
- Coal mining JV in Muara Enim (Bukit Asam and Huadian Corp.; $14.40 million).
- Coal mining cooperation in East Kalimantan (PT Budi Dharma Kencana and Lark Guangdong Power Resources Inc.; $350 million).
- Power plant financing in Pelabuhan Ratu (Exim Bank of China and PT PLN; $481.94 million).
- Power plant financing in Pacitan (Exim Bank of China and PT PLN; $293.23 million).
- Power plant construction in Cilacap (PLN, CNTIE, and Shanghai Electric; $605.29 million).
- Power purchase agreement for Muara Enim (PT PLN and PT GH EMM Indonesia; $330 million).

Exploration and Development

Makassar Strait

The government approved in third quarter 2008 a plan by Chevron Indonesia Co. to develop natural gas reserves in five offshore fields in Makassar Strait: Gendalo, Maha, Gandang, Gehem, and Ranggas.

Gendalo, Maha, and Gandang fields form part of the Ganal Block concession operated by Chevron Ganal Ltd.

Ranggas and Gehem fields are on the Rapak Block operated by Chevron Rapak Ltd.

Officials called the project the country's first deep-sea drilling project.

West Timor Block

In mid-2008, Indonesia awarded a new exploration block to Eni SPA (Rome).

The West Timor Block (Timor Island) encompasses 2,725 sq km (1,052 sq mi) offshore and 1,350 sq km (521 sq mi) onshore. Eni will be the operator and the sole license holder.

Galan Block

Indonesia approved in mid-2008 a proposal by Chevron Corp. (San Ramon, CA) to develop natural gas fields on the deepwater Galan Block off East Kalimantan.

Chevron committed to spend $311 million. The block is believed to have the potential to produce an average of 800 MMcfd.

Husky-CNOOC joint development

Husky Energy Inc. (Calgary) and a subsidiary of China's CNOOC Ltd. (Beijing) agreed in early 2008 to jointly develop the Madura BD gas and natural gas liquids field located offshore East Java.

The agreement covers development and further exploration of the Madura Straits Production Sharing Contract (PSC).

The PSC is located about 40 km (25 mi) north of East Java and covers 2,794 sq km (1,079 sq mi) of exploration acreage.

There have been two discoveries on the block since 1984: the Madura BD and MDA fields.

Husky has plans for a two well exploration program in 2009.

Kambuna field

Operator Serica Energy PLC (London) completed the drilling phase of the Kambuna gas field development program in early 2008 and planned to start production by yearend.

The field is in the Malacca Straits, 40 km (25 mi) off North Sumatra.

Serica operates the 380-sq km (147-sq mi) Glagah Kambuna Technical Assistance Contract, has installed a wellhead platform in 40 m (131 ft) of water, and was completing production tests in mid-2008 on three development wells.

Serica expects plateau production of 1.4 MMscmd (50 MMscfd) of gas and 5,000 b/d of condensate from proved reserves estimated at 19 million boe.

A pipeline to be laid in fourth quarter will carry production to an onshore gas plant.

Lease Sales

E&P investments

Indonesian officials said in late 2008 that they expect investments to reach $375 million for the 22 oil and gas blocks awarded in October. A total of 25 blocks had been offered.

Chevron Corp. (San Ramon, CA), ConocoPhillips CO. (Houston), and Husky Energy Inc. (Calgary)—all of which won oil and gas exploration rights for four blocks—have pledged total investment of at least $91 million for 2008–2011. More specifically:

- Chevron Indonesia Ventures Ltd.: $24.5 million in each of two fields, West Papua I and West Papua II Blocks in West Papua.
- ConocoPhillips: $30 million in the first three years of exploration of Arafura Sea Block.
- Husky Energy: $12.4 million for three years of exploration on North Sumbawa II Block.

In addition to the three IOCs, 17 other companies or consortia won exploration rights.

New bidding round

In November 2008, the government opened a new tender for 31 oil and gas blocks onshore and offshore. The deadline for bids is January 2009 for some areas, and March for others.

Blocks are available in the following areas:

- Sumatra. One onshore-offshore block and four blocks off northern Sumatra; two onshore blocks in central Sumatra; and two onshore blocks in southern Sumatra.
- Kalimantan. Two onshore blocks in southern Kalimantan; and one block onshore and two off eastern Kalimantan;
- Papua. One onshore-offshore block in northern Papua, and five off western Papua.

Other offshore blocks include the N. Surumana Block in the Makassar Strait; the Sermata Block in the West Arafura Sea; two blocks off southern Sulawesi; the Karaeng Block off northern Selayar; the SE Halmahera-SW Kofiau Block off South Halmahera; the East Bula Block off Seram; the Penyu block off West Natuna; and the Sokang Block off East Natuna.

Other onshore blocks include the South East Madura onshore-offshore block in East Java; and the Terumbu Block, partly onshore-offshore Madura.

Drilling and Production

Cepu Block: Production, storage

In mid-December 2008, ExxonMobil Corp. (Irving, TX) announced production of first oil from the Cepu Block, located on the borders of Central Java and East Java.

The block is said to hold 600 million bbl of recoverable oil and 48 bcm (1.7 tcf) of natural gas.

Initial output volumes reached 5,600 b/d of oil, according to Pertamina.

Peak production is expected to reach 165,000 b/d by 2012.

ExxonMobil said that it expects output to increase steadily to 100,000b/d by 2010. Peak production of 165,000 b/d is anticipated in 2012.

Also in late 2008, BPMigas approved construction of a floating oil storage facility for the block with a capacity of 2 million bbl. Tender offers were in preparation.

PT Medco: 2009 view

PT Medco Energi Internasional Tbk (Jakarta) predicted in early 2009 that its oil and gas production for the year could drop by as much as 10% due to asset sales and aging fields.

The company said that installation of new gas facilities on South Sumatra's Lematang Block would be completed in June 2009, and that other projects would be completed in 2011–2012, including:

- Block A gas field in Aceh
- Block Rimau in South Sumatra (enhanced recovery)
- Area 47 in Libya
- Senoro gas field in Central Sulawesi
- a power project in Sarulla, North Sumatra.

PT Medco said its success rate for confirming feasible reserves in Indonesia was only 10–15% and that it would intensify its search for overseas oil and gas fields.

Oil, gas production outlook

PT Chevron Pacific Indonesia (CPI; Jakarta)—a unit of Chevron Corp. (San Ramos, CA)—projected in late 2008 that production from its concessions will decline due to aging fields and seasonal floods.

CPI said its Duri and Minas concessions in Sumatra might produce just 405,000 b/d in 2008, missing the 408,000-b/d target set by the government and down from the 425,000 b/d produced in 2007.

The company produces about 40% of Indonesia's total crude output.

CPI is one of several IOCs summoned in 2008 by the government's budget committee to explain reasons behind the country's falling production of crude oil.

The committee also summoned PT ConocoPhillips Indonesia, PT Pertamina EP, PT Medco E&P Indonesia, and PT ExxonMobil Oil Indonesia.

Only Pertamina EP has estimated that it would produce more oil in 2009 than in 2008 (about 6% more).

North Duri Area 12 field

PT Chevron Pacific Indonesia reported in November 2008 the start of oil production from Area 12 of the North Duri field, where Chevron produces nearly half the nation's oil.

The firm is planning expanded steamflood operations for 2009 and expects oil production from Area 12 to rise to 34,000 b/d by 2012.

Duri field, discovered in 1941 on Sumatra, currently produces 200,000 b/d of oil.

Chevron said that steamflooding to date had more than tripled oil production from the field and has resulted in recovery of more than 2 billion bbl of oil.

Pertamina: Pushing for control

PT Pertamina was vying in late 2008 with partner Korea Development Corp. (Kodeco; Seoul) to become operator of the offshore Madura oil and gas block.

Pertamina holds a 50% stake in the block, with the remaining 50% jointly controlled by Kodeco and CNOOC (Beijing).

Pertamina also has been working to obtain a 20% interest in other oil and gas concessions, including the Mahakam Block (Total SA; Paris) in East Kalimantan; the offshore Masela Block (Inpex; Tokyo) in the Timor Sea; and deepwater fields (Chevron; San Ramos, CA) off East Kalimantan.

Drilling services outlook

The Indonesian Oil & Gas Drilling Contractors Association projected in late 2008 that demand for drilling services in the country would jump by 20% in 2009.

The group sees domestic and international oil companies seeking to boost their production in light of government incentives and rising crude prices.

Government plans to tap marginal wells also could help boost drilling.

Also, the Energy and Mineral Resources Ministry issued regulations in 2008 allowing domestic oil and gas companies to exploit wells formerly managed by IOCs.

Mengoepeh field

Continental Energy Corp. (Dallas) was planning in mid-2008 to buy a 30% working interest in the 565,500-acre (883-sq mi) Tungkal PSC in Central Sumatra.

Operator Pearl Energy Ltd. (Abu Dhabi) produces 1,000 b/d from Mengoepeh field and was planning to drill 16 development wells on a southern extension of the field starting in late 2008.

Its goal is to boost output to 4,500 b/d.

EOR: Mahakam River Delta

Total SA (Paris) and partner Inpex (Tokyo) said in mid-2008 that they will invest a combined $2 billion in 2009 to enhance recovery in their oil and gas fields in offshore and in the Mahakam River Delta.

Total said it has already invested $11 billion in the fields.

The group produces 90,000 b/d of oil and condensates and 73.6 MMcmd (2.6 bcfd) of gas.

Deepwater drilling planned

StatoilHydro AS (Stavanger, Norway) was proceeding with plans under which it would drill three ultradeep exploration wells on the Karama Block in the Makassar Strait off West Sulawesi in late 2009.

StatoilHydro is in the Makassar Strait Explorers Consortium, which will drill 12 exploration wells.

The first of the three wells in the Karama license is scheduled to be drilled in late 2010.

LNG

Decline in 2009 LNG output

Indonesian officials said that LNG output likely will fall in 2009, largely due to a decline in the supply of natural gas to the country's Bontang LNG facility.

Expected in 2009 are 349 cargoes, compared to 359 in 2008. One cargo contains 125,000 cu m (4.4 MMcf). The gas fields facing output declines were not identified.

An official at the Bontang LNG liquefaction plant expected 307 cargoes in 2009, down from 317 in 2008. At the Arun LNG plant, officials said that they expect no change from the 42 cargoes of 2008.

The country is struggling to meet its LNG commitments to offshore buyers—especially Asian clients such as Japan and South Korea.

In late 2008, Indonesia said it would seek nine cargoes of LNG in 2009 from the spot market to meet those contractual commitments.

BPMigas said Indonesia would likely divert the gas it had previously intended for South Korean buyers to a state-owned fertilizer company.

INDONESIAN LNG PRODUCTION

	Total	Bontang	Arun
	Number of cargoes		
2009*	349	307	42
2008	359	317	42
2007	372	320	52
2006	394	335	59

*Estimated.

First Tangguh exports

Indonesian officials said in late 2008 that the first LNG exports from the Tangguh project will be shipped in February 2009 as scheduled.

There had been earlier reports that the shipment would be delayed until mid-2009.

BP PLC (London) is leading the $5-billion Tangguh project, which will process gas from new fields in Papua Bintuni Bay.

Though Indonesia and China continued to negotiate LNG price, shipments from Tangguh to the Fujian terminal of CNOOC are still scheduled to start in first quarter 2009.

Offshore LNG terminal

Inpex Corp. (Tokyo) said in late 2008 that it likely will construct an offshore export terminal at a cost of more than ¥1 trillion. The new terminal is expected to provide 4.5 mtpy of LNG by 2015—all of it bound for Japan.

An agreement with Indonesia on terminal location and construction schedule was expected by yearend.

Gas for the terminal would come from the Masela block, currently being developed in the Timor Sea, 400 km (249 mi) from Darwin.

In July 2008, Inpex confirmed sufficient gas reserves to start production of 4.5 mtpy of LNG in the Masela block in 2016.

In August, the company also said it will build a regasification terminal in northwestern Japan. Its operation would start in 2014, handling 0.5–1.0 million tonnes of LNG imports in its first year.

Seroro start-up delay

In early 2008, the consortium slated to develop Indonesia's planned fourth LNG plant, in Senoro, Central Sulawesi, said that startup will be delayed from 2011 to 2012.

Disagreement over pricing gas for the facility was said to be the cause of the delay.

Project partners are Mitsubishi Corp. (Tokyo), PT Pertamina, and Medco E&P Indonesia (Jakarta).

Supply gas will come from two fields in Senoro, on Toili Block and Donggi Block.

Indonesia operates two LNG plants:

- Arun LNG plant, Nanggroe Aceh Darussalam, has a total capacity of 12.5 mtpy
- Bontang plant, East Kalimantan, has a capacity of 18.5 million tpy.

A third plant, the Tangguh facility, expects to export first LNG in early 2009.

New deal with Japan

Japanese buyers in early 2008 forgave Indonesia's failures to supply 72 cargoes of LNG in 2007 as part of a 10-yr extension to supply contracts that were due to expire in 2010 and 2011.

The new contracts are for 3 mtpy over the first five years of the contract and 2 mtpy over the next five. This compares to the 12 mtpy called for in current contracts.

The Indonesian government had announced plans to reduce LNG exports to Japan in order to have more gas available for domestic markets.

Refining

Greater Java project

Planning was under way in late 2008 for the $4-billion Greater Java refinery project with completion of a feasibility study.

JV partners are Pertamina (Jakarta), National Iranian Oil Refining and Distribution Co. (Tehran), and Petrofield Refining Co. Ltd. (Kuala Lumpur).

The refinery is to be built in Bojonegoro and is expected to go on stream in 2013.

Phase I of the refinery is to have a processing capacity of 150,000 b/d of crude oil, subsequently ramping up to 300,000 b/d.

Iran and Libya each will supply 150,000 b/d of crude for the plant.

Refinery output would be 200,000 b/d of fuel, about one-sixth of Indonesia's total requirements.

Indonesia has nine oil refineries with a total processing capacity of nearly 1.2 million b/d of crude.

Cilacap upgrade

In 2007, Pertamina and Mitsui & Co. (Tokyo) announced plans to establish a joint venture to build a $1-billion, 45,000-b/d gasoline cracker at the Cilacap refinery in Central Java.

In May 2008, Pertamina said it would soon seek financial support for the $1.9-billion upgrade. A final investment decision was expected in late 2008.

The refinery processes 350,000 b/d of crude to produce fuel, asphalt and lubricants.

Alternative Fuels

Biodiesel plant

Samsung Group (Seoul) said in mid-2008 that it intends to invest $1.63 billion to develop a 25,000-hectare (61,775-acre) oil-palm plantation and biodiesel plant in Riau province.

The facility is expected to go online in 2009, producing 50,000 kl./yr (13,200 gal/yr) of biodiesel.

Indonesia currently produces both bioethanol (from cassava, sugarcane, and sorghum) and biodiesel (from castor and crude palm oil).

Output of both is expected to increase greatly in late 2008 when Indonesia has said it plans to impose a new regulation requiring that at least 2.5% of fuel consumed by manufacturers be comprised of biofuel.

JAPAN

Capital: Tokyo
Monetary unit: Yen
Refining capacity: 4,690,700 b/cd
Oil production: 16,500 b/d
Oil reserves: 44.115 million bbl
Gas reserves: 738.0 bcf

International Activities

Aid to Iraq oil, gas industry

Japanese and Iraqi officials met in 2008 to discuss cooperation in rebuilding Iraq's oil and natural gas industry. This included reconstruction of existing facilities as well oil and gas development.

Details are presented in the Iraq section of the Middle East chapter of this book.

CHINA
RUSSIA
NORTH
KOREA
SOUTH
KOREA
JAPAN
Sea of Japan
Pacific Ocean
HOKKAIDO
HONSHU
SHIKOKU
KYUSHU
Wakkanai
Soya
Katsu
Menas
Koetoi
Toyotomi
Kitatoyotomi
Teshio Fields
Iskikari Bay
Ishikari
Atsuta
Barato
Tokachi Fields
Sapporo
Azuma
Fureoi
Kinausu
Tomakomai
133
180
Muroran
Oshamambe
Mori
Hakodate
Tsugaru Strait
Aomori
Nichi-Hachimori
Higashi-Hachimori
Sawame
Iwako
Sakake
Akita Fields
Surukawa
Kurokawa
Funakawa
Asakawa
Morioka
Akita
Yabase
Yuri
Kamekawa
Shonai
Innai
Kisagata
Okuoguni
Yokuka
Yamagata Fields
Chokaizan
Fukura
Sakata
Nogouchi
Mogaami
Shonai
Yamagata
Amarume
145
Sendai
Sendai Bay
Iwafune
Aga
Niitsu
Shibata
Niigata
Niigata
Niigata-Plain
Higashi
Okawatsu
Mitsuke
Echigo Fields
Higashiyama
Nishiyama
Nagaoka
Kashiwazaki
Sekihara
Ojiya-Sanjo
Iwaki
Iwaki
Toyama Bay
Kubiki
Ubase
Gozu
Kubiki
Nadachi
Menji
Maki
Asakawa
Odagiri
Toyama
57
Kanazawa
13in. 31cm
Kashima
181
Choshi
Fukui
Wakasa Bay
Tokyo
Kawasaki
Chiba
Yokoshiba
Shirasato
Mobara
Yokohama
Goi
Negishi
Sodegaura
Chiba Fields
110
296
65
228
175
209
192
340
Kyoto
Nagoya
Shimizu
Shimizu
Yaizu
Sagara
Chita
152
Yokkaichi
147
205
Ise Bay
Himeyi
115
Hyogo
Osaka
Sakai
250
195
76
155
Wakayama
Kainan
Owase
160
50
Shimotsu
Kii Strait
Mizushima
Hiroshima
Takamatsu
Marifu
Sakaide
114
Iwakuni
Yamaguchi
127
111
114
Shimonoseki
Onoda
Ube
Tokuyama
Kikuma
Kudamatsu
Matsuyama
Yahata
Ehime
120
Fukuoka
Tsushima Channel
Tosa Bay
Sasebo
Oita
160
Bungo Strait
Kashima
Nagasaki
Kumamoto
Miyazaki
Osumi Strait
Okinawa
9000
100
Off Map
Vladivostok
Nakhodka
Peter the Great Bay
Chongjin
42
Pohang
Ulsan
817
Pusan
10
Korean Channel
JAPAN
0
50
100 Mi.
0
100 Km.

Japan-Vietnam cooperation

Japanese firms are stepping up their involvement in Vietnam's oil industry, including both upstream and downstream sectors. Details are given in the Vietnam section of this chapter.

Agreement with Brunei

Japan took action to reduce dependency on Middle Eastern supplies and shorten its long and expensive supply chain from the Persian Gulf.

It signed a free-trade agreement with Brunei, effective July 2008, that will substantially increase investment in Brunei, with Japan being the sultanate's largest export destination.

Brunei sent shipments to Japan worth ¥293.8 billion in 2007, mainly oil and natural gas. About 70% of Japan's total imports of LNG coming from Brunei.

Saudi Arabia, the UAE, and Qatar were the top three crude exporters to Japan in mid-2008.

Petrobras refinery help

Brazil's Petroleo Brasileiro SA (Petrobras; Rio de Janeiro) acquired in early 2008 an 87.5% stake in the 100,000-b/d Nishihara refinery in Okinawa.

Sumitomo Corp. (Tokyo) will retain its 12.5% stake in the company.

Sumitomo and Petrobras are expected to spend as much as ¥100 billion to upgrade the plant and boost its role as an export base.

Petrobras will build facilities for producing high-value gasoline, kerosene and petrochemicals.

It called the agreement a key first step in its strategy to launch downstream operations in Asia.

Takaishi refinery

Nippon Oil Corp. and China National Petroleum Corp. (CNPC; Beijing) agreed in mid-2008 to jointly manage the existing 115,000-b/d Takaishi refinery in Osaka province and to export its products to China and other regional markets.

Though demand was slumping in Japan for gasoline, gas oil, and fuel oil, demand for these products was surging in China.

Australian pipeline deal

In October 2008, Osaka Gas Co. Ltd. and Marubeni Corp. (Tokyo) said they would spend about $500 million for an 80.1% stake in a gas and pipeline operating unit of APA Group. That group comprises Australian Pipeline Trust Ltd. and APT Investment Trust (Mascot, New South Wales).

The two Japanese firms will jointly operate gas pipelines, power generation plants and processing plants in Australia with APA through its wholly owned unit GasNet Australia Investment Ltd.

LNG

New import terminal at Sendai

Tohoku Electric Power Co. (Tepco; Miyagi) announced plans in early 2008 to build a new LNG import terminal at Sendai.

The aim is to disperse its LNG sites to minimize the risk of supply disruptions from earthquakes and other problems. Tepco currently receives all of its LNG at its Niigata terminal.

The new facility, with storage capacity of 1 million tonnes, is expected to begin operations in 2016.

Power plant sale

In late 2008, Osaka Gas Co. Ltd. was negotiating to sell a stake in one of its electric power plants, now under construction, to Oman in return for LNG supplies.

Four units are being built by Senboku Natural Gas Power Generation Co., each with approximately 277 MW of capacity. All are expected to come online between April and November 2009.

Osaka Gas buys 7.3 mtpy of LNG from ten countries.

Settlement with Indonesia

In early 2008, Japanese buyers agreed to forgive Indonesia's failures to supply 72 cargoes of LNG as part of a 10-year extension to supply contracts that were due to expire in 2010 and 2011.

Details are given in the Indonesia section of this chapter.

Japan: Leading 2007 LNG Importer

Japan remained the world's largest LNG importer in 2007, at 65.7 million tonnes.

This was an increase of 5.7% over 2006, according to Pan EurAsian Enterprises (Raleigh, NC).

South Korea was second, at 25.6 million tonnes, up 1.2% over 2006.

The next six largest importers in 2007 were Spain, the United States, France, Taiwan, India, and Turkey.

Refining

Products output cut

The three Japanese refiners who control some 35% of the nation's refining capacity were planning to cut their output by yearend 2008 due to reduced demand for fuels.

Showa Shell Sekiyu KK said it would cut production by 4.4%; Cosmo Oil Co. expected a cut of up to 2% in fourth quarter.

Idemitsu Kosan Co. said it also would reduce its crude refining output by about 14% compared to a year earlier for the October-December period.

Gas oil to Mexico

Idemitsu Kosan Co. (Tokyo) signed an initial contract in late 2008 with a subsidiary of Petroleos Mexicanos (Pemex; Mexico City) to supply it with 200,000 kl/yr (52.8 million gal/yr) of gas oil.

The company took the action as domestic and Asian demand fell for its gasoline, diesel, and kerosene.

For the first half of 2008, Mexico increased its imports of gasoline by 17.6% over January-July 2007.

Upgrades to four refineries

Idemitsu Kosan Co. (Tokyo) said in mid-2008 that it will spend ¥16 billion to upgrade its 220,000-b/d Chiba refinery, east of Tokyo, by 2011.

After the upgrade, production of heavy fuel oil will drop by 200,000 kl./yr (52.8 million gal/yr), while production of naphtha and propylene will increase by 260,000 kl./yr (68.6 million gal/yr).

In November 2007, the firm had announced plans to triple its capacity to export refined products at Chiba and at its 160,000-b/d Aichi refinery.

Reports said the refiner wants to raise its combined exports of jet fuel, gas oil and gasoline to 3 million kl./yr (729 million gal/yr) from the 2008 level of 1 million kl./yr (264 million gal/yr).

After revamping Chiba and Aichi, the company plans to expand the export capacity of its 140,000-b/d Hokkaido refinery and 120,000-b/d Tokuyama refinery.

Nippon Oil Corp. (Tokyo) announced plans in mid-2008 to increase exports and to begin mass-producing ethyl tertiary butyl ether at yearend 2009.

At its Negishi refinery, the firm will build a facility capable of turning out 100,000 kl./yr (26.4 million gal/yr) of ETBE.

Toyama refinery conversion

Nippon Oil Corp., Japan's largest refiner, said in early 2008 that is will convert its smallest refinery, the 60,000-b/d Toyama facility, to a site for importing oil products from other Nippon refineries in Japan.

The shutdown of Toyama is scheduled for March 2009.

Alternative Fuels

Wider biofuels use by 2013

The Petroleum Association of Japan said in early 2008 that Japanese refiners will aim to substitute biofuels for 500,000 kl./yr (132 million gal/yr) of oil equivalent of fuels by first quarter 2013.

This would more than double the initial target that was set for 2010.

It comes in response to government requests that the industry reduce carbon dioxide emissions.

Japanese automakers Nissan and Toyota have decided to develop cars that use bioethanol fuel to help reduce gasoline use.

Brazilian ethanol venture

Brazil's national oil company Petrobras said in early 2008 that it will form a 50:50 JV with Mitsui & Co. Ltd. (Tokyo) to pursue investments in renewable energy.

The new venture, called Participacoes Nippo Brasileira em Complexos Bioenergeticos SA, will provide ethanol for the Japanese market, Petrobras said.

Petrobras had previously said that it would join a JV to build the world's first ethanol-only pipeline. Details are given in the Brazil section of the Latin America chapter of this book.

Company News

Nippon Oil, Nippon Mining

In late 2008, Japan's leading refiner, Nippon Oil Corp., and sixth-ranked Nippon Mining Holdings Inc. announced plans to merge. The merger is expected to be completed in late 2009.

The resulting holding company will have subsidiaries in refining and marketing, field development, and metals.

MALAYSIA

Capital: Kuala Lumpur
Monetary unit: Ringgit
Refining capacity: 514,832 b/cd
Oil production: 750,000 b/d
Oil reserves: 4.000 billion bbl
Gas reserves: 83.000 tcf

Exploration and Development

Shelf exploration costs

State oil company Petronas (Kuala Lumpur) reported in mid-2008 that the cost of an exploration well on the country's continental shelf had increased by more than 200% since 2005, to $7 million.

Industry officials agreed and said a number of deepwater projects were also being delayed or phased differently due to the rising costs.

Drilling and Production

Blocks PM303 and PM324

Total E&P Malaysia (a unit of Total SA; Paris) signed a PSC in early 2008 with Petronas. The deal covers Blocks PM303 and PM324, some 100 km (62 mi) offshore in 50–80 m (164–262 ft) of water.

The initial work period will include seismic surveying and exploratory drilling.

Newfield reports

Newfield Exploration Co. (Houston) reported in mid-2008 the start of production from two offshore fields: East Belumut/Chermingat fields in PM 323, and the Puteri field in PM 318.

The firm also announced the signing of a new PSC in shallow water on license area SK 310.

The area covers nearly 1.2 million acres (1,875 sq mi), located in about 300 ft (91 m) of water offshore Sarawak.

Newfield and its partners have committed to shoot 500 sq km (193 sq mi) of new 3D seismic data and to drill three exploration wells. Seismic acquisition was slated to start by late 2008, with drilling to begin by early 2010.

International Activities

Iran, Malaysia agreements

In December 2008, Iran and Malaysia signed three agreements for cooperation in the oil and gas sector State-owned NIOC (Teheran) and Petrofield LNG Co. (a subsidiary of private Malaysian firm, SKS Ventures) signed a framework agreement and a gas sales and purchase agreement.

Petrofield LNG also entered into a project development agreement with a subsidiary of NIOC.

Gas fields

In one deal, SKS Ventures will develop the Golshan and Ferdousi gas fields in Iran. Golshan has estimated gas reserves of 1.19–1.59 tcm (42–56 tcf) and Ferdousi field 255–368 bcm (9–13 tcf).

LNG

Petrofield LNG will invest in and develop an LNG facility with a total capacity of 10 mtpy near the port of Deir in southern Iran. Golshan field will supply the facility. NIOC and Petrofield also will set up a 50:50 JV to transport LNG.

Refining

The two nations also signed agreements to export 250,000 b/d of heavy and ultra-heavy crude oil from Iran to the Kedah refinery being developed in Malaysia by SKS Ventures. Iran also will sell gas condensates to Malaysia.

Malaysia has offered a minority stake to Iran in the $2.2-billion refinery.

Petrochemical/refinery complex

In a separate development, Hampa Engineering Corp. (Tehran) signed an agreement with a Malaysian firm to develop a $6-billion, 250,000-b/d petrochemical and refinery complex in Terengganu state, Malaysia, to be operational in 2014.

U.S. officials expressed concern over the growing interaction between Iran and Malaysia.

The two nations are members of the Group of Eight (D-8), which aims to boost economic and trade cooperation that will enhance the position of developing Muslim nations.

Territorial dispute with Indonesia

Late in the year, the Malaysian government downplayed concerns about potential military conflict with Indonesia regarding rights to and development of the Bukat Block.

Officials stated that the block, operated by Eni SPA (Rome), falls outside the territory disputed with Indonesia.

They also said that some parties had wrongly referred to the block as part of the maritime area of Ambalat, off eastern Kalimantan, which is thought to contain huge reserves of oil and gas.

Eni has announced plans to build a floating LNG plant on Bukat Block.

Indonesia and Malaysia have long been involved in a dispute over the Ambalat territory, which consists of the Ambalat Block and the East Ambalat Block.

As recently as 2007, Malaysia sent naval ships and warplanes into Indonesian waters to protect its ND-6 and ND-7 Blocks, which overlap the Ambalat Block.

Refining

Oil and gas complex in Perak

Gulf Petroleum Ltd. (Doha, Qatar) said that it expects to start building a $5-billion integrated oil and gas complex at Manjung in the northern state of Perak in Malaysia by early 2009.

A feasibility study was to be completed by yearend 2008.

The complex, to cover almost 1,000 acres, will consist of an oil refinery, a petrochemical unit and storage facilities. It will serve as Gulf Petroleum's regional hub in the Asia-Pacific region.

In Phase I of the development program, an oil refinery with a capacity of 150,000 b/d will be constructed, using as its supply oil from Gulf's fields in the Middle East.

Phase II will develop a petrochemical plant, and Phase III will see construction of storage facilities.

Gulf also plans to produce LNG at the complex.

It is not clear if the Gulf Perak complex will connect with the major pipeline and related refinery projects announced by Malaysia in 2007.

The trans-peninsular pipeline—an alternative transport route to the busy Straits of Malacca—will cross the states of Kedah, Perak, and Kelantan to carry oil from West Asia to East Asia.

MYANMAR

Capital: Naypyitaw
Monetary unit: Kyat
Refining capacity: 57,000 b/cd
Oil production: 19,000 b/d
Oil reserves: 50.000 million bbl
Gas reserves: 10.000 tcf

Block M-2

Media reports in October 2008 said that Myanmar's national oil company had concluded licensing agreements with Vietnamese companies for an offshore block in the Andaman Sea.

The agreements were production sharing contracts granted by Myanmar Oil and Gas Enterprise (MOGE) with Petrovietnam Exploration & Production Corp. and with the joint venture of Vietsovpetro and Eden Group.

The parties are to explore Myanmar's Block M-2 in the Mottama offshore region.

It is the first venture into Myanmar exploration and production by Vietnamese companies.

Disputed waters

In November 2008, Bangladesh lodged a formal complaint with the Myanmar government for alleged incursion in a disputed maritime area between the two neighbors.

The disputed sector of the Bay of Bengal was being explored for oil and gas from the Myanmar side, said Bangladesh officials, who also stated that Bangladesh reserved the right to protect its interests.

The complaint called upon Myanmar to enforce an end to the incursion until a maritime boundary can be delineated and sovereignty over the area resolved.

Exploration vessels were observed in or near Bangladesh's Block DS-08-13.

Block A-6

In early 2008, MPRL Exploration & Production began a three-year exploration period offshore the Rakhine coast of Myanmar in the Bay of Bengal.

MPRL (formerly Myanmar Petroleum Resources Ltd.) was planning a seismic survey over the concession area, Block A-6, which it shares with project partner Burma Oil & Gas.

The field has some 250 wells in production out of about 650 on the field.

Cyclone Nargis

Cyclone Nargis struck Myanmar on May 3, 2008. The death toll was initially estimated at 22,500 with more than 40,000 unaccounted for, but some aid organizations said that 100,000 were believed dead.

A tidal wave compounded the destructive force of the 130-mph winds of the Category 4 storm.

Residents, foreign diplomats and others criticized what they called a slow and ineffectual government response to the storm.

The storm caused a delay in some regions—including the Irrawaddy Delta and much of the city of Yangon—in voting on a constitutional referendum related to continue rule by the military junta that took power in 2007.

Blocks A-1 and A-3

In mid-2008, Myanmar Oil and Gas Enterprise (MOGE) announced its intention to exercise an option to take a 15% stake in two large offshore blocks.

Blocks A-1 and A-3, in the Bay of Bengal off Myanmar's Rahkine coast, hold an estimated 128.3 bcm (4.53-tcf) of natural gas.

Others with stakes in those blocks are Daewoo Group (Seoul), Oil and Natural Gas Corp. (ONGC; Dehradun, India), Gas Authority of India Ltd. (GAIL), and Korean Gas Corporation (Kogas; Sungnam).

The two concessions are expected to produce 12.7–15.9 MMcmd (450–560 MMcfd) of gas, which is to be transported by pipeline to China beginning in 2012.

NEW ZEALAND

Capital: Wellington
Monetary unit: Dollar
Refining capacity: 107,000 b/cd
Oil production: 56,000 b/d
Oil reserves: 60.000 million bbl
Gas reserves: 1.200 tcf

Exploration and Development

Maari field

Maari field in the Taranaki basin is New Zealand's largest undeveloped offshore field, with estimated total reserves of 68 million boe.

In early 2008, Horizon Oil Ltd. (Sydney) said that the field could have 12 million bbl recoverable in the M2A sands above the main Miocene Moki sands structure.

Production is expected to begin in February 2009 and to rise to 35,000 b/d by August 2009.

Horizon Oil also estimates that there could be 25 million bbl recoverable in the Manaia structure 10 km (6 mi) southwest of Maari.

Tui reserves

Australian Worldwide Exploration Ltd. (AWE; Sydney) finalized its review of the reserves of Tui field in mid-2008 and increased the initial proven and probable (2P) oil reserves to 50.1 MMbbl from 47 MMbbl.

The Tui Area Oil Project is in PMP38158 in the Taranaki Basin, 50 km (31 mi) off the coast in 120 m (394 ft) of water. It is the country's first stand-alone offshore oil development.

Total field production was expected to reach 14.2 MMbbl by mid-2008.

D'Urville Block

Australian Worldwide Exploration Ltd. (AWE; Sydney) began in late 2008 the first seismic survey in the offshore D'Urville Block (PEP 38524).

The company will acquire 400 km (249 mi) of 2D infill seismic to advance the Malvern prospect to drill-ready status. PEP 38524 is interpreted to form an extension of the South Taranaki basin, which includes the Kupe South and Maari oil and gas fields.

Results from the seismic survey were slated to be available in early 2009, to support a drilling decision by mid-2009.

MALAYSIA-INDONESIA
MYANMAR (BURMA)
THAILAND
LAOS
CAMBODIA
VIETNAM
MALAYSIA
MALAYA
SABAH
SARAWAK
BRUNEI
SINGAPORE
SUMATRA
KALIMANTAN (BORNEO)
INDONESIA
JAVA
MADURA
BALI
SUMBAWA
BANGKA
BELITUNG
KEPULAUAN LINGGA
NATUNA ISLAND (INDONESIA)
HAINAN DAO
ANDAMAN ISLANDS
NICOBAR ISLANDS
SIMEULUE
NIAS
SIBERUT
SIPORA
PAGAI UTARA
PAGAI SELATAN
ENGGANO
PALAWAN
South China Sea
Andaman Sea
Indian Ocean
Bay of Bengal
Gulf of Martaban
Gulf of Thailand
Gulf of Tonkin
Java Sea
Sulu
Flores
Strait of Malacca
Strait of Makassar
Sunda Strait
Noakhali
Patuakhali
Chittagong
Shahbazpur
Patiya
Mandalay
Thanbayakan
Ayadaw
Chauk/Lanywa
Yenangyaung
Mann
Htaukshabin
Paduakpin
Kyaukoyauk
Yenandaung
Pyaye
Pyato
Ledang
Prome
Yenanchaung
Myanaung
Shwepyitha
Yangon (Rangoon)
Mae Sot
Martaban
Yadana
Yetagun
Mae Fang
Mae Suhn
Chiang Mai
Nan
Vientiane
Uttaradit
Loei
KDN-1
Mae Sot
Sirikit
Pru Krathiarn
Bung Ya
Phu Horm
Nam Phong
Chonnabot
Khon Kaen
Sisaket
Sawng Neung
Saraburi
Bangkok
Kanchanaburi
Sri Racha
Rayong
Phnom Penh
Sihanoukville
Nha Trang
Ho Chi Minh City
Vung Tau
Long Xuyen
Sutu Den
Rang Dong
Bach Ho
Rong Dong
Dai Hung
Dua-1X
Hanoi
Hai Phong
Tien Hai
Weizhou 11-1
Weizhou 10-3
Weizhou 6-1
Haik'ou
Jinfeng
Wenchang 9-2
Yacheng 13-1
Hue
Da Nang
Macau
Hong Kong
Moaming
Lufeng 13-1
Lufeng 15-1
Xijiang 24-3
Xijiang 30-2
Huizhou 26-1
Huizhou 21-1
Liuhua 4-1
Liuhua 11-1
Yungan
Subic Bay
Calauit
Galoc
Octon
West Linapacan
Malampaya
Camago
Destacado
Nido
Matinloc
Cadlao
Linapacan
Sampaguita
Nang Nuan
Plamuk
Kapong
Pakarang
Surat
Chaba
Dara
Erawan
Baanpot
Pailin
Pladang
Tantawan
Platong
Platong S.
Trat
Sabun
Jakrawan
Gomin
Punan
Khanom
Si Thammarot
Arthit
Bongkot
17-B-1
Songkhla
Bua Ban
Pilong
Lawit
Jernch
Sepat
Damar
Bintang
Noring
Bedong
Tujoh
Inas
Bujang
Tangga
Larut
Bubu
Lerek
Dulang
Tabu
Barat
Tapis
Kakap
Besar
Angsi
Duyong
Belumut
N. Belut
W. Belut
Malong
Sotong
Mamung
Kerisi
Belanak
Terebuk
Sepat
Keong
Udang
Tembang
Belida
Kijing
Buntal
Kepiting
Natuna
Bursa
Pinang
Kerteh
Kuala Lumpur
Port Dickson
Dumai
Singapore
Batam
Kutaradja
Julu Rayeu
Perlak
Serang Jaya
Pangkalan Brandan
Belawan
Medan
Meulabah
Sibolga
Pinang
Bangko
Buaya
Menggala
Mandar
Duri
Mellbur
Libo
Minas
Minik
Langgak
Topaz
Zamrud
Lirik Fields
Pekanbaru
N. Palai
Palai
Andan
Lirik
Sago
Molek
Binio
Bukittinggi
Pontianak
Jambi
Setiti
Bedjubang
Tempino
Limau Fields
Sumpal
Suban
Burong
Bentayan
Dayung
Keban
Sambar
Sungei Gerong
Tanjung Laban
Benakat
Djimar
Palembang
Muraraenim Fields
Sambar
Ogan Timur
Kampong Minyak
Taham
T. Tiga
Pagardewa
Batukeras
Intan
Indri
Winduri
Sundari
Krisna
Karmila
Cinta
AV
Nora
AA
Rama
KN
Arjuna
HZE
Cilegon
Arimbi
Jakarta
MB
Sidong
Cirebon
Jatibarang
Randegan
Bandung
Telukbetung
Semarang
Rembang
Ledok
Nglobo
Petak
Camar
Candi
Banyubang
Cepu
Gresik
Kawengan
Kendangan
Waron
Maleo
Wonokromo
Cilacap
Jogjakarta
Singaradja
Raba
Kuching
Bintulu
Miri
Kuala Belait
Bandar Seri Begawan
Labuan
Kota Kinabalu
Sandakan
Tawau
Kudat
S.W. Emerald
Tembungo
Erb W.
Barton
S. Furious
St. Joseph
Kinarut
Kikeh
Samarang
Kecil
Iron Duke
Glayzer
Magpie
Seria
Jintan
M1
M3
M4
F6
F28
Serai
Bijan
F14
F27
B23
B11
B12
E8
D12
J4
W. Bayan
C2
Patricia
Temana
Mamburungan
Bunyu Nibung
Tarakan
Batuputih
Sangkulirang
Kerindingon
Panyilatan
Semberah
Badak
Serang
Attaka
Kerendan
Sanga-Sanga
Pamaguan
Tunu
Balikpapan
Handil
Louise
Bekapi
Sanggur
Mutiara
Sepinggan
Bangkerak
Wailawi
Kambatin
Tanjung
TapianTimur
Warukin Tengah
Warukin Selatan
Palangkaraya
Bandjarmasin
Palu
Ujung Pand
0 50 100 150 200 250 Miles
0 100 200 300 400 Km
20°
15°
10°
5°
0°
95°
100°
105°
110°
115°
120°

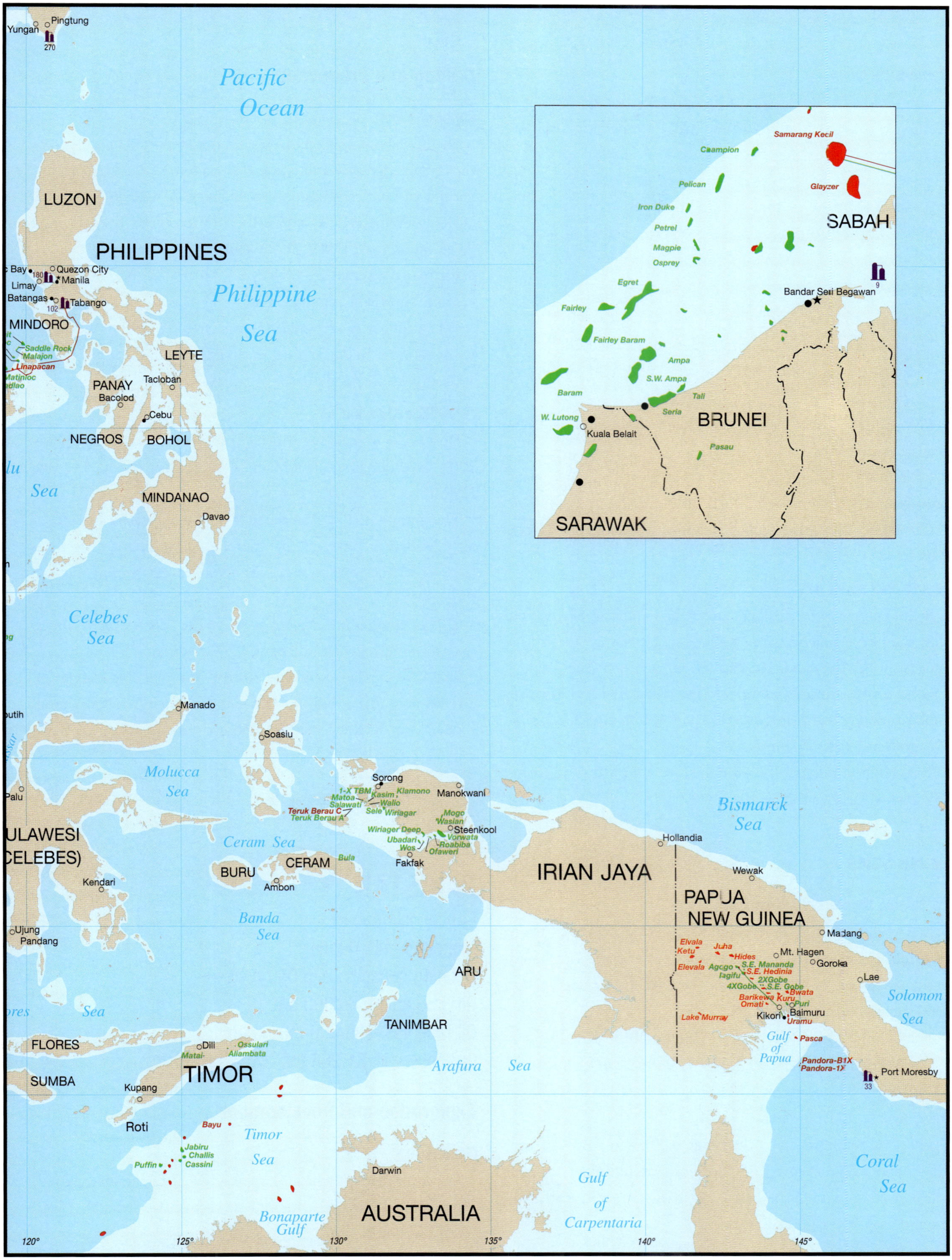

Yungan
Pingtung
270
Pacific Ocean
LUZON
PHILIPPINES
Quezon City
Manila
Limay
Batangas
Tabango
MINDORO
Saddle Rock
Malajon
Linapacan
Matinloc
Philippine Sea
LEYTE
PANAY
Tacloban
Bacolod
Cebu
NEGROS
BOHOL
Sea
MINDANAO
Davao
Celebes Sea
Manado
Soasiu
Molucca Sea
Palu
SULAWESI (CELEBES)
Kendari
Ujung Pandang
Sorong
1-X TBM
Matoa
Salawati
Kasim
Klamono
Walio
Sele
Wiriagar
Manokwari
Teruk Berau C
Teruk Berau A
Mogo
Wasian
Steenkool
Wiriager Deep
Ubadari
Wos
Vorwata
Roabiba
Ofaweri
Ceram Sea
BURU
CERAM
Bula
Ambon
Fakfak
Banda Sea
ARU
TANIMBAR
FLORES
Dili
Ossulari
Matai-Aliambata
SUMBA
TIMOR
Kupang
Roti
Bayu
Jabiru
Challis
Cassini
Puffin
Timor Sea
Bonaparte Gulf
Darwin
AUSTRALIA
Arafura Sea
Gulf of Carpentaria
IRIAN JAYA
Hollandia
Bismarck Sea
Wewak
PAPUA NEW GUINEA
Madang
Mt. Hagen
Goroka
Lae
Elvala
Ketu
Juha
Hides
Elevala
Agogo
Iagifu
S.E. Mananda
S.E. Hedinia
2XGobe
4XGobe
S.E. Gobe
Bwata
Barikewa
Kuru
Omati
Puri
Lake Murray
Kikori
Baimuru
Uramu
Gulf of Papua
Pasca
Pandora-B1X
Pandora-1X
Port Moresby
33
Solomon Sea
Coral Sea
120°
125°
130°
135°
140°
145°
Samarang Kecil
Champion
Glayzer
Pelican
Iron Duke
Petrel
Magpie
Osprey
SABAH
Egret
Bandar Seri Begawan
Fairley
Fairley Baram
Ampa
S.W. Ampa
Baram
Tali
Seria
W. Lutong
Kuala Belait
BRUNEI
Pasau
SARAWAK

Canterbury basin

Origin Energy Resources (NZ) Ltd. (Wellington) was proceeding with plans in mid-2008 to drill an exploration well on a prospect on Canterbury basin permit PEP 38262.

The Carrack/Caravel prospect complex is about 65 km (40 mi) east of Dunedin in around 1,000 m (3,281 ft) of water.

It is estimated to contain a mean of 750 MMbbl of recoverable oil or 2.7 tcf (76 bcm) of recoverable gas and 500 MMbbl of associated condensate.

Deepwater Taranaki basin

Hyundai Hysco (Seoul) announced plans in early 2008 to participate in PEP 38451 in the undrilled deepwater Taranaki basin.

PEP 38451 covers nearly the entire basin and is New Zealand's largest exploration permit.

GNS Science (Wellington) said it has identified three giant prospects in 225 m (738 ft), 1,411 m (4,629 ft), and 1,540 m (5,052 ft) of water, overlying formations of Cretaceous North Cape sandstone.

Huntly coal seam gas

Solid Energy New Zealand Ltd. (Christchurch) was testing flow rates from four appraisal wells at its Huntly coal seam gas project on North Island, 100 km (62 mi) south of Auckland.

Solid Energy drilled the wells to the Kupakupa and Renown subbituminous coal seams at a depth of 400 m (1,312 ft), some 10 km (6 mi) west of Huntly.

The Huntly resource is estimated at 300 PJ (9.48 bcm; 335 bcf), as large as that at Kupe gas field.

Lease Sales

Northland, Raukumara blocks

In December 2009, the government's New Zealand Crown Minerals agency offered six exploration blocks in the Northland basin northwest of Auckland and two off East Cape in the Raukumara basin off the North Island.

The bidding round will close in January 2010.

The Northland blocks cover 3,900–11,000 sq km (1,506–4,247 sq mi) each. Three are on the shelf and three are adjacent in intermediate to deep water.

The two Raukumara blocks cover 9,907 and 7,381 sq km (3,825 and 2,850 sq mi). The unexplored Raukumara basin is at the northern end of the East Coast basin, where the water is more than 3,000 m (9,842 ft) deep.

Drilling and Production

Pohokura field

Shell Exploration NZ Ltd. (Wellington) reported in mid-2008 the completion of Pohokura field, New Zealand's biggest natural gas and condensate field, offshore New Plymouth.

Field development is a $1-billion JV of Shell, Todd Energy (Wellington), and OMV AG (Vienna).

Estimated reserves are 884 PJ (27.9 bcm; 985 bcf) of natural gas and 44 MMbbl of condensate.

Pohokura gas reserves are approximately twice the volume of the remaining Maui field gas reserves, enough to supply about 40% of New Zealand's natural gas production.

Tui FPSO expansion

Partners in developing Tui oil field were considering in late 2008 expanding the fluid-handling capacity of the *Umuroa* FPSO vessel to 180,000 b/d.

The group—led by Australian Worldwide Exploration Ltd.—also decided to defer the drilling of a third horizontal well in favor of further exploration within the larger mining license PMP 38158.

Tui is gradually increasing its water cut, which limits maximum oil recovery to about 32,000 b/d.

The field has an estimated proven and probable reserve of around 50 million bbl. At the end of November 2007, total production had reached 19 million bbl.

PAKISTAN

Capital: Islamabad
Monetary unit: Rupee
Refining capacity: 276,050 b/cd
Oil production: 66,000 b/d
Oil reserves: 339.000 million bbl
Gas reserves: 31.266 tcf

Exploration and Development

Dhodak well

Oil & Gas Development Corp. (OGDC; Islamabad) reported a natural gas discovery in mid-2008 with its Dhodak Deep No. 1 exploration well.

The well was drilled in Dera Ghazi Khan district in Punjab province, to a target depth of 4,150 m (13,616 ft).

After tests, a Chiltan limestone formation produced 156,000 cu m/d (5.5 MMcfd) of gas.

Drilling and Production

Badhra field

A joint venture led by Eni SPA (Rome) began gas production in early 2008 from Badhra field, southeast of Bhit gas field and 250 km (155 mi) northeast of Karachi.

The group also commissioned the third train at Bhit gas treatment plant, which will process gas from nearby Badhra field, following the Badhra development and Bhit acceleration projects.

The $50-million Badhra development and Bhit acceleration project boosted the existing capacity of the Bhit plant capacity to 8.9 MMscmd (315 MMscfd) from 7.6 MMscmd (270 MMscfd).

Pipelines

TAPI project

Turkmenistan and Afghanistan signed agreements in 2007 to cooperate on the Turkmenistan-Afghanistan-Pakistan-India (TAPI) pipeline project.

In April 2008, a tenth meeting of the steering committee was held to discuss progress, during which India formally joined the consortium.

Earthquake kills 170

As reported by various media, a strong earthquake struck southwestern Pakistan on October 29, 2008, killing at least 170 people, injuring 375, and destroying many homes in the impoverished region.

An estimated 15,000 people were left homeless in Baluchistan where the magnitude 6.4 quake struck.

In the town of Ziarat and about eight surrounding villages, hundreds of houses were destroyed.

Pakistan natural gas consumption

	2000	2001	2002	2003	2004	2005	2006
	% by sector						
Residential	20	18	17	18	15	15	14
Commercial	3	3	3	3	2	2	2
Cement	1	1	1	0.3	1	1	1
Fertilizer	25	23	22	21	18	16	16
Power	32	37	38	38	45	44	40
Industrial	19	18	18	19	18	20	23
Transport-CNG	0.3	0.6	0.9	1.3	1.5	2.1	3.2

Source: Hydrocarbon Development Institute of Pakistan

Pakistan primary energy mix

	Actual				Projections			
	2003	2004	2005	2006	2010	2015	2020	2025
	% share							
Oil	38.3	30.2	29.8	28.7	26.1	27.1	25.6	22.7
Gas	23.8	49.5	51.3	50.1	49.1	44.1	43.9	45.0
Coal	5.4	6.5	6.0	7.0	9.0	12.0	14.0	15.0
Hydropower	11.3	12.7	11.3	12.7	13.9	13.6	12.1	11.9
Nuclear	0.9	0.8	1.2	1.0	0.9	1.9	2.7	3.2

Source: Government of Pakistan, Planning Commission, 2005

Pakistan crude oil statistics

Year	1997	1998	1999	2000	2001	2002	2003	2004	2005	2006
	1,000 b/d									
Production	57	54.9	53	54.4	59.9	64.3	60	62	65.6	65.7
Consumption	333	346.8	368.6	365	360.1	355.9	336.6	326.8	345	345
Import	−274	−289	−312	−308	−297	−288	−272	−260	−277	−277
Proved reserves, billion bbl	0.2	0.2	0.2	0.2	0.2	0.3	0.3	0.3	0.3	0.3

Source: US Energy Information Administration

Pakistan natural gas statistics

Year	1997	1998	1999	2000	2001	2002	2003	2004	2005	2006
	bcf									
Production	699	710	784	856	775	809	890	968	1,088	1,112
Consumption	699	710	784	856	775	809	890	968	1,088	1,112
Import	—	—	—	—	—	—	—	—	—	—
Proved reserves, tcf	22	21	22	22	22	25	26	27	27	28

Source: US EIA

Pakistan natural gas supply-demand

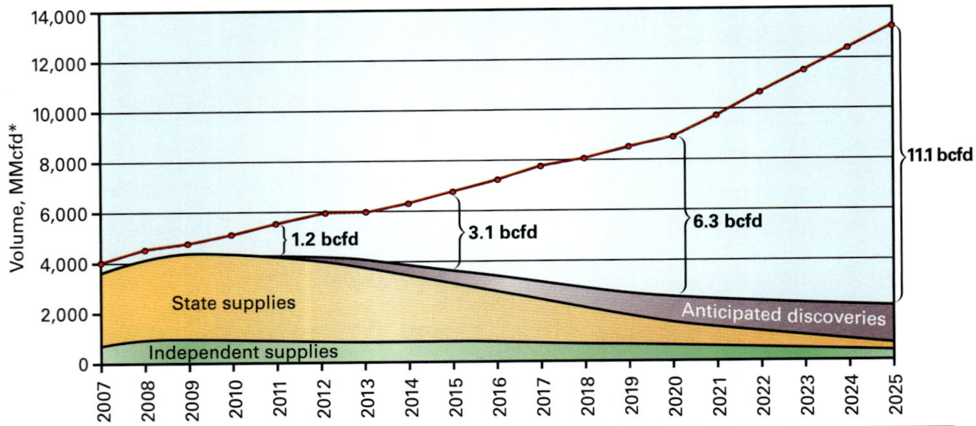

	2007	2008	2010	2011	2012	2013	2014	2015	2020	2025
Demand, MMcfd	4,013	4,492	5,086	5,516	5,921	5,912	6,305	6,763	8,854	13,259
Supplies, MMcfd	3,666	4,184	4,308	4,313	4,181	4,065	3,815	3,674	2,521	2,167
Shortfall	**(346)**	**(308)**	**(778)**	**(1,203)**	**(1,740)**	**(1,847)**	**(2,490)**	**(3,089)**	**(6,333)**	**(11,092)**

*At 950 btu/scf.

Source: PriceWaterhouseCoopers-Hagler Bailly Pakistan

The $3.3-billion project would build a 1,680-km (1,044-mi) gas line across Afghanistan to Pakistan and then to India.

With a capacity of 33 bcm (1.17 tcf), it would deliver of up to 90.6 MMcmd (3.2 bcfd) for 30 yrs from Turkmenistan to Pakistan.

The line would run 145 km (90 mi) from the Turkmen Dauletabad fields to the Afghan border, then 145 km (90 mi) to Kandahar, Afghanistan, then 800 km (497 mi) to Multan in central Pakistan.

A further extension of 220 km (137 miles) would run to the Indian town of Fazilka, near the border between Pakistan and India.

Later in 2008, some energy analysts in the region expressed uncertainty about the project's future. The pipeline would pass through regions of strong Taliban influence; also, some ask if Turkmenistan has sufficient gas reserves to meet its obligations to China and Russia as well as Pakistan and India.

Gas storage for TAPI

Pakistan it would require $1 billion to build underground storage for gas to be imported from Iran and Turkmenistan, according to the country's petroleum ministry and the Asian Development Bank.

The parties were reporting in late 2008 the results of a joint study conducted with Sui Northern Gas Pipeline Ltd. (SNGPL; Lahore) and Sui Southern Gas Co. Ltd. (SSGCL; Karachi).

SNGPL, SSGCL, and Inter-State Gas Systems (Islamabad), responsible for building the TAPI pipeline, would build three or four underground gas storage tanks.

Pakistan plans to import 62 MMcmd (2.2 bcfd) of gas from Iran, of which Pakistani share would be 30 MMcmd (1.05 bcfd).

If India does not participate, Pakistan would take the entire volumes.

From Turkmenistan, Pakistan will import 91 MMcmd (3.2 bcfd) of gas, to be shared equally with India. Pakistan also would need underground storage for gas supplied to different areas of the country.

IPI Pipeline

In October 2008, Iran and Pakistan decided to go ahead with the $7.4-billion Iran-Pakistan-India (IPI) gas pipeline project bilaterally, saying India could join the project later if it chose to.

Details are presented in the Iran section of the Middle East chapter of this book.

Qatar-Oman-Pakistan pipeline

Still in the very early stages of consideration is a pipeline that would deliver gas from the Qatar North Field through a subsea line from Oman to Pakistan.

As of mid-2008, Pakistan had signed a preliminary agreement to buy gas in the future from Qatar, but the pipeline project itself remained undefined.

LNG

LNG as supply alternative

A researcher at the University of Dundee (Dundee, UK) concluded in a 2008 study that LNG will likely play a large role in supplying Pakistan's emerging gas market.

National leaders are increasingly interested in importing natural gas and in 2008 were considering construction of pipelines from Iran, Qatar, or Turkmenistan.

Pakistan Potential Import Pipelines

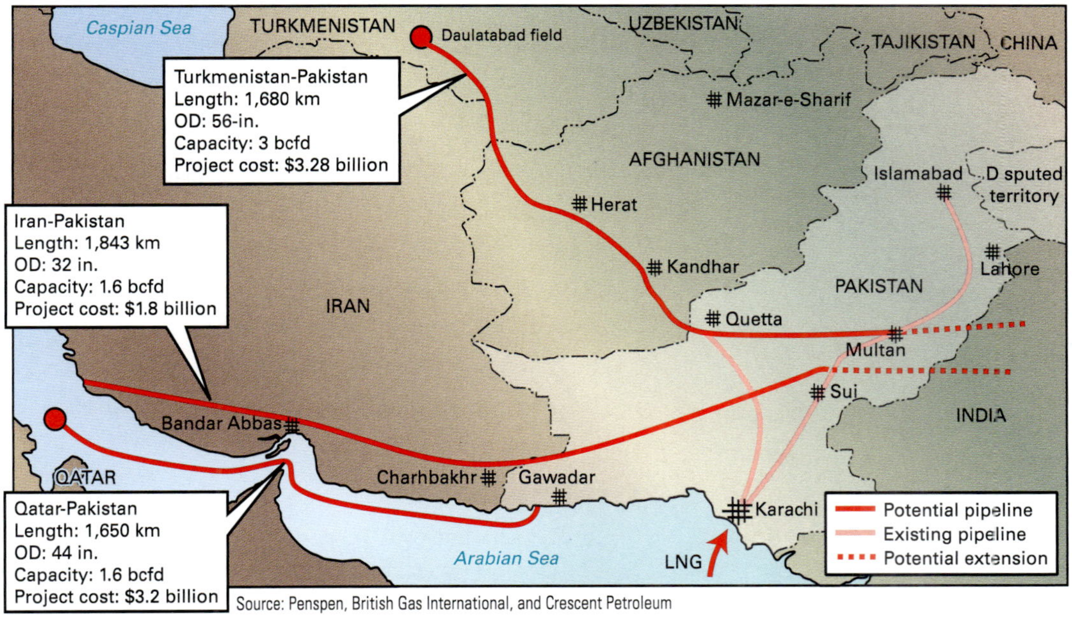

Source: Penspen, British Gas International, and Crescent Petroleum

The study suggests that Pakistan also must consider LNG if it is to meet its import needs.

The government is implementing an LNG investment policy friendly to private-sector participation and has established regulations covering LNG terminal construction and operation.

Refining

Khalifa refinery

The United Arab Emirates agreed in late 2008 to invest more than $5 billion to develop the Khalifa Coastal Refinery at Khalifa Point in Baluchistan's Hub district, on the coast of the Arabian Sea.

The investment was called the single largest foreign direct investment ever made in Pakistan.

Production is expected to begin by December 2012, refining 200,000 b/d of diesel and other petroleum products. Feedstock crude will be supplied from Iran and other Arabian nations.

Khalifa Point is to be developed into a full-fledged industrial township.

The refinery will be developed by state-owned International Petroleum Investment Co. (IPIC; Abu Dhabi, UAE) and by Pakistan Arab Refinery Co. (PARCO; Karachi).

Details of the investment plan are to be finalized in first quarter 2009.

Export duties cut

The fixed regulatory duty imposed by Pakistan on diesel and kerosene exports to Afghanistan was cut by almost 50% in November 2008.

Price differential claims came down because of a sharp fall in product prices in the international market, paving the way for the duty cuts.

Pakistan had imposed the duty in October 2008 because Afghanistan was subsidizing Kabul consumers.

Total exports of petroleum products to Afghanistan stood at more than $300 million during FY2007–08.

Refinery relocation

TransAsia Refinery Ltd. (Karachi) said in mid-2008 that it plans to relocate a 100,000-b/d refinery from Naples to Port Qasim, Karachi.

Details are given in the Italy section of the Europe chapter of this book.

Refinery output cuts

Pakistan refineries reduced gasoline production in September 2008 following drastic changes in the pricing formula, the availability of enough stocks, and a slight drop in consumption.

According to the Pakistan Oil Companies Advisory Committee (OCAC), four refineries cut production by 20–38%.

OCAC ruled out any immediate impact of the cuts on consumers because gasoline stocks were well above the consumption level, but said that its impact may be felt in the future.

PAPUA NEW GUINEA

Capital: Port Moresby
Monetary unit: Kina
Refining capacity: 32,500 b/cd
Oil production: 41,000 b/d
Oil reserves: 88.000 million bbl
Gas reserves: 8.000 tcf

Exploration and Production

Deal with Eni

In late 2008, Eni SPA (Rome) signed an agreement with Papua New Guinea for a long-term partnership to develop the latter's untouched hydrocarbon resources.

Eni will run a detailed exploration study and outline a data acquisition program across all regions in the country that are either not currently under permit or are completely unexplored wildcat areas.

The aim is to better delineate remaining hydrocarbon potential.

Eni will have first option to identify areas of interest for possible exploration.

The company sees significant potential for large gas finds and estimates that as many as four LNG trains could operate in PNG.

The company also will upgrade the technical and managerial skills of local workers in oil and gas.

PPL 235

Rift Oil PLC (London) reported flows of 29.2 and 20.85 MMcfd of gas in late 2008 from two pay intervals during a test at its Puk Puk-1 wildcat on PPL 235 in the Western Province.

Testing was to be completed in October. The company is also was planning 2D seismic in the area.

Rift is aiming to produce enough gas from the discovery to supply a floating LNG plant in the Gulf of Papua. The firm has signed a heads of agreement with Flex LNG Ltd. (London) to jointly develop Rift's onshore gas reserves in PNG using a floating facility.

PPL 238

InterOil Corp. (Cairns, Queensland) said in mid-2008 that its Elk-4 well established the presence of gas-condensate on the Antelope structure beneath the Elk structure on PPL 238.

The well intersected the Antelope structure at 2,256 m (7,402 ft) and found gas condensate.

Plans were under way to deepen the well.

Elk-Antelope field

In mid-2008, InterOil Corp. sold its leases in western PNG to Horizon Oil Ltd. (Sydney). InterOil will concentrate on developing the Elk-Antelope field in the eastern region.

The leases involved are PRL4, which contains the undeveloped Stanley gas and condensate field, and PRL5, which contains the undeveloped Elevara and Ketu gas and condensate fields.

In early 2008, Elk-4 well flowed at 14 MMcfd. InterOil said the structure is potentially 13 km (8 mi) long and 5 km (3 mi) wide with gross reservoir thickness of 500 m (1,640 ft).

The Antelope-1 appraisal well was spudded in October.

The planned PNG LNG plant (Port Moresby) likely will use Elk-Antelope gas as feedstock.

Stanley condensate project

Horizon Oil Ltd. was preparing to drill a second appraisal well in late 2008 at the Stanley gas-condensate reservoir on PRL 4 in western PNG in late 2008.

Results will guide a decision on investment in a condensate-stripping and gas-recycling facility at the site. A stripping plant could start up in 2010 if Stanley-2 were drilled in 2009.

The Stanley gas resource has been estimated independently at 7.4 bcm (260 bcf) proved and probable.

In the longer term, produced gas likely would serve power generation markets in PNG's western province and in West Papua, Indonesia.

Papuan and North New Guinea basins

In late 2008, LNG Energy Ltd. (Vancouver, BC) was exploring the Papuan basin and the North New Guinea basin. The company estimates gas reserves in the Kuru reservoir at 2.2–6.4 bcm (77–227 bcf) of gas in place.

LNG

Port Moresby LNG project

The $11-billion Port Moresby LNG liquefaction project entered the FEED stage in May 2008. Bechtel Corp. (San Francisco) will manage the work, which is expected to last until third quarter 2009.

The project is viewed as a major milestone in the country's petrochemical sector, indicating the feasibility of setting up other large-scale projects in PNG.

Planned capacity of the plant—to be located near the Napa Napa oil refinery of InterOil Corp. (Cairns, Queensland) in Port

Moresby—is 6.3 mtpy from two processing trains. Start-up is planned for 2014.

Natural gas in the Elk/Antelope structure is expected to underpin the project, but feedstock also will come from recently discovered and established fields in the Southern Highlands and Western provinces.

The project is being led by ExxonMobil Corp. (Irving, TX) in a JV with several other parties. Its subsidiary, Esso Highlands Ltd., is serving as operator.

Overall project leadership is provided by PNG LNG Inc., the holding company for Liquid Niugini Gas Ltd. (Melbourne).

Liquid Niugini Gas, in turn, is a three-way JV of InterOil Corp. (Cairns, Queensland), investment firm Pacific LNG—an investment vehicle created and controlled by Clarion Finanz AG (Zurich)—and Merrill Lynch Commodities, a unit of Bank of America Merrill Lynch.

In November 2008, the PNG LNG project concluded a $1-billion loan agreement with state-owned International Petroleum Investment Co. (IPIC; Abu Dhabi).

PHILIPPINES

Capital: Manila
Monetary unit: Peso
Refining capacity: 282,000 b/cd
Oil production: 15,800 b/d
Oil reserves: 138.500 million bbl
Gas reserves: 3.480 tcf

Galoc field

Otto Energy (West Perth) reported that oil production began in late 2008 from the Galoc 3 and Galoc 4 wells on the Galoc field, offshore northwest Palawan.

Combined production from the two wells was expected to average 17,000 b/d for the rest of 2008.

The reserves estimate in Galoc is 10 MMbbl, based on a 2006 assessment.

The development by Galoc Production Co. (GPC) is the first in the Philippines since 1992.

The first cargo of just under 300,000 bbl of Galoc crude was delivered by tanker in November to the Petron refinery at Limay, near Manila.

Tindalo-1, Yakal-1 wells

Nido Petroleum Ltd. (Perth) reported in late 2008 that the Tindalo-1 well in the offshore NW Palawan basin proved oil. The well is in service contract SC54.

The well was drilled to TD 1,875 m (6,152 ft). Logs confirmed a 119-m (390-ft) oil column. Good oil shows while drilling support the estimation that the total column could be at least 134 m (440 ft) thick, the company said.

The company's Yakal-1 wildcat drilled earlier in SC54—the first exploration well drilled in the area in the past decade—went to TD 1,975 m (6,480 ft) and encountered a 66-m (217-ft) gross oil column.

Nido said that it found more than 20 additional prospects—beyond Yakal-1 and Tindalo-1—in the SC54 inboard area covered by its 3D seismic survey data.

SC50 and SC51

NorAsian Energy Ltd. (Perth) started site surveys in mid-2008 in offshore service contracts SC50 and SC51. Three well sites will be surveyed to identify potential shallow gas drilling hazards.

The company anticipates drilling two or three wells at the Calauit-2 location in SC50, and the Argao-1 and Bahay-1 locations in SC51.

Natural gas to fuel small vehicles

Energtek Inc. (New York City) signed a purchase contract in mid-2008 with the Philippine National Oil Co. (PNOC) for an unusual gas production project.

The firm plans to extract and purchase the remaining gas from the nearly depleted San Antonio well (Isabela province) for use in fueling small vehicles.

The company will use its adsorbed-natural-gas (ANG) technology for packing additional gas into vehicle fuel cylinders.

Initial plan calls for converting 3,000 vehicles by yearend 2009 and then providing fuel for them for seven years.

The government wants to convert more than 500,000 three-wheel vehicles by 2012 to reduce pollution from their two-cycle, liquid-fuel engines. The country has about 1.5 million such vehicles.

Lumba Lumba-1 well

Salamander Energy PLC (London) reported that exploration well Lumba Lumba-1 was spudded in mid-2008 in permit SC41.

The well, in the Sandakan Basin offshore northeast Borneo, was to be drilled to a planned depth of 3,000 m (9,842 ft) subsea in 1,100 m (3,609 ft) of water.

A resource potential of about 200 MMboe is estimated for the targeted sandstone structure.

Sulu Sea

Documents from the Philippines Department of Energy in early 2009 indicated that ExxonMobil Corp. (Irving, TX) was planning to invest as much as $100 million exploring for oil and gas in the Sulu Sea, off the southwestern Philippine coast.

The company would work with private Mitra Energy Ltd. (Hamilton, Bermuda) on Service Contract 56 to drill two deep-water exploration wells in 2009.

Independent Tap Oil Ltd. (West Perth) also was planning in mid-2008 to start exploration drilling on its Service Contract 41 off the southern Philippines in the Sulu Sea.

Tap acquired 750 sq km (290 sq mi) of 3D seismic over the offshore block in 2007 and was projecting reserves of 50–150 million bbl of oil.

SINGAPORE

Capital: Singapore
Monetary unit: Dollar
Refining capacity: 1,344,000 b/cd
Oil production: 9,835 b/d
Oil reserves: 0
Gas reserves: 0

LNG

LNG aggregator

In April 2008, Singapore's Energy Market Authority signed a memorandum of agreement for BG Asia Pacific Pte Ltd.—a unit of BG Group PLC (London)—to be the LNG aggregator for Singapore.

In that role, BG will aggregate demand for regasified LNG from all end-users of gas in Singapore and procure LNG supply to meet demand.

BG will have an exclusive license to import LNG and sell regasified LNG in Singapore for 20 years, at a level of up to 3 mtpy starting in early 2012.

Part of BG's role will be to supply the country's planned LNG regasification facility, on which construction is slated to begin on Jurong Island in early 2009. First imports of LNG are slated for 2012, according to the Ministry of Trade and Industry.

BG will source the LNG from its global portfolio. PowerGas Ltd., a wholly owned subsidiary of Singapore Power Ltd., was contracted to build, own, and operate the terminal.

The plant's capacity could be expanded to 10 mtpy by 2025.

Singapore also has expressed a desire to create an LNG trade hub with spot cargoes.

Refining

New Jurong Island complex

In November 2007, ExxonMobil Chemical Co. (Houston) said it would build a second complex on Jurong Island in a petrochemical project that will produce ethylene, polyethylene, polypropylene, specialty elastomers, and benzene as well as a 220-MW power cogeneration unit.

The $4-billion chemical complex is scheduled to come on stream in 2011. The complex, together with ExxonMobil's 605,000 b/d refinery, makes ExxonMobil Singapore's single largest foreign manufacturing investor.

Petrochemicals

Aromatics plant

Jurong Aromatics Corp. reported in early 2008 that UOP LLC (Des Plaines, IL) was moving ahead with the basic engineering and design of Jurong's planned aromatics plant in Singapore.

Construction was expected to begin by third quarter 2008, with completion slated for 2011.

The complex will produce about 1.5 million tonnes of petrochemicals, including paraxylene, orthoxylene, benzene, and petroleum products such as jet fuel and kerosene.

It will be one of the largest privately owned petrochemical facilities in Singapore.

Company News

Halliburton technology center

Halliburton Co. (Houston) opened in late 2008 a manufacturing and technology center in Singapore.

The center will produce electronic sensors used in the company's logging-while-drilling tools and openhole wireline logging services.

It also will repair and maintain drilling tools used throughout the region.

At peak, said Halliburton, the facility in Jurong industrial area will employ about 400 people.

SOUTH KOREA

Capital: Seoul
Monetary unit: Won
Refining capacity: 2,606,500 b/cd
Oil production: 20,970 b/d
Oil reserves: 0
Gas reserves: 0

Refining

Two refinery upgrades

SK Energy Corp. (Seoul) announced plans in mid-2008 to build a 40,000-b/d hydrocracking unit at its refinery in Inchon, 40 km (25 mi) west of Seoul.

The project aims to increase exports of lighter, higher-value products to China and Southeast Asia.

The new unit will produce naphtha, diesel, and kerosene using low-priced bunker-C fuel as feedstock.

Construction was scheduled to start in third quarter 2008, with commercial production to start in 2011.

The facility will eventually increase SK Energy's overall heavy oil-processing capacity to 202,000 b/d from the current 162,000 b/d.

A new residual fluid catalytic cracker at the firm's 840,000-b/d refinery in Ulsan, 414 km (257 mi) southeast of Seoul was poised to begin operation in mid-2008.

The 60,000 b/d capacity unit is set to launch operations in June.

Transportation/Pipelines

Siberian gas to South Korea

The South Korean energy ministry announced in October 2008 that the country plans to import up to 10 bcm/yr (353 bcf/yr) of gas from Russia for 30 years, beginning in 2015.

The gas would move through a $3-billion pipeline, to run from Vladivostok to South Korea through North Korea.

The line is part of a $90-billion deal under which South Korea would import Siberian gas, providing Russia with another market in addition to Europe.

Gazprom (Moscow) and Korea Gas Corp. (Sungnam) signed the agreement.

The arrangement represents a renewed attempt to complete the Trans-Korean Peninsula Pipeline project, first discussed more than ten years ago. Negotiations now are slated to begin by 2010.

If pipeline talks with North Korea are not successful, Russia will supply South Korea with LNG or compressed natural gas.

It is expected that after a pipeline-route study is complete, a final agreement could be signed between the countries in 2011.

South Korea and Russia also could provide North Korea with energy assistance through the pipeline.

South Korea consumes about 37 bcm/yr of gas, mostly as LNG that is transported in tankers.

Korea Gas estimates that domestic gas demand will reach 27.7 million tons in 2008.

Single-hull tanker ban

The government announced in early 2008 that it will ban entry of single-hulled tankers into its waters starting in 2011, about five years earlier than originally planned.

The decision was triggered by the worst-ever oil spill in South Korea (10,500 tonnes; 76,955 gal) on December 7, 2007,when the VLCC *Hebei Spirit* collided with a barge off the country's west coast.

Some 60% of the crude oil imported to South Korea is shipped on single-hull tankers.

International Activities

Oil diplomacy with Iraq

South Korea, the world's tenth largest energy consumer, relies on foreign suppliers for 97% of its oil.

As a result, the government has been exploring diplomatic initiatives to seek energy resources from outside the country.

On one front, talks were held in early 2008 with the Kurdish Regional Government (KRG) of northern Iraq, resulting in an MOU to explore a reserve of some 2 billion bbl in that region.

That effort could complicate talks between SK Energy Corp. (Seoul) and Baghdad regarding resumption of Iraq's crude exports to South Korea. Iraqi leaders oppose the signing of contracts between the KRG and international oil firms.

LNG

Second on LNG import list

South Korea was the second largest LNG importer in 2007 at 25.6 million tonnes, up by 1.2% over 2006. According to Pan EurAsian Enterprises (Raleigh, NC), Japan remained the world's largest LNG importer in 2007, at 65.7 million tonnes of LNG, up 5.7% over 2006.

Petrochemicals

Downsizing seen

The government's Financial Supervisory Service (Seoul) announced in late 2008 a decision to downsize the construction and shipbuilding sector in early 2009, and industry experts were expecting that the petrochemical sector could be next.

Weakening demand and resultant slumping prices for products have hurt the sector, and investments in new production facilities or capacity expansion are likely to be reduced or dropped.

The Samsung Economic Research Institute (SERI) projected that domestic demand for petrochemicals in Korea is likely to fall by 1.6% from 2008, to about 9.35 million tons in 2009.

Compounding the problem: an expected slump in demand in downstream industries such appliances and automobiles.

SERI predicted a marginal growth of 5.3% in South Korea's exports of petrochemicals to 11.2 million tons in 2009 versus 2008. From 2007 to 2008, such exports declined by 36.6%.

TAIWAN

Capital: Taipei
Monetary unit: New Taiwan dollar
Refining capacity: 1,290,000 b/cd
Oil production: 1,000 b/d
Oil reserves: 2.380 million bbl
Gas reserves: 220.000 bcf

Canadian oil sands interest

In mid-2008, state-owned Chinese Petroleum Corp. (CPC; Taipei) signed an MOU with Indian Oilsands Corp. (Saskatchewan), regarding a possible partnership on oil sands development in Alberta.

CPC plans to budget $792 million for a five-year exploration program. Company officials say the project is part of an effort to diversify energy sources and stabilize domestic energy supplies.

Formosa Petrochemical Corp. (Taipei), a member of the Formosa Plastics Group, also was studying a plan to tap Canada's oil sands and was considering sending an inspection delegation to Alberta.

BTX extraction unit

In mid-2008, CPC said it will use the GT-BTX extraction distillation technology of GTC Technology GP Inc. (Houston) as part of its No. 6 naphtha cracker and No. 7 BTX project in Lin Yuan, Kaohsiung.

GTC's technology extracts high-purity benzene, toluene, and mixed xylenes (BTX) from hydrotreated pyrolysis gas.

Lummus Technology (Bloomfield, NJ) won the overall contract for the project, which will produce 600,000 tpy of ethylene.

THAILAND

Capital: Bangkok
Monetary unit: Baht
Refining capacity: 729,100 b/cd
Oil production: 229,000 b/d
Oil reserves: 441.0 million bbl
Gas reserves: 11.198 tcf

Exploration and Development

Bongkot concession

Total SA (Paris) hit gas and condensate pay in late 2008 with three exploration wells on the Bongkot offshore concession in the Gulf of Thailand.

The Greater Bongkot North field currently produces approximately 600 MMcf/d of gas and 20,000 b/d of condensates.

Two of the Total discoveries, Ton Sak 7 and Ton Sak 8, have been sanctioned for development. Two more wellhead platforms were being constructed and should go into production in 2009.

The third discovery, Ton Son 2, will be developed in a subsequent phase.

Block B

Singapore Petroleum Co. Ltd. and its partners began drilling in mid-2008 Block B's first exploration well, Vimean Morodok MahaNorkor-1.

The exploration well was to be drilled to an anticipated TVD of 2,917 m (9,570 ft).

The exploration prospect is offshore Cambodia in 75 m (246 ft) of water, located 160 km (99 mi) southwest of Sihanoukville.

Songkhla field

Coastal Energy Co. (Surrey, UK) received government approval in mid-2008 of the Production Area Application (PAA) for development of its Songkhla field in the Gulf of Thailand.

The PAA covers approximately 75 sq km (29 sq mi) and includes not only the Songkhla field but several surrounding satellite features as well.

The approved production area contains proved, probable, and possible reserves of 6.4 MMbbl of oil.

It also contains eight identified oil prospects with potential oil in place of more than 200 MMbbl of oil.

Production from Songkhla field was expected to begin by yearend 2008.

Drilling and Production

Bauluang field

Salamander Energy PLC (London) estimated gross proved reserves in mid-2008 of 11 million bbl, up 52%, at Bualuang field in the Gulf of Thailand after development wells found better quality reservoir than expected.

Probable and possible reserves were elevated 34% to 20 million bbl.

Salamander operates the field in Block B8/38 in partnership with SOCO International PLC (London).

First oil was produced in late August 2008, and all five production wells were brought onstream successfully in September.

The field, on Block B8/38, was producing at maximum rates of over 17,000 b/d of oil.

Kambuna field

Salamander Energy PLC also completed an initial assessment in mid-2008 of progress on the Kambuna field development plan. Successful drilling, completion, and testing of the Kambuna development wells were conducted in the first half of 2008.

Initial gas is expected by mid-2009, with a phased increase to full production by third quarter.

The energy ministry expects output from the field of as much as 10,000 b/d.

Platong platform

Chevron Thailand Ltd. (Bangkok) let a contract in early 2008 for building a central production-processing platform and associated facilities for the $3.1-billion Platong Gas II project, Gulf of Thailand.

Installation is expected by the end of 2010, with production to start in 2011.

The Platong Gas II project, in shallow water 200 km (124 mi) offshore, is designed to add 420 MMcfd of gas processing capacity to meet growing domestic gas demand.

Arthit North field

PTT Exploration & Production PLC (PTTEP; Bangkok) said in early 2008 that plans to ramp up its natural gas output from Arthit gas field in the Gulf of Thailand by 120 MMcfd, or 36%, to meet rising domestic gas demand.

In a 3-yr program set to start in August, the extra production will come from sister field Arthit North.

Arthit North's output will supplement delivery from the main field, with production of 330 MMcfd of gas and 22,000 b/d of condensate.

An FPSO vessel will support production startup at Arthit North. Development will include the installation of three platforms and the drilling of 27 development wells.

Refining

Jet fuel plant

Star Petroleum Refining Co. Ltd. (Bangkok)—a joint venture of Chevron Corp. and PTT Public Company Ltd.—let a contract to CB&I (The Woodlands, TX) to build a jet fuel processing plant and related storage and shipping systems in Rayong Province.

The resulting plant will treat and sweeten 20,000 b/d of jet fuel. Storage tanks, pipeline segments and marine facilities are included in the project.

CB&I also is contracted to build an oil and diesel storage facility and associated infrastructure.

The full project is scheduled for completion by early 2009.

Star Petroleum is a joint venture of Chevron Corp. and PTT PCL.

LNG

First LNG terminal

PTT PLC (Chatuchak, Thailand) awarded a contract to GS Engineering & Construction Corp. (Seoul) to build Thailand's first LNG receiving terminal on the eastern coast.

Construction of the 5-mtpy facility in Rayong, about 200 km (124 mi) southeast of Bangkok, was slated to start by mid-2008, aiming for completion in mid-2011.

PTT said it plans to double the terminal's capacity to 10 mtpy in a later stage.

The company has already signed an agreement to import 1 mtpy of LNG from Qatargas Operating Co. (Doha) for 10 yrs starting in early 2011.

PTT also has been in talks with other LNG suppliers in Indonesia, Australia, and the Middle East as discussions with bogged down on arrangement of a 20-yr supply of 3 mtpy from Iran's Pars LNG project.

VIETNAM

Capital: Hanoi
Monetary unit: Dong
Refining capacity: 0
Oil production: 275,000 b/d
Oil reserves: 600.000 million bbl
Gas reserves: 6.800 tcf

Exploration and Production

Blocks on continental shelf

State oil firm Petrovietnam signed a 30-yr oil and gas contract in late 2008 for Blocks 129, 130, 131, and 132 on the continental shelf.

The contract covers exploration, development, and production of hydrocarbons in the blocks. Vietgazprom, the joint operating company of Gazprom (Moscow) and Petrovietnam, will act as the project operator.

Petrovietnam also was carrying out geological exploration in Block 112.

Deal challenged

The Chinese government warned ExxonMobil Corp. (Irving, TX) in mid-2008 to pull out of an exploration deal off Vietnam, calling the project a breach of Chinese security, said a Hong Kong newspaper.

The protest was over a preliminary cooperation agreement between Petrovietnam and ExxonMobil covering exploration in the South China Sea off Vietnam's south and central coasts.

China has voiced a historical claim to virtually all of the South China Sea, while both nations claim the entire Paracels and Spratlys archipelagos and much of the surrounding waters.

Vietnam bases its claims on its extensive continental shelf and exclusive economic zone.

Block 15-2/01

Talisman Energy Inc. (Calgary) tested peak rates of 13,450 b/d of light oil and 6.87 MMcfd of natural gas from its second offshore exploration well, Hai Su Den (HSD), on Block 15-2/01.

HSD was drilled to a TVD of 3,404 m (11,168 ft), encountering a hydrocarbon-bearing interval of about 732 m (2,400 ft).

Another four exploration wells were planned for 2008.

Block 15-2/01 is 80 km (50 mi) off eastern Vietnam and is on trend with large oil and gas discoveries in the Cuu Long basin.

Block 16-1

According to SOCO International PLC (London), its Te Giac Trang appraisal well (TGT-7X) offshore on Block 16-1 in the Cuu Long basin flowed at 8,100 b/d of oil from two drillstem tests in mid-2008.

TGT-7X was drilled to test a separate, previously untested fault block in the TGT field, designated H3N, and was drilled to improve knowledge of the reservoir distribution.

Activities were ongoing to evaluate options for the TGT-8X appraisal well.

SOCO also announced a discovery of gas and condensate in mid-2008 on the Te Giac Den 1X sidetrack well on Block 16-1.

To date, seven exploration/appraisal wells have been drilled in the TGT field, with an average oil and gas flow of 11,300 boe/d from each well.

The company was expecting approval of its development plan for the field by late 2008.

Drilling and Production

Song Doc, Rang Dong fields

Talisman Energy Inc. (Calgary) reported first oil production in late 2008 from the Song Doc field in offshore Block 46/02.

Gross production from five pre-drilled wells was expected to reach 25,000 b/d of oil by early 2009.

Three more development wells were being drilled.

The firm's share of proved and probable reserves in the field is estimated at 6 MMbbl of oil, with proved reserves of 3 MMbbl of oil.

Also in late 2008, first oil was produced from the Rang Don offshore field, according to MODEC Inc. (Tokyo).

Refining and Products

Nam Van Phong complex

In early 2009, Vietnam National Petroleum Corp. (Petrolimex) got approval from central Khanh Hoa province to build the $4.8-billion Nam Van Phong Petrochemical Complex about 1,200 km (745 mi) south of Hanoi to process 10 mtpy (78 MMbbl/yr) of oil.

China Petroleum and Chemical Corp. (Sinopec; Beijing) will be a partner in the project, which is scheduled for completion in 2013.

Output will include polypropylene, benzene, LP gas, unleaded gasoline, kerosene and diesel.

Dung Quat refinery

Petrovietnam said in early 2009 that it plans to sell a 49% stake in its Dung Quat refinery, scheduled to go online in February.

Preference likely will be given to international partners committed to supplying oil to the refinery. Initial discussions were planned with BP PLC (London).

Oil for the refinery will be imported. Vietnam's own oil reserves are limited and could earn the country more revenue as an export because they are of higher quality than needed at Dung Quat.

By yearend, the plant it is scheduled to process some 6.5 million tonnes (48 MMbbl) of oil.

Nghi Son refinery

Construction began in early 2008 on the $6-billion Nghi Son refinery and petrochemical project in Thanh Hoa province, 180 km (112 mi) south of Hanoi.

The Dung Quat refinery (see above) and the Nghi Son complex are expected to jointly meet some 60% of the country's demand for petroleum products.

The Nghi Son project will process oil imported from Kuwait. Output will be 200,000 b/d, including several gasoline formulations, diesel, kerosene, and paraxylene. Completion is scheduled for 2011.

Vung Ro refinery

Central Phu Yen province granted a license in early 2008 to a joint venture of Techno Star Management Ltd. (UK) and Telloil (Russia) to build a 4-mtpy refinery.

The $1.7-billion Vung Ro refinery in the province's Dong Hoa district will be the country's first wholly foreign-owned refinery. It is to be operational in 2012.

Alternative Fuels

Biofuels production

The Vietnam Oil and Gas Corporation (PetroVietnam) said in late 2008 that it will invest $120 million to set up a bioethanol plant in the Dung Quat economic zone in central Ngai province.

The project is a JV of PetroVietnam Oil Corp. and three other firms.

The facility is to begin producing 100 mtpy of ethanol in 2010 for addition to gasoline, using as feedstock 220,000 tpy of the tropical cassava plant

Sugarcane, jatropha and castor oil trees also could be used to produce biofuels, and the government has set aside land to cultivate such crops in the provinces of Vinh Phu, Phu Tho, Yen Bai, Son La, Hao Binh and Tuyen Quang.

By 2015, the goal is to produce 250,000 tpy of ethanol to meet 1% of the country's gasoline demand, and 1.8 mtpy by 2025 to meet 5% of demand.

E5 gasoline (5% bioethanol by volume) and B5 oil (5% biodiesel) is to be available nationwide by 2025.

International Activities

Exploration on Block M-2

Media reports in October 2008 said that Myanmar's national oil company had concluded licensing agreements with Vietnamese companies for an offshore block in the Andaman Sea.

Details are given in the Myanmar section of this chapter.

Japan-Vietnam cooperation

In a late-2008 analysis, *Oil & Gas Journal* noted that Japanese firms are stepping up their involvement in Vietnam's oil industry, entering both upstream exploration and production and downstream refining and marketing.

Phuong Dong field

State-owned Petrovietnam and Nippon Oil Corp. subsidiary Japan Vietnam Petroleum Co. have begun producing crude oil at offshore Phuong Dong oil field.

Production from Block 15.2, about 120 km (75 mi) off Ba Ria-Vung Tau province, is expected to increase to 15,000 b/d in 2009 from the 2008 level of 10,000 b/d.

The field has reserves estimated at 36.3 MMbbl of crude, 5.3 MMbbl of condensate, and 3.16 bcm (116 bcf) of natural gas.

Block 09-3

VRJ Petroleum—a JV of Idemitsu Kosan subsidiary Idemitsu Cuu Long Petroleum Co., Petrovietnam, and Zarubezneft—plans to develop Block 09-3, in water 50 m (164 ft) deep some 135 km (84 mi) off southern Vietnam.

VJR Petroleum announced a find at Block 09-3 in early 2008; one well had an oil flow of 4,150 b/d. The company estimates that Block 09-3 holds 50–70 million tonnes (367–513 MMbbl) of oil. Production of 20,000 b/d slated to start in the latter half of fiscal 2009.

Downstream activities

Idemitsu Kosan and Mitsui Chemicals announced in April 2008 that they would take part in a 200,000-b/d, $5.8-billion oil refinery and petrochemical complex project in Vietnam.

The facility would be built in the Nghi Son economic zone in Thanh Hoa Province, 180 km (112 mi) south of Hanoi.

In September, Petrovietnam and Marubeni Corp. signed a contract under which Marubeni will buy 75,000–150,000 tpy of propylene products from Vietnam's Dung Quat oil refinery.

The 140,000 b/d refinery is scheduled to start operations in early 2009.

Several other nations in the Africa-Pacific region saw action in 2008 and early 2009 on hydrocarbon exploration, development and processing.

Sri Lanka

Capital: Colombo

The government awarded an oil and gas exploration license in the Mannar basin to Cairn India (Gurgaon, India) in mid-2008.

Block SL 2007-01-001, off the northwest coast, covers approximately 3,400 sq km (1,313 sq mi) in water depths of up to 1,800 m (5,905 ft).

Iranian officials launched a $1.2-billion project in early 2008 to upgrade Sri Lanka's only refinery, at Sapugaskande, outside Colombo. The 4-yr project will triple the country's refinery capacity to 150,000 b/d.

Mongolia

Capital: Ulaanbaata

In late 2008, Mongolsekiu LLC (Ulaanbaatar) and Toyo Engineering Corp. (Tokyo) announced plans to build Mongolia's first oil refinery, at Darkhan-Uul aimag near the Russian border.

The $1.2-billion facility, with a total capacity of 44,000 b/d, is expected to be operational by 2012.

Earlier in the year, Kazakhstan signed an agreement for delivery of oil for the planned refinery.

Japanese officials also were discussing the potential development of Mongolia's uranium deposits.

Timor Leste

Capital: Dili

MEO Australia Ltd. (Melbourne) was evaluating a natural gas find made by its Blackwood-1 wildcat in the Timor Sea in early 2008. Blackwood is within the Epenarra structure, whose gas resource is estimated at 170 bcm (6 tcf).

At the same time, the Australian subsidiary of Eni SPA (Rome) was evaluating a significant oil discovery in the Joint Petroleum Development Area (Timor Gap) administered by Australia and Timor Leste.

The Kitan-1 wildcat is on Permit JPDA 06-105 about 500 km (311 mi) north of Darwin. Drilled to 3,568 m (1,088 ft), it flow-tested at 6,100 b/d of oil.

Key Stats

Section 1:
Future energy supply

Oil and Gas

IEA: Oil production outlook

For several years, the International Energy Agency (IEA; Paris) has predicted that supplies of crude and other liquid fuels will rise smoothly to meet rising demand, topping 116 million b/d by 2030, up from about 87 million b/d in 2008.

However, in its *World Energy Outlook* for 2008, published in November, IEA expressed concern about aging oil fields and reduced investment.

The agency concluded that it could be a challenge to produce more than 100 million b/d over the next two decades.

For the first time, IEA studied the world's top 800 oil fields that supply more than two-thirds of daily world production. Some big producers, including Venezuela, Iran and China, were not cooperative, and others (e.g., Saudi Arabia) treat production data as state secrets.

Nonetheless, the agency used computer modeling as well as information from IHS Inc. (Englewood, CO), the U.S. Geological Survey (USGS), some oil and gas service companies, and national petroleum councils to develop its analysis.

In the past, the IEA focused chiefly on assessing future demand, and then analyzed how much non-OPEC countries were likely to produce to meet that demand. It assumed that any gap would be met by big OPEC members such as Saudi Arabia or Kuwait.

A study released in 2008 by Cambridge Energy Research Associates (CERA; Cambridge, MA; a unit of IHS) concluded that the depletion rate of the world's 811 biggest fields is around 4.5% a year.

Geopolitical issues

In its *World Energy Outlook* for 2008, IEA also identified several geopolitical factors that it said were delaying oil development projects. These included:

- Continued fighting in Iraq has stymied efforts to revive aging fields.
- International sanctions on Iran have kept investments there from moving forward.
- Rebel attacks in Nigeria and political turmoil in Venezuela have cut into both countries' output.
- Big non-OPEC producers such as Mexico and Russia, which have either barred or sidelined international operators, are seeing production slump.
- A legal moratorium has barred exploration in 85% of U.S. offshore waters.

EIA: World oil output by 2020

The Energy Information Administration—part of the U.S. Department of Energy—released preliminary results in mid-2008 of its global oil-supply study.

EIA suggests that output of conventional crude alone—about 73 million b/d in 2008—will plateau at 84 million b/d, and that significant production of nonconventional fuels such as ethanol will be needed to push global fuel supplies over 100 million b/d by 2030.

Like IEA, the DOE agency expressed concern about the scale and timing of investments in development of needed resources.

Total: World oil output by 2020

In a long-term, internal study released in mid-2008, Total SA (Paris) projected that worldwide oil production will stabilize at about 95 million b/d before 2020, including extra heavy crude from Venezuela and Canada. Some other highlights:

- Energy savings and efficiency are "absolutely necessary" to limit increasing demand—driven by emerging countries and transportation fuel-use.
- Refining should place a stronger focus on light products.
- A further 5 million b/d might be added as biofuels, GTL, CTL, condensates, LPG, and refining gains.
- The 116-million b/d supply assumption for 2030 by the IEA is "too optimistic."
- Production, not reserves, is failing, limited by geological and geopolitical uncertainties.
- By 2030, the share of fossil energies in the energy mix would still be about 75%. Oil will account for 30% and gas for 22%.

BMI: North America, U.S. view

Analyst Business Monitor International (BMI; London) predicted in late 2008 that the United States will contribute about 65% of North American regional oil supply by 2012.

In 2007, North America produced 25.8 tcf (730 bcm) of natural gas in 2007. That level should rise to 25.9 tcf (733 bcm) in 2012, said BMI.

By 2012, the U.S. share of North American gas production is projected to be about 74%.

Regarding oil prices, BMI estimated an OPEC basket price average of $106/bbl for 2008.

For 2007–18, BMI forecasts that U.S. oil production will increase by 14.% and output will peak at 7.92 million b/d in 2017.

Global Primary Energy Reserves

Resource	Proved reserves	Energy, quadrillion btu	Million tons of oil equivalent	Years remaining at current production
Oil*	1,372 × 10^9 bbl 191 × 10^9 tons	7,600	191,000	41
Natural gas	6,405 tcf	6,600	165,000	63
Coal	479 × 10^9 tons	8,500	213,000	147

* Includes Canadian oil sands.

Source: BP Statistical Review of World Energy, 2007

U.S. gas production is expected to rise to 20.1 tcf (570 bcm) in 2018 from the 2007 level of 15.5 tcf (546 bcm).

Offshore Europe outlook

Douglas-Westwood Ltd. (Canterbury, UK) in mid-2008 released its annual summary of Europe's offshore oil production status.

Nearly 600 offshore fields have been developed off Europe, involving a similar number of platforms plus some 380 subsea wells, more than 200 subsea templates, and nearly 1,000 pipelines.

However, declining production and high costs have forced the oil majors to turn their attention to other regions, opening the way for smaller players. These newcomers are better positioned to economically extract remaining reserves from a multitude of small fields and from depleted existing fields.

Looking worldwide, Douglas-Westwood expects global offshore oil and gas production to average grow to 56.7 million boe/d by 2012.

Greatest growth is expected in the Middle East. Offshore Europe production is expected to decline by a nearly 1 million boe/d from its 2008 forecast level of 8.6 million boe/d.

The Global Context

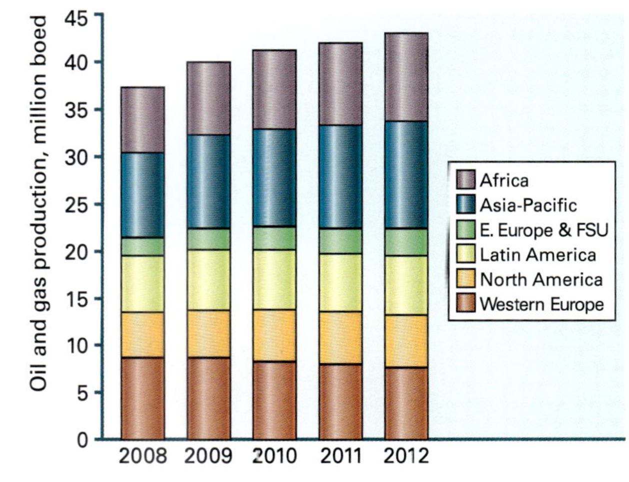

Offshore Europe Oil and Gas Production

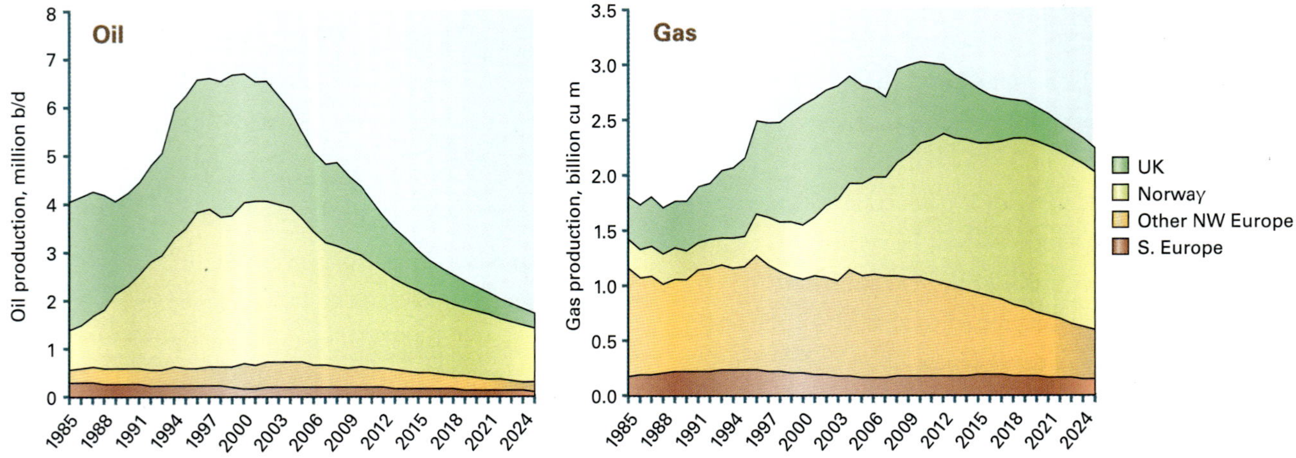

Source: Energyfiles Ltd.

Barents Sea off Norway

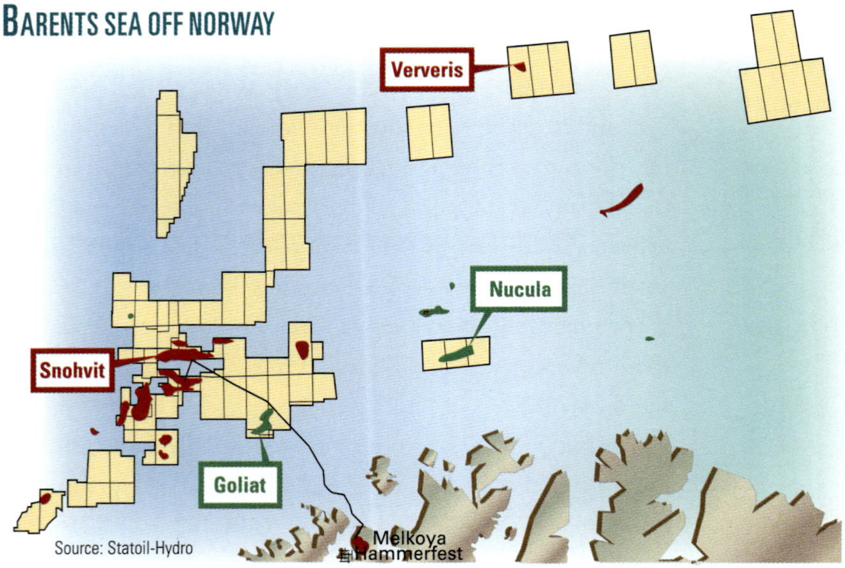

Source: Statoil-Hydro

FPS Prospects off Western Europe, 2008-12

Country	FPSO	FPSS	TLP	Total
	Installations			
Ireland	1	—	—	1
Norway	4	1	—	5
UK	9	1	2	12
Total	**14**	**2**	**2**	**18**

Source: The World Floating Production Report 2008-12

Aramco views

Abdallah S. Jum'ah, president and CEO of Saudi Aramco, said in early 2008 that there are enough conventional and unconventional petroleum resources to satisfy global demand for liquid fuels "for many decades."

However, he noted that it will require better planning and cooperation between industry and governments to accomplish that task.

Mr. Jum'ah spoke at a Houston energy conference sponsored by CERA.

Other highlights from his opening address:

- Even without CTL, GTL and biofuels, the global in-place endowment of conventional oil and nonconventional fuels ranges from 13–16 trillion bbl.
- Analysts estimate recoverable conventional and unconventional resources at 3–6 trillion bbl.
- New tools and techniques can boost oil-field recovery rates; Saudi Aramco has achieved 70% recovery of original oil-in-place for some fields.
- There has been "considerable confusion over what is realistic" regarding the future energy mix of alternative and conventional resources.
- The U.S. DOE and the IEA predict that fossil fuels will continue to meet 82–87% of total energy demand through 2030.
- Consensus is needed on: clean, efficient development of fossil energy resources; rational development of alternative energy resources; appropriate efforts on conservation and efficient energy use; and development of carbon sequestration technologies.

Oil shale promise

A late-2008 symposium at the Colorado School of Mines (Golden) highlighted projects and technologies for exploiting oil shale resources in various countries.

The global oil shale resource base is estimated to be at least 2.6 trillion bbl, of which 2 trillion bbl are in the United States. Regarding resource development:

- Estonia, China, and Brazil have operated commercial-scale oil shale projects for many years.
- Jordan, Morocco, and Australia all are planning oil shale projects.
- Several U.S. demonstration projects are under way in Western Colorado and Eastern Utah.

Developing shale gas plays

Source: Tristone Capital

Resource estimates

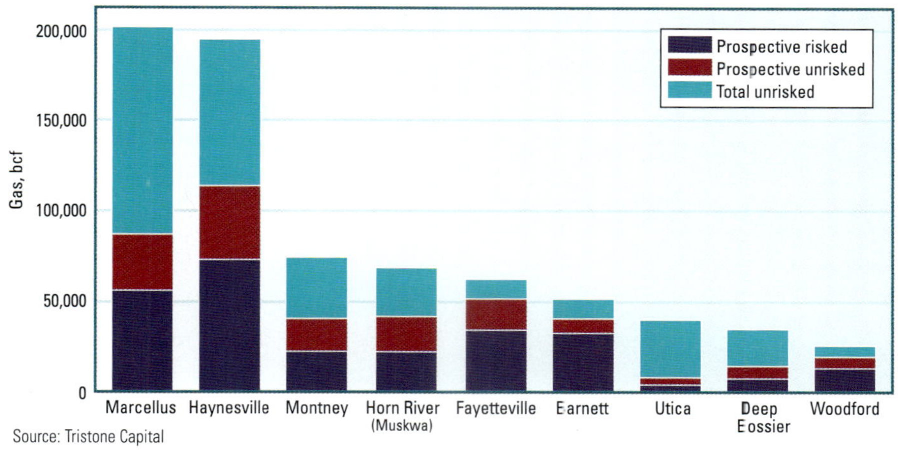

Source: Tristone Capital

Canada, U.S. shale gas

In a late-2008 study, Tristone Capital Inc. (Calgary) estimated that nine Canadian and U.S. shale-gas plays may produce as much as 24 bcfd (680 MMcm) by 2018.

The firm evaluated the Barnett (Fort Worth basin), Deep Bossier, Haynesville, Fayetteville, Woodford, and Marcellus shales in the United States; and the Montney, Hom River (Muskwa), and Utica shales in Canada.

The study expects ultimate recovery from these resources to be 261 tcf (7.4 tcm) of gas, based on various risk factors applied and a long-term average gas price of $8.50/MMbtu.

Without the risk factors, Tristone says these shales have a 743-tcf (7.4-tcm) recovery potential.

The study also identified three emerging shale plays: Pearsall shales in the Maverick basin of South Texas, the Niobrara shales of Western Colorado, and the Barnett shale in the Delaware basin of West Texas.

Risked production additions

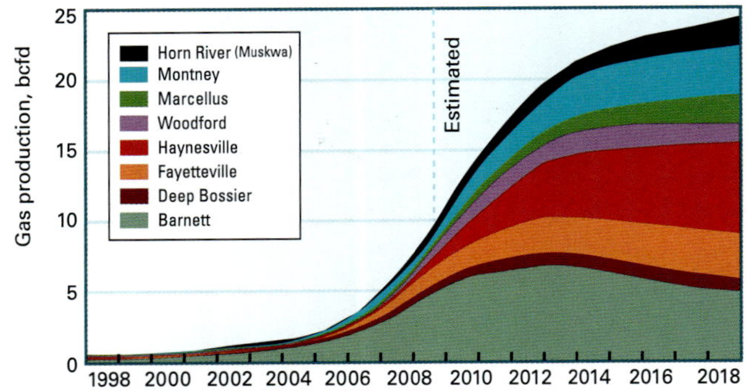

Source: Tristone Capital

US shale completions

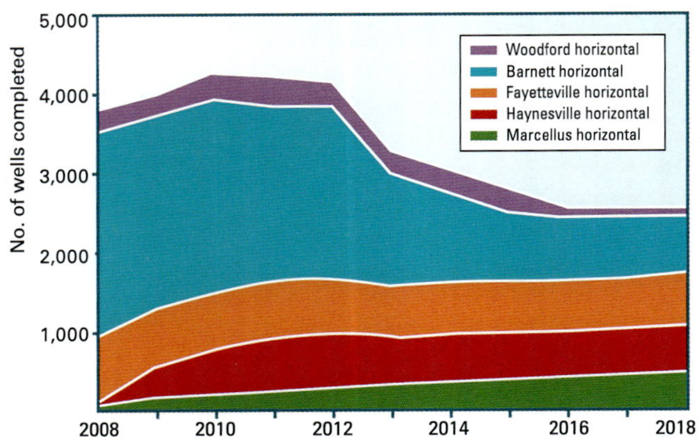

Source: Tristone Capital

Typical lateral length, frac stimulation

	Lateral, ft	Frac size, tons/frac	No. of frac stages	Frac fluid
Barnett	2,500-3,000	100	4-6	Slick water
Fayetteville	+3,000	100	6-8	Slick water
Haynesville	3,000-6,000	100	9-11	Gel cross-linked oil
Horn River	+2,000	100	6-8	Slick water
Marcellus	2,000-6,500	100	6-8	CO_2 polymer
Montney	1,700-3,000	100	6-8	Slick water
Utica		100	8	Slick water
Woodford	+3,000	100	6-8	Gel cross-linked oil

Sources: Tristone Capital, company reports, Halliburton

USGS: Arctic oil, gas potential

The area north of the Arctic Circle contains an estimated 90 billion bbl of undiscovered, technically recoverable crude oil, said the USGS in a July 2008 estimate—its first for the region.

The Arctic, especially offshore, is essentially unexplored with respect to petroleum, it noted.

The region also contains an estimated 1,670 tcf (47.3 tcm) of technically recoverable gas and 44 million bbl of technically recoverable NGLs in 25 geologically defined areas.

More than half of the undiscovered oil resources are believed to be in just three provinces: Arctic Alaska, the Amerasia basin, and the East Greenland Rift basins.

Gas is estimated to be three times more abundant than oil in the Arctic on an energy equivalency basis, with more than 70% of it in the West Siberian basin, the East Barents basins, and Arctic Alaska.

Unconventional resources

Researchers from Texas A&M University said at a September 2008 international conference that sedimentary basins in the United States appear to contain a volume of technically recoverable unconventional gas that is 10 times the ultimately recoverable volume of conventional gas.

That positive outlook was echoed at the annual conference of the Society of Petroleum Engineers (September 2008; Denver). SPE speakers included oil sands and oil shale in their commentary, as well as gas from shales, tight sands, and coal seams.

Spurring development of unconventional resources in recent years were innovations in rig and bit design, new drilling fluids, directional drilling, fracturing, coring, microseismic, and *in situ* retorting.

Barriers to development include rising infrastructure costs, limited human capabilities, and fluctuating world oil prices.

Geologic attributes

	Depth (m)	Net thickness (m)	Total organic content (%)	Total porosity (%)	Pressure gradient, psi/ft	Gas-in-place, bcf/section
Barnett	2,000-2,800	75-200	3.5-8.0	3.0-4.8	0.46-0.50	35-185
Fayetteville	450-2,000	15-105	4.0-9.5	2.0-8.0		25-65
Haynesville	3,200-4,000	60-80	3.0-5.0	8.0-12.0		150-250
Horn River	2,000-3,000	150-175	0.5-10.0	2.0-4.5	0.65	130-320
Marcellus	1,000-2,000	15-75	1.0-12.0	1.6-7.0	0.45-0.60	20-100
Montney	900-3,000	150-300	2.5-6.0	2.0-8.0	0.44-0.70	55-90
Utica	500-3,500	90-300	0.3-3.1	2.2-3.7		25-160
Woodford	1,800-2,600	15-60	2.0-10.0	6.0-8.0		40-120

Note: One section = 640 acres.

Sources: Company reports, Tristone Capital estimates

IEA: Gas supply security

In its *2008 Natural Gas Market Review*, released in October, the IEA noted a steady rise in natural gas prices since mid-2007, in both established and new markets.

IEA saw this as "not only a reflection of higher demand, but also of a delayed supply response."

The agency concluded that investments uncertainties, cost increases, and delays continue to be a major problem in most gas markets and continue to constitute a threat to long-term security of supply.

These factors will be compounded, said IEA, by the world financial turmoil unfolding in late 2008.

To meet this growing demand, LNG trade is likely to play a stronger role in regional markets within the OECD countries in the short and medium term, said the agency.

LNG is expected to reach 20% in Europe, where imports will account for over half of total supplies.

In North America, LNG imports are expected to more than double by 2015 from levels seen in 2007.

All told, gas production is expected to increase in all major regions except OECD Europe, where North Sea output is declining. North American production growth should slow after 2015.

Coalbed methane, shale gas and other nonconventional supplies are expected to play a growing role in some regions, notably North America.

Shell: TANIA scenarios

Royal Dutch Shell PLC (The Hague, Netherlands) unveiled new scenarios in early 2008 for studying future energy use that reflect growing concerns about climate issues and increased energy demand from emerging economies.

Shell officials say that energy producers, consumers, and governments could either cooperate or "scramble" to solve major energy and environment problems.

Both scenarios, under the banner, TANIA ("There Are No Ideal Answers"), run to 2050.

In the "Scramble" scenario, policymakers simply react to problems, developing supply-side responses that eventually hit constraints. Growth of wind, solar, and other alternates could take place locally but would be limited.

In the second scenario, called "Blueprints:"

- New coalitions emerge within cities or regions and sometimes across national borders.
- The growth of alternates and carbon capture-and-storage could result in most new power plants—in both emerging economies and industrialized nations—incorporating carbon capture and storage (CCS) by 2012.
- By 2020, 50% of vehicle miles traveled could still be with liquid fuels but using much more efficient engines.

The company concludes that the "blueprints" strategy would be difficult but preferable.

Renewable Resources

CERA: Renewables rising

High energy prices, concerns about climate change and energy security, and a major shift in public opinion are converging to drive the development of renewable energy, according to Cambridge Energy Research Associates (CERA).

A 2008 CERA study—which looked at biofuels, renewable power, carbon capture and storage, nuclear power, and hydropower—found that renewable resources could be supplying 7–16% of the world's electric needs by 2030.

Clean energy investment—including in nuclear and hydroelectric systems as well as renewables—could reach a cumulative total of $7 trillion by 2030. Other key points:

- Wind power will make the largest gains, followed by solar power and biomass.
- Key policy drivers will be: putting a price on CO_2 emissions, setting mandates, and providing subsidies.
- Nuclear and hydroelectric systems will account for almost half the gross clean power additions by 2030.
- The size and use of the coal resource base in the United States and China will create a powerful drive to develop "clean coal" technologies.

Countries expected to have a major impact on development of clean energy are the United States, European Union member, China, Japan, India, and Brazil.

Corporate, national strategies

Consulting firm Accenture (London) said in late 2008 that oil companies are re-examining their alternative energy assets to determine which ones may be viable alternates to fossil fuels.

Several firms have conducted initial studies of biofuels and other resources such as wind or solar energy systems.

Accenture sees a move now from investigative to pilot stage, with oil companies preparing for full-scale ramp-up into some subset of the businesses that they have evaluated.

Focused strategies rather than broad-based portfolios appear to be the pathway of choice.

The firm pointed to a range of recent decisions by BP PLC (London) regarding wind power projects, carbon capture and sequestration, solar photovoltaics, and biofuels.

Regarding national oil company actions, Accenture noted that, in early 2008, Abu Dhabi announced plans to invest $15 billion in the Masdar Initiative. That initiative will focus on developing and commercializing renewable, alternative, and sustainable energy.

In China, CNPC and Sinopec have chosen to focus on ethanol distribution.

Wind power: U.S and Europe

The American Wind Energy Association (AWEA; location? Washington, DC?) reported in early 2008 that Texas is at the forefront in redeveloping wind energy resources.

The state has the most cumulative wind-power capacity installed—4,356 MW.

Other states with large installed wind-power capacity (in MW) include:

- California: 2,439
- Minnesota: 1,299
- Iowa: 1,273
- Washington: 1,163.

The U.S. wind energy industry installed a record 5,244 MW in 2007, said AWEA, expanding the nation's total wind power generating capacity by 45% in a single calendar year.

Global Offshore Wind Power Capex

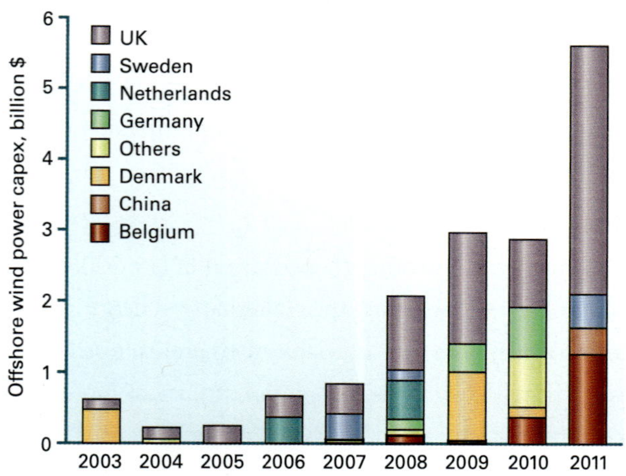

Source: Douglas-Westwood

Wind power: Europe

Looking worldwide, Douglas-Westwood Ltd. (Canterbury, UK) forecasts that 1,157 offshore wind turbines with a total capacity of 3.6 GW will be installed 2008–12.

Annual capital expenditure—about $1 billion in 2007—is expected to grow to near $6.5 billion by 2012. From 2008–12, capital expenditures are projected to total $11.8 billion.

Areas with greatest wind-power spending are expected to be northwestern Europe, followed by the United States and China.

The UK is expected to become the largest offshore wind power market, with 1.6 GW of new capacity forecast for installation 2008–12.

World biofuels demand

World biofuels demand is expected to grow at 19.5% per year to 92 million tonnes in 2011, despite concerns about the impact of biofuels on the environment and food supplies, Freedonia Group Inc. (Cleveland, OH) said in an early-2008 report.

The 2006 world biofuels demand was 37.7 million tonnes, and the annual growth rate was 19.9% during 2001–06.

Future market expansion will come chiefly from a more-than-doubling of ethanol demand and even faster increases in global biodiesel demand.

The annual growth rate for biofuels demand from 2006–11 is forecast at 35.3% in Asia Pacific, 22.1% in Western Europe, and 19.5% across North America.

At the same time, criticism was growing on several fronts regarding biofuels, particularly ethanol.

For example, the United Nations called biofuels a "crime against humanity" for diverting food crops toward fuel production as a global scarcity deepens and food prices rise.

Second-generation biofuels

Brazil's Petroleo Brasileiro SA (Petrobras; Rio de Janeiro) and biotechnology firm Kior Inc. (Menlo Park, CA) signed a cooperation agreement in early 2008 regarding development of second-generation biofuels.

Petrobras wants to use plant waste materials to produce bio-oil using Kior's biomass catalytic cracking technology. Second-generation biofuels don't compete with food crops.

Kior has been studying biofuel production from sugarcane bagasse since 2006. Brazil is a leading producer of biofuels from sugarcane. It is the world's second-largest producer of ethanol and its biggest exporter.

In parallel, Chevron Corp. (San Ramos, CA) and Weyerhaeuser Co. (Federal Way, WA) have launched Catchlight Energy LLC, a 50-50 JV to develop the next generation of renewable transportation fuels from nonfood sources. Feedstock will be timber and wood waste.

Ethanol from wood waste

A mid-2008 prefeasibility study by Sklar & Associates (Murrells Inlet, SC) explored wood waste-to-ethanol as a significant, ready source of second-generation biofuels and the opportunities for refiners and marketers to participate with pulp and paper mills in producing such fuels.

The study suggested that if all forestry wastes, mill wastes, and cellulosic wastes taken to landfills were instead used as feedstock, there would be enough additional wood waste in the U.S. Southeast to support 172 new, 20-million-gal/yr ethanol plants.

These plants could produce 3.4 billion gal/yr of cellulosic ethanol.

Two basic technologies are used to convert cellulosic material into renewable liquid fuels. The first is "two-staged thermal with gasification"; the other is "two-staged concentrated acid hydrolysis."

Other Energy Sources

IFP: CTL a 'credible' option

The Institut Francais du Petrole (IFP; Rueil-Malmaison, near Paris) said in early 2008 that coal-to-liquids (CTL) technology appears to be one of the more credible alternatives to oil "for the long-term supply of transport fuels."

IFP noted that coal-rich countries from China, the United States and Australia, to Indonesia, India, and Germany are pursuing CTL ventures, and that South Africa and China are already engaged in industrial CTL production.

The greatest hurdle facing CTL development, said IFP, is the need to combine it with carbon capture and sequestration (CCS) technology because CTL systems emit carbon dioxide.

Cost is also a key issue. At the Sasol plant in South Africa, CTL was evaluated at $67–82/bbl, excluding the cost of CO_2 capture.

GTL, CTL moving forward

Interest in Fischer-Tropsch (F-T) chemistry and associated technologies such as gas-to-liquids (GTL) and coal-to-liquids (CTL) continues to grow worldwide.

According to an early-2008 evaluation by E-MetaVenture Inc. (Houston), in the last few years, one new world-scale unit came on line, three others are in construction, and many more are planned.

South Carolina Wood Waste

Source, dry tons/year	
Logging residues	2,205,750
Precommercial biomass (<5 in. DBH)	4,277,898
Commercial biomass (5-8.9 in. DBH)	2,662,951
Southern scrub oak	917,440
Mill residue	2,452,866
Urban wood waste	310,726
Potential additional wood waste for ethanol	12,827,630
Midsized cellulosic ethanol plant feedstock requirements = 1,000 dry tons/day × 330 days/year = 330,000 dry tons/year	
Plants that could be supported by wood waste = 12,827,630/330,000= 39	
Potential ethanol output in South Carolina = 39 × 20 million gal/year/plant = 780 million gal/year	
Mills in Southeast US	22
Mills in South Carolina	5
Potential annual ethanol output in Southeast US = 22/5 × 780 million gal/year = 3.432 billion gal/year	

Economies of Scale

	Plant size		
	Small	Med.	Large
Feedstock capacity, dry tons/day	500	1,000	2,560
Ethanol production, million gal/year	10.8	22.1	55.2
Site footprint requirements, acres	<2	2-3	4
Steam requirements (150 psi), lb/hr	11,328	22,656	58,000
Water requirements, million gal/year	84	168	431
Capital costs, million $	85	142	275
Capital cost/installed production, $/gal	7.88	6.41	4.99

Financial Summary

	Plant size		
	Small	Med.	Large
Wood waste used, dry tons/day	500	1,000	2,500
Capital costs, million $	85	142	275
Debt, million $	59.5	99.4	192.5
Equity, million $	25.5	42.6	82.5
Mill equity, million $	6.7	11.5	26.5
Mill equity, % of total	26	27	32
Annual revenue, million $	25.1	50.3	128.7
Annual cash flows, million $	8.9	24.7	53.1
Cash flow to mill, million $	2.3	6.7	16.4
Annual mill subsidies, million $	1.1	2.1	5.1
Cash-on-cash return (cash flow/subsidies)	2.1	3.2	3.2
Ethanol sales, million gal/year	10.8	22.1	55.2
Average selling price, $/gal	2.23	2.23	2.23
Direct cost, $/gal	1.01	0.86	1.1
Operating income, $/gal	1.32	1.47	1.23
After-tax income, $/gal	0.4	0.57	0.51

Commercial products have been tested, certified, shipped, and used. However, several factors continue pose challenges. These include technology development and demonstration, costs of construction and operation, and varied regulatory approaches in different jurisdictions.

Fischer-Tropsch-Based Processes Diagram

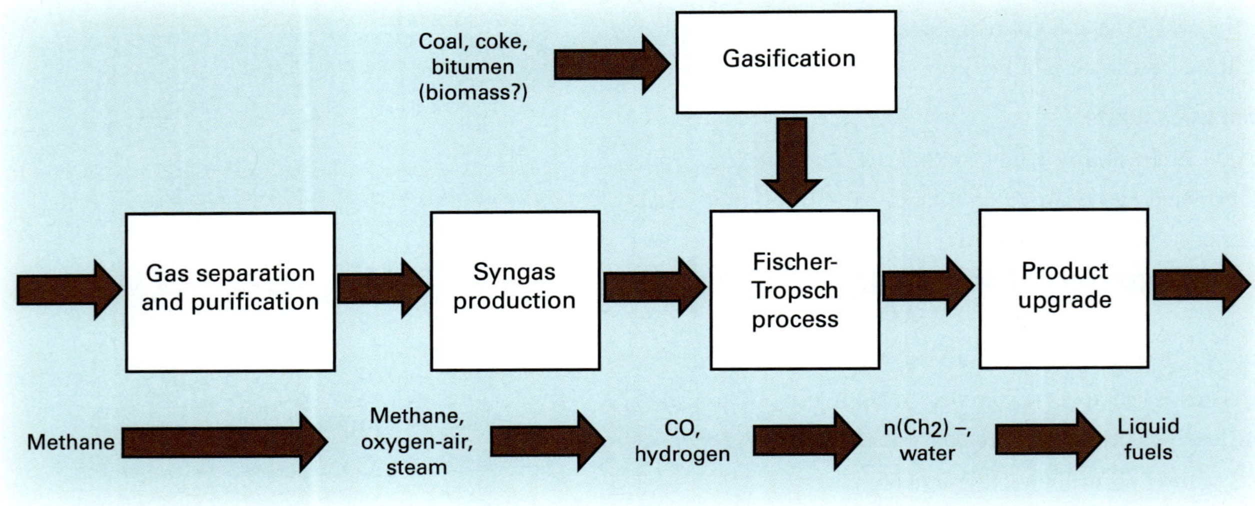

Block Flow Diagram of a CTL Process

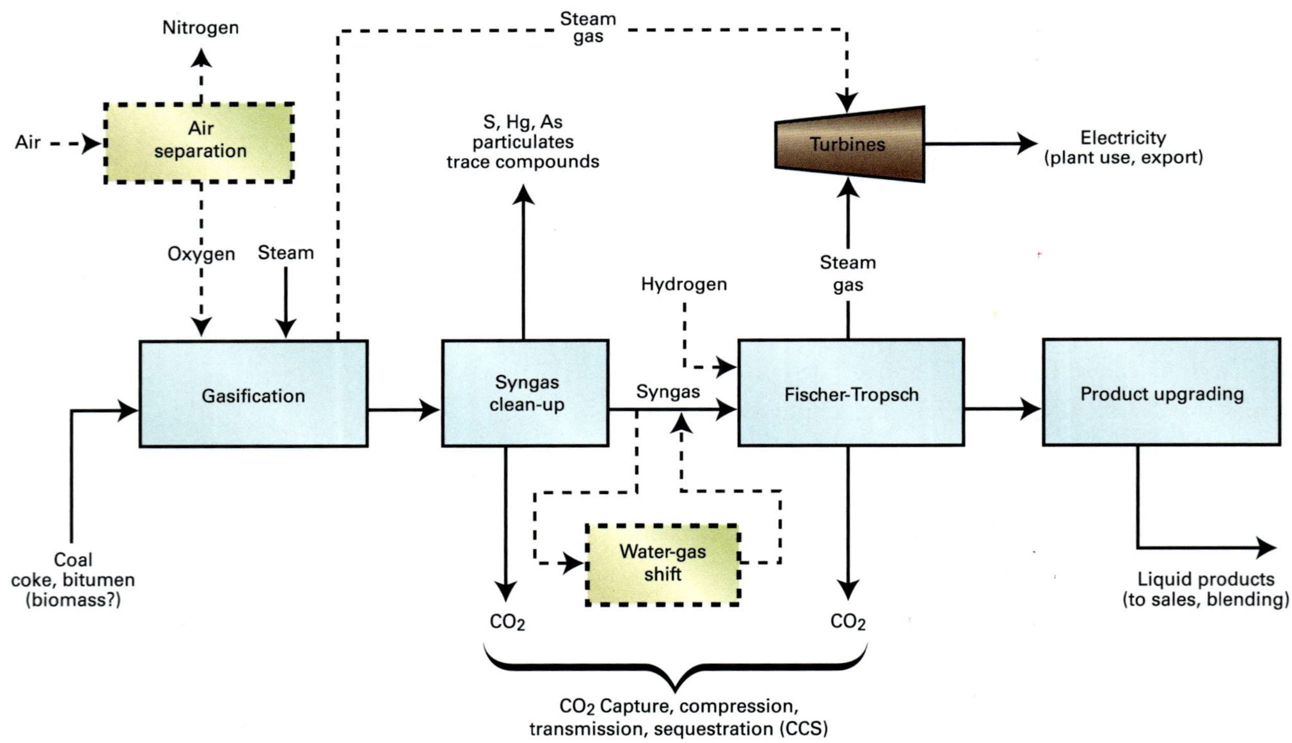

Gasification involves pyrolysis, combustion, and gasification chemistries:

$$2C\text{–}H + 3/2\ O_2 \rightarrow 2CO + H_2O + \text{Heat}$$

$$C\text{–}H + H_2O \rightarrow CO + 1.5\ H_2$$

Also, some water-gas shift: $CO + H_2O \leftrightarrows CO_2 + H_2$

F–T converts syngas to hydrocarbons:

$$CO + ?H_2 \rightarrow \text{–}CH_2\text{–} + CO_2 + H_2O + \text{heat}$$

(long chain)

COMMERCIAL GTL-CTL UNITS*

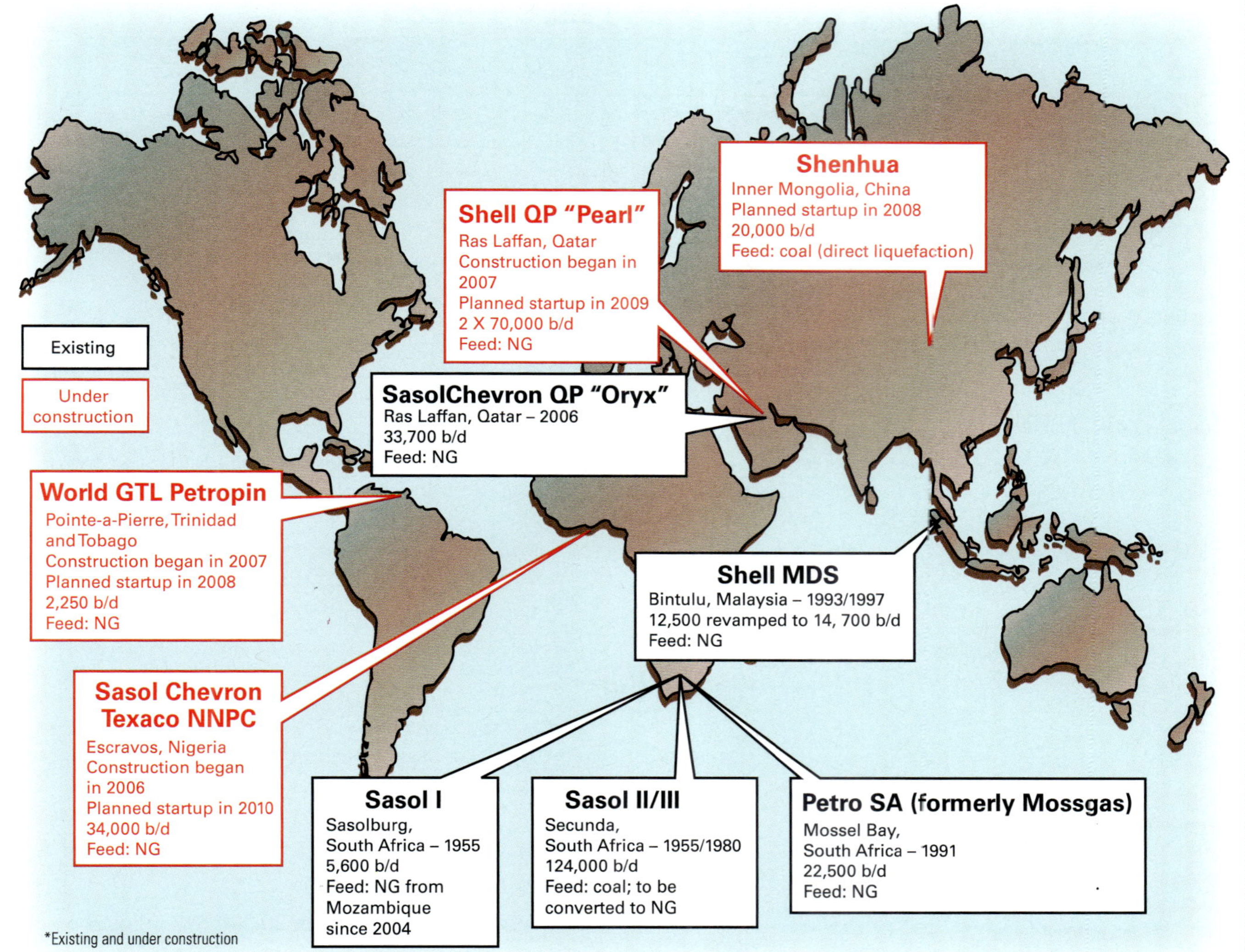

*Existing and under construction

GTL

Construction of the 140,000 b/d Shell-Qatar Petroleum (QP) Pearl plant continued in 2008, although with major cost overruns.

Several GTL projects in Qatar were cancelled or placed on hold until at least 2009. Of particular note was cancellation of the ExxonMobil and QP Palm GTL plant with a planned capacity of 154,000 b/d.

Other facilities put on hold included plants planned with ConocoPhillips and with Marathon/Petro-Canada as Qatar assesses North field reserves.

The Escravos GTL plant in Nigeria has a design capacity of 34,000 b/d and would use more than 300 MMcfd of gas. First production is slated for 2010.

The 2,250-b/d World GTL plant being built in Trinidad and Tobago was slated to begin operation in 2008.

Several other projects are at various stages of study, planning, and design around the world, including plants in Australia, Egypt, Thailand, and Papua New Guinea, among others.

Proved Worldwide Coal Reserves, Million Tons

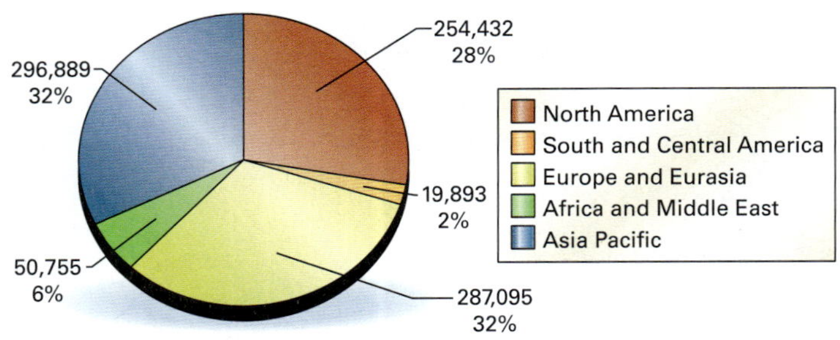

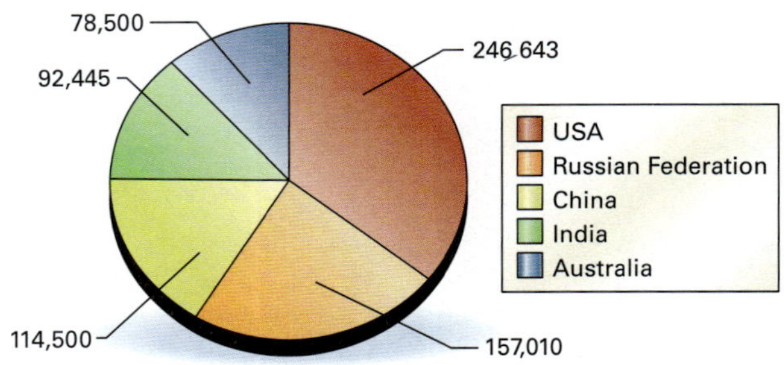

Source: Energy Information Administration, US Department of Energy

Partial List of US CTL Projects

Project lead	Project partners	Location	Feedstock	Status	Capacity, b/d	Cost, $ billion
American Clean Coal Fuels	None cited	Oakland, Ill.	Bituminous, biomass	Feasibility	25,000	—
Synfuels Inc.	GE, Haldor-Topsoe, NACC, ExxonMobil	Ascension Parish, La.	Lignite	Feasibility	—	5
DKRW Advanced Fuels	Rentech, GE	Medicine Bow, Wyo.	Bituminous	Design (2011)	15,000-20,000	1.4 (?)
DKRW Advanced Fuels	Rentech, GE, Bull Mountain Land Co.	Roundtop, Mont.	Sub-bituminous, lignite	Feasibility	22,000	1-5
AIDEA	ANTRL, CPC	Cook Inlet, Alas.	Sub-bituminous	Feasibility	80,000	5-8
Mingo County	Rentech	WV	Bituminous	Planning	25,000	2 (?)
WMPI	Sasol, Shell, DOE	Gilberton, Pa.	Anthracite culm	Design	5,000	0.612
Rentech/Peabody	N/A	Montana	Sub-bituminous, lignite	Feasibility	10,000-30,000	—
Rentech/Peabody	N/A	Illinois, SW Indiana, Kentucky	Bituminous	Feasibility	10,000-30,000	—
Rentech		Natchez, Miss.	Coal, petcoke, biomass	Planning	1,600 (Ph. I)	—
Baard Energy	AMEC Paragon	Wellsville, Ohio	Sub-bituminous, lignite	Feasibility	35,000	4

Overall CTL Balance

Total liquid product capacity	11,000 b/d[1]	50,000 b/d[2]
Coal (Illinois #6, bituminous), tons/day	4,891	24,533
Other feeds: air, water,...		
Diesel, b/d	7,500	27,819
Naphtha, b/d	3,509	22,173
Co_2, tons/d	6,035	32,481
Net Power, Mw	9.7	124.3
Other products: S, slag, fuel gas, ...		
Bbl liquid/Ton coal	2.25	2.04
Ton CO_2/Ton coal (carbon /carbon)	0.53	0.57
Ton CO_2/Ton coal	1.23	1.32
Overall thermal efficiency, % HHV	[3]51	[3]47

[1]NETL study for DOD/Air Force Apr. 9, 2007. [2]NETL/DOE study Aug. 24, 2007. [3]Not verified. Does not include all energy recovered in process.
Sources: NETL/DOD feasibility study, DOE/NETL feasibility study.

CTL

Under construction and slated for start-up in 2008 is the Shenhua Inner Mongolia facility.

The plant, with a capacity of 20,000 b/d of liquid products, uses direct liquefaction technology rather than the classical CTL route that includes manufacture of syngas.

Other CTL projects are at various stages in China, Australia, Indonesia, India, South Africa, Botswana, and New Zealand.

Typical CTL Capital Cost for Nominal 50,000 b/d Unit

	Capital cost, $ million[1]	
Coal and slurry prep	425	67%
Gasification	1,150	
Air separation unit	425	
Syngas clean-up	850	
WGS + F.T	510	12%
Product upgrading	210	5%
Power generation	255	16%
Other	425	
TIC	4,250	
~$85,000/bbl installed capacity		

[1]Excludes CO_2 compression, transportation, sequestration costs.

Global CTL, GTL Liquid Fuels Production

	1992	2004	2005	2010	2015	2020	2025	2030
GTL liquid products				1,000 b/d				
High case	—	—	—	20	60	100	110	140
Low case	—	—	—	20	30	40	50	60
Reference case	—	—	—	20	50	90	100	120
CTL liquid products								
High case	100	100	100	300	600	1,800	3,000	3,900
Low case	100	100	100	300	300	500	800	900
Reference case	100	100	100	300	600	1,200	1,700	2,400

Source: Energy Information Administration, International Energy Outlook 2007

Progress on gas hydrates

At a May 2008 conference, U.S. DOE officials noted that research on methane hydrates has made unexpected strides since 2000, but that technical and economic obstacles must be resolved before hydrates production can be proved feasible.

DOE also spotlighted the value of international collaboration in hydrates research. Japan has one of the most active programs; others are India, China, South Korea, Canada, and the United States.

The USGS has said the world's gas hydrates accumulations exceed the volume of known conventional gas resources. The U.S. gas hydrates resources could total 200,000 tcf (5,664 tcm) of gas in place, and world hydrates resources could total about 700,000 tcf (19,824 tcm) of gas in place.

DOE is optimistic that hydrates production can be demonstrated on a commercial scale by 2020.

PROJECTED US CTL DIESEL

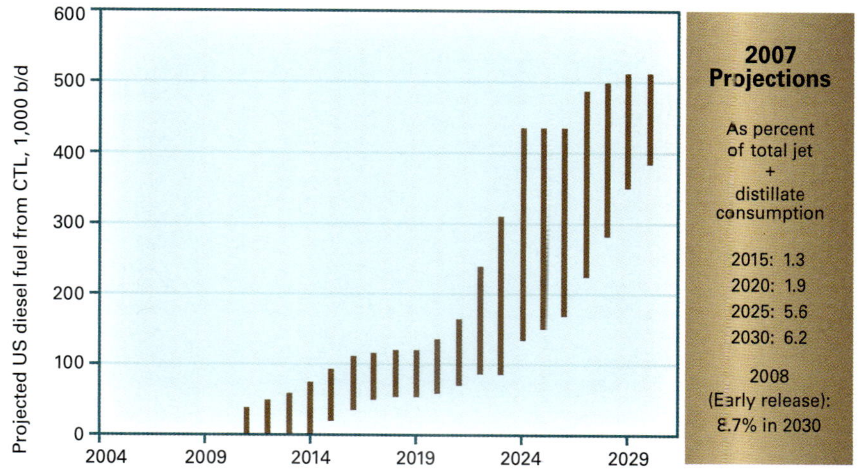

Sources: Energy Information Administration, International Energy Outlook 2007, International Energy Outlook 2008 (Early release)

A DECADE OF RESERVES CHANGES

Jan. 1	World (1,000 bbl)	OPEC (1,000 bbl)	World gas, bcf
2009	1,342,207,320	944,017,000	6,254,363
2008	1,331,698,077	931,999,000	6,185,693
2007	1,317,447,415	902,343,000	6,182,692
2006	1,293,344,534	901,659,000	6,101,158
2005	1,277,701,992	885,188,000	6,040,208
2004	1,265,811,583	869,521,000	6,068,302
2003	1,212,880,852	819,007,000	5,501,424
2002	1,031,100,681	818,842,000	5,451,332
2001	1,028,457,585	814,398,710	5,278,484
2000	1,016,041,221	802,479,710	5,146,207

Source: OGJ Worldwide Production Reports

RUSSIAN PRODUCTION

	1,000 b/d
2008	9,760*
2007	9,830
2006	9,498
2005	9,190
2004	8,887
2003	8,216
2002	7,405
2001	6,781
2000	6,325
1999	5,930

*Estimate.

Section 2:
Reserves and production

New estimates boost worldwide oil, gas reserves

Larger estimates for Venezuela increased worldwide oil and gas reserves from previous estimates, according to the annual survey of proved reserves, published in late 2008, by *Oil & Gas Journal.*

Worldwide oil production declined slightly in 2007. But preliminary estimates indicate that worldwide crude and condensate production increased 1.1% in 2008, spurred by robust demand in late 2007 and early 2008, and record high oil prices that peaked in July.

New estimates of the world's oil reserves total 1.34 trillion bbl, up 10.5 billion bbl from those reported a year ago.

Latest estimates of gas reserves total 6.254 quadrillion cu ft (177.1 tcm), up 68.67 tcf (1.94 tcm) from those reported a year ago.

OGJ compiles the estimates of proved reserves from an annual survey of official sources, including government agencies and ministries.

For additional details for individual countries, see Table 1: World Reserves and Production, in the Statistical Tables chapter of this book.

Production, reserves leaders

Worldwide production of crude oil, condensate, and NGLs in 2007 averaged 85.6 million b/d, according to the International Energy Agency (IEA; Paris), up just 100,000 b/d from the 2006 average.

Topping *OGJ's* OGJ100 list of production leaders is Saudi Aramco, with total production of 3.1 billion bbl in 2007. This compares to its total output in 2006 of 3.25 billion bbl.

OGJ100: OIL PRODUCTION AND RESERVE LEADERS

Rank	Company	Production, million bbl
1	Saudi Arabian Oil Co.	3,102.5
2	National Iranian Oil Co.	1,429.7
3	Petroleos Mexicanos	1,124.9
4	Petroleos de Venezuela SA	1,060.0
5	BP PLC	881.1
6	Abu Dhabi National Oil Co.	846.8
7	PetroChina Co. Ltd.	838.8
8	Nigerian National Petroleum Corp.	791.0
9	Kuwait Petroleum Corp.	788.4
10	Iraq National Oil Co.	759.2
11	OAO Rosneft	740.0
12	OAO Lukoil	713.0
13	Petroleo Brasileiro SA	700.2
14	Royal Dutch Shell	663.6
15	National Oil Corp. (Libya)	620.5
16	Sonangol	618.7
17	Total SA	550.8
18	Sonatrach	494.6
19	OJSC Surgutneftegas	474.3
20	StatoilHydro	391.0
Total		**17,589.1**

Rank	Company	Reserves, million bbl
1	Saudi Arabian Oil Co.	259,900.0
2	National Iranian Oil Co.	138,400.0
3	Iraq National Oil Co.	115,000.0
4	Kuwait Petroleum Corp.	101,500.0
5	Petroleos de Venezuela SA	99,377.0
6	Abu Dhabi National Oil Co.	92,200.0
7	National Oil Corp. (Libya)	41,464.0
8	Nigerian National Petroleum Corp.	36,220.0
9	OAO Rosneft	17,513.0
10	OAO Lukoil	15,715.0
11	Qatar Petroleum Corp.	15,207.0
12	Sonatrach	12,200.0
13	PetroChina Co. Ltd.	11,706.0
14	Petroleos Mexicanos	11,047.6
15	Petroleo Brasileiro SA	9,613.0
16	Sonangol	9,035.0
17	Total SA	5,778.0
18	Petroleum Development Oman LLC	5,500.0
19	BP PLC	5,492.0
20	Petronas	5,360.0
Total		**1,008,227.6**

TOP 20 IN SPENDING AND US NET WELLS DRILLED

Rank	Company	Capital, exploratory spending, $1,000
1	ExxonMobil Corp.	20,853,000
2	Chevron Corp.	16,678,000
3	ConocoPhillips	11,791,000
4	Devon Energy Corp.	6,158,000
5	Chesapeake Energy Corp.	5,305,000
6	Apache Corp.	4,802,343
7	Marathon Oil Corp.	4,466,000
8	Anadarko Petroleum Corp.	4,246,000
9	Hess Corp.	3,578,000
10	Occidental Petroleum Corp.	3,497,000
11	EOG Resources Inc.	3,401,986
12	Exco Resources Inc.	2,847,000
13	XTO Energy Inc.	2,668,000
14	El Paso Corp.	2,495,000
15	Pioneer Natural Resources Co.	2,067,648
16	Murphy Oil Corp.	1,949,219
17	Newfield Exploration Co.	1,930,000
18	Southwestern Energy Co.	1,519,433
19	Noble Energy Inc.	1,414,515
20	Questar Corp.	1,398,300
	Total	**103,065,444**

Rank	Company	US net wells drilled
1	Chesapeake Energy Corp.	1,919.0
2	XTO Energy Inc.	1,073.3
3	Devon Energy Corp.	1,015.1
4	EOG Resources Inc.	938.3
5	Anadarko Petroleum Corp.	926.5
6	Williams Cos. Inc.	904.0
7	Chevron Corp.	892.0
8	ConocoPhillips	823.0
9	Dominion Energy Inc.	814.0
10	Noble Energy Inc.	803.9
11	Range Resources Corp.	698.2
12	Pioneer Natural Resources Co.	619.0
13	Quest Resource Inc.	571.0
14	ExxonMobil Corp.	486.0
15	Occidental Petroleum Corp.	476.2
16	Atlas America Inc.	473.8
17	El Paso Corp.	465.0
18	Equitable Supply	456.3
19	Quicksilver Resources Inc.	432.8
20	Exco Resources Inc.	405.2
	Total	**15,192.6**

TOP 20 IN LIQUIDS RESERVES

Rank	Company	US liquids reserves, million bbl
1	ConocoPhillips	2,242.0
2	ExxonMobil Corp.	1,851.0
3	Occidental Petroleum Corp.	1,707.0
4	Chevron Corp.	1,624.0
5	Anadarko Petroleum Corp.	662.0
6	Apache Corp.	551.6
7	Devon Energy Corp.	452.0
8	Pioneer Natural Resources Co.	451.1
9	Plains Exploration & Production Co.	436.5
10	Cano Petroleum Inc.	423.3
11	XTO Energy Inc.	308.0
12	Noble Energy Inc.	207.4
13	Hess Corp.	204.0
14	Whiting Petroleum Corp.	196.3
15	Encore Acquisition Co.	188.6
16	Marathon Oil Corp.	166.0
17	EOG Resources Inc.	160.0
18	Denbury Resources Inc.	135.0
19	Kinder Morgan CO2 Co. LP	132.5
20	Chesapeake Energy Corp.	123.6
	Total	**12,221.9**

Rank	Company	Worldwide liquids reserves, million bbl
1	ExxonMobil Corp.	7,744.0
2	Chevron Corp.	7,087.0
3	ConocoPhillips	6,320.0
4	Occidental Petroleum Corp.	2,228.0
5	Apache Corp.	1,133.7
6	Anadarko Petroleum Corp.	1,014.0
7	Devon Energy Corp.	998.0
8	Hess Corp.	885.0
9	Marathon Oil Corp.	650.0
10	Pioneer Natural Resources Co.	469.7
11	Plains Exploration & Production Co.	436.5
12	Noble Energy Inc.	329.0
13	XTO Energy Inc.	308.0
14	Whiting Petroleum Corp.	196.3
15	Encore Acquisition Co.	188.6
16	EOG Resources Inc.	179.3
17	Murphy Oil Corp.	178.2
18	Denbury Resources Inc.	135.0
19	Kinder Morgan CO2 Co. LP	132.5
20	Chesapeake Energy Corp.	123.6
	Total	**30,736.4**

Earnings mixed for 2007

The OGJ200 group of companies—a compilation by *OGJ*—recorded a decline in earnings in 2007 as greater expenses ate into returns.

The collective 2.5% dip in net income accompanied a 7% swell in revenues.

High costs for labor, equipment, and supplies underpinned the group's capital and exploration spending.

Top 20 in total revenue

Rank	Company	Total revenue, $1,000
1	ExxonMobil Corp.	404,552,000
2	Chevron Corp.	220,904,000
3	ConocoPhillips	194,495,000
4	Marathon Oil Corp.	65,207,000
5	Hess Corp.	31,924,000
6	Occidental Petroleum Corp.	20,013,000
7	Murphy Oil Corp.	18,439,098
8	Anadarko Petroleum Corp.	15,892,000
9	Devon Energy Corp.	11,362,000
10	Apache Corp.	9,977,858
11	Chesapeake Energy Corp.	7,815,000
12	XTO Energy Inc.	5,513,000
13	El Paso Corp.	4,648,000
14	EOG Resources Inc.	4,190,791
15	Dominion Energy Inc.	3,527,000
16	Noble Energy Inc.	3,272,030
17	Questar Corp.	2,740,900
18	Williams Cos. Inc.	2,093,000
19	Pioneer Natural Resources Co.	1,833,349
20	Newfield Exploration Co.	1,783,000
	Total	**1,030,182,026**

Top 20 in liquids production

Rank	Company	US liquids production, million bbl
1	Chevron Corp.	168.0
2	ConocoPhillips	166.0
3	ExxonMobil Corp.	111.0
4	Occidental Petroleum Corp.	95.0
5	Anadarko Petroleum Corp.	64.0
6	Devon Energy Corp.	41.0
7	Apache Corp.	35.9
8	Marathon Oil Corp.	23.0
9	XTO Energy Inc.	22.1
10	Plains Exploration & Production Co.	18.1
11	Noble Energy Inc.	15.5
12	Hess Corp.	15.0
13	Kinder Morgan CO2 Co. LP	15.0
14	Pioneer Natural Resources Co.	13.6
15	EOG Resources Inc.	13.0
16	Dominion Energy Inc.	11.6
17	Denbury Resources Inc.	10.2
18	Chesapeake Energy Corp.	9.9
19	Whiting Petroleum Corp.	9.6
20	Encore Acquisition Co.	9.5
	Total	**867.1**

Rank	Company	Worldwide liquids production, million bbl
1	ExxonMobil Corp.	801.0
2	Chevron Corp.	641.0
3	ConocoPhillips	506.0
4	Occidental Petroleum Corp.	166.0
5	Hess Corp.	100.0
6	Anadarko Petroleum Corp.	96.0
7	Apache Corp.	95.6
8	Devon Energy Corp.	81.0
9	Marathon Oil Corp.	72.0
10	Noble Energy Inc.	30.8
11	Murphy Oil Corp.	28.7
12	XTO Energy Inc.	22.1
13	Plains Exploration & Production Co.	18.1
14	Pioneer Natural Resources Co	16.2
15	EOG Resources Inc.	15.8
16	Kinder Morgan CO2 Co. LP	15.0
17	Dominion Energy Inc.	12.2
18	Newfield Exploration Co.	10.4
19	Denbury Resources Inc.	10.2
20	Chesapeake Energy Corp.	9.9
	Total	**2,747.9**

Top 20 in gas production

Rank	Company	US gas production, bcf
1	ConocoPhillips	948.0
2	Anadarko Petroleum Corp.	698.0
3	Chesapeake Energy Corp.	655.0
4	ExxonMobil Corp.	641.0
5	Devon Energy Corp.	635.0
6	Chevron Corp.	620.0
7	XTO Energy Inc.	532.1
8	EOG Resources Inc.	438.9
9	Williams Cos. Inc.	334.0
10	Apache Corp.	280.9
11	El Paso Corp.	238.0
12	Occidental Petroleum Corp.	216.0
13	Dominion Energy Inc.	206.0
14	Newfield Exploration Co.	185.2
15	Marathon Oil Corp.	174.0
16	Noble Energy Inc.	150.5
17	Pioneer Natural Resources Co.	132.8
18	Questar Corp.	121.9
19	Cimarex Energy Co.	119.9
20	Southwestern Energy Co.	109.9
	Total	**7,437.1**

Rank	Company	Worldwide gas production, bcf
1	ExxonMobil Corp.	2,683.0
2	ConocoPhillips	2,110.0
3	Chevron Corp.	1,832.0
4	Devon Energy Corp.	863.0
5	Anadarko Petroleum Corp.	698.0
6	Chesapeake Energy Corp.	655.0
7	EOG Resources Inc.	637.7
8	XTO Energy Inc.	532.1
9	Williams Cos. Inc.	334.0
10	Marathon Oil Corp.	319.0
11	Occidental Petroleum Corp.	261.0
12	Noble Energy Inc.	250.9
13	El Paso Corp.	242.0
14	Hess Corp.	241.0
15	Dominion Energy Inc.	214.0
16	Apache Corp.	204.9
17	Newfield Exploration Co.	185.2
18	Pioneer Natural Resources Co.	151.1
19	Questar Corp.	121.9
20	Cimarex Energy Co.	119.9
	Total	**12,655.7**

Top 20 in Gas Reserves

Rank	Company	US gas reserves, bcf
1	ExxonMobil Corp.	13,172.0
2	ConocoPhillips	12,634.0
3	Chesapeake Energy Corp.	10,137.3
4	XTO Energy Inc.	9,441.1
5	Anadarko Petroleum Corp.	8,504.0
6	Devon Energy Corp.	7,143.0
7	EOG Resources Inc.	5,180.2
8	Williams Cos. Inc.	4,143.0
9	Chevron Corp.	3,677.0
10	Pioneer Natural Resources Co.	2,903.1
11	Ultra Petroleum	2,842.7
12	Apache Corp.	2,699.0
13	Occidental Petroleum Corp.	2,672.0
14	Equitable Supply	2,669.9
15	El Paso Corp.	2,248.0
16	Noble Energy Inc.	1,840.4
17	Range Resources Corp.	1,832.8
18	Newfield Exploration Co.	1,810.0
19	Questar Corp.	1,668.5
20	Cabot Oil & Gas Corp.	1,560.0
	Total	**98,777.8**

Rank	Company	Worldwide gas reserves, bcf
1	ExxonMobil Corp.	32,610.0
2	ConocoPhillips	25,438.0
3	Chevron Corp.	22,140.0
4	Chesapeake Energy Corp.	10,137.3
5	XTO Energy Inc.	9,441.1
6	Devon Energy Corp.	8,994.0
7	Anadarko Petroleum Corp.	8,504.0
8	EOG Resources Inc.	7,745.1
9	Williams Cos. Inc.	4,143.0
10	Occidental Petroleum Corp.	3,843.0
11	Marathon Oil Corp.	3,450.0
12	Noble Energy Inc.	3,307.5
13	Pioneer Natural Resources Co.	2,964.4
14	Ultra Petroleum	2,842.7
15	Equitable Supply	2,669.9
16	Hess Corp.	2,668.0
17	Apache Corp.	2,445.8
18	El Paso Corp.	2,299.0
19	Range Resources Corp.	1,832.8
20	Newfield Exploration Co.	1,810.0
	Total	**159,285.5**

Comparison of Heavy Oil Distillation, Thermal Cracking

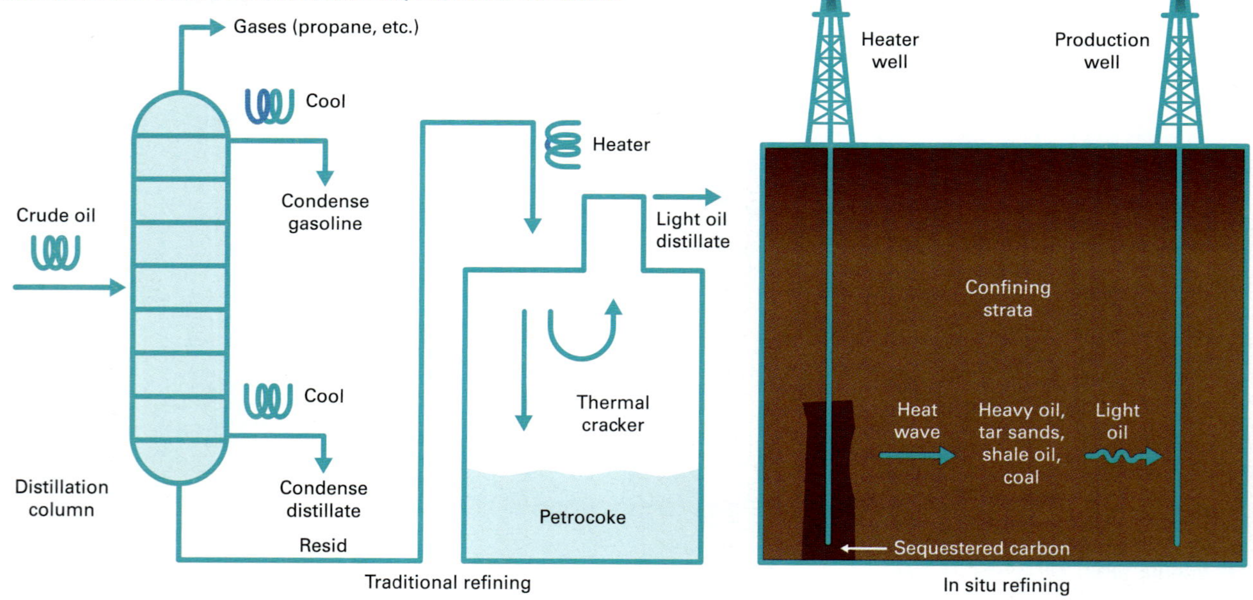

Nuclear heat for in *situ* refining

At a mid-2008 technical conference, an engineer from Massachusetts Institute of Technology proposed a novel use of heat from a nuclear reactor to refine hydrocarbon feedstocks *in situ* (underground).

In addition to processing heavy oils, tar sands, oil shale and coal to produce light distillates, underground high-temperature heating (near 700° C) might also recover remaining oil in depleted fields.

Floating production systems

International Maritime Associates (IMA; Washington, DC) noted the growing use of floating production systems for oil and gas production from fields otherwise not commercially feasible.

Such facilities can handle 250,000 bo/d of production from a complex subsea network of wells in water deeper than 3,281 ft (1,000 m).

They have enabled such countries as Angola, Nigeria, and Brazil to become major global oil producers and have kept alive production in the North Sea and Gulf of Mexico.

Oil shale heating underground with nuclear heat*

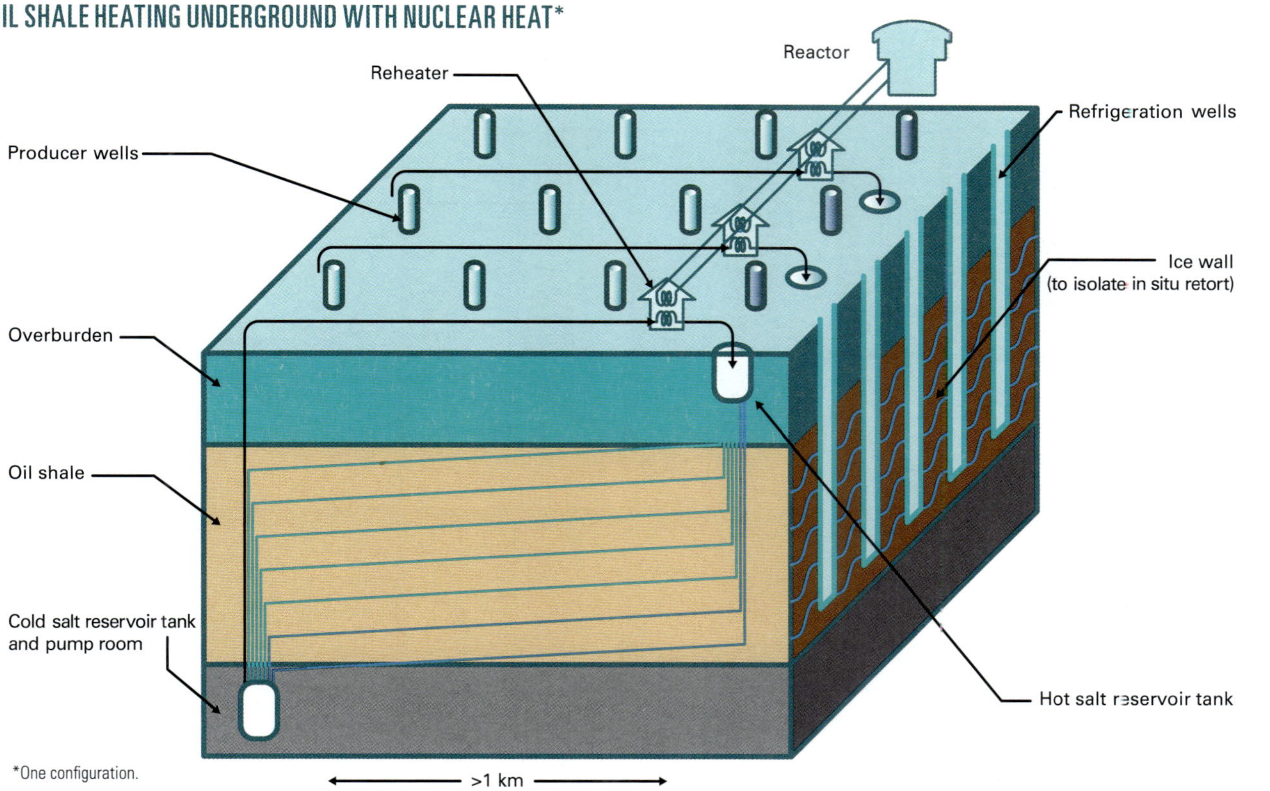

*One configuration.

Operating production, storage floaters

In service as of March 2008 are 212 floating production systems and 86 storage vessels.

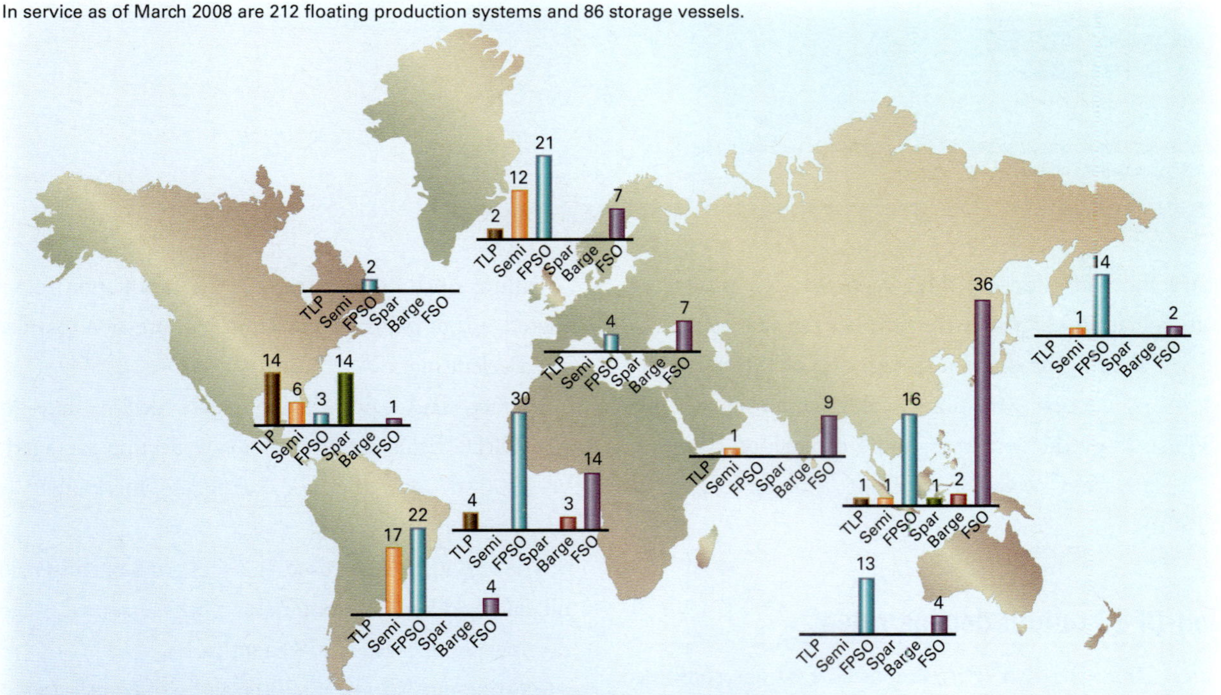

Note: Not shown in the bar charts are six FPSOs, two production semis, two FSOs that are off site and available for reuse. The three FPSOs in the Gulf of Mexico are off Mexico and two of these are early well test units.
Source: IMA, Floating Production Systems, March 2008

Planned, under study floater projects

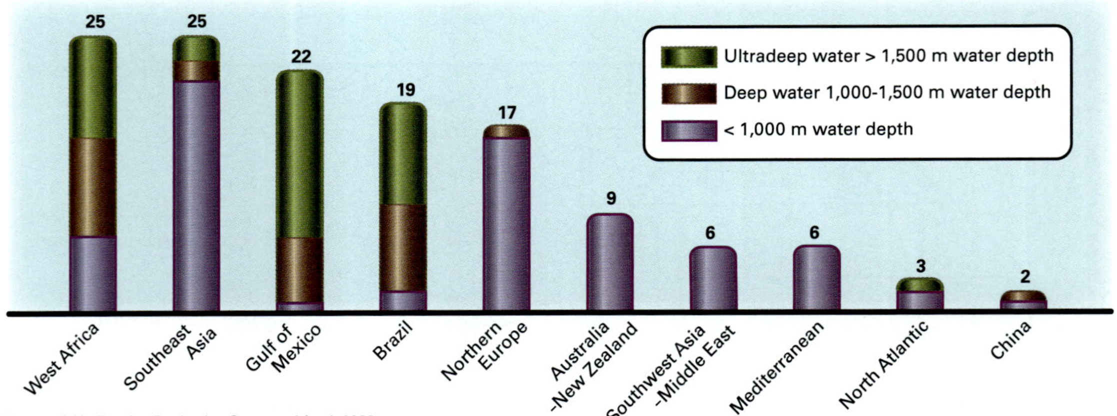

Source: IMA, Floating Production Systems, March 2008

Average non-OPEC decline rates, 2000-07*

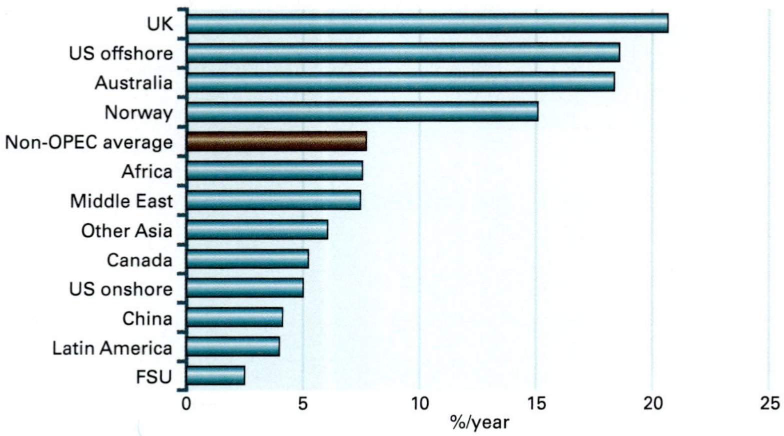

*Conventional crude and condensate production from mature fields in sustained decline.
Source: International Energy Agency

In a March 2008 update, IMA reported 212 production floaters in operation, 2.5 times the number ten years ago. This included 131 floating production storage and offloading vessels (FPSOs), 40 production semisubmersibles, 21 tension leg platforms (TLPs), 15 production spars, and 5 production barges.

Also in operation were 86 floating storage offloading vessels (FSOs).

IEA: Non-OPEC output decline rates

In March 2008, the IEA (Paris) reported that the combined production decline rate for countries outside OPEC was steadier and less steep than generally believed.

Failure since 2004 of non-OPEC production to achieve predicted output gains has created the impression that decline rates are accelerating.

In fact, IEA said, the aggregate decline rate for mature non-OPEC fields already in steady decline averaged 7.7% per yr 2000–07 and did not "accelerate markedly" during the study period.

The IEA analysis focused on production of conventional crude oil and condensate from fields with sustained, yearly output declines for at least 12–18 months.

Decline-rate surges in 2005 and 2006 reflected distortions such as those caused by hurricane damage in the Gulf of Mexico, extended shut-ins in the North Sea, and asset divestments in Russia.

Observed production decline rates*

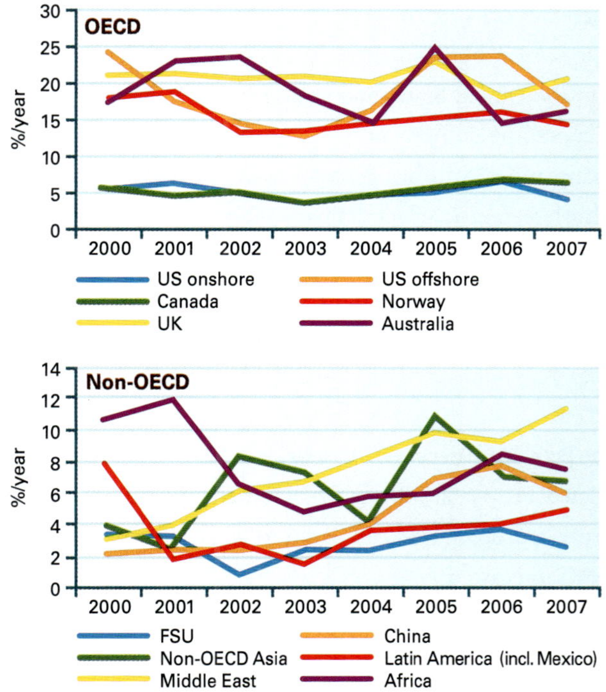

*Conventional crude and condensate from mature fields in steady decline.
OECD is Organization for Economic Cooperation and Development.
Source: International Energy Agency

CERA-IHS: Global oil field decline rate

The aggregate global decline rate for producing oil fields is 4.5% per yr rather than the 8% per yr sometimes cited in other studies, said CERA and IHS Inc.

The estimate was contained in their January 2008 report, *"Finding the Critical Numbers."*

The analysis was based on production characteristics of 811 oil fields that account for about two-thirds of current global production and half of the total estimated proved and probable conventional oil reserves base.

Annual field decline rates are not increasing with time, said CERA, which "provides the basis for more confidence about the future availability of oil."

Offshore fields held by producers outside the Organization of Petroleum Exporting Countries decline 5% per yr compared with 12% per yr for offshore fields belonging to OPEC members.

CERA said global liquids capacity of about 91 million b/d in 2007 could climb to 112 million b/d by 2017.

2008 OPEC production capacity

Production projects due on stream in 2008 by OPEC members represent peak gross capacity additions of 3.1 million b/d of crude oil and other liquids, said IEA in February 2008.

In 2007, OPEC members started up projects with peak total-liquids capacities of 1.25 million b/d.

IEA estimated sustainable OPEC capacity to produce crude oil—the output level that members can reach with 30 days and maintain for at least 90 days—at 35.04 million b/d.

IEA said it expected OPEC output of NGLs in 2008 to average 5.18 million b/d, up 365,000 b/d.

Main OPEC capacity projects due on line in 2008

Country	Project	Peak crude	Peak NGL	Peak condensate
		1,000 b/d		
Angola	Kizomba C-Mondo	100		
Angola	Kizomba C-Saxi/Batuque	100		
Indonesia	Kerisi-Hiu			20
Iran	Darkhovin II	110		
Iran	Khesht	35		
Iran	Jufeyr I	25		
Iran	Salman	50		
Iran	Masjid e Suleiman expansion	15		
Iran	Azadegan I	50		
Iran	Pars 6-8			120
Iran	Pars 9-10			80
Iran	Pars 9-10		16	
Iraq	Taq Taq	20		
Iraq	Majnoon increase	50		
Kuwait	Burgan water treatment, etc.	40		
Kuwait	Sabriyah			50
Libya	NC 186 ramp-up	20		
Libya	Waha EOR	20		
Libya	El Shahara ramp-up	10		
Libya	Elephant ramp-up	25		
Nigeria	Akpo			180
Nigeria	Agbami	250		
Qatar	Al Shaheen increments	100		
Qatar	RasGas Train 6			50
Qatar	RasGas Train 6		25	
Qatar	Qatargas Train 4			160
Saudi Arabia	Khursaniyah	500		
Saudi Arabia	Khursaniyah			80
Saudi Arabia	Khursaniyah		220	
Saudi Arabia	Shaybah	200		
Saudi Arabia	Hawiyah NGL		300	
Venezuela	Corocoro	75		
Total		**1,795**	**561**	**740**

Source: International Energy Agency

IOR-EOR: Lowest-cost reserves additions

IPC Petroleum Consultants Inc. (Tulsa) said in April 2008 that improved or enhanced oil recovery (IOR-EOR) is now the least costly method for adding reserves, unlike in past years.

The firm estimated that IOR-EOR could add reserves for a capital expenditure of $2–4/bbl.

In contrast, the firm noted that $4.30–6.25/bbl capex is needed for heavy oil production, $4–6/bbl for deepwater, and $12.86/bbl for acquisitions.

Global finding and development costs were pegged at $14.42/bbl.

IPC estimated that the $200–400 billion capital spending needed for adding 100 billion bbl of IOR-EOR reserves is in the same range as the industry's current global spending of $260 billion/year.

Section 3:
Drilling and exploration

Beyond North America

In mid-2008, *OGJ* highlighted trends and expectations for oil and gas drilling in areas other than North America.

The pace of drilling in the rest of the world was up about 7% from January through May 2008, according to international statistics on active rigs.

Year-over-year gains for January through May were posted in all geographic regions.

The rest of the world excluding the United States and Canada had 1,058 rigs at work at the beginning of June 2008.

Not a single country outside the United States and Canada averaged even 100 rigs working in January through May 2008, according to Baker Hughes. The closest were: Mexico with 99 active rigs; Indonesia with 83; Argentina with 82; Venezuela with 81; and Saudi Arabia with 78.

Canadian drilling

Canadian drilling continued to decline in late 2008, said *Oil & Gas Journal*, but Canadian drillers were diversifying, sending rigs to the United States and abroad, purchasing U.S. rig fleets, and upgrading equipment.

Dampening drilling activity were a higher government royalty rate taking effect January 2009 in Alberta, falling oil prices, and the crash of financial and credit markets.

At the same time, Canada's oil exports increased to nearly $31 billion (Can.) in first-half 2008. Some observers said that oil exports could reach an all-time high of around $60 billion for 2008.

The Petroleum Services Association of Canada (PSAC; Calgary) predicted in mid-2008 that 16,500 wells would be drilled for the year, a drop of 11% compared to 2007 (18,557 wells).

PSAC forecasted:

- 11,815 wells in Alberta
- 820 wells in British Columbia
- 3,445 wells in Saskatchewan
- 360 wells in Manitoba
- 40 wells in Ontario
- 14 wells in eastern Canada (Quebec, New Brunswick, Nova Scotia, Prince Edward Island, and Newfoundland-Labrador)
- 6 wells in northern Canada (Yukon, Northwest Territories, Nunavut).

Western Canada Drilling Activity

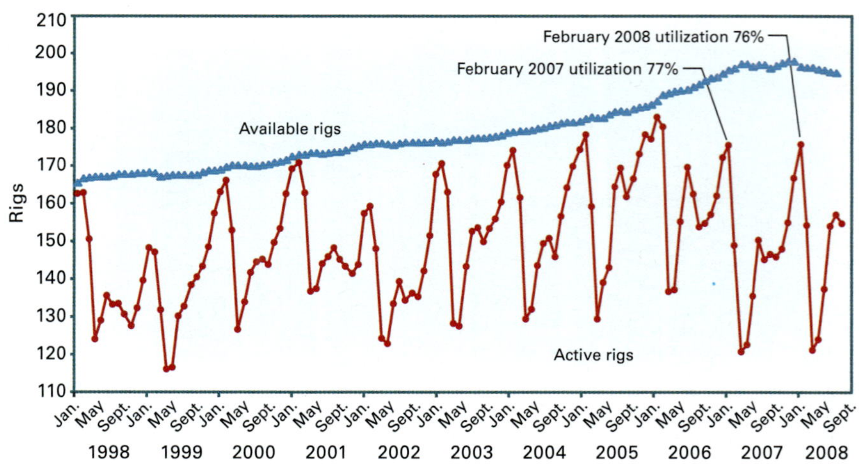

Source: Nickle's Energy Group, Rig Locator

Nickle's Rig Locator reported that the busiest operators drilling in Canada in mid-October 2008 were:

- Husky Energy Inc., 32 rigs
- EnCana Corp., 32 rigs
- Talisman Energy Inc., 25 rigs
- ConocoPhillips Canada Ltd., 20 rigs
- Canadian Natural Resources Ltd., 19 rigs.

Oil, gas price impact

Crude oil prices hovered just below $100/bbl in fourth-quarter 2007, spurring drilling as well as investment in capital-intensive projects.

As the prices rose through mid-2008, more new rigs were ordered and day rates continued to increase.

However, prices dropped precipitously through year end, to a level of about $35/bbl in late December.

Turmoil in the U.S. financial system drove speculation about a looming recession and slowed demand for energy. On September 29, Deutsche Bank lowered its 2009 New York oil price forecast by 23% to $92.50/bbl because of the economic crisis.

Driven by high natural gas prices, operators pursued hydraulic fracturing of new U.S. and Canadian shale plays.

Western Canadian oil, gas wells

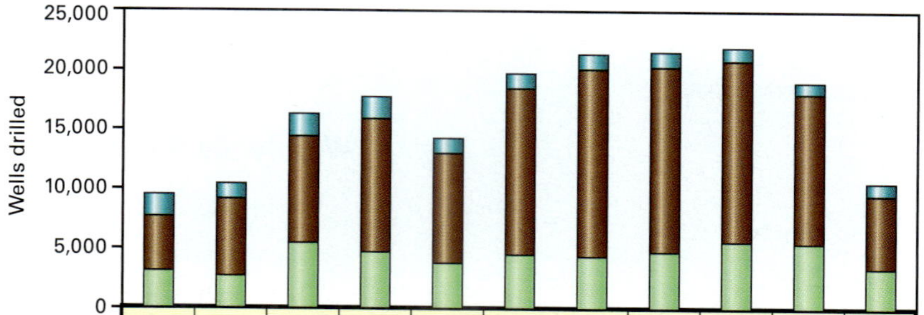

	1998	1999	2000	2001	2002	2003	2004	2005	2006	2007	*2008
% DRY	19.3	13.6	11.7	10	9.1	6.3	5.9	6.5	4.9	4.9	9.7
DRY	1,850	1,427	1,903	1,759	1,289	1,233	1,266	1,414	1,072	935	1,015
GAS	4,585	6,309	8,934	11,177	9,073	13,944	15,645	15,359	15,289	12,621	6,114
OIL	3,142	2,730	5,462	4,689	3,832	4,473	4,427	4,822	5,599	5,429	3,332

*Through September 2008.
Sources: Nickle's Daily Oil Bulletin; Canadian Assoc. of Oilwell Drilling Contractors

World rig activity

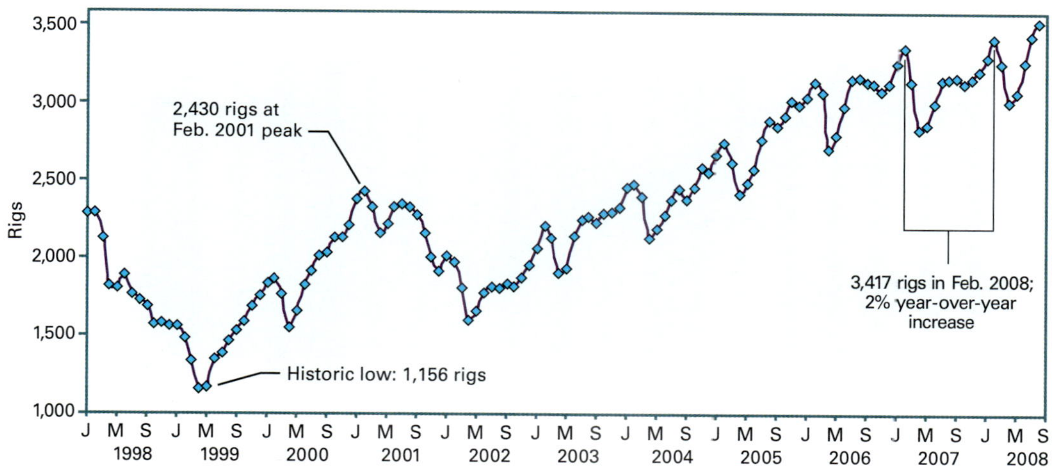

Source: Baker Hughes Inc.

North American, international rig count

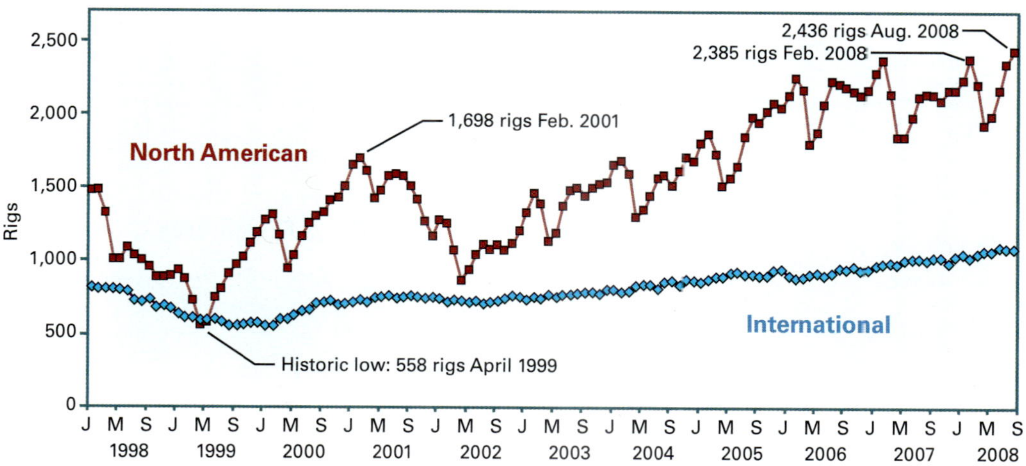

Source: Baker Hughes Inc.

International Rig Count

Average number of active rigs per month, year to date

	Jan. 2007– May 2007	Jan. 2008– May 2008
WESTERN HEMISPHERE		
Argentina	86	82
Bolivia	3	2
Brazil	39	46
Chile	1	1
Colombia	30	40
Ecuador	11	7
Mexico	89	99
Peru	7	7
Trinidad	7	6
Venezuela	77	81
Other	2	1
Total Western Hemisphere*	**353**	**373**
ASIA-PACIFIC		
Australia	20	26
Brunei	4	3
China-offshore	18	20
India	84	83
Indonesia	51	65
Japan	2	3
Malaysia	16	11
Myanmar (Burma)	10	8
New Zealand	4	5
Papua New Guinea	3	4
Philippines	0	1
Taiwan	0	0
Thailand	11	11
Vietnam	7	7
Other	4	3
Total Asia-Pacific	**234**	**249**
AFRICA		
Algeria	25	27
Angola	5	5
Congo	3	2
Gabon	3	2
Kenya	0	0
Libya	12	15
Nigeria	7	9
South Africa	0	0
Tunisia	3	4
Other	6	2
Total Africa	**64**	**67**
MIDDLE EAST		
Abu Dhabi	13	12
Dubai	1	1
Egypt	42	52
Iran	0	0
Iraq	0	0
Jordan	1	0
Kuwait	13	12
Oman	45	54
Pakistan	19	20
Qatar	11	11
Saudi Arabia	76	78
Syria	22	21
Yemen	14	14
Other	1	1
Total Middle East	**258**	**275**
EUROPE		
Croatia	1	0
Denmark	3	2
France	1	1
Germany	5	8
Hungary	2	3
Italy	4	5
Netherlands	5	3
Norway	19	18
Poland	2	2
Romania	2	20
Turkey	5	5
United Kingdom	27	20
Other	5	7
Total Europe	**80**	**94**
Total World	**989**	**1,058**

*Excludes US and Canada. Source: Baker Hughes Inc.

In late 2008, operators of new pipeline projects in Australia and elsewhere were increasing drilling in order to maintain a steady supply of natural gas to feed LNG trains.

Worldwide drilling activity

Worldwide drilling set a record in August 2008, with 3,523 rigs operating, based on the Baker Hughes Inc. (Houston) international rig count.

The North American rig count reached a historic high of 2,436 rigs in August 2008, up 13% from a year earlier and diverging from a pattern set over the last five years.

The seasonal cycle in which drilling spikes in February is driven by Canadian operations.

The Baker Hughes international rotary rig count (excluding North America, Iran, and Sudan) reached a record high of 1,102 rigs in June 2008, up 8.6% from 1,015 rigs a year earlier.

MODUs* under construction

	Jack ups	Semisubmersibles	Drillships	Tenders	Total
US Gulf of Mexico	12	—	—	—	12
Latin America	2	2	2	—	6
North Sea	0	2	—	—	2
Europe - other	0	3	—	—	3
Mediterranean	0	1	—	—	1
Middle East	15	3	—	2	20
India	3	—	—	—	3
Southeast Asia	35	25	—	4	64
Far East	16	18	41	—	75
Caspian Sea	—	1	—	—	1
Russian Arctic	1	1	—	—	2
Total	**84**	**56**	**43**	**6**	**189**
MODUs under construction by delivery date					
2008	17	9	1	—	27
2009	39	19	6	2	66
2010	23	13	15	3	31
2011	5	10	17	1	28
Undetermined	—	5	4	—	—
Total	**84**	**56**	**43**	**6**	**189**
MODUs under construction with no drilling contract					
2008	3	—	—	—	3
2009	28	—	—	2	30
2010	19	4	5	2	30
2011	4	5	12	1	22
Undetermined	—	3	—	—	3
Total	**54**	**12**	**17**	**5**	**88**

*MODU: mobile offshore drilling unit. Source: ODS-Petrodata Inc., Offshore Rig Locator, Sept. 1, 2008

Offshore rig fleets: October 2008

Rig type	Current fleet size	Units under construction	Future fleet size
Jack ups	379	84	463
Semisubmersibles	160	56	216
Drillships	38	43	81
Tenders	26	6	32
Drill barges	11	—	11
Submersibles	6	—	6
Total	**620**	**189**	**809**

Sources: Rigzone; ODS-Petrodata Inc.; shipyard and operator announcements

International Rig Count by Region

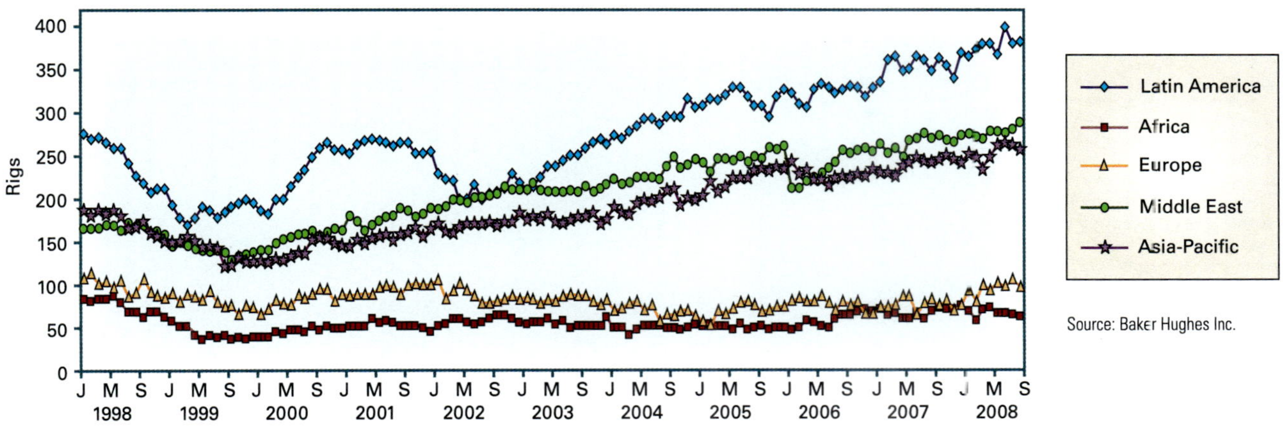

Source: Baker Hughes Inc.

Top 50 US Operators in 2008

Rank[1]	Company	Footage, drilled	% of total	Avg. footage	Well starts[2]	% of total	Dir. wells
1	Chesapeake Operating Inc.	12,653,336	5.4	9,250	1,371	4.0	922
2	XTO Energy Inc.	9,139,878	3.9	9,828	932	2.7	388
3	EOG Resources Inc.	8,985,167	3.8	9,311	970	2.8	549
4	Devon Energy Production Co. LP	7,214,821	3.1	9,121	797	2.3	622
5	EnCana Oil & Gas (USA) Inc.	6,461,157	2.7	10,388	624	1.8	406
6	Pioneer Natural Resources USA Inc.	4,154,350	1.8	9,684	435	1.3	77
7	Noble Energy Inc.	4,103,167	1.7	5,912	697	2.0	147
8	BP America Production Co.	3,935,948	1.7	9,327	436	1.3	298
9	Williams Production RMT Co.	3,748,284	1.6	6,429	585	1.7	418
10	Kerr-McGee Oil & Gas Onshore LP	3,707,453	1.6	8,369	444	1.3	239
11	Apache Corp.	3,245,238	1.4	9,864	370	1.1	56
12	Burlington Resources Oil & Gas Co. LP	3,243,868	1.4	9,951	335	1.0	171
13	SandRidge Energy Inc.	2,850,113	1.2	11,681	244	0.7	42
14	Forest Oil Corp.	2,681,748	1.1	11,036	244	0.7	31
15	Questar E&P Co.	2,637,255	1.1	12,210	218	0.6	101
16	Samson Lone Star LLC	2,571,075	1.1	12,361	209	0.6	54
17	ConocoPhillips Co.	2,544,455	1.1	6,858	383	1.1	97
18	Atlas Resources Inc.	2,441,997	1.0	6,326	397	1.2	9
19	HighMount E&P Texas LLC	2,357,455	1.0	7,938	297	0.9	1
20	Ultra Resources Inc.	1,781,680	0.8	13,498	133	0.4	112
21	Anadarko E&P Co. LP	1,706,947	0.7	8,286	206	0.6	74
22	Quicksilver Resources Inc.	1,582,400	0.7	8,462	187	0.5	182
23	El Paso E&P Co. LP	1,551,868	0.7	5,705	278	0.8	61
24	Cimarex Energy Co.	1,428,592	0.6	11,710	122	0.4	52
25	Cabot Oil & Gas Production Corp.	1,420,500	0.6	8,211	178	0.5	30
26	SEECO Inc.	1,393,796	0.6	4,052	345	1.0	318
27	Range Production Co.	1,388,155	0.6	8,024	173	0.5	66
28	Endeavor Energy Resources LP	1,358,766	0.6	9,119	149	0.4	3
29	Petroleum Development Corp.	1,327,375	0.6	6,412	207	0.6	49
30	St. Mary Land & Exploration Co.	1,316,097	0.6	9,401	142	0.4	40
31	Chevron USA Inc.	1,310,164	0.6	7,799	475	1.4	69
32	Bill Barrett Corp.	1,303,320	0.6	6,062	215	0.6	135
33	Henry Petroleum LP	1,286,258	0.5	10,809	119	0.3	0
34	Berry Petroleum Co.	1,257,522	0.5	3,881	329	1.0	95
35	Newfield Production Co.	1,250,367	0.5	6,946	183	0.5	58
36	Mariner Energy Inc.	1,201,875	0.5	10,926	123	0.4	14
37	Energen Resources Corp.	1,199,199	0.5	4,835	252	0.7	31
38	Whiting Oil & Gas Corp.	1,165,943	0.5	8,511	138	0.4	56
39	Marathon Oil Co.	1,142,143	0.5	12,551	149	0.4	137
40	Swift Energy Operating LLC	1,125,589	0.5	10,327	109	0.3	47
41	Cimarex Energy Co. of Colorado	1,110,088	0.5	8,741	134	0.4	44
42	Unit Petroleum Co.	1,096,770	0.5	9,706	113	0.3	23
43	Newfield Explor. Mid-Continent Inc.	1,080,198	0.5	10,802	101	0.3	91
44	Penn Virginia Oil & Gas LP	1,078,900	0.5	13,157	82	0.2	44
45	Shell Rocky Mountain Prod. LLC	1,056,240	0.4	13,370	79	0.2	76
46	Continental Resources Inc.	1,032,276	0.4	10,120	159	0.5	143
47	ExxonMobil Development Co.	1,025,291	0.4	12,658	83	0.2	63
48	KCS Resources Inc.	1,015,247	0.4	9,146	111	0.3	34
49	BP Exploration (Alaska) Inc.	933,801	0.4	8,413	140	0.4	129
50	Goodrich Petroleum Co.	929,300	0.4	12,069	77	0.2	7
Total		**54.4**			**44.2**	**6,911**	

[1]Rank according to number of wells drilled. [2]Wells deeper than 2,500 ft.

Source: RigData; data from January-September 2008

US ONSHORE AND OFFSHORE DRILLING ACTIVITY

	Baker Hughes's rig count				RigData's working rigs				
	4-week avg.		Week, as of		4-week avg.		Week, as of		Waiting
State, district	2007	2008	9/14/08	9/21/08	2007	2008	9/14/08	9/21/08	to spud
Alabama	4	5	4	4	4	4	4	3	1
Offshore	1	1	1	1	—	—	—	—	—
Total	**5**	**6**	**5**	**5**	**4**	**4**	**4**	**3**	**1**
Alaska	5	8	8	8	12	14	15	15	1
Offshore	—	3	3	3	—	2	2	2	—
Total	**5**	**11**	**11**	**11**	**12**	**16**	**17**	**17**	**—**
Arkansas	49	59	59	59	52	67	66	64	4
California	34	47	46	46	39	43	42	40	1
Offshore	2	1	1	1	—	—	—	—	—
Total	**36**	**48**	**47**	**47**	**39**	**43**	**42**	**40**	**—**
Colorado	116	115	116	109	117	130	129	126	4
Florida	—	1	1	1	—	1	1	1	—
Offshore	1	2	2	2	—	—	—	—	—
Total	**1**	**3**	**3**	**3**	**—**	**1**	**1**	**1**	**—**
Illinois	1	—	—	1	7	10	10	11	—
Indiana	2	2	2	2	4	2	1	2	—
Kansas	14	11	10	12	50	59	55	59	4
Kentucky	11	12	12	12	11	14	13	17	—
Louisiana									
North, land	62	85	89	85	71	105	102	108	3
South, land	25	26	25	26	42	36	36	34	5
Offshore	56	55	56	51	46	59	61	58	4
Inland waters	25	19	21	14	39	36	37	35	—
Total	**168**	**185**	**191**	**176**	**198**	**236**	**236**	**235**	**12**
Michigan	2	2	2	2	11	11	10	10	—
Mississippi	11	16	17	16	24	20	19	19	1
Montana	13	11	10	10	18	14	15	11	2
Nebraska	—	—	—	—	5	1	—	1	—
Nevada	3	3	3	4	2	1	1	1	—
New Mexico	74	91	91	91	67	93	95	93	—
New York	6	7	7	8	6	9	9	8	—
North Dakota	43	74	73	75	44	88	85	88	4
Ohio	14	10	10	10	12	14	14	14	—
Oklahoma	195	212	211	202	218	246	240	244	12
Oregon	—	1	1	1	—	—	—	—	—
Pennsylvania	16	27	27	27	43	59	63	61	—
South Dakota	2	1	1	2	2	1	1	1	—
Tennessee	5	2	2	2	4	2	1	1	1
Texas	168								
District 1	26	27	27	25	27	28	30	28	3
District 2	33	36	35	38	42	42	44	44	—
District 3	56	62	59	61	63	77	76	79	10
District 4	86	91	93	87	93	98	99	100	12
District 5	186	187	185	189	186	188	186	186	13
District 6	122	132	135	141	128	146	148	146	9
District 7b	37	31	29	30	51	50	46	51	4
District 7c	59	71	71	69	62	73	69	69	3
District 8	110	132	130	129	126	150	149	149	9
District 8a	19	27	29	26	24	32	30	32	4
District 9	37	42	43	44	44	57	56	58	1
District 10	60	99	100	100	70	96	96	94	3
Inland waters	1	1	—	—	2	4	4	4	—
Offshore	6	11	11	11	10	12	12	13	—
Total	**831**	**937**	**936**	**939**	**916**	**1037**	**1029**	**1036**	**71**
Utah	42	45	42	43	53	61	62	61	2
Virginia	3	6	6	6	3	4	4	4	—
Washington	1	1	1	1	1	1	1	1	—
West Virginia	33	28	28	28	33	35	33	33	1
Wyoming	78	80	83	80	135	146	142	148	2
US deep water	—	—	—	—	37	42	43	41	1
Total	**1,787**	**2,018**	**2,018**	**1,995**	**2,140**	**2,483**	**2,457**	**2,468**	**124**

Source: RigData; a summary of data presented in the Sept. 26, 2008 edition of "LOCATION & OPERATORS" report

2008 Utilization of Top 25 US Drilling Contractors—Onshore and Offshore*

Rank	Company	Jan.	Feb.	Mar.	Apr.	May	June	July	Aug.	Avg. monthly rig use	2008 fleet	2007 fleet
1	Patterson-UTI Drilling Co. LP	242	243	247	250	245	251	264	269	251	331	324
2	Nabors Drilling USA LP	234	223	229	234	245	250	254	263	242	334	319
3	Helmerich & Payne IDC	159	160	165	168	173	174	180	185	171	194	164
4	Grey Wolf Drilling Co. LP	101	100	102	101	102	105	110	111	104	119	123
5	Unit Drilling Co.	73	71	69	68	71	71	72	74	71	88	81
6	Nomac Drilling LLC	1	2	3	4	7	18	62	62	20	64	58
7	Pioneer Drilling Co.	57	56	57	61	60	59	63	63	60	64	64
8	Ensign United States Drilling Inc.	53	52	51	51	51	51	51	51	51	56	56
9	Union Drilling Inc.	35	33	35	38	38	42	44	45	39	57	52
10	Trinidad Drilling LP	38	41	44	44	44	44	44	45	43	46	39
11	Bronco Drilling Co. Inc.	35	34	37	39	41	41	39	38	38	54	56
12	Cactus Drilling Co. LLC	39	40	40	40	41	41	41	41	40	43	38
13	Lariat Services Inc.	36	37	38	38	38	39	40	40	38	47	43
14	Hercules Drilling Co. LLC	27	23	29	26	31	34	35	35	30	56	55
15	Goober Drilling LLC	34	33	33	34	35	35	35	36	34	37	31
16	Key Energy Services Inc.	15	15	21	21	25	26	17	19	20	44	16
17	Unit Texas Drilling LLC	28	27	26	28	29	30	30	31	29	35	38
18	Rowan Drilling Co. Inc.	25	25	25	26	27	28	28	29	27	31	28
19	Cyclone Drilling Inc.	19	18	20	21	24	23	24	25	22	27	27
20	Precision Drilling Oilfield Services Inc.	11	13	14	14	17	18	21	25	17	27	7
21	Capstar Drilling LP	24	25	25	26	26	26	26	26	26	27	24
22	Nabors Well Services Co.	13	14	14	16	19	19	17	16	16	26	73
23	Big Dog Drilling Co.	23	23	23	22	23	24	24	25	23	25	23
24	Nabors Offshore Corp.	13	13	14	13	18	17	17	17	15	44	42
25	Felderhoff Brothers Drilling LLC	18	21	20	20	19	20	22	22	20	23	22
	Active rigs	**1,353**	**1,342**	**1,381**	**1,403**	**1,449**	**1,486**	**1,560**	**1,593**	**1,446**	**1,899**	**1,803**

*Some rigs that worked earlier in the year may no longer be marketed.

Source: RigData; based on data Jan. 1-Aug. 31, 2008

Offshore Drilling Rigs Under Construction*

	Rigs
Drillships	39
Semisubmersibles	53
Drilling tenders	7
Jack ups	77
Inland barges	3
Total	**177**

*As of Aug. 4, 2008.

Work-class ROV operations market

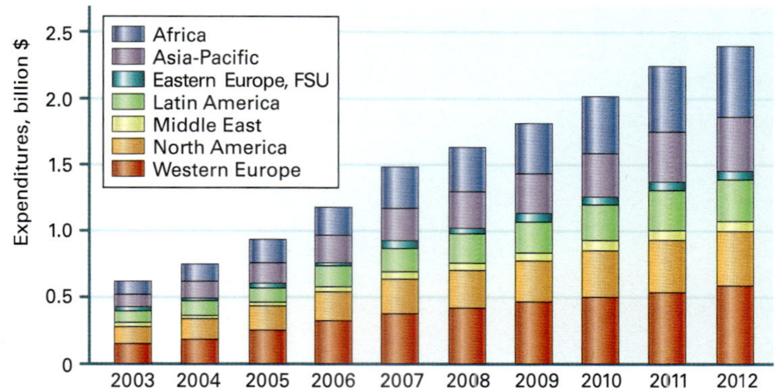

Source: Douglas Westwood

ROV market

Oil and gas industry spending on remotely operated underwater vehicle (ROV) operations is likely to total $1.6 billion in 2008, and the market could reach $2.4 billion by 2012.

The projection comes from a market study by Douglas-Westwood (Canterbury, UK).

In deep water, ROVs are a key enabling technology for drilling and field development, and service thousands of existing underwater installations worldwide.

The analysts conclude that from 2008–12, some 655 new work-class ROVs must be built to satisfy the dual demands of market growth and attrition within the existing fleet.

Billion-dollar drillship

In May 2008, Stena Drilling, a subsidiary of Stena AB (Göteborg, Sweden), announced that it was building a new Arctic-class drillship at Samsung Heavy Industries shipyard in South Korea.

The estimated total delivery price for the dynamically positioned *Stena DrillMAX IV* is $1.15 billion.

With a displacement of 97,000 tons, the ship will have a thickened hull for drilling in harsh, high-latitude environments and will be capable of drilling in ultradeep water.

The scheduled delivery date is December 2011.

U.S. EOR project survey

In April 2008, *OGJ* released its exclusive biennial enhanced oil recovery (EOR) survey.

The data show that the number of U.S. EOR projects increased compared with the last survey, taken in 2006.

However, total production from all U.S. EOR projects is less than in the 2006 survey.

Production decline in California steam injection projects mostly accounts for the lower production, while new carbon dioxide floods mostly account for the increase in projects.

Outside of the United States, the number of projects continued to increase.

Projects listed are those that involve injecting fluids, other than water or methane, into a reservoir to improve or enhance oil recovery.

US EOR PRODUCTION

	1986	1988	1990	1992	1994	1996	1998	2000	2002	2004	2006	2008
Thermal												
Steam	468,692	455,484	444,137	454,009	415,801	419,349	439,010	417,675	365,717	340,253	286,668	273,947
Cumbustion in situ	10,272	6,525	6,090	4,702	2,520	4,485	4,760	2,781	2,384	1,901	13,260	17,025
Hot water	705	2,896	3,985	1,980	250	250	2,200	306	3,360	3,360	4,370	1,776
Total thermal	**479,669**	**464,905**	**454,212**	**460,691**	**418,571**	**424,084**	**445,970**	**417,675**	**371,461**	**345,514**	**304,298**	**292,748**
Chemical												
Micellar-polymer	1,403	1,509	617	254	64	0	0	0				
Polymer	15,313	20,992	11,219	1,940	1,828	139	139	1,598				
Caustic/alkaline	185											
Surfactant			20					60	60	60		
Total chemical	**16,901**	**22,501**	**11,856**	**2,194**	**1,892**	**139**	**139**	**1,658**	**60**	**60**	**0**	**0**
Gas												
Hydrocarbon miscible/immiscible	33,767	25,935	55,386	113,072	99,693	96,263	102,053	124,500	95,300	97,300	95,800	81,000
CO_2 miscible	28,440	64,192	95,591	144,973	161,486	170,715	179,024	189,493	187,410	205,775	235,344	240,313
CO_2 immiscible	1,349	420	95	95				66	66	102	2,698	9,350
Nitrogen	18,510	19,050	22,260	22,580	23,050	28,017	28,117	14,700	14,700	14,700	14,700	19,700
Flue gas (miscible and immiscible)	26,150	21,400	17,300	11,000	—	—	—	—				
Other				6,300	4,400	4,350	4,350	—				
Total gas	**108,216**	**130,997**	**190,632**	**298,020**	**288,629**	**299,345**	**313,544**	**328,759**	**297,476**	**317,877**	**348,542**	**350,363**
Other												
Carbonated waterflood												
Microbial				2	2							
Total other			**2**	**2**								
Grand total	**604,786**	**618,403**	**656,700**	**760,907**	**709,094**	**723,568**	**759,653**	**748,092**	**668,997**	**663,451**	**652,840**	**643,111**

ACTIVE US EOR PROJECTS

	1988	1990	1992	1994	1996	1998	2000	2002	2004	2006	2008
Thermal											
Steam	133	137	119	109	105	92	86	55	46	40	45
Cumbustion in situ	9	8	8	5	8	7	5	6	7	12	12
Hot water	10	9	6	2	2	1	1	4	3	3	3
Total thermal	**152**	**154**	**133**	**116**	**115**	**100**	**92**	**65**	**56**	**55**	**60**
Chemical											
Micellar-polymer	9	5	3	2							
Polymer	111	42	44	27	11	10	10	4	4	0	1
Caustic/alkaline	4	2	2	1	1	1					
Surfactant		1									1
Total chemical	**124**	**50**	**49**	**30**	**12**	**11**	**10**	**4**	**4**	**0**	**2**
Gas											
Hydrocarbon miscible/immiscible	22	23	25	15	14	11	6	7	8	13	13
CO_2 miscible	49	52	52	54	60	66	63	66	70	79	100
CO_2 immiscible	8	4	2	1	1		1	1	1	2	5
Nitrogen	9	9	7	8	9	10	4	4	4	3	4
Flue gas (miscible and immiscible)	2	3	2								
Other			1	1							
Total gas	**90**	**91**	**89**	**79**	**84**	**87**	**74**	**78**	**83**	**97**	**122**
Other											
Microbial			2	1	1	1					
Total other	**0**	**0**	**2**	**1**	**1**	**1**	**0**	**0**	**0**	**0**	**0**
Grand total	**366**	**295**	**273**	**226**	**212**	**199**	**176**	**147**	**143**	**152**	**184**

Denbury CO_2 project potential

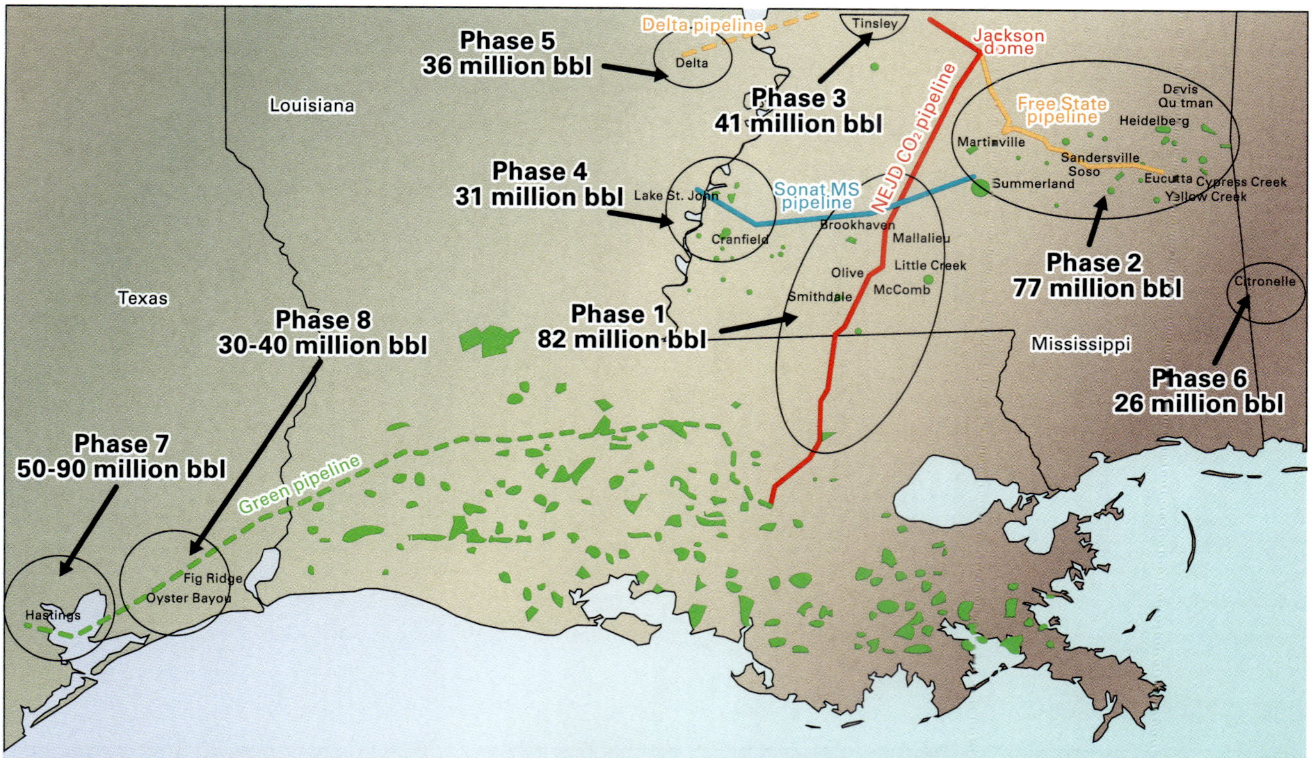

Note: For Phases 7 and 8 the tertiary reserves are based on 10-17% recovery factors as of Dec. 31, 2006, estimates. Hastings field is under contract but not owned by Denbury.
Source: Denbury Resources Inc., The future in CO_2, January 2008.

Current CO_2 sources

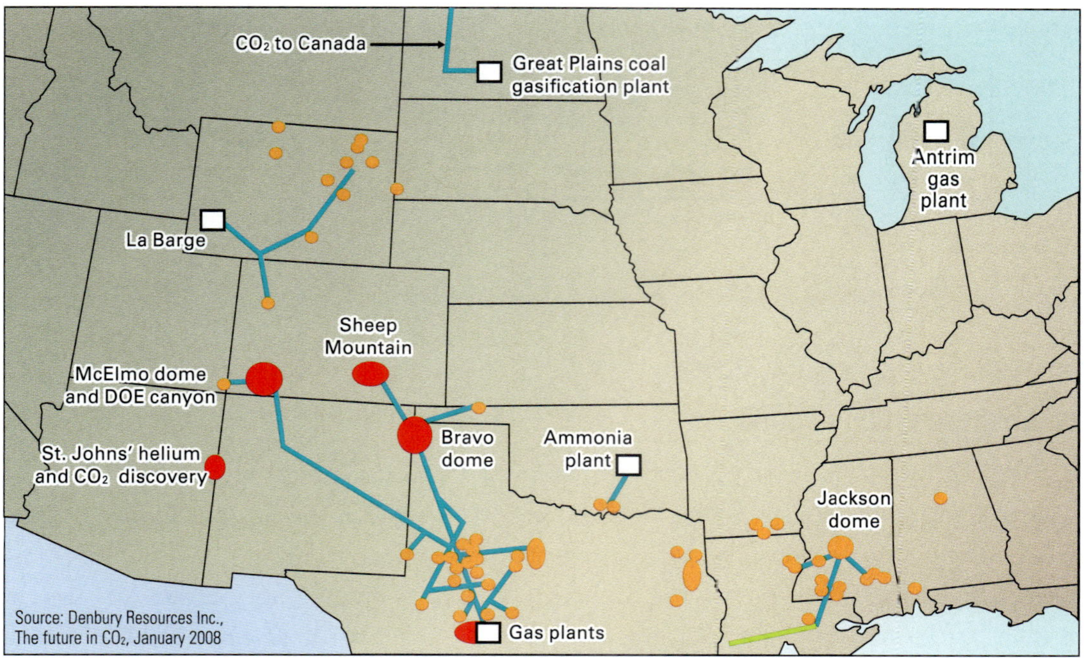

Source: Denbury Resources Inc., The future in CO_2, January 2008

Spend for shallow and deepwater offshore wells

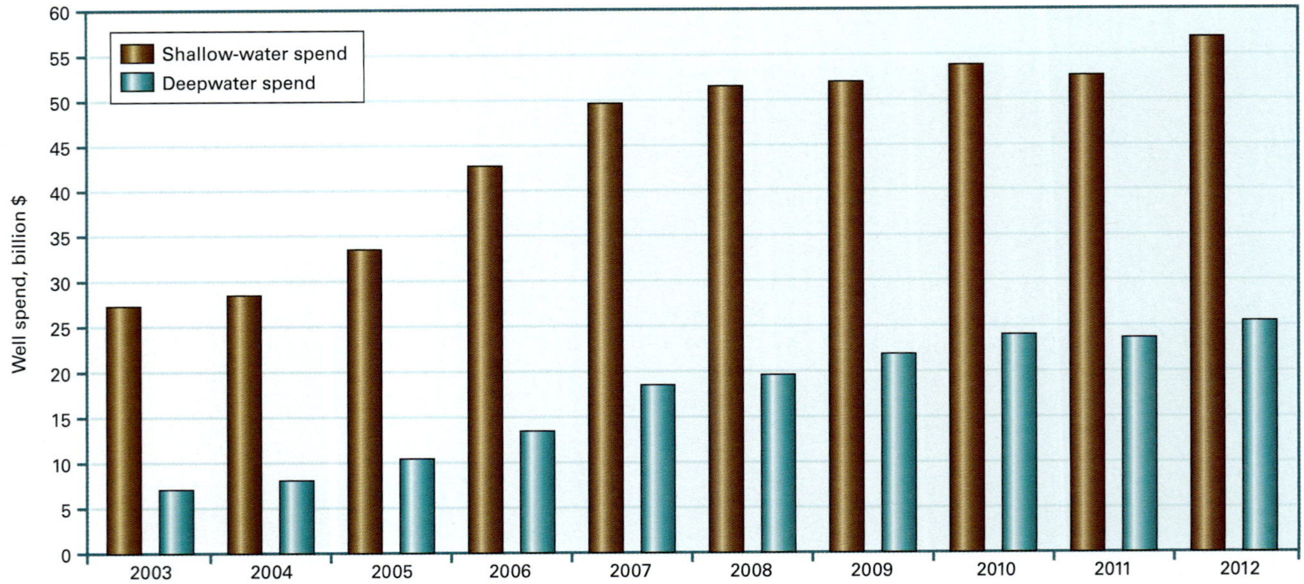

Source: Douglas-Westwood and Energyfiles

Offshore capex plus opex: total spend

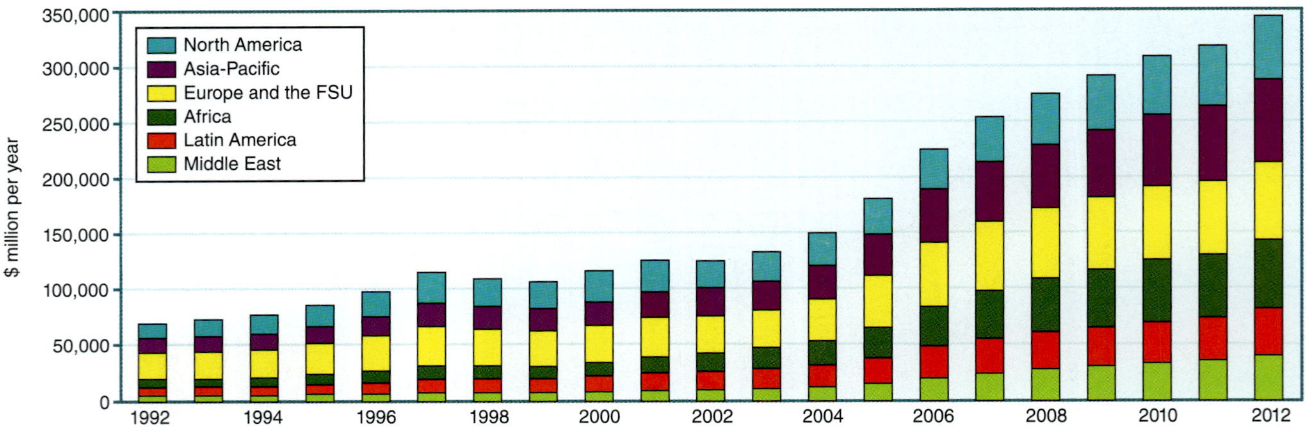

Source: Energyfiles Ltd

Global offshore drilling spend

From 2008–12, high oil prices will continue to drive oil and gas industry spending on offshore drilling to a total of $380 billion, according to Douglas-Westwood Ltd. (Canterbury, UK) and Energyfiles (Bucks, UK).

This is an increase of nearly 60% compared with the $240 billion spent in the previous five years.

The figures are reported in the latest edition of the *"World Offshore Drilling Spend Forecast 2008–12,"* released in April 2008 by the firms.

By 2012, they expect that the global drilling market will be worth an estimated $80 billion, more than doubling since 2003.

Offshore wells: total spend and wells drilled

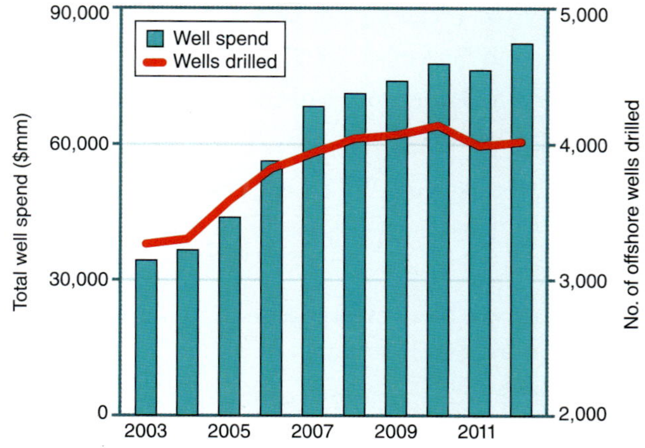

Source: Energyfiles Ltd

Offshore production: deep water oil from 1990

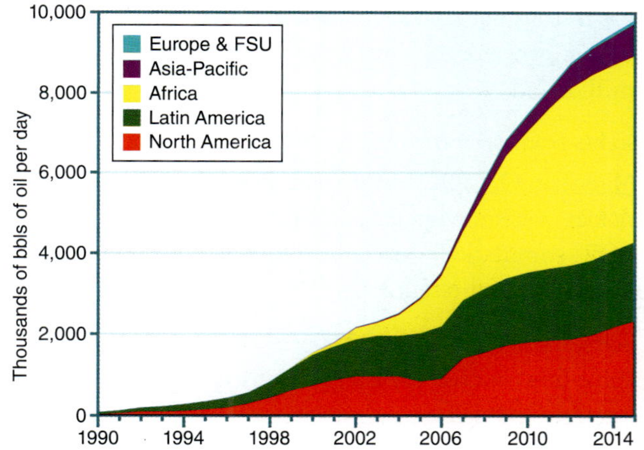

Source: Energyfiles Ltd

Deep waters: oil forecast

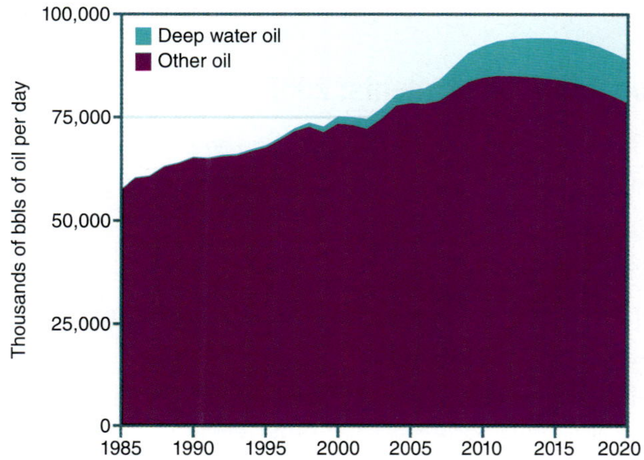

Source: Energyfiles Ltd

Ultradeepwater challenges

Officials from Petrobras America Inc. (Houston) and Chevron North America E&P Co. (Houston) told a committee of the U.S. Minerals Management Service in early 2008 about the technical challenges of ultradeepwater operations in the Gulf of Mexico.

- It can cost $250 million to drill a single well (not including expenses for equipment, services and crews).
- It can cost $250,000/day just to rent a deepwater drilling rig.
- Drilling must penetrate a salt canopy and then high-pressure, high-temperature rock formations.
- For some wells, a year or more of work is expended before drilling even begins.
- Undersea cliffs and canyons make it very difficult to place a platform and production system.
- Wells planned by Petrobras in the Cascade and Chinook fields will be in 8,000 ft (2,438 m) of water and reach as far as 28,000 ft (8,535 m).

Section 4:
Investment and markets

Two related factors roiled the waters in 2008 for those in the oil and gas industry charged with making decisions about markets and investments.

The first was the wild swing in oil prices—from about $100/bbl at the start of the year, to just over $147/bbl in July, and back down to the $40/bbl range by year end.

The second was the rapid and widespread collapse of financial markets that began in the autumn of 2008, triggering a sharp global recession and extreme tightening of credit markets.

The industry struggled with question such as:

- How deep will the recession be and how long will it last?
- What actions will governments and the private sector take to limit its impact?
- Will some regions or industries escape with relatively little damage?

A major concern among both private-sector and governmental organizations was that the worsening economy and attendant financial difficulties would reduce demand for oil and gas, depress prices, and discourage investment in exploration and development.

Lower 2009 E&P spending

Global spending for exploration and production is expected to decline 12% to $400 billion in 2009—"a reversal after six years of global growth," said analysts at Barclays Capital Resources (New York City).

The late-2008 projection was based on a semiannual survey of 357 oil and gas companies.

Driving budget cuts were a significant decline in commodity prices, constrained cash flow, and the tight credit markets.

Surveyed companies based their 2009 budgets on average prices of $58/bbl for oil and $6.35/Mcf for gas.

Barclays said that the sharpest decline in spending is expected to be in the United States. A drop of 26% (to $79 billion) is seen, ending a four-year upturn.

They expect Canadian budgets to fall 23% to $22 billion in 2009.

Natural gas prices remain the leading driver of 2009 E&P budget decisions, but 49% of respondents listed oil prices among key determinants.

Exploration investment trend

In late 2008, falling oil prices and difficulty in obtaining credit were prompting cuts and delays in oil exploration investment—a trend, said the International Energy Agency (IEA; Paris), that was putting the world "on a bad path."

In 2007, some $390 billion was invested in oil and gas exploration and production, said IEA, one of the highest levels in recent years. Yet it still fell short of the $450 billion that the IEA said would be needed in both sectors.

IEA said that developing countries are expected to be the only source of growth in oil demand until 2030, including China, India, the Middle East and other emerging economies in Asia.

But meeting demand growth, said the agency, is secondary to the big challenge of compensating for the fast-declining production from the world's older fields.

Slump: Mixed opportunities

PFC Energy Inc. (Washington, DC) shared its views with a Houston forum in November 2008:

- Most Persian Gulf members of OPEC can weather the slump in crude prices.
- National "break-even" WTI crude prices in 2010 are $54.26/bbl in Saudi Arabia, $52.07 in Kuwait, and $45.59 in the UAE.
- In contrast, the break-even price in 2010 is $102.68 in Venezuela and $83.31 in Iran.
- National oil companies (NOCs) likely to encounter project delays are those in Saudi Arabia, Abu Dhabi, and China.
- NOCs least likely to encounter delays are in Nigeria, Angola, Malaysia, Thailand, India (ONGC), Iran, and Kazakhstan.
- For reasons not related to "peak-oil pessimism", the world cannot generate the capacity to produce more than 100 million b/d.

Insights from other firms:

- Marathon Oil: IOCs must concentrate on technology—seismic imaging, reservoir characterization, EOR, stimulation, gas conversion—to gain access to unconventional resources.
- Software firm Paradigm: Major energy innovations will come from outside the energy industry (e.g., the medical, satellite, and defense industries).
- Banc of America Securities: Trends in OPEC production, U.S. rig count, and earnings of service companies suggest possible reductions in oil and gas supply, with attendant increases in price.

Global E&P investments: 2007

Global investment by oil and gas companies in 2007 for exploration and development totaled $402 billion—unchanged from 2006.

That was the finding of IHS Herold Inc. (Englewood, CO) and Harrison Lovegrove & Co. Ltd. (London) in their annual upstream performance review (September 2008).

However, capital spending varied greatly by region. Lower levels of investment in Canada and the United States were offset by gains in regional spending elsewhere.

Regional upstream spending, 2005-07*

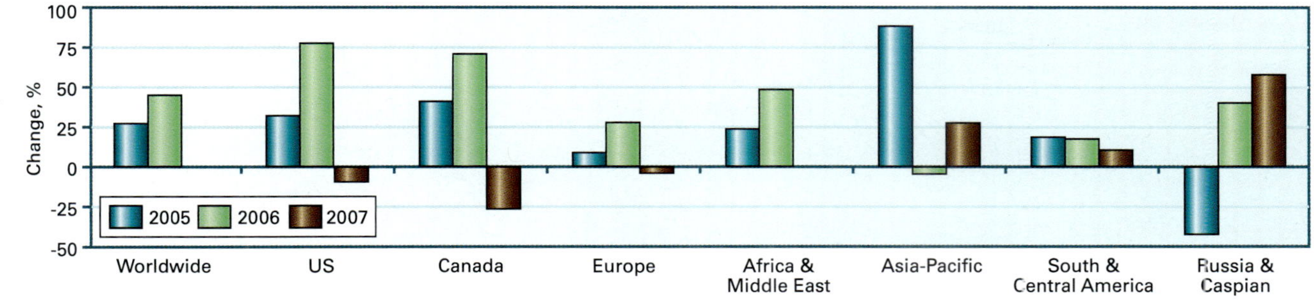

The 2008 Global Upstream Performance Review shows the 2007 investment of 232 oil and gas companies was unchanged at $402 billion in 2007 as compared with 2006.
Source: IHS Herold Inc. and Harrison Lovegrove & Co. Ltd.

Rising royalty rates in Alberta and new royalty trust legislation contributed to lower overall upstream investments in Canada by more than 25%. Conventional oil and gas spending in Canada plunged by nearly $30 billion from 2006.

In the United States, substantially lower acquisition activity contributed to a 9% decline in 2007 upstream spending compared with 2006.

Worldwide acquisition spending was $90 billion, down 30% from 2006 record levels.

Where funds will go for US projects

	2008, million $	Change 2008-2007, %	2007, million $	Change 2007-2006, %	2006, million $
Exploration-production					
Drilling-exploration	130,200	4.2	125,010	0.3	124,000
Production	24,750	4.2	23,760	0.8	23,560
OCS lease bonus	5,250	87.8	2,795	205.8	914
Subtotal	**160,200**	**5.7**	**151,565**	**2.1**	**148,474**
Other					
Refining	13,000	57.0	8,280	–8.0	9,000
Petrochemicals	1,000	19.0	840	7.7	780
Marketing	3,000	20.0	2,500	0.0	2,500
Crude and products pipelines	6,629	269.1	1,796	1,173.8	141
Natural gas pipelines	5,710	30.8	4,367	94.5	2,245
Other transportation	1,200	23.7	970	14.1	850
Mining, other energy	1,200	20.0	1,000	0.0	1,000
Miscellaneous	5,000	22.0	4,100	10.8	3,700
Subtotal	**36,739**	**54.0**	**23,853**	**18.0**	**20,216**
Total	**196,939**	**12.3**	**175,418**	**4.0**	**168,690**

2008 capital budgets: U.S. and Canada

An April 2008 evaluation by *OGJ* projected that capital spending for oil and gas projects in the United States would increase in 2008, buoyed by growth in upstream, midstream, and downstream developments.

In Canada, total project spending was expected to decline, but by a smaller margin than in 2007. Oil sands expenditures were projected to increase, with outlays for most other types of projects expected to shrink.

Total U.S. budgets for 2008 capital spending were estimated at $197 billion, up from $175 billion in 2007 and $169 billion in 2006.

Canadian spending plans

	2008, million $ (Can.)	Change 2008-2007, %	2007, million $ (Can.)	Change 2007-2008, %	2006, million $ (Can.)
Exploration-production					
Drilling-exploration	17,085	–15.0	20,100	–28.0	27,926
Production	6,515	–15.0	7,665	–28.0	10,648
Subtotal	**23,600**	**–15.0**	**27,765**	**–28.0**	**38,574**
Oil sands*	21,000	16.7	18,000	25.5	14,337
Other					
Refining	2,000	–42.3	3,465	–1.0	3,500
Petrochemicals	200	–39.9	333	11.0	300
Marketing	550	–18.8	677	–3.3	700
Crude and products pipelines	336	300.0	84	–84.6	546
Natural gas pipelines	470	—	0	–100.0	17
Other transportation	300	4.5	287	4.4	275
Miscellaneous	650	2.2	636	7.8	590
Subtotal	**4,506**	**–17.8**	**5,482**	**–7.5**	**5,928**
Total	**49,106**	**–4.2**	**51,247**	**–12.9**	**58,839**

*In situ, mining, and upgrading.

Capital budgets, operating companies*

	2008 Billion $	2007
OAO Gazprom	28.8	31.6
Chevron Corp.	22.9	19.9
Royal Dutch Shell PLC	24-25	23.8
ConocoPhillips	15.3	13.5
StatoilHydro	13.7	12.0
EnCana Corp.	6.9	6.1
Petro-Canada	5.34	4.17
Hess Corp.	4.4	3.6
Husky Energy Inc.	3.7	2.89
Nexen	2.4	3.6
Pioneer Natural Resources	1.0	1.5
W&T Offshore	0.8	0.45

*From company announcements.

Total spending in Canada was expected to be $49 billion (Can.) in 2008. In 2007 and 2006, the figures were $51 billion and nearly $59 billion, respectively.

Countries implementing the EITI*

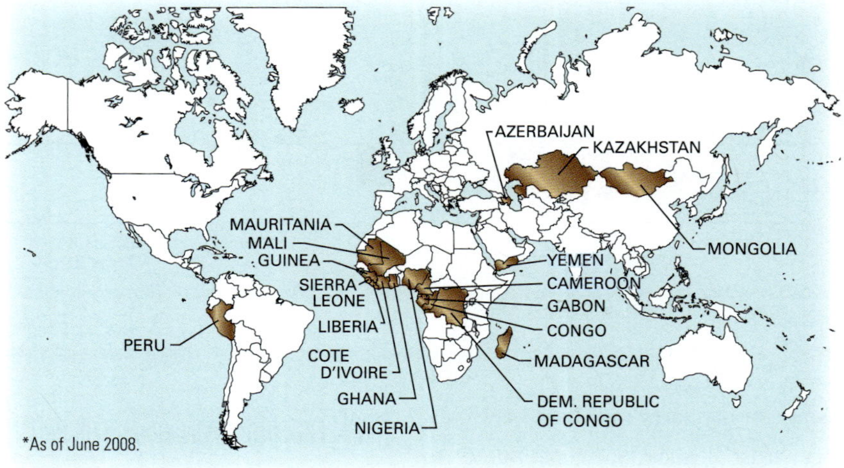

*As of June 2008.

EITI guide

In May 2008, the Extractive Industries Transparency Initiative (EITI) published the "EITI Business Guide" to encourage oil and gas companies, as well as companies in other extractive industries, to support implementation of the EITI in countries that are developing or operating oil and gas assets.

The multi-nation EITI program (launched in 2002) aims to publish government revenues generated by extractive industries and to compare those revenues with payments reported by private companies in order to improve transparency in countries rich in oil, gas, and mineral resources.

It is hoped that making this data public will allow in-country civil society organizations to hold their governments accountable for any discrepancies.

This in turn should minimize corruption in the countries that adopt the EITI.

Currently, companies in 23 countries are subject to EITI requirements. Of these, 16 are in Africa. The others are Azerbaijan, Kazakhstan, Kyrgyzstan, Mongolia, Peru, Timor-Leste, and Yemen.

Government oil funds

The accumulation of large oil revenues, particularly since the early 2000s, has placed oil exporting countries among the world's largest sources of capital.

The bulk of these assets are invested by government-owned and government-managed oil funds, which constitute a large component of broader sovereign wealth funds (SWFs).

A 2008 study by Morgan Stanley (New York City) estimated that by 2015, the financial assets of oil funds and other SWFs will be about $6 trillion each.

McKinsey Global Institute (New York City) estimates that at the end of 2006, oil exporters collectively owned $3.4–3.8 trillion in foreign financial assets.

The Gulf Cooperation Council (GCC) states—Bahrain, Kuwait, Oman, Qatar, Saudi Arabia, and the UAE—own $1.8 trillion of these assets, according to a late-2007 report by the Institute of International Finance (Washington, DC).

Selected OPEC Members' Net Oil Export Revenues

	Iran	Iraq	Kuwait	Qatar	Saudi Arabia	UAE	Total OPEC Nominal	Total OPEC Real (2000 $)
	Billion $							
1975	23.5	9.8	9.7	2.1	32.7	7.9	121.8	317.1
1976	25.0	10.1	9.7	2.3	38.3	8.9	133.7	329.0
1977	25.6	10.6	9.6	2.3	44.6	9.8	146.5	339.3
1978	23.3	11.6	10.3	2.5	39.4	8.8	138.0	298.6
1979	20.9	24.9	18.4	4.0	69.3	13.8	219.1	435.8
1980	13.0	27.7	18.6	6.0	113.5	20.3	296.3	544.1
1981	10.9	10.4	13.4	5.6	122.0	18.1	264.9	444.9
1982	18.7	9.8	8.4	4.0	69.7	13.5	191.3	302.2
1983	17.5	7.8	9.8	3.1	44.9	10.6	150.5	228.4
1984	14.7	9.9	10.6	4.1	39.3	10.4	147.8	216.5
1985	14.1	11.1	8.6	2.8	23.4	9.8	123.3	175.1
1986	6.0	6.6	6.2	1.4	18.6	5.5	71.2	99.0
1987	9.0	11.0	9.2	1.8	20.7	8.6	93.8	126.8
1988	6.8	11.4	6.8	1.7	20.2	6.7	82.2	107.6
1989	11.5	15.6	10.3	2.3	25.6	10.3	113.6	143.2
1990	16.5	9.9	6.5	3.1	43.4	15.0	147.4	178.5
1991	14.8	0.5	0.8	2.6	45.7	14.0	87.4	102.3
1992	14.8	0.7	5.8	2.5	45.5	13.0	127.9	146.6
1993	14.2	0.5	9.6	2.3	38.7	10.7	115.7	129.8
1994	13.5	0.6	10.2	2.4	41.1	11.0	122.7	134.6
1995	15.3	0.6	11.7	2.9	47.2	12.6	140.6	151.3
1996	19.2	0.9	14.1	4.0	57.1	15.8	175.7	185.4
1997	16.7	4.6	12.4	4.0	53.5	14.3	165.8	172.1
1998	10.7	6.5	8.1	3.4	34.8	9.8	112.4	115.4
1999	14.8	12.5	10.6	4.7	45.6	12.4	152.6	154.3
2000	24.3	21.1	18.3	8.1	76.3	21.5	258.6	256.1
2001	21.0	15.4	15.0	7.0	62.4	18.1	210.2	203.5
2002	18.5	12.7	15.0	6.9	57.8	17.8	197.2	187.4
2003	23.4	8.7	19.1	8.5	76.9	22.7	240.5	223.9
2004	32.3	18.4	26.3	12.5	103.6	29.7	336.8	304.5
2005	48.0	23.8	40.4	19.5	156.9	44.2	507.4	444.4
2006	54.6	31.8	50.0	24.4	182.8	57.5	612.7	520.8
2007	57.3	37.8	54.3	26.6	193.8	63.0	674.7	557.8

Source: Energy Information Administration, OPEC Revenues Fact Sheet, Mar. 12, 2008, (http://www.eia.doe.gov/emeu/cabs/OPEC_Revenues/Factsheet.html)

Worldwide oil, gas transaction value

The total transaction value for worldwide upstream corporate and asset deals slipped to $154 billion in 2007, compared with $166 billion in 2006.

That was a major finding of John S. Herold Inc. (Norwalk, CT) and Harrison Lovegrove & Co. Ltd. (London) in a joint report released March 2008, *"2008 Global Upstream Mergers and Acquisitions Review."*

The study analyzed more than 330 major upstream transactions announced during 2007.

The pricing for transactions declined during 2007 after hitting two consecutive records in 2005–06.

Herold said that it saw difficult conditions continuing for 2008.

The number of worldwide asset transactions surged 30% (to 240) 2007 compared with 2006. About 75% of the transactions involved properties in the United States and Canada.

Worldwide, asset transactions totaled $89 billion in 2007.

Energy security: Massive investment

In January 2008, the U.S. Department of Energy (DOE) said that billions of dollars in investments will be required annually to achieve global energy security.

Action is needed to diversify in three broad areas, said DOE: supplies, suppliers, and supply routes.

DOE pointed to the estimate of the IEA (Paris) that $22 trillion of investment will be needed between 2008 and 2030 if the world is to meet expected energy demand.

The investment must be in developed as well as developing nations and at all stages in the energy cycle.

In parallel, government and business leaders also must recognize that greater environmental care is required in all steps of the energy value chain.

Also needed will be new energy options beyond hydrocarbons, developed through public-private partnerships involving corporations, utilities and universities.

Major oil trade movements

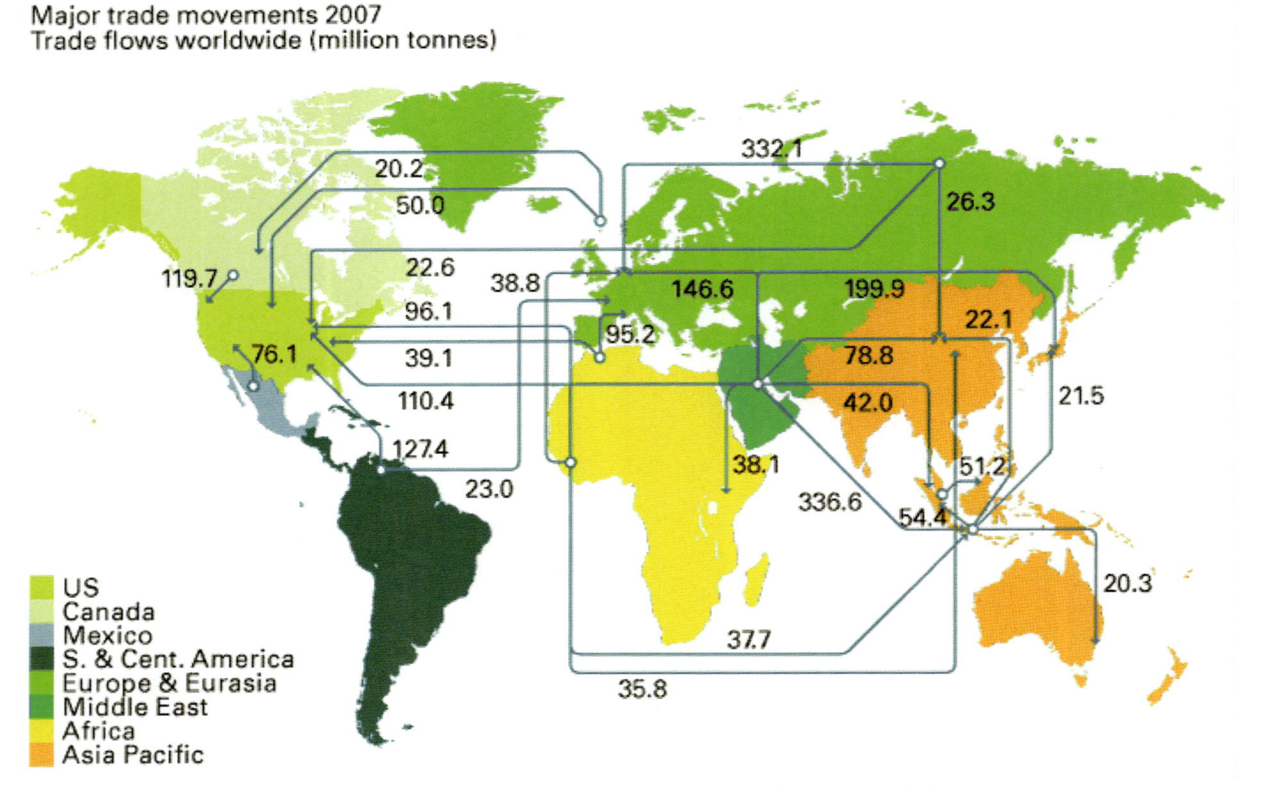

Arctic tanker tests

The American Bureau of Shipping (Houston), ConocoPhillips Co. (Houston), Sovcomflot Ltd. (London), and Samsung Heavy Industries Ltd. (SHI; Seoul) were joint participants in late 2008 in a study to measure the effect of ice loads on the performance of large Arctic-class shuttle tankers.

The study will be conducted on the 70,000-dwt *Shturman Albanov* during its initial two winter seasons operating in the Barents Sea.

The ship—third in a series of three being built by SHI—is to be delivered in February 2009.

Fiber-optic sensors in the bow and stern quarters will record ice pressures and loads and give real-time feedback to the ship operators.

Section 5:
Trade and tankers

Oil trade movements

Internationally traded volumes of oil totaled about 54.8 million b/d (2,700 mtpy) in 2007, said BP PLC. Approximately 36% originated in the Middle East.

For detailed data, see Tables 6 and 7 in the Statistical Tables section of this book.

Gas trade movements

Internationally traded volumes of gas totaled about 27.4 tcf (776 bcm) in 2007. Gas moved by pipeline accounted for 71% [19.4 tcf (550 bcm)], while gas traded as LNG represented the other 29% [7.98 tcf (226 bcm)].

For detailed data, please refer to Table 15 in the Statistical Tables section.

China: Fleet expansion

A late-2008 analysis by researchers at Shanghai Jiao Tong University (Shanghai) concluded that China's push for energy security has prompted

Major Gas Trade Movements

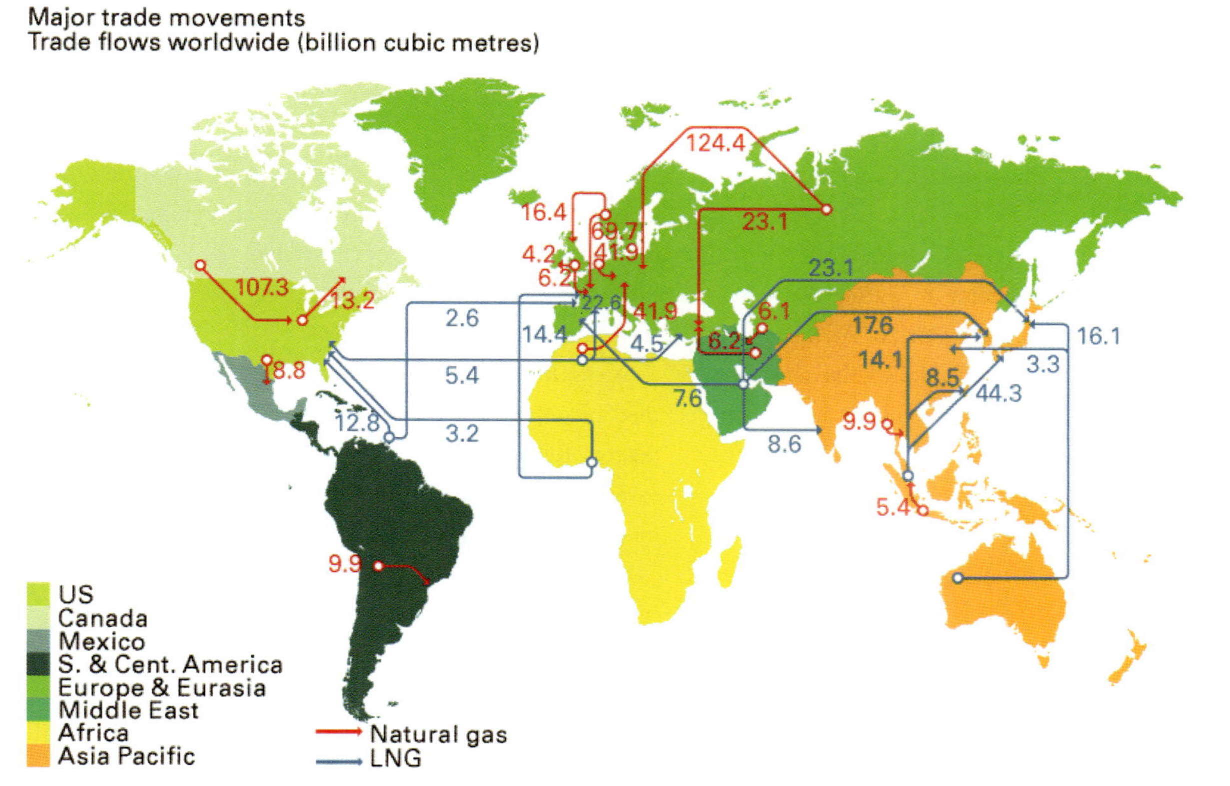

China, Imported Oil Sources

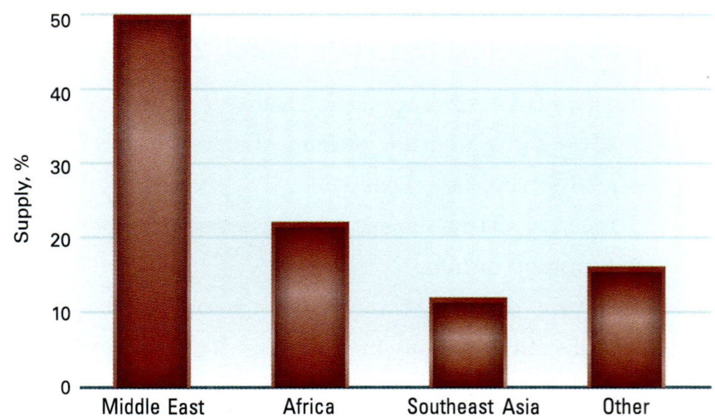

COSCO-Operated VLCCs

Name	Delivery	Displacement, dwt
Cosgreat Lake	2003	300,000
Cosgrace Lake	2006	300,000
Cosbright Lake	2003	300,000
Yuansheng Lake	2006	300,000
Cosmerry Lake	2006	300,000
Cospearl Lake	2008	300,000
Dali Lake	2004	160,000
Daming Lake	2003	160,000

China, VLCC Programs

Company	In service	Under construction
CMSNC	6	7
COSCO	8	5
CSCO	4	15
Yangtze Transportation Co. Ltd.	2	11
Petrotitan	7	2
Others	8 (est.)	10 (est.)
Total	**36**	**49**

Sources: Company web sites

a program for building 70 additional very large crude carrier (VLCC) tankers.

China is the world's most populous country and second largest oil consumer and importer, just behind the United States.

As one element of fleet expansion, China Shipping Development Co. Ltd. signed a contract with China's largest shipbuilder, Dalian Shipbuilding Industry Co. Ltd. (DSIC) in September 2008 for four oil carriers.

As of October 2008, about a half-dozen Chinese companies were operating a total of 36 VLCCs. This group included China Ocean Shipping (Group) Co. (COSCO); China Shipping (Group) Co. (CSCO); and Titan Petrochemicals Group Ltd. (Petrotitan).

COSCO also operates 11 Panamax tankers.

Piracy alters tanker routes

Between November 2007 and November 2008, Somali pirates seized three dozen ships off Africa.

On November 15, 2008, they hijacked the *Sirius Star*—a VLCC owned by Saudi Aramco (Dhahran) and operated by Vela International Ltd. (Dubai)—some 450 nautical miles off Kenya.

The ship had a crew of 25 and carried 2 million bbl of oil.

Several nations formed a naval force to patrol the waters between the Arabian Peninsula and the Horn of Africa.

In reaction to the piracy, however, AP Moller-Maersk Group (Copenhagen) decided to reroute its 83 tankers around the Cape of Good Hope rather than use the Gulf Aden and the Suez Canal.

Frontline Ltd. (Hamilton, Bermuda) and Odfjell SE (Bergen, Norway) did the same with their vessels.

Tanker Earnings

Vessel type	March-September 2007 ($/day)	September 2007-March 2008 ($/day)	Change, %
VLCC (modern)	44,494	86,009	93.3
VLCC (early 1990s)	42,398	80,384	89.6
Suezmax (modern)	36,833	53,038	44.0
Aframax (modern)	31,190	37,556	17.7
Products (dirty)	28,785	28,241	–1.9
Products (clean)	26,866	22,529	–16.1
Weighted average	30,477	34,771	14.1

VLCC Fleet Development

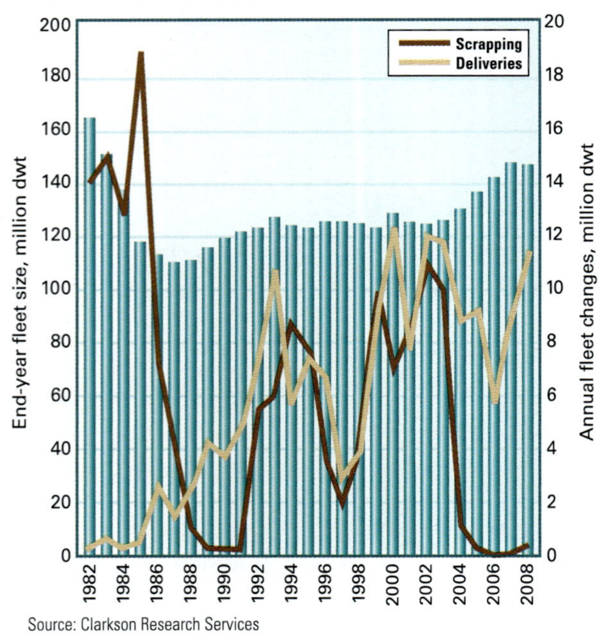

Source: Clarkson Research Services

Tighter 2008 market seen

McQuilling Services LLC (New York City) said in early 2008 that it expected a rebound in both tanker spot rates and time-charter revenues during 2008, with demand outpacing supply in most sectors.

The firm's annual *Tanker Market Outlook* saw little growth in demand, but instead expected market tightening due to accelerated exits from the fleet for conversion.

Beyond 2008, McQuilling expects the tanker market to weaken as surplus tonnage grows.

Strength once anticipated post-2010—driven by the scheduled phase-out of single-hull vessels by that year—has disappeared. Reasons include asset owners converting these vessels early, and a 600,000-b/d softening in global oil demand growth for 2010 compared with 2007 projections.

McQuilling also anticipates double-digit tonnage supply increases in 2009.

The company sees 3.3% demand growth for VLCCs in 2008, compared to no growth in 2007.

For time-charter equivalent daily revenue, McQuilling expects VLCCs, which earned $39,800/day in 2007, to peak in 2008 at $51,400/day, before slipping to an average of $38,300 through 2012.

Earnings: September 2007-March 2008

Tanker market earnings for the half-year ended March 2008 firmed by 14.1% from the previous six months, said Clarkson Research Services Ltd. (London).

Temporary increases in demand and decreases in supply drove the increase in earnings.

The firm detailed these market movements and forecast future market directions in its Spring 2008 *"Shipping Review and Outlook: A Half-Yearly Review of the Shipping Market."*

VLCC Market Summary

Worldscale rates	End 2007	March 2008	Change, %
AG-West	180	83	–54.2
AG-Far East	275	115	–58.2
Med-UK	205	133	–35.4
Revenue, $/day			
Average spot earnings, 1990s built	185,635	77,358	–58.3
1-year time-charter rate, 1990s built	52,500	45,000	–14.3
Asset values, $ million			
Newbuild price, 300,000 dwt	146	148.5	1.7
5-year old, 300,000 dwt	135	140	3.7
Tonnage supply, million dwt			
Fleet	148.3	146,8	–1.0
Order book	54.9	57.8	5.3

Aframax fleet development

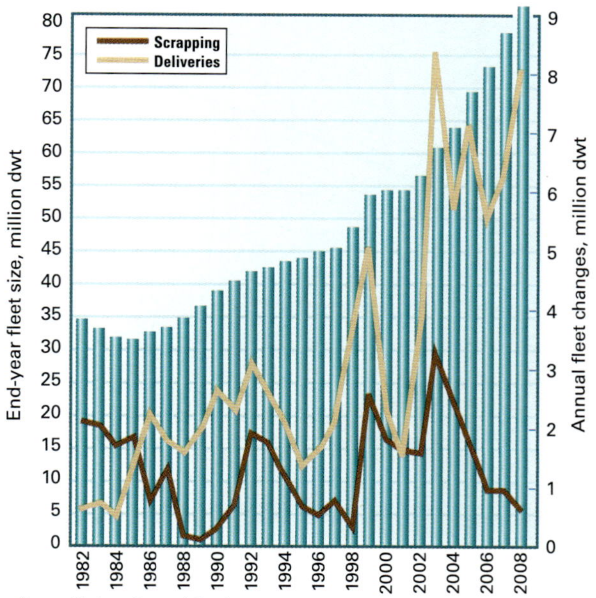

Source: Clarkson Research Services

VLCC market summary

	End 2006	Sept. 2007	Change, %
	World scale rates		
AG-West	55	43	−22.7
AG-Far East	60	53	−12.5
Med-UK	105	55	−47.6
	Revenue $/day		
Average spot earnings, 1990s built	39,147	23,212	−40.7
1-year time-charter rate, 1990s built	44,000	40,000	-9.1
	Asset values, million $		
Newbuild price, 300,000 dwt	129	140	8.5
5-year old, 300,000 dwt	118	134	13.6
	Tonnage supply, million dwt		
Fleet	142.3	148.2	4.1
Orderbook	53.3	53.3	—

Tanker earnings

Vessel type	Sept. 2006-Mar. 2007 $/day	March-Sept. 2007 $/day	Change, %
VLCC (modern)	53,172	45,149	-15.1
VLCC (early 1990s)	49,452	43,006	-13.0
Suezmax (modern)	52,769	36,833	-30.2
Aframax (modern)	44,060	31,190	-27.6
Products (dirty)	31,737	28,785	-9.3
Products (clean)	25,267	26,866	6.3
Weighted average	34,025	30,620	-10.0

Suezmax rates, early-1990s built

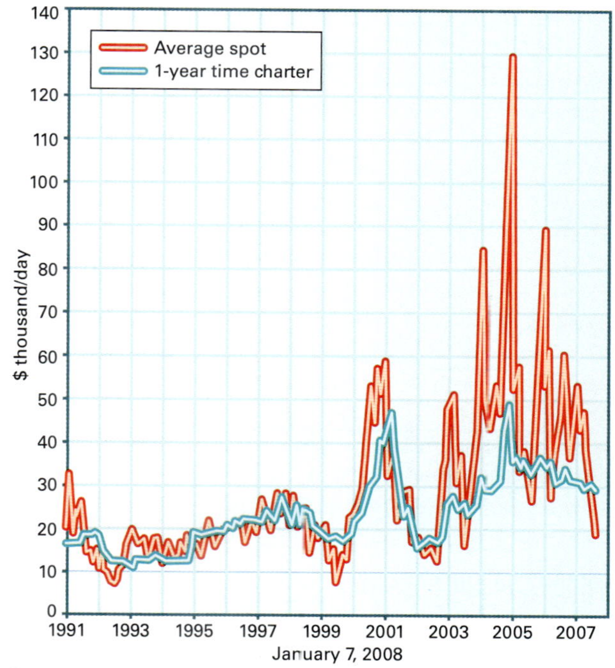

Source: Clarkson Research Services.

Clean product rates

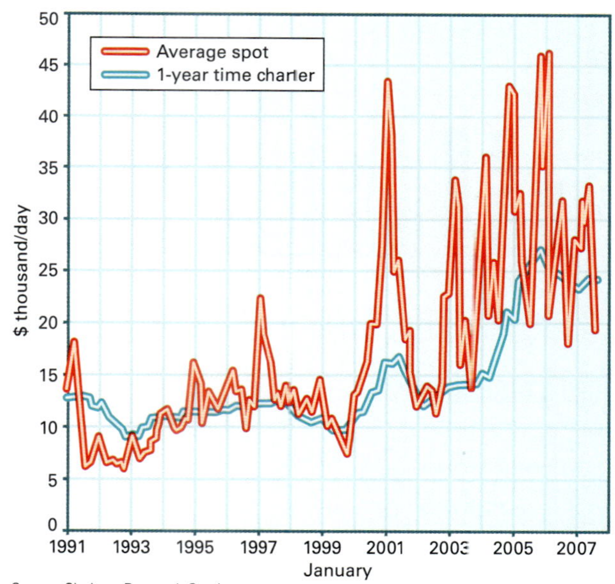

Source: Clarkson Research Services.

Earnings: March-September 2007

Clarkson Research Services Ltd. (London) reported in early 2008 on tanker-market economics for the period March-September 2007.

Tanker market earnings for that period softened by 25% from the same period in 2006, and by 10% from the prior six months.

The details behind these market movements as well as offering forecasts of future market direction in its Autumn 2007 *"Shipping Review and Outlook: A Half-Yearly Review of the Shipping Market."*

One Projection of World LNG Trade*

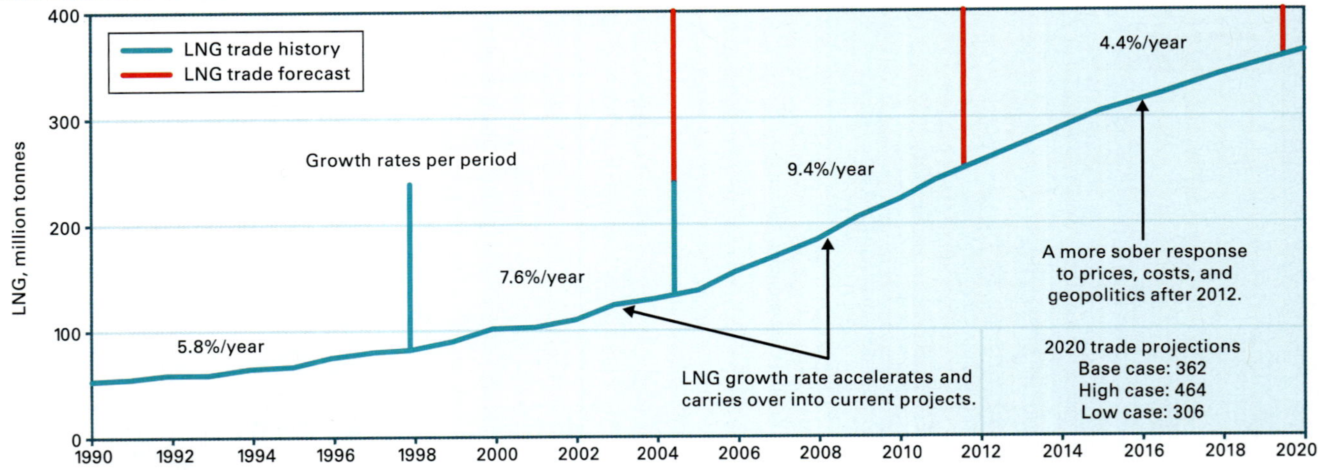

*Jensen Associates base-case estimates

Order Book's Rising Pressure

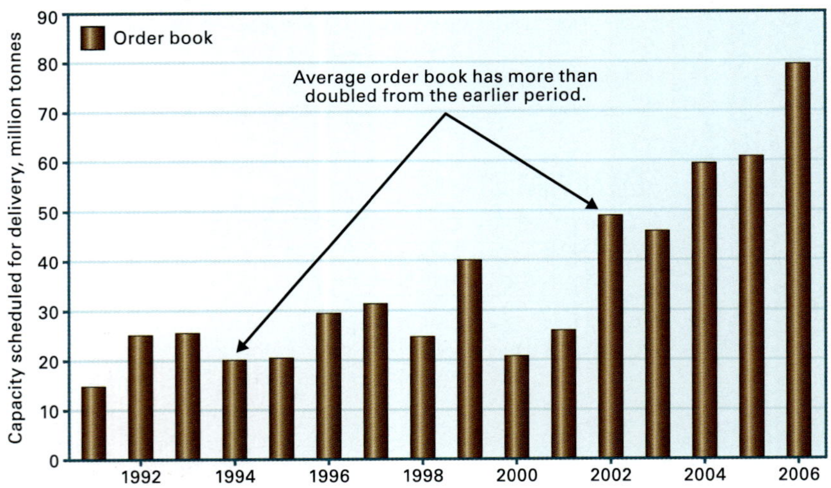

LNG Growth Scenarios

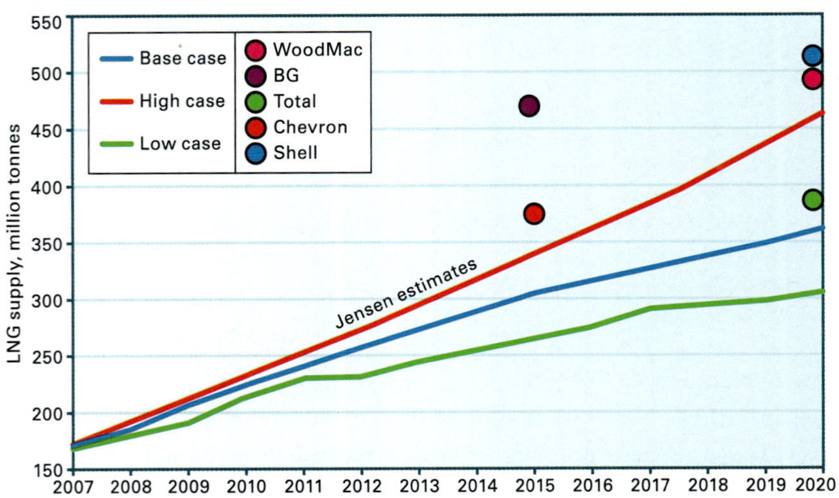

Section 6:
LNG—Industry outlook

Global trade to 2020

In an early 2008 analysis for the California Energy Commission, Jensen Associates (Weston, MA) developed scenarios for evaluating global LNG trade patterns to 2020.

Jensen concluded that a number of factors make future projections uncertain:

- The rapid growth of LNG demand, particularly in North America, Spain, and the UK.
- The inherently slow response time of LNG supply to sharply increased demand signals
- A shortage of LNG supply relative to expectations, due to supply lags.
- A burst in demand for new plant capacity, taxing the capabilities of design and construction contractors and machinery suppliers, leading to increased capital costs.
- A sharp increase in world energy prices [as of early 2008], complicating the analysis of gas demand and interfuel competition.
- Environmental concerns that in some situations favor natural gas over coal.
- Persistent and difficult geopolitical issues surrounding the natural gas export policies of countries such as Bolivia, Nigeria, Iran, Russia, and Venezuela.
- Inherent sensitivity of LNG demand to small changes in world gas supply-demand balances.

Also of key importance is evaluation of where gas will come from to feed LNG liquefaction plants.

Base case projections of world LNG demand

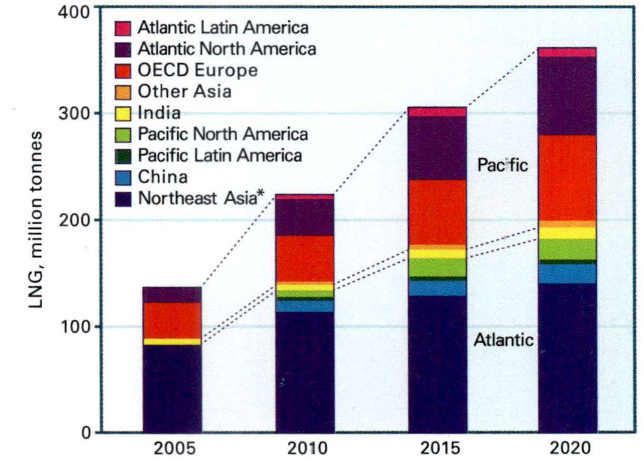

*Northeast Asia, which once dominated LNG trade, is now growing less rapidly than Atlantic.

Base case projections of world LNG supply

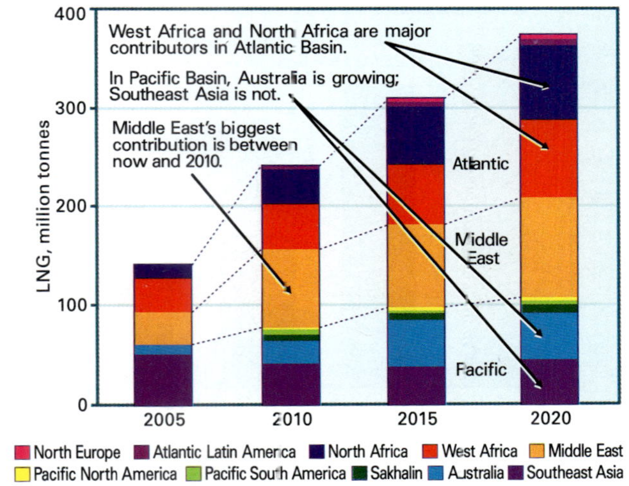

Major uncommitted FSU gas resources: yearend 2005[1]

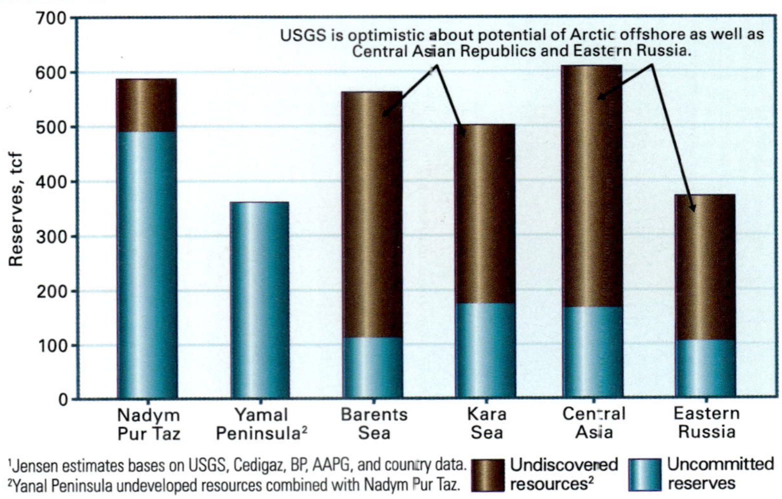

[1]Jensen estimates bases on USGS, Cedigaz, BP, AAPG, and country data.
[2]Yanal Peninsula undeveloped resources combined with Nadym Pur Taz.

World's Proved Gas Reserves: Yearend 2005

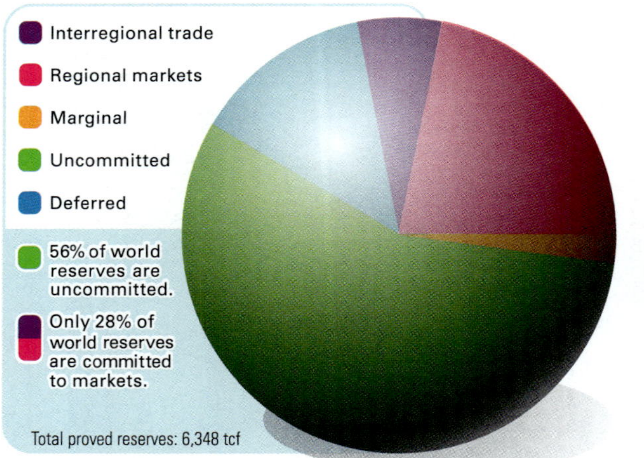

Source: Jensen Associates estimates

World's Uncommitted Gas: Yearend 2005

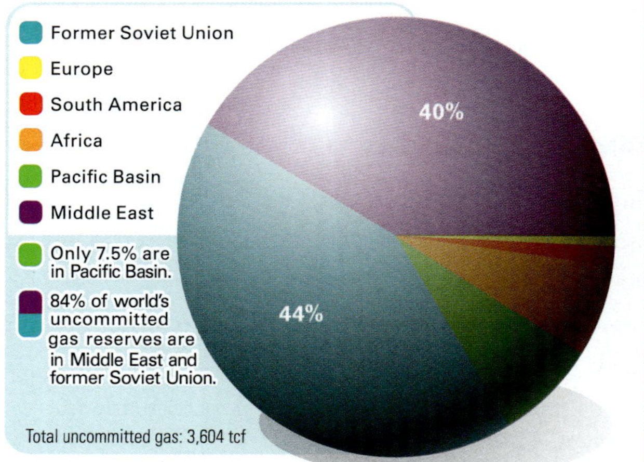

Source: Jensen Associates estimates

Uncommitted Middle East Gas Resources: Yearend 2005[1]

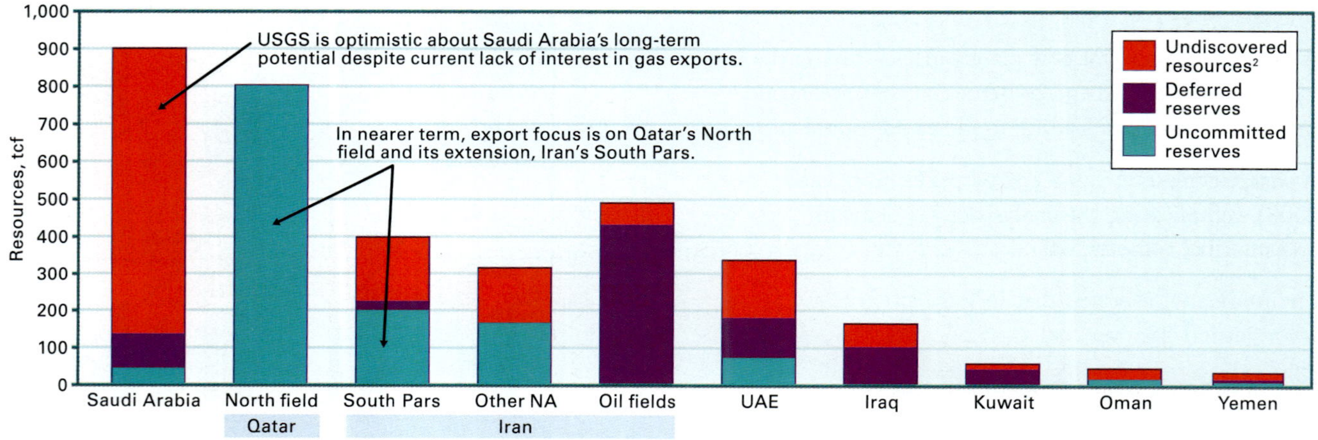

[1]Jensen estimates bases on USGS, Cedigaz, BP, AAPG, and country data. [2]Includes undeveloped reserves.

Uncommitted Gas Resources[1]

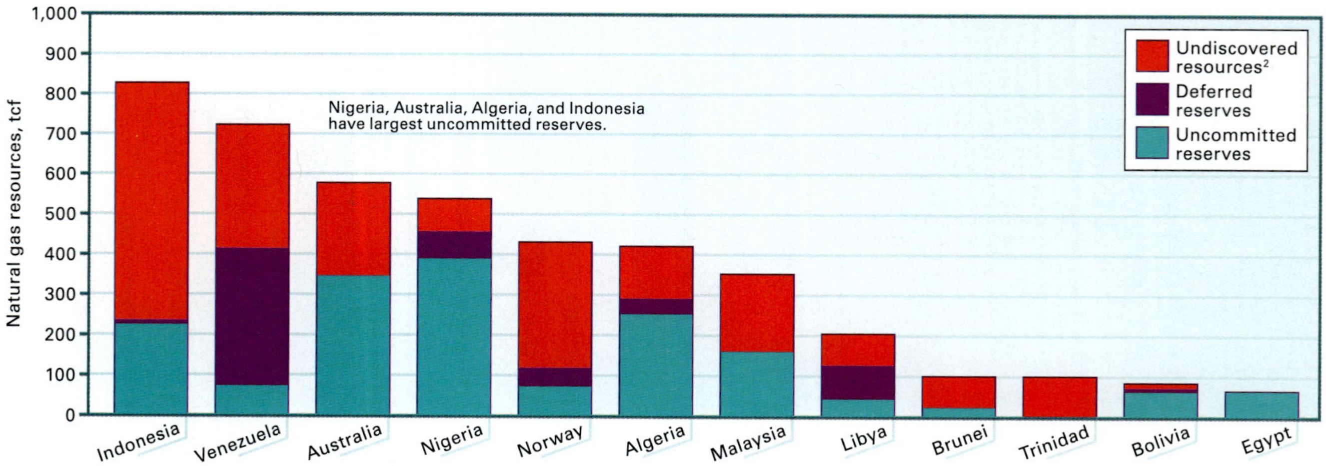

[1]Jensen estimates bases on USGS, Cedigaz, BP, AAPG, and country data. [2]Includes undeveloped reserves.

LNG boosts gas role

Cambridge Energy Research Associates (CERA; Cambridge, MA) said that an expanding LNG industry was making natural gas more of a global energy commodity in 2008 and 2009 than ever before.

The remarks came in early 2008 at the firm's annual conference in Houston.

Global liquefaction capacity is expected to increase to 12.0 tcf (341 bcm) by mid-2010 [from 9.25 tcf (262 bcm) in 2008], due to investments at Qatar's Ras Laffan along with projects in Russia, Yemen, Australia, and Indonesia.

Countries considering building LNG import facilities include Brazil and the Netherlands to Pakistan and New Zealand.

World LNG shipping capacity is expected to increase by more than 50% by 2010, said CERA.

The firm is less certain about the pace of LNG growth in the years beyond 2010.

LNG project financing

LNG became largely a sellers' market since 2004 due to strong gas demand and delays in sanctioning and constructing new liquefaction capacity. Many analysts see a sellers' LNG market prevailing well beyond 2010.

Upward pressures on LNG project lending margins

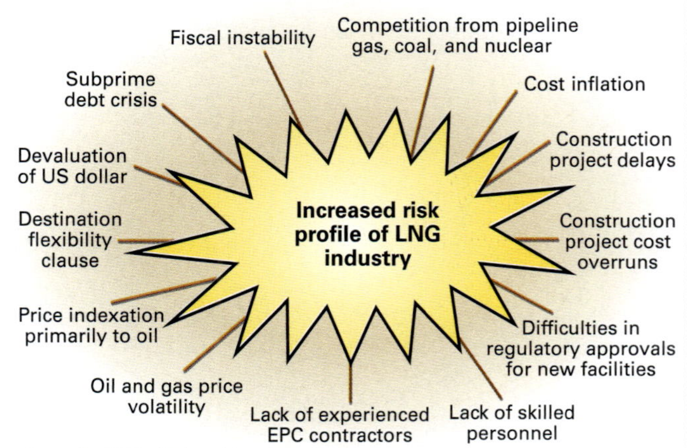

Source: David Wood & Associates

However, David Wood & Associates (Lincoln, UK) concluded in an early-2008 study that recent events seemed to conspire to make financing more problematic. These include:

- The U.S. subprime debt crisis of August 2007 and resultant global tightening of debt markets.
- Fiscal instability and toughening terms for IOCs in upstream gas supply contracts.
- Rampant oil and gas industry cost inflation, 2005–07.
- Devaluation of the U.S. dollar by 67% against the euro since 2002.
- Volatility and uncertainty in gas markets, in terms of both supply-demand fundamentals and price.
- Substantial delays and major cost overruns in some large liquefaction projects (e.g., Sakhalin II).
- Lack of skilled personnel and the shortage of experienced engineering contractors.
- Difficulties gaining regulatory approvals to build new LNG receiving terminals in key markets (e.g., California, New England).

Bank lending margin trend to large LNG projects

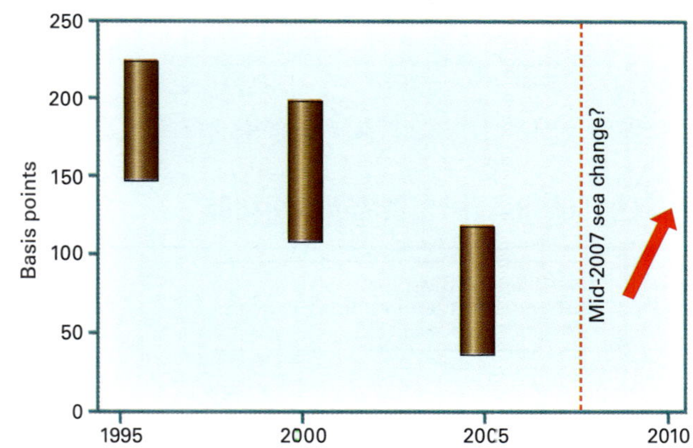

Source: David Wood & Associates

Asia-Pacific outlook

Preliminary data for 2007, said FACTS Global Energy (Singapore), show that worldwide LNG trade grew by 7.9% to 172 million tonnes.

The Asia-Pacific market accounted for about 65% of worldwide LNG trade, or 112 million tonnes. LNG imports into Asia increased by 9.6% because of strong demand from both established markets (Japan, Korea, and Taiwan) and emerging markets (India and China).

Potential new buyers include Singapore, Hong Kong, and Thailand.

Asia-Pacific LNG demand outlook

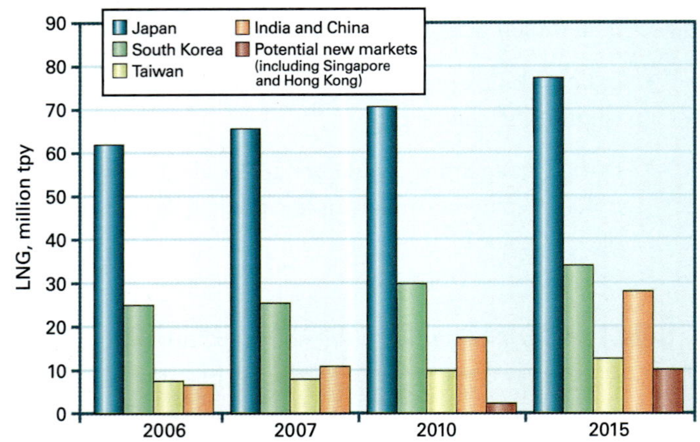

India's LNG contracts vs. demand outlook

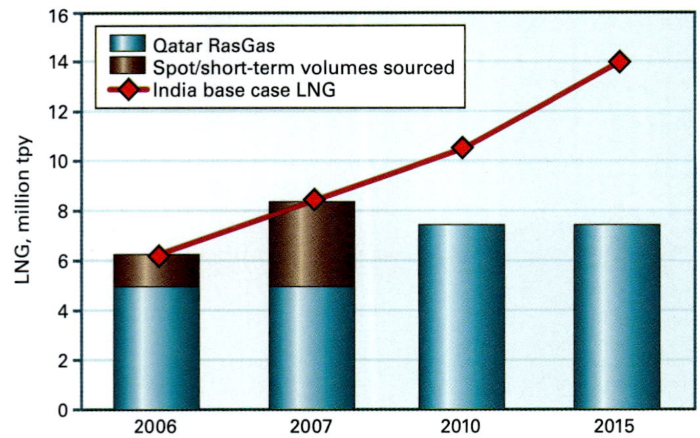

China's existing LNG contracts

Exporter	Contract volume	Start-up date	2006	2007	2010	2015
	Million tonnes/year					
NWS	3.3	2006	0.7	2.4	3.7	3.7
Tangguh	2.6	2009	—	—	1.5	2.6
MLNG	3.0	2010	—	—	1.0	3.0
Total	**8.9**	**–**	**0.7**	**2.4**	**6.2**	**9.3**
FGE base-case LNG demand forecast			0.7	2.9	7.0	14.3
China uncommitted demand			—	0.5	0.8	5.0

U.S. CO_2 policy, LNG import link

Potential regulation to cut carbon dioxide emissions in the United States could increase demand for natural gas and pit the United States against Europe in a bid to secure future supplies.

That was the finding in a mid-2008 report by Booz & Co. (Washington, DC). Key points:

- By 2015, U.S. gas demand will increase by up to 2.97 tcf (84 bcm), most of which will be met by LNG imports.
- The United States may use more clean-burning gas for power generation if nuclear or clean-coal technology does not become available.
- If its domestic gas production cannot meet growing demand, the United States will compete more strongly for international gas supplies.
- LNG redirected to U.S. markets could force Europe to rely more heavily on pipeline gas from Russia.

Section 7:
LNG—Liquefaction and receiving

Global snapshot: 2008

At the beginning of 2008, worldwide there were 26 liquefaction plants in 15 countries.

In contrast, there were 60 import (regasification) marine terminals, on or offshore, in 18 countries.

More than 270 LNG tankers deliver LNG to more than 40 ports around the world.

In addition to these terminals, 73 liquefaction projects and 182 regasification projects were either proposed or under construction around the world (though not all will be constructed).

First LNG-FPSO vessel

In September 2008, Samsung Heavy Industries Co. Ltd. (Seoul) secured an order from Flex LNG Ltd. (London) to supply the world's first liquefied natural gas floating production, storage and offloading (LNG-FPSO) vessel.

The $568-million unit is to be delivered in late 2011, for offshore operations in Nigeria. The 220,000-cu-m vessel, designed to operate for ten years, has a nameplate liquefaction capacity of 1.7 mtpy.

Samsung said it is also designing LNG floating storage and regasification units (FSRUs) and Arctic ice-breaking LNG vessels.

U.S. LNG terminals: 2008 snapshot

According to a compilation by Technip USA Inc. (Houston), as of May 2008, there were five import terminals and one export plant operating in the United States, at these locations:

- Everett, MA
- Cove Point, MD
- Elba Island, GA
- Lake Charles, LA
- Gulf Gateway Energy Bridge, Gulf of Mexico
- Penuelas, Puerto Rico
- Sabine, LA
- Freeport, TX.
- Kenai, AK (the only export plant). Kenai exports LNG to Japan.

Eight of these are under jurisdiction of the U.S. Federal Energy Regulatory Commission (FERC).

Gulf Gateway Energy Bridge is under U.S. Maritime Administration jurisdiction.

Record LNG trade in 2007

FACTS Global Energy (FGE; Honolulu) reported in early 2008 on the LNG trade in the previous year.

In 2007, global LNG demand reached about 172.2 million tonnes, a rise of roughly 13 million tonnes (8%) compared with 2006.

On a percentage basis, largest growth occurred in the Americas, followed by Asia.

Global LNG Trade[1]

Exporting countries[2]	Importing countries, territories[2]	Future exporting countries	Future importing countries
Algeria (1971)	Belgium (1987)	Angola	Brazil
Australia (1989)	China (2006)	Bolivia	Canada
Brunei (1972)	Dominican Republic (2003)	East Timor	Germany
Egypt (2004)	France (1972)	Iran	Hong Kong
Equatorial Guinea (2007)	Greece (2000)	Papua New Guinea	Jamaica
Indonesia (1977)	India (2004)	Russia	Netherlands
Libya (1970)	Italy (1971)	Venezuela	New Zealand
Malaysia (1983)	Japan (1969)	Yemen	Pakistan
Nigeria (1999)	Mexico (2006)		Singapore
Norway (2007)	Portugal (2003)		Thailand
Oman (2000)	Puerto Rico (2000)		
Qatar (1997)	South Korea (1986)		
Trinidad & Tobago (1999)	Spain (1969)		
United Arab Emirates (1977)	Taiwan (1990)		
US (1969)	Turkey (1992)		
	UK (1965)		
	US (1971)		

[1]As of Jan. 1, 2008. [2]Start-up year of earliest facility in parentheses.

LNG IMPORTS IN ASIA

Importers	2006 – Million tonnes –	2007	Change, %
Japan	62.1	66.8	7.6
Korea	25.3	25.6	1.2
Taiwan	7.8	8.3	7.0
India	6.2	8.4	35.9
China	0.7	2.9	324.0
Total	**102.0**	**112.1**	**9.9**

Source: FGE, Honolulu

LNG IMPORTS IN EUROPE

Importers	2006 – Million tonnes –	2007	Change, %
Spain	19.0	18.9	–0.4
France	11.2	9.9	–13.0
Turkey	3.4	4.2	24.6
Italy	2.4	2.1	–10.9
Belgium	3.2	2.0	–36.0
Portugal	1.7	2.1	21.6
Greece	0.4	0.7	71.6
UK	2.6	1.1	–55.6
Norway	0.1	—	–100.0
Total	**43.9**	**41.1**	**–6.6**

Source: FGE, Honolulu

LNG IMPORTS IN AMERICAS

Importers	2006 – Million tonnes –	2007	Change, %
US	12.2	16.2	32.1
Puerto Rico	0.5	0.6	2.9
Dominican Republic	0.2	0.5	92.0
Mexico	0.6	2.0	237.8
Total	**13.6**	**19.2**	**41.0**

Source: FGE, Honolulu

GLOBAL LNG SUPPLY

Exporter	2006 – Million tonnes –	2007	Change, %
Alaska	1.2	0.9	–23.5
Australia	13.9	15.1	8.5
Brunei	7.4	7.6	2.8
Indonesia	22.4	20.7	–7.3
Malaysia	21.0	22.0	4.7
Total Pacific Basin	65.8	66.2	0.7
Abu Dhabi	5.3	5.7	8.3
Qatar	24.4	29.4	20.6
Oman	8.5	8.9	5.3
Total Middle East	38.1	44.0	15.7
Algeria	18.0	18.6	3.8
Libya	0.5	0.6	18.4
Egypt	10.8	10.4	–3.9
Trinidad	12.8	14.8	12.0
Nigeria	13.5	16.1	25.0
Equatorial Guinea	—	1.5	NA
Norway	—	0.1	NA
Total Atlantic Basin	55.6	62.2	11.6
Total world	**159.4**	**172.2**	**8.0**

Source: FGE, Honolulu

US LNG TERMINALS, FERC-REGULATED

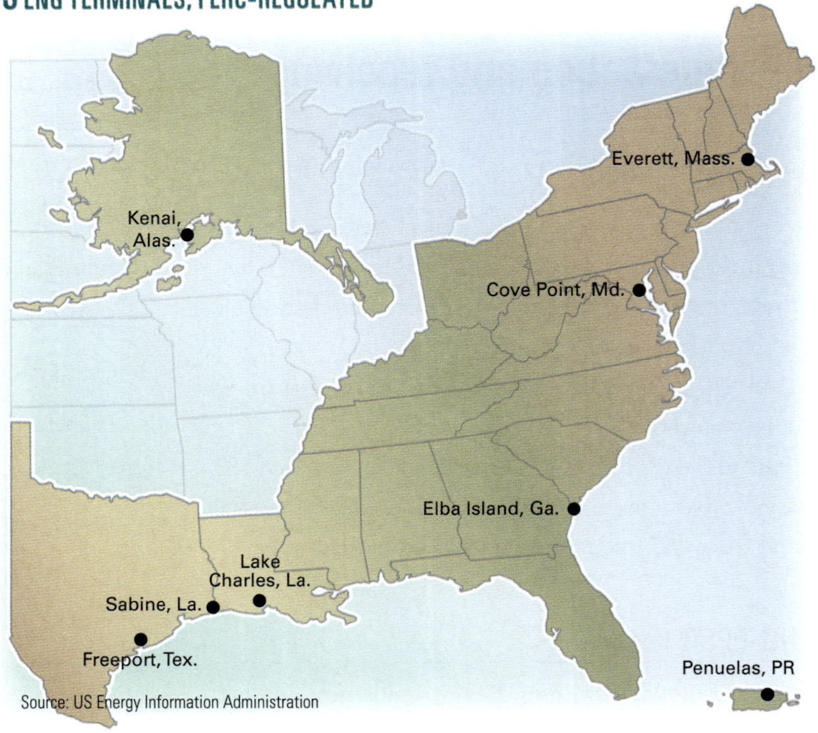

Source: US Energy Information Administration

LNG DEMAND GROWTH: 2006-07

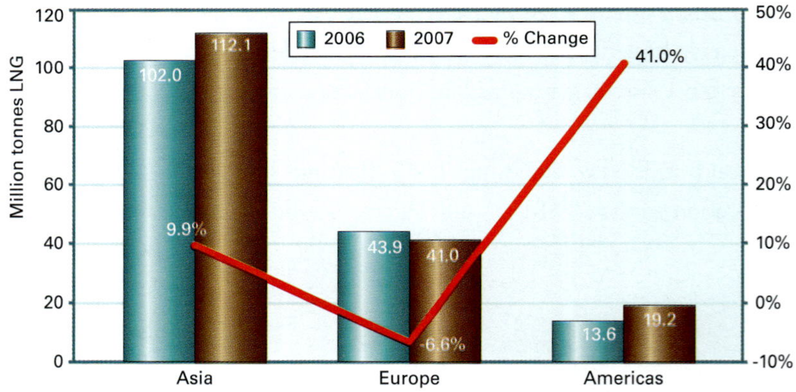

LNG SUPPLY GROWTH: 2006-07

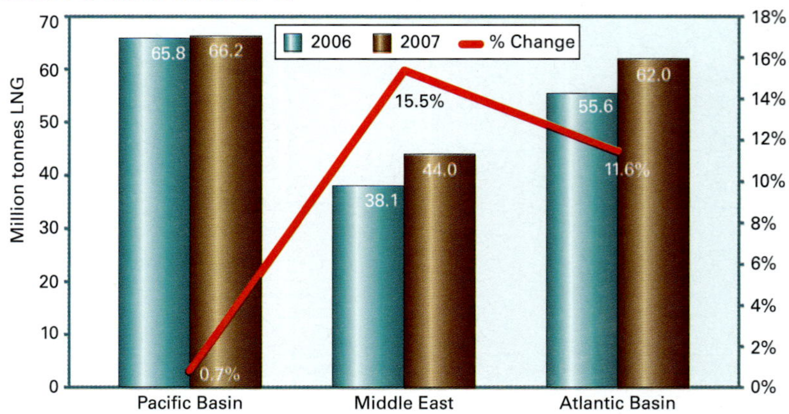

Section 8:

LNG—Carrier fleet

2008 snapshot

As of early August 2008, there were 275 LNG carriers in service, according to a survey compiled by Technip USA Inc. (Houston).

The largest commissioned vessels at that time were Teekay LNG's *Al Huwaila* and its sister ships, with a capacity of 217,000 cu m.

About 102 carriers were on order, with 28 planned for delivery during the balance of 2008.

Vessels with capacities as large as 267,000 cu m are under construction, measuring 990–1,150 ft (302–350 m) long and 164–180 ft (50–55 m) wide.

A 138,000-cu-m vessel would have the following delivery schedule, depending on terminal capacity:

- 5.8 mtpy (0.75 bcfd): one vessel about every 4 days.
- 10 mtpy (1.3 bcfd): one vessel every 2 days.
- 20 mtpy: (2.6 bcfd): one vessel every day.

Chinese-built carriers

The North West Shelf Joint Venture partners in Australia took delivery in early 2008 of the first of three Chinese-built LNG carriers.

The vessels will transport LNG from Australia's Burrup Peninsula plant to the Dapeng receiving terminal in Guandong Province, China.

The *Dapeng Sun* was delivered at Hudong-Zhonghua shipyard in Shanghai.

The second vessel, *Dapeng Moon*, was scheduled to be completed by year end 2008.

The NWS group has a 25-yr supply deal with the Guangdong consortium buyer to deliver 3.3 mtpy of LNG to Dapeng on a fob basis.

2009: More short-haul voyages seen

In late 2008, Waterborne Energy (Houston) projected that the global LNG market will start slowly in 2009, but will likely see a 30% rise in LNG production by year end.

The firm expects that 2.8 tcf (79.2 bcm) of new LNG production capacity will come online in first-half 2009, if all projects remain on schedule.

By mid-year, excess global LNG likely will begin to move toward U.S. import terminals, with Lake Charles and Cove Point being the first to receive LNG cargoes.

Waterborne Energy also expects that the LNG tanker market will face some challenges in 2009, with 45 new tankers to be delivered in 2009. The firm expects an increase in shorter voyages and fewer long-haul voyages, with some of the new capacity dedicated to new projects.

LNG CARRIERS

Capacity, cu m	266,000	216,000	155,000	138,000	137,500	125,000	87,600
Containment system	Membrane	Membrane	Membrane	Membrane	Spherical	Spherical	Spherical
Type	Qmax	Qflex	—	Repsol	Adgas	North West Shelf	Norman Lady
Deadweight, tonnes	125,700	101,000	75,000	68,200	71,543	66,875	50,922
Length overall, ft	1,032	1,033	935	933.1	961.3	892.4	818.6
Beam, ft	180	164	142.4	139.4	150.3	154.8	131.3
Maximum draft, ft	39.4	39.4	38.7	37.4	37.1	36.1	34.8
Air draft (ballast), ft	174	174	137.5	147.3	—	—	—

Section 9:
LNG—Risks and opportunities

Global market evolution

At a mid-2008 conference, BG Group Ltd. (London) said that the wave of U.S. LNG import contracts signed in 2003–05 laid the foundation for evolution to a "more globally connected, flexible, and integrated market from largely regional, bilateral trade."

Rapidly increasing commodity prices since then have increased uncertainty in both supply and demand as LNG starts to drive globalization of natural gas in a high-priced, supply-tight world. And as the industry has become more interconnected, predicting how the global system will function has become more difficult.

BG believes that three fundamentals will define the future LNG industry: price, market efficiency, and volatility.

Price

Economic growth, new players, supply security, and environmental concerns are boosting LNG demand. But cost escalation, construction capacity, and domestic priorities are slowing the arrival of new supplies.

This discrepancy has driven up prices for both term and spot volumes, which is good for sellers. However, its effect on the LNG industry is less certain.

Market efficiency

Increasing trade flexibility is allowing LNG in many cases to respond to price signals and flow to the highest-value markets worldwide. Spare transport and import infrastructure are emerging to enable this.

But inhibiting market efficiency are constrained market and resource access, infrastructure costs, existing long-term agreements, and lack of price discovery.

Volatility

While increasingly flexible volume should smooth out fluctuations of individual markets, the U.S. market will likely take on the task of balancing seasonal demand for the rest of the globe. Global volatility may be transmitted to the NYMEX, rather than the NYMEX smoothing global volatility.

Cargo-swapping

Wood Mackenzie (Edinburgh) proposed wider use of LNG "cargo-swapping." Instead of two ships embarking from different sides of the world and passing each other on the way to supply remote markets, the parties agree to supply the markets closest to them, saving time, fuel, and money.

This approach could make sense, said the firm, in view of increased LNG traffic between the Atlantic and Pacific basins.

WoodMac said that swappable Atlantic-Pacific volumes increased to 3.4 mtpy (7.8 MMcm) in 2006 from 0.3 mtpy (688,000 cu m) in 2005, and were estimated to top 6.0 mtpy (13.8 MMcm) in 2007.

Atlantic-Sourced Cargoes in Pacific Basin

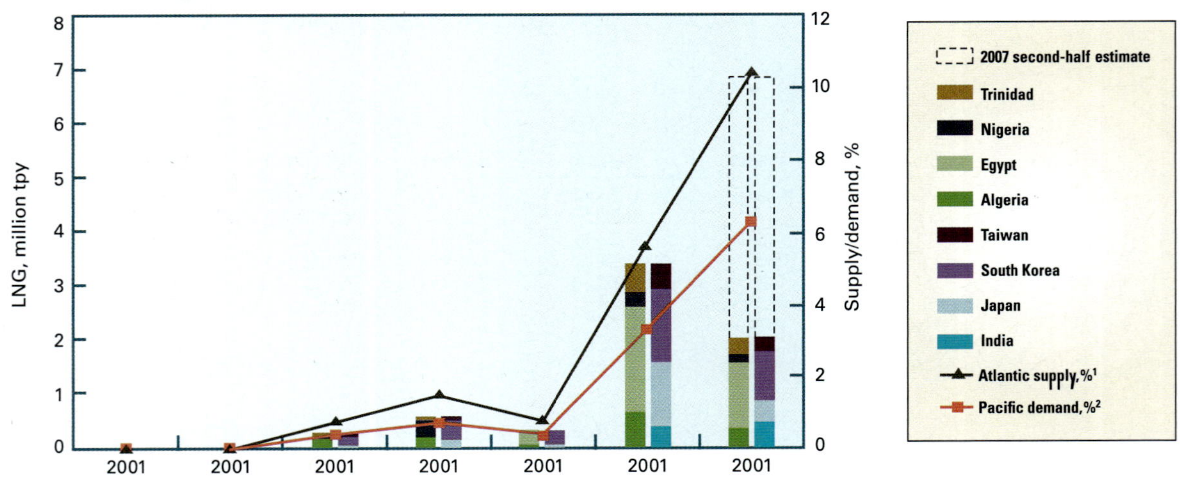

[1]Atlantic supply indicates the proportion of Atlantic supply going into the Pacific Basin–based on supply from Algeria, Egypt, Equatorial Guinea, Nigeria, and Trinidad.
[2]Pacific demand indicates the proportion of Pacific Basin imports from the Atlantic Basin–based on imports into China, India, Japan, South Korea, and Taiwan.

Approximately 120 cargoes could have been swapped in 2006. Doing so would have saved more than 700,000 nautical miles, 1,900 days (5 yrs) of sailing time, 260,000 tonnes of fuel-oil equivalent, and about $240 million in costs (about $4 million/swap, $2 million/cargo).

Factors limiting swaps are contractual constraints, execution details, and a lack of transparency across the industry. WoodMac suggested formation of a forum—an "LNG Swap Club" of suppliers and buyers within each basin—that could expedite the process.

Summary of Trans-Basin LNG Flows: 2006[1]

Flow	Fig. 2 key	Volume, million tpy[2]	No. of cargoes	Distance, nautical miles[2]	Time, days[2]	Fuel consumption, tonnes fuel-oil equivalent[2]	Cost, million $[2]
Pacific ➤ Atlantic	Blue dotted	5.3	89	838,498	2,389	313,414	296
Pacific ➤ Atlantic (swapable[3])	Blue dotted	3.4	60	575,242	1,636	214,972	201
Atlantic ➤ Pacific	Green solid	3.4	60	1,006,386	2,540	370,978	285
Swaps	Red dashed	3.4	120	866,100	2,289	327,041	246
Savings				715,528	1,887	258,908	240

[1]See shipping assumptions section for assumptions used. FOE = fuel-oil equivalent; figure includes an assumption for fuel oil consumed and boiloff in transit converted to a FOE figure. [2]Based on estimated figures only. Tonnes of FOE based upon fuel oil consumed in transit and assumed boiloff of LNG converted to FOE. [3]Swappable amount is constrained by the smallest directional flow.

Assumptions used to create the table titled, "Summary of Trans-Basin LNG Flows: 2006:"

- Shipping journeys for each cargo are based on return journeys from and to the supply project.
- Trade benefits are based on a simple annual (not monthly) calculation.
- The maximum swappable volume is based on the lower of the two flows between the two basins. The trade from the Atlantic to the Pacific is the smaller of the two. This figure was 3.4 mtpy (7.8 million cu m) of LNG in 2006.
- The estimated 60 Atlantic-to-Pacific shipments in 2006 are swapped for 60 Atlantic-to-Atlantic journeys. And 60 Pacific-to-Atlantic journeys are swapped for 60 Pacific-to-Pacific journeys.
- Shipping costs are based on a vessel of 145,000 cu m.

Trans-Basin LNG Flows: 2006

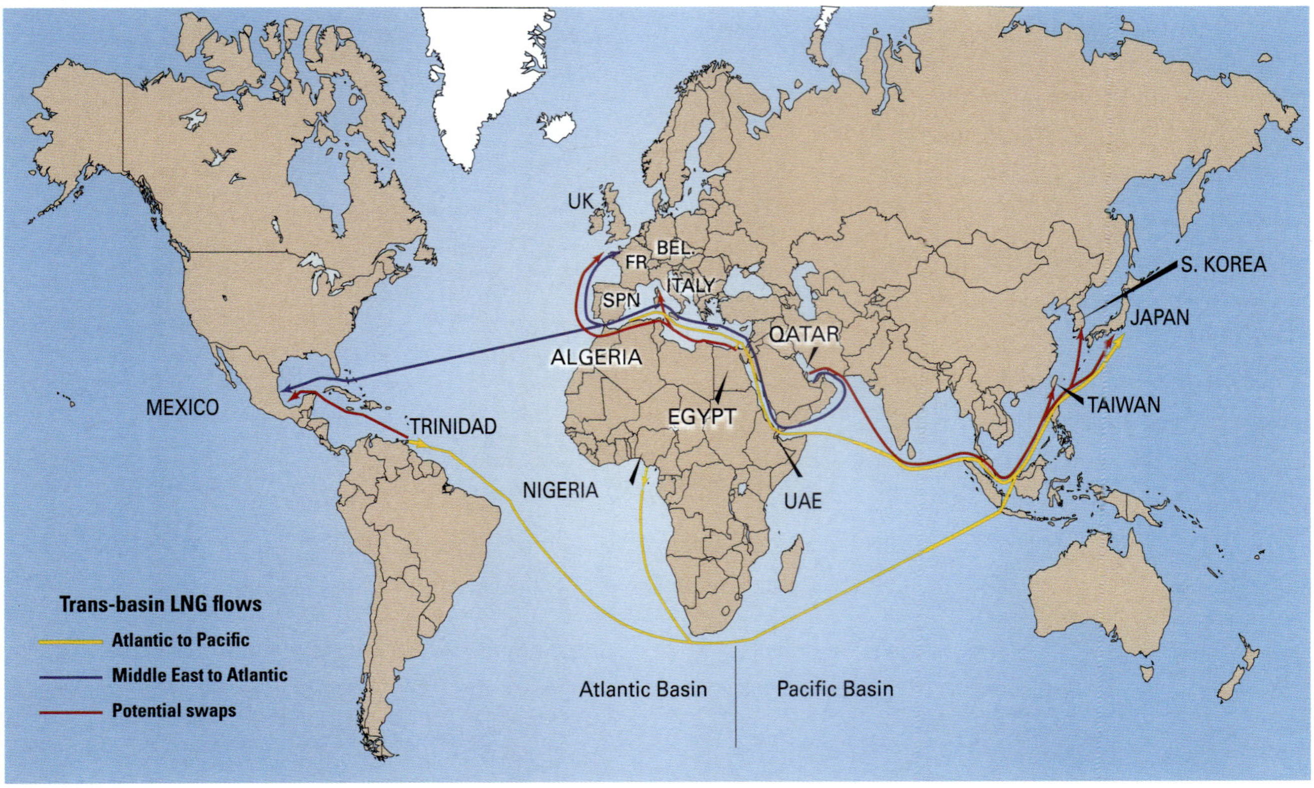

Section 10:
Refining and products

Global refining capacity growth

In 2008, growth in crude distillation capacity continued to build on the momentum of past years.

Refiners added more capacity than in the past two years, according to a year-end survey published by *Oil & Gas Journal.* For the seventh year running, worldwide capacity was at a record high.

A new refinery start-up in China, expansions in other refineries, and capacity creep that offset losses were the main reasons for the capacity increases in the 2008 survey.

Asia saw the largest increase—up about 269,000 b/cd, or 1.2%. Western Europe added about 34,000 b/cd, or 0.2%. North America showed a slight net decrease of 7,500 b/cd. Other regions had no net gain or loss in stated capacity.

The survey lists one new refinery. In mid-June, Sinopec (Beijing) started up a $1.8-billion refinery in Qingdao, China. The 200,000-b/cd facility will produce transportation fuels and petrochemical products.

No refineries shut down in 2008. High margins early in the year made most refineries profitable.

MOL Hungarian Oil & Gas Co.'s refinery in Tiszaujvaros, and Koramo Kolin's refinery in the Czech Republic were removed from the survey because they process no raw crude.

Worldwide refining

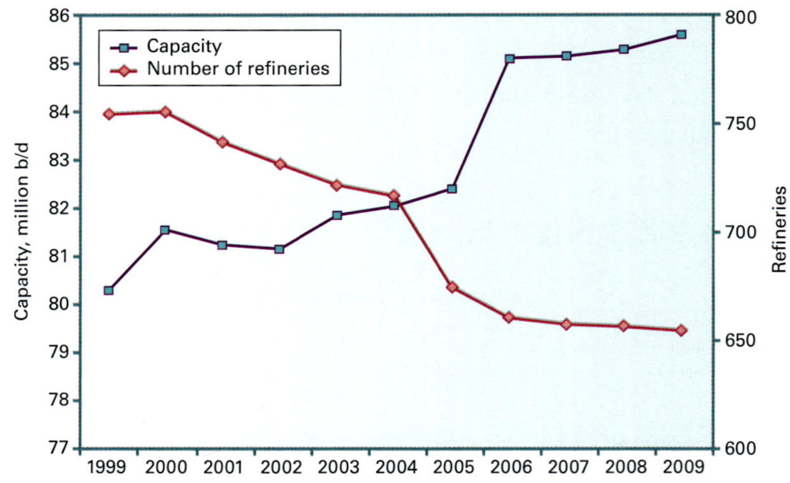

How the world's largest refiners rank

Rank Jan. 1, 2009	Rank Jan. 1, 2008	Company	Crude capacity, b/cd[1]
1	1	ExxonMobil Corp.	5,632,000
2	2	Royal Dutch Shell PLC	4,599,000
3	3	Sinopec	3,811,000
4	4	BP PLC	3,328,000
5	6	ConocoPhillips	2,696,000
6	7	Petroleos de Venezuela SA	2,678,000
7	5	Total SA	2,655,000
8	8	Valero Energy Corp.	2,596,000
9	9	China National Petroleum Corp.	2,440,000
10	10	Saudi Aramco	2,433,000
11	12	Petroleo Brasileiro SA	1,997,000
12	11	Chevron Corp.[2]	1,981,000
13	13	Petroleos Mexicanos	1,703,000
14	14	National Iranian Oil Co.	1,451,000
15	17	Nippon Oil Co. Ltd.	1,317,000
16	15	Rosneft	1,293,000
17	16	OAO Lukoil	1,217,000
18	18	Repsol YPF SA	1,105,000
19	19	Kuwait National Petroleum Co.	1,085,000
20	21	Marathon Oil Corp.	1,016,000
21	20	Pertamina	993,000
22	22	Agip Petroli SPA	904,000
23	23	Sunoco Inc.	880,000
24	24	Flint Hills Resources	817,000
25	25	SK Corp.	817,000

[1]Includes partial interests in refineries not wholly owned by the company. [2]Includes holdings in Caltex.

Statistical Tables

See Tables 9–12 in the Statistical Tables chapter of this book for more details about worldwide refining operations.

Also, note that subscribers to *Oil & Gas Journal* can download, free of charge, the text version of the *OGJ Worldwide Refining Report 2008* tables from www.ogjonline.com.

Click on the Resource Center tab, then the Surveys and OGJ Subscriber Surveys links.

Companies with 200,000+ b/cd refining capacity in Asia, the US, Western Europe

Rank	Company	No. of refineries	Crude capacity, b/cd[1]
Asia[2]			
1	Sinopec	27	3,811,000
2	China National Petroleum Corp.	24	2,425,000
3	ExxonMobil Corp.	10	1,337,507
4	Nippon Oil Co. Ltd.	7	1,317,000
5	Royal Dutch Shell PLC	13	1,271,875
6	Pertamina	8	992,745
7	SK Corp.	1	817,000
8	Indian Oil Co. Ltd.	10	787,290
9	Chinese Petroleum Corp.	3	770,000
10	LG-Caltex Corp.	1	[3]680,000
11	Chevron Corp.	6	668,667
12	Reliance Petroleum Ltd.	1	660,000
13	Tonen/General Sekiyu Seisei KK	4	[4]633,750
14	Idemitsu Kosan Co. Ltd.	4	608,000
15	Hyundai Oil Refinery Co.	3	589,500
16	Cosmo Oil Co. Ltd.	4	565,250
17	S-Oil Corp.	1	[5]520,000
17	Formosa Petrochemical Co.	1	520,000
19	BP PLC	4	348,254
20	Saudi Aramco	6	327,652
21	Hindustan Petroleum Corp. Ltd.	2	296,250
22	Showa Yokkaichi Sekiyu Co. Ltd.	1	[6]205,010
US			
1	Valero Energy Corp.	13	2,065,660
2	ConocoPhillips	13	2,000,200
3	ExxonMobil Corp.	7	1,965,750
4	BP PLC	6	1,401,525
5	Marathon Oil Corp.	7	1,016,000
6	Royal Dutch Shell PLC	8	[7]979,250
7	Chevron Corp.	5	914,000
8	Sunoco Inc.	5	880,000
9	Petroleos de Venezuela SA	4	[8]849,400
10	Flint Hills Resources (Koch Industries)	3	817,475
11	Motiva Enterprises LLC[9]	3	730,000
12	Tesoro Corp.	7	658,000
13	Saudi Aramco	3	[10]370,000
14	LyondellBasell	1	268,000
15	Husky Energy Inc.	2	237,500
16	Total SA	1	231,452
17	EnCana Corp.	2	226,000
18	Alon USA	3	206,000
Western Europe[11]			
1	Total SA	15	2,267,116
2	ExxonMobil Corp.	9	1,636,856
3	Royal Dutch Shell PLC	11	1,551,801
4	BP PLC	8	900,736
5	AgipPetroli SPA	10	876,117
6	Petroplus International NV	7	792,000
7	Repsol YPF SA	5	709,200
8	Turkish Petroleum Refineries Corp.	4	613,275
9	ConocoPhillips	4	610,125
10	Compania Espanola de Petroles SA (CEPSA)	3	427,000
11	Ineos Group Holdings Inc.	2	402,800
12	OMV AG	3	398,635
13	ERG Group	4	396,214
14	Preem Raffinaderi AB	2	316,000
15	Hellenic Petroleum SA	3	313,000
16	Neste Oil	6	[12]306,100
17	Statoil AS	3	304,210
18	Galp Energia SA	2	304,172
19	Saras SPA	1	300,000
20	Petroleos de Venezuela SA	8	294,550
21	Chevron Corp.	1	210,000

[1] Includes partial interest in refineries not wholly owned by the company.
[2] Asia includes Australia, Bangladesh, Brunei, China (and Taiwan), India, Indonesia, Japan, Malaysia, Myanmar, New Zealand, North Korea, Pakistan, Papua New Guinea, the Philippines, Singapore, South Korea, Sri Lanka, and Thailand.
[3] Includes Caltex's 50% stake.
[4] Includes ExxonMobil Corp.'s 50% stake.
[5] Includes Saudi Aramco's 35% stake.
[6] Includes Royal Dutch Shell's 50% stake.
[7] Includes Shell's stakes in Motiva and its 50% stake in the Deer Park, Tex., refinery.
[8] Consists of PDVSA's ownership of Citgo and its 50% stake in the ExxonMobil Chalmette, La., refinery.
[9] 50/50 joint venture between Shell and Saudi Aramco.
[10] Consists of 50% stake in Motiva.
[11] Western Europe includes Austria, Belgium, Denmark, Finland, France, Germany, Greece, Ireland, Italy, the Netherlands, Norway, Portugal, Spain, Sweden, Switzerland, Turkey, and the UK.
[12] Includes 50% stake in AB Nynas refineries.

Asia Processing Capability

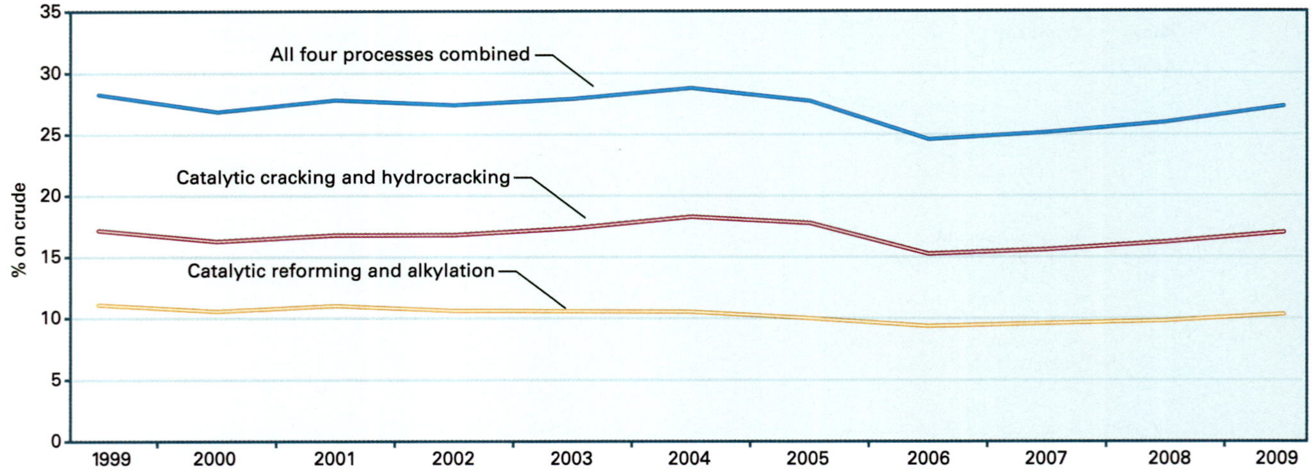

US Processing Capability

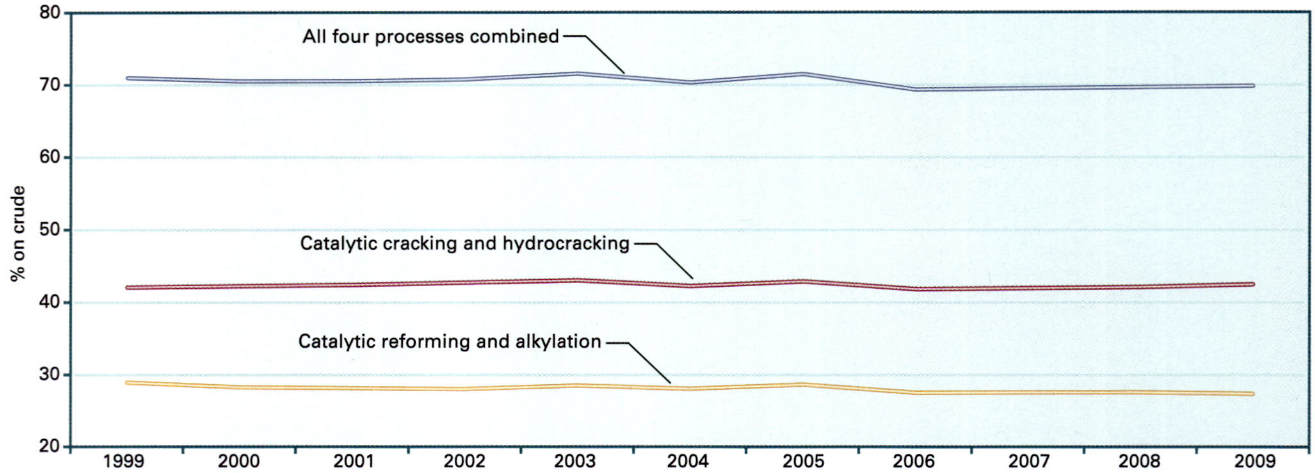

EU Processing Capability

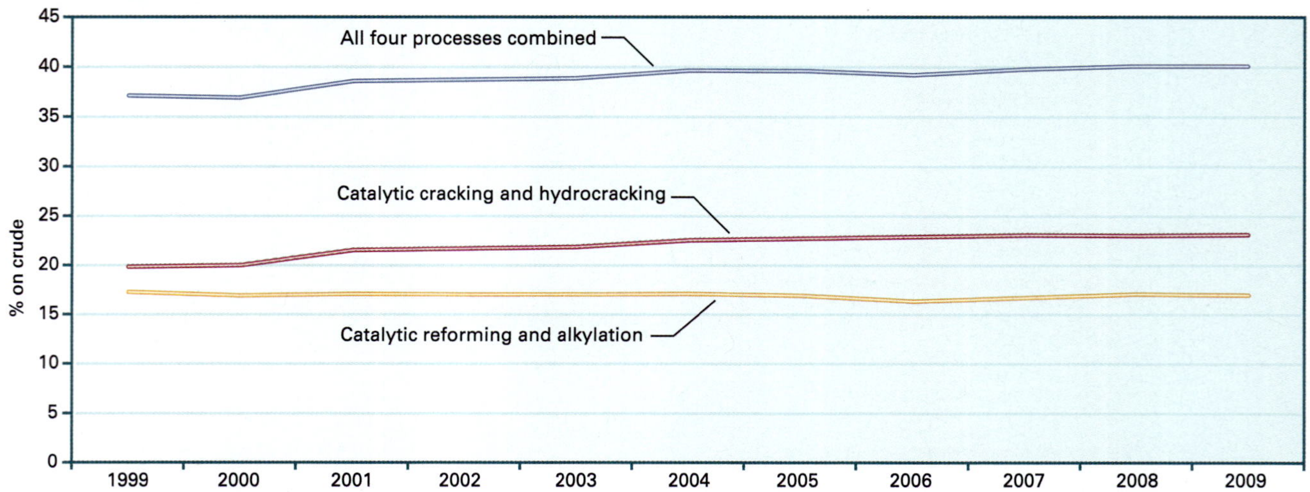

Regional look at worldwide refining operations

Region	No. of refineries	Crude distillation	Vacuum distillation	Catalytic cracking	Catalytic reforming	Catalytic hydrocracking	Catalytic hydrotreating	Coke, tonnes/day
		b/cd						
Africa	45	3,278,382	507,904	205,765	458,427	61,754	830 876	1,841
Asia	157	22,482,913	4,388,751	2,799,675	2,140,776	1,017,535	9,386 913	20,110
Eastern Europe	89	10,343,580	3,903,261	877,089	1,474,382	330,392	4,273 893	12,570
Middle East	42	7,036,215	1,983,905	364,450	652,897	602,021	2,043 563	3,300
North America	154	20,949,112	9,140,631	6,669,034	4,175,803	1,827,780	16,080 695	128,763
South America	66	6,602,903	2,845,885	1,309,507	402,195	132,400	1,904 061	24,640
Western Europe	102	14,910,507	5,785,898	2,191,672	2,183,338	1,172,716	9,879 931	11,534
Total	**655**	**85,603,612**	**28,557,235**	**14,417,192**	**11,487,818**	**5,144,598**	**44,399.932**	**202,758**

The refinery of Suncor Energy Inc. (Denver) was removed from the survey; the Denver and Commerce City refineries were operating as a single, integrated plant.

Growth in Asian refining capacity

A late-2008 evaluation by Purvin & Gertz Inc. (Singapore) showed that Asian refining capacity continues to grow rapidly despite the effects of a slumping global economy.

Fast-growing China and India are leading demand growth for transportation fuels.

World's largest refineries

	Company	Location	Crude capacity, b/cd
1	Paraguana Refining Center	Cardon/Judibana, Falcon, Venezuela	940,000
2	SK Corp.	Ulsan, South Korea	817,000
3	LG-Caltex	Yosu, South Korea	680,000
4	Reliance Industries Ltd.	Jamnagar, India	660,000
5	ExxonMobil Refining & Supply Co.	Jurong/Pulau Ayer Chawan, Singapore	605,000
6	ExxonMobil Refining & Supply Co.	Baytown, Tex.	572,500
7	Saudi Arabian Oil Co. (Saudi Aramco)	Ras Tanura, Saudi Arabia	550,000
8	Formosa Petrochemical Co.	Mailiao, Taiwan	520,000
8	S-Oil Corp.	Onsan, South Korea	520,000
10	ExxonMobil Refining & Supply Co.	Baton Rouge, La.	503,000
11	Hovensa LLC	St. Croix, Virgin Islands	500,000
12	BP PLC	Texas City, Tex.	451,250
13	Shell Eastern Petroleum (Pte.) Ltd.	Pulau Bukom, Singapore	449,000
14	Kuwait National Petroleum Co.	Mina Al-Ahmadi, Kuwait	442,700
15	Citgo Petroleum Corp.	Lake Charles, La.	440,000
16	Shell Nederland Raffinaderij BV	Pernis, Netherlands	406,000
17	Sinopec	Zhenhai, China	403,000
18	Saudi Arabian Oil Co. (Saudi Aramco)	Rabigh, Saudi Arabia	400,000
19	Saudi Aramco-Mobil	Yanbu, Saudi Arabia	400,000

Asia petroleum demand

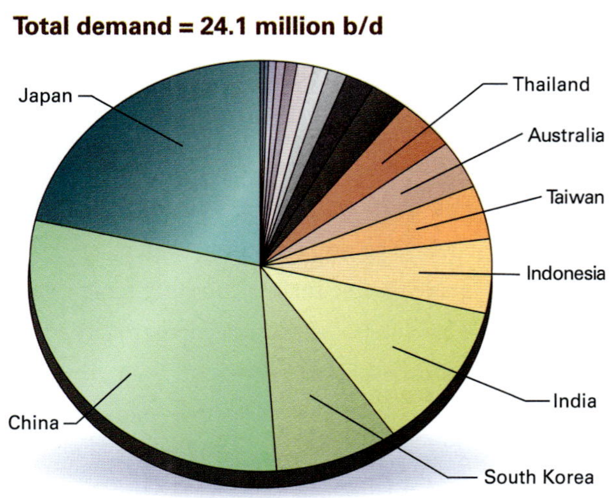

Asia diesel demand*

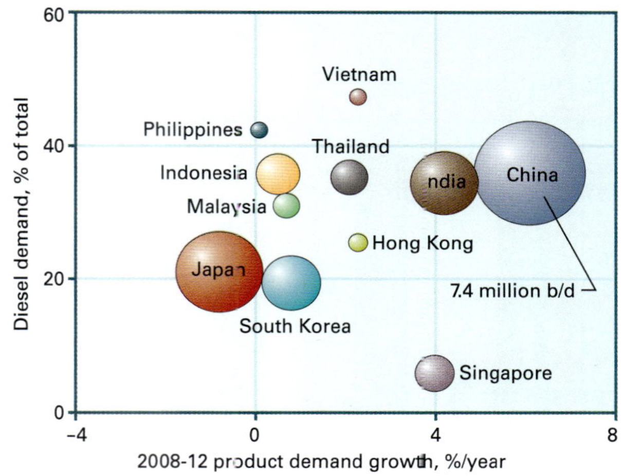

*Size of each bubble refers to total product demand.

South China gasoline

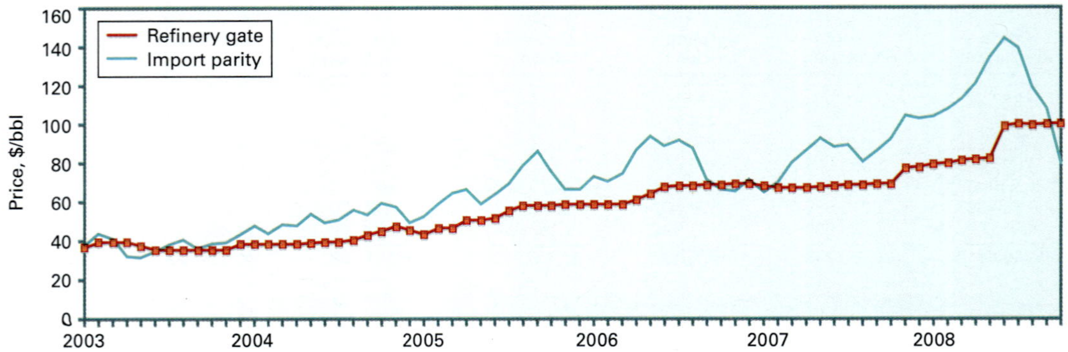

China capacity additions*

Year	Company	Location	Crude distillation capacity, 1,000 b/d
2008	CNPC Fushun Petrochemical	Fushun	160
	CNPC Dushanzi Refining	Karamay	200
	Sinopec Group	Qingdao	200
	Qinzhou Refinery	Guangxi	200
2009	CNOOC	Huizhou	240
	Sinopec Tianjin Petrochemical	Tianjin	200
	Fujian Oil Refining	Quanzhou	160

*Includes only projects or expansions of more than 100,000 b/d of capacity.

India capacity additions*

Year	Company	Location	Crude distillation capacity, 1,000 b/d
2008	Reliance Petroleum Ltd.	Jamnagar	540
2011	Essar Oil	Vadinar	110
	Mangalore Refining	Mangalore	106
	Nagarjuna Group, TIDCO	Cuddalore	125
	Bharat Oman Refineries Ltd.	Bina	125

*Includes only projects or expansions of more than 100,000 b/d of capacity.

Bunker fuel sulfur limits

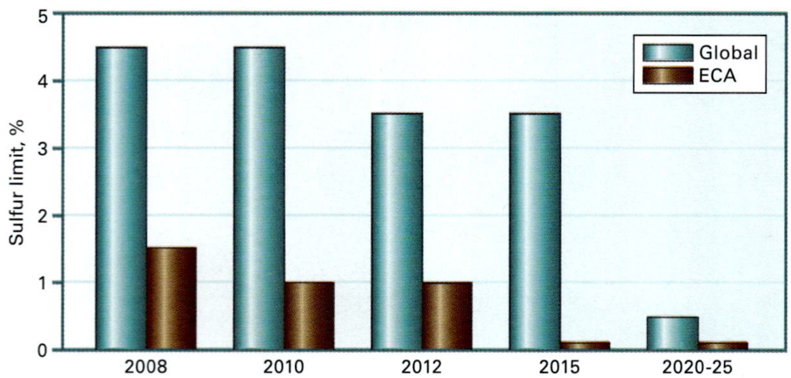

Annex VI amendments NOx

Tier	1	2	3
Application	Ships built after Jan. 1, 2000	Ships built after Jan. 1, 2011	Ships built after Jan. 1, 2016, and operating in NOx ECAs
NOx limit, g/kw-hr	17	14.4	3.4

New bunker fuel regs

Purvin & Gertz (Singapore) noted in late 2008 the recent ratification of International Marine Organization regulations that call for a decrease in the sulfur level of marine bunker fuels.

The new rules may cause refiners to invest in more desulfurization and conversion capacity.

In the past, refiners have traditionally used the unregulated marine bunker fuel pool as a way to dispose of bottom-of-the-barrel streams high in sulfur.

Annex VI Amendments also call for the control of oxides of nitrogen (NOx).

NOx controls apply to large marine engines with power output of as little as 130 kW.

The refining industry faces major obstacles in producing higher-quality marine fuels. The average sulfur content of IFO-grade marine fuels is currently more than 2.5%. Refiners may need to reduce the sulfur level to 0.5%.

Purvin & Gertz conclude that refiners require a clear signal from the shipping industry as to the quality of fuel required. The schedule is the most daunting obstacle, providing 11–16 yrs for the refining industry to respond. Only limited progress is possible until key decisions are made.

Downstream manpower shortage

In late 2008, Transmar Consult Inc. (Houston) completed a biennial survey of the global EPC industry that serves refiners and petrochemical producers.

A key finding: The global engineering contractor industry is experiencing a shortage of qualified manpower.

Transmar interviewed 109 key decision makers in refining and chemical companies in North America, China, Europe, Southeast Asia, and most of the producing countries in the Middle East.

Global refining trends

In a late 2008 study, Wood Mackenzie (Edinburgh) found that more than four of five refinery construction projects face cancellation because of the worldwide collapse in fuel demand.

Just 30 of 160 refining projects announced since 2005, scheduled for completion from 2010–15, would now go ahead. Almost all of those 30 are backed by large national oil companies.

In addition, the global refining industry is moving east after a century of western dominance.

WoodMac found that two-thirds of new global refining capacity to be built by 2015 is expected to be sited in the Middle East and Asia.

Refineries in Europe and the United States were suffering in late 2008 from a collapse in fuel demand. At the same time, Asia and the Middle East are emerging as new refining centers.

Short-term supply, demand

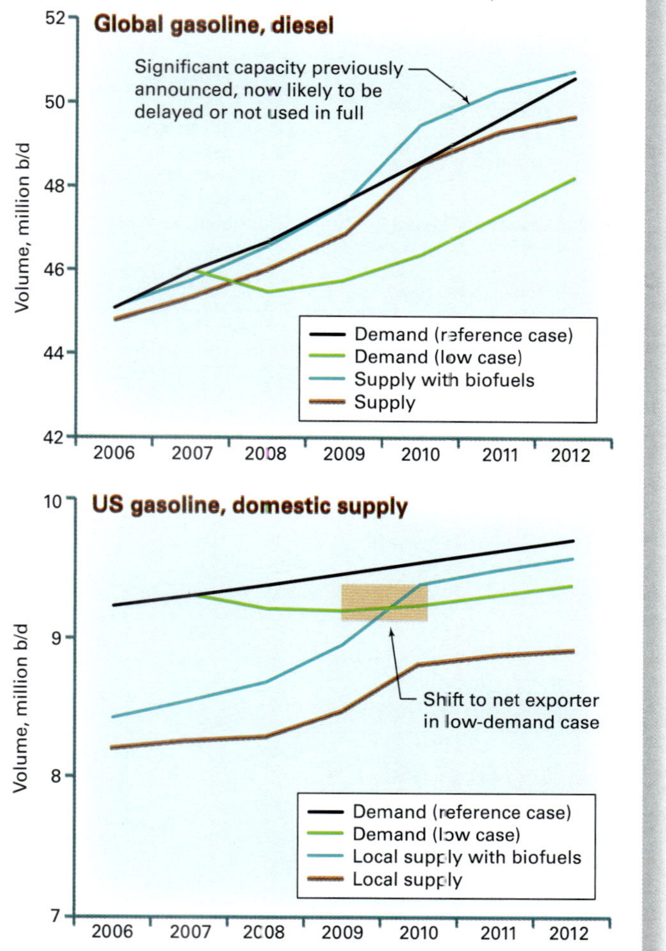

Supply is domestic only; rest of demand is fulfilled with imports.
Source: Energy Information Administration, International Energy Agency, Booz & Co.

US gasoline demand

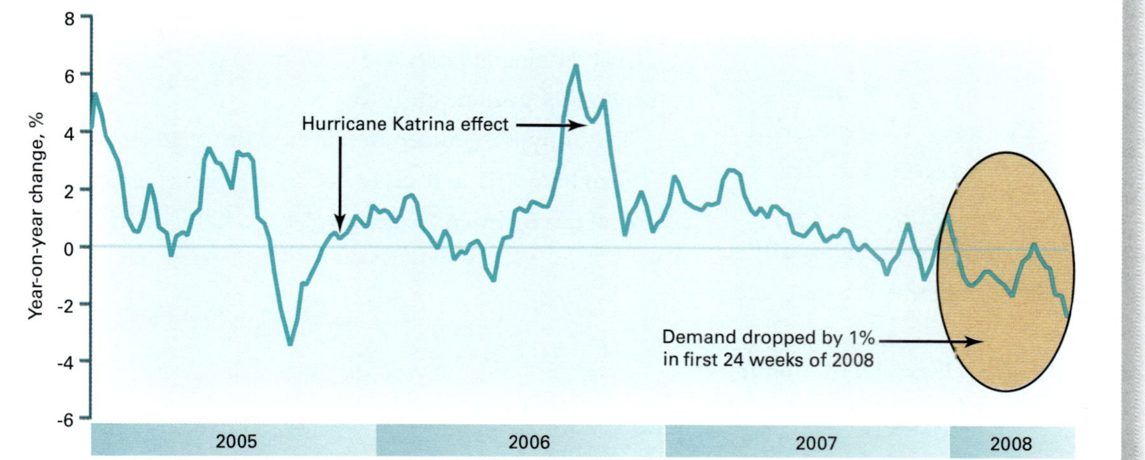

Source: Energy Information Administration, Booz & Co.

LONG-TERM DEMAND SCENARIOS

Scenario			Alternative technologies	
Growth		Economy	Hybrids	Plug-in hybrids
Robust	Continuous fleet efficiency improvement	Strong real GDP growth globally: US: 2.4%/year Europe: 1.9%/year China and India: 8.0%/year fading down to 3.0%/year Other: 1-4%/year	Continuous limited penetration	Continuous limited penetration
	High penetration of alternative vehicle technologies		From 1% of new vehicles in 2008 to 25% of new vehicles by 2032	From 0.1% of new vehicles in 2012 to 25% of new vehicles by 2042
Moderate	Continuous fleet efficiency improvement	75% of Robust growth GDP scenario	Continuous limited penetration	Continuous limited penetration
	High penetration of alternative vehicle technologies		From 1% of new vehicles in 2008 to 25% of new vehicles by 2032	From 0.1% of new vehicles in 2012 to 25% of new vehicles by 2042

Note: All cases include the effect of ultracheap cars similarly.

Source: Booz & Co.

ADDITIONAL REQUIRED REFINING CAPACITY BY 2025*

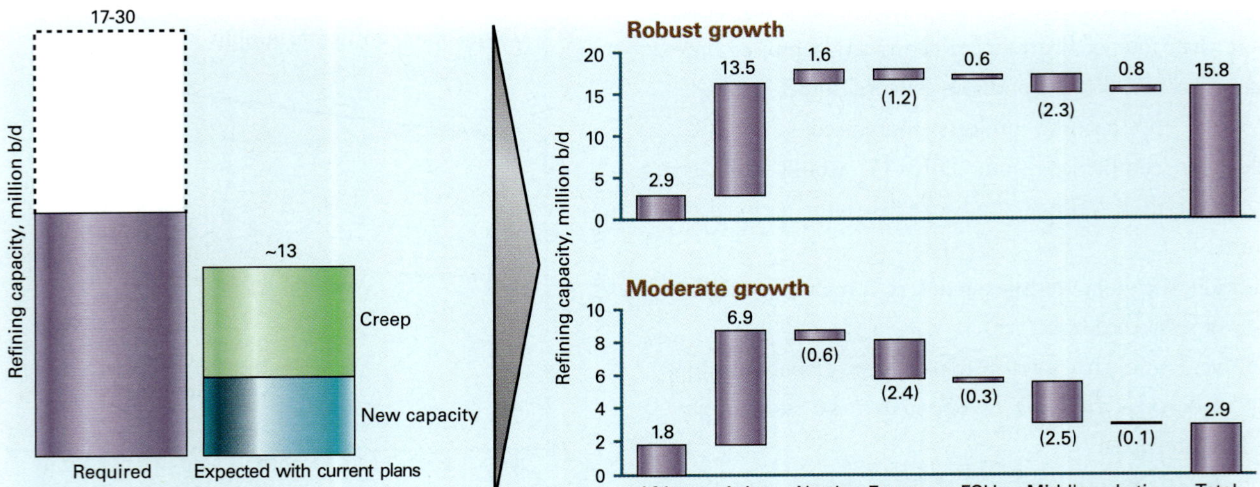

*Assumes EU and US biofuels mandates are met.
Source: Booz & Co.

The biggest projects will be led by state-owned national oil companies such as Saudi Aramco (Dhahran), Sinopec (Beijing), and Kuwait Petroleum Corp. (Kuwait City).

A big reason for the shift can be found in the *World Energy Outlook*, published in late 2008 by the International Energy Agency (IEA; Paris). The study concluded that all the growth in oil demand to 2030 is expected to come from developing countries.

China will contribute 43% to that growth. India will contribute 20%; the Middle East 20%; and other emerging economies in Asia 17%.

The rich-country share of global oil demand will drop from 59% in 2007 to less than 50% in 2030.

U.S. a gasoline exporter by 2010?

A late-2008 study by Booz & Co. (Bethesda, MD) found that shifts in energy demand and biofuels capacity will disrupt the global refining industry and could make the United States a net exporter of gasoline by 2010.

The analysis examined the effect of rising gasoline and diesel demand in Asia and other countries, biofuel mandates, and alternative fuel vehicles on the global refining industry.

US Gulf Coast Coking Refinery Netbacks*

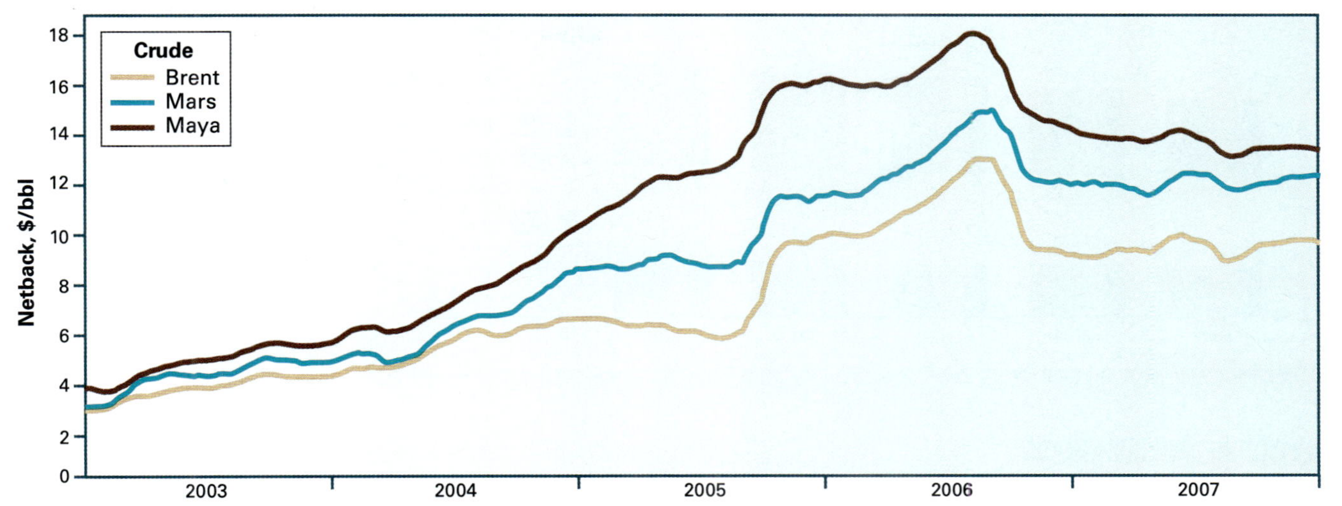

*Based on data provided to Turner, Mason & Co. and published by Platt's Oilgram. Rolling 12-month average.

US Refining, 2000s

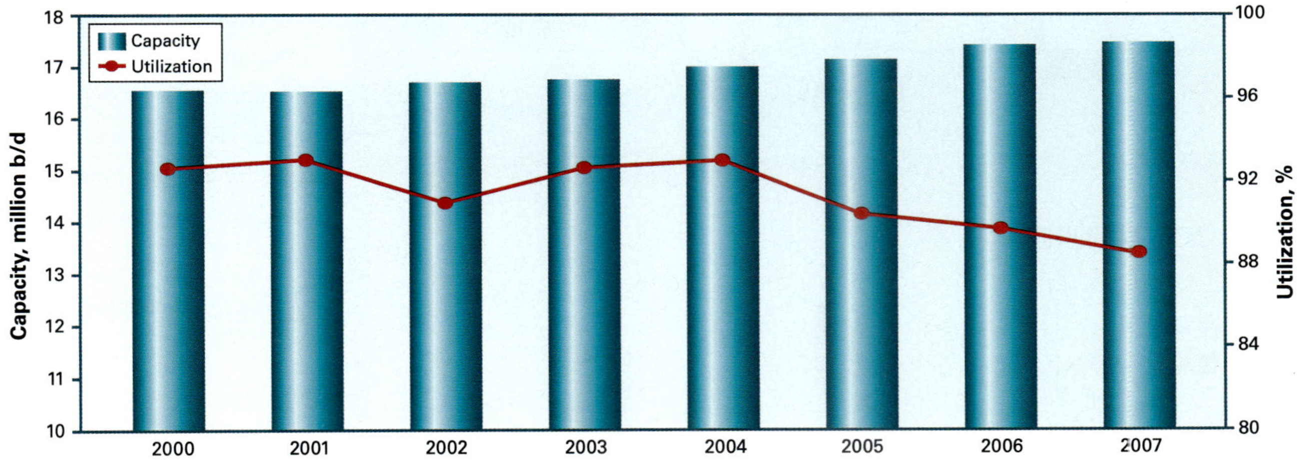

Source: US Energy Information Administration

Overall Demand Scenarios

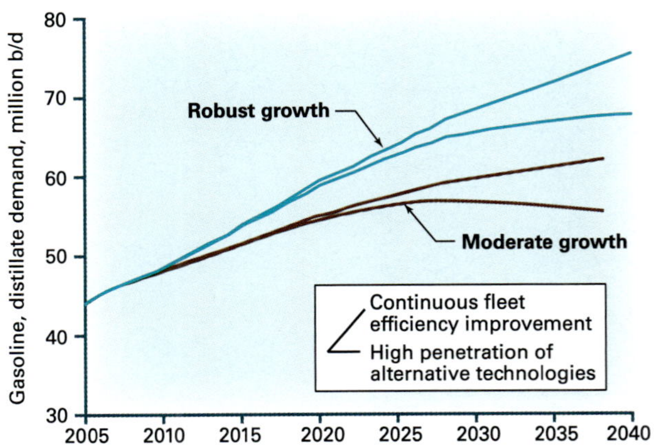

Source: OGJ, Booz & Co.

Key Regulatory Changes

Year	Requirement
1990	rvp reduction
1993	Low-sulfur diesel
1995	Reformulated gasoline
2004	Ultralow-sulfur gasoline
2006	Ultralow-sulfur diesel

Middle East Product Demand

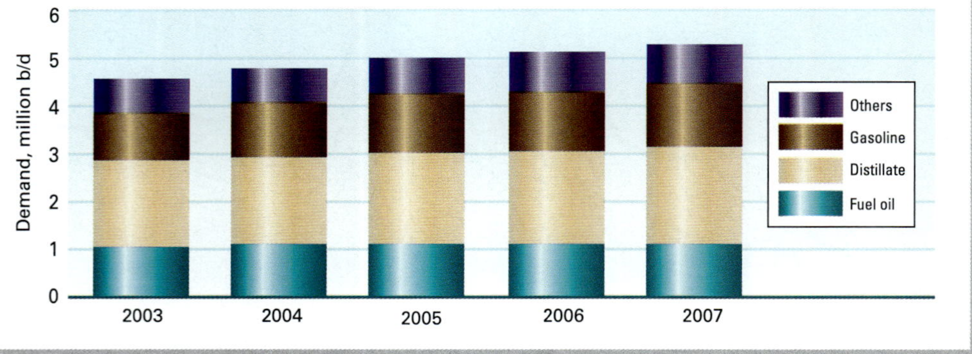

Middle East Net Product Trade

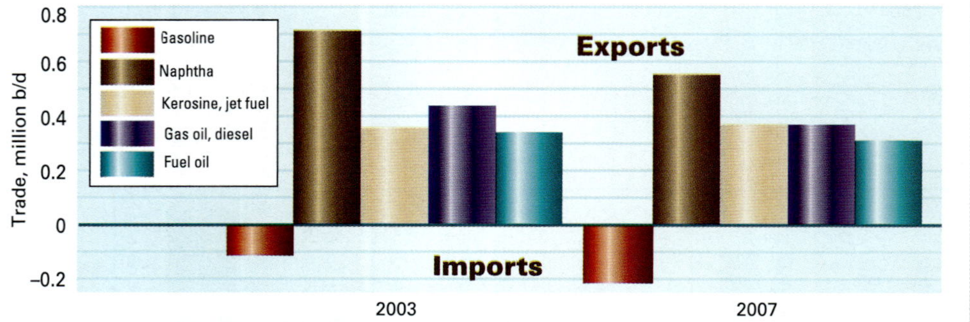

Middle East Capacity Additions

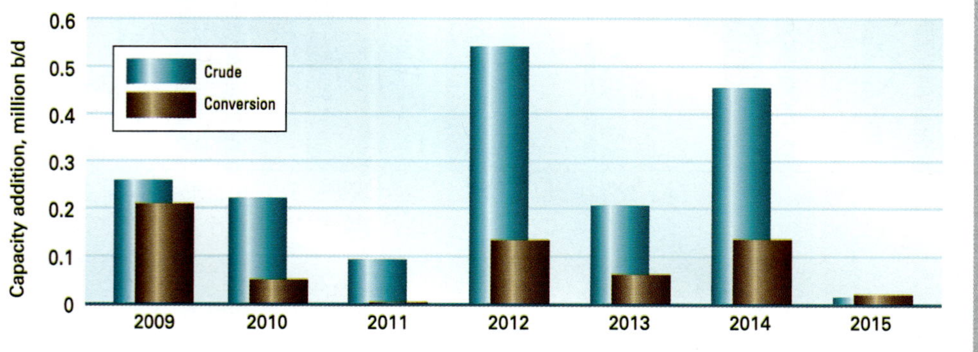

Refining Margins

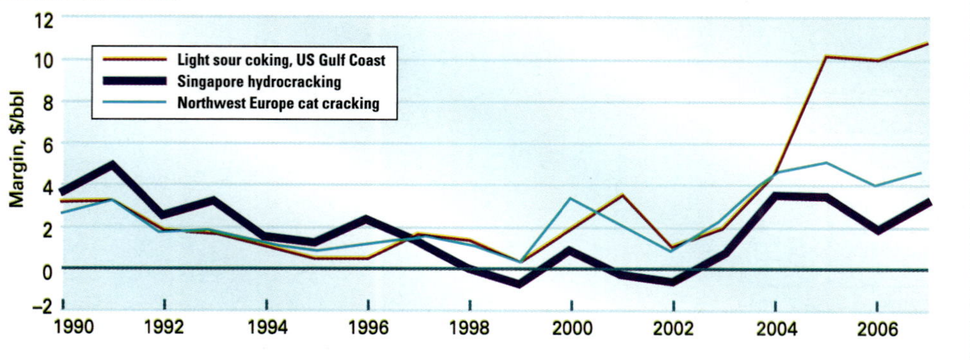

Middle East refinery investments

Purvin & Gertz Inc. (Dubai) noted in mid-2008 that development of the Middle East's oil industry during the next few years will be crucial to the outlook for the global oil industry.

Though the region likely will increase its current 31% share of global crude oil and condensate production, it also is a substantial net exporter of refined products.

In 2007, refineries produced about 6.5 million b/d compared with regional consumption of about 5.2 million b/d.

Because regional oil product consumption is increasing steadily and numerous refinery developments are proposed, potential exists for a rapid and substantial change in this position.

Middle East Refining Capacity

Country	Refineries	Capacity, 1,000 b/d	% of total
Saudi Arabia	8	2,090	28.2
Iran	9	1,661	22.4
Kuwait	3	898	12.1
UAE	5	723	9.7
Iraq	12	694	9.4
Others	11	1,352	18.2
Total	**48**	**7,418**	**100.0**

EU rules chafe

The 17-member European Petroleum Industry Association (Europia; Brussels) expressed concern in early 2008 regarding European Union environmental and energy policies.

Current EU policies, said the refining trade group, are not being matched elsewhere in the world and are increasing constraints and uncertainties as well as creating cost burdens. Of particular concern:

- Cutting greenhouse gas emissions (GHG) by at least 20% of 1990 levels by 2020.
- Mandating that 20% renewable resources be used for power production.
- Requirements that 10% biofuels be incorporated in transport fuels by 2020.
- Compulsory auctioning of CO_2 emission permits by 2013 that will trim refining margins.
- Pollution prevention rules that will make industrial operations more costly "with negligible additional human health benefits."
- Proposed global marine fuel specifications with questionable public health benefits.

The group also points out that growing refinery investments in the Middle East, India, and China are leading to increased imports of diesel into Europe.

At the same time, independents, Russian oil majors, and petrochemical firms are challenging the group's members in Europe itself.

Trade in high-acid crudes

Refiners looking for discounted crude supplies will import and use greater volumes of high total acid number (TAN) crudes, said an early-2008 study by Asia Pacific Energy Consulting (APEC; Houston).

High-TAN crudes contain substantial naphthenic acid.

The study states that incremental high-TAN crude output should rise by 1.8 million b/d during 2005–10, with 61% of that gain expected in West-of-Suez production regions.

High-TAN crude production, trade, and use will increase globally and Asia-Pacific will be the focus of inter-regional acidic crude trade through the next decade, according to the study.

The study did not include acidic crudes that have no effect on markets or trade balances.

High-TAN crude exporters

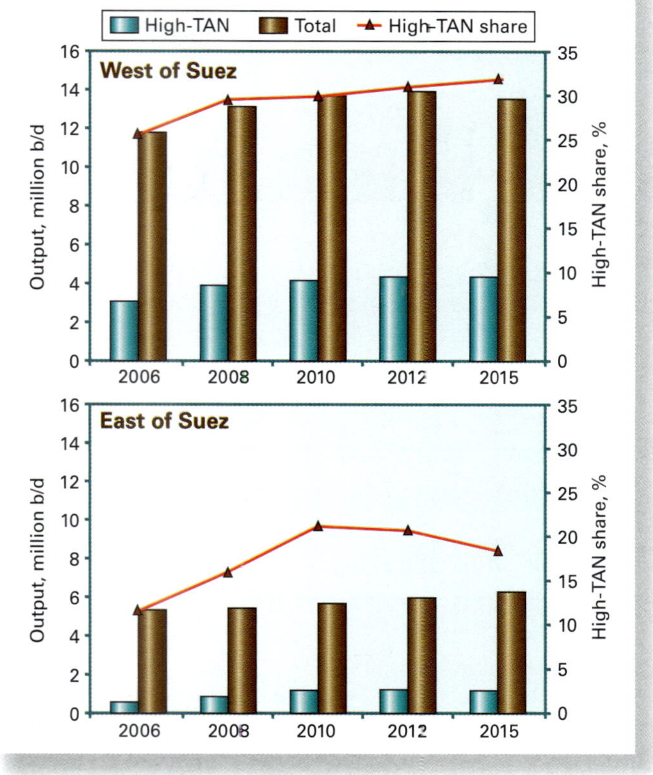

Regional production, high-TAN producers only

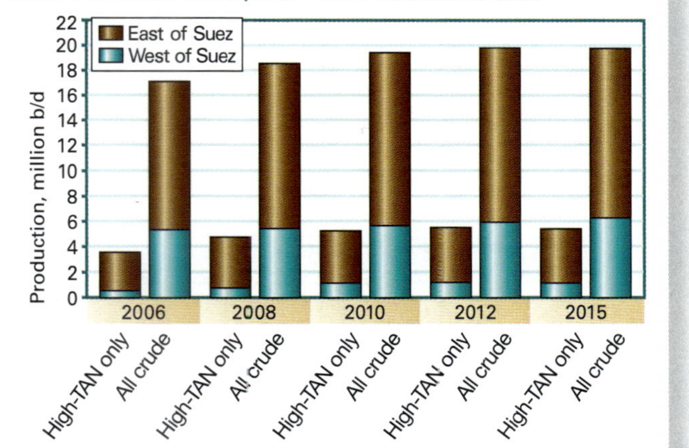

Global crude capacity: 2012

The International Energy Agency (IEA; Paris) projected that global crude distillation capacity will increase 10.6 million b/d during 2007–12.

Of this increase, 9.1 million b/d will be new capacity and 1.5 million b/d capacity creep.

The projection was part of the agency's *"Medium-Term Oil Market Report"* of early 2008.

The capacity expansion will be in response to an increase in oil demand of 9.6 million b/d by 2012.

Demand growth

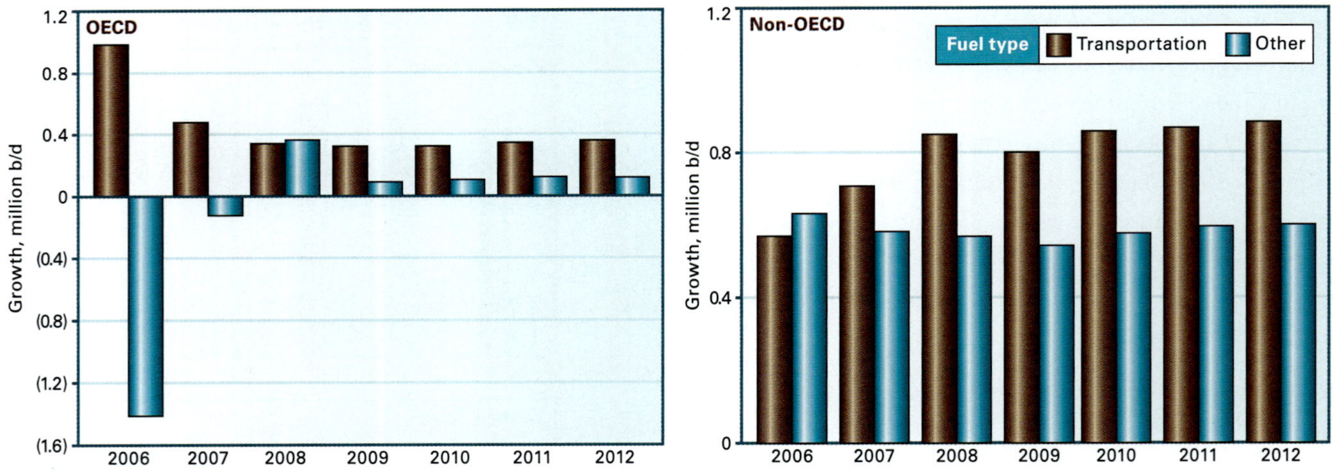

Crude distillation additions

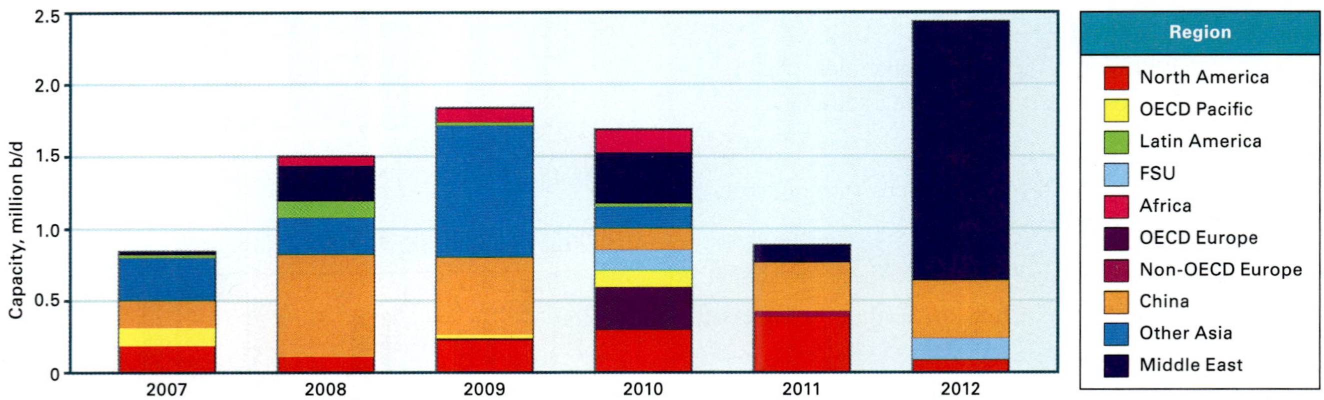

Global cumulative capacity additions*

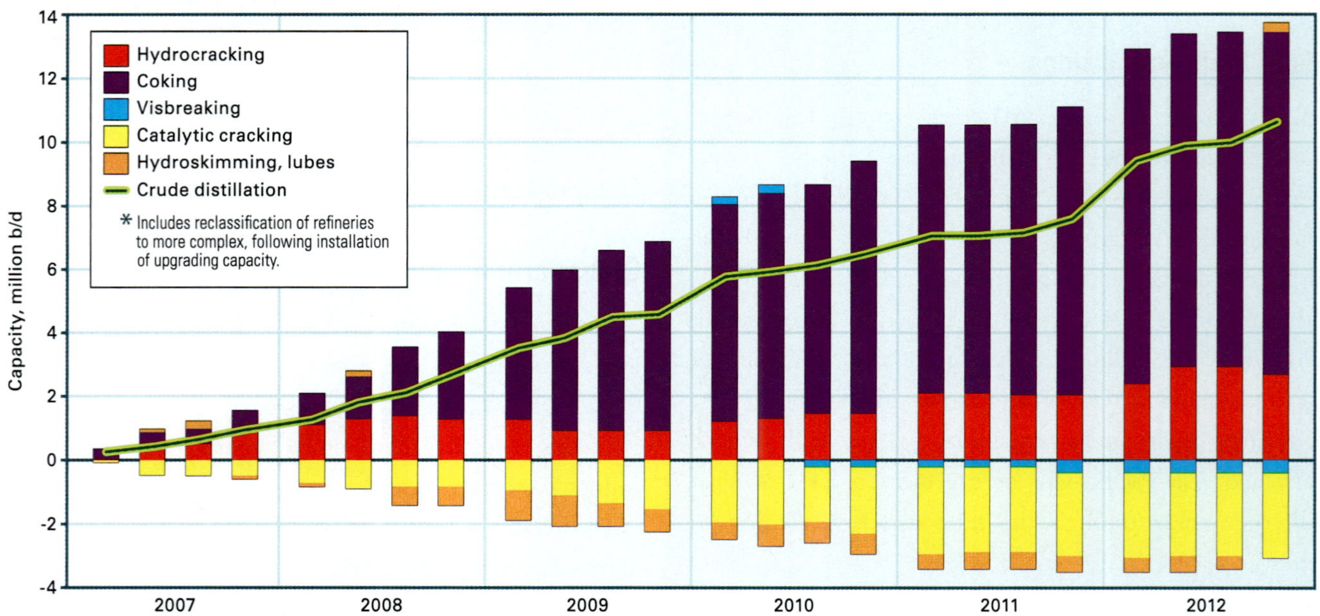

North America capacity additions

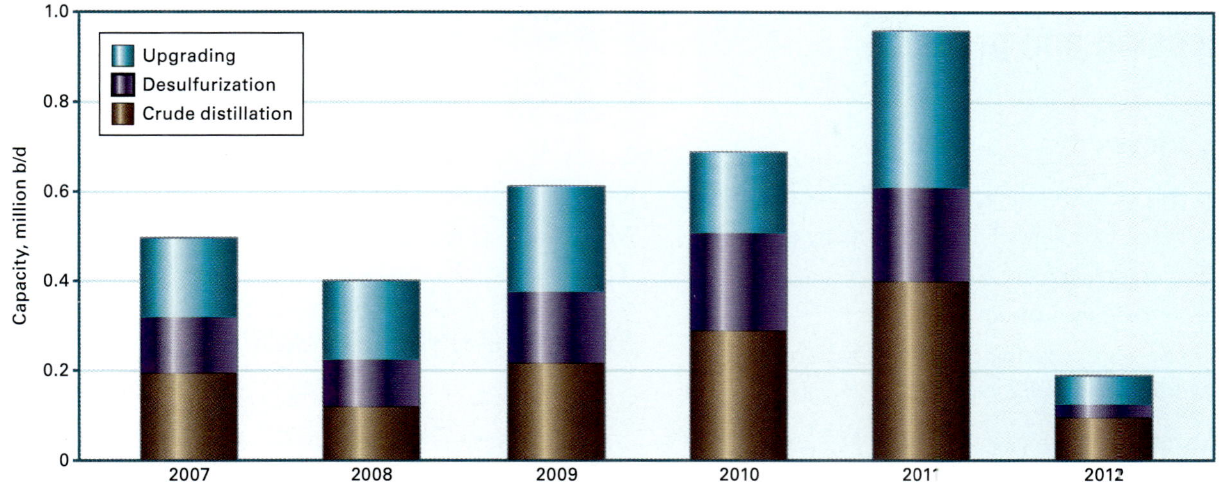

China capacity additions

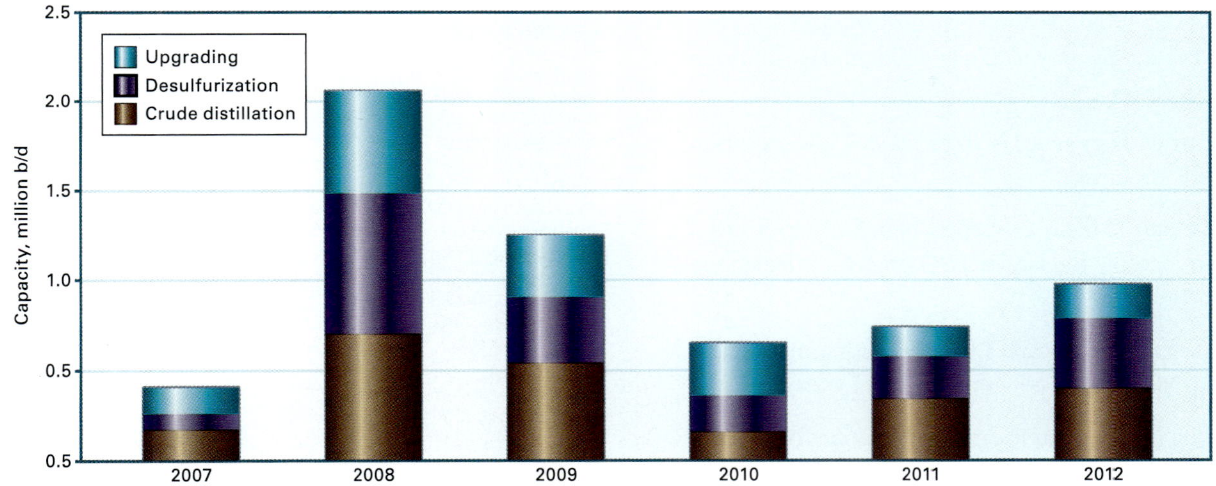

Liquids supply composition

The refining industry faces new reconfiguration and investment challenges to avoid shortages of diesel, heating oil, and jet fuel while minimizing the risk of a decline in refining capacity utilization.

That was the conclusion of an early 2008 evaluation by Cambridge Energy Research Associates (Cambridge, MA; a unit of IHS).

Growth in liquids supply capacity will be "more than sufficient" to match the volumetric increase in demand, but the "cocktail" of hydrocarbons in the liquids supply will change.

Light liquids will likely account for 32% of the total supply in 2020, up from 19% in 2007, CERA said.

Beyond 2010, the key challenge will be adding the appropriate type of conversion capacity, particularly hydrocracking, not necessarily adding more volumes of simple crude distillation capacity.

Section 11:

Gas processing and products

LPG supply, prices

Worldwide supply of LPG will expand rapidly from 2008–12, but record high prices have slowed demand in many developing markets, said Purvin & Gertz Inc. (Houston) in early 2008.

Key points from their evaluation:

- High LPG prices are reducing demand growth in residential and commercial sectors.
- A large expansion of petrochemical capacity is under way in the Middle East, which will increase base-load demand for LPG.
- Global LPG supply will increase by 43 million tonnes during 2007–12, increasing to about 272 million tonnes from 229 million tonnes.
- Overall, LPG supply growth likely will average 3.5%/year through 2012.
- LPG production is growing in most regions, the exception being North America.
- The Middle East is likely to surpass North America in a few years to become the largest LPG producing region in the world.
- The Middle East will expand LPG production more than 45% during 2007–12.

Factors affecting LPG markets in early 2008 included high global oil prices, regional supply disruptions, project delays, regional demand increases, and slowing economies in many regions.

Global processing capacity trends

Oil & Gas Journal released its global look at gas processing markets in mid-2008.

In the United States, capacity grew by barely 1%, and Canada was essentially flat.

For the entire world, Middle East processing advanced by slightly more than 2%.

Worldwide, total natural gas processing capacity outside the United States and Canada continued to outpace combined capacities in the world's two largest gas processing countries, a trend that emerged in 2005.

For 2007, U.S. and Canadian gas-plant capacities reached slightly less than 50% of world capacity. That trend was pushed by plants' capacities in the United States advancing by only 1.2%, while Canada's plants increased capacities by only 0.2%.

Note: *OGJ* subscribers can download, free, the *2008 Worldwide Gas Processing Survey* tables at www.ogjonline.com. Click on Resource Center, Surveys, OGJ Subscriber Surveys, then Worldwide Gas Processing, and choose June 23, 2008.

Worldwide Gas Processing Roundup

Region	[1]2006	[2]2007	Change	Change, %
US				
Gas capacity, MMcfd	70,218.2	71,062.7	844.5	1.2
Gas throughput, MMcfd	45,537.4	45,501.7	–35.7	–0.1
NGL production, 1,000 gpd	75,468.2	76,227.6	759.4	1.0
Proved reserves, bcf	204,385.0	211,085.0	6,700.0	3.3
CANADA				
Gas capacity, MMcfd	52,900.8	53,010.0	109.2	0.2
Gas throughput, MMcfd	29,515.9	29,946.6	430.7	1.5
NGL production, 1,000 gpd	32,330.9	32,521.8	190.9	0.6
Proved reserves, bcf	57,946.0	58,200.0	254.0	0.4
WESTERN EUROPE				
Gas capacity, MMcfd	24,349.0	24,495.0	146.0	0.6
Gas throughput, MMcfd	9,466.7	9,501.2	34.5	0.4
NGL production, 1,000 gpd	9,484.0	9,689.6	205.6	2.2
Proved reserves, bcf	168,349.0	160,132.0	–8,217.0	–4.9
EASTERN EUROPE				
Gas capacity, MMcfd	2,796.0	2,796.0	—	—
Gas throughput, MMcfd	1,426.5	1,426.5	—	—
NGL production, 1,000 gpd	10,142.4	10,142.4	—	—
Proved reserves, bcf	2,026,753.0	2,026,709.0	–44.0	—
LATIN AMERICA				
Gas capacity, MMcfd	18,448.2	18,448.2	—	—
Gas throughput, MMcfd	12,475.3	12,427.3	–48.0	–0.4
NGL production, 1,000 gpd	[3]37,202.9	35,929.9	–1,273.0	–3.4
Proved reserves, bcf	255,302.0	260,860.0	5,558.0	2.2
MIDDLE EAST				
Gas capacity, MMcfd	38,404.2	39,264.2	860.0	2.2
Gas throughput, MMcfd	26,898.0	27,318.0	420.0	1.6
NGL production, 1,000 gpd	84,358.4	86,323.1	1,964.7	2.3
Proved reserves, bcf	2,566,038.0	2,548,900.0	–17,138.0	–0.7
AFRICA				
Gas capacity, MMcfd	17,254.2	17,254.2	—	—
Gas throughput, MMcfd	9,642.4	9,642.4	—	—
NGL production, 1,000 gpd	16,777.3	16,777.3	—	—
Proved reserves, bcf	484,433.0	489,630.0	5,197.0	1.1
ASIA-PACIFIC				
Gas capacity, MMcfd	24,960.2	24,960.2	—	—
Gas throughput, MMcfd	20,173.6	20,290.8	117.2	0.6
NGL production, 1,000 gpd	25,764.7	25,799.1	34.4	0.1
Proved reserves, bcf	419,487.0	415,393.0	–4,094.0	–1.0
TOTAL-excl. US				
Gas capacity, MMcfd	179,112.6	180,227.8	1,115.2	0.6
Gas throughput, MMcfd	109,598.4	110,552.8	954.4	0.9
NGL production, 1,000 gpd	216,060.6	217,183.2	1,122.6	0.5
Proved reserves, bcf	5,978,308.0	5,959,824.0	–18,484.0	–0.3
TOTAL-incl. US				
Gas capacity, MMcfd	249,330.8	251,290.5	1,959.7	0.8
Gas throughput, MMcfd	155,135.8	156,054.5	918.7	0.6
NGL production, 1,000 gpd	291,528.8	293,410.8	1,882.1	0.6
Proved reserves, bcf	6,182,693.0	6,170,909.0	–11,784.0	–0.2

[1]Proved gas reserve totals for 2006 are as of Jan. 1, 2007 (OGJ, Dec. 18, 2006, p. 22).
[2]Proved gas reserve totals for 2007 are as of Jan. 1, 2008 (OGJ, Dec. 24, 2007, p. 24).
[3]Corrected from 2007 report.

WORLD LPG DEMAND

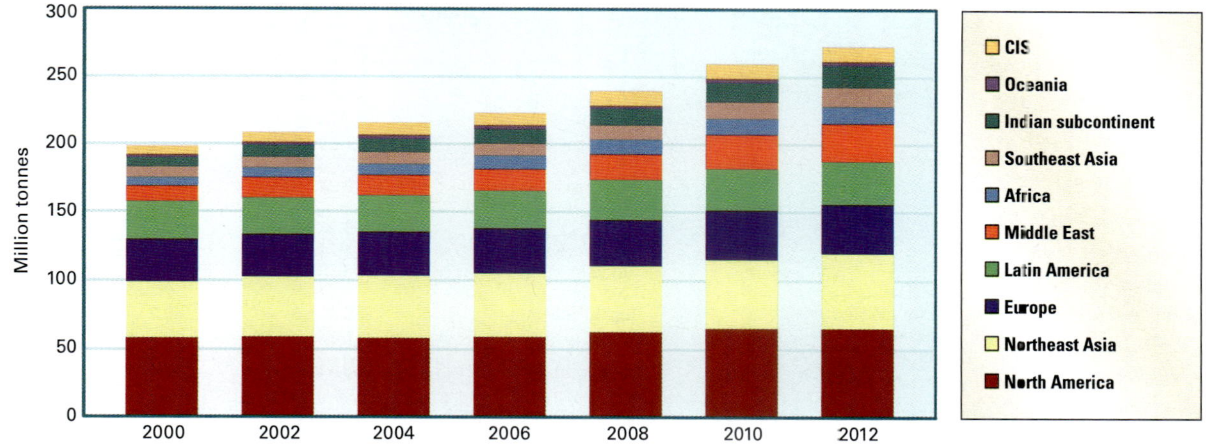

WORLD LPG DEMAND BY SECTOR

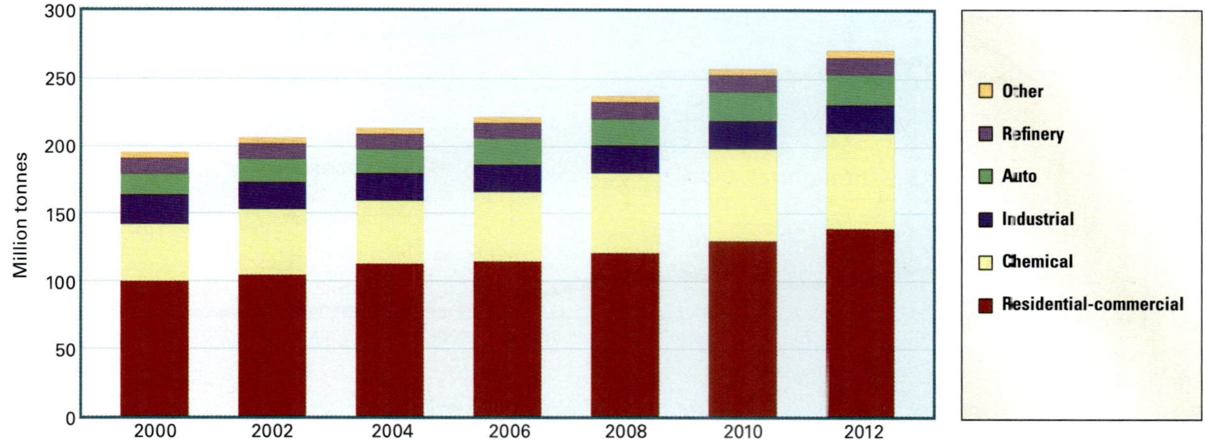

WORLD LPG SUPPLY

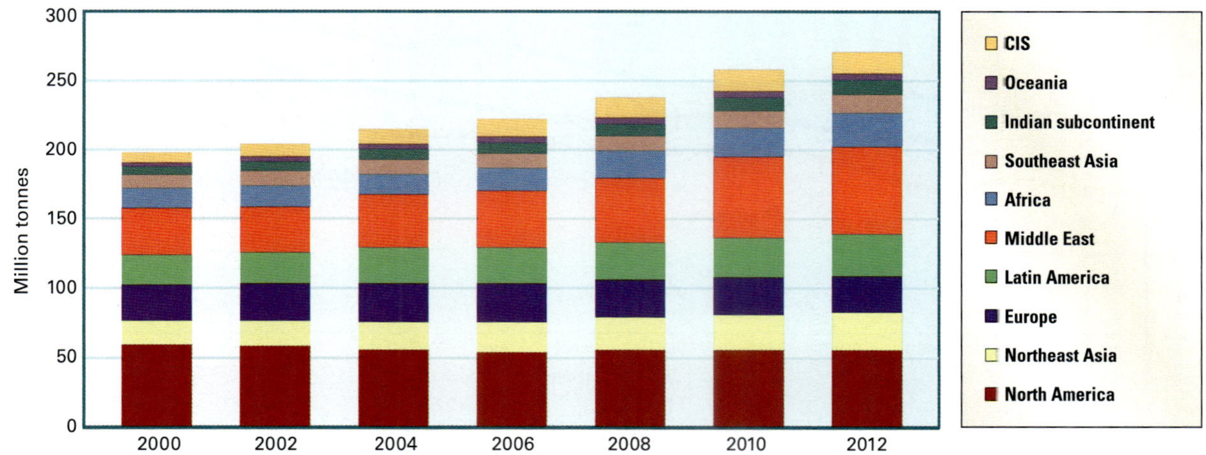

Worldwide Gas Plant Construction*

Country	2008	2007
Africa	2	2
Asia-Pacific	3	1
Canada	1	—
Eastern Europe	2	1
Middle East	4	2
Latin America	5	2
US	7	7
Western Europe	—	—
Total	**24**	**15**

*Includes new plants, expansions, and modifications scheduled for completion in the year listed. 2008 data are from OGJ, Apr. 7, 2008. 2007 data are from OGJ, Apr. 16, 2007.

World's Top 10 NGL Producing Countries: 2007

Country	Production, bcf	Share, %
US	76,227.6	26.0
Saudi Arabia	37,044.0	12.6
Canada	32,521.8	11.1
Mexico	21,649.0	7.4
Kuwait	18,684.8	6.4
Australia	10,939.3	3.7
Russia/FSU	9,127.9	3.1
UAE	8,978.6	3.1
Venezuela	7,312.9	2.5
India	6,534.5	2.2
Norway	6,201.2	2.1
Top 10 subtotal	**235,221.6**	**80.2**
Rest of world	58,189.1	19.8
Total worldwide	**293,410.7**	**100.0**

US Gas Processing Trends

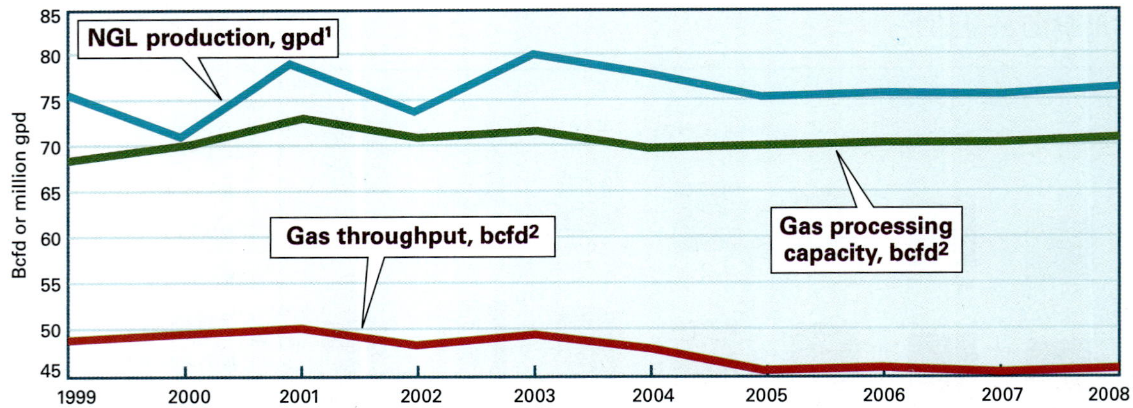

[1]Based on a 12-month average for previous year. [2]As of Jan. 1 of each year.

US LPG Differential, Price Trends: 2007

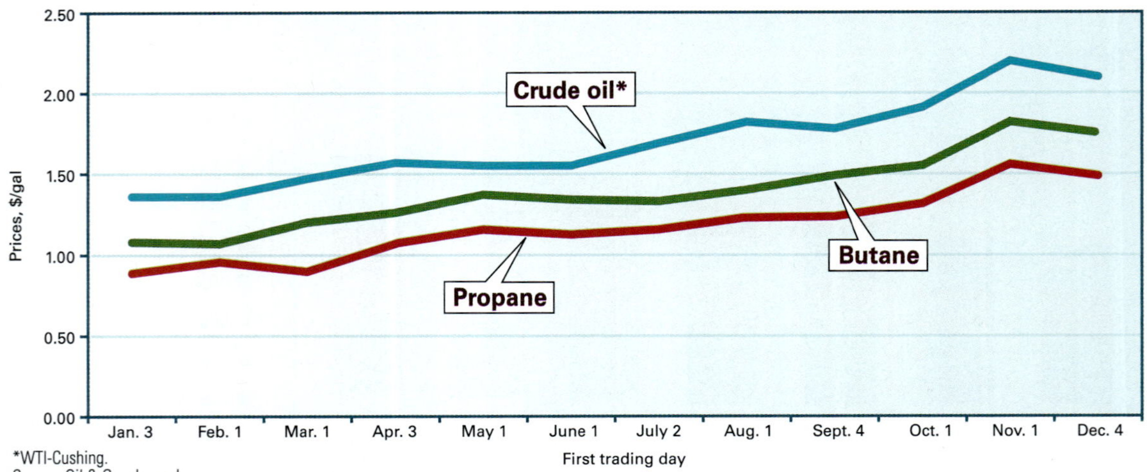

*WTI-Cushing.
Source: Oil & Gas Journal

World's Major Gas Reserves

Country	Est. proved reserves, tcf	Share, %
Russia	1,680.0	27.2
Iran	948.2	15.3
Qatar	905.3	14.6
Saudi Arabia	252.6	4.1
US	211.1	3.4
Abu Dhabi	198.5	3.2
Nigeria	184.0	3.0
Venezuela	166.3	2.7
Algeria	159.0	2.6
Iraq	112.0	1.8
Kazakhstan	100.0	1.6
Turkmenistan	100.0	1.6
Indonesia	94.0	1.5
Malaysia	83.0	1.3
China	80.0	1.3
Norway	79.1	1.3
Uzbekistan	65.0	1.1
Egypt	58.5	0.9
Canada	58.2	0.9
Kuwait	55.5	0.9
Libya	50.1	0.8
Netherlands	50.0	0.8
Ukraine	39.0	0.6
Australia	30.0	0.5
Oman	30.0	0.5
Argentina	15.8	0.3
UK	14.5	0.2
Subtotal	**5,819.7**	**94.1**
Others	366.0	5.9
Total	**6,185.7**	**100.0**

Source: OGJ, Dec. 24, 2007, p. 24

World's 20 Top Gas Producing Countries: 2007

Country	Production, bcf	Share, %
Russia/FSU	27,940.0	27.7
US	19,951.0	19.8
Canada	5,934.8	5.9
Algeria	3,310.0	3.3
Norway	3,161.8	3.1
Iran	2,970.0	2.9
Netherlands	2,705.0	2.7
UK	2,696.5	2.7
China	2,432.9	2.4
Indonesia	2,350.0	2.3
Mexico	2,211.2	2.2
Saudi Arabia	1,955.0	1.9
Qatar	1,825.0	1.8
Malaysia	1,713.0	1.7
UAE	1,577.0	1.6
Argentina	1,517.7	1.5
Pakistan	1,414.0	1.4
Australia	1,407.6	1.4
India	995.7	1.0
Venezuela	867.0	0.9
Top 20 total	**88,935.2**	**88.1**
Other	12,002.3	11.9
Total	**100,937.5**	**100.0**

Source: OGJ, Mar. 12, 2007, p. 62

Global, Chinese trade patterns

In late 2008, Poten & Partners (New York City) reported on its study of demand patterns for global LPG trade.

They concluded that those patterns will change in the near term, and also shed light on recent unexpected patterns in Chinese LPG supply and demand.

The firm noted the reversal in West-to-East LPG trade patterns that had dominated global trade for the last five years; trade is now East-to-West.

Until 2008, targeted LPG retail growth markets in the East have been the large populations of China, India, and Indonesia. But China's LPG demand has stalled, said Poten.

LPG imports to China 2002–04 once underpinned demand growth in the Far East. However, Chinese LPG imports for 2008 just 2.4–2.6 million tons, compared with 8 million tons for 2004.

The company said that high and rising international prices since 2004 forced China out of the market, and that the downward trend in imports to that nation is not likely to end soon.

Section 12:
Petrochemicals outlook

Sharp price drops

In late 2008, the price of petrochemicals fell to multi-year lows, said a report in the *Financial Times*, providing further evidence of the slump in global manufacturing activity.

The drop was sharper than the fall in oil prices, suggesting that demand for plastics and synthetic textiles was very weak as export-oriented firms in the Asia-Pacific region saw reduced overseas orders.

The price of naphtha—a key feedstock—fell in late October to a five-year low of $284/tonne in the far-east Asia market, down 76% from the July record high of $1,200/tonne, according to Platts *Petrochemical Report*.

Benzene—derived from naphtha and used for plastics and dyes—fell in price below the cost of naphtha for the first time since the 1980s.

The price of polyvinyl chloride (PVC) fell in China to a five-year low of $635/tonne, down from a record $1,320/tonne in July.

Ethylene fell in price to $780/tonne, down from almost $2,000/tonne in July.

U.S. propane snapshot

Statistics from the Energy Information Administration (EIA) of the U.S. Department of Energy showed that gas plant propane production averaged 530,000 b/d for second-quarter 2008, and was an estimated 530,000 b/d during July and August.

Gas-plant production in fourth-quarter 2008 was expected to average 515,000 b/d.

In second-quarter 2008, propane production from refineries (net of propylene for propylene chemicals markets) averaged 327,000 b/d.

In the third quarter, propane supply from refineries averaged 290,000–310,000 b/d, and was expected to average 320,000 b/d in the fourth quarter.

US PROPANE INVENTORY

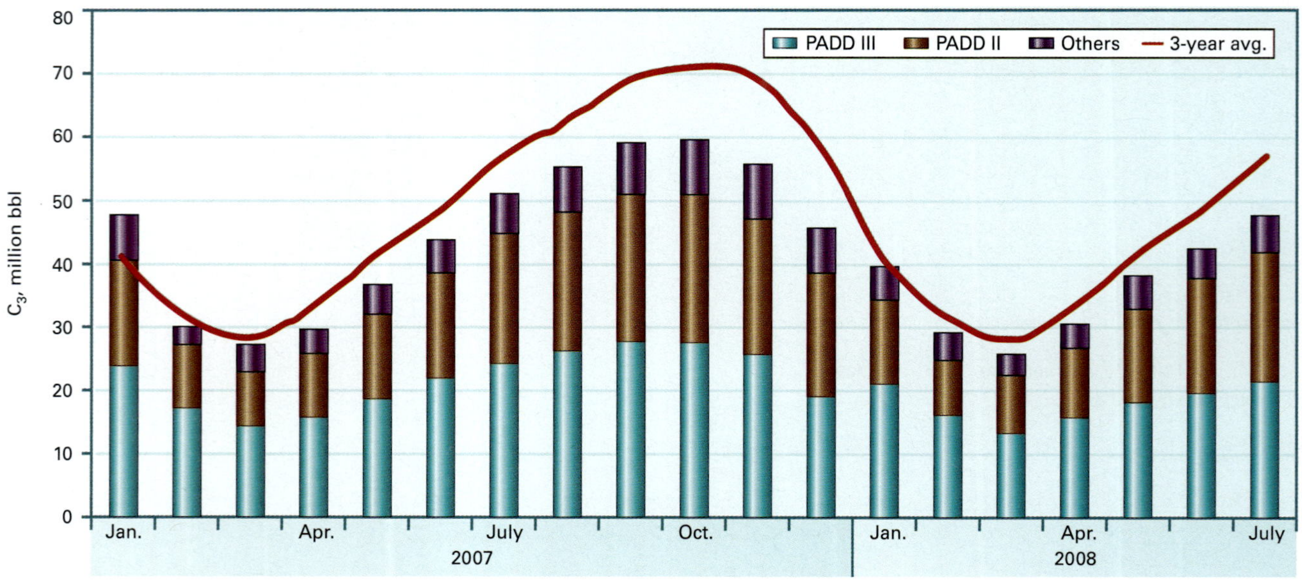

Source: US Energy Information Administration

US GAS PLANT PROPANE PRODUCTION

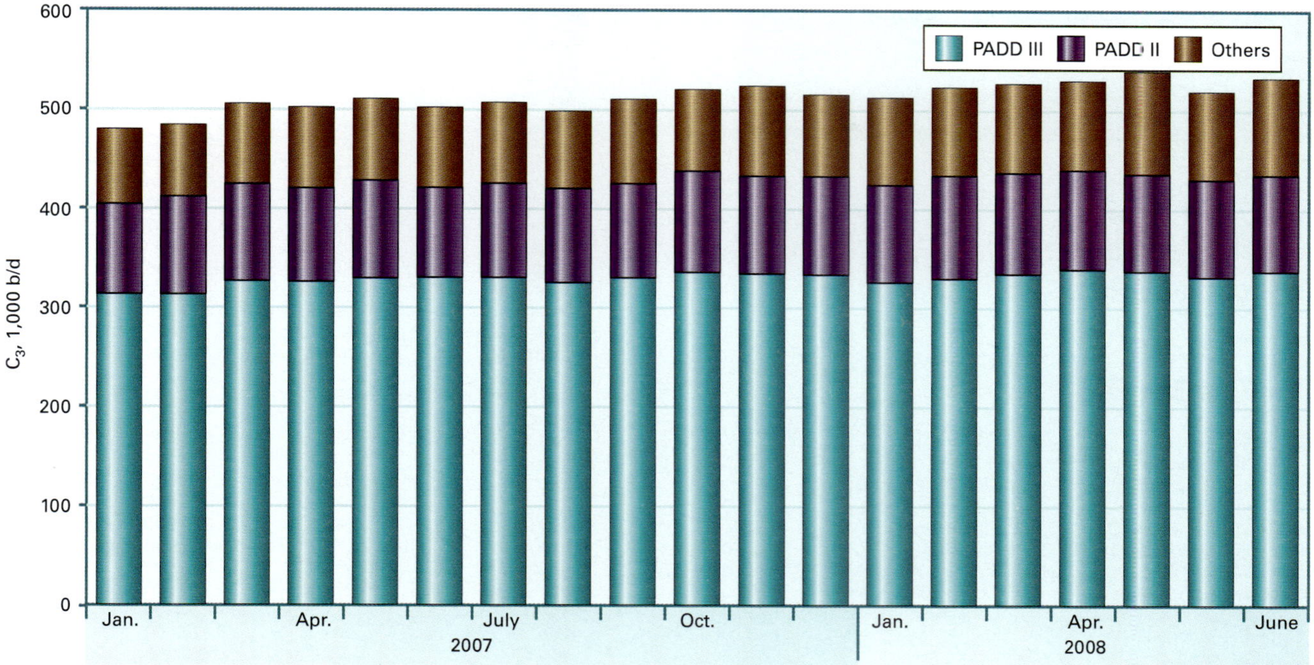

Source: US Energy Information Administration

TOTAL US PROPANE PRODUCTION

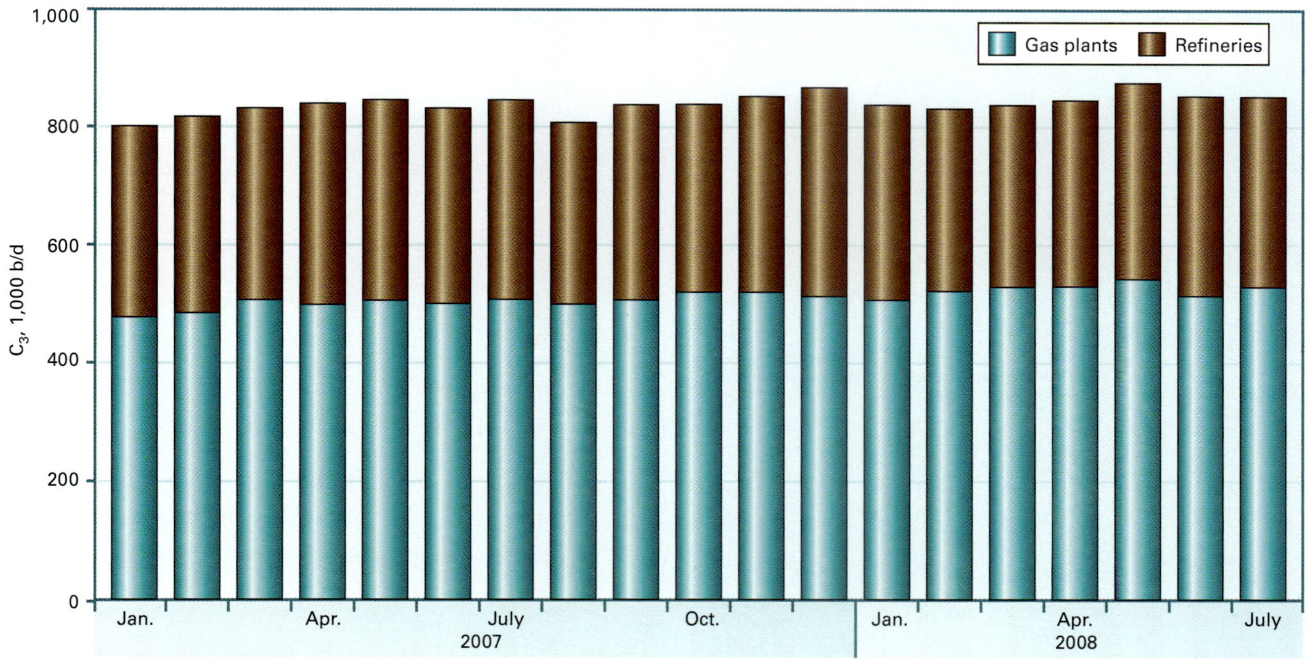

Source: US Energy Information Administration

Worldwide ethylene capacity

Ethylene producers added a net 2 million tonnes/year (mtpy) of capacity in 2007, according to the annual ethylene survey conducted by *Oil & Gas Journal* that was released in mid-2008.

This was a significant increase from the 2006 addition rate of 245,000 tpy.

Capacity as of January 1, 2008, was 119.6 million tpy, an increase of 1.7% from the 117.6 million tpy of capacity reported in OGJ's mid-2007 survey.

Top 10 Ethylene Complexes*

	Company	Location	Capacity, tpy
1	Nova Chemicals Corp.	Joffre, Alta.	2,812,000
2	Formosa Petrochemical Corp.	Mailiao, Taiwan, China	2,550,000
3	Arabian Petrochemical Co.	Jubail, Saudi Arabia	2,250,000
4	ExxonMobil Chemical Co.	Baytown, Tex.	2,197,000
5	ChevronPhillips Chemical Co.	Sweeny, Tex.	1,868,000
6	Dow Chemical Co.	Terneuzen, Netherlands	1,800,000
7	Ineos Olefins & Polymers	Chocolate Bayou, Tex.	1,752,000
8	Equistar Chemicals LP	Channelview, Tex.	1,750,000
9	Yanbu Petrochemical Co.	Yanbu, Saudi Arabia	1,705,000
10	Dow Chemical Co.	Freeport, Tex.	1,640,000

*As of Jan. 1, 2008.

Regional Capacity Breakdown

	Ethylene capacity, tpy		Change	
	Jan. 1, 2008	Jan. 1, 2007	tpy	%
Asia-Pacific	33,002,000	31,602,000	1,400,000	4.43
Eastern Europe	8,512,000	8,462,000	50,000	0.59
Middle East, Africa	12,342,000	12,367,000	–25,000	–0.20
North America	35,707,700	35,687,700	20,000	0.06
South America	5,083,500	5,018,500	65,000	1.30
Western Europe	24,928,000	24,438,000	490,000	2.01
Total capacity	**119,575,200**	**117,575,200**	**2,000,000**	**1.70**

Top 10 Ethylene Producers[1]

	Company	No. of sites	Capacity, tpy: Of entire complexes	Capacity, tpy: With only company partial interests
1	Dow Chemical Co.	14	13,155,000	10,369,500
2	ExxonMobil Corp.	15	11,470,000	8,352,000
3	Saudi Basic Industries Corp.	7	8,940,000	7,165,000
4	Royal Dutch Shell PLC	10	8,965,000	6,841,000
5	LyondellBasell[2]	7	5,200,000	5,200,000
6	Ineos	4	4,656,000	4,286,000
7	Formosa Petrochemical Corp.	2	4,091,000	4,091,000
8	Sinopec	9	4,375,000	4,075,000
9	Chevron Phillips Chemical Co. LP[3]	4	3,956,000	3,701,000
10	Total AS	9	5,713,000	3,421,800

[1]As of Jan. 1, 2008. [2]Includes subsidiary Equistar Chemicals LP. [3]Ownership: ChevronTexaco Corp. 50%, ConocoPhillips 50%.

National Ethylene Capacities

Country	Jan. 1, 2008 — Ethylene capacity, tpy	Jan. 1, 2007 — Ethylene capacity, tpy	Change, tpy
Algeria	133,000	133,000	0
Argentina	838,500	838,500	0
Australia	502,000	532,000	–30,000
Austria	500,000	500,000	0
Azerbaijan	330,000	330,000	0
Belarus	193,000	193,000	0
Belgium	2,460,000	2,180,000	280,000
Brazil	3,500,000	3,435,000	65,000
Bulgaria	400,000	400,000	0
Canada	5,531,000	5,531,000	0
Chile	45,000	45,000	0
China	6,988,000	6,988,000	0
China, Taiwan	3,621,000	2,421,000	1,200,000
Colombia	100,000	100,000	0
Croatia	90,000	90,000	0
Czech Republic	485,000	485,000	0
Egypt	330,000	300,000	30,000
Finland	330,000	330,000	0
France	3,373,000	3,373,000	0
Germany	5,757,000	5,557,000	200,000
Greece	20,000	20,000	0
Hungary	660,000	620,000	40,000
India	2,515,000	2,515,000	0
Indonesia	520,000	520,000	0
Iran	1,214,000	1,214,000	0
Israel	200,000	200,000	0
Italy	2,170,000	2,170,000	0
Japan	7,265,000	7,265,000	0
Kazakhstan	130,000	130,000	0
Kuwait	800,000	800,000	0
Libya	350,000	350,000	0
Malaysia	1,649,000	1,649,000	0
Mexico	1,384,000	1,384,000	0
Netherlands	3,975,000	3,965,000	10,000
Nigeria	300,000	300,000	0
North Korea	60,000	60,000	0
Norway	550,000	550,000	0
Poland	700,000	700,000	0
Portugal	330,000	330,000	0
Qatar	1,030,000	1,030,000	0
Romania	844,000	844,000	0
Russia	3,490,000	3,670,000	–180,000
Saudi Arabia	6,800,000	6,855,000	–55,000
Serbia and Montenegro	200,000	200,000	0
Singapore	1,980,000	1,940,000	40,000
Slovakia	220,000	210,000	10,000
South Africa	585,000	585,000	0
South Korea	5,630,000	5,440,000	190,000
Spain	1,430,000	1,430,000	0
Sweden	625,000	625,000	0
Switzerland	33,000	33,000	0
Thailand	2,272,000	2,272,000	0
Turkey	520,000	520,000	0
Ukraine	630,000	450,000	180,000
UAE	600,000	600,000	0
United Kingdom	2,855,000	2,855,000	0
United States	28,792,700	28,772,700	20,000
Uzbekistan	140,000	140,000	0
Venezuela	600,000	600,000	0
Total	**119,575,200**	**117,575,200**	**2,000,000**

Global ethylene capacity additions

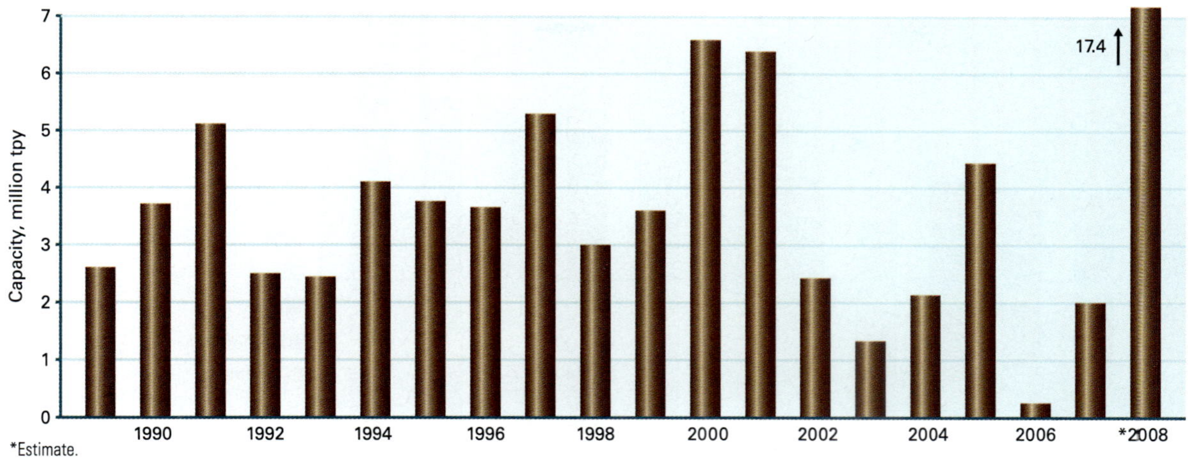

*Estimate.
Sources: ChemSystems, White Plains, NY (1988-99 data). OGJ Energy Database (2000-08 data)

Ethylene expansions, 2008-12

Location	Company	2008	2009	2010	2011	2012
		Ethylene capacity, tpy				
Arzew, Algeria	Total/Sonatrach					1,100,000
Santo Andre, Sao Paulo, Brazil	Petroquimica Uniao SA	200,000				
Nanjing, China	BASF-YPC Co. Ltd.		750,000			
Heilongjian Province, China	Daqing Petroleum & Chemical Co.					600,000
Fujian Province, China	Fujian Petrochemical Co. Ltd.	800,000				
Lanzhou, China	Lanzhou Petrochemical Co.	360,000				
Ningbo, China	PetroChina				1,200,000	
Dushanzi, Xinjang, China	PetroChina Dushanzi Petrochemical Co.	1,000,000				
Maoming, Guangdong, China	Sinopec			320,000		
Quanzhou City, China	Sinopec				800,000	
Ningbo, Zhejiang, China	Zhenhai Refining & Chemical Co. Ltd.			1,000,000		
Koln, Germany	Ineos	100,000				
Dibrugarh, Assam, India	GAIL India Ltd.				220,000	
Panipat, India	Indian Oil Co. Ltd.		800,000			
Dahej, India	Oil & Natural Gas Corp.			1,100,000		
Bandar Imam, Iran	Arvand Petrochemical Co.				1,000,000	
Assaluyeh Bushehr, Iran	Arya Sasol Polymer Co.	1,000,000				
Assaluyeh Bushehr, Iran	Jam Petrochemical Co.	1,320,000				
Kharg Island, Iran	Kharg Petrochemical Co.	500,000				
Bandar Assaluyeh, Iran	Marun Petrochemical Co.	1,200,000				
Bandar Assaluyeh, Iran	Petrochemical Industries Dev. Mgmt. Co.		1,200,000			
Shuaiba, Kuwait	Petrochemical Industries Co., Dow Chemical Co.	850,000				
Sohar, Oman	Dow Chemical Co.		850,000			
Ras Laffan, Qatar	Qatar Chemical Co. Ltd., ChevronTexaco Corp., Total AS	1,300,000				
Ras Laffan, Qatar	Qatar Petroleum Co., ExxonMobil Corp.	1,600,000				
Mesaieed, Qatar	Qatar Petroleum Co., Honam Petrochemical Co.		900,000			
Al-Jubail, Saudi Arabia	Eastern Petrochemical Co.	1,300,000				
Al-Jubail, Saudi Arabia	National Petrochemical Industrialization	1,008,000				
Rabigh, Saudi Arabia	Saudi Aramco, Sumitomo Chemical Co.	1,500,000				
Yanbu, Saudi Arabia	Saudi Basic Industries Corp.	1,300,000				
Al-Jubail, Saudi Arabia	Tasnee Petrochemicals	1,000,000				
Singapore	ExxonMobil Chemical Corp.				1,000,000	
Singapore	Shell Eastern Petroleum Ltd.			800,000		
Kaoshiung Linyuan, Taiwan	Chinese Petroleum Corp.				1,000,000	
Map Ta Phut, Thailand	PTT Polyethylene Co. Ltd.		1,000,000			
Rayong, Thailand	Siam Cement PLC, Dow Chemical Co.			900,000		
Trinidad & Tobago	Westlake Chemical Corp.			570,000		
Ruwais, Abu Dhabi, UAE	Abu Dhabi Polymers Co. Ltd.			1,500,000		
El Tablazo, Venezuela	Polinter					1,000,000
Jose, Anzoategui, Venezuela	Pequiven	1,050,000				
Total		**17,388,000**	**5,500,000**	**6,190,000**	**5,220,000**	**2,700,000**

Source: Oil & Gas Journal construction survey

Ethylene trade

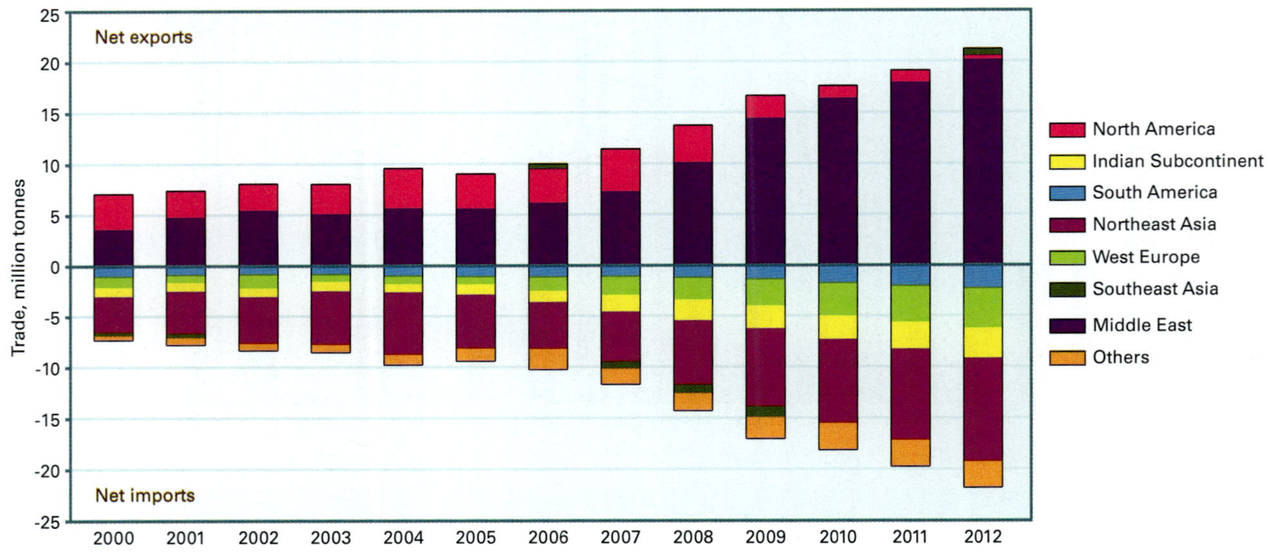

Source: CMAI

Regional capacity

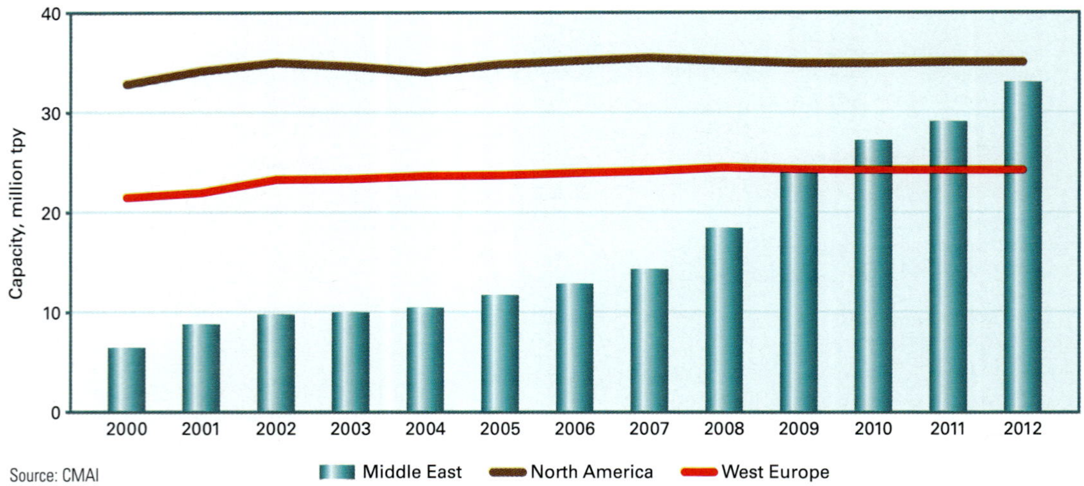

Source: CMAI

The 2008 survey showed that one new train—capacity 1.2 mtpy, in an existing plant—started up in 2007. None of the producers surveyed reported any shutdown capacity.

The net additional 800,000 tpy of capacity resulted from expansions and debottlenecking at existing sites.

More than 17 mtpy of capacity is slated for start-up in 2008, chiefly in the Middle East.

Asia-Pacific ethylene capacity

Country	Australia	China	India	Indonesia	Japan	Malaysia	Singapore	S. Korea	Taiwan	Thailand	Total
	Capacity, 1,000 tpy										
Jan. 1, 2008	500	10,115	3,009	700	7,346	1,670	1,980	8,305	4,100	2,436	40,161
Additions											
2008			247					350			597
2009		4,950					800			1,100	6,850
2010		1,120	2,477						770	900	5,267
2011		3,800					1,000				4,800
2012		2,000									2,000
2013		1,000	800						600		2,400
2014		1,200									1,200
2015									1,200		1,200
Total additions		14,070	3,524				1,800	350	2,570	2,000	24,314
Dec. 31, 2015	500	24,185	6,533	700	7,346	1,670	3,780	8,655	6,670	4,436	64,475

Asia petrochem outlook

FACTS Global Energy (FGE; Singapore) in late 2008 projected that a wave of ethylene investment will take place in Asia from 2009–11, with peaks in early and late 2009. A wave of aromatics investments will take place in 2008–09.

The firm said that the petrochemical sector in Asia is undergoing fundamental change due to three key forces.

- Massive expansion of the Asian petrochemicals sector.
- The rapid emergence of China as the largest petrochemical producer in Asia.
- The sharp rise in production capacity in the Middle East.

Ethylene production capacity

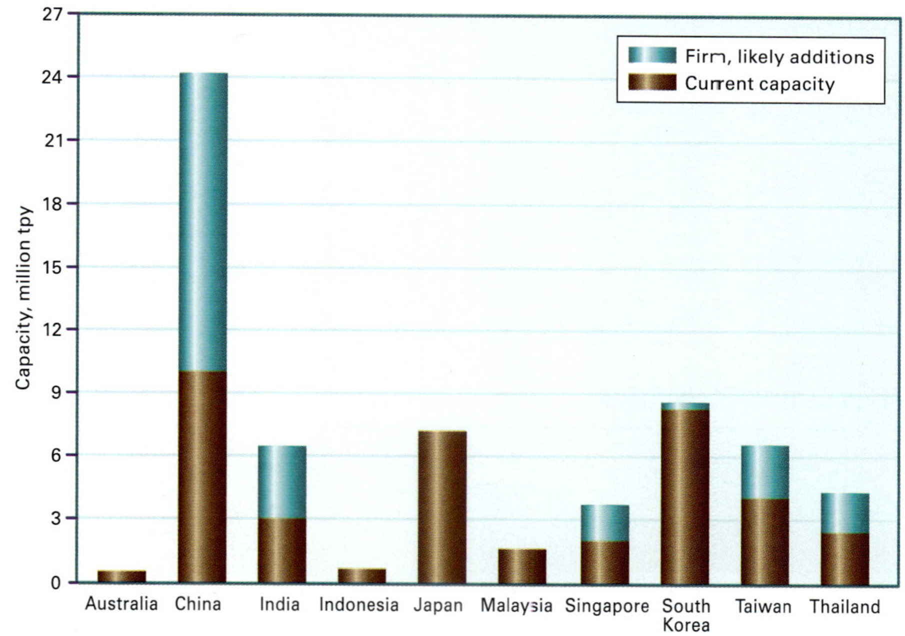

Ethylene capacity additions

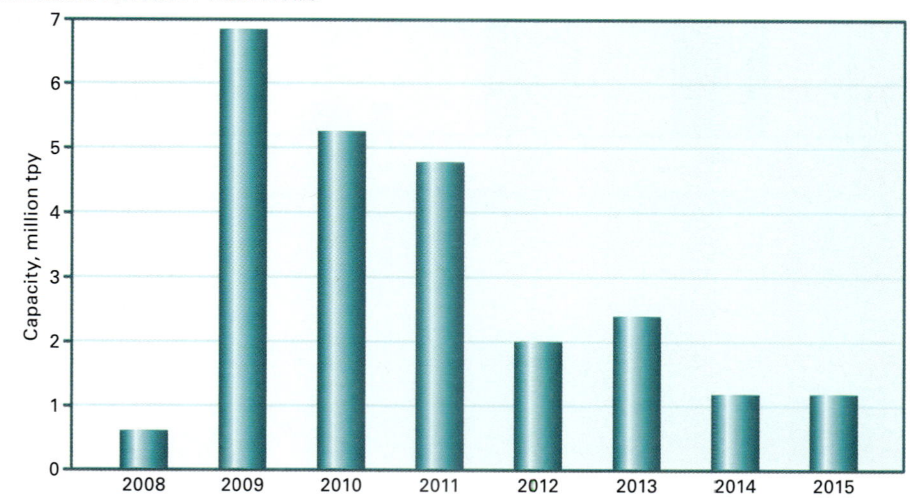

Asia-Pacific Aromatics Capacity

Country	Australia	China	India	Indonesia	Japan	Malaysia	Singapore	S. Korea	Taiwan	Thailand	Total
	Capacity, 1,000 tpy										
Jan. 1, 2008		9,800	3,527	1,287	14,960	765	1,691	8,727	1,525	2,534	44,816
Additions											
2008		2,287			580					1,519	4,386
2009		4,391						350	90		4,831
2010		400									400
2011		800					1,840				2,640
2012								1,180			1,180
2013											
2014											
2015									1,550		1,550
Total additions		7,878	731		580		1,840	1,530	1,640	1,519	15,718
Dec. 31, 2015		17,679	4,258	1,287	15,540	765	3,531	10,257	3,165	4,053	60,535

China Petchem Joint Ventures

Plant	Companies	Province	Completion	Ethylene	Aromatics
				Capacity, 1,000 tpy	
Shanghai	Secco (Sinopec, BP)	Shanghai	Existing	900	480
Yangzi	Sinopec, BP	Jiangsu	Existing	600	300
Huizhou	CNOOC, Shell		Existing	800	460
Qingdao Lidong	GS-led joint venture		Existing		340
Fujian	Sinopec, ExxonMobil, Saudi Aramco	Fujian	First-quarter 2009	800	700
Tianjin	Sinopec, SABIC	Tianjin	Third-quarter 2009	1,000	
Shanghai	Secco (Sinopec, BP)	Shanghai	Fourth-quarter 2009	190	
Nansha/Guanzhou	Sinopec, KPC	Guangdong	Fourth-quarter 2012	1,000	
Ningbo	FPC	Zhejiang	Second-quarter 2014	1,200	

Aromatics Capacity Additions

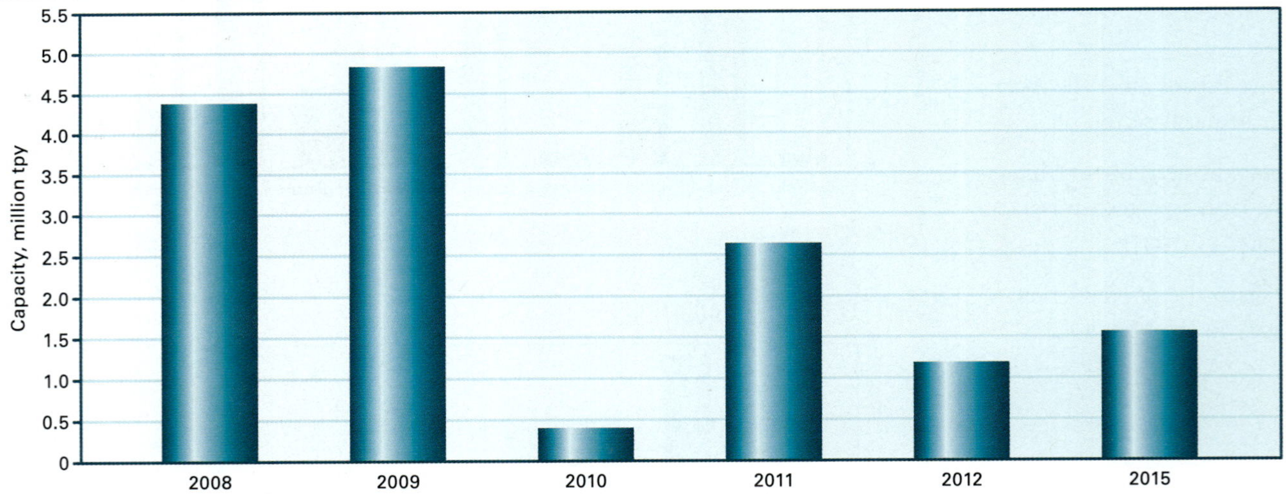

Aromatics production capacity

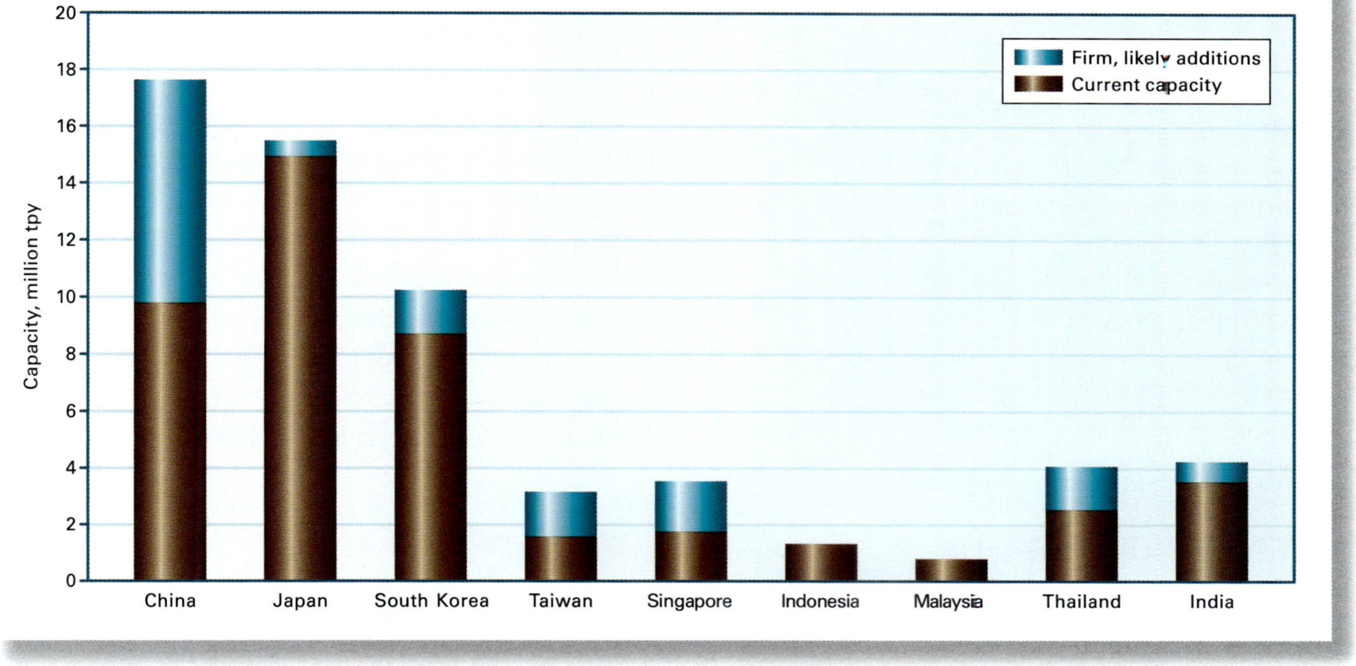

Naphtha consumption

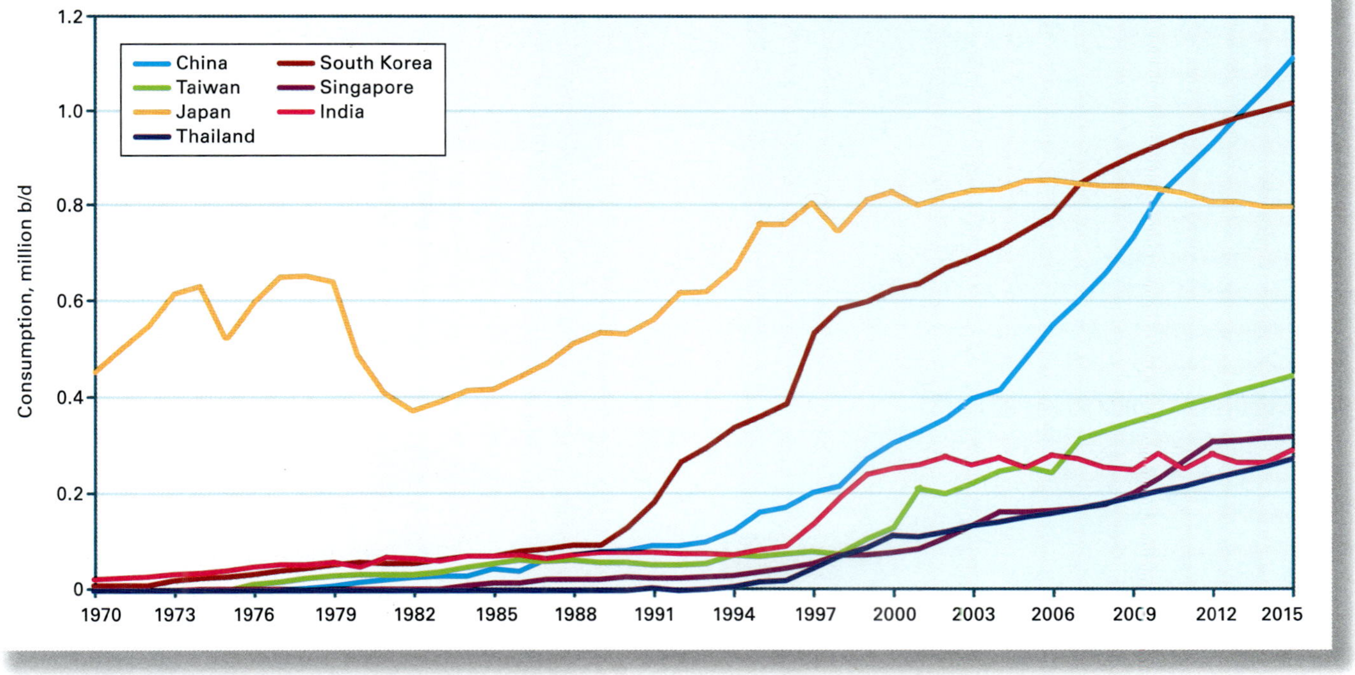

Middle East Ethylene

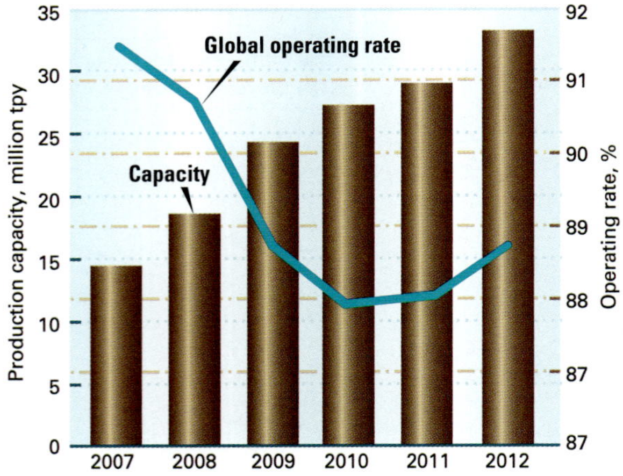

Mideast olefins capacity growth

An "unprecedented" wave of new olefins capacity is starting up in the Middle East, said Chemical Market Associates Inc. (CMAI; Houston) in a March 2008 report.

An additional 20 million tonnes of new ethylene capacity is expected by 2012, more than doubling current capacity in this region, said the firm.

CMAI sees 2008 as a pivotal year because 9 million tonnes of capacity will start up; this will lower global operating rates towards bottom-of-cycle conditions in 2009.

In addition, a "staggering" 9 million tpy of new ethylene capacity will start up by second quarter 2009, said CMAI, noting that global ethylene demand growth is around 5 million tpy.

Middle East Ethylene Projects*

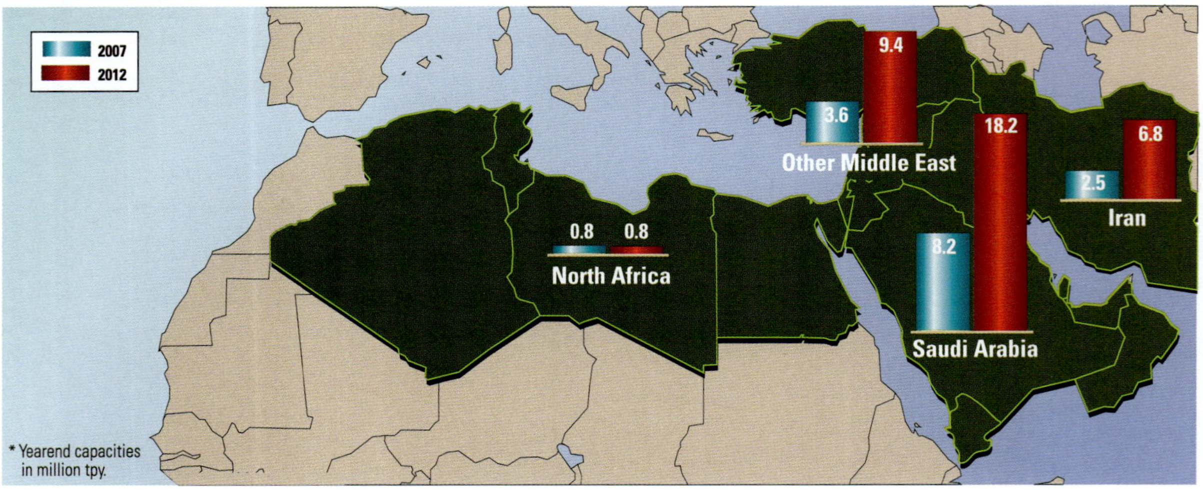

Middle East Propylene Projects*

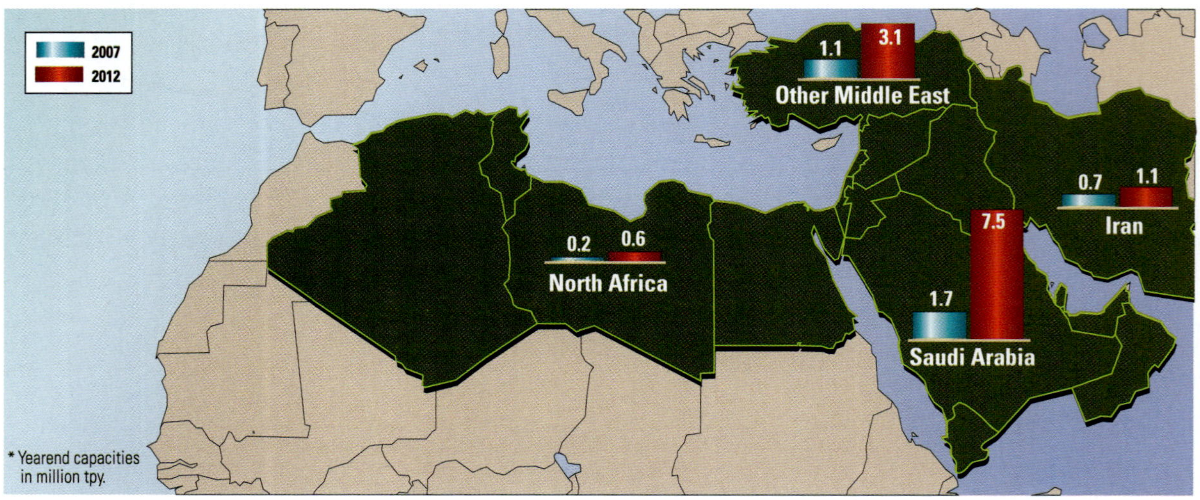

Propylene production

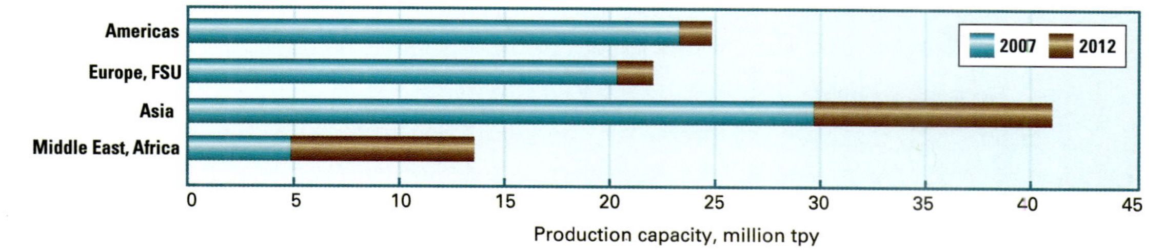

In 2008, two world-scale crackers in Iran started up and several are imminent in Saudi Arabia, Kuwait, and Qatar. Other than Iran, most petchem projects are on or ahead of schedule.

By 2010, Middle East capacity will surpass that of Western Europe, and by 2012, will nearly match capacity in North America.

By year end 2012, CMAI expects total ethylene capacity in the Middle East to be about 31 mtpy.

Saudi Arabia propylene projects

Unit	Capacity, 1,000 tpy	Start-up	Comment
Jubail ChevronPhillips	150	Second-quarter 2008	Condensate feed
SEPC	240	Third-quarter 2008	Ethane-propane feed
Yansab	360	Third-quarter 2008	Ethane-propane feed
Petro-Rabigh	100	Third-quarter 2008	Ethane-offgas feed
SHARQ	200	Fourth-quarter 2008	Ethane-propane feed
Kayan	660	Fourth-quarter 2009	Ethane-butane feed
Advanced PP	450	First-quarter 2008	Propane dehydrogenation
NATPET	420	Second-quarter 2008	Propane dehydrogenation
Ibn Zahr	250	Third-quarter 2008	Metathesis
Petro-Rabigh	800	Fourth-quarter 2008	High-severity FCC
Al-Waha	450	First-quarter 2009	Propane dehydrogenation

Regional polymer trading

Accenture (New York City) found in an early-2008 study that polymer consumption was growing in all regions around the world.

Developed nations are striving to add value to existing products, create new value-added products, and continue to lead in developing new services. Developing economies are spending on the basics; exporting is their value-added activity.

The firm used a "triangle of trade" to illustrate global competition and the forces for investment in polymer manufacturing.

Triangle of trade

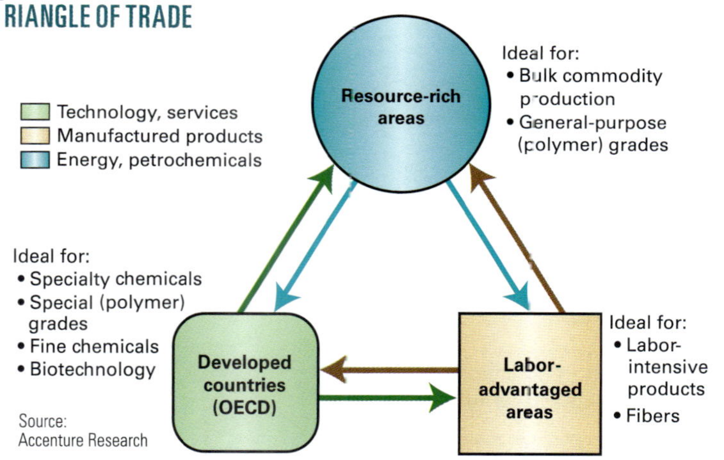

Source: Accenture Research

Polymer net trade 'heat' map, 2006

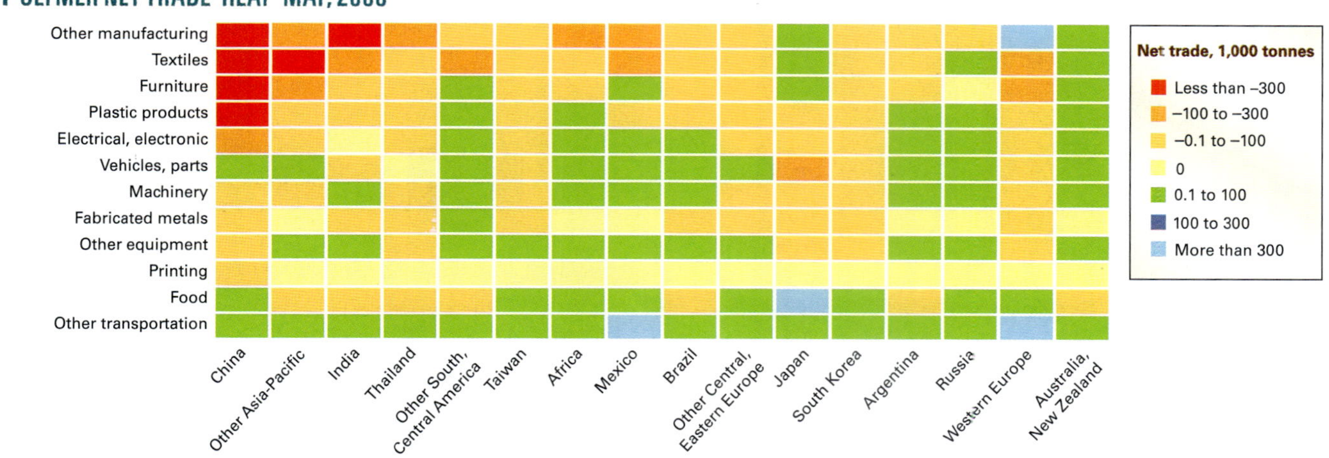

Evolving Feedstock Regions

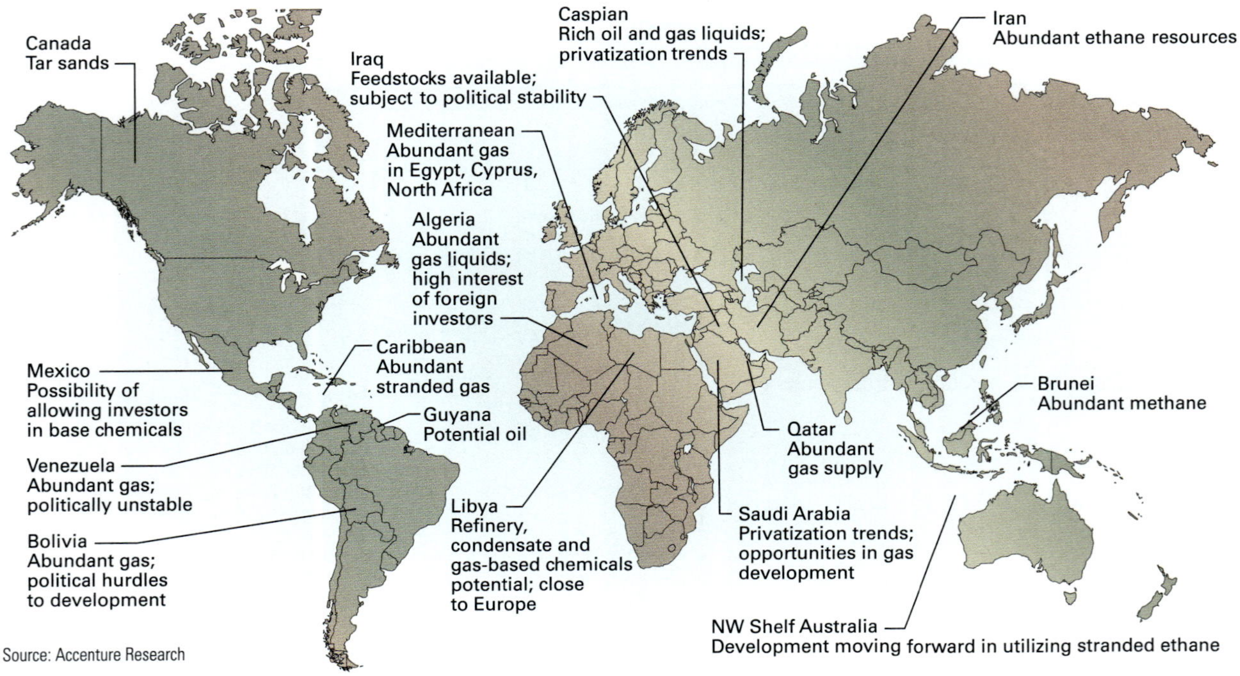

Source: Accenture Research

China Polymer Balance

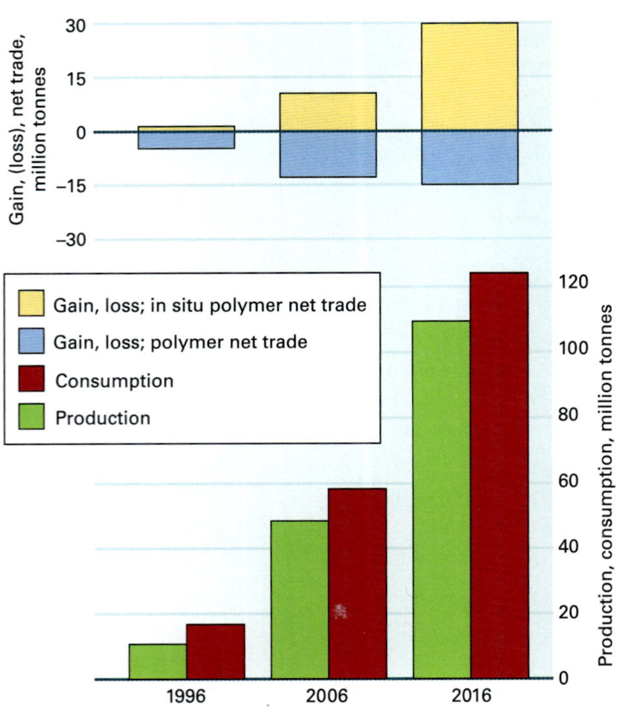

Middle East, Africa Polymer Balance

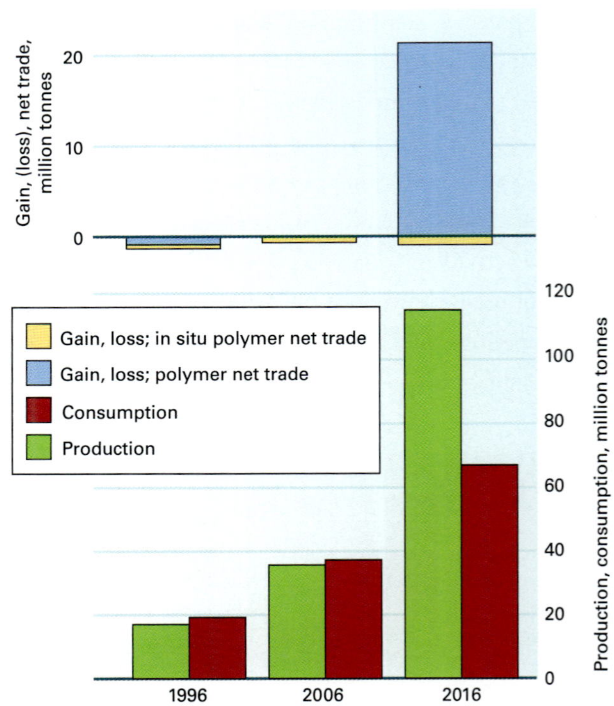

Section 13:
Pipelines

Global pipeline plans expand

Planned pipeline construction to be completed in 2008 rose by more than one-third from the previous year, said *Oil & Gas Journal* in early 2008.

The increase is driven by large crude oil transportation projects in both the United States and Asia-Pacific. Plans for 2008 construction of both natural gas and products pipelines also expanded from those for 2007.

Operators plan to install nearly 13,500 mi (21,722 km) in 2008 alone, with natural gas construction making up 56% [more than 7,500 mi (12,070 km)] of the plans, based on reports from the world's pipeline operating companies and data collected by *OGJ*.

PIPELINE CONSTRUCTION IN 2008[1]

Area	4-10 in.	12-20 in.	22-30 in.	32+ in.	Total
	Miles				
Gas pipelines					
US	—	86	1,183	1,513	2,782
Canada	31	120	90	—	241
Latin America	308	730	—	327	1,365
Asia-Pacific[2]	29	288	25	1,509	1,851
Europe[3]	—	58	—	66	124
Middle East	—	—	—	817	817
Africa	—	—	—	364	364
Total gas	**368**	**1,282**	**1,298**	**4,596**	**7,544**
Crude pipelines					
US	65	748	355	321	1,489
Canada	—	—	143	273	416
Latin America	10	—	—	—	10
Asia-Pacific[2]	29	29	—	1,500	1,558
Europe[3]	—	—	—	—	—
Middle East	—	—	15	160	175
Africa	—	—	445	—	445
Total crude	**104**	**777**	**958**	**2,254**	**4,093**
Product pipelines					
US	—	1,380	—	—	1,380
Canada	—	—	—	—	—
Latin America	—	175	—	—	175
Asia-Pacific[2]	172	—	—	—	172
Europe[3]	—	—	—	—	—
Middle East	—	106	—	—	106
Africa	—	—	—	—	—
Total product	**172**	**1,661**	**—**	**—**	**1,833**
World totals					
Gas	368	1,282	1,298	4,596	7,544
Crude	104	777	958	2,254	4,093
Product	172	1,661	—	—	1,833
Total	**644**	**3,720**	**2,256**	**6,850**	**13,470**

[1]Projects planned to be completed in 2008. [2]Regions east of the Ural Mountains and south of the Caucasus Mountains, excluding the Middle East. [3]Regions west of the Ural Mountains and north of the Caucasus Mountains.

PIPELINE CONSTRUCTION IN 2008 AND BEYOND[1]

Area	4-10 in.	12-20 in	22-30 in.	30+ in.	Total
	Miles				
Gas pipelines					
US	—	236	920	8,191	9,347
Canada	—	—	1,145	331	1,476
Latin America	458	374	2,035	1,403	4,270
Asia-Pacific[2]	—	936	108	20,289	21,333
Europe[3]	—	411	898	7,566	8,875
Middle East	—	—	56	5,877	5,933
Africa	—	—	—	2,810	2,810
Total gas	**458**	**1,957**	**5,162**	**46,467**	**54,044**
Crude pipelines					
US	—	165	2,125	1,590	3,880
Canada	—	—	1,033	1,064	2,097
Latin America	—	—	—	—	—
Asia-Pacific[2]	25	—	1,655	1,911	3,591
Europe[3]	—	83	—	1,374	1,457
Middle East	—	676	—	225	901
Africa	—	—	—	—	—
Total crude	**25**	**924**	**4,813**	**6,164**	**11,926**
Product pipelines					
US	—	2,384	314	500	3,198
Canada	281	872	—	—	1,153
Latin America	—	330	—	—	330
Asia-Pacific[2]	113	—	—	—	113
Europe[3]	—	—	—	—	—
Middle East	—	1,352	—	—	1,352
Africa	—	—	—	—	—
Total product	**394**	**4,938**	**314**	**500**	**6,146**
World totals					
Gas	458	1,957	5,162	46,467	54,044
Crude	25	924	4,813	6,164	11,926
Product	394	4,938	314	500	6,146
Total	**877**	**7,819**	**10,289**	**53,131**	**72,116**

[1]Projects under way at the start of or set to begin in 2008 and be completed in 2009 or later. Includes some probable major projects whose installation will begin in 2009 or later. [2]Regions east of the Ural Mountains and south of the Caucasus Mountains, excluding the Middle East. [3]Regions west of the Ural Mountains and north of the Caucasus Mountains.

FORECAST PIPELINE CONSTRUCTION

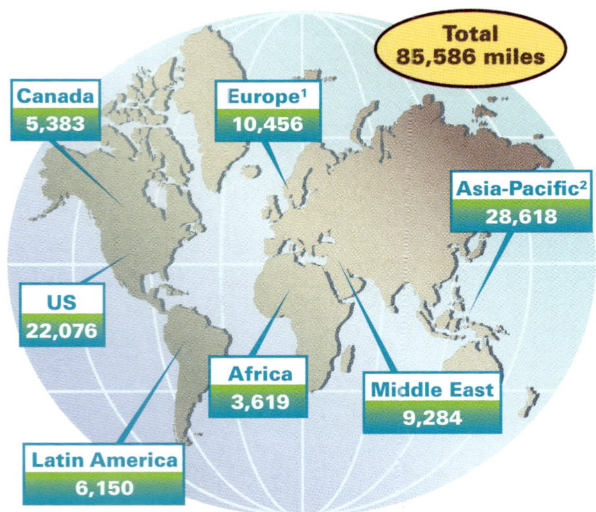

[1]Including Russia and former Soviet republics west of the Ural Mountains.
[2]Including Russia and former Soviet republics east of the Ural Mountains.

Rockies Express Pipeline

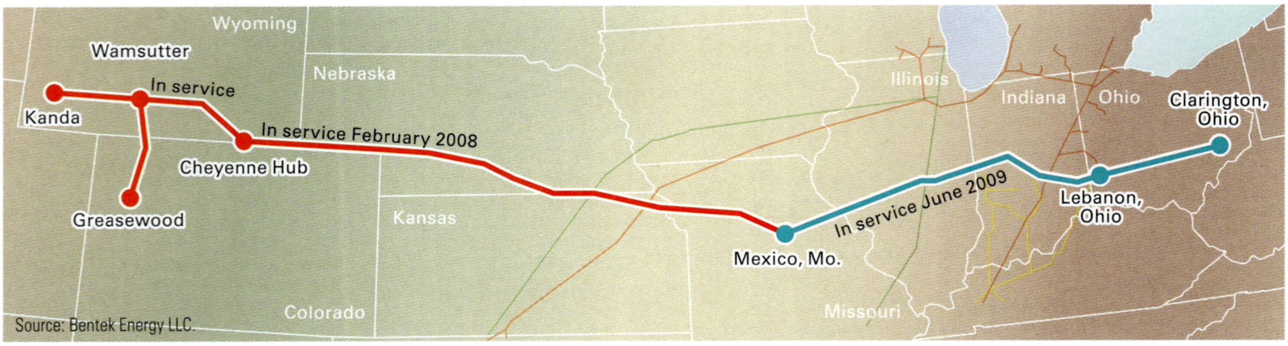

Source: Bentek Energy LLC.

Nabucco Pipeline

Source: Nabucco Gas Pipeline International GMBH

Looking to 2009 and beyond, greater mileage is planned for both natural gas and products pipelines than had been the case in 2007, with a slight downturn in crude lines stemming from the large quantity of mileage expected to be completed in 2008.

Shown in this section are maps of several major pipeline projects under way in 2008.

Note that details about other projects may be presented in the regional chapters of this book.

West Ethylene Pipeline

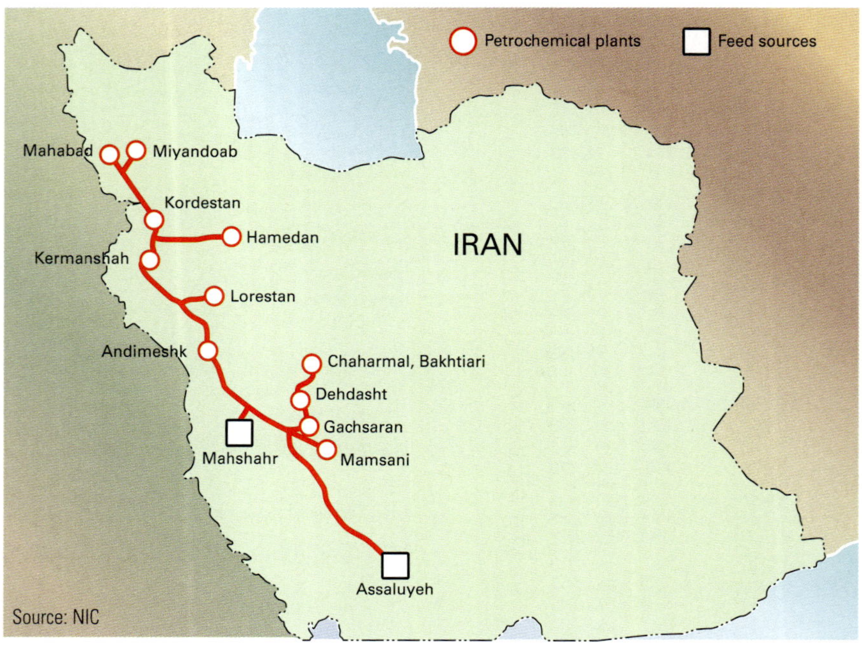

Source: NIC

US Southeast-Gulf Natural Gas Transportation, Storage Projects*

Category	No. of projects	Capacity, bcfd	Description
Feeder pipelines, outside southeast-gulf region	13	7.4	Primarily designed to move gas out of the field and into pipelines delivering gas into new or existing southeast-gulf pipelines.
Inbound pipelines	7	6.6	Some of the region's largest inbound pipeline system additions bring gas from Carthage, Tex., to Perryville, La. and beyond, including Gulf South's Southeast Expansion, Boardwalk's Gulf Crossing, and Kinder Morgan's Midcontinent Express.
Storage facilities	11	6.3	All but one of these storage projects are high-deliverability salt-dome facilities, mostly located within 200 miles of the gulf coast.
LNG terminals	4	7.1	LNG terminals should be viewed similar to storage facilities in that they can deliver large volumes of gas quickly into the pipeline grid.
Total inbound	**22**	**20.0**	
Outbound pipelines	5	4.2	Three new pipeline systems will move gas from supply areas in northeast Texas and Oklahoma to Transco's Station 85 in Choctaw County, Ala.: Gulf South's Southeast Expansion, Boardwalk's Gulf Crossing, and Kinder Morgan's Midcontinent Express. Spectra-CenterPoint's South-east Supply Header will move gas from Perryville into Florida.
Total, all projects	**40**	**31.6**	

*Planned for completion by mid-2009.

Cumulative Pipeline Projects

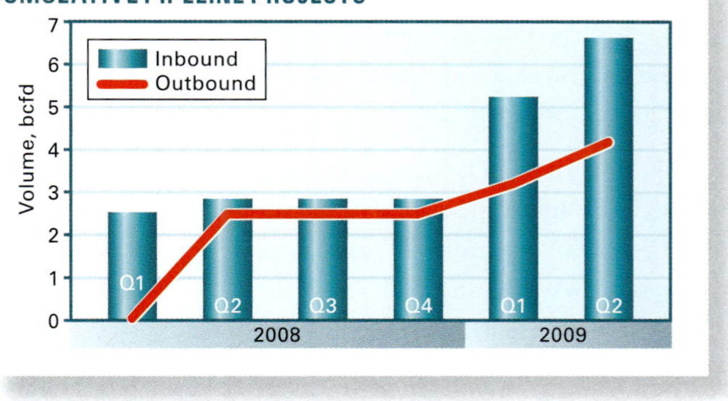

Southeast-Gulf Pipeline Projects, Cumulative

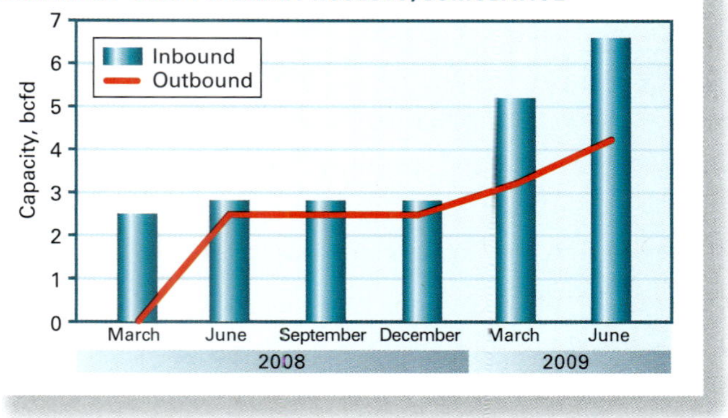

Cumulative Input Expansion

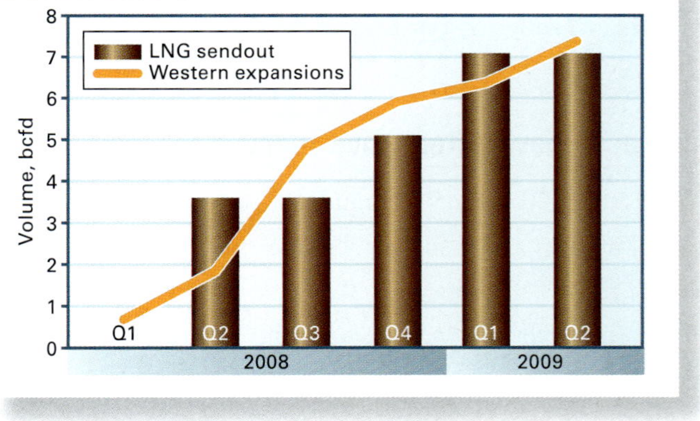

Natural Gas Flow into Southwest-Gulf Region

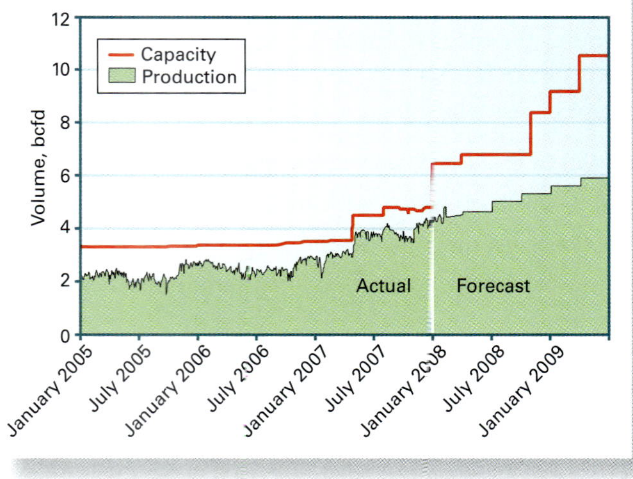

U.S. Southeast: Gas transportation

Between January 2008 and mid-2009, some 29 pipeline and LNG terminal projects are scheduled to be completed in and around the southeastern United States and Gulf Coast region.

They represent pipeline-capacity and LNG-sendout additions totaling 25.4 bcfd (719.3 MMcmd)

Bentek Energy LLC (Golden, CO) in mid-2008 concluded that this unprecedented infrastructure investment has major implications not only for that region, but also for gas markets across North America, and ultimately the world.

The significant new pipeline, LNG, and gas storage projects will affect flow patterns, commodity values, asset values, trading economics, and hedging strategies in both the near and long term.

Section 14:
Climate change issues

Government and industry officials around the world in 2008 were exploring ways to address concerns about global climate change, as affected by hydrocarbon development, delivery and use.

Presented here are highlights of some of those initiatives.

Poznan conference

In December 2008 in Poznan, in western Poland, representatives of 190 countries met to discuss approaches to mitigation of climate change. It was the 14th Conference of the Parties (COP) to the UN Framework Convention on Climate Change (UNFCCC).

Among the key actions was a decision by European Union leaders to keep their tough emissions reduction target intact while also allowing some opt-out exceptions (see box). In other actions:

- Mexico vowed to cut its greenhouse gas (GHG) emissions in half by 2050.
- Brazil said it would cut the rate of forest loss by 70% over the next decade.
- Peru said that with help, it could reduce deforestation to zero.
- Guyana presented a plan for a low-carbon economy designed to suit the interests of all countries at an early stage of developing their natural resources.

The 15th COP is scheduled for November 2009 in Copenhagen. Attendees will continue to work toward a new global climate change policy regime going beyond 2012, which is the final year of coverage of the Kyoto Protocol's first commitment period.

IEA: Energy security, climate linked

In November 2008, the International Energy Agency (IEA; Paris) released its *World Energy Outlook 2008* report (WEO2008). A key message: Mitigating climate change will substantially improve energy security.

IEA said that the global economic slowdown cannot distract policymakers worldwide from progressing on climate change mitigation efforts through international accords and related national energy policies.

The report estimated that cumulative investment of more than $26 trillion (in 2007 dollars) in energy projects will be needed during 2007–30.

Just over half of the projected 2007–30 global energy investment goes simply to maintain the current level of supply capacity.

Policy clarity sought

In an early-2008 review of the issue, *O&GJ* found that corporate pressure is building globally for lawmakers to address climate change.

- A study by Accenture (New York City) found that more than 75% of companies surveyed in worldwide energy, utilities, chemical, forest products, metals, and mining industries are looking to policymakers for a clean, consistent framework on global emission targets.
- In a U.S. survey, Deloitte LLP (New York City) found that 82% of consumers believe it is important for regulators to impose laws reducing GHG emissions.
- The Investor Network on Climate Risk released a letter in May 2008 calling on the U.S. Congress to set a target and timetable to reduce emissions. The network represents some 60 institutional investors focused on the impacts of climate change on business.

Oil execs' view: Alternative fuels

More than 50 senior oil and gas professionals took part in a November 2008 survey by the oil and gas group of Deloitte LLP.

More than half said they are thinking more about how their companies can participate in a U.S. transition to renewable energy and alternative fuels.

The phone survey involved professionals at petroleum companies with annual revenues of at least $100 million.

About 75% believe that a transition from reliance on fossil fuels for transportation is an appropriate U.S. goal and 56% believe this is also an appropriate goal for oil and gas companies.

More than half said they expect natural gas to figure more in the future as a transportation fuel.

Deloitte said 56% agreed that "world will run out of reasonably priced oil in the next 50 years."

Worldwide large stationary CO_2 sources

Process	No. of sources	Emissions, Million tons CO_2/year
Power	4,942	10,539
Cement production	1,175	932
Refineries	638	798
Iron and steel industry	269	646
Petrochemicals industry	470	379
Oil and gas processing	Not available	50
Other fossil fuels	90	33
Bioethanol and bioenergy	303	91
Total	**7,887**	**13,466**

Source: Intergovernmental Panel on Climate Change Special Report, Cambridge University Press, 2005

Technology for climate change mitigation

The IEA stated in November 2008 that restricting the rise in global temperatures to just 2°C will require (a) the use of new technologies not yet widely available, and (b) a price-tag for carbon dioxide emissions well above current levels.

Continued growth in energy use with "business as usual" policies would have shocking consequences, said IEA, raising average temperatures by up to 6°C, which could cause the melting of ice at both poles.

IEA has said that it would be "possible, but very hard" to cut carbon dioxide emissions to keep atmospheric concentration of carbon dioxide no more than 450 ppm—a level consistent with a rise of just 2°C in average temperatures.

To reach a condition in which CO_2 levels stabilize at 450ppm scenario, the following technology actions would be needed:

- wind power capacity would have to increase more than 13-fold from its present level by 2030
- use of nuclear power would have to double
- coal- and gas-fired power stations equivalent to 15% of today's global capacity would need to be fitted with equipment to capture and store CO_2 emissions
- "second-generation" biofuels made from algae or plant waste also would also be needed.

Also, under a global emissions-reduction framework, said IEA, the price of emissions would have to be on the order of $180/tonne of CO_2 in order to get to 450 ppm in the atmosphere.

In November 2008, the price was about $23 in the European Union's emissions trading scheme.

Emissions mitigation: Challenge and opportunity

A carbon cap-and-trade system and other initiatives to curb greenhouse gas (GHG) emissions will hurt the petroleum industry overall, but also will create financial opportunities for some oil companies.

That was the conclusion of the Energy Policy Research Foundation Inc. (Washington, DC) in a mid-2008 analysis.

US CO_2 emissions

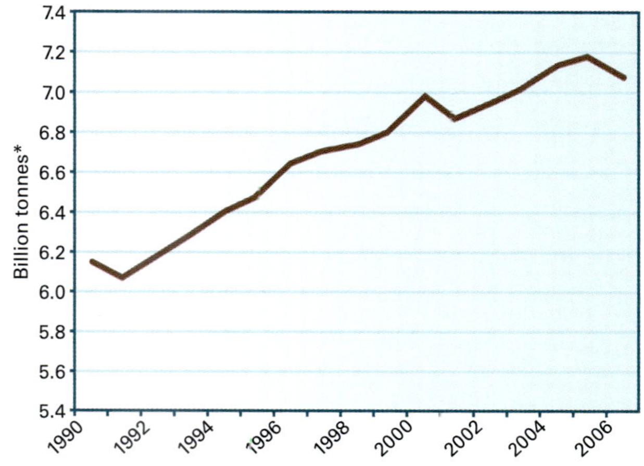

*Estimated.

Source: Energy Policy Research Foundation Inc.

World CO_2 emissions from fossil fuels, 1980-2005

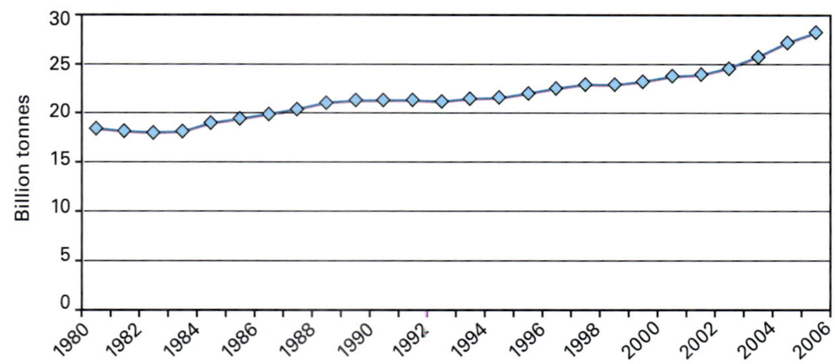

Source: Energy Policy Research Foundation Inc.

CCS: A question of scale

Carbon capture and storage (CCS) is seen as a promising method of reducing carbon dioxide emissions into the atmosphere. A typical coal-fired power plant emits 5–10 million tonnes of CO_2 each year.

But attendees at a mid-2008 conference concluded that researchers have yet to demonstrate that CCS can be accomplished on a sufficiently large scale to stabilize CO_2 concentration levels.

CCS technology would capture CO_2 from power plants, petrochemical plants, and other facilities and sequestering it underground in depleted oil and gas fields, saline aquifers, and unmineable coal seams.

Some science and technology exists for CCS. In fact, major sequestration projects are already operational in Norway, Saskatchewan, and Algeria.

But obstacles to commercialization of CCS include the following:

- It is likely that thousands of projects will be needed to reduce CO_2 emissions.
- Each of those projects will likely need to be 5–10 times larger than any existing project, to capture hundreds of megatonnes and even gigatonnes of CO_2 each year.
- Deep-sea sediments might provide offshore storage, but the concept has not been field tested.
- Legal liability issues abound, such as: who will be responsible for long-term monitoring of CCS sites, and who pays to remediate them if they start to leak in 100 years?

EU CLIMATE CHANGE ACTIONS

In 2007, the European Union committed to achieve three goals by the year 2020:

- cut overall greenhouse gas emissions by 20% over 1990 levels
- obtain 20% of overall EU energy from renewables like wind, waves and plant waste
- achieve efficiency savings of 20% over forecast consumption.

At a December 2008 conference in Poznan, Poland, the EU kept these targets, but offered some concessions.

- Struggling FSU countries that rely on coal for power will have greater leeway in distribution of carbon allowances needed by power firms.
- Heavy industries facing global competition will also get up to 100% of their allowances free, at least initially, if they use the cleanest available emission control technologies.
- EU nations will be allowed to buy credits for emissions reductions far from Europe, and count them against as much as 90% of their national reduction targets.

French CCS pilot project

Total SA agreed in early 2008 to pass on to Indonesia the results of its carbon dioxide capture and storage (CCS) pilot project near Pau in southwestern France.

The agreement gives the Indonesian Agency for the Research and Development of Energy and Mining Resources access to the main results of Total's pilot project in the Lacq basin.

This project involves extracting CO_2 from gases emitted by a boiler at a Lacq factory and reinjecting it 14,765 ft (4,500 m) underground in an old gas field.

Total was planning to make the first injection of CO_2 in late 2008, and to inject a further 150,000 tonnes over the following two years.

The project aims to prove the technical feasibility of an integrated carbon capture and storage scheme.

Table 1: World Reserves and Production

	ESTIMATED PROVED RESERVES				OIL PRODUCTION			
	Jan. 1, 2009		Jan. 1, 2008					
COUNTRY	Oil, 1,000 bbl	Gas, bcf	Oil, 1,000 bbl	Gas, bcf	Producing oil wells, Dec. 31, 2008*	Estimated 2008, 1,000 b/d	Change from 2007, %	Actual 2007, 1,000 b/d
Canada	178,092,000	57,906	178,592,000	58,200	62,519	2,570.0	-1.8	2,618.1
US	21,317,000	237,726	20,972,000	211,085	498,619	4,900.0	-3.2	5,064.5
North America	**199,409,000**	**295,632**	**199,564,000**	**269,285**	**561,138**	**7,470.0**	**-2.8**	**7,682.6**
Argentina	2,616,000	15,600	2,586,750	15,750	18,814	610.0	-2.2	623.5
Bolivia	465,000	26,500	465,000	26,500	361	40.0	-9.1	44.0
Brazil	12,623,900	12,890	12,181,620	12,280	11,995	1,815.0	3.8	1,747.8
Chile	150,000	3,460	150,000	3,460	315	2.3	-11.8	2.6
Colombia	1,355,000	3,739	1,506,000	4,342	7,600	578.0	8.9	531.0
Cuba	124,000	2,500	124,000	2,500	251	50.0	—	50.0
Ecuador	4,660,000	315	4,517,000	—	1,296	500.0	—	500.0
Guatemala	83,070	—	83,070	—	20	14.0	-12.5	16.0
Mexico	10,501,200	13,162	11,650,000	13,850	3,052	2,800.0	-9.2	3,082.6
Peru	415,769	11,842	382,866	11,928	5,035	75.0	-3.8	78.0
Trinidad & Tobago	728,300	18,770	728,300	18,770	3,903	113.0	-5.8	120.0
Venezuela	99,377,000	170,920	87,035,000	166,260	15,669	2,350.0	-1.7	2,390.0
Other Latin America	88,471	5	96,900	5	1,269	16.8	7.0	15.7
Latin America	**133,187,710**	**279,703**	**121,506,506**	**275,645**	**69,580**	**8,964.1**	**-2.6**	**9,201.2**
Austria	50,000	570	50,000	570	905	16.0	-8.0	17.4
Denmark	1,060,000	2,165	1,188,000	2,490	232	286.0	-8.2	311.7
France	103,300	245	119,800	257	462	19.8	0.8	19.6
Germany	276,000	6,200	367,000	9,000	1,141	60.0	-12.1	68.3
Greece	10,000	70	10,000	70	12	1.3	-13.3	1.5
Ireland	—	350	—	350	—	—	—	—
Italy	406,500	3,325	406,500	3,325	205	99.0	-8.2	107.9
Netherlands	100,000	50,000	100,000	50,000	203	35.0	-12.7	40.1
Norway	6,680,000	81,680	6,865,325	79,130	801	2,125.0	-6.4	2,271.0
Spain	150,000	90	150,000	90	16	2.6	-10.3	2.9
Turkey	300,000	300	300,000	300	871	41.2	1.6	40.6
United Kingdom	3,410,000	12,110	3,600,000	14,550	1,360	1,400.0	-8.1	1,524.1
Western Europe	**12,545,800**	**157,105**	**13,156,625**	**160,132**	**6,208**	**4,085.9**	**-7.2**	**4,405.0**
Abu Dhabi	92,200,000	198,500	92,200,000	198,500	1,200	2,450.0	2.9	2,380.0
Bahrain	124,560	3,250	124,560	3,250	496	170.0	0.1	169.9
Dubai	4,000,000	4,000	4,000,000	4,000	200	110.0	10.0	100.0
Iran	136,150,000	991,600	138,400,000	948,200	1,128	3,900.0	-2.0	3,980.0
Iraq	115,000,000	111,940	115,000,000	111,940	1,685	2,365.0	13.2	2,090.0
Israel	1,940	1,075	1,940	1,075	6	—	—	—
Jordan	1,000	213	1,000	213	4	—	—	—
Kuwait	101,500,000	62,860	101,500,000	55,515	790	2,320.0	7.4	2,160.0
Neutral Zone	5,000,000	1,000	5,000,000	1,000	578	570.0	1.8	560.0
Oman	5,500,000	30,000	5,500,000	30,000	2,298	715.0	0.7	710.0
Qatar	15,210,000	891,945	15,207,000	905,300	421	855.0	6.9	800.0
Ras al Khaimah	100,000	1,200	100,000	1,200	7	1.0	42.9	0.7
Saudi Arabia	264,210,000	257,970	264,251,000	252,607	1,560	8,900.0	8.5	8,200.0
Sharjah	1,500,000	10,700	1,500,000	10,700	49	55.0	10.0	50.0
Syria	2,500,000	8,500	2,500,000	8,500	132	385.0	-1.1	389.2
Yemen	3,000,000	16,900	3,000,000	16,900	1,649	305.0	-4.8	320.3
Middle East	**745,997,500**	**2,591,653**	**748,285,500**	**2,548,900**	**12,203**	**23,101.0**	**5.4**	**21,910.2**
Algeria	12,200,000	159,000	12,200,000	159,000	1,285	1,375.0	1.1	1,360.0
Angola	9,040,000	9,530	9,035,000	9,530	1,064	1,870.0	10.1	1,698.2
Cameroon	200,000	4,770	200,000	4,770	255	85.0	-0.2	85.2
Chad	1,500,000	—	1,500,000	—	449	145.0	—	145.0
Congo (former Zaire)	180,000	35	180,000	35	150	25.0	—	25.0
Congo (Brazzaville)	1,600,000	3,200	1,600,000	3,200	460	240.0	—	240.0
Egypt	3,700,000	58,500	3,700,000	58,500	1,491	680.0	5.4	645.0
Equatorial Guinea	1,100,000	1,300	1,100,000	1,300	67	320.0	—	320.0
Gabon	2,000,000	1,000	2,000,000	1,000	395	235.0	2.2	230.0
Ivory Coast	100,000	1,000	100,000	1,000	9	30.0	—	30.0
Libya	43,660,000	54,380	41,464,000	50,100	1,543	1,720.0	0.6	1,710.0

COUNTRY	ESTIMATED PROVED RESERVES Jan. 1, 2009 Oil, 1,000 bbl	Jan. 1, 2009 Gas, bcf	Jan. 1, 2008 Oil, 1,000 bbl	Jan. 1, 2008 Gas, bcf	OIL PRODUCTION Producing oil wells, Dec. 31, 2008*	Estimated 2008, 1,000 b/d	Change from 2007, %	Actual 2007, 1,000 b/d
Mauritania	100,000	1,000	100,000	1,000	—	—	—	—
Morocco	752	53	836	55	7	0.2	-25.0	0.2
Mozambique	—	4,500	—	4,500	—	—	—	—
Nigeria	36,220,000	184,160	36,220,000	183,990	2,524	1,940.0	-8.9	2,130.0
South Africa	15,000	—	15,000	—	28	14.0	-2.8	14.4
Sudan	5,000,000	3,000	5,000,000	3,000	9	490.0	3.8	472.0
Tanzania	—	230	—	230	—	—	—	—
Tunisia	425,000	2,300	400,000	2,300	237	85.0	-10.1	94.6
Other Africa	23,428	6,120	23,428	6,120	11	6.0	—	6.0
Africa	117,064,180	494,078	114,838,264	489,630	9,984	9,260.2	0.6	9,205.6
Albania	199,140	30	199,140	30	1,945	9.9	3.1	9.6
Azerbaijan	7,000,000	30,000	7,000,000	30,000	62	925.0	11.4	830.0
Belarus	198,000	100	198,000	100	—	35.0	-0.6	35.2
Bulgaria	15,000	200	15,000	200	100	1.0	—	1.0
Croatia	79,300	1,080	79,150	1,008	846	14.8	-6.9	15.9
Czech Repub.	15,000	140	15,000	140	—	4.5	—	4.5
Georgia	35,000	300	35,000	300	283	1.0	—	1.0
Hungary	20,180	286	20,180	286	875	14.4	-8.9	15.8
Kazakhstan	30,000,000	85,000	30,000,000	100,000	1,006	1,385.0	25.9	1,100.0
Kyrgyzstan	40,000	200	40,000	200	—	1.0	—	1.0
Lithuania	12,000	—	12,000	—	—	3.0	-3.2	3.1
Poland	96,375	5,820	96,375	5,820	512	16.5	-6.3	17.6
Romania	600,000	2,225	600,000	2,225	6,000	92.5	-5.2	97.6
Russia	60,000,000	1,680,000	60,000,000	1,680,000	100,637	9,760.0	-0.7	9,830.0
Serbia	77,500	1,700	77,500	1,700	646	15.0	—	15.0
Slovakia	9,000	500	9,000	500	—	0.1	—	0.1
Tajikistan	12,000	200	12,000	200	—	—	—	—
Turkmenistan	600,000	94,000	600,000	100,000	2,460	220.0	11.1	198.0
Ukraine	395,000	39,000	395,000	39,000	2,487	75.0	-6.3	80.0
Uzbekistan	594,000	65,000	594,000	65,000	2,190	105.0	-7.9	114.0
Eastern Europe & FSU	99,997,495	2,005,781	99,997,345	2,026,709	120,049	12,678.7	2.5	12,369.4
Afghanistan	—	1,750	—	1,750	—	—	—	—
Australia	1,500,000	30,000	1,500,000	30,000	1,317	453.0	0.4	451.4
Bangladesh	28,000	5,000	28,000	5,000	41	4.0	—	4.0
Brunei	1,100,000	13,800	1,100,000	13,800	779	157.0	-12.3	179.0
China	16,000,000	80,000	16,000,000	80,000	71,542	3,800.0	1.6	3,739.4
China, Taiwan	2,380	220	2,380	220	71	1.0	—	1.0
India	5,624,640	37,960	5,624,640	37,960	3,686	668.0	-2.8	687.3
Indonesia	3,990,000	106,000	4,370,000	93,900	8,331	855.0	1.8	840.0
Japan	44,115	738	44,115	738	145	16.5	-1.8	16.8
Malaysia	4,000,000	83,000	4,000,000	83,000	788	750.0	-1.3	760.0
Myanmar	50,000	10,000	50,000	10,000	450	19.0	46.2	13.0
New Zealand	60,000	1,200	55,000	1,048	72	56.0	37.6	40.7
Pakistan	339,000	31,266	289,202	28,000	204	66.0	-3.5	68.4
Papua New Guinea	88,000	8,000	88,000	8,000	46	41.0	-12.2	46.7
Philippines	138,500	3,480	138,500	3,480	11	15.8	-2.9	16.3
Thailand	441,000	11,198	460,000	11,697	1,180	229.0	7.9	212.2
Vietnam	600,000	6,800	600,000	6,800	28	275.0	-11.3	310.0
Asia-Pacific	34,005,635	430,412	34,349,837	415,393	88,691	7,406.3	0.3	7,386.1
Total World	1,342,207,320	6,254,363	1,331,698,077	6,185,693	867,853	72,966.1	1.1	72,160.0
Total OPEC	944,017,000	3,216,020	931,999,000	3,151,742	39,330	32,136.0	3.8	30,948.9

Source: *Oil & Gas Journal* survey data as of Jan. 1, 2009

NOTES: All reserves figures are reported as proved reserves recoverable with current technology and prices except Former Soviet Union, some eastern European countries, and Canada's gas reserves; these figures include proved plus some probable.

Other Latin America: Barbados, Belize, Suriname. Other Africa: Benin, Ethiopia, Ghana, Namibia, Rwanda, Somalia.

Ecuador rejoined OPEC Nov. 2007. 'Total OPEC' line does include Ecuador data.

* Does not include shut-in, injection, or service wells.

Some totals might not compute exactly due to rounding and other factors.

Table 2: World Oil Consumption

	2005	2006	2007	2005	2006	2007	Change 2007 over 2006	2007 share of total
COUNTRY	million tonnes per year (mtpy)			thousand barrels per day (1,000 b/d)			%	
North America								
US	951.4	943.8	943.1	20,802	20,687	20,698	-0.1	23.9
Canada	100.3	99.6	102.3	2,247	2,246	2,303	2.7	2.6
Mexico	87.7	86.9	89.2	1,974	1,970	2,024	2.8	2.3
Total	1,139.4	1,130.2	1,134.7	25,023	24,904	25,024	0.4	28.7
Latin America								
Argentina	20.1	21.0	23.5	421	440	492	12.0	0.6
Brazil	90.1	91.2	96.5	2,048	2,064	2,192	5.8	2.4
Chile	11.9	12.3	16.0	256	265	342	29.9	0.4
Colombia	10.1	10.6	10.3	225	235	228	-2.4	0.3
Ecuador	7.5	7.8	8.1	168	175	181	4.0	0.2
Peru	7.8	6.4	6.6	165	140	145	2.9	0.2
Venezuela	25.8	26.3	26.8	547	582	596	1.9	0.7
Other Latin America	62.6	64.2	64.1	1,290	1,323	1,317	-0.2	1.6
Total	236.0	239.9	252.0	5,147	5,225	5,493	5.0	6.4
Europe/Eurasia								
Austria	14.2	14.2	13.5	294	294	281	-4.8	0.3
Azerbaijan	5.3	4.9	4.5	108	99	93	-7.1	0.1
Belarus	6.9	6.8	7.0	143	142	145	1.9	0.2
Belgium/ Luxembourg	39.9	41.1	41.2	815	839	839	0.4	1.0
Bulgaria	4.9	5.4	5.4	109	120	120	*	0.1
Czech Republic	9.9	9.8	9.9	211	208	210	0.8	0.3
Denmark	9.2	9.3	9.3	195	197	197	0.4	0.2
Finland	11.0	10.6	10.6	233	225	226	0.6	0.3
France	93.1	92.8	91.3	1,960	1,953	1,919	-1.7	2.3
Germany	122.4	123.6	112.5	2,605	2,624	2,393	-9.0	2.8
Greece	21.2	22.1	21.6	434	453	443	-2.2	0.5
Hungary	7.5	7.8	7.6	163	169	168	-1.4	0.2
Iceland	1.0	1.0	1.0	21	20	21	5.5	*
Republic of Ireland	9.4	9.3	9.4	196	195	198	1.0	0.2
Italy	86.7	86.7	83.3	1,819	1,812	1,745	-3.9	2.1
Kazakhstan	10.0	11.0	10.6	207	227	219	-3.2	0.3
Lithuania	2.8	2.8	2.9	58	59	61	3.5	0.1
Netherlands	49.6	49.0	48.5	1,070	1,043	1,044	-0.9	1.2
Norway	9.7	10.0	10.1	212	217	221	1.2	0.3
Poland	21.9	23.3	24.3	479	516	532	4.0	0.6
Portugal	16.0	14.4	14.4	331	300	302	0.2	0.4
Romania	10.5	10.3	10.8	223	219	229	4.7	0.3
Russian Federation	121.9	127.1	125.9	2,601	2,709	2,699	-0.9	3.2
Slovakia	3.8	3.4	3.8	81	73	80	10.3	0.1
Spain	78.8	78.1	78.7	1,619	1,602	1,615	0.8	2.0
Sweden	15.1	16.6	16.8	315	353	364	1.1	0.4
Switzerland	12.2	12.6	11.3	262	269	243	-9.9	0.3
Turkey	30.0	30.7	31.1	650	655	666	1.5	0.8
Turkmenistan	4.4	4.5	4.7	100	103	107	3.6	0.1
Ukraine	13.9	15.0	15.3	294	318	325	2.2	0.4
United Kingdom	83.0	82.3	78.2	1,802	1,785	1,696	-5.0	2.0
Uzbekistan	5.5	5.6	5.8	114	117	119	2.1	0.1
Other Europe/Eurasia	26.2	26.9	27.7	548	563	581	3.3	0.7
Total	957.8	969.0	949.4	20,274	20,477	20,100	-2.0	24.0

COUNTRY	2005	2006	2007	2005	2006	2007	Change 2007 over 2006	2007 share of total
	million tonnes per year (mtpy)			thousand barrels per day (1,000 b/d)			%	
Middle East								
Iran	74.8	76.7	77.0	1,578	1,625	1,621	0.4	1.9
Kuwait	15.6	14.0	14.0	302	275	276	0.3	0.4
Qatar	2.8	3.4	4.1	65	79	95	21.7	0.1
Saudi Arabia	87.2	92.6	99.3	1,891	2,005	2,154	7.2	2.5
UAE	18.3	20.5	22.0	376	419	450	7.7	0.6
Other Middle East	72.8	74.0	77.0	1,519	1,546	1,608	4.0	1.9
Total	271.5	281.2	293.5	5,731	5,949	6,203	4.4	7.4
Africa								
Algeria	11.0	11.5	12.0	251	260	270	3.9	0.3
Egypt	29.8	28.7	30.6	629	610	651	6.8	0.8
South Africa	24.8	25.3	25.8	526	537	549	2.1	0.7
Other Africa	64.3	66.7	69.8	1,368	1,418	1,485	4.7	1.8
Total	129.9	132.1	138.2	2,773	2,824	2,955	4.6	3.5
Asia-Pacific								
Australia	39.8	41.4	42.2	886	918	935	1.9	1.1
Bangladesh	4.6	4.8	5.0	96	99	102	3.5	0.1
China	327.8	353.3	368.0	6,984	7,530	7,855	4.1	9.3
China Hong Kong	13.8	14.8	16.9	285	303	341	14.1	0.4
India	119.6	120.4	128.5	2,569	2,580	2,748	6.7	3.3
Indonesia	58.2	53.4	54.4	1,232	1,139	1,157	1.9	1.4
Japan	244.1	237.1	228.9	5,358	5,224	5,051	-3.5	5.8
Malaysia	22.0	22.8	23.6	477	496	514	3.4	0.6
New Zealand	7.2	7.2	7.0	155	154	151	-2.5	0.2
Pakistan	15.3	17.4	17.9	312	352	362	2.8	0.5
Philippines	14.7	13.2	13.9	314	283	298	5.3	0.4
Singapore	40.9	44.0	47.4	794	853	917	7.7	1.2
South Korea	105.4	105.6	107.6	2,308	2,318	2,371	1.9	2.7
Taiwan	51.3	51.7	52.5	1,090	1,097	1,123	1.5	1.3
Thailand	44.5	43.8	43.0	929	918	911	-1.8	1.1
Other Asia-Pacific	27.2	27.5	28.4	580	587	608	3.2	0.7
Total	1,136.4	1,158.5	1,185.1	24,368	24,851	25,444	2.3	30.0
Total World	**3,871.0**	**3,910.9**	**3,952.8**	**83,317**	**84,230**	**85,220**	**1.1**	**100.0**

Source: Adapted from *BP Statistical Review of World Energy 2008* (June 2008).

NOTES: "Oil Consumption" includes inland demand plus international aviation and marine bunkers and refinery fuel and loss. Consumption of fuel ethanol and biodiesel is also included.

Some totals might not compute exactly due to rounding and other factors.

Growth rates are adjusted for leap years.

Differences between these world consumption figures and world production statistics are accounted for by stock changes, consumption of non-petroleum additives and substitute fuels, and unavoidable disparities in the definition, measurement or conversion of oil supply and demand data.

Annual changes and shares of total are calculated using million tonnes per annum figures.

* Less than 0.05%

Table 3: World Petroleum Product Consumption

REGION/PRODUCT	2002	2003	2004	2005	2006	2007	Change 2007 over 2006	2007 share of total
	thousand barrels per day (1,000 b/d)						%	
US								
Light distillates	9,167	9,275	9,518	9,548	9,599	9,601	*	46.4
Middle distillates	5,735	5,886	6,116	6,198	6,226	6,222	-0.1	30.1
Fuel oil	686	763	858	914	683	727	6.4	3.5
Others	4,172	4,109	4,239	4,143	4,178	4,147	-0.7	20.0
Total	19,761	20,033	20,731	20,802	20,687	20,698	*	100.0
North America (including US)								
Light distillates	10,523	10,675	10,979	11,020	11,131	11,207	0.7	44.8
Middle distillates	6,655	6,861	7,133	7,249	7,297	7,339	0.6	29.3
Fuel oil	1,209	1,271	1,339	1,396	1,055	1,095	3.7	4.4
Others	5,278	5,243	5,446	5,359	5,420	5,383	-0.7	21.5
Total	23,665	24,050	24,897	25,023	24,904	25,024	0.5	100.0
South & Central America								
Light distillates	1,506	1,496	1,452	1,549	1,591	1,657	4.1	30.2
Middle distillates	1,650	1,650	1,775	1,822	1,861	2,001	7.5	36.4
Fuel oil	789	755	745	779	767	800	4.3	14.6
Others	1,028	925	973	997	1,006	1,036	3.0	18.9
Total	4,974	4,826	4,944	5,147	5,225	5,493	5.1	100.0
Europe								
Light distillates	4,025	3,928	3,888	3,770	3,627	3,506	-3.3	21.7
Middle distillates	6,935	7,138	7,356	7,591	7,850	7,758	-1.2	48.0
Fuel oil	1,996	1,954	1,888	1,854	1,785	1,668	-6.6	10.3
Others	3,113	3,154	3,217	3,289	3,293	3,246	-1.4	20.1
Total	16,069	16,174	16,349	16,503	16,555	16,177	-2.3	100.0
Former Soviet Union								
Light distillates	882	937	918	935	990	1,043	5.3	26.6
Middle distillates	1,048	1,085	1,108	1,151	1,205	1,239	2.8	31.6
Fuel oil	670	646	578	555	607	517	-14.8	13.2
Others	1,066	1,080	1,157	1,130	1,120	1,125	0.4	28.7
Total	3,667	3,748	3,762	3,771	3,923	3,923	*	100.0
Middle East								
Light distillates	1,011	1,059	1,177	1,241	1,322	1,319	-0.2	21.3
Middle distillates	1,666	1,723	1,766	1,850	1,956	2,042	4.4	32.9
Fuel oil	1,320	1,335	1,394	1,487	1,488	1,597	7.3	25.7
Others	1,013	1,112	1,170	1,153	1,182	1,245	5.3	20.1
Total	5,011	5,229	5,507	5,731	5,949	6,203	4.3	100.0
Africa								
Light distillates	591	600	612	630	651	683	4.8	23.1
Middle distillates	1,075	1,107	1,133	1,185	1,230	1,297	5.4	43.9
Fuel oil	416	422	444	482	462	478	3.4	16.2
Others	428	438	455	477	480	497	3.5	16.8
Total	2,510	2,567	2,644	2,773	2,824	2,955	4.6	100.0
China								
Light distillates	1,352	1,489	1,673	1,776	1,926	2,051	6.5	26.1
Middle distillates	1,791	1,945	2,285	2,528	2,723	2,908	6.8	37.0
Fuel oil	792	880	956	904	938	854	-9.0	10.9
Others	1,352	1,489	1,858	1,777	1,942	2,041	5.1	26.0
Total	5,288	5,803	6,772	6,984	7,530	7,855	4.3	100.0

	2002	2003	2004	2005	2006	2007	Change 2007 over 2006	2007 share of total
REGION/PRODUCT	thousand barrels per day (1,000 b/d)						%	
Japan								
Light distillates	1,759	1,788	1,793	1,822	1,813	1,790	-1.3	35.4
Middle distillates	1,930	1,904	1,856	1,876	1,758	1,620	-7.8	32.1
Fuel oil	649	770	672	698	675	667	-1.2	13.2
Others	1,021	993	960	963	979	974	-0.4	19.3
Total	5,359	5,455	5,281	5,358	5,224	5,051	-3.3	100.0
Asia-Pacific (including China, Japan)								
Light distillates	6,022	6,283	6,642	6,781	6,917	7,193	4.0	28.3
Middle distillates	8,102	8,238	8,769	9,017	9,090	9173	0.9	36.1
Fuel oil	3,421	3,568	3,578	3,574	3,655	3,595	-1.7	14.1
Others	4,389	4,614	5,019	4,995	5,188	5,483	5.7	21.5
Total	21,934	22,702	24,008	24,368	24,851	25,444	2.4	100.0
World								
Light distillates	24,560	24,977	25,669	25,926	26,230	26,608	1.4	31.2
Middle distillates	27,133	27,802	29,039	29,866	30,490	30,848	1.2	36.2
Fuel oil	9,821	9,952	9,967	10,127	9,821	9,749	-0.7	11.4
Others	16,316	16,566	17,437	17,399	17,689	18,014	1.8	21.1
Total	77,829	79,296	82,111	83,317	84,230	85,220	1.2	100.0
European Union								
Light distillates	3,763	3,671	3,619	3,520	3,376	3,271	-3.1	22.0
Middle distillates	6,405	6,592	6,795	6,996	7,225	7,108	-1.6	47.8
Fuel oil	1,793	1,745	1,704	1,704	1,647	1,554	-5.7	10.5
Others	2,834	2,857	2,912	2,982	2,985	2,929	-1.9	19.7
Total	14,795	14,865	15,030	15,201	15,233	14,861	-2.4	100.0
OECD								
Light distillates	17,390	17,502	17,759	17,759	17,753	17,754	*	36.3
Middle distillates	16,296	16,682	17,104	17,478	17,664	17,482	-1.0	35.7
Fuel oil	4,175	4,309	4,218	4,253	3,810	3,702	-2.8	7.6
Others	9,826	9,801	10,024	10,008	10,093	9,995	-1.0	20.4
Total	47,687	48,294	49,104	49,497	49,319	48,934	-0.8	100.0
Other EMEs								
Light distillates	6,287	6,538	6,992	7,232	7,487	7,811	4.3	24.1
Middle distillates	9,789	10,034	10,827	11,236	11,621	12,127	4.4	37.5
Fuel oil	4,976	4,997	5,171	5,319	5,404	5,530	2.3	17.1
Others	5,424	5,685	6,256	6,262	6,476	6,895	6.5	21.3
Total	26,476	27,254	29,246	30,049	30,989	32,363	4.4	100.0

Source: Adapted from *BP Statistical Review of World Energy 2008* (June 2008).

NOTES: Other EMEs excludes Former Soviet Union.
Annual changes and shares of total are calculated using thousand barrels daily figures.
Some totals might not compute exactly due to rounding and other factors.
'Light distillates' consists of aviation and motor gasolines and light distillate feedstock (LDF).
'Middle distillates' consists of jet and heating kerosenes, and gas and diesel oils (including marine bunkers).
'Fuel oil' includes marine bunkers and crude oil used directly as fuel
'Others' consists of refinery gas, LPGs, solvents, petroleum coke, lubricants, bitumen, wax, other refined products and refinery fuel and loss.
* Less than 0.05%

Table 4: World Oil Production

	1980	1990	2000	2002	2003	2004	2005	2006	Actual 2007	Est. 2008
COUNTRY	1,000 b/d									
North America										
Canada	1,412.0	1,508.0	2,035.0	2,212.9	2317.4	2,417.8	2,368.8	2,517.2	2,618.1	2,570.0
US	8,569.0	7,220.0	5,822.0	5,746.0	5,681.6	5,418.7	5,178.4	5,101.7	5,064.5	4,900.0
Total	9,981.0	8,728.0	7,857.0	7,958.9	7,999.0	7,836.5	7,547.2	7,618.9	7,682.6	7,470.0
Latin America										
Argentina	487.0	473.0	750.7	738.1	707.0	738.3	645.4	640.5	623.5	610.0
Bolivia	30.0	19.2	27.6	30.9	32.2	35.0	41.6	44.8	44.0	40.0
Brazil	182.0	633.0	1,128.0	1,495.0	1,535.7	1,479.2	1,634.0	1,722.3	1,747.8	1,815.0
Chile	29.0	20.3	7.0	10.0	10.0	10.0	10.0	10.0	2.6	2.3
Colombia	125.0	445.0	687.3	578.5	541.3	527.6	526.1	527.5	531.0	578.0
Cuba	—	15.0	44.0	40.0	40.0	40.0	38.6	38.7	50.0	50.0
Ecuador	222.0	287.0	401.1	392.5	417.8	525.5	532.0	534.6	500.0	500.0
Guatemala	5.0	4.0	20.7	23.6	23.9	20.2	18.5	16.1	16.0	14.0
Mexico	1,936.0	2,633.0	3,012.0	3,177.0	3370.4	3,383.3	3,334.2	3,256.3	3,082.6	2800.0
Peru	191.0	132.0	98.0	92.8	87.3	88.6	111.2	115.6	78.0	75.0
Trinidad & Tobago	211.0	151.0	119.0	130.6	136.5	122.9	144.5	142.8	120.0	113.0
Venezuela	2,167.0	2,118.0	3,028.0	2,285.0	2,006.7	2,208.3	2,705.8	2,561.7	2,390.0	2,350.0
Other Latin America	—	—	—	—	12.8	12.3	13.0	14.1	15.7	16.8
Total	5,585.0	6,935.7	9,323.4	9,007.6	8,921.6	9,191.1	9,754.9	9,625.0	9201.2	8,964.1
Europe										
Albania	—	—	5.9	6.2	6.0	6.1	6.1	8.9	9.6	9.9
Austria	29.0	24.0	19.1	18.6	18.2	17.1	17.3	17.4	17.4	16.0
Bulgaria	6.0	5.0	1.0	1.0	1.0	1.0	1.0	1.0	1.0	1.0
Croatia	—	—	23.0	20.7	20.7	19.1	17.5	16.7	15.9	14.8
Czech Republic	1.9	3.0	3.2	4.8	5.8	5 .9	4.8	4.9	4.5	4.5
Denmark	6.0	118.0	363.0	370.8	367.7	389.1	377.3	342.2	311.7	286.0
France	26.0	62.0	28.4	26.4	24.5	22.9	21.7	21.4	19.6	19.8
Germany	100.0	78.0	62.5	74.1	76.0	70.3	71.5	70.3	68.3	60.0
Greece	—	15.0	5.1	3.3	2.4	2.3	1.8	1.8	1.5	1.3
Hungary	40.6	40.0	25.8	22.0	23.7	22.5	19.0	16.4	15.8	14.4
Italy	39.0	97.0	90.8	84.4	95.8	115.0	115.0	110.4	107.9	99.0
Lithuania	—	—	—	—	—	—	4.3	3.6	3.1	3.0
Netherlands	25.0	71.0	29.3	43.7	45.8	44.0	29.0	26.3	40.1	35.0
Norway	528.0	1,574.0	3,212.0	3,148.8	3,065.0	2,955.1	2,697.6	2,492.0	2,271.0	2,125.0
Poland	6.6	3.0	14.3	16.6	17.6	20.1	19.9	17.5	17.6	16.5
Romania	230.9	160.0	122.9	116.1	115.2	103.0	100.0	98.2	97.6	92.5
Serbia	—	—	—	—	15.5	15.0	15.0	15.0	15.0	15.0
Slovakia	—	—	—	—	1.0	1.0	1.0	0.3	0.1	0.1
Spain	32.0	14.3	4.6	6.5	6.6	5.3	3.3	2.9	2.9	2.6
Turkey	44.0	70.0	56.0	46.9	44.4	42.0	42.0	41.5	40.6	41.2
UK	1,619.0	1,860.0	2,513.7	2,305.8	2,103.2	1,888.3	1,678.0	1,518.2	1,524.1	1,400.0
Total	2,734.0	4,194.3	6,580.6	6,333.8	6,056.1	5,745.1	5,243.1	4,826.9	4,585.3	4,257.6
Middle East										
Bahrain	48.0	42.0	102.3	174.1	33.8	174.2	174.4	171.9	169.9	170.0
Iran	1,467.0	3,120.0	3,681.7	3,430.0	3,784.2	3,931.7	3,890.8	3,890.8	3,980.0	3,900.0
Iraq	2,638.0	2,083.0	2,566.7	2,014.2	1,328.3	2,002.5	1,810.0	1,900.0	2,090.0	2,365.0
Israel	—	—	—	—	0.1	0.1	0.1	0.1	—	—
Kuwait	1,382.0	1,080.0	1,765.0	1,600.0	1,870.0	2,050.0	2,130.0	2,210.0	2,160.0	2,320.0
Neutral Zone	540.0	315.0	630.0	538.0	610.0	600.0	580.0	580.0	560.0	570.0
Oman	283.0	658.0	933.3	904.2	828.3	764.2	757.5	740.0	710.0	715.0
Qatar	472.0	387.0	688.3	644.2	736.7	782.5	798.3	822.5	800.0	855.0
Saudi Arabia	9,630.0	6,215.0	7,995.0	6,850.0	8,480.0	8,750.0	9,060.0	8,930.0	8,200.0	8,900.0
Syria	165.0	385.0	522.5	508.3	527.8	503.3	460.0	422.5	389.2	385.0
UAE	1,709.0	2,101.0	2,230.5	1,873.5	2,287.2	2,353.7	2,453.0	2,605.7	2,530.7	2,616.0
Yemen	0	179.0	354.0	350.0	350.0	350.0	412.8	354.3	320.3	305.0
Total	18,335	16,566	21,469	18,886.6	20,836.4	22,262.2	22,526.9	22,627.8	21,910.2	23,101.1

COUNTRY	1980	1990	2000	2002	2003	2004	2005	2006	Actual 2007	Est. 2008
	1,000 b/d									
Africa										
Algeria	1,016.0	797.0	808.0	850.0	1,110.8	1,206.7	1,351.7	1,347.5	1,360.0	1,375.0
Angola	150.0	480.0	740.0	894.2	879.7	986.2	1,238.1	1,410.0	1,698.2	1,870.0
Cameroon	58.0	164.0	85.0	69.0	65.0	74.0	82.5	87.4	85.2	85.0
Chad	—	—	—	—	—	175.4	180.0	150.0	145.0	145.0
Congo (Zaire)	20.0	28.0	25.0	22.7	25.5	20.0	20.0	20.0	25.0	25.0
Congo (Brazzaville)	56.0	161.0	265.0	254.1	240.3	240.0	240.0	240.0	240.0	240.0
Egypt	596.0	873.0	810.0	751.3	750.1	709.2	695.0	670.0	645.0	680.0
Equatorial Guinea	—	—	167.5	200.0	241.6	320.0	350.0	320.0	320.0	320.0
Gabon	180.0	289.0	325.0	248.0	240.4	235.0	234.2	235.0	230.0	235.0
Ivory Coast	—	2.1	12.0	10.0	25.8	30.0	30.0	30.0	30.0	30.0
Libya	1,785.0	1,369.0	1,410.0	1,315.8	1,427.5	1,546.7	1,640.0	1,708.3	1,710.0	1,720.0
Morocco	0	0	0	0	0.2	0.2	0.2	0.2	0.2	0.2
Nigeria	2,057.0	1,808.0	2,030.0	1,945.0	2,145.0	2,342.5	2,406.7	2,219.2	2,130.0	1,940.0
South Africa	—	—	25.9	20.1	16.5	26.2	24.0	20.0	14.4	14.0
Sudan	—	—	185.0	200.0	255.0	286.7	290.0	414.2	472.0	490.0
Tunisia	110.0	93.0	75.9	72.2	65.8	69.7	71.1	68.2	94.6	85.0
Other Africa	—	—	—	—	6.0	6.0	6.0	6.0	6.0	6.0
Total	6,032.0	6,048.4	6,971.6	6,858.6	7,495.2	8,274.5	8,859.9	8,946.0	9,205.6	9,260.2
Former Soviet Union										
Azerbaijan	—	—	275.0	300.0	306.3	300.0	443.3	642.5	830.0	925.0
Belarus	—	—	35.0	35.0	35.0	37.0	35.5	36.0	35.2	35.0
Georgia	—	—	2.0	2.0	2.8	3.0	2.0	2.0	1.0	1.0
Kazakhstan	—	—	675.0	818.3	888.0	970.0	994.2	1,064.2	1,100.0	1,385.0
Kyrgyzstan	—	—	1.0	2.0	2.0	2.0	1.0	1.0	1.0	1.0
Russia	—	—	6,325.0	7,404.6	8,240.0	8,886.7	9,190.0	9,498.3	9,830.0	9,760.0
Turkmenistan	—	—	140.0	180.0	210.0	200.0	195.0	165.0	198.0	220.0
Ukraine	—	—	73.0	85.0	80.0	84.0	98.0	95.0	80.0	75.0
USSR	12,031.0	11,500.0	—	—	—	—	—	—	—	—
Uzbekistan	—	—	155.0	149.0	144.0	150.0	113.0	108.0	114.0	105.0
Total	12,031.0	11,500.0	7,896.2	8,975.9	9,908.1	10,632.7	11,072.0	11,612.0	12,189.2	12,507.0
Asia-Pacific										
Australia	379.0	582.0	700.0	622.6	518.2	440.4	471.5	469.8	451.4	453.0
Bangladesh	—	0.9	3.3	5.0	3.7	4.0	4.0	3.6	4.0	4.0
Brunei	230.0	143.0	177.9	188.4	193.4	189.2	186.5	203.5	179.0	157.0
China	2,119.0	2,755.0	3,237.3	3,405.7	3,422.9	3,485.2	3,627.1	3,684.4	3,739.4	3,800.0
China, Taiwan	5.0	2.5	0.6	0.9	0.8	0.9	0.9	1.0	1.0	1.0
India	182.0	679.0	646.0	661.8	669.4	687.7	659.1	680.2	687.3	668.0
Indonesia	1,576.0	1,274.0	1,267.3	1,116.3	1,013.3	971.7	945.0	892.5	840.0	855.0
Japan	10.0	10.5	12.7	12.5	14.2	14.3	15.6	15.5	16.8	16.5
Malaysia	288.0	605.0	690.0	770.0	800.8	859.2	770.0	750.8	760.0	750.0
Myanmar	30.0	13.0	8.0	10.0	13.0	13.0	13.0	13.0	13.0	19.0
New Zealand	7.0	39.0	36.0	30.0	23.9	21.0	13.9	15.6	40.7	56.0
Pakistan	10.0	60.0	53.3	60.9	61.1	62.1	64.5	64.8	68.4	66.0
Papua New Guinea	—	—	69.7	47.0	48.2	46.0	46.0	49.0	46.7	41.0
Philippines	15.0	5.0	1.1	14.4	14.8	12.4	15.3	15.2	16.3	15.8
Thailand	—	41.0	111.0	128.7	159.7	154.1	186.0	186.9	212.2	229.0
Vietnam	—	40.0	304.0	330.7	333.2	340.0	340.0	344.6	310.0	275.0
Total	4,851.0	6,249.9	7,318.2	7,404.9	7,290.6	7,301.3	7,358.5	7,390.4	7,386.1	7,406.3
Total World	**59,684**	**60,317**	**67,234**	**65,426.3**	**68,507.0**	**71,243.4**	**72,362.1**	**72,647.0**	**72,160.0**	**72,966.1**

Table 4: World Oil Production, cont'd.

	1980	1990	2000	2002	2003	2004	2005	2006	Actual 2007	Est. 2008
COUNTRY	Million tonnes per year (mtpy)									
North America										
Canada	70.3	75.1	101.4	110.2	115.4	120.4	118.0	125.4	130.4	128.0
US	426.7	359.6	289.5	286.1	282.9	269.8	257.9	254.1	252.2	244.0
Total	497.0	434.7	390.9	396.3	398.3	390.2	375.9	379.4	382.6	372.0
Latin America										
Argentina	24.3	23.6	37.4	36.8	35.2	36.8	32.1	31.9	31.1	30.4
Bolivia	1.5	1.0	1.4	1.5	1.6	1.7	2.1	2.2	2.2	2.0
Brazil	9.1	31.5	56.2	74.4	76.5	73.7	81.4	85.8	87.0	90.4
Chile	1.4	1.0	0.3	0.5	0.5	0.5	0.5	0.5	0.1	0.1
Colombia	6.2	22.2	34.2	28.8	27.0	26.3	26.2	26.3	26.4	28.8
Cuba	—	0.7	2.2	2.0	2.0	2.0	1.9	1.9	2.5	2.5
Ecuador	11.1	14.3	20.0	19.5	20.8	26.2	26.5	26.6	24.9	24.9
Guatemala	0.2	0.2	1.0	1.2	1.2	1.0	0.9	0.8	0.8	0.7
Mexico	96.4	131.1	150.0	158.2	167.8	168.5	166.0	162.2	153.5	139.4
Peru	9.5	6.6	4.9	4.6	4.3	4.4	5.5	5.8	3.9	3.7
Trinidad & Tobago	10.5	7.5	5.9	6.5	6.8	6.1	7.2	7.1	6.0	5.6
Venezuela	107.9	105.5	150.8	113.7	99.9	110.0	134.7	127.6	119.0	117.0
Other Latin America	—	—	—	—	0.6	0.6	0.6	0.7	0.8	0.8
Total	278.1	345.5	464.3	448.5	444.3	457.7	485.8	479.3	458.2	446.4
Europe										
Albania	—	—	0.3	0.3	0.3	0.3	0.3	0.4	0.5	0.5
Austria	1.4	1.2	1.0	0.9	0.9	0.9	0.9	0.9	0.9	0.8
Bulgaria	0.3	0.2	—	—	—	—	—	—	—	—
Croatia	—	—	1.1	1.0	1.0	1.0	0.9	0.8	0.8	0.7
Czech Republic	0.1	0.1	0.2	0.2	0.3	0.3	0.2	0.2	0.2	0.2
Denmark	0.3	5.9	18.1	18.5	18.3	19.4	18.8	17.0	15.5	14.2
France	1.3	3.1	1.4	1.3	1.2	1.1	1.1	1.1	1.0	1.0
Germany	5.0	3.9	3.1	3.7	3.8	3.5	3.6	3.5	3.4	3.0
Greece	—	0.7	0.3	0.2	0.1	0.1	0.1	0.1	0.1	0.1
Hungary	2.0	2.0	1.3	1.1	1.2	1.1	0.9	0.8	0.8	0.7
Italy	1.9	4.8	4.5	4.2	4.8	5.7	5.7	5.5	5.4	4.9
Lithuania	—	—	—	—	—	—	0.2	0.2	0.2	0.1
Netherlands	1.2	3.5	1.5	2.2	2.3	2.2	1.4	1.3	2.0	1.7
Norway	26.3	78.4	160.0	156.8	152.6	147.2	134.3	124.1	113.1	105.8
Poland	0.3	0.1	0.7	0.8	0.9	1.0	1.0	0.9	0.9	0.8
Romania	11.5	8.0	6.1	5.8	5.7	5.1	5.0	4.9	4.9	4.6
Serbia	—	—	—	—	0.8	0.7	0.7	0.7	0.7	0.7
Slovakia	—	—	—	—	—	—	—	—	—	—
Spain	1.6	0.7	0.2	0.3	0.3	0.3	0.2	0.1	0.1	0.1
Turkey	2.2	3.5	2.8	2.3	2.2	2.1	2.1	2.1	2.0	2.1
UK	80.6	92.6	125.2	114.8	104.7	94.0	83.6	75.6	75.9	69.7
Total	136.0	208.7	327.8	315.4	301.6	286.1	261.1	240.4	228.3	212.0
Middle East										
Bahrain	2.4	2.1	5.1	8.7	1.7	8.7	8.7	8.6	8.5	8.5
Iran	73.1	155.4	183.4	170.8	188.4	195.8	193.8	193.8	198.2	194.2
Iraq	131.4	103.7	127.8	100.3	66.1	99.7	90.1	94.6	104.1	117.8
Israel	—	—	—	—	—	—	—	—	—	—
Kuwait	68.8	53.8	87.9	79.7	93.1	102.1	106.1	110.1	107.6	115.5
Neutral Zone	26.9	15.7	31.4	26.8	30.4	29.9	28.9	28.9	27.9	28.4
Oman	14.1	32.8	46.5	45.0	41.2	38.1	37.7	36.9	35.4	35.6
Qatar	23.5	19.3	34.3	32.1	36.7	39.0	39.8	41.0	39.8	42.6
Saudi Arabia	479.6	309.5	398.2	341.1	422.3	435.7	451.2	444.7	408.4	443.2
Syria	8.2	19.2	26.0	25.3	26.3	25.1	22.9	21.0	19.4	19.2
UAE	85.1	104.6	111.1	93.3	113.9	117.2	122.2	129.8	126.0	130.3
Yemen	—	8.9	17.6	17.4	17.4	17.4	20.6	17.6	16.0	15.2
Total	913.1	825.0	1,069.2	940.5	1,037.7	1,108.6	1,121.8	1,126.9	1,091.1	1,150.4

COUNTRY	1980	1990	2000	2002	2003	2004	2005	2006	Actual 2007	Est. 2008
	Million tonnes per year (mtpy)									
Africa										
Algeria	50.6	39.7	40.2	42.3	55.3	60.1	67.3	67.1	67.7	68.5
Angola	7.5	23.9	36.8	44.5	43.8	49.1	61.7	70.2	84.6	93.1
Cameroon	2.9	8.2	4.2	3.4	3.2	3.7	4.1	4.4	4.2	4.2
Chad	—	—	—	—	—	8.7	9.0	7.5	7.2	7.2
Congo (Brazzaville)	2.8	8.0	13.2	12.7	12.0	12.0	12.0	12.0	12.0	12.0
Congo (Zaire)	1.0	1.4	1.2	1.1	1.3	1.0	1.0	1.0	1.2	1.2
Egypt	29.7	43.5	40.3	37.4	37.4	35.3	34.6	33.4	32.1	33.9
Equatorial Guinea	—	—	8.3	10.0	12.0	15.9	17.4	15.9	15.9	15.9
Gabon	9.0	13.4	16.2	12.3	12.0	11.7	11.7	11.7	11.5	11.7
Ivory Coast	—	0.1	0.6	0.5	1.3	1.5	1.5	1.5	1.5	1.5
Libya	88.9	68.2	70.2	65.5	71.1	77.0	81.7	85.1	85.2	85.7
Morocco	—	—	—	—	—	—	—	—	—	—
Nigeria	102.4	90.0	101.0	96.9	106.8	116.6	119.9	110.5	106.1	96.6
South Africa	—	—	1.3	1.0	0.8	1.3	1.2	1.0	0.7	0.7
Sudan	—	—	9.2	10.0	12.7	14.3	14.4	20.6	23.5	24.4
Tunisia	5.5	4.6	3.8	3.6	3.3	3.5	3.5	3.4	4.7	4.2
Other Africa	—	—	—	—	0.3	0.3	0.3	0.3	0.3	0.3
Total	300.5	301.2	347.2	341.5	373.3	412.0	441.2	445.5	458.4	461.2
Former Soviet Union										
Azerbaijan	—	—	13.7	14.9	15.3	14.9	22.1	32.0	41.3	46.1
Belarus	—	—	1.7	1.7	1.7	1.8	1.8	1.8	1.8	1.7
Georgia	—	—	0.1	0.1	0.1	0.1	0.1	0.1	—	—
Kazakhstan	—	—	33.6	40.7	44.2	48.3	49.5	53.0	54.8	69.0
Kyrgyzstan	—	—	0	0.1	0.1	0.1	0.1	0.1	0.1	0.1
Russia	—	—	315.0	368.7	410.4	442.5	457.7	473.0	489.5	486.0
Turkmenistan	—	—	7.0	9.0	10.5	10.0	9.7	8.2	9.9	11.0
Ukraine	—	—	3.6	4.2	4.0	4.2	4.9	4.7	4.0	3.7
USSR	599.1	572.7	—	—	—	—	—	—	—	—
Uzbekistan	—	—	7.7	7.4	7.2	7.5	5.6	5.4	5.7	5.2
Total	599.1	572.7	393.2	447.0	493.4	529.5	551.4	587.3	607.0	622.8
Asia-Pacific										
Australia	18.9	29.0	34.9	31.0	25.8	21.9	23.5	23.4	22.5	22.6
Bangladesh	—	—	0.2	0.2	0.2	0.2	0.2	0.2	0.2	0.2
Brunei	11.5	7.1	8.9	9.4	9.6	9.4	9.3	10.1	8.9	7.8
China	105.5	137.2	161.2	169.6	170.5	173.5	180.6	183.5	186.2	189.2
China, Taiwan	0.2	0.1	—	—	—	—	—	—	—	—
India	9.1	33.8	32.2	33.0	33.3	34.2	32.8	33.9	34.2	33.3
Indonesia	78.5	63.4	63.1	55.6	50.5	48.4	47.1	44.4	41.8	42.6
Japan	0.5	0.5	0.6	0.6	0.7	0.7	0.8	0.8	0.8	0.8
Malaysia	14.3	30.1	34.4	38.3	39.9	42.8	38.3	37.4	37.8	37.4
Myanmar	1.5	0.6	0.4	0.5	0.6	0.6	0.6	0.6	0.6	0.9
New Zealand	0.3	1.9	1.8	1.5	1.2	1.0	0.7	0.8	2.0	2.8
Pakistan	0.5	3.0	2.7	3.0	3.0	3.1	3.2	3.2	3.4	3.3
Papua New Guinea	—	—	3.5	2.3	2.4	2.3	2.3	2.4	2.3	2.0
Philippines	0.7	0.2	—	0.7	0.7	0.6	0.8	0.8	0.8	0.8
Thailand	—	2.0	5.5	6.4	8.0	7.7	9.3	9.3	10.6	11.4
Vietnam	—	2.0	15.1	16.5	16.6	16.9	16.9	17.2	15.4	13.7
Total	241.5	310.9	364.4	368.7	363.1	363.6	366.5	368.0	367.8	368.8
Total World	**2,972.0**	**3,003.4**	**3,348.2**	**3,257.9**	**3,411.6**	**3,547.6**	**3,603.6**	**3,617.8**	**3,593.6**	**3,633.7**

Source: *Oil & Gas Journal* survey data as of Jan. 1, 2009.

NOTES: Some totals might not compute exactly due to rounding and other factors.
Other Latin America includes Barbados, Suriname. Other Africa includes Benin, Ethiopia, Ghana, Namibia, Rwanda, Somalia.
Slovakia production is included in Czech Republic's total before 2000.
USSR includes total of all Former Soviet Union countries before 2000.

Table 5: Historical Oil Reserves

	January 1										
	1980	1990	2000	2002	2003	2004	2005	2006	2007	2008	2009
COUNTRY	billion bbl										
North America											
Canada	6.4	6.1	4.9	4.9	180.0	178.9	178.8	178.8	179.2	178.6	178.1
US	26.4	25.9	21.0	22.0	22.4	22.7	21.9	21.4	21.6	21.0	21.3
Latin America											
Argentina	2.5	2.3	2.8	3.0	2.9	2.8	2.7	2.3	2.5	2.6	2.6
Brazil	1.3	2.8	7.4	8.5	8.3	8.5	10.6	11.2	11.8	12.2	12.6
Colombia	1.0	2.0	2.6	1.8	1.8	1.8	1.5	1.5	1.5	1.5	1.4
Ecuador	1.1	1.5	2.1	2.1	4.6	4.6	4.6	4.6	4.5	4.5	4.7
Mexico	44.0	56.4	28.4	26.9	12.6	15.7	14.6	12.9	12.3	11.7	10.5
Venezuela	17.9	58.5	72.6	77.7	77.8	77.8	77.2	79.7	80.0	87.0	99.4
Europe											
Norway	5.5	11.5	10.8	9.4	10.3	10.4	8.5	7.7	7.9	6.9	6.7
UK	14.8	4.2	5.2	4.9	4.7	4.7	4.5	4.0	3.9	3.6	3.4
Middle East											
Iran	57.5	92.8	89.7	89.7	89.7	125.8	125.8	132.5	136.3	138.4	136.2
Iraq	30.0	100.0	112.5	112.5	112.5	115.0	115.0	115.0	115.0	115.0	115.0
Kuwait	64.9	94.5	94.0	94.0	94.0	96.5	99.0	101.5	99.0	101.5	101.5
Neutral Zone	6.0	5.2	5.0	5.0	5.0	5.0	5.0	5.0	5.0	5.0	5.0
Oman	2.3	4.2	5.3	5.5	5.5	5.5	5.5	5.5	5.5	5.5	5.5
Qatar	3.6	4.5	3.7	15.2	15.2	15.2	15.2	15.2	15.2	15.2	15.2
Saudi Arab.	165.0	254.9	261.0	259.3	259.3	259.4	259.4	264.3	259.8	264.3	264.2
Syria	1.3	1.7	2.5	2.5	2.5	2.5	2.5	2.5	2.5	2.5	2.5
UAE	30.4	96.2	96.2	97.8	97.8	97.8	97.8	97.8	97.8	97.9	97.9
Yemen	—	3.0	4.0	4.0	4.0	4.0	4.0	4.0	3.0	3.0	3.0
Africa											
Algeria	8.2	9.2	9.2	9.2	9.2	11.3	11.8	11.4	12.3	12.2	12.2
Angola	1.2	2.0	5.4	5.4	5.4	5.4	5.4	5.4	8.0	9.0	9.0
Egypt	2.9	4.5	2.9	2.9	3.7	3.7	3.7	3.7	3.7	3.7	3.7
Gabon	—	0.7	2.5	2.5	2.5	2.5	2.5	2.5	2.0	2.0	2.0
Libya	23.0	22.8	29.5	29.5	29.5	36.0	39.0	39.1	41.5	41.5	43.7
Nigeria	16.7	16.0	22.5	24.0	24.0	25.0	35.3	35.9	36.2	36.2	36.2
Former Soviet Union											
Azerbaijan	—	—	—	1.2	7.0	7.0	7.0	7.0	7.0	7.0	7.0
Kazakhstan	—	—	—	5.4	9.0	9.0	9.0	9.0	30.0	30.0	30.0
Russia	—	—	—	48.6	60.0	60.0	60.0	60.0	60.0	60.0	60.0
USSR	63.0	24.0	59.0	—	—	—	—	—	—	—	—
Asia-Pacific											
Australia	2.4	1.7	2.9	3.5	3.5	3.5	1.5	1.4	1.6	1.5	1.5
Brunei	1.7	1.4	1.4	1.4	1.4	1.4	1.4	1.4	1.1	1.1	1.1
China	—	24.0	24.0	24.0	18.3	18.3	18.3	18.3	16.0	16.0	16.0
India	2.5	7.5	4.8	4.8	5.4	5.4	5.4	5.8	5.6	5.6	5.6
Indonesia	9.5	8.2	4.9	5.0	5.0	4.7	4.7	4.3	4.3	4.4	4.0
Malaysia	3.0	3.0	3.9	3.0	3.0	3.0	3.0	3.0	3.0	4.0	4.0
Total World	**648.5**	**1,001.6**	**1,016.0**	**1,031.6**	**1,212.9**	**1,265.8**	**1,277.7**	**1,292.5**	**1,309.3**	**1,331.7**	**1,342.2**

Source: *Oil & Gas Journal* data as of Jan. 1, 2009.

NOTES: Data are for nations with major reserves. Total World includes other nations as well. Some totals might not compute exactly due to rounding and other factors. Canada's large increase in 2002-03 was due to including its oil sands.

Table 6: World Oil Imports and Exports

COUNTRY/REGION	2001	2002	2003	2004	2005	2006	2007	Change 2007 over 2006	2007 share of total
	thousand barrels per day (1,000 b/d)							%	
IMPORTS									
US	11,618	11,357	12,254	12,898	13,525	13,612	13,632	0.1	24.9
Europe	11,531	11,895	11,993	12,538	13,261	13,461	13,953	3.6	25.4
Japan	5,202	5,070	5,314	5,203	5,225	5,201	5,032	-3.2	9.2
Rest of World	16,436	16,291	17,191	18,651	19,172	20,287	22,207	9.5	40.5
Total World	44,787	44,613	46,752	49,290	51,182	52,561	54,824	4.3	100.0
EXPORTS									
US	910	904	921	991	1,129	1,317	1,439	9.3	2.6
Canada	1,804	1,959	2,096	2,148	2,201	2,330	2,457	5.5	4.5
Mexico	1,882	1,966	2,115	2,070	2,065	2,102	1,975	-6.0	3.6
South & Central America	3,143	2,965	2,942	3,233	3,528	3,681	3,570	-3.0	6.5
Europe	1,947	2,234	2,066	1,993	2,149	2,173	2,273	4.6	4.1
Former Soviet Union	4,679	5,370	6,003	6,440	7,076	7,115	8,334	16.5	15.2
Middle East	19,098	18,062	18,943	19,630	19,821	20,204	19,680	-2.6	35.9
North Africa	2,724	2,620	2,715	2,917	3,070	3,225	3,336	3.4	6.1
West Africa	3,182	3,134	3,612	4,048	4,358	4,704	4,830	2.7	8.8
Asia-Pacific	3,914	3,848	3,978	4,189	4,243	4,312	5,274	22.3	9.6
Rest of World	1,506	1,551	1,361	1,631	1,542	1,359	1,656	21.8	3.0
Total World	44,789	44,613	46,752	49,290	51,182	52,561	54,824	4.3	100.0

Source: Adapted from *BP Statistical Review of World Energy 2008.*

NOTES: Some totals might not compute exactly due to rounding and other factors.
'Rest of World' includes unidentified trade.
'Asia-Pacific' excludes Japan.
Annual changes and shares of total are calculated using thousand barrels daily figures.

Table 7: World Oil Trade Movements

	2007 IMPORTS TO												
	North America		Latin America				Asia-Pacific						
	US	Canada	Mexico	S. & Central America	Europe	Africa	Australasia	China	Japan	Singapore	Other Asia-Pacific	Rest of World	Total
2007 IMPORTS FROM	million tonnes												
US	—	10.9	11.1	21.6	15.8	1.4	0.4	0.3	4.2	0.7	1.1	1.6	69.1
Canada	119.7	—	0.1	*	0.5	—	—	0.5	0.4	0.1	*	—	121.2
Mexico	76.1	1.3	—	9.6	8.8	—	—	—	—	0.2	1.9	0.2	98.1
S. & Central America	127.4	5.3	2.9	—	23.0	2.0	*	13.7	0.4	0.1	0.3	0.1	175.3
Europe	50.0	20.2	5.1	7.5	—	14.4	0.2	0.5	0.7	0.4	1.6	9.4	109.9
FSU	22.6	2.1	—	1.7	332.1	0.2	*	26.3	8.2	0.2	10.9	6.8	411.1
Middle East	110.4	7.0	0.8	4.5	146.6	38.1	7.7	78.8	199.9	42.0	336.6	3.0	975.3
N. Africa	39.1	9.1	0.2	5.6	95.2	4.3	—	4.6	0.3	0.1	6.4	—	164.9
W. Africa	96.1	4.0	0.1	21.1	38.8	3.8	0.1	35.8	2.2	0.2	37.7	—	240.2
E. & S. Africa	—	—	—	*	0.1	—	—	12.7	5.1	0.9	1.4	—	20.2
Australasia	0.2	—	—	*	*	—	—	1.4	3.3	9.2	12.5	*	26.6
China	0.6	*	0.1	2.4	0.9	0.2	0.2	—	1.2	3.7	9.7	0.2	19.2
Japan	2.6	0.4	0.7	*	0.9	*	1.5	3.1	—	1.1	1.2	*	11.5
Singapore	0.7	*	0.2	0.3	1.2	1.0	9.2	3.3	1.3	—	51.2	0.5	68.9
Other Asia-Pacific	11.4	0.2	0.5	4.8	4.1	0.6	20.3	22.1	21.5	54.4	—	0.3	140.1
Unidentified	14.9	6.2	—	*	20.8	—	1.2	*	—	—	5.8	*	49.0
Total	671.9	66.7	21.6	79.2	688.8	66.1	40.8	203.1	248.8	113.4	478.3	22.1	2,700.6
	1,000 b/d												
US	—	226	231	450	329	30	9	6	88	15	22	33	1,439
Canada	2,426	—	1	*	10	—	—	9	7	3	*	—	2,457
Mexico	1,533	26	—	194	176	—	—	—	—	4	38	4	1,975
S. & Central America	2,592	109	60	—	469	41	*	278	9	3	7	3	3,570
Europe	1,038	408	106	156	—	300	3	9	14	8	34	195	2,273
FSU	467	42	—	35	6,726	4	*	532	166	4	220	137	8,334
Middle East	2,218	140	16	91	2,957	772	154	1,587	4,032	844	6,806	63	19,680
N. Africa	795	182	3	113	1,923	87	—	93	6	3	129	—	3,336
W. Africa	1,933	81	3	425	781	77	2	719	45	4	758	—	4,830
E. & S. Africa	—	—	—	*	3	—	—	255	103	19	27	—	407
Australasia	4	—	—	*	*	—	—	29	67	191	252	*	543
China	13	1	1	50	18	3	3	—	26	77	202	5	399
Japan	54	8	14	*	19	1	32	64	—	22	26	1	241
Singapore	15	*	4	6	25	21	192	69	27	—	1,069	11	1,440
Other Asia-Pacific	235	5	11	99	85	12	408	458	441	1,132	—	5	2,892
Unidentified	308	125	—	*	432	—	26	*	—	—	117	*	1,008
Total	13,632	1,354	451	1,620	13,953	1,350	830	4,111	5,032	2,329	9,705	457	54,824

Source: Adapted from *BP Statistical Review of World Energy 2008.*

NOTES: Some totals might not compute exactly due to rounding and other factors.
'Unidentified' includes changes in the quantity of oil in transit, movements not otherwise shown, and unidentified military use.
* Less than 0.05.

Table 8: World Oil Balance

	Avg. 2004	Avg. 2005	Avg. 2006	2007 1Q07	2007 2Q07	2007 3Q07	2007 4Q07	Avg. 2007
	million b/d							
SUPPLY								
OECD								
US	8.70	8.32	8.33	8.38	8.50	8.36	8.58	8.46
Canada	3.14	3.09	3.29	3.45	3.37	3.48	3.39	3.42
Mexico	3.85	3.78	3.71	3.59	3.61	3.46	3.35	3.50
North Sea	5.62	5.17	4.77	4.80	4.49	4.28	4.57	4.54
Other OECD	1.51	1.52	1.49	1.50	1.54	1.56	1.57	1.54
Total OECD	22.81	21.88	21.59	21.72	21.51	21.15	21.45	21.46
Non-OECD								
OPEC								
OPEC Crude								
Algeria	1.68	1.80	1.81	1.83	1.82	1.83	1.85	1.83
Angola	1.05	1.25	1.41	1.61	1.68	1.74	1.94	1.74
Ecuador	0.53	0.53	0.54	0.50	0.51	0.51	0.52	0.51
Indonesia	1.10	1.07	1.02	0.98	0.96	0.95	0.96	0.96
Iran	4.00	4.14	4.03	3.95	3.90	3.90	3.90	3.91
Iraq	2.01	1.88	2.00	1.93	2.07	2.05	2.29	2.09
Kuwait	2.38	2.53	2.54	2.43	2.42	2.48	2.52	2.46
Libya	1.52	1.63	1.68	1.68	1.68	1.71	1.74	1.70
Nigeria	2.33	2.63	2.44	2.34	2.29	2.38	2.39	2.35
Qatar	0.78	0.84	0.85	0.83	0.83	0.87	0.88	0.85
Saudi Arabia	9.10	9.55	9.15	8.65	8.60	8.67	8.97	8.72
UAE	2.48	2.54	2.64	2.60	2.61	2.66	2.54	2.60
Venezuela	2.56	2.56	2.51	2.40	2.44	2.44	2.44	2.43
OPEC NGL	2.71	2.94	3.02	3.09	3.10	3.10	3.09	3.10
Refinery proc. gain	0.06	0.06	0.06	0.06	0.06	0.06	0.06	0.06
Total OPEC	34.45	36.09	35.83	34.98	35.07	35.44	35.18	35.42
Non-OPEC								
Former Soviet Union	11.34	11.77	12.15	12.61	12.60	12.55	12.66	12.60
China	3.66	3.78	3.84	3.92	3.96	3.87	3.86	3.90
Other Non-OECD	10.84	11.04	11.10	10.70	11.04	11.21	11.13	11.02
Total Non-OECD	60.29	62.68	62.93	62.21	62.66	63.08	63.82	62.95
Total World Supply	**83.10**	**84.56**	**84.52**	**83.93**	**84.17**	**84.23**	**85.28**	**84.40**

Table 8: World Oil Balance, cont'd.

	Avg. 2004	Avg. 2005	Avg. 2006	2007 1Q07	2007 2Q07	2007 3Q07	2007 4Q07	Avg. 2007
	million b/d							
DEMAND								
OECD								
OECD North America								
US	20.73	20.80	20.69	20.79	20.63	20.73	20.58	20.68
US Territories	0.35	0.34	0.35	0.30	0.32	0.33	0.32	0.32
Canada	2.31	2.34	2.30	2.38	2.29	2.40	2.39	2.36
Mexico	2.00	2.07	2.08	2.12	2.14	2.06	2.16	2.12
Total OECD N. America	25.39	25.55	25.42	25.59	25.38	25.52	25.45	25.48
OECD Asia-Pacific								
Australia & New Zealand	1.06	1.09	1.11	1.12	1.10	1.12	1.15	1.12
Japan	5.32	5.32	5.20	5.43	4.64	4.70	5.25	5.01
South Korea	2.16	2.19	2.18	2.36	2.12	2.06	2.31	2.21
Total OECD Asia-Pacific	8.54	8.60	8.49	8.91	7.86	7.88	8.71	8.34
OECD Europe								
France	2.01	1.99	1.98	1.98	1.86	1.94	2.02	1.95
Germany	2.67	2.65	2.69	2.37	2.37	2.55	2.54	2.46
Italy	1.79	1.75	1.74	1.72	1.69	1.65	1.75	1.70
UK	1.79	1.83	1.81	1.80	1.78	1.75	1.73	1.76
Other OECD Europe	7.27	7.44	7.45	7.36	7.25	7.52	7.60	7.43
Total OECD Europe	15.52	15.67	15.68	15.23	14.95	15.41	15.62	15.30
Total OECD	49.44	49.83	49.58	49.74	48.20	48.82	49.78	49.13
Non-OECD								
Former Soviet Union	4.04	4.07	4.21	4.25	4.32	4.22	4.32	4.28
Non-OECD Europe	0.74	0.76	0.78	0.85	0.78	0.73	0.79	0.79
China	6.44	6.72	7.20	7.33	7.52	7.59	7.87	7.58
Other Asia	8.37	8.53	8.66	8.74	8.83	8.64	8.93	8.78
Other Non-OECD	13.38	13.91	14.54	14.94	15.22	15.54	15.26	15.24
Total Non-OECD	32.97	33.99	35.38	36.11	36.68	36.72	37.16	36.67
Total World Demand	82.41	83.82	84.95	85.84	84.88	85.54	86.94	85.80
Supply/Demand Balance	0.69	0.74	- 0.43	-1.91	-0.71	-1.31	-1.66	-1.40

Source: U.S. Energy Information Administration, www.eia.doe.gov/emeu/ipsr/balance.html, November 2008 *International Petroleum Monthly*. Data as of Dec. 20, 2008.

NOTES: Some totals might not compute exactly due to rounding.
OECD = Organization for Economic Cooperation and Development.

North Sea includes the UK Offshore, Norway, Denmark, Netherlands Offshore, and Germany Offshore.

Table 9: Summary of Operating Refineries Worldwide

COUNTRY	Number of refineries	Charge capacity, b/cd		
		Crude	Catalytic cracking	Catalytic reforming
Albania	2	26,300	0	3,500
Algeria	4	450,000	0	88,900
Angola	1	39,000	0	1,900
Argentina	10	626,075	141,180	56,630
Aruba	1	270,750	0	0
Australia	7	696,346	228,200	165,403
Austria	1	208,600	26,250	32,725
Azerbaijan	2	398,978	57,750	24,466
Bahrain	1	262,000	42,000	12,000
Bangladesh	1	33,000	0	1,800
Belarus	2	493,323	44,000	92,000
Belgium	4	797,629	132,684	104,517
Bolivia	2	41,200	0	12,100
Brazil	13	1,908,278	505,287	24,386
Brunei	1	8,600	0	5,700
Bulgaria	1	115,240	23,300	4,060
Cameroon	1	37,000	0	6,500
Canada	18	2,029,450	504,500	372,100
Chile	3	226,800	50,540	26,460
China	53	6,446,000	588,000	178,000
China, Taiwan	4	1,290,000	217,900	115,000
Colombia	5	285,850	90,000	0
Congo, Republic of	1	21,000	0	2,000
Costa Rica	1	24,000	0	1,200
Croatia	3	250,317	51,002	49,368
Cuba	4	301,400	14,700	20,000
Czech Republic	3	183,000	0	27,470
Denmark	2	174,400	0	21,990
Dominican Republic	2	47,000	0	8,200
Ecuador	3	176,000	18,000	12,800
Egypt	9	726,250	0	62,240
El Salvador	1	22,000	0	3,200
Eritrea	1	14,564	0	1,465
Finland	2	255,600	51,400	45,730
France	13	1,986,197	388,559	289,299
Gabon	1	24,000	0	1,400
Germany	15	2,417,523	350,171	405,850
Ghana	1	45,000	14,000	65,000
Greece	4	423,000	75,550	49,200
Hungary	1	161,000	24,000	29,600
India	17	2,255,540	303,305	41,673
Indonesia	8	992,745	101,450	92,970
Iran	9	1,451,000	35,000	164,700
Iraq	8	597,500	0	82,000
Ireland	1	71,000	0	11,000
Israel	2	220,000	49,500	26,500
Italy	17	2,337,229	313,500	286,569
Ivory Coast	1	63,990	0	12,330
Jamaica	1	36,000	0	3,700
Japan	31	4,690,700	932,630	782,825
Jordan	1	90,400	4,000	10,900
Kazakhstan	3	345,093	38,356	51,586
Kenya	1	90,000	0	8,260
Kuwait	3	889,200	36,000	46,620
Kyrgyzstan	1	10,000	0	0
Liberia	1	15,000	0	2,000
Libya	5	378,000	0	20,250
Lithuania	1	190,000	43,200	45,900
Macedonia	1	50,000	0	10,860
Malaysia	6	514,832	42,700	86,570

Table 9: Summary of Operating Refineries Worldwide, cont'd.

COUNTRY	Number of refineries	Charge capacity, b/cd		
		Crude	Catalytic cracking	Catalytic reforming
Martinique	1	17,000	0	3,019
Mexico	6	1,540,000	380,500	279,300
Morocco	2	154,901	5,040	24,359
Myanmar	3	57,000	0	0
Netherlands	6	1,207,714	49,000	149,753
Netherlands Antilles	1	320,000	50,000	20,000
New Zealand	1	107,000	0	25,840
Nicaragua	1	20,000	0	2,900
Nigeria	4	505,000	82,700	70,070
North Korea	2	71,000	0	7,300
Norway	2	315,000	49,000	36,400
Oman	1	85,000	0	16,000
Pakistan	5	276,050	0	26,365
Papua New Guinea	1	32,500	0	0
Paraguay	1	7,500	0	0
Peru	6	192,950	32,000	2,100
Philippines	3	282,000	19,000	51,000
Poland	4	492,950	32,985	67,514
Portugal	2	304,172	40,500	50,182
Puerto Rico	1	73,000	0	21,000
Qatar	1	200,000	60,000	29,400
Romania	10	516,557	109,478	61,763
Russia	40	5,428,497	330,817	745,735
Saudi Arabia	7	2,080,000	103,600	193,160
Senegal	1	25,030	0	1,590
Serbia & Montenegro	2	214,826	18,950	18,822
Sierra Leone	1	10,000	0	0
Singapore	3	1,344,000	80,000	167,970
Slovakia	1	115,000	18,000	21,000
Slovenia	1	13,500	0	0
South Africa	4	485,297	104,025	77,143
South Korea	6	2,606,500	195,000	261,470
Spain	9	1,271,500	191,300	196,750
Sri Lanka	1	50,000	0	5,300
Sudan	3	121,700	0	1,900
Suriname	1	7,000	0	0
Sweden	5	437,000	29,700	70,660
Switzerland	2	132,000	20,400	28,000
Syria	2	239,865	0	31,242
Tanzania	1	14,900	0	2,500
Thailand	4	729,100	91,490	125,590
Trinidad & Tobago	1	168,000	24,000	18,000
Tunisia	1	34,000	0	3,300
Turkey	6	714,275	28,935	65,662
Turkmenistan	2	236,970	15,151	52,540
Ukraine	6	879,759	70,100	144,711
United Arab Emirates	5	781,250	34,350	25,875
United Kingdom	11	1,857,668	444,723	339,051
United States	130	17,379,662	5,784,034	3,524,403
Uruguay	1	50,000	12,000	12,000
Uzbekistan	3	222,271	0	23,487
Venezuela	5	1,282,100	231,800	49,500
Virgin Islands	1	500,000	140,000	105,000
Yemen	2	140,000	0	14,500
Zambia	1	23,750	0	5,320
Total	655	85,603,612	14,417,192	11,487,818

Source: *Oil & Gas Journal* survey data as of Jan. 1, 2009.

NOTES: Some totals might not compute exactly due to rounding and other factors.
b/cd = barrels per calendar-day.

Table 10: World Refining Capacity

COUNTRY	1980	1990	2000	2002	2003	2004	2005	2006	2007	2008	2009
	1,000 b/cd										
North America											
Canada	2,222	1,852	1,912	1,944	1,983	1,988	2,017	2,017	2,041	1,969	2,029
US	17,720	16,244	16,541	16,564	16,623	16,698	16,775	17,126	17,273	17,447	17,380
Total	19,942	18,096	18,453	18,508	18,606	18,686	18,792	19,143	19,314	19,416	19,409
Latin America											
Argentina	676	689	662	639	639	639	625	625	625	626	626
Bolivia	74	58	48	63	63	63	47	47	41	41	41
Brazil	1,205	1,397	1,783	1,756	1,865	1,914	1,920	1,908	1,908	1,908	1,908
Chile	139	147	205	205	205	205	227	227	227	227	227
Colombia	193	227	286	286	286	286	286	286	286	286	286
Cuba	160	160	301	301	301	301	301	301	301	301	301
Ecuador	86	145	176	176	176	176	176	176	176	176	176
Guatemala	16	16	20	16	16	16	—	—	—	—	—
Jamaica	—	—	—	—	—	—	—	36	36	36	36
Mexico	1,394	1,514	1,525	1,525	1,684	1,684	1,684	1,684	1,540	1,540	1,540
Peru	170	172	182	182	191	191	193	193	193	193	193
Trinidad & Tobago	456	300	160	160	160	165	165	175	168	168	168
Venezuela	1,446	1,201	1,239	1,282	1,282	1,282	1,282	1,282	1,282	1,282	1,282
Other Caribbean	—	—	—	—	—	1,125	1,045	1,224	1,229	1,227	1,228
Other Latin America	—	—	—	—	—	—	—	131	131	131	131
Total	8,620	7,315	7,973	7,942	6,868	8,047	7,951	8,295	8,143	8,142	8,143
Europe											
Albania	—	—	26	26	26	26	26	26	26	26	26
Austria	280	204	209	209	209	209	209	209	209	209	209
Belgium	1,064	614	719	791	791	793	803	858	791	798	798
Bulgaria	—	—	115	115	115	115	115	115	115	115	115
Croatia	—	—	236	293	260	250	250	250	250	250	250
Czech Republic	455	455	186	198	198	198	198	198	198	183	183
Denmark	214	187	135	176	176	176	176	176	176	174	174
Finland	336	241	200	239	252	252	252	252	252	252	256
France	3,385	1,820	1,902	1,896	1,903	1,951	1,951	1,979	1,959	1,932	1,986
Germany	2,986	1,507	2,275	2,259	2,267	2,289	2,323	2,344	2,417	2,417	2,417
Greece	431	385	383	407	407	401	401	413	413	423	423
Hungary	290	220	232	161	161	161	161	161	161	161	161
Ireland	56	56	71	71	71	71	71	71	71	71	71
Italy	4,131	2,804	2,341	2,283	2,300	2,313	2,321	2,324	2,337	2,337	2,337
Netherlands	1,828	1,381	1,188	1,206	1,207	1,222	1,228	1,222	1,212	1,227	1,208
Norway	264	295	358	310	310	310	310	310	310	315	315
Poland	385	385	382	382	350	350	350	467	497	493	493
Portugal	378	313	304	304	304	304	304	304	304	304	304
Romania	608	617	499	504	501	501	517	517	517	517	517
Slovakia	—	—	115	115	115	115	115	115	115	115	115
Spain	1,455	1,293	1,316	1,294	1,322	1,272	1,272	1,272	1,272	1,227	1,272
Sweden	458	428	427	424	424	424	434	434	434	437	437
Switzerland	137	132	132	132	132	132	132	132	132	132	132
Turkey	356	725	691	719	719	714	714	714	714	714	714
UK	2,527	1,831	1,785	1,784	1,789	1,817	1,825	1,877	1,887	1,858	1,858
Other Europe	—	—	—	—	—	—	—	278	279	278	278
Total	4,933	4,409	16,227	16,298	16,309	16,366	16,458	17,018	17,048	16,965	17,049
Middle East											
Bahrain	250	243	249	249	249	249	249	255	249	262	262
Iran	921	530	1,474	1,484	1,474	1,474	1,474	1,451	1,451	1,451	1,451
Iraq	168	319	348	418	418	588	598	598	598	598	598
Israel	195	180	220	220	220	220	220	220	220	220	220
Kuwait	645	819	865	773	889	889	889	889	889	889	889
Oman	—	—	—	85	85	85	85	85	85	85	85
Qatar	11	62	58	58	200	200	200	200	200	200	200
Saudi Arabia	487	1,484	1,710	1,745	1,745	1,745	1,745	2,095	2,095	2,080	2,080

Table 10: World Refining Capacity, cont'd.

COUNTRY	1980	1990	2000	2002	2003	2004	2005	2006	2007	2008	2009
	1,000 b/cd										
Syria	223	244	242	242	240	240	240	240	240	240	240
UAE	14	180	429	515	514	514	514	781	781	781	781
Yemen	175	162	120	130	130	130	130	130	140	140	140
Other Middle East	—	—	—	—	—	—	—	90	90	90	90
Total	3,163	4,447	5,928	5,532	6,164	6,334	6,344	7,034	7,038	7,036	7,036
Africa											
Algeria	122	465	503	450	450	450	450	450	450	450	450
Angola	36	32	39	39	39	39	39	39	39	39	39
Cameroon	—	43	35	42	42	42	42	42	42	37	37
Congo (Republic of)	—	21	21	21	21	21	21	21	21	21	21
Congo (Zaire)	16	17	17	15	15	—	—	—	—	—	—
Egypt	234	489	578	726	726	726	726	726	726	726	726
Gabon	20	24	17	17	17	17	17	17	17	24	24
Ivory Coast	50	69	69	65	65	65	65	65	65	64	64
Libya	138	329	348	343	343	343	380	380	378	378	378
Madagascar	—	—	—	—	15	15	15	—	—	—	—
Morocco	72	155	157	155	155	155	155	155	155	155	155
Nigeria	160	433	439	439	439	439	439	439	439	505	505
South Africa	478	434	467	469	490	520	490	505	488	485	485
Sudan	26	25	82	122	122	122	122	122	122	122	122
Tanzania	—	—	—	—	—	—	—	15	15	15	15
Tunisia	34	34	34	34	34	34	34	34	34	34	34
Other Africa	—	—	—	—	—	—	—	221	222	223	223
Total	1,665	2,824	3,046	3,169	2,973	2,988	2,995	3,231	3,213	3,278	3,278
Former Soviet Union											
Azerbaijan	—	—	—	442	442	399	399	399	399	399	399
Belarus	—	—	—	493	493	493	493	493	493	493	493
Georgia	—	—	—	106	106	106	0	0	0	0	0
Kazakhstan	—	—	—	427	427	427	345	345	345	345	345
Kyrgyzstan	—	—	—	10	10	10	10	10	10	10	10
Lithuania	—	—	—	—	—	263	263	190	190	190	190
Russia	—	—	—	5,435	5,435	5,435	5,433	5,341	5,339	5,429	5,429
Turkmenistan	—	—	—	237	237	237	237	237	237	237	237
Ukraine	—	—	—	1,206	1,025	1,025	880	880	880	880	880
USSR	10,950	12,300	9,762	—	—	—	—	—	—	—	—
Uzbekistan	—	—	—	222	222	222	222	222	222	222	222
Total	10,950	12,300	9,762	8,578	8,397	8,617	8,282	8,117	8,115	8,205	8,205
Asia-Pacific											
Australia	725	675	812	848	848	755	755	702	705	707	696
Bangladesh	31	31	33	33	33	33	33	33	33	33	33
Brunei	—	—	—	9	9	9	9	9	9	9	9
China	1,600	2,200	4,347	4,528	4,528	4,528	4,650	6,246	6,246	6,246	6,446
India	557	1,080	1,858	2,135	2,135	2,135	2,255	2,255	2,256	2,256	2,256
Indonesia	528	714	993	993	993	993	993	993	993	993	993
Japan	5,509	4,198	4,998	4,786	4,767	4,703	4,707	4,672	4,677	4,651	4,691
Malaysia	—	—	—	515	516	516	545	545	545	515	515
Myanmar	26	32	32	32	57	57	57	57	57	57	57
New Zealand	74	95	98	106	106	104	104	104	104	107	107
Pakistan	98	121	143	239	234	269	269	269	269	267	276
Papua New Guinea	—	—	—	—	—	—	33	33	33	33	33
Philippines	253	284	401	420	420	333	333	333	333	282	282
Singapore	1,093	1,049	1,788	1,259	1,259	1,319	1,337	1,337	1,337	1,344	1,344
South Korea	601	867	2,540	2,560	2,560	2,544	2,577	2,577	2,577	2,577	2,607
Taiwan	425	570	770	920	920	920	1,220	1,220	1,290	1,290	1,290
Thailand	186	215	713	682	703	703	703	703	729	729	729
Other Asia-Pacific	—	—	—	—	—	—	—	121	121	124	121
Total	11,986	12,181	19,645	20,185	20,088	19,921	20,595	22,209	22,314	22,220	22,485
Total World	**79,514**	**74,535**	**81,556**	**81,166**	**81,878**	**82,055**	**82,409**	**85,043**	**85,185**	**85,262**	**85,604**

Table 10: World Refining Capacity, cont'd.

COUNTRY	1980	1990	2000	2002	2003	2004	2005	2006	2007	2008	2009
	million tonnes per year (mtpy)										
North America											
Canada	110.7	92.2	95.2	96.8	98.8	99.0	100.4	100.4	101.6	98.1	101.0
US	882.5	809.0	823.7	824.9	827.8	831.5	835.3	852.8	860.0	868.7	865.4
Total	993.2	901.2	918.9	921.7	926.6	930.5	935.7	953.2	961.6	966.8	966.4
Latin America											
Argentina	33.7	34.3	33.0	31.8	31.8	31.8	31.1	31.1	31.1	31.2	31.2
Bolivia	3.7	2.9	2.4	3.1	3.1	3.1	2.3	2.4	2.0	2.1	2.0
Brazil	60.0	69.6	88.8	88.9	92.9	95.3	95.6	95.0	95.0	95.0	95.0
Chile	6.9	7.3	10.2	10.2	10.2	10.2	11.3	11.3	11.3	11.3	11.3
Colombia	9.6	11.3	14.2	14.2	14.2	14.2	14.2	14.2	14.2	14.2	14.2
Cuba	8.0	8.0	15.0	15.0	15.0	15.0	15.0	15.0	15.0	15.0	15.0
Ecuador	4.3	7.2	8.8	8.8	8.8	8.8	8.8	8.8	8.8	8.8	8.8
Guatemala	0.8	0.8	1.0	0.8	0.8	0.8	—	—	—	—	—
Jamaica	—	—	—	—	—	—	—	1.8	1.8	1.8	1.8
Mexico	69.4	75.4	75.9	75.9	83.9	83.9	83.9	83.9	76.7	76.7	76.7
Peru	8.5	8.6	9.1	9.1	9.5	9.5	9.6	9.6	9.6	9.6	9.6
Trinidad & Tobago	22.7	14.9	8.0	8.0	8.0	8.2	8.2	8.7	8.4	8.4	8.4
Venezuela	72.0	59.8	61.7	63.8	63.8	63.8	63.8	63.8	63.8	63.8	63.8
Other Caribbean	—	—	—	—	—	56.0	52.0	61.0	61.2	61.1	61.1
Other Latin America	—	—	—	—	—	—	—	6.5	6.5	6.5	6.5
Total	429.2	364.0	396.9	395.2	342.0	400.6	395.8	413.1	405.4	405.4	405.4
Europe											
Albania	—	—	1.3	1.3	1.3	1.3	1.3	1.3	1.3	1.3	1.3
Austria	13.9	10.2	10.4	10.4	10.4	10.4	10.4	10.4	10.4	10.4	10.4
Belgium	53.0	30.6	35.8	39.4	39.4	39.5	40.0	42.7	39.4	39.7	39.7
Bulgaria	—	—	5.7	5.7	5.7	5.7	5.7	5.7	5.7	5.7	5.7
Croatia	—	—	11.8	14.6	12.9	12.4	12.4	12.5	12.5	12.5	12.5
Czech Republic	22.7	22.7	9.3	9.9	9.9	9.9	9.9	9.9	9.9	9.1	9.1
Denmark	10.7	9.3	6.7	8.8	8.8	8.8	8.8	8.8	8.8	8.7	8.7
Finland	16.7	12.0	10.0	11.9	12.4	12.4	12.4	12.5	12.5	12.5	12.8
France	168.6	90.6	94.7	94.4	94.8	97.2	97.2	98.6	97.5	96.2	98.9
Germany	148.7	75.0	113.3	112.5	112.9	114.0	115.7	116.7	120.3	120.3	120.3
Greece	21.5	19.2	19.1	20.2	20.2	20.0	20.0	20.6	20.6	21.1	21.1
Hungary	14.4	11.0	11.6	8.0	8.0	8.0	8.0	8.0	8.0	8.0	8.0
Ireland	2.8	2.8	3.5	3.5	3.5	3.5	3.5	3.5	3.5	3.5	3.5
Italy	205.7	139.6	116.6	113.7	114.5	115.2	115.6	115.7	116.4	116.4	116.4
Netherlands	91.0	68.8	59.2	60.1	60.1	60.8	61.1	60.8	60.3	61.1	60.2
Norway	13.1	14.7	17.8	15.4	15.4	15.4	15.4	15.4	15.4	15.7	15.7
Poland	19.2	19.2	19.0	19.0	17.4	17.4	17.4	23.2	24.7	24.5	24.5
Portugal	18.8	15.6	15.1	15.1	15.1	15.1	15.1	15.1	15.1	15.1	15.1
Romania	30.3	30.7	24.9	25.1	24.9	24.9	25.7	25.7	25.7	25.7	25.7
Slovakia	—	—	5.7	5.7	5.7	5.7	5.7	5.7	5.7	5.7	5.7
Spain	72.5	64.4	65.5	64.4	65.8	63.3	63.3	63.3	63.3	63.6	63.3
Sweden	22.8	21.3	21.3	21.1	21.1	21.1	21.6	21.6	21.6	21.8	21.8
Switzerland	6.8	6.6	6.6	6.6	6.6	6.6	6.6	6.6	6.6	6.6	6.6
Turkey	17.7	36.1	34.4	35.8	35.8	35.6	35.6	35.6	35.6	35.6	35.6
UK	125.8	91.2	88.9	88.8	89.1	90.5	90.9	93.5	94.0	92.5	92.5
Other Europe	—	—	—	—	—	—	—	13.9	13.9	13.9	13.9
Total	245.6	761.0	808.2	811.4	811.7	814.7	819.3	847.3	848.8	847.2	848.9
Middle East											
Bahrain	12.4	12.1	12.4	12.4	12.4	12.4	12.4	12.7	12.4	13.0	13.0
Iran	45.9	26.4	73.4	73.9	73.4	73.4	73.4	72.3	72.3	72.2	72.2
Iraq	8.4	15.9	17.3	20.8	20.8	29.3	29.8	29.8	29.8	29.7	29.7
Israel	9.7	9.0	11.0	11.0	11.0	11.0	11.0	11.0	11.0	11.0	11.0
Kuwait	32.1	40.8	43.1	38.5	44.3	44.3	44.3	44.3	44.3	44.3	44.3
Oman	—	—	—	4.2	4.2	4.2	4.2	4.2	4.2	4.2	4.2
Qatar	0.5	3.1	2.9	2.9	10.0	10.0	10.0	10.0	10.0	10.0	10.0
Saudi Arabia	24.3	73.9	85.2	86.9	86.9	86.9	86.9	104.3	104.3	103.6	103.6
Syria	11.1	12.2	12.1	12.1	12.0	12.0	12.0	11.9	11.9	11.9	11.9
UAE	0.7	9.0	21.4	25.6	25.6	25.6	25.6	38.9	38.9	38.9	38.9
Yemen	8.7	8.1	6.0	6.5	6.5	6.5	6.5	6.5	7.0	7.0	7.0
Other Middle East	—	—	—	—	—	—	—	4.5	4.5	4.5	4.5
Total	151.4	221.6	295.4	275.6	307.0	315.6	316.1	350.4	350.4	350.3	350.3

Table 10: World Refining Capacity, cont'd.

COUNTRY	1980	1990	2000	2002	2003	2004	2005	2006	2007	2008	2009
	million tonnes per year (mtpy)										
Africa											
Algeria	6.1	23.2	25.0	22.4	22.4	22.4	22.4	22.4	22.4	22.4	22.4
Angola	1.8	1.6	1.9	1.9	1.9	1.9	1.9	1.9	1.9	1.9	1.9
Cameroon	—	2.1	1.7	2.1	2.1	2.1	2.1	2.1	2.1	1.8	1.8
Congo (Republic of)	—	1.0	1.0	1.0	1.0	1.0	1.0	1.0	1.0	1.0	1.0
Congo (Zaire)	0.8	0.8	0.8	0.7	0.7	—	—	—	—	—	—
Egypt	11.7	24.4	28.8	36.2	36.2	36.2	36.2	36.2	36.2	36.2	36.2
Gabon	1.0	1.2	0.8	0.8	0.8	0.8	0.8	0.9	0.9	1.2	1.2
Ivory Coast	2.5	3.4	3.4	3.2	3.2	3.2	3.2	3.2	3.2	3.2	3.2
Libya	6.9	16.4	17.3	17.1	17.1	17.1	18.9	18.9	18.8	18.8	18.8
Madagascar	—	—	—	—	0.7	0.7	0.7	—	—	—	—
Morocco	3.6	7.7	7.8	7.7	7.7	7.7	7.7	7.7	7.7	7.7	7.7
Nigeria	8.0	21.6	21.9	21.9	21.9	21.9	21.9	21.8	21.8	25.1	25.1
South Africa	23.8	21.6	23.3	23.3	24.4	25.9	24.4	25.1	24.3	24.2	24.2
Sudan	1.3	1.2	4.1	6.1	6.1	6.1	6.1	6.1	6.1	6.1	6.1
Tanzania	—	—	—	—	—	—	—	0.7	0.7	0.7	0.7
Tunisia	1.7	1.7	1.7	1.7	1.7	1.7	1.7	1.7	1.7	1.7	1.7
Other Africa	—	—	—	—	—	—	—	11.0	11.1	11.1	11.1
Total	82.9	140.4	151.2	157.1	147.3	148.0	148.3	160.7	160.0	163.1	163.2
Former Soviet Union											
Azerbaijan	—	—	—	22.0	22.0	19.9	19.9	19.9	19.9	19.9	19.9
Belarus	—	—	—	24.6	24.6	24.6	24.6	24.6	24.6	24.6	24.6
Georgia	—	—	—	5.3	5.3	5.3	0.0	0.0	0.0	0.0	0.0
Kazakhstan	—	—	—	21.3	21.3	21.3	17.2	17.2	17.2	17.2	17.2
Kyrgyzstan	—	—	—	0.5	0.5	0.5	0.5	0.5	0.5	0.5	0.5
Lithuania	—	—	—	—	—	13.1	13.1	9.5	9.5	9.5	9.5
Russia	—	—	—	270.7	270.7	270.7	270.5	266.0	265.8	270.3	270.3
Turkmenistan	—	—	—	11.8	11.8	11.8	11.8	11.8	11.8	11.8	11.8
Ukraine	—	—	—	60.1	51.0	51.0	43.8	43.8	43.8	43.8	43.8
USSR	545.3	612.5	486.1	—	—	—	—	—	—	—	—
Uzbekistan	—	—	—	11.1	11.1	11.1	11.1	11.1	11.1	11.1	11.1
Total	545.3	612.5	486.1	427.4	418.2	429.3	412.5	404.4	404.2	408.7	408.5
Asia-Pacific											
Australia	36.1	33.6	40.4	42.2	42.2	37.6	37.6	34.9	35.1	35.2	34.7
Bangladesh	1.5	1.5	1.6	1.6	1.6	1.6	1.6	1.6	1.6	1.6	1.6
Brunei	—	—	—	0.4	0.4	0.4	0.4	0.4	0.4	0.4	0.4
China	79.7	109.6	216.5	225.5	225.5	225.5	231.5	311.0	311.0	311.0	320.9
India	27.7	53.8	92.5	106.3	106.3	106.3	112.3	112.3	112.3	112.3	112.3
Indonesia	26.3	35.6	49.5	49.5	49.5	49.5	49.5	49.4	49.4	49.4	49.4
Japan	274.3	209.1	248.9	238.3	237.4	234.2	234.4	232.6	232.9	231.6	233.6
Malaysia	—	—	—	25.6	25.7	25.7	27.1	27.1	27.1	25.6	25.6
Myanmar	1.3	1.6	1.6	1.6	2.8	2.8	2.8	2.8	2.8	2.8	2.8
New Zealand	3.7	4.7	4.9	5.3	5.3	5.2	5.2	5.2	5.2	5.3	5.3
Pakistan	4.9	6.0	7.1	11.9	11.7	13.4	13.4	13.4	13.4	13.3	13.7
Papua New Guinea	—	—	—	—	—	—	1.6	1.6	1.6	1.6	1.6
Philippines	12.6	14.1	20.0	20.9	20.9	16.6	16.6	16.6	16.6	14.0	14.0
Singapore	54.4	52.2	89.0	62.7	62.7	65.7	66.6	66.6	66.6	66.9	66.9
South Korea	29.9	43.2	126.5	127.5	127.5	126.7	128.3	128.3	128.3	128.3	129.8
Taiwan	21.2	28.4	38.3	45.8	45.8	45.8	60.8	60.8	64.2	64.2	64.2
Thailand	9.3	10.7	35.5	34.0	35.0	35.0	35.0	35.0	36.3	36.3	36.3
Other Asia-Pacific	—	—	—	—	—	—	—	6.0	6.0	6.2	6.0
Total	596.9	606.6	978.2	1,005.1	1,000.4	992.0	1,025.4	1,105.6	1,111.0	1,106.0	1,119.5
Total World	3,959.5	3,711.5	4,060.8	4,041.4	4,077.5	4,086.0	4,103.6	4,234.7	4,241.4	4,250.6	4,262.3

Source: *Oil & Gas Journal* survey data as of Jan. 1, 2009.

NOTES: Some totals might not compute exactly due to rounding and other factors.

b/cd = barrels per calendar-day.

Other Caribbean includes Aruba, Dominican Republic, Martinique, Netherlands Antilles, Puerto Rico, and Virgin Islands.

Other Latin America includes Costa Rica, El Salvador, Nicaragua, Paraguay, Suriname, and Uruguay.

Other Europe includes Macedonia, Serbia & Montenegro, and Slovenia.

Other Middle East is Jordan.

Other Africa includes Eritrea, Ghana, Kenya, Liberia, Senegal, Sierra Leone, and Zambia.

USSR includes total of all Former Soviet Union countries before 2002.

Brunei and Malaysia were totaled with Singapore before 2002.

Other Asia-Pacific includes North Korea and Sri Lanka.

Table 11: World Refinery Throughput

REGION	2002	2003	2004	2005	2006	2007	Change 2007 over 2006	2007 share of total
	thousand barrels per day (1,000 b/d)						%	
US	14,947	15,304	15,475	15,220	15,242	15,148	-0.6	20.1
Canada	1,862	1,880	1,957	1,882	1,843	1,869	1.4	2.5
Mexico	1,390	1,436	1,436	1,415	1,407	1,395	-0.8	1.8
South & Central America	4,941	5,040	5,430	5,402	5,354	5,448	1.8	7.2
Europe & Eurasia	19,485	19,988	20,441	20,815	20,877	20,829	-0.2	27.6
Middle East	5,482	5,566	5,805	5,934	6,321	6,301	-0.3	8.3
Africa	2,377	2,301	2,304	2,468	2,374	2,470	4.0	3.3
Australasia	855	823	820	757	749	767	2.4	1.0
China	4,395	4,823	5,382	5,916	6,155	6,563	6.6	8.7
Japan	3,986	4,118	4,038	4,136	4,026	3,994	-0.8	5.3
Other Asia-Pacific	8,827	9,298	9,944	10,060	10,396	10,761	3.5	14.2
Total World	**68,547**	**70,576**	**73,032**	**74,005**	**74,743**	**75,545**	**1.1**	**100.0**

Source: Adapted from *BP Statistical Review of World Energy 2008.*
Includes data from Energy Security Analysis, Inc. (ESAI).

NOTES: Input to primary distillation units only.
Some totals might not compute exactly due to rounding and other factors.
Annual changes and shares of total are calculated using thousand barrels daily figures.

Table 12: World Refining Margins

MARKET	1998	1999	2000	2001	2002	2003	2004	2005	2006	Dec. 2007	Dec. 2008	2007-2008 Change, %
	$/bbl											
US Gulf Coast												
West Texas sour	3.31	2.15	5.54	2.91	3.81	3.83	5.67	12.05	9.69	9.53	6.75	-29.2
Composite US gulf refinery	—	—	—	—	4.54	4.16	7.87	13.42	10.33	13.01	8.93	-31.3
Arabian light	4.04	3.14	7.27	5.02	5.45	5.51	8.81	13.22	10.58	15.39	6.58	-57.3
Bonny light	2.83	1.74	6.40	0.97	1.54	2.89	5.27	7.74	1.87	3.39	-0.55	-116.1
US PADD II Region												
Chicago (WTI)	3.17	2.47	4.60	2.52	4.30	3.18	3.31	7.90	7.58	6.87	2.42	-64.8
US East Coast												
NY Harbor (Arabian medium)	3.62	2.50	6.38	4.15	3.38	3.45	5.90	9.14	4.48	14.29	9.94	-30.5
East Coast composite-RFG	—	—	—	2.88	4.33	4.15	6.53	10.06	6.42	12.35	7.89	-36.1
US West Coast												
Los Angeles (ANS)	5.60	6.55	12.48	4.33	3.57	6.95	10.03	8.60	19.56	12.31	8.19	-33.4
NW Europe												
Rotterdam (Brent)	3.26	2.19	4.71	0.88	0.76	2.21	0.52	0.65	-0.57	4.02	3.14	-21.8
Mediterranean												
Italy (Urals)	2.73	1.67	3.28	0.05	0.74	1.88	7.38	5.60	7.34	3.95	4.91	24.5
Far East												
Singapore (Dubai)	1.61	0.78	3.89	0.63	1.98	3.43	4.24	1.46	-1.39	4.65	4.13	-11.2

Source: Jacobs Consultancy, Inc.; in *Oil & Gas Journal*, January 5, 2009.

NOTES: WTI = West Texas Intermediate.
ANS = Alaska North Slope.

Table 13: World Natural Gas Consumption

	2005	2006	2007	2005	2006	2007	Change 2007 over 2006	2007 share of total
COUNTRY	billion cubic meters (bcm)			bcfd			%	
North America								
US	623.3	613.1	652.9	60.3	59.3	63.2	6.5	22.6
Canada	98.1	96.9	94.0	9.5	9.4	9.1	-3.1	3.2
Mexico	46.2	51.4	54.1	4.5	5.0	5.2	5.3	1.8
Total	767.6	761.4	801.0	74.3	73.7	77.5	5.2	27.6
South & Central America								
Argentina	40.4	41.8	44.1	3.9	4.0	4.3	5.5	1.5
Brazil	19.3	20.6	22.0	1.9	2.0	2.1	7.0	0.8
Chile	8.3	7.9	4.4	0.8	0.8	0.4	-43.6	0.2
Colombia	6.8	7.3	7.7	0.7	0.7	0.7	5.6	0.3
Ecuador	0.3	0.3	0.3	*	*	*	-1.9	**
Peru	1.5	1.8	2.7	0.1	0.2	0.3	50.7	0.1
Venezuela	28.1	27.9	28.5	2.7	2.7	2.8	1.9	1.0
Other S/C America	19.5	23.8	24.9	1.9	2.3	2.4	4.5	0.8
Total	124.3	131.3	134.5	12.0	12.7	13.0	2.5	4.6
Europe and Eurasia								
Austria	10.0	9.4	8.9	1.0	0.9	0.9	-5.3	0.3
Azerbaijan	8.9	9.4	8.3	0.9	0.9	0.8	-12.0	0.3
Belarus	18.9	19.6	19.4	1.8	1.9	1.9	-1.0	0.7
Begium & Luxembourg	16.6	17.0	16.9	1.6	1.6	1.6	-0.6	0.6
Bulgaria	2.9	3.0	3.1	0.3	0.3	0.3	2.0	0.1
Czech Republic	9.6	9.8	8.9	0.9	1.0	0.9	-9.7	0.3
Denmark	5.0	5.1	4.6	0.5	0.5	0.4	-10.5	0.2
Finland	4.0	4.2	4.1	0.4	0.4	0.4	-4.0	0.1
France	45.8	44.1	41.9	4.4	4.3	4.1	-5.0	1.4
Germany	86.2	87.2	82.7	8.3	8.4	8.0	-5.1	2.8
Greece	2.8	3.2	4.0	0.3	0.3	0.4	24.3	0.1
Hungary	13.2	12.5	11.8	1.3	1.2	1.1	-5.8	0.4
Republic of Ireland	3.9	4.5	4.8	0.4	0.4	0.5	6.8	0.2
Italy	79.1	77.4	77.8	7.7	7.5	7.5	0.5	2.7
Kazakhstan	19.4	20.9	19.8	1.9	2.0	1.9	-5.4	0.7
Lithuania	3.3	3.2	3.8	0.3	0.3	0.4	17.3	0.1
Netherlands	39.5	38.3	37.2	3.8	3.7	3.6	-2.9	1.3
Norway	4.5	4.4	4.3	0.4	0.4	0.4	-2.9	0.1
Poland	13.6	13.7	13.7	1.3	1.3	1.3	-0.1	0.5
Portugal	4.2	4.1	4.3	0.4	0.4	0.4	6.6	0.1
Romania	17.6	18.2	16.4	1.7	1.8	1.6	-9.9	0.6
Russian Federation	405.1	432.1	438.8	39.2	41.8	42.5	1.6	15.0
Slovakia	6.6	6.4	5.9	0.6	0.6	0.6	-7.8	0.2
Spain	32.4	33.7	35.1	3.1	3.3	3.4	4.3	1.2
Sweden	0.8	1.0	1.0	0.1	0.1	0.1	3.5	**
Switzerland	3.1	3.0	2.9	0.3	0.3	0.3	-2.4	0.1
Turkey	26.9	30.5	35.1	2.6	3.0	3.4	15.0	1.2
Turkmenistan	16.6	18.9	21.9	1.6	1.8	2.1	15.8	0.7
Ukraine	73.0	67.1	64.6	7.1	6.5	6.3	-3.6	2.2
United Kingdom	94.9	90.0	91.4	9.2	8.7	8.8	1.5	3.1
Uzbekistan	44.0	43.2	45.6	4.3	4.2	4.4	5.6	1.6
Other Europe/Eurasia	16.1	16.4	16.8	1.6	1.6	1.6	2.3	0.6
Total	1,128.3	1,151.5	1,155.7	109.2	111.4	111.8	0.4	39.4
Middle East								
Iran	102.4	108.7	111.8	9.9	10.5	10.8	2.9	3.8
Kuwait	12.3	12.9	12.6	1.2	1.2	1.2	-2.3	0.4
Qatar	18.7	19.6	20.5	1.8	1.9	2.0	4.6	0.7
Saudi Arabia	71.2	73.5	75.9	6.9	7.1	7.3	3.3	2.6
United Arab Emirates	41.3	41.7	43.2	4.0	4.0	4.2	3.7	1.5
Other Middle East	30.9	35.0	35.3	3.0	3.4	3.4	1.0	1.2
Total	276.8	291.4	299.4	26.8	28.2	29.0	2.7	10.2

COUNTRY	2005	2006	2007	2005	2006	2007	Change 2007 over 2006	2007 share of total
	billion cubic meters (bcm)			bcfd			%	
Africa								
Algeria	23.2	23.7	24.4	2.2	2.3	2.4	2.7	0.8
Egypt	25.8	29.2	32.0	2.5	2.8	3.1	9.9	1.1
Other Africa	24.3	25.0	27.1	2.3	2.4	2.6	8.2	0.9
Total	73.3	77.9	83.5	7.1	7.5	8.1	7.2	2.8
Asia-Pacific								
Australia	21.9	24.0	25.1	2.1	2.3	2.4	4.9	0.9
Bangladesh	14.5	15.3	16.3	1.4	1.5	1.6	6.2	0.6
China	46.8	56.1	67.3	4.5	5.4	6.5	19.9	2.3
China Hong Kong SAR	2.7	2.9	3.0	0.3	0.3	0.3	0.8	0.1
India	35.7	37.3	40.2	3.5	3.6	3.9	7.6	1.4
Indonesia	37.2	34.9	33.8	3.2	3.4	3.3	-3.3	1.2
Japan	78.6	83.7	90.2	7.6	8.1	8.7	7.8	3.1
Malaysia	28.9	29.6	28.3	2.8	2.9	2.7	-4.4	1.0
New Zealand	3.4	3.5	3.7	0.3	0.3	0.4	6.2	0.1
Pakistan	30.2	30.5	30.8	2.9	3.0	3.0	1.0	1.1
Philippines	3.3	2.9	3.4	0.3	0.3	0.3	19.4	0.1
Singapore	6.6	6.6	6.6	0.6	0.6	0.6	—	0.2
South Korea	33.7	35.6	37.0	3.3	3.4	3.6	4.0	1.3
Taiwan	10.3	11.1	11.8	1.0	1.1	1.1	6.1	0.4
Thailand	32.5	33.3	35.4	3.1	3.2	3.4	6.2	1.2
Other Asia-Pacific	13.1	13.4	15.0	1.3	1.3	1.4	11.4	0.5
Total	395.0	420.9	447.8	38.2	40.7	43.3	6.4	15.3
Total World	**2,765.2**	**2,834.4**	**2,921.9**	**267.5**	**274.2**	**282.7**	**3.1**	**100.0**

Source: Adapted from *BP Statistical Review of World Energy 2008.*

NOTES: Annual changes and shares of total are calculated using million tonnes of oil equivalent figures.

As far as possible, the data above represent standard cubic meters (at 15 deg C and 1,013 mbar). Because it is derived directly from tonnes of oil equivalent using an average conversion factor, it does not necessarily equate with gas volumes expressed in specific national terms.

The difference between these world consumption figures and the world production figures in Table 14 is due to variations in stocks at storage facilities and liquefaction plants, together with unavoidable disparities in the definition, measurement, or conversion of gas supply and demand data.

* Less than 0.05

** Less than 0.05%

Table 14: World Natural Gas Production

COUNTRY	2005	2006	2007	2005	2006	2007	Change 2007 over 2006	2007 share of total
	billion cubic meters (bcm)			bcfd			%	
North America								
US	511.1	523.2	545.9	49.5	50.6	52.8	4.3	18.8
Canada	187.4	188.4	183.7	18.1	18.2	17.8	-2.5	6.2
Mexico	38.9	42.8	46.2	3.8	4.1	4.5	8.0	1.6
Total	737.4	754.4	775.8	71.3	73.0	75.1	2.9	26.6
South & Central America								
Argentina	45.6	46.1	44.8	4.4	4.5	4.3	-2.8	1.5
Bolivia	11.9	12.7	13.5	1.2	1.2	1.3	6.5	0.5
Brazil	11.0	11.3	11.3	1.1	1.1	1.1	-0.5	0.4
Colombia	6.8	7.3	7.7	0.7	0.7	0.7	5.6	0.3
Trinidad & Tobago	30.3	36.4	39.0	2.9	3.5	3.8	7.0	1.3
Venezuela	28.1	27.9	28.5	2.7	2.7	2.8	1.9	1.0
Other S/C America	4.8	5.3	6.1	0.5	0.5	0.6	13.2	0.2
Total	138.6	147.2	150.8	13.4	14.2	14.6	2.5	5.1

Table 14: World Natural Gas Production, cont'd.

	2005	2006	2007	2005	2006	2007	Change 2007 over 2006	2007 share of total
COUNTRY	billion cubic meters (bcm)			bcfd			%	
Europe and Eurasia								
Azerbaijan	5.3	6.3	10.3	0.5	0.6	1.0	63.2	0.3
Denmark	10.4	10.4	9.2	1.0	1.0	0.9	-11.4	0.3
Germany	15.8	15.6	14.3	1.5	1.5	1.4	-8.4	0.5
Italy	11.1	10.1	8.9	1.1	1.0	0.9	-11.5	0.3
Kazakhstan	23.3	24.6	27.3	2.3	2.4	2.6	10.8	0.9
Netherlands	62.9	65.3	64.5	6.1	6.3	6.2	-1.2	2.2
Norway	85.0	87.6	89.7	8.2	8.5	8.7	2.4	3.0
Poland	4.3	4.3	4.3	0.4	0.4	0.4	-0.8	0.1
Romania	12.4	11.9	11.6	1.2	1.2	1.1	-3.1	0.4
Russian Federation	598.0	612.1	607.4	57.9	59.2	58.8	-0.8	20.6
Turkmenistan	58.8	62.2	67.4	5.7	6.0	6.5	8.4	2.3
Ukraine	19.4	19.1	19.0	1.9	1.8	1.8	-0.3	0.6
United Kingdom	88.2	80.0	72.4	8.5	7.7	7.0	-9.5	2.5
Uzbekistan	55.0	55.4	58.5	5.3	5.4	5.7	5.6	2.0
Other Europe/Eurasia	10.8	11.5	11.0	1.0	1.1	1.1	-4.0	0.4
Total	1,060.6	1,076.3	1,075.7	102.6	104.1	104.1	-0.1	36.5
Middle East								
Bahrain	10.7	11.1	11.5	1.0	1.1	1.1	4.0	0.4
Iran	100.9	108.6	111.9	9.8	10.5	10.8	3.0	3.8
Kuwait	12.3	12.9	12.6	1.2	1.2	1.2	-2.3	0.4
Oman	19.8	23.7	24.1	1.9	2.3	2.3	1.7	0.8
Qatar	45.8	50.7	59.8	4.4	4.9	5.8	17.9	2.0
Saudi Arabia	71.2	73.5	75.9	6.9	7.1	7.3	3.3	2.6
Syria	5.4	5.5	5.3	0.5	0.5	0.5	-5.0	0.2
United Arab Emirates	47.0	47.4	49.2	4.5	4.6	4.8	3.8	1.7
Other Middle East	4.4	5.6	5.5	0.4	0.5	0.5	-2.8	0.2
Total	317.5	339.0	355.8	30.7	32.8	34.4	4.9	12.1
Africa								
Algeria	88.2	84.5	83.0	8.5	8.2	8.0	-1.7	2.8
Egypt	34.6	44.7	46.5	3.4	4.3	4.5	4.2	1.6
Libya	11.3	14.8	15.2	1.1	1.4	1.5	2.7	0.5
Nigeria	22.4	28.4	35.0	2.2	2.8	3.4	23.0	1.2
Other Africa	9.0	9.2	10.7	0.9	0.9	1.0	15.4	0.4
Total	165.6	181.6	190.4	16.0	17.6	18.4	4.8	6.5
Asia-Pacific								
Australia	37.1	38.9	40.0	3.6	3.8	3.9	2.8	1.4
Bangladesh	14.5	15.3	16.3	1.4	1.5	1.6	6.2	0.6
Brunei	12.0	12.6	12.3	1.2	1.2	1.2	-2.5	0.4
China	49.3	58.6	69.3	4.8	5.7	6.7	18.4	2.4
India	29.6	29.3	30.2	2.9	2.8	2.9	2.8	1.0
Indonesia	68.7	69.3	66.7	6.6	6.7	6.5	-3.7	2.3
Malaysia	59.9	60.2	60.5	5.8	5.8	5.9	0.5	2.1
Myanmar	13.0	13.4	14.7	1.3	1.3	1.4	9.7	0.5
New Zealand	3.5	3.6	4.0	0.3	0.4	0.4	10.1	0.1
Pakistan	30.2	30.5	30.8	2.9	3.0	3.0	1.0	1.0
Thailand	23.7	24.4	25.9	2.3	2.4	2.5	6.4	0.9
Vietnam	6.9	7.0	7.7	0.7	0.7	0.7	10.1	0.3
Other Asia-Pacific	7.2	10.7	13.1	0.7	1.0	1.3	23.4	0.4
Total	355.8	373.7	391.5	34.4	36.2	37.9	4.8	13.3
Total World	**2,775.5**	**2,872.2**	**2,940.0**	**268.5**	**277.9**	**284.5**	**2.4**	**100.0**

Source: Adapted from *BP Statistical Review of World Energy 2008.*

NOTE: These data exclude gas flared or recycled.
Annual changes and shares of total are calculated using million tonnes of oil equivalent figures.
As the data above are derived from tonnes of oil equivalent using an average conversion factor, they do not necessarily equate with gas volumes expressed in specific national terms.
The difference between these world production figures and the world consumption figures in Table 13 is due to variations in stocks at storage facilities and liquefaction plants, together with unavoidable disparities in the definition, measurement, or conversion of gas supply and demand data.

Table 15a: World Natural Gas Trade Movements: Pipelines

2007 Pipeline Exports To	2007 Pipeline Exports From									
	North America			Latin America		Europe/Former Soviet Union				
	US	Canada	Mexico	Bolivia	Other LA	Belgium	Germany	Netherlands	Norway	UK
	billion cubic meters (bcm)									
North America										
US	—	107.3	1.60	—	—	—	—	—	—	—
Canada	13.20	—	—	—	—	—	—	—	—	—
Mexico	8.81	—	—	—	—	—	—	—	—	—
South & Central America										
Argentina	—	—	—	1.85	—	—	—	—	—	—
Brazil	—	—	—	9.88	0.12	—	—	—	—	—
Chile	—	—	—	—	2.38	—	—	—	—	—
Other S/C America	—	—	—	—	0.19	—	—	—	—	—
Europe and Former Soviet Union										
Austria	—	—	—	—	—	—	1.10	—	0.78	—
Belgium	—	—	—	—	—	—	1.60	7.10	9.50	0.64
Bulgaria	—	—	—	—	—	—	—	—	—	—
Croatia	—	—	—	—	—	—	—	—	—	—
Czech Republic	—	—	—	—	—	—	—	—	2.20	
Finland	—	—	—	—	—	—	—	—	—	—
France	—	—	—	—	—	1.90	0.10	8.92	15.11	0.10
Germany	—	—	—	—	—	—	—	19.13	23.74	2.90
Greece	—	—	—	—	—	—	—	—	—	—
Hungary	—	—	—	—	—	—	0.83	—	—	—
Ireland	—	—	—	—	—	—	—	—	—	4.15
Italy	—	—	—	—	—	—	1.50	6.11	8.99	0.75
Latvia	—	—	—	—	—	—	—	—	—	—
Lithuania	—	—	—	—	—	—	—	—	—	—
Luxembourg	—	—	—	—	—	0.80	0.70	—	—	—
Netherlands	—	—	—	—	—	—	5.50	—	7.00	1.82
Poland	—	—	—	—	—	—	0.80	—	—	—
Portugal	—	—	—	—	—	—	—	—	—	—
Romania	—	—	—	—	—	—	1.30	—	—	—
Serbia	—	—	—	—	—	—	—	—	—	—
Slovakia	—	—	—	—	—	—	—	—	—	—
Slovenia	—	—	—	—	—	—	—	—	—	—
Spain	—	—	—	—	—	—	—	—	2.15	
Sweden	—	—	—	—	—	—	0.15	—	—	—
Switzerland	—	—	—	—	—	—	1.20	0.60	0.18	—
Turkey	—	—	—	—	—	—	—	—	—	—
United Kingdom	—	—	—	—	—	1.80	1.60	8.20	16.40	—
Other Europe/FSU	—	—	—	—	—	—	—	—	—	
Middle East										
Iran	—	—	—	—	—	—	—	—	—	—
Jordan	—	—	—	—	—	—	—	—	—	—
UAE	—	—	—	—	—	—	—	—	—	—
Africa										
Tunisia	—	—	—	—	—	—	—	—	—	—
Asia-Pacific										
Singapore	—	—	—	—	—	—	—	—	—	—
Thailand	—	—	—	—	—	—	—	—	—	—
Total Exports	22.01	107.30	1.60	11.73	2.69	4.50	16.38	50.06	86.05	10.36

Table 15a: World Natural Gas Trade Movements: Pipelines, cont'd.

2007 Pipeline Exports To	2007 Pipelines Exports From: Europe/Former Soviet Union			Middle East			Africa			Asia-Pacific			Total
	Russian Fed.	Turkmenistan	Other Europe & Eurasia	Iran	Oman	Qatar	Algeria	Egypt	Libya	Indonesia	Malaysia	Myanmar	
	billion cubic meters (bcm)												
North America													
US	—	—	—	—	—	—	—	—	—	—	—	—	108.90
Canada	—	—	—	—	—	—	—	—	—	—	—	—	13.20
Mexico	—	—	—	—	—	—	—	—	—	—	—	—	8.81
South & Central America													
Argentina	—	—	—	—	—	—	—	—	—	—	—	—	1.85
Brazil	—	—	—	—	—	—	—	—	—	—	—	—	10.00
Chile	—	—	—	—	—	—	—	—	—	—	—	—	2.38
Other S/C America	—	—	—	—	—	—	—	—	—	—	—	—	0.19
Europe and Former Soviet Union													
Austria	5.60	—	—	—	—	—	—	—	—	—	—	—	7.48
Belgium	0.50	—	—	—	—	—	—	—	—	—	—	—	19.34
Bulgaria	3.10	—	—	—	—	—	—	—	—	—	—	—	3.10
Croatia	0.70	—	0.10	—	—	—	—	—	—	—	—	—	0.80
Czech Republic	6.43	—	—	—	—	—	—	—	—	—	—	—	8.63
Finland	4.30	—	—	—	—	—	—	—	—	—	—	—	4.30
France	7.63	—	—	—	—	—	—	—	—	—	—	—	33.76
Germany	35.55	—	2.40	—	—	—	—	—	—	—	—	—	83.72
Greece	2.89	—	—	—	—	—	—	—	—	—	—	—	2.89
Hungary	7.85	—	1.80	—	—	—	—	—	—	—	—	—	10.48
Ireland	—	—	—	—	—	—	—	—	—	—	—	—	4.15
Italy	23.80	—	—	—	—	—	22.10	—	9.20	—	—	—	72.45
Latvia	1.60	—	—	—	—	—	—	—	—	—	—	—	1.60
Lithuania	3.40	—	—	—	—	—	—	—	—	—	—	—	3.40
Luxembourg	—	—	—	—	—	—	—	—	—	—	—	—	1.50
Netherlands	2.30	—	2.24	—	—	—	—	—	—	—	—	—	18.86
Poland	6.20	—	2.30	—	—	—	—	—	—	—	—	—	9.30
Portugal	—	—	—	—	—	—	1.39	—	—	—	—	—	1.39
Romania	2.50	—	1.00	—	—	—	—	—	—	—	—	—	4.80
Serbia	1.90	—	—	—	—	—	—	—	—	—	—	—	1.90
Slovakia	5.80	—	—	—	—	—	—	—	—	—	—	—	5.80
Slovenia	0.56	—	0.10	—	—	—	0.44	—	—	—	—	—	1.10
Spain	—	—	—	—	—	—	8.80	—	—	—	—	—	10.95
Sweden	—	—	0.96	—	—	—	—	—	—	—	—	—	1.11
Switzerland	0.35	—	0.65	—	—	—	—	—	—	—	—	—	2.98
Turkey	23.15	—	1.28	6.16	—	—	—	—	—	—	—	—	30.59
UK	—	—	—	—	—	—	—	—	—	—	—	—	28.00
Other Europe/FSU	1.42	—	—	—	—	—	—	—	—	—	—	—	1.42
Middle East													
Iran	—	6.10	—	—	—	—	—	—	—	—	—	—	6.10
Jordan	—	—	—	—	—	—	—	2.35	—	—	—	—	2.35
UAE	—	—	—	—	0.95	0.80	—	—	—	—	—	—	1.75
Africa													
Tunisia	—	—	—	—	—	—	1.30	—	—	—	—	—	1.30
Asia-Pacific													
Singapore	—	—	—	—	—	—	—	—	—	5.39	1.78	—	7.17
Thailand	—	—	—	—	—	—	—	—	—	—	—	9.89	9.89
Total Exports	147.53	6.10	12.82	6.16	0.95	0.80	34.03	2.35	9.20	5.39	1.78	9.89	549.67

Source: Adapted from *BP Statistical Review of World Energy 2008.*

NOTES: Flows are on a contractual basis and may not correspond to physical gas flows in all cases.
Data exclude trade within the Former Soviet Union and United Arab Emirates (UAE).
Data from Cedigaz (provisional).

Table 15b: World Natural Gas Trade Movements: LNG

2007 LNG Exports To	2007 LNG Exports From															
	Americas		Europe	Middle East			Africa					Asia-Pacific				
	US	Trinidad & Tobago	Norway	Oman	Qatar	UAE	Algeria	Egypt	Equatorial Guinea	Libya	Nigeria	Australia	Brunei	Indonesia	Malaysia	Total
	Billion cubic meters (bcm)															
Americas																
US	—	12.76	—	—	0.52	—	2.11	3.24	0.50	—	2.69	—	—	—	—	21.82
Dominican Republic	—	0.36	—	—	—	—	—	—	—	—	—	—	—	—	—	0.36
Puerto Rico	—	0.74	—	—	—	—	—	—	—	—	—	—	—	—	—	0.74
Mexico	—	0.62	—	—	—	—	—	0.99	—	—	0.56	—	—	—	—	2.17
Europe																
Belgium	—	0.07	—	—	2.75	—	0.35	—	—	—	—	—	—	—	—	3.17
France	—	0.06	0.07	—	—	—	7.85	1.21	—	—	3.78	—	—	—	—	12.97
Greece	—	—	—	—	—	—	0.50	0.31	—	—	—	—	—	—	—	0.81
Italy	—	—	—	—	—	—	2.43	—	—	—	—	—	—	—	—	2.43
Portugal	—	—	—	—	—	—	—	—	—	—	2.31	—	—	—	—	2.31
Spain	—	2.09	0.07	0.12	4.45	—	4.32	4.04	—	0.76	8.33	—	—	—	—	24.18
Turkey	—	0.06	—	—	—	—	4.45	0.08	—	—	1.42	—	—	—	—	6.01
UK	—	0.39	—	—	0.27	—	0.64	0.16	—	—	—	—	—	—	—	1.46
Asia-Pacific																
China	—	—	—	0.07	—	—	0.42	—	—	—	0.08	3.30	—	—	—	3.87
India	—	0.21	—	0.21	8.27	0.07	0.44	0.07	—	—	0.64	—	—	—	0.07	9.98
Japan	1.18	0.57	—	4.81	10.87	7.41	0.78	1.62	0.36	—	0.88	16.05	8.57	18.07	17.65	88.82
South Korea	—	0.22	—	6.74	10.79	0.07	0.24	1.48	—	—	0.24	0.56	0.78	5.12	8.15	34.39
Taiwan	—	—	—	0.21	0.57	—	0.14	0.41	0.56	—	0.23	0.33	—	4.55	3.92	10.92
Total	1.18	18.15	0.14	12.17	38.48	7.55	24.67	13.61	1.42	0.76	21.16	20.24	9.35	27.74	29.79	226.41

Source: Adapted from *BP Statistical Review of World Energy 2008*. Data from Cedigaz (provisional)

Table 16: Crude Oil Prices

	US refiner acquisition cost			US landed cost of imports							US f.o.b. cost	
	Domestic	Imported	Composite	Total	Canada	Mexico	UK	Nigeria	Saudi Arabia	Venezuela	OPEC	Non-OPEC
YEAR	$/bbl											
1985	26.66	26.99	26.75	26.67	25.71	25.63	28.36	28.96	24.72	24.43	25.67	25.96
1986	14.82	14.00	14.55	13.49	13.43	12.17	14.63	15.29	12.84	11.52	12.21	12.87
1987	17.76	18.13	17.90	17.65	17.04	16.69	18.78	19.32	16.81	15.76	16.43	16.99
1988	14.74	14.56	14.67	14.08	13.50	12.58	15.82	15.88	13.37	13.66	13.43	13.05
1989	17.87	18.08	17.97	17.68	16.81	16.35	18.74	19.19	17.34	16.78	17.06	16.72
1990	22.59	21.76	22.22	21.13	20.48	19.64	22.65	23.33	21.82	20.31	20.40	20.32
1991	19.33	18.70	19.06	18.02	17.16	15.89	21.37	21.39	17.22	15.92	16.99	16.77
1992	18.63	18.20	18.43	17.75	17.04	15.60	20.63	20.78	17.48	15.13	16.87	16.66
1993	16.67	16.14	16.41	15.72	15.27	14.11	17.92	18.73	15.40	13.39	14.78	14.65
1994	15.67	15.51	15.59	15.18	14.83	14.09	16.64	17.21	15.11	13.12	14.00	14.34
1995	17.33	17.14	17.23	16.78	16.65	16.19	17.91	18.25	15.84	14.81	15.36	16.02
1996	20.77	20.64	20.71	20.31	19.94	19.64	20.88	21.95	20.49	18.59	18.94	19.65
1997	19.61	18.53	19.04	18.11	17.63	17.30	20.64	20.64	17.52	16.35	16.26	17.51
1998	13.18	12.04	12.52	11.84	11.62	11.04	13.55	14.14	11.16	10.16	10.20	11.21
1999	17.90	17.26	17.51	17.23	17.54	16.12	18.26	17.63	17.48	15.58	15.90	16.84
2000	29.11	27.70	28.26	27.53	26.69	26.03	29.26	30.04	26.58	26.05	25.56	26.77
2001	24.33	22.00	22.95	21.82	20.72	19.37	25.32	26.55	20.98	19.81	19.73	21.04
2002	24.65	23.71	24.10	23.91	22.98	22.09	26.35	26.45	24.77	21.93	22.18	22.93
2003	29.82	27.71	28.53	27.69	26.76	25.48	30.62	31.07	27.50	25.70	25.36	26.21
2004	38.97	35.90	36.98	36.07	34.51	32.25	39.28	40.95	37.11	33.79	33.95	33.58
2005	52.94	48.86	50.24	49.29	44.73	43.47	55.28	57.55	50.31	47.87	49.60	45.79
2006	62.62	59.02	60.24	59.11	53.90	53.76	67.44	68.26	59.19	57.37	59.18	55.35
2007	69.65	67.04	67.94	67.35	60.32	62.21	70.95	76.72	70.24	65.92	69.58	62.71

Source: U.S. Energy Information Administration: *Annual Energy Review*, June 23, 2008 (www.eia.doe.gov/emeu/aer/petro.html); *Monthly Energy Review*, Aug. 21, 2008 (www.eia.doe.gov/emeu/mer/contents. html); Petroleum Navigator for RAC data Aug. 29, 2008 (http: //tonto.eia.doe.gov/dnav/pet/pet_pri_rac2_dcu_nus_a.htm)

NOTES: All prices are US annual average (nominal dollars)..

Table 17: Petroleum Product Prices

	Retail unleaded motor gasoline			Refiner resale (wholesale)								
	Regular	Premium	All	Home Heating Oil	Motor Gasoline	Aviation Gasoline	Kerosene Jet Fuel	Kerosene	No. 2 Fuel Oil	No. 2 Diesel Fuel	Propane	Residual Fuel Oil
YEAR	$/gallon (nominal)			cents/gallon								
1985	1.20	1.34	1.20	77.4	83.5	113.0	79.4	87.4	77.6	77.2	39.8	57.7
1986	0.93	1.09	0.93	47.0	53.1	91.2	49.5	60.6	48.6	45.2	29.0	30.5
1987	0.95	1.09	0.96	53.1	58.9	85.9	53.8	59.2	52.7	53.4	25.2	38.5
1988	0.95	1.11	0.96	47.3	57.7	85.0	49.5	54.9	47.3	47.3	24.0	30.0
1989	1.02	1.20	1.06	56.6	65.4	95.0	58.3	66.9	56.5	56.7	24.7	36.0
1990	1.16	1.35	1.22	69.5	78.6	106.3	77.3	83.9	69.7	69.4	38.6	41.3
1991	1.14	1.32	1.20	61.8	69.9	100.1	65.0	72.2	62.2	61.5	34.9	31.4
1992	1.13	1.32	1.19	58.5	67.7	99.1	60.5	63.2	57.9	59.1	32.8	30.8
1993	1.11	1.30	1.17	55.9	62.6	96.5	57.7	60.4	54.4	57.0	35.1	29.3
1994	1.11	1.31	1.17	52.2	59.9	93.3	53.4	61.8	50.6	52.9	32.4	31.7
1995	1.15	1.34	1.21	53.0	62.6	97.5	53.9	58.0	51.1	53.8	34.4	36.3
1996	1.23	1.41	1.29	65.3	71.3	105.5	64.6	71.4	63.9	65.9	46.1	42.0
1997	1.23	1.42	1.29	60.2	70.0	106.5	61.3	65.3	59.0	60.6	41.6	38.7
1998	1.06	1.25	1.12	43.9	52.6	91.2	45.0	46.5	42.2	44.4	28.8	28.0
1999	1.17	1.36	1.22	53.6	64.5	100.7	53.3	55.0	49.3	54.6	34.2	35.4
2000	1.51	1.69	1.56	89.6	96.3	133.0	88.0	96.9	88.6	89.8	59.5	56.6
2001	1.46	1.66	1.53	77.9	88.6	125.6	76.3	82.1	75.6	77.5	54.0	47.6
2002	1.36	1.56	1.44	71.8	82.8	114.6	71.6	75.2	69.4	72.4	43.1	53.0
2003	1.59	1.78	1.64	88.2	100.2	128.8	87.1	95.5	88.1	88.3	60.7	66.1
2004	1.88	2.07	1.92	117.8	128.8	162.7	120.8	127.1	112.5	118.7	75.1	68.1
2005	2.30	2.49	2.34	172.0	167.0	207.6	172.3	175.7	162.3	173.7	93.3	97.1
2006	2.59	2.81	2.64	199.1	196.9	249.0	196.1	200.7	183.4	201.2	103.1	113.6
2007	2.80	3.03	2.85	219.0	218.2	274.6	216.4	224.9	206.3	220.4	119.4	135.0

Source: U.S. Energy Information Administration, *Annual Energy Review 2007*, posted June 23, 2008 (www.eia.doe.gov/emeu/aer/petro.html).

NOTE: All prices are US annual average (nominal dollars).

Table 18: Comparative Energy Prices

	Natural gas delivered to consumers, $/1000 cu ft				Retail electricity, cents/kWh			
YEAR	Residential	Commercial	Industrial	Electric Power	Residential	Commercial	Industrial	All
1985	6.12	5.50	3.95	3.55	7.39	7.27	4.97	6.44
1986	5.83	5.08	3.23	2.43	7.42	7.20	4.93	6.44
1987	5.54	4.77	2.94	2.32	7.45	7.08	4.77	6.37
1988	5.47	4.63	2.95	2.33	7.48	7.04	4.70	6.35
1989	5.64	4.74	2.96	2.43	7.65	7.20	4.72	6.45
1990	5.80	4.83	2.93	2.39	7.83	7.34	4.74	6.57
1991	5.82	4.81	2.69	2.18	8.04	7.53	4.83	6.75
1992	5.89	4.88	2.84	2.36	8.21	7.66	4.83	6.82
1993	6.16	5.22	3.07	2.61	8.32	7.74	4.85	6.93
1994	6.41	5.43	3.05	2.28	8.38	7.73	4.77	6.91
1995	6.06	5.05	2.71	2.02	8.40	7.69	4.66	6.89
1996	6.34	5.40	3.42	2.69	8.36	7.64	4.60	6.86
1997	6.94	5.80	3.59	2.78	8.43	7.59	4.53	6.85
1998	6.82	5.48	3.14	2.40	8.26	7.41	4.48	6.74
1999	6.69	5.33	3.10	2.62	8.16	7.26	4.43	6.64
2000	7.76	6.59	4.45	4.38	8.24	7.43	4.64	6.81
2001	9.63	8.43	5.24	4.61	8.58	7.92	5.05	7.29
2002	7.89	6.63	4.02	3.68	8.44	7.89	4.88	7.20
2003	9.63	8.40	5.89	5.57	8.72	8.03	5.11	7.44
2004	10.75	9.43	6.53	6.11	8.95	8.17	5.25	7.61
2005	12.70	11.34	8.56	8.47	9.45	8.67	5.73	8.14
2006	13.75	11.99	7.86	7.11	10.40	9.46	6.16	8.90
2007	13.01	11.31	7.58	7.30	10.64	9.67	6.36	9.14

Source: U.S. Energy Information Administration: http://tonto.eia.doe.gov/dnav/ng/ng_pri_sum_dcu_nus_a.htm;
Electric Power Monthly: http://www.eia.doe.gov/fuelelectric.html

NOTE: All prices are U.S. annual average.

Table 19: Price History of Oil, Gas, and Gasoline

	Actual prices					Inflation-adjusted prices (1982=100)			
	Petroleum		Natural gas			Petroleum		Natural gas	
YEAR	Wellhead, $/bbl	Retail gasoline, $/gal	Wellhead $/1000 cu ft	Residential $/1000 cu ft	Producer price index	Wellhead, $/bbl	Retail gasoline, $/gal	Wellhead $/1000 cu ft	Residential $/1000 cu ft
1940	1.02	0.184	0.045	N/A	13.5	7.556	1.364	0.333	NA
1945	1.22	0.205	0.049	N/A	18.2	6.703	1.126	0.269	NA
1950	2.51	0.268	0.065	N/A	27.3	9.194	0.980	0.238	NA
1955	2.77	0.291	0.104	N/A	29.3	9.454	0.992	0.355	NA
1960	2.88	0.311	0.140	N/A	31.7	9.085	0.982	0.442	NA
1965	2.86	0.312	0.156	N/A	32.3	8.854	0.966	0.483	NA
1970	3.18	0.357	0.171	1.09	36.9	8.618	0.967	0.463	2.954
1975	7.67	0.567	0.445	1.71	58.4	12.945	0.971	0.762	2.928
1980	21.59	1.191	1.59	3.68	89.7	24.069	1.328	1.773	4.103
1981	31.77	1.311	1.98	4.29	98.0	32.418	1.338	2.020	4.378
1982	28.52	1.296	2.46	5.17	100.0	28.520	1.296	2.460	5.170
1983	26.19	1.241	2.59	6.06	101.3	25.879	1.226	2.559	5.982
1984	25.88	1.212	2.66	6.12	103.7	24.981	1.170	2.568	5.902
1985	24.09	1.202	2.51	6.12	103.2	23.366	1.166	2.435	5.930
1986	12.51	0.927	1.94	5.83	100.2	12.498	0.926	1.938	5.818
1987	15.40	0.948	1.67	5.54	102.8	14.981	0.922	1.625	5.389
1988	12.58	0.946	1.69	5.47	106.9	11.768	0.885	1.581	5.117
1989	15.86	1.021	1.69	5.64	112.2	14.135	0.910	1.506	5.027
1990	20.03	1.164	1.71	5.80	116.3	17.223	1.001	1.470	4.987
1991	16.54	1.140	1.64	5.82	116.5	14.197	0.979	1.408	4.996
1992	15.99	1.127	1.74	5.89	117.2	13.643	0.962	1.485	5.026
1993	14.25	1.108	2.04	6.16	118.9	11.985	0.932	1.716	5.181
1994	13.19	1.112	1.85	6.41	120.5	10.955	0.924	1.537	5.320
1995	14.62	1.205	1.55	6.06	124.8	11.724	0.966	1.243	4.856
1996	18.46	1.288	2.17	6.34	127.7	14.456	1.009	1.699	4.965
1997	17.23	1.291	2.32	6.94	127.6	13.503	1.012	1.818	5.439
1998	10.87	1.115	1.96	6.82	124.4	8.738	0.896	1.559	5.482
1999	15.56	1.221	2.19	6.69	125.5	12.398	0.973	1.657	5.331
2000	26.72	1.563	3.68	7.76	132.7	20.143	1.178	2.773	5.848
2001	21.84	1.531	4.00	9.63	134.2	16.274	1.141	3.070	7.183
2002	22.51	1.441	2.95	7.89	131.1	17.170	1.099	2.250	6.018
2003	27.56	1.638	4.88	9.63	138.1	19.957	1.186	3.533	6.973
2004	36.77	1.923	5.46	10.75	146.7	25.065	1.311	3.742	7.321
2005	50.28	2.338	7.33	12.70	157.4	31.944	1.485	4.657	8.069
2006	59.69	2.635	6.40	13.75	164.8	36.220	1.599	3.883	8.343
2007	66.52	2.849	6.39	13.01	172.7	38.518	1.650	3.700	7.533

Sources: U.S. Energy Information Administration:
Monthly Energy Report (Aug. 2008): http://www.eia.doe.gov/emeu/mer/contents.html
Petroleum Navigator: http://tonto.eia.doe.gov/dnav/pet/hist/f000000__3A.htm
Capital District Regional Planning Commission, www.cdrpc.org/CPI_PPI.html.

NOTES: All prices are U.S. annual averages.
Gasoline prices: 1975 and prior for leaded regular; 1976-94 for unleaded regular; 1995 forward for all grades.

Table 20: International Rig Count

REGION	2002	2003	2004	2005	Dec. 2006			Dec. 2007			Dec. 2008		
					Land	Off.	Total	Land	Off.	Total	Land	Off.	Total
North America													
Canada	277	332	440	575	452	4	456	358	2	360	359	2	361
US	862	1,126	1,246	1,470	1,634	84	1,718	1,749	62	1,811	1,716	66	1,782
Subtotal	1,139	1,458	1,686	2,045	2,086	88	2,174	2,107	64	2,171	2,075	68	2,143
Latin America													
Argentina	56	60	74	81	81	0	81	87	0	87	70	0	70
Bolivia	5	5	4	5	3	0	3	2	0	2	3	0	3
Brazil	26	24	26	28	14	19	33	21	25	46	29	30	59
Chile	1	0	0	0	1	0	1	1	1	2	1	0	1
Colombia	10	11	13	21	24	0	24	40	0	40	42	0	42
Ecuador	8	7	12	14	12	0	12	9	0	9	13	0	13
Mexico	80	107	103	94	57	27	84	68	30	98	83	28	111
Peru	1	2	2	3	5	1	6	6	2	8	4	3	7
Trinidad	2	3	3	5	3	4	7	0	5	5	0	4	4
Venezuela	30	45	69	75	58	18	76	58	13	71	66	12	78
Other	—	—	—	1	2	0	2	2	0	2	1	0	1
Subtotal	219	264	306	327	260	69	329	294	76	370	312	77	389
Europe													
Croatia	3	3	3	3	1	0	1	1	0	1	0	0	0
Denmark	4	4	3	2	0	0	0	0	2	2	0	3	3
France	0	1	1	1	0	0	0	1	0	1	1	0	1
Germany	4	2	2	3	4	0	4	4	1	5	10	0	10
Hungary	2	2	2	2	3	0	3	2	0	2	3	0	3
Italy	5	7	5	3	4	1	5	4	1	5	4	0	4
Netherlands	5	5	4	5	1	2	3	0	2	2	0	5	5
Norway	15	17	14	15	0	15	15	0	19	19	0	25	25
Poland	12	11	3	2	2	0	2	2	0	2	1	0	1
Romania	2	2	2	2	2	0	2	2	1	3	16	2	18
Turkey	4	4	4	5	4	0	4	5	0	5	6	0	6
UK	23	21	16	27	1	21	22	2	20	22	4	18	22
Other	4	4	4	4	4	0	4	8	0	8	3	0	3
Subtotal	83	83	63	74	26	39	65	31	46	77	48	53	101
Middle East													
Iran*	34	39	40	38	0	0	0	0	0	0	0	0	0
Jordan	—	—	—	1	1	0	1	0	0	0	1	0	1
Kuwait	7	8	12	13	16	0	16	11	0	11	11	0	11
Oman	32	37	36	36	41	0	41	53	0	53	52	0	52
Qatar	9	8	11	14	2	7	9	3	13	16	1	9	10
Saudi Arabia	33	31	34	44	66	9	75	65	11	76	64	10	74
Syria	24	24	23	23	25	0	25	19	0	19	20	0	20
UAE	16	16	14	15	10	5	15	10	4	14	11	3	14
Yemen	11	7	12	14	15	0	15	16	0	16	15	0	15
Other	3	1	5	2	2	0	2	0	0	0	1	0	1
Subtotal	169	171	187	200	178	21	199	177	28	205	176	22	198

Table 20: International Rig Count, cont'd.

REGION	2002	2003	2004	2005	Dec. 2006			Dec. 2007			Dec. 2008		
					Land	Off.	Total	Land	Off.	Total	Land	Off.	Total
Africa													
Algeria	22	20	20	21	25	0	25	29	0	29	27	0	27
Angola	2	7	3	3	0	4	4	2	3	5	0	6	6
Congo (Brazzaville)	1	4	2	1	1	0	1	2	1	3	1	2	3
Egypt	24	27	31	32	31	8	39	39	11	50	49	11	60
Gabon	2	3	2	3	2	1	3	3	1	4	1	0	1
Libya	10	11	10	9	12	1	13	14	1	15	14	1	15
Nigeria	12	9	11	9	2	9	11	2	7	9	3	4	7
South Africa	0	1	0	0	0	0	0	0	0	0	0	1	1
Sudan*	5	7	17	19	0	0	0	0	0	0	0	0	0
Tunisia	3	2	2	0	3	2	5	4	1	5	2	1	3
Other	4	5	3	3	2	3	5	2	2	4	3	1	4
Subtotal	85	96	101	100	78	28	106	97	27	124	100	27	127
Asia-Pacific													
Australia	10	13	14	14	11	11	22	10	11	21	15	11	26
Brunei	3	2	2	2	1	3	4	1	2	3	1	3	4
China: Offshore	9	10	12	17	0	17	17	0	22	22	0	20	20
India	58	62	76	83	54	31	85	55	29	84	55	27	82
Indonesia	53	42	51	60	35	18	53	44	22	66	46	15	61
Japan	9	4	2	3	1	0	1	1	1	2	2	0	2
Malaysia	15	14	11	12	0	14	14	0	11	11	0	19	19
Myanmar	7	10	9	9	8	2	10	7	0	7	1	1	2
New Zealand	1	2	3	6	4	2	6	4	1	5	3	1	4
Pakistan	13	12	12	11	17	0	17	19	0	19	21	0	21
Papua New Guinea	0	1	1	3	1	0	1	1	0	1	2	0	2
Philippines	2	2	3	2	1	0	1	0	1	1	3	0	3
Thailand	7	5	6	9	3	8	11	3	8	11	2	10	12
Vietnam	9	9	6	10	0	6	6	0	4	4	0	4	4
Other	1	1	4	6	2	2	4	1	2	3	0	1	1
Subtotal	197	189	212	247	138	114	252	146	114	260	151	112	263
Total World	1,892	2,261	2,555	2,993	2,766	359	3,125	2,852	355	3,207	2,862	359	3,221

Source: Baker Hughes Inc. in *Oil & Gas Journal.*

NOTES: U.S. land count includes inland water rigs.
China count includes offshore rigs only.
No Iraq rig report received since the Iraq Invasion of Kuwait.
Iran and Sudan rig counts discontinued in Jan. 2006..

Information Sources

National Oil Companies and Energy Ministries

A

Albania

Ministry of Economy, Trade and Energy
Tirana, ALBANIA
Telephone: 355 42 32746
Fax: 355 42 26717

Albpetrol Sh.A.
c/o Geological Institute for Oil & Gas
Patos, Fieri District, ALBANIA
Telephone: 355 423 4542
Fax: 355 423 4204

National Agency of Natural Resources
Agjensia Kombetare e Burimeve Natyrore (AKBN)
Boulevard Bajram Curri
Vasil Shanto, Tirana, ALBANIA
Telephone: 355 422 57117; 355 422 58830
Fax: 355 422-57382
Web site: www.akbn.gov.al

Algeria

Ministry of Energy & Mining
Tour A, Val d'Hydra.
B.P 677 AlgerGare
80 Av. Ghermoul
Algiers, ALGERIA
Telephone: 213 21 48 85 26
Fax: 213 21 48 85 57
E-mail: info@mem-algeria.org
Web site: www.mem-algeria.org

Sonatrach
80 Av. Ghermoul
Algiers, ALGERIA
Telephone: 213 21 54 80 11; 213 21 54 70 00
Fax: 213 21 54-77 00
E-mail: sonatrach@sonatrach.dz
Web site: www.sonatrach-dz.com

Angola

Sonangol (Sociedad Nacional de Combustiveis de Angola)
Rua 1° Congresso do MPLA, N.° 8-16
Caixa Postal 1316
Luanda, ANGOLA
Telephone: 244 2 63 2103
Fax: 244 2 39 1782
E-mail: CGI@sonangol.co.ao
Web site: www.sonangol.co.ao

Argentina

Secretary of Energy
Av. Paseo Colón 171
Capital Federal – CP (C1063ACB)
ARGENTINA
Telephone: 54 11 4349 5000
E-mail: energia@minplan.gov.ar
Web site: http://energia3.mecon.gov.ar/

Repsol YPF SA
Av. Pres. Roque Sáenz Peña #777
CP 1364 Buenos Aires, ARGENTINA
Telephone: 54 11 43 29 20 00
Fax: 54 11 43 29 20 01
Web site: www.repsolypf.com

Armenia

Ministry of Energy
Government House 2
Republic Square
Yerevan 375010, ARMENIA
Telephone: 37 410 52 19 64
Fax: 37 410 52 63 65
E-mail: minenergy@minenergy.am
Web site: www.minenergy.am

Australia

Commonwealth of Australia
Department of Resources, Energy and Tourism
Industry House
10 Binara Street
Mail: GPO Box 1564
Canberra City, ACT 2601, AUSTRALIA
Telephone: 61 2 6276 1000
E-mail: ret@ret.gov.au
Websites: www.australia.gov.au
www.ret.gov.au

Australia, New South Wales

Department of Water and Energy
Level 17, 227 Elizabeth St.
Mail: GPO Box 3889
Sydney, New South Wales 2001, AUSTRALIA
Telephone: 02 8281 7777
Fax: 02 8281 7799
E-mail: information@dwe.nsw.gov.au
Web site: www.deus.nsw.gov.au

Australia, Northern Territory

Department of Minerals and Energy
Geological Survey: Darwin
Centrepoint Building, 3rd Floor
Smith Street Mall
Darwin 0800, Northern Territory, AUSTRALIA
Telephone: 618 8999 5313
Fax: 618 8999 6824
Website: www.nt.gov.au/d/Minerals_Energy
www.minerals.nt.gov.au

Australia, Queensland

Department of Natural Resources and Water
Mineral House
41 George St.
GPO Box 2454
Brisbane 4001, AUSTRALIA
Telephone: 07 3896 3111
Web site: www.nrm.qld.gov.au

Australia, South Australia

Department of Primary Industries and Resources
Petroleum and Geothermal Group
Level 6, 101 Grenfell St.
GPO Box 1671
Adelaide South Australia 5001, AUSTRALIA
Telephone: 61 8 8463 3204 (international)
Fax: 61 8 8463 3229 (international)
Web site: http://www.pir.sa.gov.au/petroleum

Australia, Tasmania

Department of Infrastructure, Energy and Resources
10 Murray St.
Mail: GPO Box 936
Hobart, Tasmania 7001, AUSTRALIA
Telephone: 1300 135 513
E-mail: webmaster@dier.tas.gov.au
Web site: www.dier.tas.gov.au

Australia, Victoria

Department of Primary Industries
Minerals and Petroleum Section
1 Spring St.
Mail: GPO Box 4440
Melbourne, Victoria 3001, AUSTRALIA
Telephone: 61 3 5332 5000 (international)
Web site: www.dpi.vic.gov.au

Australia, Western Australia

Office of Energy
Level 9
Governor Stirling Tower
197 St Georges Terrace
Perth, Western Australia 6000, AUSTRALIA
Telephone: 61 8 9420 5600
Fax: 61 8 9420 5700
E-mail: enquiries@energy.wa.gov.au
Web site: www.energy.wa.gov.au

Austria

Federal Ministry of Economics and Labour
Energy and Mining Division
A-1010 Wien (Vienna), Stubenring 1, AUSTRIA
Telephone: 43 1 711 00 0
E-mail: service@bmwa.gv.at
Web site: www.bmwa.gv.at

OMV AG (Aktiengesellschaft)
Otto-Wagner-Platz 5
A-1090 Wien (Vienna), AUSTRIA
Ph.: 43 1 40 440 0
Fax: 43 1 40 440 20091
E-mail: info@omv.com
Web site: www.omv.com

Azerbaijan

AIOC (Azerbaijan International Operating Co.)
Villa Petrolea, 2 Neftchilar Prospect (Bayil)
Baku, AZERBAIJAN
Telephone: 994 12 97 90 00; 994 12 97 97 54
Fax: 994 97 96 02; 994 12 97 97 37
E-mail: bayatltf@bp.com
Web site: www.azer.com

SOCAR (State Oil Co. of Azerbaijan)
73 Neftchilar Prospect (Bayil)
Baku 1000, AZERBAIJAN
Telephone: 994 12 492 1789; 994 12 497 5347
Fax: 994 12 497 1167
E-mail: info@socar.az
Web site: http://socar.az/index-en.html

B

Bahrain

NOGA (National Oil and Gas Authority)
P.O. Box 1435, Manama, BAHRAIN
Telephone: 973 1731 2644
Fax: 973 1729 3007
E-mail: info@noga.gov.bh
Web site: http://www.noga.gov.bh

Banagas (Bahrain National Gas Co.)
P.O. Box 29099, Manama, BAHRAIN
Telephone: 973 17 756222
Fax: 973 17 756 991
E-mail: bng@banagas.com.bh
Web site: www.banagas.com.bh

Bapco (Bahrain Petroleum Company)
P.O. Box 25555
Awali, BAHRAIN
Telephone: 973 1770 4040
Fax: 973 1770 4070

Bangladesh

Petrobangla (Bangladesh Oil, Gas & Mineral Corp.)
Petrocentre
3 Kawranbazar
Dhaka 1215, BANGLADESH
Fax: 880 2 912 0224
Web site: www.petrobangla.bd.org

Belize

Ministry of Natural Resources and the Environment
Market Square, Belmopan, BELIZE
Telephone: 501 822 2226; 501 822 2249; 501 822 2711
Fax: 501 822 2333
E-mail: info@mnrei.gov.bz
Web site: www.governmentofbelize.gov.bz

Bolivia

Ministry of Hydrocarbons and Energy
Superintendent of Hydrocarbons
Av. 20 de Octubre #2685 esq. Campos
La Paz, BOLIVIA
Telephone: 591 2 374 050; 591 2 374 053
Fax: 591 2 392-758
Web site: www.hidrocarburos.gov.bo

YPFB (Yacimientos Petroliferos Fiscales Bolivianos)
Calle Bueno No. 185
Box 401
La Paz, BOLIVIA
Telephone: 591 2 237 0210
Web site: www.ypfb.gov.bo

Brazil

ANP (Agência Nacional do Petróleo,
Gas Natural e Biocombustiveis)
Av. Rio Branco #65 – del 12 al 22 piso
Rio de Janeiro, BRAZIL 20090-004
Telephone: 55 21 2112 8100; 55 21 2112 8331
Fax: 55 21 2112 8129
E-mail: imprensa@anp.gov.br
Web site: www.anp.gov.br

Ministry of Mines and Energy
Esplanada dos Ministerios Block "U"
Brasilia, DF, BRAZIL
Telephone: 61 3319 5555
Web site: www.mme.gov.br

Petrobras (Petroleo Brasilerio SA)
Av. República do Chile 65, sala 401 E
20031-912 Rio de Janeiro, BRAZIL 20031-912
Telephone: 55 21 2534 4477
Fax: 55 21 2534 3247
Web Site: www.petrobras.com.br

Brunei

Office of the Prime Minister
Petroleum Unit
Bahirah Building, Jalan Menteri Besar
Darussalam BB3910, BRUNEI
Telephone: 673 2 387 102
Fax: 673 2 383 004
E-mail: brupet@brunet.bn
Web site: www.petroleum-unit.gov.bn

Bulgaria

Energy Committee
1169 Sofia, 1 Kniaz Alexander I Square
Sofia 1040, BULGARIA
Telephone: 12 939 32 01
Fax: 02 987 50 13
E-mail: kom_energy@parliament.bg
Web site: www.parliament.bg

C

Cameroon

Societe Nationale des Hydrocarbures (SNH)
Box 955
Yaounde, CAMEROON
Telephone: 237 22 1910; 237 23 2530
Fax: 237 22 3253

Canada

National Energy Board
444 Seventh Avenue SW
Calgary, Alberta
T2P 0X8 CANADA
Telephone: 403-292-4800;
1 800 899 1265
Fax: 403 292 5503
Toll-free fax: 1 877 288 8803
E-mail: info@neb-one.gc.ca
Web site: www.neb-one.gc.ca

Natural Resources Canada
Energy Sector
580 Booth St.
Ottawa, Ontario
ON K1A CANADA
Telephone: 613 995 0947
Web site: www.nrcan-rncan.gc.ca

Canada, Alberta

Alberta Department of Energy (Edmonton)
North Petroleum Plaza
7th Floor, 9945-108 St
Edmonton, Alberta
T5K 2G6 CANADA
Telephone: 780 427 0265
Fax: 780 422 0698
Web site: www.energy.gcv.ab.ca

Alberta Department of Energy (Calgary)
300, 801-6 Avenue SW
Calgary, Alberta
T2P 3W2 CANADA
Telephone: 403 297 8955
Web site: www.energy.gov.ab.ca

ERCB (Alberta Energy Resources Conservation Board)
Calgary Office
640 - 5 Avenue SW
Calgary, Alberta
T2P 3G4 CANADA
Telephone: 403 297 8311
Fax: 403 297 7336
E- mail: Inquiries@ercb.ca
Web site: www.ercb.ca

Alberta Utilities Commission
Calgary Office
Fifth Avenue Place
4th Floor, 425 First St. SW
Calgary, Alberta
T2P 3L8 CANADA
Telephone: 780 427 9362
Email: Info@auc.ab.ca
Web site: www.auc.ab.ca

Alberta Utilities Commission
Edmonton Office
10th Floor, 10055, 106 St.
Edmonton, Alberta
T5J 2Y2 CANADA
Telephone: 780 427 9362
Email: Info@auc.ab.ca
Web site: www.auc.ab.ca

Canada, British Columbia

Ministry of Energy, Mines and Petroleum Resources
Public Affairs Bureau
PO Box 9318, Stn Prov Govt
Victoria, British Columbia
V8W 9N3 CANADA
Phone: 250 952 0241
E-mail: EnquiryBC@gov.bc.ca/empr
Web site: www.gov.bc.ca/ministries

Canada, Manitoba

Department of Science, Technology, Energy and Mines
Petroleum Branch
360-195 Ellice Avenue
Winnipeg, Manitoba
R3B 3P2 CANADA
Telephone: 204 945 6577
Fax: 204 945 0586
E-mail: mgi@gov.mb.ca
Websites: www.gov.mb.ca
www.gov.mb.ca/stem/petroleum

Canada, New Brunswick

Brunswick Square
1 Germain Street – Suite M-100
St. John, New Brunswick
E2L 4V1 CANADA
Telephone: 506 658 3180
Fax: 506 658 3191
Web site: www.gnb.ca/0085/index-e.asp

Canada, Newfoundland & Labrador

Department of Natural Resources
7th Floor, Natural Resources Building
50 Elizabeth Avenue
P.O. Box 8700
St. John's, Newfoundland
A1B 4J6 CANADA
Telephone: 709 729 2920
Fax: 709 729 0059
Websites: www.gov.nf.ca
www.nr.gov.nl.ca/nr/

Canada, Northwest Territories

Environment and Natural Resources Department
PO Box 1320
Yellowknife, Northwest Territories
X1A 2L9 CANADA
Telephone: 867 873 7532
Fax: 867 920 2756
Web site: www.gov.nt.ca

Canada, Nova Scotia

Department of Energy
Bank of Montreal Building, Suite 400
5151 George St.
P.O. Box 2664
Halifax, Nova Scotia
B3J 3P7 CANADA
Telephone: 902 424 4575
Fax: 902 424 0528
Web site: www.gov.ns.ca/energy

Canada, Nunavut

Department of Environment
P.O.Box 1000, Station 1300
Iqaluit, Nunavut, CANADA
X0A 0H0
Telephone: 867 975 7700
Fax: 867 975 7740
E-mail: environment@gov.nu.ca

Canada, Ontario

Ministry of Energy and Infrastructure
900 Bay Street, 4th Floor
Hearst Block
Toronto, Ontario
M7A 2E1 CANADA
Telephone: 1 888 668 4636
Web site: www.energy.gov.on.ca

Canada, Prince Edward Island

Department of Environment, Energy and Forestry
Jones Building
11 Kent Street
P.O. Box 2000
Charlottetown, Prince Edward Island
C1A 7N8 CANADA
Telephone: 902-368-5000
Fax: 902 368 5830
E-mail: Island@gov.pe.ca
Web site: www.gov.pe.ca

Canada, Quebec

Department of Natural Resources
5700, 4e Avenue Ouest
B401 (Energy & Mines)
Charlesbourg, Québec
G1H 6R1 CANADA
Telephone: 418 627 6385
E-mail: hydrocarbures@mrnf.gouv.qc.ca
Web site: www.mrnf.gouv.qc.ca/english/energy

Canada, Saskatchewan

Ministry of Energy and Resources
300-2103 11th Avenue
Regina, Saskatchewan
S4P 3Z8 CANADA
Telephone: 306 787 2528
Web site: www.er.gov.sk.ca

Canada, Yukon Territory

Energy, Mines and Resources
Box 2703
Whitehorse, Yukon
Y1A 2C6 CANADA
Telephone: 867 667 3130
Fax: 867 667 8601
E-mail: emr@gov.yk.ca
Web site: www.emr.gov.yk.ca

Chile

Enap (Empresa Nacional del Petróleo)
Todos los derechos reservados
Av. Vitacura 2736 piso 10 Las Condes
Santiago, CHILE
Telephone: 56 2 280 3000
Fax: 56 2 280 3199
Web site: www.enap.cl

China

CNOOC Ltd. (China National Offshore Oil Corp. Ltd.)
Box 4705
No. 25 Chao Yangmen North Street
Dongcheng District
Beijing 100010, PEOPLE'S REPUBLIC OF CHINA
Telephone: 010 845 21010
Fax: 010 646 02600
E-mail: webmaster@cnooc.com.cn
Web site: www.cnooc.com.cn/yyww

CNPC (China National Petroleum Corp.)
6, Liu Pu Kang Jie, Xicheng District
Beijing, 100724, PEOPLE'S REPUBLIC OF CHINA
Telephone: 86 10 6209 4114
Fax: 86 10 6209 5148
E-mail: admin_eng@cnpc.com.cn
Web site: www.cnpc.com.cn/eng

Sinopec Corp. (China Petroleum & Chemical Corp.)
6A Huixindong Street, Chaoyang District
Beijing 100029, CHINA
Telephone: 86 10 649 90060
Fax: 86 10 649 90022
E-mail: ir@sinopec.com
Web site: www.sinopec.com

Colombia

Ecopetrol SA (Empresa Colombiana del Petroleos)
Edificio Principal Cr 13 No. 36 – 24
Bogota, COLOMBIA
Telephone: 57 1 234 4000; 57 1 234 4444
Fax: 57 1 234 4099
E-mail: prozo@infantas.ecp.com
Web site: www.ecopetrol.com.co

Ministerio de Minas y Energia de Colombia
Transversal 45, No. 26-86, CAN
Bogota, 80319 COLOMBIA
Telephone: 57 1 220 0300
Fax: 57 1 222 3656
Web site: www.minminas.gov.co
E-mail: minas.energia@minminas.gov.co

Congo (Brazzaville)

Ministry of Mines and Energy - S.P. 2120
Brazzaville, CONGO
Telephone: 242 81 10 85
Fax: 242 81 10 86

Hydro-Congo
B.P. 742
Pointe Noire, Brazzaville, CONGO
Telephone: 242 94 04 04
Fax: 242 83 12 38

Congo (former Zaire)

Petrozaire
B.P. 7617
Kinshasa/Gombe, CONGO ZAIRE
Telephone: 243 12 20 344; 243 12 20 492

Croatia

INA-Industrija nafte, dd
Avenija Veceslava Holjevca 10, p.p. 555
10002 Zagreb, CROATIA
Telephone: 00385 1 6450 000
E-mail: ina-besplatni.telefon@ina.hr
Web site: www.ina.hr

Cuba

Comercial Cupet SA
km 37.5 Via Blanca
Caribe, Sta. Cruz del Norte
Havana 32900, CUBA
Telephone: 53 7 62 40 11
Fax: 53 7 33 34 60; 53 7 33 80 27; 53 7 33 30 72

Ministerio de la Industria Basica
Empedarado 113
Havana, CUBA
Web site: www.cubagob.cu

Czech Republic

Urad Vlady CR
Nabrezi Edvarda Benese 4
Mala Strana, PSC 118 01
Praha 1,CZECH REPUBLIC
Telephone: 420 224 002 111
Fax: 420 224 002 502
E-mail: sedlackova.silvie@vlada.zc
Web site: www.vlada.cz

D

Denmark

Ministry of the Environment
Frontlinien
Rentemestervej 8
2400 Copenhagen NV DENMARK
Telephone: 45 70 12 02 11
E-mail: mim@mim.dk
frontlinien@frontlinien.dk
Web site: www.mim.dk

E

Ecuador

Petroecuador
Calle Alpallana E8-86 y
Av. 6 de Diciembre
Edificio Alpallana
P.O. Box 17-11-5007
Quito, ECUADOR
Telephone: 593 2 222 9043; 593 2 250 3770
Fax: 593 2 256 9738
Web site: www.petroecuador.com.ec

Ministerio de Minas y Petroleos
Av. Orellana N26-220 y Juan León Mera (esquina) Edificio MOP
PBX: 593 2 2550 018; 593 2 2550 041
Quito, ECUADOR
Web site: www.menergia.gov.ec

OLADE (Organizacion LatinoAmericana de Energia)
Av. Mariscal Antonio José de Sucre N 58-63 y
Fernández Salvador Edificio
Sector San Carlos
P.O. Box 17-11-6413 CCNU
Quito, ECUADOR
Telephone: 593 2 259 8122; 593 2 259 8280
Fax: 593 2 253 1691
E-mail: olade@olade.org.ec
Web site: www.olade.org.ec

Egypt

Arab Republic of Egypt
Ministry of Petroleum
1A, Ahmed al Zomor Street
8th District, Nasr City
Cairo, EGYPT
Telephone: 00 202 670 64 01; 00 202 670 64 02
Fax: 00 202 670 64 19
E-mail: contactus@petroleum.gov.eg
Web site: www.petroleum.gov.eg

EGPC (Egyptian General Petroleum Corp.)
Corporation Building
Palestine Street, Sector 4, New Maadi
P.O. Box 2130
Cairo 11742, EGYPT
Telephone: 20 2 2703 1438; 20 2 2703 1439
Fax: 20 2 2703 1457
E-mail: info@egpc.com.eg
Web site: www.egpc.com.eg

EGAS (Egyptian Natural Gas Holding Co.)
Company Building
Al Nasr Road, Abbas Al Akkad Street
Nasr City
P.O. Box 8064
Cairo 11371, EGYPT
Telephone: 20 2 2405 5865; 20 2 2405 5846
Fax: 20 2 2405 5876
Email: egas@egas.com.eg
Website: www.egas.com.eg

Equatorial Guinea

Ministry of Mines, Industry and Energy
David Shaw, Technical Advisor
Carretera de Punta Europa
Malabo, EQUATORIAL GUINEA
Telephone: 240 09 3567; 27 7502
Fax: 240 09 3353
E-mail: shawd@rpsgroup.com
Web site: www.equatorialoil.com

F

Faroe Islands

Faroese Earth and Energy Directorate (Jardfeingi)
Brekkutún 1
Postmoga 3059
FO-110 Tórshavn, FAROE ISLANDS
Telephone: 298 357000
Fax: 298 357001
Email: jardfeingi@jardfeingi.fo
Web site: www.oms.fo

Finland

Ministry of Employment and the Economy
P.O. Box 32
FI-00023 Government
Aleksanterinkatu 4, FI-00170
Helsinki, FINLAND
Telephone: 358 10 606 000
E-mail: kirjaamo@tem.fi
Web site: www.tem.fi

France

Ministry of the Economy, Industry and Employment
139, rue de Bercy 75012
Paris, FRANCE
Web site: www.minefi.gouv.fr

G

Gabon

Ministry of Mines, Energy, Oil and Hydraulic Resources
B.p 576/874
Libreville, GABON
Telephone: 241 72 22 39/ 40/ 45
Fax: 241 72 49 90
E-mail: mcn@internetgabon.com

Petrogab (Societe Nationale Petroliere Gabonaise)
Box 564
Libreville, GABON

Greece

Hellenic Petroleum S.A.
17th km., Athens-Corinth National Hwy.
GR-193 00 Aspropyrgos
Attikí, GREECE
Telephone: 30 210 55 33 000
Fax: 30 210 55 39 298
Web site: www.hellenic-petroleum.gr

Ministry of Development
Energy and Natural Rescurces
119 Messoghion Ave.
GR-101 92
Athens 115 28, GREECE
Telephone: 30 210 696 9823
E-mail: grammatia@ypan.gr
Web site: www.ypan.gr/fysikoi_poroi/diarthrosi_eng.htm

Guatemala

Ministerio de Energia y Minas
Diagonal 17 29-78, Zona 11
Las Charcas, GUATEMALA
Telephone: 502 2477 0382; 502 2477 0901
Fax: 502 2476 8506
E-mails: informatica@mem.gob.gt
relapubli@mem.gob.gt
Web site: www.mem.gob.gt

H

Hungary

MOL plc (Hungarian Oil & Gas)
Október huszonharmadika u. 18
H-1117 Budapest, HUNGARY
Telephone: 36 20 9665 665
Fax: 36 1 46 40 794
Web site: www.mol.hu

I

Iceland

Orkustofnun (National Energy Authority)
Orkugardur, Grensasvegur 9
Reykjavík 108, ICELAND
Telephone: 354 569 6000
Fax: 354 568 8896
E-mail: os@os.is
Web site: www.os.is/page/english/

India

Ministry of Petroleum & Natural Gas
Rajendra Prasad Road
New Delhi 110 001, INDIA
Telephone: 91 233 83692
Fax: 91 233 83535
E-mail: dspdi.png@sb.nic.in
Web site: www.petroleum.nic.in

ONGC (Oil & Natural Gas Corp. Ltd.)
Tel Bhavan
Dehradun 248 003, INDIA
Telephone: 0135 275 9561 67
Web site: www.ongcindia.com

Indonesia

Pertamina
(Perusahaan Pertambangan Minyak Dan Gas Bumi Negara)
Jalan Merdeka Timur No. 1 A
Jakarta-10110, INDONESIA
Telephone: 62 21 7917 3000
Fax. 62 21 7972 177
Web site: www.pertamina.com

Iran

NIOC (National Iranian Oil Co.)
Hafez Crossing, Taleghani Avenue
P.O. Box 1863
Tehran, IRAN
Telephone: 98 21 615 2929
Fax: 98 21 615 3886
E-mail: managing.director@nioc.com
Web site: www.nioc.com

Iraq

Ministry of Oil
Oil Complex Building
Port Saeed St.
Baghdad, IRAQ
Telephone: 946 1 8177 000; international 946 1 2863 103
Fax: 946 1 8869 432; international 946 1 2869 432
E-mails: oil@uruklink.net
oilcompo@uruklink.net
Web site: www.uruklink.net/oil

SOMO (State Oil Marketing Organization)
Al Muthanah District
P.O. Box 5118
Baghdad, IRAQ
Telephone: 946 1 7742 040; 964 1 8869 561
Fax: 964 1 7742 040 ext. 297; 964 1 8853 925
Cable: Inocmart
E-Mail: somo@uruklink.net

Ireland

Department of Communications,
Marine and Natural Resources
29-31 Adelaide Rd.
Dublin 2, IRELAND
Telephone: 353 1 678 2000
Fax: 353 1 678 2449
Web site: www.dcmnr.gov.ie

Israel

Israel Government Ministries
Web site: www.gov.il/firstGov/topNavEng/Engoffices/EngMinistries

Italy

Eni SpA (Ente Nazionale Idrocarburi SpA)
Piazzale Mattei, 1
00144 Rome, ITALY
Telephone: 39 06 59 82 1
Fax: 39 06 59 82 21 41
Web site: www.eni.it

Ministero dello Sviluppo Economico
Direzione Generale dell'Energia e delle Risorse Minerarie
Via Molise 2
00187 Roma, ITALY
Telephone: 39 06 4705 2531 / 2639
Fax: 39 06 4705 2847
E-mail: martino@mica-dgfe.casaccia.enea.it
Web site: http://sviluppoeconomico.gov.it

Ivory Coast

Director of the Office of the Minister
Ministere de Mines
Box V 50
Abidjan, COTE D'IVOIRE
Telephone: 225 21 50 03
Fax: 225 21 53 20

Petroci (Societe Nationale d'Operations Petrolieres de la Cote d'Ivoire)
Box V 194
Abidjan, COTE D'IVOIRE
Telephone: 225 20 20 25 00
Fax: 225 21 68 24

J

Japan

Japan Ministry of Economy, Trade and Industry
1-3-1 Kasumigaseki, Chiyoda-ku
Tokyo 100-8901, JAPAN
Telephone: 81 3 3501 1511
Fax: 81 3 3501 6942
E-mail: webmail@meti.go.jp
Web site: www.meti.go.jp/english/index.html

JOGMEC (Japan Oil, Gas and Metals National Corp.)
1310 Omiya-cho, Saiwai-ku, Kawasaki-shi
Kanagawa 212-8554 JAPAN
Telephone: 81 44 520 8600
Fax: 81 44 520 8710
Web site: www.jogmec.go.jp/english/index.html

K

Kazakhstan

JSC KazMunaiGaz Exploration Production
2 Tauelsizdik Street
Astana 010000, KAZAKHSTAN
Telephone: 7 7172 977 427
Fax: 7 7172 977 426
E-mail: info@kmg.kz
Web site: www.kmg.kz/eng/

Ministry of Energy and Mineral Resources
Astana, KAZAKHSTAN
Telephone: 7 7172 317 172
Fax: 7 7172 102 474
E-mail: bertisbayev@minenergo.kegoc.kz

Kuwait

KOC (Kuwait Oil Co.)
P.O. Box 9758 Ahmadi
61008 Ahmadi, KUWAIT
Telephone: 00965 398 9111
Fax: 00965 398 3661
E-mail: kocinfo@kockw.com
Web site: www.kockw.com

KPC (Kuwait Petroleum Co.)
P.O. Box 26565
Postal Code 13126
Safat, KUWAIT
Telephone: 965 1 85 85 85
Fax: 965 2499 4991
E-mail: info@kpc.com.kw
Web site: www.kpc.com.kw

Ministry of Oil
P.O.Box 5077
13051, KUWAIT
E-mail: alnaft@moo.gov.kw
Web site: www.moo.gov.kw

L

Libya

General Department of Technical Affairs
Ministry of Petroleum
PO Box 256
Tripoli, LIBYA

NOC (National Oil Corp)
Bashir Saadawi St.
Tripoli, LIBYA
Phone: 218 333 7141
Fax: 218 333 1930
E-mail: info@www.noclibya.com
Web site: www.noclibya.com

RASCO (Ras Lanuf Oil and Gas Processing Company)
PO Box 1971
Ras Lanuf Terminal
Benghazi, GSPLAJ LIBYA
Telephone: 218 21 360 5177
Fax: 218 21 360 5174
E-mail: info@raslanuf.com
Web site: http://www.raslanuf.com

M

Malaysia

Petronas (Petroliam Nasional Berhad)
Tower 1, Petronas Twin Towers
Kuala Lumpur City Centre
Kuala Lumpur 50088 MALAYSIA
Telephone: 603 2051 5000
Fax: 603 2026 5050
E-mail: webmaster@petronas.com.my
Web site: www.petronas.com.my

Mexico

Pemex (Petroleos Méxicanos)
Marina Nacional 329, Col. Huasteca
C.P. 11311
México D.F., MEXICO
Telephone: 52 55 1944 2500
Web site: www.pemex.com

Mongolia

Mongol Petroleum Co.
Uildverchnii Street 37
Ulaanbaatar, MONGOLIA
Telephone: 976 1 38 34 83
Fax: 976 1 33 11 76

Mineral Resources and Petroleum Authority of Mongolia
Builders' Square 3, State Property
Building #5, West Leaf, 1st Building of MPRAM
Ulaanbaatar 211239 Post Office 38 MONGOLIA
Telephone: 976 11 327180, 976 11 263701
Fax: 976 11 327180, 976 11 310370
E-mail: mram@mram.mn
Web site: www.mrpam.gov.mn

Myanmar

Myanmar Ministry of Energy
Director General, Energy Planning Department
No. 23, Pyay Road, Lanmadaw Township
Yangon, MYANMAR
Telephone: 95 1 222 545; 95 1 220 010
Fax: 95 1 222 960
E-mail: myanmoe@mptmail.net.mm
Web site: www.energy.gov.mm

N

Netherlands

SenterNovem
Juliana van Stolberglaan 3
Postbus 93144
2509 AC Den Haag, NETHERLANDS
Telephone: 31 70 3735000
Fax: 31 70 3735100
E-mail: info@senternovem.nl
Web site: http://www.senternovem.nl/english/

New Zealand

Ministry of Economic Development
New Zealand Crown Minerals
PO Box 1473
33 Bowen St. – Level 5
Wellington, NEW ZEALAND
Telephone: 64 3 962 6179
Fax: 64 4 499 0968
E-mail: crown.minerals@med.govt.nz
Web site: www.crownminerals.govt.nz

Nicaragua

INE (Nicaraguan Institute of Energy)
Petronic Building, 2nd Floor
Post Box 3226
Managua, NICARAGUA
Telephone: 505 222 7527
Fax: 505 228 2049
Web site: www.ine.gob.ni

Nigeria

Ministry of Energy
Block D, NNPC Towers
Herbert Macaulay Way
Central Business District
PMB 449
Abuja, NIGERIA
Telephone: 234 09 780 5085; 234 09 673 7815
Fax: 234 09 523 6652
Web site: www.nigeria.gov.ng/

NNPC (Nigerian National Petroleum Corp.)
NNPC Towers
Herbert Macaulay Way
Central Business District
PMB 190
Garki, Abuja, NIGERIA
Telephone: 09 2008 1133
Fax: 09 234 0029
E-mail: www.nnpcgroup.com

Norway

Ministry of Petroleum and Energy
Einar Gerhardsens plass 1 (R4)
P.O. Box 8148 Dep
Oslo 0033, NORWAY
Telephone: 47 22 24 90 90
Fax: 47 22 24 95 65
E-mail: postmottak@oed.dep.no
Web site: www.regjeringen.no/en/dep/oed/The-Ministey.html?id=755

StatoilHydro AS
Forusbeen 50
N-4035 Stavanger, NORWAY
Telephone: 47 51 99 00 00
Fax: 47 51 99 00 50
E-mail: statoilhydro@statoilhydro.com
Web site: www.statoilhydro.com

O

Oman

Ministry of Oil & Gas
Director General of Petroleum and Gas Affairs
Sultanate of Oman
P.O. Box 600
P.C. 113 Muscat, OMAN
Telephone: 968 2460 3222
Fax: 968 2469 3770
E-mail: webmaster@omanet.om
Web site:
http:www.omanet.om/english/government/ministers.asp?cat=gov

PDO (Petroleum Development Oman)
Muscat 113 OMAN
Telephone: 968 678 111
Fax: 968 677 106
E-mail: external-affairs@pdo.co.om
Web site: www.pdo.co.om

P

Pakistan

Ministry of Petroleum and Natural Resources
3rd Floor, A Block, Room 301, Pak Secretariat
Islamabad, PAKISTAN
Telephone: 92 51 921 0220; 92 51 920 6416
Fax: 92 51 921 3180
E-mail: info@mpnr.gov.pk
Web site: www.mpnr.gov.pk

Papua New Guinea

Department of Petroleum and Energy
P O Box 1993
Port Moresby, PAPUA NEW GUINEA
Telephone: 675 322 4200
Fax: 675 322 4222
E-mail: Bernard_Pawih@petroleum.gov.pg
Web site: www.petroleum.gov.pg

Peru

Ministerio de Energia & Minas
Av. Las Artes Sur 260 San Borja
Lima, PERU
Telephone: 511 618 8700
Web site: www.minem.gob.pe

Petroperu (Petroleos del Peru SA)
Av. Paseo de la Republica 3361
Lima 27, PERU
Telephone: 511 211 7800; 641 5000
Web site: www.petroperu.com.pe

Philippines

Department of Energy
Energy Center, Merritt Road
Fort Bonifacio
Taguig, NCR 1634
Metro Manila, PHILIPPINES
Telephone: 63 2 840 1401 to 21
Fax: 840 61 94 (Office of the Secretary)
E-mail: info@doe.gov.ph
Web site: www.doe.gov.ph

PNOC (Philippine National Oil Co.)
PNOC Bldg. 6, Energy Center
Merritt Road
Fort Bonifacio
Makati NCR
Metro Manila, PHILIPPINES
Telephone: 840 63 2 555 0321
Web site: www.pnoc.com.ph

Poland

Ministry of the Environment
Ul. Wawelska 52/54
Warsaw 00-922, POLAND
Telephone: 48 22 57 92 900
Web site: www.mos.gov.pl

PGNiG (Polskie Górnictwo Naftowe i Gazownictwo SA)
Ul. Kasprzaka 25
Warsaw 01-224, POLAND
Ph.: 48 22 589 45 55; 48 22 691 79 00
Fax: 48 22 691 82 73
E-mail: pr@pgnig.pl
Web site: www.pgnig.pl

Portugal

Galp Energia SA
Edifício Galp Energia
Rua Tomás da Fonseca
Lisboa 1600-209, PORTUGAL
Telephone: 351 21 724 25 00
Fax: 351 21 724 29 65
E-mail: comunicacao@galpenergia.com
Web site: www.galpenergia.com

Q

Qatar

Ministry of Foreign Affairs
Web site: http://english.mofa.gov.qa

Qatar Petroleum
P.O. Box 3212
Doha, QATAR
Telephone: 974 440 2000
Fax: 974 483 1125
Web site: www.qp.com.qa

R

Romania

Ministry of Economy and Finance
1 Victoriei Square, District 1
Bucharest 010366, ROMANIA
Telephone: 4021 314 34 00
Fax: 4021 230 36 60
Web site: www.mefromania.ro

National Agency for Mineral Resources
36-38, Mendeleev St., Sector 1
Bucharest 010366, ROMANIA
Phone: 4021 313 22 04
Fax: 4021 210 74 40
Web site: www.namr.ro

Petrom SA
239 Calea Dorobantilor, District 1
Bucharest 010567 ROMANIA
Telephone: 40 21 4060040
Fax: 40 21 4060420
Web site: www.petrom.ro
E-mail: crem.services@petrom.com
E-mail: press.office@petrom.com

Russia

OAO Gazprom
16 Nametkina St., 117997
Moscow, V-420, GSP-7, RUSSIA
Telephone: 7 495 719 30 01
Fax: 7 495 719 83 33
E-mail: gazprom@gazprom.ru
Web site: www.gazprom.ru

OAO Rosneft Oil Co.
26/1 Sokolniki
Moscow 115035, RUSSIA
Telephone: 7 495 777 44 22
Fax: 7 495 777 44 44
E-mail: postman@rosneft.ru
Web site: www.rosneft.ru

S

Saudi Arabia

Saudi Aramco (Saudi Arabian Oil Co.)
P.O. Box 5000
Dhahran 31311, SAUDI ARABIA
Telephone: 966 3 872 0115
Fax: 966 3 873 8190
E-mail: webmaster@aramco.com
Web site: www.saudiaramco.com

South Africa

Department of Minerals and Energy
Mineralia Centre
234 Visagie Streets
Pretoria 001, SOUTH AFRICA
Telephone: 27 12 317 8000
Fax: 27 12 322 3416
Web site: www.dme.gov.za

Petroleum Agency SA
PO Box 1174
151 Frans Conradie Drive
Parow, Capetown 7499, SOUTH AFRICA
Telephone: 27 21 938 3500
Fax: 27 21 938 3520
E-mail: info@petroleumagencysa.com
Web site: www.petroleumagencysa.com

South Korea

KOGAS (Korea Gas Corp.)
215 Jeongja-dong
Bundang-gu, Seongnam
Gyeonggi-do, 463-754, KOREA
Telephone: 82 31 710 0070 2
Fax: 82 31 710 0079
E-mail: kogasmaster@kogas.or.kr
Web site: www.kogas.or.kr

KNOC (Korea National Oil Corp.)
1588-14, Gwanyang-dong, Dongan-gu,
Anyang-si, Gyeonggi-do KOREA
Telephone: 82 31 380 2114
Fax : 82 31 387 9321
Web site: www.knoc.co.kr

Spain

Comisión Nacional de Energía
Dirección de Relaciones Externas y Documentación
Calle Alcalá, 47
28014 Madrid, SPAIN
Telephone: 91 432 96 00
Fax: 91 577 62 18
E-mail: dre@cne.es
Web site: www.cne.es

Enagás SA
Paseo de los Olmos, 19
Madrid 28005, SPAIN
Telephone: 34 902 443 700
Fax: 34 91 709 93 44
E-mail: dircom@enagas.es
Web site: www.enagas.es

Repsol YPF SA
Paseo de la Castellana, 278-280
Madrid 28046, SPAIN
Telephone: 34 91 348 81 00; 34 91 348 80 00
Fax: 34 91 348 28 21; 34 91 348 94 94
E-mail: info@repsol.com
Web site: www.repsol.com

Sweden

Ministry of Enterprise, Energy and Communications
Jakobsgatan 26
103 33 Stockholm, SWEDEN
Telephone: 46 8 405 10 00
Web site: www.sweden.gov.se/sb/d/2067

Syria

Ministry of Petroleum and Mineral Resources
Al-Adawi Khateeb Street
P.O.Box 31483
Damascus, SYRIA
Telephone: 963 11 445 9355; 963 11 445 5972
Fax: 963 11 446 3942
E-mail: mopmr@net.sy

Syrian Petroleum Co.
Al-Mutanabbi St.
P.O.Box: 2849
Damascus, SYRIA
Telephone: 963 11 222 7007; 963 11 222 4272
Fax: 963 11 222 5648; 963 11 221 0953
E-mail: spccom1@scs-net.org; spccom2@scs-net.org
Web site: www.spc-sy.com

T

Taiwan

Chinese Petroleum Corp.
No. 3 Songren Rd., Sinyi District
Taipei City 110, TAIWAN
Telephone: 02 8789 8989
Fax: 02 8789 9000
E-mail: ir@cpc.com.tw
Web site: www.cpc.com.tw

Tanzania

Ministry of Energy and Minerals
Permanent Secretary
Sokoine/Mkwepu St.
P.O. Box 2000/9152
Dar es Salaam, TANZANIA
Telephone: 255 22 211 7153 9
Fax: 255 22 211 6719
E-mail: madini@africaonline.co.tz
Web site: www.tanzania.go.tz/government

Thailand

Ministry of Energy
17 Rama I Road
Pathumwan
Bangkok 10330, THAILAND
Telephone: 66 0 2223 3344
Fax: 66 0 2245 9855
Web site: www.energy.go.th/en/index.asp

PTT plc (formerly Petroleum Authority of Thailand)
555 Vibhavadi Rangsit Rd., Chatuchak
Bangkok 10900, THAILAND
Telephone: 66 0 2537 2000
Fax. 66 0 2537 3498 9
E-mail: corporate@pttplc.com
Web site: www.pttplc.com

Trinidad & Tobago

Ministry of Energy and Energy Industries
Level 9, Riverside Plaza
3 Besson Street
Port of Spain, TRINIDAD
Telephone: 868 623 6708; 868 623 6719
Fax: 868 625 0306
Web site: www.energy.gov.tt

National Gas Co. of Trinidad and Tobago Ltd.
Orinoco Drive
Point Lisas Industrial Estate
P.O Box 1127
Port of Spain, TRINIDAD
Telephone: 868 636 4662; 868 636 4680
Fax: 868 679 2384
E-mail: ngc@ngc.co.tt
Web site: www.ngc.co.tt

Petrotrin (Petroleum Co. of Trinidad & Tobago)
Pointe-a-Pierre, TRINIDAD
Telephone: 868 658 4200; 868 658 4210
Fax: 868 624 6461
E-mail: petrochair@petrotrin.com
Web site: www.petrotrin.com

Tunisia

Etap (Entreprise Tunisienne d'Activites Petrolieres)
27 bis, Ave. Khereddine Pacha
BP 83-1073 Montplaisir
1002 Tunis, TUNISIA
Telephone: 216 71 782 288
Fax: 216 71 780 123; 216 71 786 141
E-mail: dexprom@etap.com.tn
Web site: www.etap.com.tn

Turkey

BOTAS Petroleum Pipeline Corp.
Bilkent Plaza A, II Blok, Bilkent
06800 Ankara, TURKEY
Telephone: 90 312 297 20 00
Fax: 90 312 266 07 33; 90 312 266 07 34
Web site: www.botas.gov.tr

TPAO Turkish Petroleum Corp.
Sogutozu Mahallesi 2
Cadde No. 86
06100 Cankaya/Ankara, TURKEY
Telephone: 90 312 207 2000
Fax: 90 312 286 9000
E-mail: tpaocc@petrol.tpao.gov.tr
Web site: www.tpao.gov.tr

Turkmenistan

Ministry of Energy and Industry
6 Nurberdy Pomma St.
Ashgabat 744000, TURKMENISTAN
Telephone: 9 9312 35 49 21; 9 9312 51 08 82
Fax: 9 9312 29 06 82

U

Ukraine

Ukrtransnafta
60 Artema St.
Kyiv 04050, UKRAINE
Telephone: 044 246 80 07
Fax: 044 246 83 77

State Committee on Geology and Utilization of Mineral Resources
Volodymyrska 34
252601, Kyiv 34, UKRAINE
Telephone: 044 228 60 51; 044 228 32 43
Fax: 044 228 60 51
E-mail: admin@brama.com
Web site: www.brama.com/ua-geology

United Arab Emirates

ADNOC (Abu Dhabi National Oil Co.)
P.O. Box 898
Abu Dhabi, UNITED ARAB EMIRATES
Telephone: 971 2 602 0000
Fax: 971 2 602 3389
Web site: www.adnoc.ae

Ministry of Energy
Petroleum & Minerals Sector
P.O. Box 59
Abu Dhabi, UNITED ARAB EMIRATES
Telephone: 971 2 626 2288; 971 2 612 6500
Fax: 971 2 627 2291; 971 2 626 0037
E-mail: mopmr@uae.gov.ae
Web sites: www.moenr.gov.ae
www.government.ae/gov/en/gov/federal/index.jsp

United Kingdom

DBERR (Department for Business Enterprise & Regulatory Reform)
Energy Group: Oil and Gas Directorate
1 Victoria Street
London SW1H 0ET, UNITED KINGDOM
Web site: www.og.berr.gov.uk/
Note: The new Department of Energy and Climate Change was established in October 2008.
Telephone: 020 7215 3505.

Uzbekistan

Republic of Uzbekistan
Telephone: 998 71 138 4148; 998 71 138 4249
E-mail: info@uzinfocom.uz
Web site: www.gov.uz/

Uzneftegaz
Amir Temir Ulitsa 66
700084 Tashkent, UZBEKISTAN
Telephone: 371 2 35 83 23
Fax: 371 2 35 83 37

V

Venezuela

Ministerio del Poder Popular
para la Energia y Petróleo
Edificio Petroleos de Venezuela, Torre Oeste
Av. Libertador con Av. Empalme Urb.
La Campiña, Parroquia El Recreo
Distrito Capital
Caracas, VENEZUELA
Telephone: 58 212 708 7001; 58 212 708 7045
E-mail: consultasopap@menpet.gob.ve
Web site: www.menpet.gob.ve

PDVSA (Petroleos de Venezuela)
Av. Libertador La Campina, Torre Este
Edificio Petróleos de Venezuela
Apartado 169
Caracas, 1050-A, VENEZUELA
Telephone: 58 212 708 4111
Fax: 58 212 708 4660
E-mail: webmaster@pdvsa.com
Web site: www.pdvsa.com.ve/

Vietnam

Ministry of Petroleum & Gas
193/6A Nam Ky Khoi Nghia
Telephone: 080 48137; 080 48354
Fax: 080 48299
Ho Chi Minh City 3, VIET NAM

PetroVietnam (Vietnam Oil and Gas Corp.)
18 Lang Ha
Thanh Cong, Ba Dinh
Hanoi, VIET NAM
Telephone: 84 4 772 5761
Fax: 84 4 772 5903
E-mail: webmaster@hn.pv.com.vn
Web site: www.petrovietnam.com.vn

Y

Yemen

Ministry of Oil and Minerals
Al-Zubairy Street
PO Box 81
Sana'a, YEMEN
Telephone: 967 1 202 309
Fax: 967 1 202 314
Web site: www.mom.gov.ye/eng/contacte.htm

YPC (Yemen Petroleum Company)
YPC Building
P.O. Box 187
Sana'a, YEMEN
Telephone: 967 1 704 322; 967 1 704 320
Fax: 967 1 209 598
Web site: www.ypcye.com/english

GLOSSARY

Abbreviations and Acronyms

A

ACG	Azeri-Chirag-Guneshli (Azerbaijan oil fields)
Adma Opco	Abu Dhabi Marine Operations Co.
Adnoc	Abu Dhabi National Oil Co.
AED	Alberta Economic Development
AEP	American Electric Power Co.
A.G.A.	American Gas Association
Agip KCO	Agip Kazakhstan North Caspian Operating Co., formerly OKIOC
AGPA	Alaska Gasline Port Authority
AIOC	Azerbaijan International Operating Co.
altfuels	alternative fuels including wind power and solar energy
AMBO	Albanian Macedonian Bulgarian Oil Co.
AMPCO	Atlantic Methanol Production Co. LLC
ANP	Agencia Nacional de Petroleo, or National Petroleum Agency (Brazil)
ANWR	Arctic National Wildlife Refuge (Alaska, US)
AOSP	Alberta Oil Sands Project; Alberta Oil Sands Pipeline
APG	Aboriginal Pipeline Group (Inuvik, NT)
API	American Petroleum Institute
Aramco	(Saudi) Arabian Oil Co.
ASEAN	Association of Southeast Asian Nations
ASPO	Association for the Study of Peak Oil
ATHEER	Abu Dhabi Gas Co. (United Arab Emirates)

B

BALAK	national oil and gas agency's Indonesian-language acronym
Banagas	Bahrain National Gas Co.
Banco	Bahrain National Oil Co.
Bapco	Bahrain Petroleum Co.
bbl	barrels
b/cd	barrels per calendar day
bcf	billion cubic feet
bcfd	billion cubic feet per day
bcfe	billion cubic feet equivalent
bcm	billion cubic meters
bcmd	billion cubic meters per day
bcme	billion cubic meters equivalent
b/d	barrels per day
bbl-d	barrels daily
BEDB	Brunei Economic Development Board
BGL	Bhagyanagar Gas Ltd. (India)
b/h	barrels per hour
BLM	Bureau of Land Management (US)
BNA	Bundesnetzagentur (Germany)
BNDES	National Economic & Social Development Bank (Brazil)
bo/d	barrels of oil per day
boe	barrels of oil equivalent
boe/d	barrels of oil equivalent per day
BOOT	build, own, operate, and transfer
BOOTS	Bulk Oil Offshore Transfer System (US)
BOTAS	Turkey government natural gas and pipeline company
bscf	billion standard cubic feet
bscfd	billion standard cubic feet per day
bscm	billion standard cubic meters
b/sd	barrels per stream day
BTC	Baku-Tbilisi-Ceyhan oil pipeline (Azerbaijan-Georgia-Turkey)
Btu	British thermal unit
bw/d	barrels of water per day

C

CAA	commercial arrangement area (Malaysia/Viet Nam)
CAC	Central Asia-Center (Former Soviet Union pipeline system)
CAGC	Canadian Association of Geophysical Contractors
CAPP	Canadian Association of Petroleum Producers
Caricom	Caribbean Community
CBM	coalbed methane
CBNG	coalbed natural gas
CBOT	Chicago Board of Trade
CD	compact disk
CEO	chief executive officer
CEPA	Canadian Energy Pipeline Association
Cepsa	Compania Espanola de Petroleos SA (Spain)
CERA	Cambridge Energy Research Associates (CERA)
CERI	Canadian Energy Research Institute
cf	cubic feet
cfd	cubic feet per day
CFE	Comision Federal de Electricidad (Federal Electricity Commission, Mexico)
cfed	cubic feet of equivalent per day
CFTC	Commodity Futures Trading Commission (US)
CGES	Centre for Global Energy Studies (London)
CHDP	cricondentherm hydrocarbon dew point
Chinaoil	China National United Oil Co.
CICESE	Centro de Investigación Científica y de Educación Superior de Enseñada (Mexico)

CIS	Commonwealth of Independent States (Former Soviet Union)
CITIC	China International Trust & Investment Corporation
CLJOC	Cuu Long Joint Operating Co. (Viet Nam)
C-NLOPB	Canada-Newfoundland Labrador Offshore Petroleum Board
CNG	compressed natural gas
CNOOC	China National Offshore Oil Corporation
CNPC	China National Petroleum Corporation
CNR	Canadian Natural Resources Ltd.
CO	carbon monoxide
CO_2	carbon dioxide
COS	China Oilfield Services Ltd.
cP	centi-Poise (viscosity measurement)
CPC	Caspian Pipeline Consortium
CPC	Chinese Petroleum Corp. (Taiwan)
CPOT	Constanta-Pancevo-Omisalj-Trieste pipeline (Romania-Croatia-Italy)
CPU	central processing unit
CPUC	California Public Utilities Commission
CRE	Energy Regulatory Commission (Mexico)
CRP	Paraguana Refining Complex (Venezuela)
CSO	Coflexip Stena Offshore
CTL	coal to liquids
cu m	cubic meters (metres)
CUCBM	China United Coalbed Methane Corp. Ltd.
CV	core venture (Saudi Arabia gas initiative)

D

DDCV	deep-draft caisson vessel
DEPA	Public Gas Co. (Greece)
DIGP	Dauphin Island Gathering Partners
dilbit	diluted bitumen (heavy crude oil and bitumen blended with condensate diluent)
DOE	Department of Energy
DONG	Dansk Olie ag Naturgas AS (Denmark)
DUC	Dansk Undergrunds Consortium (Denmark)
dwt	deadweight tons

E

E&D	exploration and development
E&P	exploration and production
EBRVs	Energy Bridge regasification vessels
EC	European Commission
ECO	Economic Cooperation Organization (Central Asia)
EdF	Electricité de France
EDP	Electricidade de Portugal SA
EGAS	Egyptian Natural Gas Holding Co.
EGPC	Egyptian General Petroleum Corp.
EIA	Energy Information Administration (US)
EIA	environmental impact assessment
ELNG	Egyptian LNG joint venture
ENAP	Empresa Nacional de Petroleo (Chile)
Enarsa	Energía Argentina SA
Enel	Ente Nazionale Energia Elettrica SpA (Italy)
Eni	Ente Nazionale Idrocarburi SpA (Italy)
EPA	Environmental Protection Agency
EPC	engineering, procurement, and construction
EPCA	Energy Policy & Conservation Act (US)
EPDC	European Pipeline Development Co. BV
EPSA	exploration and production-sharing agreement
ERC	Energy Regulatory Commission (France)
ESP	electrical submersible pump
ETAP	Enterprises Tuniseienne d'Activites Petrolieres
ETLP	extended tension leg platform
EU	European Union
EUB	Energy and Utilities Board (Alberta, Canada)
EVE	Basque Energy Authority (Spain)
EWT	extended well test
Ex-Im Bank	Export-Import Bank of the US

F

FAWAG	foam-assisted water alternating gas
FCC	fluid catalytic cracking
FCCU	fluid catalytic cracking unit
FDI	foreign direct investment
FEED	front-end engineering design
FERC	Federal Energy Regulatory Commission
FNA	Frigg Norwegian Association (Norway)
fob	free on board
FPC	Formosa Petrochemical Corp.
fph	feet per hour
FPSO	floating production, storage, and off-loading
FPU	floating production units
FSRU	floating storage and regasification unit
FSU	Former Soviet Union
ft	foot
FTC	Federal Trade Commission
FTIR	Fourier transform infrared

G

GAIL	Gas Authority of India Ltd.
GASCO	Abu Dhabi Gas Industries Ltd. (United Arab Emirates)
GBS	gravity-based structure
GC	Green Canyon (US Gulf of Mexico)

GCC	Gulf Cooperation Council
GCDS	Gasoducto Cruz del Sur (Argentina-Uruguay)
GdF	Gaz de France SA
GDP	gross domestic product
GEA	Gas Energy Adria
GFU	Gas Negotiating Committee (Norway)
GNA	Gasoducto del Noreste Argentino (Bolivia-Argentina)
GNPOC	Greater Nile Petroleum Operating Co. (Sudan)
GOM	Gulf of Mexico
GOR	gas-oil ratio
GOSP	gas-oil separation plant
gpd	gallons per day
GSCP	Georgia Strait Crossing Project (Canada-US)
GTI	Gas Technology Institute
GTL	gas-to-liquids
GTLB	GTL Bolivia SA
GWh	gigawatt-hours

H

H&P	Helmerich & Payne Inc.
HCU	Hydrocarbon Unit of the Bangladesh Energy and Mineral Resources Division
HDPE	high-density polyethylene
HDS	hydroskimming
HOA	heads-of-agreement
hp	horsepower
HP	Hellenic Petroleum (Greece)
HPCL	Hindustan Petroleum Corp Ltd. (India)
HPHT	high-pressure, high-temperature
HSE	health, safety, and environment

I

IADB	Inter-American Development Bank
IADC	International Association of Drilling Contractors
IAEA	International Atomic Energy Agency (Vienna)
ICO	Interconexion Centro-Occidente pipeline (Venezuela)
ID	inside diameter
IDB	Inter-American Development Bank
IEA	International Energy Agency
IEC	Israel Electric Co.
IET	integrated exploration technology
IFP	Institut Français du Pétrole (France)
IGAT	Iran Gas Trunkline
IGCC	integrated gasification combined cycle
ILNGA	International LNG Alliance (US)
IMF	International Monetary Fund
IMO	International Maritime Organization
INA	Industrija nafte dd (Croatia)
INAC	Indian and Northern Affairs Canada
INGAA	Interstate Natural Gas Association of America
INOC	Iraq National Oil Co.
IOC	Indian Oil Corp.
IOC	international oil company
IPAA	Independent Petroleum Association of America
IPO	initial public offering
IRNA	Iranian news agency
ISAC	Information Sharing & Analysis Center (US)
IUA	International Unitization Agreement

J

Japex	Japan Petroleum Exploration Corp.
JDA	Joint Development Area (Malaysia-Thailand, Australia-East Timor)
JIPA	joint investment production activity
JNOC	Japan National Oil Company
JPSC	Joint Pipeline Security Commission (Azerbaijan, Georgia, and Turkey)
JV	joint venture

K

KGOC	Kuwait Gulf Oil Co.
KIO	Karachaganak Integrated Organization (consortium)
KKPL	Kenai Kachemak pipeline (Alaska)
km	kilometer
KNOC	Korean National Oil Corporation
Kogas	Korea Gas Corporation
KPC	Kuwait Petroleum Corp.
KVA	kilo-volt-amperes
KZM	Ku-Zaap-Maloob complex (Mexico)

L

LAB	linear alkylbenzene
LION	large increase in octane number
LLDPE	linear low-density polyethylene
LNG	liquefied natural gas
LOGGS	Lincolnshire offshore gas gathering system (UK)
Logic	leading oil and gas industry competitiveness (UK)
LPG	liquefied petroleum gas
LWD	logging-while-drilling

M

m	meter
M&A	merger and acquisition
M&NP	Maritimes and Northeast Pipeline (Canada-US)
Marad	Maritime Administration (US)
MC	Mississippi Canyon (US Gulf of Mexico)
Mcf	thousand cubic feet
MD	measured depth
MEG	monoethylene glycol
MFL	magnetic flux leakage
MHP	Market Hub Partners (US)
MIBEL	Mercado Ibérico de Electricidade (Portugal, Spain)
MIDOR	Middle East Oil Refinery Ltd.
Migas	Directorate General of Oil and Gas (Indonesia)
MMBtu	million British thermal units
MMcf	million cubic feet
MMcfd	million cubic feet per day
MMcfe	million cubic feet of equivalent
MMcfed	million cubic feet of equivalent per day
MMcm	million cubic meters
MMcmd	million cubic meters per day
MMS	Minerals Management Service
MODU	mobile offshore drilling unit
MOL plc	Hungarian Oil and Gas Co.
MSC	multiple-service contract
MTBE	methyl tertiary butyl ether
mtoe	million tonnes of oil equivalent
mtpy	million tonnes per year
MW	megawatt
MWD	measurement-while-drilling
MWh	megawatt-hour

N

NAM	Nederlandse Aardolie Maatschappij BV (Netherlands)
NARUC	National Association of Regulatory Utility Commissioners (US)
NCMA	North Coast Marine Area (Trinidad & Tobago)
NEGP	North European Gas Pipeline (Russia-Germany)
NELP IV	New Exploration Licensing Policy IV (India)
Nemed	North East Mediterranean deepwater concession (Egypt)
NETA	New Electricity Trading Arrangements (UK)
NGO	non-governmental organization
NGSA	Natural Gas Supply Association (US)
NIMBY	not in my back yard
NIOC	National Iranian Oil Company
NLNG	Nigeria Liquefied Natural Gas Co.
NNCL	Bangladesh Business Development Corp. Ltd.
NNPC	Nigerian National Petroleum Co.
NO_x	nitrogen oxides
NOIA	Newfoundland Ocean Industries Association
NPD	Norwegian Petroleum Directorate
NPPAP	New Players Petroleum Association of the Philippines
NPR-A	National Petroleum Reserve-Alaska
NPRA	National Petrochemical & Refiners Association (US)
NTN	Nova Transportadora do Nordeste SA (Brazil)
NTS	Nova Transportadora do Sudeste SA (Brazil)
NWS	North West Shelf (Australia)
NYBOT	New York Board of Trade
Nymex	New York Mercantile Exchange
NYSE	New York Stock Exchange

O

OCA	Overlapping Claims Area (Thailand-Cambodia)
OCP	Oleoducto de Crudos Pesados Ecuador SA (consortium and pipeline)
OCTG	oil country tubular goods
OD	outer diameter
OECD	Organization for Economic Cooperation and Development
OFAC	Office of Foreign Assets Control (US Department of Treasury)
OGDC	Oil & Gas Development Co. Ltd. (Pakistan)
OGIP	original gas in place
OGJ	Oil & Gas Journal
OKIOC	Offshore Kazakhstan International Operating Company, re-named Agip KCO
OLNGC	Oman Liquefied Natural Gas Co.
OML	oil mining lease (Nigeria)
Omnis	Office of National Mines and Strategic Industries (Madagascar)
ONAREP	Office National de Recherches et d'Explorations Pétrolières (Morocco)
ONGC	Oil & Natural Gas Corporation (India)
Onhym	National Office of Hydrocarbons and Mining (Morocco)
OPEC	Organization of Petroleum Exporting Countries
OPIC	Overseas Private Investment Corp. (US)
OPL	oil prospecting license (Nigeria)
OPS	Office of Pipeline Safety (US)
OSA	operating service agreement (Venezuela)

P

PADD Petroleum Administration for Defense District (US)
PALs Plantwide Applicability Limits (US)
Parco Pak-Arab Refinery Co. Ltd.
Parsep Proyecto de Asistencia para La Reglamentación del Sector Energético del Perú
PCO-W Iraq Project and Contracting Office (Washington, DC), US Army
PCR production change rate
PDET Oil Outflow and Treatment Plan (Campos basin, Brazil)
PDO Petroleum Development Oman LLC
PDO plan for development and operation (Norway)
PdVSA Petróleos de Venezuela SA
PEGASO Excellence in Environmental Management & Operational Safety Program (Brazil)
Pemex Petroleos Mexicanos
PEP Petroleum Exploration Permit (New Zealand)
Pertamina Perusahaan Pertambangan Minyak Dan Gas Bumi Negara (Indonesia)
PET polyethylene terephthalate
Petrobangla Bangladesh Oil, Gas & Mineral Corp.
Petrobras Petroleo Brasileiro SA
Petroecuador Petroleos del Ecuador
Petrogal Petroleos de Portugal EP
Petronas Petroliam Nasional Sdn. Berhad (Malaysia)
Petroperu Petroleos del Peru SA
PetroSA Petroleum Oil and Gas Corp. of South Africa
Petrotrin Petroleum Co. of Trinidad and Tobago Ltd.
PetroVietnam Vietnam Oil & Gas Co.
PG&E Pacific Gas & Electric Co.
PGC Potential Gas Committee (US)
PGC Process Gas Consumers (US)
PGN PT Perusahaan Gas Negara (Persero) Tbk (Indonesia)
PGNiG Polskie Gomictwo Naftowe I Gazownictow SA (Polish Oil & Gas Co.)
PIC Petrochemical Industries Co. (Kuwait)
PIO plan for installation and operation (Norway)
PIRINC Petroleum Industry Research Foundation Inc.
PJ petajoule (approx. = 1 trillion Btu)
PKN Polski Koncern Naftowy (Poland)
PL Production License (Norway)
PLEM pipeline end manifold
PMP Petroleum Mining Permit (New Zealand)
PNSC Pakistan National Shipping Corporation
Poas Petrol Ofisis (Turkey)
PoG Perseus-over-Goodwyn (Australia)
POGC Polish Oil and Gas Co.
PPL petroleum prospecting license (New Zealand)
PSA production-sharing agreement (or arrangement)
PSA pressure swing adsorption
PSC production-sharing contract
PSO Pakistan State Oil
PTEN PT Exspan Nusantara (Indonesia)
PTT polytrimethylene terephthalate
PTT plc Thailand government oil and gas firm (formerly Petroleum Authority of Thailand)
PTTEP PTT Exploration and Production plc (Thailand)

R

RasGas Ras Laffan Liquefied Natural Gas Co. Ltd. (Qatar)
RFCC residual fluid catalytic cracking
RFG reformulated gasoline
RFR right of first refusal
RIK royalty-in-kind
ROSBOS Rosetti Marino SPA-Bouygues Offshore joint venture
ROV remotely operated vehicle
RR railroad
RTO regional transmission organization
RTP rapid thermal conversion

S

Sabic Saudi Basic Industries Corp.
SADC South African Development Community
SAGD steam-assisted gravity drainage
SAGE Scottish Area Gas Evacuation
SAP solvent-aided process
SAPREF South African Petroleum Refineries (Pty) Ltd.
SARS severe acute respiratory syndrome
scfd standard cubic feet per day
SCO synthetic crude oil
SCP South Caucasus Pipeline
SDFI state's direct financial interest (Norway), renamed Petoro
SEA State Energy Administration (China)
SEEL South-East European Line
semi semisubmersible offshore production platform
Sinochem China National Chemical Import & Export Company
SIR Ivorian Refining Company (Ivory Coast)
SNG synthetic natural gas
SNGPL Sui Northern Gas Pipelines Ltd. (Pakistan)
SNH National Hydrocarbon Corp. (Cameroon)
SNPC Société Nationale des Pétroles du Congo (Brazzaville)
SO_2 sulfur dioxide
SO_x sulfur oxides
SOCAR State Oil Company of Azerbaijan Republic

SOEP	Sable Offshore Energy Project (Canada)
SOMO	State Oil Marketing Organization (Iraq)
Sonangol EP	Sociedade Nacionale de Combustiveis de Angola
SOTE	Trans-Ecuadorian pipeline's Spanish-language acronym
SPA	sale and purchase agreement
SPE	Society of Petroleum Engineers (US)
SPLA	Sudan People's Liberation Army
SPM	single-point mooring
SPP	Slovensky plynarensky priemysel (Slovakia)
sq km	square kilometer
sq mi	square mile
SPR	Strategic Petroleum Reserve
SRS	Schmierstoff Raffinerie Salzbergen (German refinery)
SRU	sulfur-recovery unit
SSGC	Sui Southern Gas Co. Ltd. (Pakistan)
SSP	Sevan Stabilized Platform (cylindrical monohull FPSO vessel design)
Statoil	Den norske stats oljeselskap AS (Norway)
STL	submerged turret loading
SVA	security vulnerability assessment (US)
synbit	syncrude-bitumen blend (raw bitumen diluted with synthetic crude oil)

T

TAC	technical assistance contract
TAG	Trans Austria Gasleitung pipeline
TAP	Trans-Alpine Pipeline
TAP	Turkmenistan-Afghanistan-Pakistan pipeline
TAWN	Tariki, Ahuroa, Waihapa, and Ngaere area (New Zealand)
tcf	trillion cubic feet
tcfe	trillion cubic feet equivalent
tcm	trillion cubic meters
tcme	trillion cubic meters equivalent
TCP	Trans-Caspian Pipeline
TD	total depth
TGS	Transportadora de Gas del Sur (South American pipeline operator)
TJ	terajoules
TMD	total measured depth
TNK	Tyumen Oil Co. (Russia)
toe	tons or tonnes of oil equivalent
TPAO	Turkiye Petrolleri Anonim Ortakligi (Turkey)
tpd	tons or tonnes per day
tpy	tons or tonnes per year
TRG	Rompetrol Group NV (Netherlands)
TSB	Transportadora Sulbrasileira de Gas pipeline (Brazil)
tscm	trillion standard cubic meters
Tupras	Turkish Petroleum Refineries Corp.
TVD	true vertical depth
TVDS	true vertical depth subsea
TVK	Tiszai Vegyi Kombinat Rt. (Hungary)
TW	terawatt
TWh	terawatt-hour

U

UAE	United Arab Emirates
UDS	Ultramar Diamond Shamrock Corp.
UES	Unified Energy Systems (Russia)
UKCS	UK Continental Shelf
UKOOA	UK Offshore Operators Association Ltd.
ULCC	ultra large crude carrier
UN	United Nations
Unipec	United International Petroleum & Chemicals Company Ltd. (China)
UOG	United Arab Emirates Offsets Group
URR	ultimate resource recovery
USACE	US Army Corps of Engineers
UXO	unexploded ordnance

V

Vapex	vapor extraction
VGO	vacuum gas oil
VLCC	very large crude carrier
VOCs	volatile organic compounds
VP	Vice President
VSS	vortex separation system

W

WAG	water-alternating-gas
WAGP	West Africa Gas Pipeline (Nigeria. Ghana, Togo, and Benin)
WDDM	West Delta Deep Marine (Egypt)
WLGP	Western Libyan Gas Project
WNTS	West Natuna Transportation System (Indonesia to Singapore)
Wocap	World Oil Production Capacity (model)
WOPP	White Oil Pipeline Project (Pakistan)
WTO	World Trade Organization

Z

ZIC	Zone d'Interet Commun (Angola/Congo-Brazzaville)

North American

Geographical Abbreviations

U.S. States

AL	Alabama
AK	Alaska
AR	Arkansas
AZ	Arizona
CA	California
CO	Colorado
CT	Connecticut
DC	District of Columbia
DE	Delaware
FL	Florida
GA	Georgia
HI	Hawaii
ID	Idaho
IL	Illinois
IN	Indiana
IA	Iowa
KS	Kansas
KY	Kentucky
LA	Louisiana
MA	Massachusetts
MD	Maryland
ME	Maine
MI	Michigan
MN	Minnesota
MO	Missouri
MS	Mississippi
MT	Montana
NC	North Carolina
ND	North Dakota
NE	Nebraska
NH	New Hampshire
NJ	New Jersey
NM	New Mexico
NV	Nevada
NY	New York
OH	Ohio
OK	Oklahoma
OR	Oregon
PA	Pennsylvania
RI	Rhode Island
SC	South Carolina
SD	South Dakota
TN	Tennessee
TX	Texas
UT	Utah
VA	Virginia
VT	Vermont
WA	Washington
WI	Wisconsin
WV	West Virginia
WY	Wyoming

Canadian Provinces

AB	Alberta
BC	British Columbia
MB	Manitoba
NB	New Brunswick
NL	Newfoundland & Labrador
NS	Nova Scotia
NT	Northwest Territories
NU	Nunavut
ON	Ontario
PE	Prince Edward Island
QC	Quebec
SK	Saskatchewan
YT	Yukon Territory

Index

Advertisers

Contents

A

B

C

D

E

F

G

H

I

J

K

L

M

N

O

P

Q

R

S

T

U

V

W

X

Y

Z

Fold-out Map Legend

Use this legend for all IPE 2009 maps

Legend

Oil field

Oil sand

Gas field

Crude oil pipeline

Natural gas pipeline

Products pipeline

Pipeline planned or under construction

600

Refinery in operation
Refinery capacity in 1,000 b/d

Tanker terminal

Cities

Capital

International boundary

Water depth

0 to 200 m

200 m and deeper

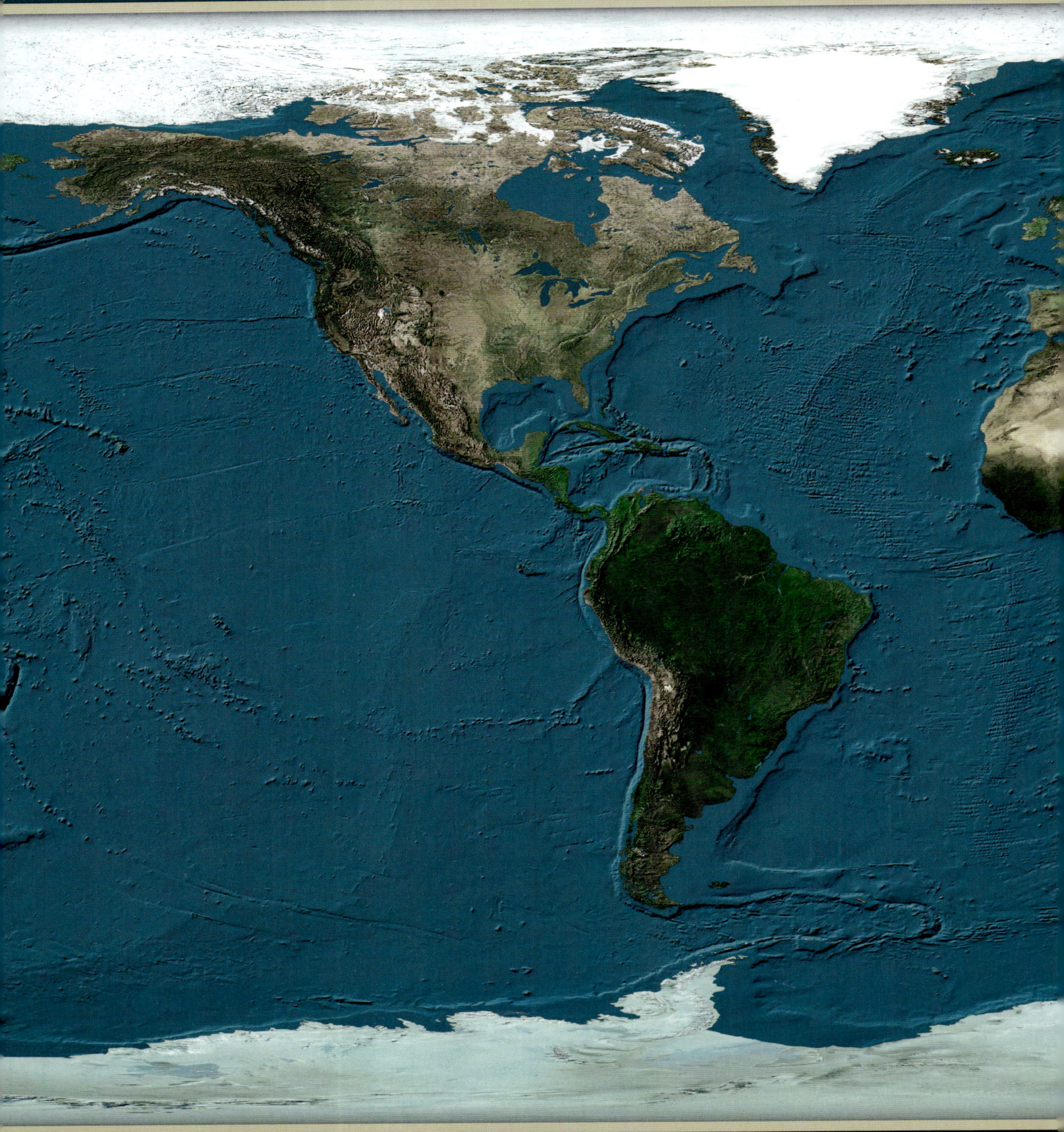